Lecture Notes in Computer Science   6809

Commenced Publication in 1973
Founding and Former Series Editors:
Gerhard Goos, Juris Hartmanis, and Jan van Leeuwen

Judith Bayard Cushing   James French
Shawn Bowers (Eds.)

Scientific and Statistical Database Management

23rd International Conference, SSDBM 2011
Portland, OR, USA, July 20-22, 2011
Proceedings

 Springer

Volume Editors

Judith Bayard Cushing
The Evergreen State College, Olympia, WA 98505, USA
E-mail: judyc@evergreen.edu

James French
CNRI and University of Virginia, Charlottesville, VA 22908-0816, USA
E-mail: jfrench@cnri.reston.va.us

Shawn Bowers
Gonzaga University, Spokane, WA 99258, USA
E-mail: bowers@gonzaga.edu

ISSN 0302-9743   e-ISSN 1611-3349
ISBN 978-3-642-22350-1   e-ISBN 978-3-642-22351-8
DOI 10.1007/978-3-642-22351-8
Springer Heidelberg Dordrecht London New York

Library of Congress Control Number: 2011930828

CR Subject Classification (1998): H.3, I.2, H.4, C.2, H.2.8, H.5

LNCS Sublibrary: SL 3 – Information Systems and Application, incl. Internet/Web
and HCI

Typesetting: Camera-ready by author, data conversion by Scientific Publishing Services, Chennai, India

Printed on acid-free paper

Springer is part of Springer Science+Business Media (www.springer.com)

Message from the General Chair

Welcome to the proceedings of the 23rd International Conference on Scientific and Statistical Database Management held in Portland, Oregon, where it celebrated its 30th birthday. The first incarnation of SSDBM (then called the Workshop on Statistical Database Management) took place in Menlo Park, California, in December 1981. Since that time, SSDBM added a second 'S,' for "Scientific," then switched the orders of the 'S's, and upgraded "Workshop" to "Working Conference" to "Conference." Initially held roughly every other year, SSDBM now convenes annually, alternating between North America and Europe or Asia. It remains an intimate conference, where one has a chance to interact with all the other attendees.

This year marked the return of SSDBM to the Pacific Northwest, having previously been held in Olympia, Washington (1997). The shift of the conference time from winter to summer may have deprived attendees of enjoying the famous Portland rain, but I hope that lack was compensated by the other attractions of the city: its brewpubs and baristas; the urban parks and gardens; the wineries and farmers markets; and the nearby natural wonders of the Columbia River Gorge, the Oregon Coast and Mount St. Helens.

SSDBM is an independent conference, and succeeds by the efforts of its all-volunteer Organizing Committee. On the technical side, I thank Judy Cushing and Jim French for their recruitment of the Program Committee and oversight of the review process. Shawn Bowers excelled as Proceedings Editor and EasyChair wrangler. Thanks as well to the PC members and ancillary reviewers. In the financial realm, Len Shapiro and Michael Grossniklaus set up SSDBM 2011 as an Oregon Corporation, and Michael oversaw our budget and banking. Bill Howe ran the registration site and was instrumental in arranging our sponsorships. Locally, Laura Bright and Kristin Tufte managed our arrangements within the hotel, as well as doing tireless research on a suitable restaurant for our dinner. Dave Hansen oversaw all the student volunteers. Thanks as well to Stephanie Lewis and the staff of University Place. For conference communications, Pete Tucker maintained our website and David Chiu coordinated announcements and information for SSDBM participants. I also thank the SSDBM Steering Committee, particularly Ari Shoshani for his corporate memory, and Michael Gertz for information on the Heidelberg conference (and surplus funds!).

I thank our Gold Sponsor, Microsoft Research, and Silver Sponsors, the eScience Institute at the University of Washington, The Gordon and Betty Moore Foundation and Paradigm4 for their support of this year's conference.

Their generous contributions helped support discounted registration for students, the keynote speaker, and social events for student volunteers. Also, I express our gratitude to Springer, for their continuing role as our proceedings publisher. Finally, I thank all those who submitted papers and proposed panels. Their interest and participation is what keeps the quality of SSDBM high and the topics timely.

July 2011   David Maier

Message from the Program Co-chairs

We were pleased at the high quality of work presented at SSDBM 2011. In addition to our keynote speaker, Michael Stonebraker who inaugurated the conference with his talk "The Architecture of SciDB," we had two excellent panels "Data-Intensive Science: Moving Towards Solutions" chaired by Terence Critchlow and "Data Scientists, Data Management and Data Policy" chaired by Sylvia Spengler. These "hot topics" within the main themes of the conference were selected to promote fruitful discussion on directions of the field.

As always, however, the "main course" of the conference was in the research presented. In response to the call for papers, we received 80 submissions. As in prior years, there was solid Praxis *and* Theoria—papers ranging from the practical or applied Applications and Models, and Architectures and Privacy, Workflows and Provenance to theoretical areas that provide fundamentals for the field Clustering and Data Mining, Ranked Search, Temporal Data and Queries, and Graph Querying.

The number of paper submissions for the conference was about average for SSDBM, which given the poor global economic climate boded well for the field and future SSDBM conferences. There were 80 submissions, 67 as long papers and 13 as short papers. Each was carefully reviewed by at least three Program Committee members and, in some cases, electronically discussed by reviewers. The Program Co-chairs, after reading each review and considering the ratings, accepted 23 of the 67 (34%) long papers and 3 of the 13 (23%) short papers. We also accepted 9 long submissions as short papers, for a total of 12 short papers in the conference.

In addition to research paper publication and presentations, we had a lively poster and demo session. As Program Co-chairs, we adopted a strict policy to assure high-quality research papers, but in accepting submissions as posters sought to engage broadly. Where we saw promise, value for the field, as well as the possibility that authors would benefit from attending the conference, listening to talks, and presenting the work as a poster, we accepted the work as posters. Our objective here was to provide a mechanism for improving the overall quality of submissions to the conference in subsequent years, and to increase active participation in the present conference. We thus included 15 posters in the conference. After the conference proceedings went to press, we invited all accepted authors to present demonstrations during the poster session; a list of demos was available in the conference program and on the website.

The Program Committee thanks all those who submitted papers and posters to the conference, as the high-quality program reflects the viability of the field and this conference. As Program Co-chairs, we also express our sincere appreciation to the 65 members of the Program Committee as well as to 32 additional reviewers (recruited for specialized expertise by members of the Program Committee) for their hard work and dedication during the electronic interactions during the review process.

July 2011

Judith Bayard Cushing
James French

SSDBM 2011 Conference Organization

General Chair

David Maier   Portland State University, USA

Assistant Chair

Leonard Shapiro   Portland State University, USA

Program Committee Co-chairs

Judith Bayard Cushing   The Evergreen State College, USA
James French   CNRI and University of Virginia, USA

Proceedings Editor

Shawn Bowers   Gonzaga University, USA

Information Officer

David Chiu   Washington State University Vancouver, USA

Web and Publicity Chair

Peter Tucker   Whitworth University, USA

Registration

Bill Howe   University of Washington, USA

Treasurer

Michael Grossniklaus   Portland State University, USA

Local Arrangements

Laura Bright   McAfee, USA
Kristin Tufte   Portland State University, USA

Student Volunteers

David Hansen   George Fox University, USA

Program Committee

Ken Barker   University of Calgary, Canada
Randal Burns   Johns Hopkins University, USA
Sarah Cohen-Boulakia   University of Paris-Sud 11, France
Isabel Cruz   University of Illinois at Chicago, USA
Alfredo Cuzzocrea   University of Calabria, Italy
Nilesh Dalvi   USC Information Sciences Institute, USA
Ewa Deelman   USC Information Sciences Institute, USA
Dejing Dou   University of Oregon, USA
Amr El Abbadi   University of California, Santa Barbara, USA
Conny Franke   University of Heidelberg, Germany
Juliana Freire   University of Utah, USA
James Frew   University of California, Santa Barbara, USA
Johann Gamper   Free University of Bozen-Bolzano, Italy
Michael Gertz   University of Heidelberg, Germany
Carole Goble   University of Manchester, UK
Michael Goodchild   University of California, Santa Barbara, USA
Wilfried Grossmann   University of Vienna, Austria
Dimitrios Gunopulos   University of Athens, Greece
Amarnath Gupta   San Diego Supercomputer Center, USA
Bill Howe   University of Washington, USA
Theo Härder   University of Kaiserslautern, Germany
Ray Idaszak   RENCI, UNC, USA
H.V. Jagadish   University of Michigan, USA
Matthias Jarke   RWTH Aachen, Germany
Chris Jermaine   Rice University, USA
Matthew Jones   University of California, Santa Barbara, USA
Jessie Kennedy   Napier University, UK
Larry Kerschberg   George Mason University, USA
Martin Kersten   CWI, The Netherlands
Hans-Joachim Klein   University of Kiel, Germany
Peer Kröger   Ludwig-Maximilians-Universität München,
                                  Germany
Zoe Lacroix   Arizona State University, USA
Ulf Leser   Humboldt University of Berlin, Germany
Feifei Li   Florida State University, USA
Bertram Ludäscher   University of California, Davis, USA
Yannis Manolopoulos   Aristotle University of Thessaloniki, Greece
Claudia Medeiros   University of Campinas, Brazil
Kyriakos Mouratidis   Singapore Management University, Singapore

Wolfgang Mueller	HITS gGmbH, Germany
Silvia Nittel	University of Maine, USA
Frank Olken	University of California, Berkeley, USA
Beng Chin Ooi	National University of Singapore, Singapore
Gultekin Ozsoyoglu	Case Western Reserve University, USA
Andreas Reuter	HITS gGmbH, Germany
Philippe Rigaux	Université Paris-Dauphine, France
Kenneth Ross	Columbia University, USA
Doron Rotem	Lawrence Berkeley National Laboratory, USA
Nagiza F. Samatova	North Carolina State University, USA
Leonard Shapiro	Portland State University, USA
Linda Shapiro	University of Washington, USA
Sylvia Spengler	National Science Foundation, USA
Jianwen Su	University of California, Santa Barbara, USA
Kian-Lee Tan	National University of Singapore, Singapore
Yufei Tao	Chinese University of Hong Kong, SAR China
Dimitri Theodoratos	New Jersey Institute of Technology, USA
Shengru Tu	University of New Orleans, USA
Can Türker	Functional Genomics Center Zürich, Switzerland
Richard Weiss	The Evergreen State College, USA
Andrew Westlake	Survey and Statistical Computing, UK
Kesheng John Wu	Lawrence Berkeley National Laboratory, USA
Yan Xu	Microsoft Corporation, USA
Jeffrey Yu	Chinese University of Hong Kong, SAR China
Xiangliang Zhang	King Abdullah University of Science and Technology, Saudi Arabia
Daniel Zinn	University of California, Davis, USA

Additional Reviewers

Aksoy, Cem
Beck, David
Belhajjame, Khalid
Budak, Ceren
Cugler, Daniel
Dalamagas, Theodore
Das, Sudipto
Dey, Saumen
Elmore, Aaron
Graham, Martin
Jestes, Jeffrey
Kim, Jinoh
Koehler, Sven
Koschmieder, André
Le, Wangchao
Lin, Yimin

Liu, Haishan
Malaverri, Joana
Nguyen, Kim
Papadopoulos, Apostolos
Rheinlaender, Astrid
Scheers, Bart
Tang, Mingwang
Tiakas, Eleftherios
Valkanas, Georgios
Vilar, Bruno
Wang, Shiyuan
Wu, Xiaoying
Yao, Bin
Zhang, Chongsheng
Zhang, Jilian
Zhu, Yuanyuan

SSDBM Steering Committee

Michael Gertz	University of Heidelberg, Germany
Bertram Ludäscher	University of California, Davis, USA
Nikos Mamoulis	University of Hong Kong, SAR China
Arie Shoshani	Lawrence Berkeley National Laboratory (Chair), USA
Marianne Winslett	University of Illinois, USA

SSDBM 2011 Conference Sponsors

Microsoft Research
eScience Institute
Gordon and Betty Moore Foundation
Paradigm4 Inc.

Table of Contents

Keynote Address

The Architecture of SciDB 1
   Michael Stonebraker, Paul Brown, Alex Poliakov, and Suchi Raman

Ranked Search

Location-Based Instant Search 17
   Shengyue Ji and Chen Li

Continuous Inverse Ranking Queries in Uncertain Streams 37
   Thomas Bernecker, Hans-Peter Kriegel, Nikos Mamoulis,
   Matthias Renz, and Andreas Zuefle

Finding Haystacks with Needles: Ranked Search for Data Using
Geospatial and Temporal Characteristics 55
   V.M. Megler and David Maier

Using Medians to Generate Consensus Rankings for Biological Data 73
   Sarah Cohen-Boulakia, Alain Denise, and Sylvie Hamel

A Truly Dynamic Data Structure for Top-k Queries on Uncertain
Data 91
   Manish Patil, Rahul Shah, and Sharma V. Thankachan

Temporal Data and Queries

Efficient Storage and Temporal Query Evaluation in Hierarchical Data
Archiving Systems 109
   Hui (Wendy) Wang, Ruilin Liu, Dimitri Theodoratos, and
   Xiaoying Wu

Update Propagation in a Streaming Warehouse 129
   Theodore Johnson and Vladislav Shkapenyuk

Efficient Processing of Multiple DTW Queries in Time Series
Databases 150
   Hardy Kremer, Stephan Günnemann, Anca-Maria Ivanescu,
   Ira Assent, and Thomas Seidl

Probabilistic Time Consistent Queries over Moving Objects 168
   Xiang Lian and Lei Chen

Workflows and Provenance

Knowledge Annotations in Scientific Workflows: An Implementation in
Kepler   189
 Aída Gándara, George Chin Jr., Paulo Pinheiro da Silva,
 Signe White, Chandrika Sivaramakrishnan, and Terence Critchlow

Improving Workflow Fault Tolerance through Provenance-Based
Recovery   207
 Sven Köhler, Sean Riddle, Daniel Zinn, Timothy McPhillips, and
 Bertram Ludäscher

PROPUB: Towards a Declarative Approach for Publishing Customized,
Policy-Aware Provenance   225
 Saumen C. Dey, Daniel Zinn, and Bertram Ludäscher

Provenance-Enabled Automatic Data Publishing   244
 James Frew, Greg Janée, and Peter Slaughter

Panel I

A Panel Discussion on Data Intensive Science: Moving towards
Solutions   253
 Terence Critchlow

Querying Graphs

Querying Shortest Path Distance with Bounded Errors in Large
Graphs   255
 Miao Qiao, Hong Cheng, and Jeffrey Xu Yu

PG-Join: Proximity Graph Based String Similarity Joins   274
 Michail Kazimianec and Nikolaus Augsten

A Flexible Graph Pattern Matching Framework via Indexing   293
 Wei Jin and Jiong Yang

Subgraph Search over Massive Disk Resident Graphs   312
 Peng Peng, Lei Zou, Lei Chen, Xuemin Lin, and Dongyan Zhao

BR-Index: An Indexing Structure for Subgraph Matching in Very Large
Dynamic Graphs   322
 Jiong Yang and Wei Jin

Clustering and Data Mining

CloudVista: Visual Cluster Exploration for Extreme Scale Data in the
Cloud   332
    Keke Chen, Huiqi Xu, Fengguang Tian, and Shumin Guo

Efficient Selectivity Estimation by Histogram Construction Based on
Subspace Clustering   351
    Andranik Khachatryan, Emmanuel Müller, Klemens Böhm, and
    Jonida Kopper

Finding Closed MEMOs   369
    Htoo Htet Aung and Kian-Lee Tan

Density Based Subspace Clustering over Dynamic Data   387
    Hans-Peter Kriegel, Peer Kröger, Irene Ntoutsi, and Arthur Zimek

Hierarchical Clustering for Real-Time Stream Data with Noise   405
    Philipp Kranen, Felix Reidl, Fernando Sanchez Villaamil, and
    Thomas Seidl

Architectures and Privacy

Energy Proportionality and Performance in Data Parallel Computing
Clusters   414
    Jinoh Kim, Jerry Chou, and Doron Rotem

Privacy Preserving Group Linkage   432
    Fengjun Li, Yuxin Chen, Bo Luo, Dongwon Lee, and Peng Liu

Dynamic Anonymization for Marginal Publication   451
    Xianmang He, Yanghua Xiao, Yujia Li, Qing Wang,
    Wei Wang, and Baile Shi

Pantheon: Exascale File System Search for Scientific Computing   461
    Joseph L. Naps, Mohamed F. Mokbel, and David H.C. Du

Massive-Scale RDF Processing Using Compressed Bitmap Indexes   470
    Kamesh Madduri and Kesheng Wu

Database-as-a-Service for Long-Tail Science   480
    Bill Howe, Garret Cole, Emad Soroush, Paraschos Koutris,
    Alicia Key, Nodira Khoussainova, and Leilani Battle

Panel II

Data Scientists, Data Management and Data Policy   490
    Sylvia Spengler

Applications and Models

Context-Aware Parameter Estimation for Forecast Models in the
Energy Domain   491
   Lars Dannecker, Robert Schulze, Matthias Böhm,
   Wolfgang Lehner, and Gregor Hackenbroich

Implementing a General Spatial Indexing Library for Relational
Databases of Large Numerical Simulations   509
   Gerard Lemson, Tamás Budavári, and Alexander Szalay

Histogram and Other Aggregate Queries in Wireless Sensor Networks . . .   527
   Khaled Ammar and Mario A. Nascimento

Efficient In-Database Maintenance of ARIMA Models   537
   Frank Rosenthal and Wolfgang Lehner

Recipes for Baking Black Forest Databases: Building and Querying
Black Hole Merger Trees from Cosmological Simulations   546
   Julio López, Colin Degraf, Tiziana DiMatteo, Bin Fu,
   Eugene Fink, and Garth Gibson

CrowdLabs: Social Analysis and Visualization for the Sciences   555
   Phillip Mates, Emanuele Santos, Juliana Freire, and Cláudio T. Silva

Posters

Heidi Visualization of R-tree Structures over High Dimensional Data . . .   565
   Shraddha Agrawal, Soujanya Vadapalli, and Kamalakar Karlapalem

Towards Efficient and Precise Queries over Ten Million Asteroid
Trajectory Models   568
   Yusra AlSayyad, K. Simon Krughoff, Bill Howe,
   Andrew J. Connolly, Magdalena Balazinska, and
   Lynne Jones

Keyword Search Support for Automating Scientific Workflow
Composition   571
   David Chiu, Travis Hall, Farhana Kabir, and Gagan Agrawal

FastQuery: A General Indexing and Querying System for Scientific
Data   573
   Jerry Chou, Kesheng Wu, and Prabhat

Retrieving Accurate Estimates to OLAP Queries over Uncertain and
Imprecise Multidimensional Data Streams   575
   Alfredo Cuzzocrea

Hybrid Data-Flow Graphs for Procedural Domain-Specific Query
Languages   577
   Bernhard Jaecksch, Franz Faerber, Frank Rosenthal, and
   Wolfgang Lehner

Scalable and Automated Workflow in Mining Large-Scale Severe-Storm
Simulations   579
   Lei Jiang, Gabrielle Allen, and Qin Chen

Accurate Cost Estimation Using Distribution-Based Cardinality
Estimates for Multi-dimensional Queries   581
   Andranik Khachatryan and Klemens Böhm

Session-Based Browsing for More Effective Query Reuse   583
   Nodira Khoussainova, YongChul Kwon, Wei-Ting Liao,
   Magdalena Balazinska, Wolfgang Gatterbauer, and Dan Suciu

The ETLMR MapReduce-Based ETL Framework   586
   Xiufeng Liu, Christian Thomsen, and Torben Bach Pedersen

Top-k Similarity Search on Uncertain Trajectories   589
   Chunyang Ma, Hua Lu, Lidan Shou, Gang Chen, and Shujie Chen

Fast and Accurate Trajectory Streams Clustering   592
   Elio Masciari

Data-Driven Multidimensional Design for OLAP   594
   Oscar Romero and Alberto Abelló

An Adaptive Outlier Detection Technique for Data Streams   596
   Shiblee Sadik and Le Gruenwald

Power-Aware DBMS: Potential and Challenges   598
   Yi-cheng Tu, Xiaorui Wang, and Zichen Xu

Author Index   601

The Architecture of SciDB

Michael Stonebraker, Paul Brown, Alex Poliakov, and Suchi Raman

Paradigm4, Inc.
186 Third Avenue
Waltham, MA 02451

Abstract. SciDB is an open-source analytical database oriented toward the data management needs of scientists. As such it mixes statistical and linear algebra operations with data management ones, using a natural nested multi-dimensional array data model. We have been working on the code for two years, most recently with the help of venture capital backing. Release 11.06 (June 2011) is downloadable from our website (SciDB.org).

This paper presents the main design decisions of SciDB. It focuses on our decisions concerning a high-level, SQL-like query language, the issues facing our query optimizer and executor and efficient storage management for arrays. The paper also discusses implementation of features not usually present in DBMSs, including version control, uncertainty and provenance.

Keywords: scientific data management, multi-dimensional array, statistics, linear algebra.

1 Introduction and Background

The Large Synoptic Survey Telescope (LSST) [1] is the next "big science" astronomy project, a telescope being erected in Chile, which will ultimately collect and manage some 100 Petabytes of raw and derived data. In October 2007, the members of the LSST data management team realized the scope of their data management problem, and that they were uncertain how to move forward. As a result, they organized the first Extremely Large Data Base (XLDB-1) conference at the Stanford National Accelerator Laboratory [2]. Present were many scientists from a variety of natural science disciplines as well as representatives from large web properties. All reported the following requirements:

Multi-petabyte amounts of data. In fact a recent scientist at a major university reported that 20 research groups at his university had more than a quarter of a petabyte each [3].

A preponderance of array data. Geospatial and temporal data such as satellite imagery, oceanographic data telescope data, telematics data and most simulation data all are naturally modeled as arrays. Genomics data generated from high throughput sequencing machines are also naturally represented as arrays.

J.B. Cushing, J. French, and S. Bowers (Eds.): SSDBM 2011, LNCS 6809, pp. 1–16, 2011.

Complex analytics. Traditional business intelligence has focused on simple SQL aggregates or windowing functions. In contrast, scientists need much more sophisticated capabilities. For example, satellite imagery can be reported at various resolutions and in different co-ordinate systems. As a result, earth scientists need to regrid such imagery in order to correlate the data from multiple satellites. In addition, most satellites cannot see through cloud cover. Hence, it is necessary to find the "best" cloud-free composite image from multiple passes of the satellite. These are representative of the complex operations required in this application area.

A requirement for open source code. Every scientist we have talked to is adamant about this requirement. Seemingly, the experience of the Large Hadron Collider (LHC) project [4] with one proprietary DBMS vendor has "poisoned the well". Hence, scientists require the option of fixing bugs and adding their own features, if the vendor of their chosen solution is unable, unwilling, or just slow to respond. In effect, only open source software is acceptable.

A requirement for no overwrite. Scientists are equally adamant about never throwing anything away. For example, large portions of the earth are currently not very interesting to earth scientists. However, that could change in the future, so discarding currently uninteresting data is not an option. Also, they wish to keep erroneous data that has been subsequently corrected. The reason for this is to redo analyses on the data as it existed at the time the original analysis was done, i.e. they want auditability for their analyses. This is related to the provenance discussion below, and requires that all data be kept indefinitely.

A requirement for provenance. If a data element looks suspicious, then scientists want to be able to trace backward through its derivation to find previous data that appears faulty. In other words, trace the error back to its source. Similarly, they would then want to find all of the derived data that came from this faulty item. In other words, they want the ability to do forward and backward derivation efficiently.

One reason for this requirement is assistance in the error correction noted above. A second reason is to facilitate sharing. Different scientists generally cannot make use of derived data unless they know the algorithm that was used to create it. For example, consider the "best" cloud free image discussed above. There is no universal way to choose the best composite image, and any scientist who wants to use a composite image must know what algorithm was used to construct it. They want to find this information by exploring the provenance of the data of interest.

A requirement for uncertainty. After all, every bit of scientific data comes with error bars. Current DBMSs were written to support the business market, and assume the data is perfect. Obviously, enterprises must know accurate salaries, in order to write pay checks. Essentially all information collected from sensors (nearly 100% of science data) does not have this property. Furthermore, scientists may want to propagate error data through a sequence of calculations.

A requirement for version control. There is no universal agreement on the cooking algorithms, which turn raw data into derived data sets. Hence, scientists would like to re-cook raw data for their study areas, retaining the conventional derivations for the rest of the data set. Although they can construct a complete copy of the data, with the required characteristics, it is wildly more efficient to delta their copies off of the conventional one, so the common data only appears once. Version control software has been supporting this functionality for years.

At XLDB-1, there was a general feeling that RDBMSs would never meet the above requirements because they have:

- The wrong data model,
- The wrong operators, and
- Are missing required capabilities.

Moreover, the RDBMS vendors appear not to be focused on the science market, because the business enterprise market is perceived to be larger. Hence, there was skepticism that these shortcomings would ever be addressed.

A second theme of the meeting was the increasing difficulty of meeting big science requirements with "from the bare metal up" custom implementations. The software stack is simply getting too large. Several of the web properties indicated the scope of their custom efforts, and said "we are glad we have sufficient resources to move forward". Also, there was frustration that every big science project re-grows the complete stack, leading to limited shared infrastructure. The Sloan Digital Sky Survey [5] was also noted as a clear exception, as they made use of SQLServer.

In effect, the community was envious of the RDBMS market where a common set of features is used by nearly everybody and supported by multiple vendors. In summary, the mood was "Why can't somebody do for science what RDBMS did for business?"

As a result, Dave Dewitt and Mike Stonebraker said they would try to build a from-the-ground-up DBMS aimed at science requirements. Following XLDB-1, there were meetings to discuss detailed requirements and a collection of use cases written, leading to an overall design. This process was helped along by the LSST data management team who said, "If it works, we will try to use it".

We began writing code in late 2008, with a pick-up team of volunteers and research personnel. This led to a demo of an early version of SciDB at VLDB in Lyon, France in Sept 2009 [6]. We obtained venture capital support for the project, and additional assistance from NSF in 2010. This has allowed us to accelerate our efforts. We have recently released SciDB 11.06 and are working toward a full featured high performance system in late 2011. We have been helped along the way by the subsequent annual XLDB meetings [7, 8, 9] where SciDB issues have been discussed in an open forum.

This paper reports on the SciDB design and indicates the status of the current system.

2 SciDB Design

In this section we present the major design decisions and our rationale for making them the way we did. We start with system assumptions in Section 2.1, followed by a data model discussion in Section 2.2. The query language is treated in Section 2.3.

The optimizer and storage management components are treated respectively in Section 2.4 and 2.5. Other features, such as extensibility, uncertainty, version control and provenance are discussed at appropriate times.

2.1 System Assumptions

It was pretty obvious that SciDB had to run on a grid (or cloud) of computers. A single node solution is clearly not going to make LSST happy. Also, there is universal acceptance of Linux in this community, so the OS choice is easy. Although we might have elected to code the system in Java, the feeling was that C++ was a better choice for high performance system software.

The only point of contention among the team was whether to adopt a shared-disk or a shared-nothing architecture. On the one hand, essentially all of the recent parallel DBMSs have adopted a shared nothing model, where each node talks to locally attached storage. The query optimizer runs portions of the query on local data. In essence, one adopts a "send the query to the data" model, and strives for maximum parallelism.

On the other hand, many of the recent supercomputers have used a shared-disk architecture. This appears to result from the premise that the science workload is computation intensive, and therefore the architecture should be CPU-focused rather than data focused. Also, scientists require a collection of common operations, such as matrix multiply, which are not "embarrassingly parallel". Hence, they are not obviously faster on a shared-nothing architecture.

Since an important goal of SciDB is petabyte scalability the decision was made that SciDB would be a shared nothing engine.

2.2 Data Model

It was clear that we should select an array data model (rather than a table one) as arrays are the natural data object for much of the sciences. Furthermore, early performance benchmarks on LSST data [10] indicated that SciDB is about 2 orders of magnitude faster than an RDBMS on a typical science workload. Finally, most of the complex analytics that the science community uses are based on core linear algebra operations (e.g. matrix multiply, covariance, inverse, best-fit linear equation solution). These are all array operations, and a table model would require a conversion back and forth to arrays. As such, it makes sense to use arrays directly.

Hence, SciDB allows any number of **dimensions** for an array. These can be traditional integer dimensions, with any starting and ending points or they can be unbounded in either direction. Moreover, many arrays are more natural with non-integer dimensions. For example, areas of the sky are naturally expressed in polar co-ordinates in some astronomy projects. Hence, dimensions can be any user-defined data type using a mechanism we presently describe.

Each combination of dimension values defines a **cell** of an array, which can hold an arbitrary number of **attributes** of any user-defined data type. Arrays are **uniform** in that all cells in a given array have the same collection of values. The only real decision was whether to allow a nested array data model or a flat one. Many use cases, including LSST, require nested arrays, so the extra complexity was deemed well worth it. Also, nested arrays support a mechanism for hierarchical decomposition of cells, so that systematic refinement of specific areas of an array can be supported, a feature often cited as useful in HDF5 [11].

Hence, an example array specification in SciDB is:

```
CREATE ARRAY example <M: int, N: float> [I=1:1000, J=1000:20000]
```

Here, we see an array with attributes M and N along with dimensions I and J.

2.3 Query Language

SciDB supports both a functional and a SQL-like query language. The functional language is called AFL for array functional language; the SQL-like language is called AQL for array query language. AQL is compiled into AFL.

AFL, the functional language includes a collection of operations, such as filter and join, which a user can cascade to obtain his desired result. For example, if A and B are arrays with dimensions I and J, and c is an attribute of A, then the following utterance would be legal:

```
temp = filter (A, c = value)
result = join (B, temp: I, J)
```

Or the composite expression: `result = join (B, filter (A, c = value), I, J)`

Such a language is reminiscent of APL [12] and other functional languages and array languages [13, 14].

For commercial customers more comfortable with SQL, SciDB has created an array query language, AQL, which looks as much like SQL as possible. Hence, the above example is expressed as:

```
select *
from A, B
where A.I = B.I and A.J = B.J and A.c = value
```

We have had considerable discussion concerning two aspects of the semantics of AQL, namely joins and non-integer dimensions, and we turn to these topics at this time.

Consider the arrays A and B from above, and suppose A has attributes c and d, while B has attributes e and f. The above example illustrated a dimension join, i.e., one where the dimensions indexes must be equal. The result of this operation has dimensions I and J, and attributes c, d, e and f. In essence this is the array version of a relational natural join. It is also straightforward to define joins that match less than all dimensions.

Non equi-dimensional joins are also reasonably straightforward. For example the following join result must be defined as a three dimensional array, I (from A), I (from B) and J.

```
select *
from A, B
where A.I > B.I and A.J = B.J and A.c = value
```

The problem arises when we attempt to define attribute joins, e.g.,

```
select *
from A, B
where A.c = B.e and A.d = B. f
```

In effect, we want to join two arrays on attribute values rather than dimensions. This must be defined as a four dimensional result: I (from A), J (from A), I (from B), and J (from B).

To understand array joins, it is useful to think of an array as having a relational representation, where the dimensions are columns as are the cell values. Then any array join can be defined as a relational join on the two argument tables. This naturally defines the semantics of an array join operation, and SciDB must produce this answer. In effect, we can appeal to relational semantics to define array joins.

A second semantic issue is an offshoot of the first one. It is straightforward to change attribute values in AQL with an update command. For example, the following command increments a value for a specific cell:

```
update A set (d = d+1)
where A.I = value and A.J = value
```

Obviously, it must be possible to manipulate SciDB dimensions, and an update command is not the right vehicle. Hence, SciDB has included a new powerful command, **transform**, to change dimension values. The use cases for transform include:

- Bulk changes to dimensions, e.g. push all dimension values up one to make a slot for new data,
- Reshape an array; for example change it from 100 by 100 to 1000 by 10,
- Flip dimensions for attributes, e.g. replace dimension I in array A with a dimension made up from d.
- Transform one or more dimension, for example change I and J into polar co-ordinates.

Transform can also map multiple dimension or attribute values to the same new dimension value. In this case, transform allows an optional aggregation function to combine the multiple values into a single one for storage. In the interest of brevity we skip a detailed discussion of the transform command and the interested reader is referred to the online documentation on SciDB.org, for a description of this command.

Non-integer dimensions are supported by an index that maps the dimension values into integers. Hence, an array with non-integer dimensions is stored as an integer array mapping index.

2.4 Extensibility

It is well understood that a DBMS should not export data to an external computation (i.e. move the data to the computation), but rather have the code execute inside the DBMS (move the computation to the data). The latter has been shown to be wildly faster, and is supported by most modern day relational DBMSs. The norm is to use the extension constructs pioneered by Postgres more than 20 years ago [15].

Since science users often have their own analysis algorithms (for example examining a collection of satellite passes to construct the best cloud-free composite image) and unique data types (e.g. 7 bit sensor values), it is imperative to support user-defined extensibility. There are four mechanisms in SciDB to support user extensions.

First, SciDB supports **user-defined data types**. These are similar to Postgres user defined types as they specify a storage length for a container to hold an object of the given type. User-defined types allow a (sophisticated) user to extend the basic SciDB data types of integer, float, and string. Hence, the attribute values in a SciDB cell can be user-defined.

Second, a user must be able to perform operations on new data types. For example, a user could define arbitrary precision float as a new data type and then would want to define operations like addition and subtraction on this type. **User-defined functions** are the mechanism for specifying such features. These are scalar functions that accept one or more arguments of various data types and produce a result of some data type. Again, the specification is similar to Postgres, and right now such functions must be written in C++.

Third, SciDB supports **user-defined aggregates**, so that conventional aggregates can be written for user-defined types. As well, science-specific aggregates can be written for built-in or user-defined data types. An aggregate requires four functions, along the lines of Postgres [16]. Three of the functions are the standard Init (), Increment (), and Final () that are required for any single node user-defined aggregate calculation. Since SciDB is a multi-node system, these three functions will be run for the data at each node. Subsequently, a rollup () must be specified to pull the various partial aggregates together into the final answer.

The last extension mechanism in SciDB is **user-defined array operators**. These functions accept one or more arrays as arguments and usually produce an array as an answer. Although Join is a typical example, the real use case is to support linear algebra operations, such as matrix multiply, curve fitting, linear regression, equations solving and the like. Also in this category are data clustering codes and other machine learning algorithms.

There are two wrinkles to array functions that are not present in standard Postgres table functions. As will be discussed in Section 2.6 SciDB decomposes storage into multi-dimensional chunks, which may overlap. Some array functions are **embarrassingly parallel**, i.e. they can be processed in parallel on a collection of computing nodes, with each node performing the same calculation on its data. However, some array functions can only be run in parallel if chunks overlap by a minimum amount, as discussed in more detail in Section 2.6. Hence, a user-defined array function must specify the minimum overlap for parallel operation.

Second, many array operations are actually algorithms consisting of several steps, with conditional logic between the steps. For example, most algorithms to compute the inverse of a matrix proceed by iterating a core calculations several times. More complex operations may perform several different kinds of core operations, interspersed with conditional logic. Such logic may depend on the size or composition of intermediate results (e.g. an array column being empty). As such, a user-defined array operation must be able to run other operations, test the composition of intermediate results and control its own parallelism. To accomplish this objective, SciDB has a system interface that supports these kinds of tasks.

It should be noted that writing user-defined array operations is not for the faint of heart. We expect experts in the various science disciplines to write libraries to our interface that other scientists can easily use in AQL, without understanding their detailed composition. This is similar to ScaLAPACK [17], which was written by rocket scientists and widely used by mere mortals.

2.5 Query Processing

Users specify queries and updates in AQL, and the job of the optimizer and executor is to correctly solve such queries. We have several guiding principles in the design of this component of SciDB.

First, we expect a common environment for SciDB is to run on a substantial number of nodes. As such, SciDB must scale to large configurations. Also, many science applications are CPU intensive. Hence, the user-defined functions that perform the complex analytics usually found in this class of problems are often CPU bound. Also, many are not "embarrassingly parallel", and entail moving substantial amounts of data if one is not careful. Thus, the three guiding principles of query processing in SciDB are:

Tenet 1: aim for parallelism in all operations with as little data movement as possible.

This goal drives much of the design of the storage manager discussed in Section 2.6. Also, if an operation cannot be run in parallel because the data is poorly distributed, then SciDB will redistribute the data to enable parallelism. Hence, SciDB is fundamentally focused on providing the best response time possible for AQL utterances.

Second, the optimizers in relational DBMSs often choose poor query plans because their cost functions entail predicting the size of intermediate results. If a query has three or four cascading intermediate results, then these size estimates become wildly inaccurate, resulting in a potentially poor choice of the best query plan. Because SciDB queries are expected to be complex, it is imperative to choose a good plan.

To accomplish this goal, the SciDB optimizer processes the query parse tree in two stages. First, it examines the tree for operations that **commute**. This is a common optimization in relational DBMSs, as filters and joins are all commutative. The first step in the SciDB optimizer is to push the cheaper commuting operation down the tree. In our world, we expect many user defined array operations will not commute. For example, re-gridding a satellite imagery data set will rarely, if ever, commute with operations above or below it in the tree. Hence, this tactic may be less valuable than in a relational world.

The next step is to examine the tree looking for **blocking** operations. A blocking operation is one that either requires a redistribution of data in order to execute, or cannot be pipelined from the previous operation, in other words it requires a temporary array to be constructed. Note that the collection of blocking operations separates a query tree into sub-trees.

Tenet 2: Incremental optimizers have more accurate size information and can use this to construct better query plans.

The SciDB optimizer is **incremental**, in that it picks the best choice for the first sub-tree to execute. After execution of this sub-tree, SciDB has a perfect estimate for the size of the result, and can use this information when it picks the next sub-tree for execution.

Of course, the downside is that SciDB has a run-time optimizer. Such run-time overhead could not be tolerated in an OLTP world; however, most scientific queries run long enough that optimizer overhead is insignificant.

Tenet 3: Use a cost-based optimizer.

This third principle is to perform simple cost-based plan evaluation. Since SciDB only plans sub-trees, the cost of exhaustive evaluation of the options is not onerous.

Right now the optimizer is somewhat primitive, and focuses on minimizing data movement and maximizing the number of cores that can be put to work, according to tenet 1.

In summary, the optimizer/execution framework is the following algorithm:

```
Until no more {
        Choose and optimize next sub-plan
        Reshuffle data, if required
        Execute a sub-plan in parallel on a collection of local nodes
        Collect size information from each local node
}
```

2.6 Storage of Arrays

Basic Chunking

It is apparent that SciDB should **chunk** arrays to storage blocks using some (or even all) of the dimensions. In other words, a **stride** is defined in some or all of the dimensions, and the next storage block contains the next stride in the indicated dimensions. Multiple dimension chunking was explored long ago in [18] and has been shown to work well. Equally obviously, SciDB should chunk arrays across the nodes of a grid, as well as locally in storage. Hence, we distribute chunks to nodes using hashing, range partitioning, or a block-cyclic algorithm.

In addition, chunks should be large enough to serve as the unit of I/O between the buffer pool and disk. However, CPU time can often be economized by splitting a chunk internally into **tiles** as noted in [19]. In this way, subset queries may be able to examine only a portion of a chunk, and economize total time. Hence, we support a two level chunk/tile scheme.

One of the bread-and-butter operations in LSST is to examine raw imagery looking for interesting celestial objects (for example, stars). Effectively this is a data clustering problem; one is looking for areas of imagery with large sensor amplitude. In other areas of science, nearest neighbor clustering is also a very popular operation.

For example, looking for islands in oceanographic data or regions of snow cover in satellite imagery entails exactly the same kind of clustering.

To facilitate such neighborhood queries, SciDB contains two features. First, chunks in SciDB can be specified to **overlap** by a specific amount in each of several dimensions. This overlap should be the size of the largest feature that will be searched for. In this way, parallel feature extraction can occur without requiring any data movement. As a result, unlike parallel RDBMSs, which use non-overlapping partitions, SciDB supports the more general case.

At array creation time, stride and overlap information must be specified in the create array command. Hopefully, overlap is specified to be the largest size required by any array function that does feature extraction. Also, every user-defined array operation specifies the amount of overlap it requires to be able to perform parallel execution. If insufficient overlap is present, then SciDB will reshuffle the data to generate the required overlap.

Fixed or Variable Size Chunks

A crucial decision for SciDB was the choice of fixed or variable size chunks. One option is to fix the size of the stride in each dimension, thereby creating logically fixed size chunks. Of course, the amount of data in each chunk can vary widely because of data skew and differences in compressibility. In other words, the first option is fixed logical size but variable physical size chunks.

The second option is to support variable logical size chunks. In this case, one fills a chunk to a fixed-size capacity, and then closes it, thereby a chunk encompasses a variable amount of logical array real estate.

Variable chunk schemes would require an R-tree or other indexing scheme to keep track of chunk definitions. However, chunks would be a fixed physical size, thereby enabling a simple fixed size main memory buffer pool of chunks. On the other hand, fixed size logical chunks allow a simple addressing scheme to find their containers; however, we must cope with variable size containers.

We have been guided by [19] in deciding what to do. The "high level bit" concerns join processing. If SciDB joins two arrays, with the same fixed size chunking, then they can be efficiently processed in pairs, with what amounts to a generalization of merge-sort. If the chunking of the two arrays is different, then performance is much worse, because each chunk in the first array may join to several chunks in the second array. If chunking is different, then the best strategy may be to rechunk one array to match the other one, a costly operation as noted in [19].

This argues for fixed chunking, since frequently joined arrays can be identically chunked. That will never be the case with variable chunking. Hence, SciDB uses fixed logical size chunks. Right now, the user executing the Create Array command specifies the size of these chunks. Obviously, a good choice makes a huge difference in performance.

In summary, chunks are fixed (logical) size, and variable physical size. Each is stored in a container (file) on disk that can be efficiently addressed. The size of a chunk should average megabytes, so that the cost of seeks is masked by the amount of data returned.

There are several extensions to the above scheme that are required for good performance. These result from our implementation of versions, our desire to perform skew management, and our approach to compression. These topics are addressed in the next three sections.

Version Control

There are three problems which SciDB solves using version management. First, there is a lot of scientific data that is naturally temporal. LSST, for example, aims its telescope at the same portion of the sky repeatedly, thereby generating a time series. Having special support for temporal data seems like a good idea.

Second, scientists **never** want to throw old data away. Even when the old data is wrong and must be corrected, a new value is written and the old one is retained. Hence, SciDB must be able to keep everything.

The third problem deals with the "cooking" of raw data into derived information. In LSST, raw data is telescope imagery, and feature extraction is used to identify stars and other celestial objects, which constitute derived data. However, there is no universal feature extraction algorithm; different ones are used by different astronomers for different purposes. As such, LSST supports a "base line" cooking process, and individual astronomers can recook portions of the sky that they are interested in. Hence, astronomers want the base line derived information for the whole sky, except for the portions they have recooked. Such versioning of data should be efficiently supported.

To support the no overwrite model, all SciDB arrays are versioned. Data is loaded into the array at the time indicated in the loading process. Subsequent updates, inserts or bulk loads add new data at the time they are run, without discarding the previous information. As such, new information is written at the time it becomes valid. Hence, for a given cell, a query can scan particular versions referenced by timestamp or version number.

We now turn to the question: "How are array versions stored efficiently?" As updates or inserts occur to a chunk, we have elected to keep the most up-to-date version of the chunk stored contiguously. Then, previous versions of the chunk are available as a chain of "deltas" referenced from the base chunk. In other words, we store a given chunk as a base plus a chain of "**backwards deltas**". The rationale is that users usually want the most current version of a chunk, and retrieval of this version should be optimized. The physical organization of each chunk contains a reserved area, for example 20% additional space, to maintain the delta chain.

Arrays suffixed with a timestamp can be used in scan queries. Since we expect queries of the state of the array at a specific time to be very popular, we allow the select arrayname@T shorthand popularized in Postgres. If no specification is made, the system defaults to select arrayname@now.

We turn briefly to support for **named versions**. A user can request a named version to be defined relative to a given materialized array at time T. At this point, no storage is allocated, and the time T is noted in the system catalogs. As updates to the named version are performed, new containers for stored chunks are allocated and updates recorded in the new chunks. Multiple updates are backwards chained, just like in normal arrays. Over time, a **branch** is constructed, which is maintained as a chain of deltas based on the base array at time T. Clearly a tree of such versions is possible.

Query processing must start with the named version looking for data relevant to a given query. If no object exists in the version, its parent must be explored, ultimately leading back to the stored array from which the version was derived. This architecture looks much like configuration management systems, which implement similar functionality. A more elaborate version management solution is described in [20], and we may incorporate elements of this system into SciDB in the future.

Skew Management

Data in SciDB arrays may be extremely skewed for two reasons. As noted above, update traffic may be skewed. In addition, the density of non-null data may also be skewed. For example, consider a population database with geographic co-ordinates. The population density of New York City is somewhere around 1000000 times that of Montana.

There are two skew issues which we discuss in this section: what to do with chunks that have too little data, and what to do with chunks that have too much data.

Decades of system design experience dictates that it is advantageous to move data from disk to main memory in fixed size blocks (pages) versus variable size blocks (segments). The universal consensus was that fixed size blocks were easier to manage and performed better. Hence, SciDB has a fixed-size block model, where the main memory buffer pool is composed of a collection of fixed size slots containing "worthy" fixed size disk blocks. As noted above, this block size must be at least several megabytes.

If the user specifies a chunk size that results in a chunk containing more than B bytes, then the chunk must be split. We cycle through the chunking dimensions, splitting each in turn. As such, actual chunks will be some binary tree refinement of the user-specified chunk size. Unlike [19] which reports experiments on two chunk sizes, SciDB supports an arbitrary number of splits to keep the chunk size below B.

If a chunk is too small, because it is sparsely populated with data, then it can accommodate many updates before it fills. In the meantime, it can be co-located in a disk block of size B with neighboring sparse chunks. The storage manager current performs this "bin packing".

Compression

All arrays are aggressively compressed on a chunk-by-chunk basis. Sparse arrays can be stored as a list of non-null values with their dimension indexes, followed by prefix encoding. Additionally, value encoding of many data types is also profitable. This can include delta encoding, run-length encoding, subtracting off an average value, and LZ encoding. The idea is that the compression system will examine a chunk, and then choose the appropriate compression scheme on a chunk-by-chunk basis.

In addition, if the chunk is subject to intensive update or to small geographic queries, then it will spend much overhead decompressing and recompressing chunks to process either modest queries or updates. In this case, it makes sense to divide a chunk into **tiles**, and compress each tile independently. In this way, only relevant tiles need to be decompressed and recompressed to support these kinds of queries and updates. Hence, tiling will result in better performance on workloads with many small updates and/or small geographic queries. On a chunk-by-chunk basis, the compression system can optionally elect to tile the chunk.

Also, we have noted that some SciDB environments are CPU limited, and compressing and decompressing chunks or tiles is the "high pole in the tent". In this case, SciDB should switch to a lighter weight compression scheme.

The compression system is inside the storage manager and receives a new chunk or tile to encode. After encoding, the result is obviously variable sized, so the compression engine controls the splitting of chunks described above as well as the packing of small chunks into storage blocks mentioned in the previous section.

Uncertainty

Essentially all science data is uncertain. After numerous conversations with scientists, they pretty much all say:

Build in the common use case (normal distributions) to handle 80% of my data automatically.

My other 20% is specific to my domain of interest, and I am willing to write error analysis code in my application to handle this.

As such we have implemented both uncertain and precise versions of all of the common data types. Operating on precise data gives a precise answer; operating on uncertain data yields an uncertain answer. The uncertain versions of SciDB operations "do the right thing" and carry along errors in the internal calculations being performed. Moreover, a challenge to the compression system is to be smart about uncertainty. Specifically, most uncertain values in a chunk will have the same or similar error information. Hence, uncertainty information can be aggressively compressed

Notice that SciDB supports uncertain cell values but not uncertain dimensions. That functionality would require us to support approximate joins, which is a future extension.

Provenance

A key requirement for most science data is support for provenance. The common use case is the ability to point at a data value or a collection of values and say "show me the derivation of this data". In other words, the data looks wrong, and the scientist needs to trace backwards to find the actual source of the error. Once, the source has been identified, it should be fixed, of course using the no-overwrite processing model. Then, the scientist wants to trace forward to find all data values that are derived from the incorrect one, so they can also be repaired.

In other words, SciDB must support the ability to trace both backward and forward. Some systems support coarse provenance (for example at the array level) that allow this functionality only for arrays, not cells. Since SciDB expects some very big arrays, this granularity is unacceptable. Other systems, e.g. Trio [21] store provenance by associating with each output value the identifier of all input values that contributed to the calculation. This approach will cause the data volumes to explode. For example, matrix multiply generates a cell from all the values in a particular source row and source column. If an array is of size M, then the provenance for matrix multiply will be of size $M^{**}3$. This is obviously not an engineering solution.

Our solution is to allow database administrators to specify the amount of space they are willing to allocate for provenance data. The SciDB provenance system chooses how to best utilize this space, by varying the granularity of provenance information, on a command-by-command basis. The details of this system are discussed in [22].

Discussions with LSST personnel indicate a willingness to accept approximate provenance, if that can result in space savings or run time efficiency. For example, many LSST operations are "region constrained", i.e. the cell value that results from an operation comes from a constrained region in the input array. If true, approximate provenance can be supported by just recording the centroid of this region and its size. Often, the centroid is easily specified by a specific mapping from input to output, thereby further reducing the amount of provenance information that must be kept. The details of our approximate provenance are also discussed in [22].

In-situ Data

Most of the scientists we have talked to requested support for in-situ data. In this way, they can use some of SciDB's capabilities without having to go through the effort of loading their data. This would be appropriate for data sets that are not repeatedly accessed, and hence not worth the effort to load.

We are currently designing an interface (wrapper) that will allow SciDB to access data in other formats than SciDB natively understands. The details of how to do this as well as how to make the optimizer understand foreign data are still being worked out.

3 Summary, Status Performance, and Related Work

3.1 Related Work

SciDB is a commercial, open-source analytical database oriented toward scientific applications. As such, it differs from RDBMSs, which must simulate arrays on top of a table data model. The performance loss in such a simulation layer may be extreme [6]. The loss of performance in linear algebra operations may be especially daunting [23]. Also, most RDBMSs have trouble with complex analytics, because they are expressed on arrays, not tables. SciDB implements such operations directly, whereas RDBMSs, such as GreenPlum and Netezza, must convert a table to an array inside user-defined functions, then run the analytic code, and convert the answer back to a table to continue processing. Such out-and-back conversion costs do not need to be paid by SciDB. A similar comment can be made about interfaces between R [24] and RDBMSs. In addition, RDBMSs do not support multi-dimensional chunked storage, overlapping chunks, uncertainty, versions or provenance.

MonetDB [25] has an array layer [26], implemented on top of its column store table system. All of the comments in the previous paragraph apply to it. Similarly RasDaMan [27] is an array DBMS. However, it is implemented as an application layer that used Postgres for blob storage. As such, it is implementing multi-dimension chunking in an application layer external to the DBMS. It also lacks overlapping chunks, version control, uncertainty and provenance.

There are a myriad of statistical packages, including R [24], S [28], SAS [29], ScaLAPACK [17], and SPSS [30]. All of these perform complex analytics, often on a single node only, but perform no data management. SciDB is an integrated system to provide both data management and complex analytics.

Status and Summary

At the time of the SSDBM conference, SciDB version 11.06 will be available for download. SciDB development is backed by the commercial company Paradigm4 who will provide support as well as offer extensions for the commercial marketplace (monitoring tools, proprietary function libraries, etc.)

Development is proceeding with a global team of contributors across many time zones. Some are volunteers but at this early stage, most are employees of Paradigm4, including the engineering manager and chief architect. QA is being performed by volunteers in India and California. User-defined extensions are underway in Illinois, Massachusetts, Russia, and California.

References

1. http://arxiv.org/abs/0805.2366
2. Becla, J., Lim, K.-T.: Report from the First Workshop on Extremely Large Databases. Data Science Journal 7 (2008)
3. Szalay, A.: Private communication
4. Branco, M., Cameron, D., Gaidioz, B., Garonne, V., Koblitz, B., Lassnig, M., Rocha, R., Salgado, P., Wenaus, T.: Managing ATLAS data on a petabyte-scale with DQ2. Journal of Physics: Conference Series 119 (2008)
5. Szalay, A.: The Sloan Digital Sky Survey and Beyond. In: SIGMOD Record (June 2008)
6. Cudre-Mauroux, P., et al.: A Demonstration of SciDB: a Science-oriented DBMS. VLDB 2(2), 1534–1537 (2009)
7. Becla, J., Lim, K.-T.: Report from the Second Workshop on Extremely Large Databases, http://www-conf.slac.stanford.edu/xldb08/, http://www.jstage.jst.go.jp/article/dsj/7/0/1/_pdf
8. Becla, J., Lim, K.-T.: Report from the Third Workshop on Extremely Large Databases, http://www-conf.slac.stanford.edu/xldb09/
9. Becla, J., Lim, K.-T.: Report from the Fourth Workshop on Extremely Large Databases, http://www-conf.slac.stanford.edu/xldb10/
10. Cudre-Maroux, P., et al.: SS-DB: A Standard Science DBMS Benchmark (submitted for publication)
11. http://www.hdfgroup.org/HDF5/
12. http://en.wikipedia.org/wiki/APLprogramming_language
13. http://en.wikipedia.org/wiki/Functional_programming
14. http://kx.com/
15. Stonebraker, M., Rowe, L.A., Hirohama, M.: The Implementation of POSTGRES. IEEE Transactions on Knowledge and Data Engineering 2(1), 125–142 (1990)
16. http://developer.postgresql.org/docs/postgres/xaggr.html
17. http://www.netlib.org/ScaLAPACK/

18. Sarawagi, S., Stonebraker, M.: Efficient organization of large multidimensional arrays. In: ICDE, pp. 328–336 (1994),
    `http://citeseer.ist.psu.edu/article/sarawagi94efficient.html`
19. Soroush, E., et al.: ArrayStore: A Storage Manager for Complex Parallel Array Processing. In: Proc. 2011 SIGMOD Conference (2011)
20. Seering, A., et al.: Efficient Versioning for Scientific Arrays (submitted for publication)
21. Mutsuzaki, M., Theobald, M., de Keijzer, A., Widom, J., Agrawal, P., Benjelloun, O., Das Sarma, A., Murthy, R., Sugihara, T.: Trio-One: Layering Uncertainty and Lineage on a Conventional DBMS. In: Proceedings of the 2007 CIDR Conference, Asilomar, CA (January 2007)
22. Wu, E., et al.: The SciDB Provenance System (in preparation)
23. Cohen, J., et al.: Mad Skills: New Analysis Practices for Big Data. In: Proc. 2009 VLDB Conference
24. `http://www.r-project.org/`
25. `http://monetdb.cwi.nl/`
26. van Ballegooij, A., Cornacchia, R., de Vries, A.P., Kersten, M.L.: Distribution Rules for Array Database Queries. In: Andersen, K.V., Debenham, J., Wagner, R. (eds.) DEXA 2005. LNCS, vol. 3588, pp. 55–64. Springer, Heidelberg (2005)

Location-Based Instant Search

Shengyue Ji and Chen Li[*]

University of California, Irvine

Abstract. Location-based keyword search has become an important part of our daily life. Such a query asks for records satisfying both a spatial condition and a keyword condition. State-of-the-art techniques extend a spatial tree structure by adding keyword information. In this paper we study location-based instant search, where a system searches based on a partial query a user has typed in. We first develop a new indexing technique, called filtering-effective hybrid index (FEH), that judiciously uses two types of keyword filters based on their selectiveness to do powerful pruning. Then, we develop indexing and search techniques that store prefix information on the FEH index and efficiently answer partial queries. Our experiments show a high efficiency and scalability of these techniques.

1 Introduction

Location-based services have become an important part of our daily life. We use online maps to search for local businesses such as stores and movie theaters; we use Yelp.com to search for restaurants; and we use Twitter to search for nearby tweets. On the back-end of these systems there are spatial records with keyword descriptions. For instance, Figure 1 shows an example data set that includes business listings in Manhattan, New York, such as museums, schools, and hospitals. Each record has a *Name* value (e.g., Metropolitan Museum of Art) and a *Location* value including the latitude and longitude of the entity (e.g., ⟨40.7786, -73.9629⟩). The entities are also shown as points on the map in the figure. These applications need to answer *location-based keyword queries*, a.k.a. spatial keyword search. A query includes a location, such as the point P or the area R in Figure 1. The query also includes keywords, and asks for answers that match the keywords and are close to the spatial location. Example queries include "finding movie theaters close to downtown New York" and "finding Japanese restaurants near the Disneyland in California."

Instant keyword search has become popular in recent years. It returns the search results based on partial query keywords as a user is typing. Users of an instant search system can browse the results during typing. In this paper, we study location-based instant search, a search paradigm that combines location-based keyword search with instant search. Figure 2 shows an interface of location-based instant search on our example data set. As the user types a query letter by letter, the system responds to the partial queries and returns the results to

---

[*] The authors have financial interest in Bimaple Technology Inc., a company currently commercializing some of the techniques described in this publication.

J.B. Cushing, J. French, and S. Bowers (Eds.): SSDBM 2011, LNCS 6809, pp. 17–36, 2011.

ID	Name	Location
1	Trinity School	<40.7903, -73.9707>
2	Metropolitan Hospital Center	<40.7846, -73.9441>
3	Solomon R. Guggenheim Museum	<40.7831, -73.9596>
4	Brearly School	<40.7783, -73.9501>
5	Metropolitan Museum of Art	<40.7786, -73.9629>
6	American Museum of Natural History	<40.7791, -73.9730>
7	Manhattan Church of Christ	<40.7766, -73.9613>
8	Mt Sinai Hospital	<40.7901, -73.9538>
9	Cooper Hewitt Museum	<40.7844, -73.9580>
...	...	...

Fig. 1. Spatial keyword records of business listings. The map on the right shows the listings as well as a query area R and a query point P.

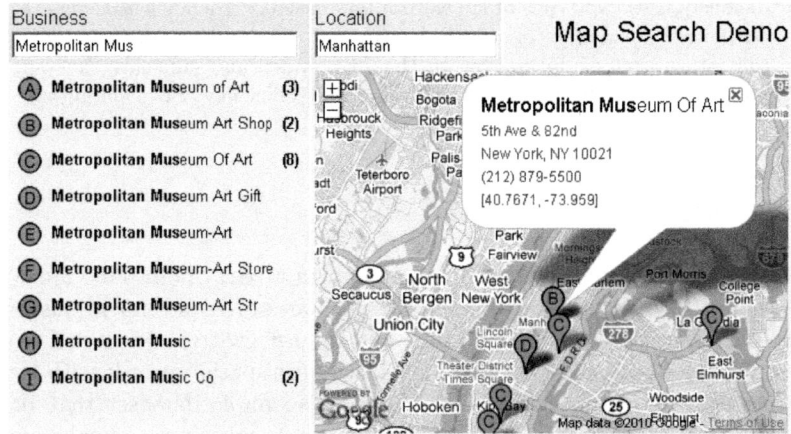

Fig. 2. Location-based instant search

the user, listing them and plotting them on the map. When the system receives the partial query "Metropolitan Mus", it returns businesses that are nearest to Manhattan (represented as a point returned by the geo-coder), having the keyword Metropolitan and a keyword with Mus as a prefix.

In these systems, it is critical to answer queries efficiently in order to serve a large amount of query traffic. For instance, for a popular map-search service provider, it is not uncommon for the server to receive thousands of queries per second. Such a high query throughput requires the search process to be able to answer each search very fast (within milliseconds). Instant search would further increase the server workload. Therefore, we focus on searching using in-memory index in this paper to achieve a high efficiency for instant search. There are recent studies on supporting location-based keyword queries. A common solution is extending a spatial tree structure such as R-tree or R*-tree by adding keyword information on the tree nodes. For example, Figure 3 in Section 2 shows such an index structure, in which each node has a set of keywords and their children with these keywords. We can use the keyword information to do efficient pruning during the traversal of the tree to answer a query, in addition to the pruning based on the spatial information in the tree.

In this paper we first present an index structure called "filtering-effective hybrid" (FEH) index. It judiciously uses two types of keyword filters in a node of a spatial tree based on the selectiveness of each keyword. One filter, called *child filter*, maps keywords and their corresponding children nodes. Another filter, called *"object filter"*, maps keywords to their corresponding records in the subtree of the node. During a traversal of the FEH index tree, the object filter at each node allows us to directly retrieve records for these keywords in the filter, thus bypassing those intermediate nodes in the subtree. Next we study how to answer location-based instant queries on spatial data, i.e., finding answers to a query as the user is typing the keywords character by character. We utilize existing indexing techniques and FEH to answer queries. We develop a technique to store prefix filters on spatial-tree nodes using a space-efficient representation. In addition, we develop a method to compress the representation in order to further reduce the index size.

We show that our techniques can be applied to efficiently support both range queries (where the user specifies a spatial area) and nearest-neighbor queries (where the user wants to find objects closest to a location). Our techniques can also reduce the index size. Such reduction can minimize the hardware cost. We have conducted a thorough experimental study to evaluate these techniques. The results show that, our techniques can support efficient instant search on large amounts of data. For instance, we are able to index a data set of 20 million records in memory on a commodity machine, and answer a location-based instant keyword search in microseconds.

The rest of the paper is organized as follows. In Section 2 we give the problem formulation of location-based instant search, and describe a state-of-the-art approach for answering location-based keyword queries. In Section 3, we present our FEH index. In Section 4, we study how to efficiently answer location-based instant queries, using existing index techniques and FEH. We report our experimental results in Section 5, and conclude in Section 6.

1.1   Related Work

Location-based keyword search received a lot of attention recently. Early studies utilize a keyword index (inverted lists) and a spatial index (such as R-tree [10] or R*-tree [4]) separately [22,7,18]. These proposed methods answer a query by using the keyword and spatial indexes separately. A main disadvantage of these approaches is that filtering on the spatial and keyword conditions is not achieved at the same time. Therefore, the pruning power cannot be fully utilized.

There are recent studies on integrating a spatial index with a keyword index [11,9,8,20,21,19,14]. The proposed methods add a keyword filter to each node in the spatial tree node that describes the keyword information in the subtree of that node. (More details are explained in Section 2.) This structure allows keyword-based pruning at the node. A filter can be implemented as an inverted list, a signature file [9], or a bitmap of the keywords in the subtree. The work in [11] constructs a global inverted index to map from keywords to tree nodes that have these keywords. These studies consider the problem of range search, nearest neighbor search [12,17], or top-k search [8]. The work in [8] proposed two ideas to improve the performance of searching with this type of indices: using object similarities to influence the structure of the tree index, and creating clusters for similar objects and indexing on them. The work in [20,21] studied how to find records that are close to each other and match query keywords. The work in [19,1] studied approximate keyword search on spatial data. The results in [11,9,8] show that these "integrated" index structures combining both conditions outperform the methods using two separate index structures. Therefore, in this paper, we focus on the "integrated" index structure as the baseline approach to show the improvement of our techniques.

Instant keyword search (a.k.a. interactive search, auto-complete search, or type-ahead search) has become in many search systems. Bast et al. [3,2] proposed indexing techniques for efficient auto-complete search. The studies in [13,6] investigated how to do error-tolerant interactive search. Li et al. [15] studied type-ahead search on relational databases with multiple tables. These studies did not focus on how to answer location-based queries. Some online map services, such as Yelp and Google Maps, recently provide location-based instant search interfaces. To our best knowledge, there are no published results about their proprietary solutions.

2   Preliminaries

In this section we formulate the problem of location-based instant search, and present a state-of-the-art approach for answering location-based keyword queries.

2.1   Data

Consider a data set of spatial keyword records. Each record has multiple attributes, including a spatial attribute A_S and a keyword attribute A_W. For simplicity, we assume that the data set has one keyword attribute, and our techniques can be extended to data sets with multiple keyword attributes. The value of the spatial attribute A_S of a record represents the geographical location of the record. The value can be a rectangle, or a point with a latitude and a longitude. The keyword attribute A_W is a textual string that can be tokenized into keywords. Figure 1 shows an example data set, and we will use it to explain the related techniques throughout this paper.

2.2   Location-Based Instant Search

A location-based instant search combines spatial search with keyword search using the AND semantics. That is, we want to retrieve records that satisfy both spatial and keyword conditions. We consider the following types of queries.

Range query: An instant range query consists of a pair $\langle R, W \rangle$, where R is a geographical area (usually represented as a rectangle or a circle), and W is a set of keywords $W = \langle w_1, w_2, \ldots, w_l \rangle$. The answer to the query is the set of records whose spatial attributes A_S geographically overlap with R, and whose keyword attributes A_W contain $w_1, w_2, \ldots, w_{l-1}$ and a keyword with w_l as a prefix[1]. For example, if we define a rectangle $R = \langle 40.776, 40.783, -73.976, -73.956 \rangle$ using latitudes and longitudes (shown in the map of Figure 1 as the dashed rectangle), the query $\langle R, \{\texttt{Christ}, \texttt{Chu}\} \rangle$ is to ask for entities located within R that have the keyword $\texttt{Christ}$ and the prefix $\texttt{Chu}$ in their name (e.g., record 7).

kNN query: An instant k-nearest-neighbor (kNN) query for a positive integer k is a pair $\langle P, W \rangle$, where P is a geographical point (e.g., the current location of the user), and W is a set of keywords. The answer to the query is the set of top-k records that are geographically closest to P, having the keywords $w_1, w_2, \ldots, w_{l-1}$ and a keyword with w_l as a prefix in their A_W value. For instance, the 2-NN query $\langle P, \{\texttt{Muse}\} \rangle$ asks for top-2 entities that are nearest to $P = \langle 40.786, -73.957 \rangle$ (shown in the map of Figure 1), having the keyword $\texttt{Muse}$ as a prefix in their name value (e.g., records 3 and 9).

2.3   Baseline Approach for Location-Based Keyword Search

We describe a baseline approach for answering location-based keyword queries presented in previous studies [11,9,8,20]. It uses an index that extends a spatial tree index such as an R*-tree, by adding keyword filters to the tree nodes. A keyword filter at a node is a summary of the keywords in the records in the subtree of the node. We say an R*-tree node n and a keyword w are *consistent* if there exists at least one record in the subtree of node n that has keyword w. The purpose of using filters is to prune *inconsistent* branches at this node when traversing the tree to answer queries with keyword conditions. We can implement filters using multi-maps[2] from keywords to their consistent children (for this purpose, the "children" of a leaf node are its records). In the literature, the multi-values associated with a keyword w are also referred as the inverted list of w or the posting list of w.

Figure 3 presents the baseline index built on our data set. The minimum bounding rectangles (MBRs) of the tree nodes and the locations of the records are shown on the top-left map. Each node stores a multi-map from the keywords to their sets of consistent children. Notice that the multi-map is not necessarily stored in the node. For instance, in our in-memory implementation, the multi-map is stored in a separate buffer, linked to the node through a pointer. When we visit the node p when answering the query $\langle R, \{\texttt{Church}\} \rangle$, we only need to traverse the consistent children nodes (leaf c), as indicated by the keyword filter on node p. Record 7 is then retrieved from the leaf c using the keyword filter on the node, as an answer to this query. The main advantage of using this index is that we can do pruning on both the spatial condition and the keyword condition simultaneously during a search.

---

[1]   Our techniques can also be extended to treat all query keywords as prefixes.

[2]   A multi-map is a generalized associative array that maps keys to values. In a multi-map, multiple values can be associated for a key.

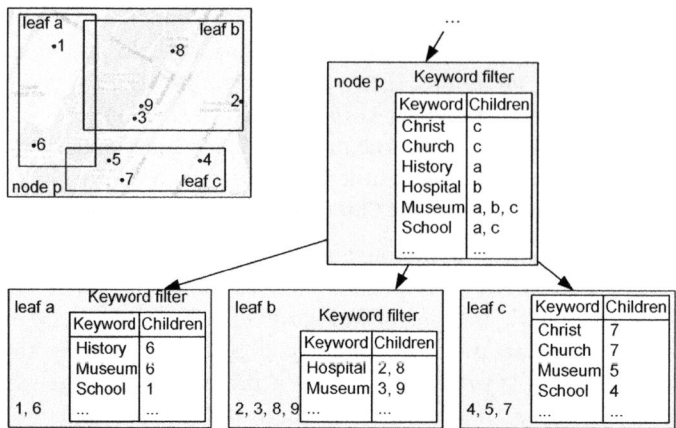

Fig. 3. A tree-based index to support location-based keyword search. The map shows the locations of the records and minimum bounding rectangles (MBRs) of the nodes.

3   Filtering-Effective Indexing

In this section we present an index that improves the baseline index for both complete queries (in which each keyword is treated as a complete keyword) and instant queries, by using more effective filtering[3]. We first show the index and search technique for complete queries in this section, and then adapt it for instant queries in Section 4. We use an R*-tree as an example to explain the algorithms, since it is one of the state-of-the-art indices for geographical data. Our techniques can also be used in other tree-based indices.

3.1   Keyword Selectiveness

Our index is called filtering-effective hybrid (FEH) index. The main idea is to treat the keywords at a tree node differently when creating the filters, so that we can do direct access to records for those very selective keywords during a traversal of the R*-tree. Specifically, we define the following selectiveness es of a keyword w on a node n:

Record selectiveness: The keyword w is *record selective* on the node n, when there are only a few records in the subtree of n that have w. We use a threshold t_r to determine whether w is considered to be record selective on n, i.e., there are at most t_r records in the subtree of n that have w. In the example of Figure 3, let t_r be 1. Since there is only one record (record 7) that has the keyword Church in the subtree of node p, Church is record selective on p.

Child selectiveness: The keyword w is *child selective* on the node n, if n has at least one child that is inconsistent with w. When n is traversed to answer a

---

[3] This technique is orthogonal to and can be applied to the work in [8]. For simplicity we present it by improving the baseline index.

query with w, only those consistent children need to be visited, and the rest can be pruned. For example, the keyword Hospital is child selective on the node p in Figure 3, because only the child node b of p is consistent with the keyword, and the other two children a and c can be pruned.

Non-selectiveness: The keyword w is *not selective* on the node n, if w is neither record nor child selective. In this case, the baseline index would gain no additional pruning power on n by adding w to the filter. For instance, in Figure 3, the keyword Museum is not selective on p, since all the children of p are consistent with the keyword (not child selective), and there are many records in the subtree of p that have the keyword (not record selective).

3.2   FEH Index

Based on the analysis above, we construct an FEH index by introducing two types of keyword filters on each tree node for record-selective keywords and child-selective keywords. An *object filter* ("O-filter" for short) on an internal node n in the R*-tree is a multi-map from some record-selective keywords directly to the records that have these keywords. We say that such keywords are *bypassing* at node n in the subtree of n. Using an O-filter we can directly retrieve records for these record-selective keywords without traversing the subtree. The multi-values of a keyword in an O-filter can be stored as an array of the record ids with this keyword. Note that we build O-filters only on those internal nodes.

A *child filter* ("C-filter" for short) on a node n is a multi-map from some child-selective keywords to their consistent children. The C-filter is used to prune inconsistent children of n for child-selective keywords. Considering that the total number of children of n is bounded by the maximum branching factor of the node, which is often relatively small, the multi-values of the multi-map can be stored as a bit vector, in which each bit in the vector of a keyword indicates whether the corresponding child is consistent with the keyword. Notice that a non-selective keyword is absent from both filters, which implies that all children of n need to be visited. Figure 4 shows an example FEH index. On the internal node p, for instance, the C-filter contains the keywords Hospital and School, which are mapped to their consistent children nodes. The O-filter has the keywords Christ, Church, and History, with lists of records that have these keywords.

In the FEH index, a keyword w is added to the O-filter and the C-filter on a node n, based on its selectiveness on n. (1) If the keyword w is record selective and not bypassing in n's ancestors, then it is added to the O-filter on n. (2) Otherwise, if w is child selective, then it is added to the C-filter on n. (3) Otherwise, w is absent in both filters on n. For instance, Museum is absent in the filters of node p in Figure 4, since it is not selective.

3.3   Searching with FEH Index

We present a search algorithm using an FEH index on a range query $\langle R, W \rangle$. The search algorithm for a kNN query $\langle P, W \rangle$ is similar by replacing the depth-first

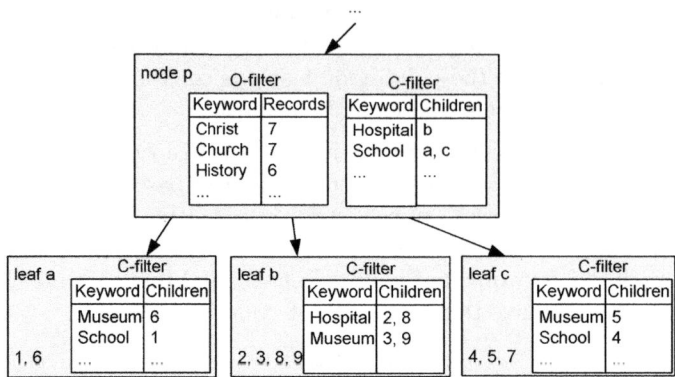

Fig. 4. An FEH index based on the tree structure of Figure 3

traversal with a priority-queue driven traversal. Using minimum distance to P as the priority.

Using a global vocabulary to check query keywords: We use a vocabulary to keep all the keywords in the data set. For each query, we first check if each of its keywords appears in the vocabulary. The search continues only if all the query keywords are found in the vocabulary. We can also use the vocabulary to map query keywords from strings to integer ids.

Range search: We show how to do a range search with the FEH index. We recursively access tree nodes and return records that satisfy the query conditions R and W using the O-filter and the C-filter on the input node. (1) We first try to find a query keyword of W in the O-filter. If such a query keyword can be found in the O-filter, we retrieve all the records on the inverted list of this keyword, and verify whether they satisfy the keyword and spatial conditions. For those that pass the verification, we add them to the result. If we can successfully use the O-filter, we prune the whole subtree of the current node without any further processing. (2) Otherwise, if the keyword appears in the C-filter, for the current node's children that are spatially consistent with R, we prune them using the keyword's consistent children. The resulting set of children are those consistent with both the spatial condition and the keyword condition. (3) If none of the keywords in W is found in the O-filter and the C-filter, we consider all children of the current node that satisfy the spatial condition R. If the current node is an internal node, we then recursively visit its survived children nodes and retrieve answers from them; otherwise, we add all the survived records to the result.

Advantages: Compared to the baseline index, FEH can not only reduce the query time, but also reduce the index size. Specifically, (1) it can effectively prune an entire subtree if a query keyword is record selective on the root of the subtree. In Figure 4, the number of churches in the subtree of p is relatively small. When processing the query $\langle R, \{\texttt{Church}\} \rangle$ on node p, with the O-filter on p we can directly answer the query from the list of record Ids of the keyword

Church (i.e., record 7) without visiting the subtree of p. (2) FEH reduces storage space at a node n by ignoring the keywords that are bypassing in an ancestor of n or are not selective. For example, on node p in Figure 4, Museum is not selective as it is consistent with all the children of p, and appears in many records in the subtree of p. We save space by not storing it in the filters on p.

4   Answering Location-Based Instant Queries

In this section we study answering location-based instant queries using the baseline index and the FEH index. We present a technique that represents the prefix filters efficiently in Section 4.1. Then we study how to compress the prefix filters in Section 4.2.

We extend the baseline index and the FEH index to answer instant queries. We first use the baseline index as an example to illustrate the techniques. The goal is to support pruning of branches that will not give answers satisfying the keyword conditions. During the search, we need to utilize a filter so that only children nodes that have a record with the query prefix are visited. We extend the baseline index by building the filters on the *consistent prefixes* as prefix filters. We say a keyword w' extends a string w, if w is a prefix of w'. A string w is a *consistent prefix* of an R*-tree node n if there exists a consistent keyword w' in the subtree of n that extends w. The prefix filter on node n maps from n's consistent prefixes to their lists of n's consistent children. For instance, on the node p in our running example, its children nodes a and c are consistent with the prefix Sch. To answer a query with Sch, only nodes a and c are traversed.

By building filters on prefixes we can use the baseline index and the FEH index to efficiently answer instant queries, since pruning on prefix and spatial conditions happens at the same time during the search. A main challenge here is that, as prefixes (in addition to complete keywords) are added to the filters, the size of the index could dramatically increase. Like most instant search systems, it is crucial to make the index reside in memory in order to achieve a high query performance. Therefore, reducing the index size is important. Next we present techniques that efficiently store the prefix filters.

4.1   Representing Prefix Filters Efficiently

In this section we present an efficient representation of the prefix filters. A straightforward approach is to add all consistent prefixes into a prefix filter. This approach requires a lot of space. For instance, the consistent prefixes of the keyword Hospital on node p in Figure 3 are H, Ho, Hos, Hosp, Hospi, Hospit, Hospita, and Hospital.

Next, we show that many prefixes can be combined with each other using a radix tree [16]. The radix tree is built on *all* the keywords in the data set, with each edge of the tree labeled with one or more characters. Each node of the radix tree represents one or more prefixes in the data set. For example, Figure 5 shows the radix tree on the keywords Christ\$, Church\$, History\$, Hospital\$, Museum\$, Scheme\$, and School\$. The character \$ is appended to the end of each

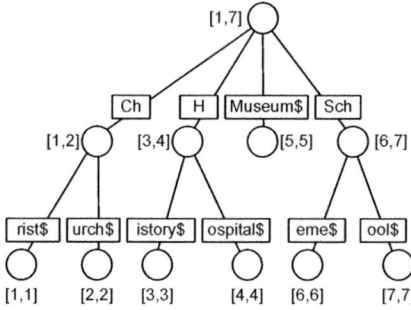

Fig. 5. Radix tree built on the global vocabulary

keyword to distinguish it from the same string as a prefix, assuming it is not in the alphabet. The leftmost leaf node represents prefixes Chr, Chri, Chris, Christ, and Christ$. We can see that all the prefixes represented by a radix tree node have the same set of extended keywords in the dataset. Therefore, they have the same set of consistent children of an R*-tree node and records. For instance, the prefixes Mus and Museu, represented by the same radix tree node, have the same set of consistent children of p (nodes a, b, and c). We first illustrate our technique using the baseline index. Instead of adding all consistent prefixes to the filter, our technique adds consistent radix tree nodes to the prefix filter, where each radix tree node can represent multiple prefixes.

To efficiently store the prefix filter on each R*-tree node, we assign numerical ids to all the keywords sorted in their lexicographical order. A prefix in the data set can then be represented as an interval $[Id_{min}, Id_{max}]$, where Id_{min} is the minimum id of the keywords that extend the prefix, and Id_{max} is their maximum id. This representation has the following benefits. (1) All the prefixes represented by a radix tree node have the same interval. For instance, prefixes C and Ch, represented by the same node in Figure 5, have the same interval $[1, 2]$. Therefore, we can use an interval to represent a radix tree node to be stored in the prefix filter, instead of storing all the prefixes that share the same interval. (2) It is easy to test whether the node represented by one interval I_1 is an ancestor of the node represented by another interval I_2, by simply checking if I_1 contains I_2. For instance, the radix tree node for $[3, 4]$ (H) is an ancestor of that for $[3, 3]$ (History$) since the former interval contains the latter. We also say "$[3, 4]$ is a prefix of $[3, 3]$", or "$[3, 3]$ extends $[3, 4]$" interchangeably. This property is used later in the compressed prefix filter, where we need to test whether a prefix in the filter is a prefix of a query string. (3) The lexicographical order of the prefixes is defined as follows. For two intervals $I_1 = [Id_{min1}, Id_{max1}]$ and $I_2 = [Id_{min2}, Id_{max2}]$, we define:

$$\begin{cases} I_1 < I_2, \text{ if } Id_{min1} < Id_{min2} \vee (Id_{min1} = Id_{min2} \\ \qquad \wedge Id_{max1} > Id_{max2}); \\ I_1 = I_2, \text{ if } Id_{min1} = Id_{min2} \wedge Id_{max1} = Id_{max2}; \\ I_1 > I_2, \text{ otherwise.} \end{cases}$$

For example, $[3,4] < [3,3]$, because they tie on Id_{min}, and $[3,4]$ has a bigger Id_{max}. This order allows us to store the intervals in a sorted array. It also allows us to use a binary search to efficiently locate the longest prefix of the query string in the array.

Within an R*-tree node, we store all its consistent prefixes in a sorted array using their interval representations and their corresponding consistent children. Figure 6 shows the prefix filter stored as sorted intervals on the R*-tree node p. The interval $[3,4]$ for the prefix H is stored in the prefix filter, with its set of consistent children (nodes b and c). To answer a prefix query using the prefix filters, we can first lookup from the global vocabulary (the radix tree) to convert the query prefix to its interval representation. When accessing an R*-tree node n, we locate the corresponding interval in the prefix filter on n by performing a binary search on the sorted array of intervals, and then retrieving the consistent children nodes to visit.

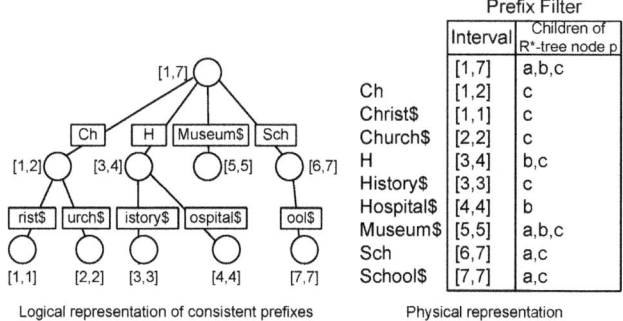

Fig. 6. The prefix filter on the R*-tree node p

FEH prefix filters: The techniques presented in Section 4.1 can be applied to the FEH index as well. Figure 7 shows the prefix filters with intervals on node p in the FEH prefix index in our running example. For instance, the record-selective prefix Ch is added to the O-filter on node p. It is stored as its corresponding interval $[1,2]$, with the list of the records that have the prefix (record 7). The intervals added to a filter on the node p form a radix forest (a set of disjoint radix trees), shown on the left side of the filter in the figure.

When traversing an R*-tree node to answer a query, we first lookup the query interval from the O-filter. If the query interval is found in the O-filter, we verify the records from its list without accessing this subtree. Otherwise, we lookup the query interval from the C-filter, and visit the consistent R*-tree children. For instance, to answer the query with the prefix Ch, we lookup its corresponding interval $[1,2]$ from the global vocabulary. When visiting the R*-tree node p, we lookup $[1,2]$ from the O-filter, and then directly retrieve the records without visiting any other children nodes of p.

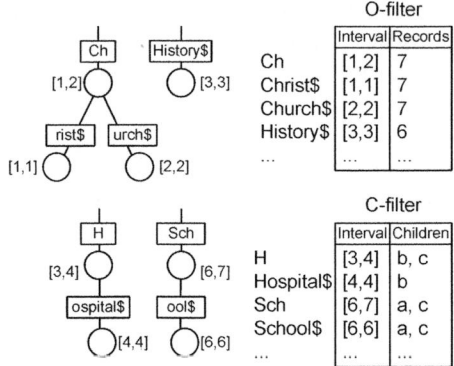

Fig. 7. Filters on node p in the FEH prefix index

4.2   Compressing Prefix Filters

In this section we present a technique for further compressing the prefix filters. An interesting property of the prefix filters is the following: For two prefixes w_1 and w_2 in the same filter, if w_1 is a prefix of w_2, then the consistent set of children or records of w_1 is a superset of that of w_2. This property allows us to compress the prefix filter by removing w_2 from the filter, since we can use the superset of w_1 to answer a query of w_2, with possible false positives. Formally, we require an interval to hold the following condition in order to remain in a compressed filter on node n: the interval does not extend another interval in the same prefix filter. Otherwise, the interval can be removed from the prefix filter. For example, Figure 8(a) shows the radix forest of record-selective prefixes. $[1,1]$ and $[2,2]$ are removed from the compressed O-filter, as they extend another interval $[1,2]$ in the same filter. The O-filter is physically stored as in Figure 8(b).

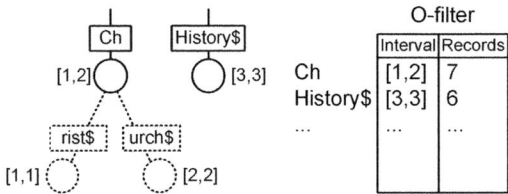

(a) Logical representation. (b) Physical represen-
Dashed nodes are removed. tation.

Fig. 8. Compressed O-filter on node p

The technique of compressing prefix filters discussed in Section 4.2 can be utilized on both C-filters and O-filters. Here we focus on compressing the O-filters, as experiments show that the size of O-filters is much larger compared to that of C-filters. Another reason is that the performance penalty is large when the compression method is applied on the C-filters, since a traversal of a query would go to branches not consistent with the query.

We can apply this technique on O-filters by enforcing the condition on some instead of all intervals to support flexible size reduction. Our experimental results show that the performance penalty on O-filters is very small even if the condition is enforced on all prefixes.

After we remove some prefixes from the filters, we need to modify the search algorithm for answering the query $\langle R, W \rangle$. The existence test of the query keyword w in the O-filter is replaced with finding a prefix w' of w in the O-filter. It can be achieved by doing a binary search on the interval array for w. Either w is returned, or the largest string w' that is smaller than w is returned. The returned w' is a prefix of w if w has a prefix in the filter. For instance, a binary search for $[2, 2]$ (Church$) on the interval array of the O-filter on node p returns $[1, 2]$ (Ch). We verify that $[1, 2]$ is a prefix of $[2, 2]$, since $[1, 2]$ contains $[2, 2]$. We use the records for $[1, 2]$ (record 7) to answer the query. The verification is performed on the records that have w' as a prefix, by checking whether they have w as a prefix. Since the set of records with w' as a prefix is a superset of the records with w as a prefix, retrieving and verifying records from the list of w' guarantees to retrieve all the answers. False positives can be eliminated by the verification. As the size of the list for w' is bounded by the record-selectiveness threshold t_r, we can answer queries using compressed O-filters with limited performance penalty, especially when t_r is selected properly. This fact is shown in our experimental evaluations.

5   Experimental Evaluation

In this section, we report our experimental results of the proposed techniques. All the algorithms (including the R*-tree) were implemented using GNU C++ and run on a Dell PC with 3GB main memory, and a 2.4GHz Dual Core CPU running a Ubuntu operating system. Our indices were created in memory using R*-trees with a branching factor of 32. The multi-map filters of all indices were implemented either using keyword ids, or prefix intervals (for instant queries). We evaluated the index size and construction time, as well as the query performances of the baseline index, and the FEH indices with various record-selectiveness thresholds t_r (denoted by FEH-t_r), for answering complete and instant queries.

5.1   Data Sets

We used two data sets in our experiments:

Business listings (Businesses). It consists of 20.4 million records of businesses in the USA from a popular directory website. We built the indices on the name, the latitude, and the longitude attributes of each business. The size of the raw data was around 4GB. The entire data set was used in the scalability tests. Since the baseline index can only handle around 5 million records in memory in some of the experiment settings, we used 5 million records of the data set for most of the experiments.

Image posts (CoPhIR) [5]. This image data set was extracted from the user posts on flickr.com. We selected 3.7 million posts that have a geographical location in the USA, and used their textual and spatial attributes. The indices were built on the title, description, tags, latitude, and longitude attributes. The size of the data was about 500MB.

We observed similar performance in our experiments on the two data sets. Due to space limitation, we mainly present the results on the Businesses data set unless otherwise noted.

5.2   Complete Keyword Search

Index Construction. We constructed the baseline index and the FEH indices with different record-selectiveness thresholds t_r. The baseline index on the Business data set took 3 minutes to build, and FEH-16 took 7 minutes. The construction of FEH-16 on the Business data set used a maximum of 400MB memory.

Figure 9 shows the sizes of indices decomposed to different components on the two data sets. Indices of the same data set had the same R*-tree size as they extend the same R*-tree (e.g., 154MB for the Businesses data set). The total sizes of the filters in the FEH indices were smaller than that of the baseline index for each data set. For instance, the total size of the baseline index for Businesses was 474MB, while the total size of FEH-16 was 279MB. We also noticed that the sizes of the C-filters on the FEH indices were much smaller compared to that of the O-filters, which included the keywords as well as their corresponding lists of record ids. For both data sets, the C-filter size of FEH decreased from 48MB to 5MB as we increased the record-selectiveness threshold t_r from 4 to 64. The reason is as we increased the threshold, more keywords were qualified as record selective and bypassing and they were not added to the C-filters.

Search Performance. We created query workloads by combining spatial locations with query keywords to evaluate the search performance of different indices.

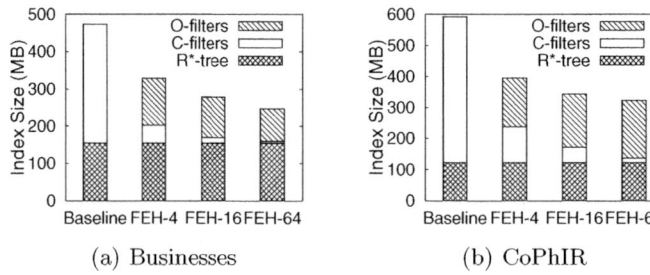

(a) Businesses   (b) CoPhIR

Fig. 9. Index size for complete keyword search

We manually selected 10 populous geographical areas. Each area spans 0.2 latitudes and longitudes, resulting in a 20km by 20km rectangle approximately (the exact size depends on the latitude). These areas were used as the rectangles in the range queries. The centers of these areas were used as the query points in the kNN queries, which ask for the top 10 results. To generate the queries, for each area we selected a set of keywords from its *local vocabulary*, a vocabulary for the businesses within the area, to avoid the situation where a range query returns no answer. To study the performance of queries with different keyword frequencies, we partitioned each local vocabulary into groups. For a frequency f, we used "G-f" to represent the group of keywords that appear in at least f records and in at most $4f - 1$ records. We randomly selected 100 keywords from each group, and generated 1000 location-based keyword queries for this group.

Figure 10 presents the query performance using different indices for answering range queries and kNN queries on the Businesses data set. Within each group, the query time is normalized against the time of using the baseline index. From the figures we can see that FEH indices can achieve a good speedup (up to 60%) over the baseline index for both range queries and kNN queries. The speedup was especially significant for range queries of all the groups, and for kNN queries with less frequent keywords. We can also see that the search became faster as we increased the record-selectiveness threshold from 4 to 16, and started to slow down from 16 to 64. This trend can be explained by the trade off between reduction in the number of visited nodes and increase in the number of candidate records to be verified. We observed similar trends for multi-keyword queries, due to space limitation we do not show the details.

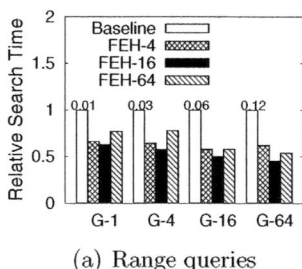

(a) Range queries

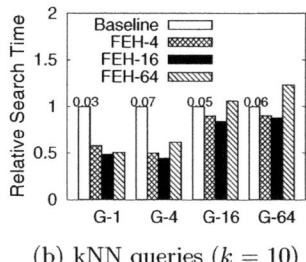

(b) kNN queries ($k = 10$)

Fig. 10. 1-keyword query performance for complete keyword search. The absolute search time using the baseline index is shown on the corresponding bar in ms.

Scalability. We used the Businesses data to study scalability. Figure 11(a) shows the index sizes of the baseline and FEH-16 indices when we increased the number of records from 5 million to 20 million. The size of both indices increased linearly, and FEH-16 had a smaller index size compared to the baseline index.

We conducted a scalability test of the query performance by selecting keywords from the local vocabulary for each geographical area. We randomly selected 100 keywords based on their frequencies for each geographical area, and generated 1000 queries for each setting. We also compared the performance of our techniques to the technique in [9]. Figure 11(b) shows the average search time on range queries, using the signature-file index, the baseline index, and the FEH-16 index, as we increased the number of records. We created the signature-file indices to have the same size as their corresponding baseline indices. We can see that the search time increased sub-linearly for all the queries. The results showed that the baseline index and FEH-16 outperformed the signature-file index. This was due to the fact that the signature-file index could identify a lot of consistent children nodes as false positives, resulting in unnecessary traversal. Further, the results showed that FEH-16 is faster than the baseline index. The average search time for the baseline index increased from 0.085 ms to 0.116 ms, while that for FEH-16 increased from 0.053 ms to 0.058 ms.

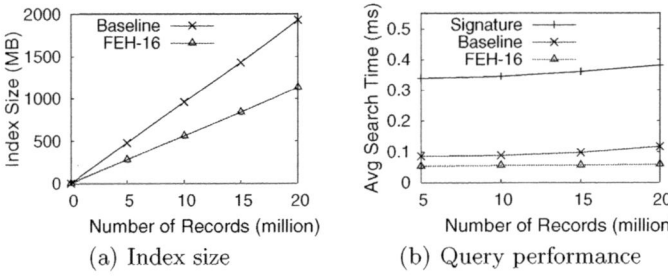

(a) Index size   (b) Query performance

Fig. 11. Scalability for complete keyword search

5.3   Instant Keyword Search

Index Construction. We evaluated the construction and update of the baseline index and FEH for instant search. The baseline prefix index on the Business data set took 3 minutes to build, and FEH-16 on prefixes took 6 minutes. The construction of FEH-16 on prefixes used a maximum of 400MB memory on the Business data set.

Figure 12 presents the sizes of the prefix indices on the two data sets. We use "FEHc-t_r" to denote the FEH index with compressed O-filters using record-selectiveness threshold t_r. Similar trends are observed as in Figure 9. In addition, the filters on the prefix baseline indices are significantly larger compared to the filters on the non-prefix baseline indices, due to the cost of adding prefixes to the filters. We noticed that FEH and FEHc indices can significantly reduce index size. For instance, on the Businesses data set, FEH-16 reduced the index size by half (from 1.6GB to 800MB), while FEHc-16 further reduced the size by roughly half from FEH-16 (to 440MB).

Search Performance 1-keyword queries: We evaluated the search performance of the prefix indices for instant queries. We used the same methods to

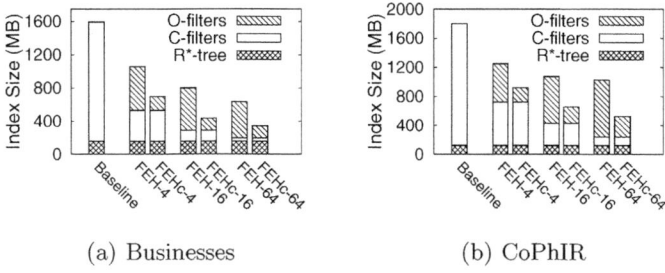

Fig. 12. Index size for instant search

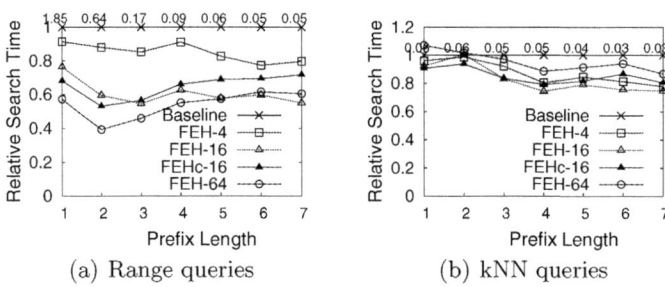

Fig. 13. 1-keyword query performance for instant search. The absolute search time using the baseline index is shown along the corresponding line in ms.

generate the query keywords as in Section 5.2. We generated 1000 keywords for each setting. To test prefix query performance, all prefixes of a query keyword were individually combined with the location to form the prefix queries, simulating the case where a user types character by character. We measured the performance of the queries with different prefix lengths. We only reported results for prefixes that appeared in at least 500 queries for each setting.

Figure 13 shows the query performance of different prefix indices for answering range queries and kNN queries. The query time was normalized against the time of using the baseline prefix index for each prefix length. The range queries with short prefix lengths were more expensive for all the indices due to the large number of results that need to be retrieved. We retrieved the top-10 results for the kNN queries. We noticed that the search-time variance for different prefix lengths was small on kNN queries compared to range queries. FEH prefix indices can improve performance of both range and kNN queries, for all the prefix lengths compared to the baseline prefix index, for up to 60%. Due to space limitation we only plotted the results FEHc-16 for FEH prefix index with compressed O-filters. The query performance penalty for compressing the O-filter was small (around 10% less improvement over the baseline index). The performance improvement of FEH indices over the baseline index on kNN queries is relatively small, due

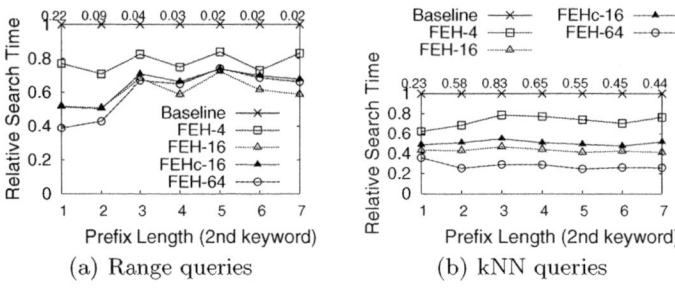

Fig. 14. 2-keyword query performance for instant search. The absolute search time using the baseline index is shown along the corresponding line in ms.

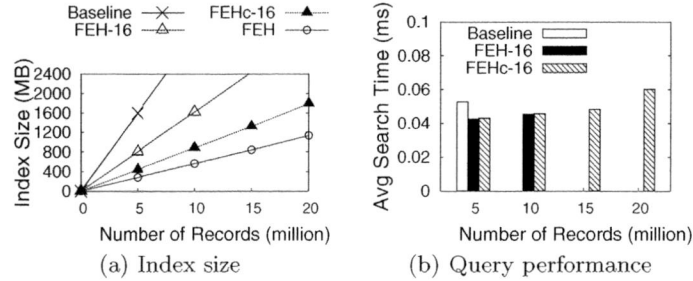

Fig. 15. Scalability for instant search. "FEH" is the index for complete keywords with $t_r = 16$.

to the fact that kNN queries only retrieve the top-k answers, where k is usually very small, leaving little room for FEH to improve.

Multi-keyword queries: We also evaluated the performance of multi-keyword instant queries. To generate multi-keyword instant queries, we first randomly selected 100 sets of keywords based on their frequencies from the local co-occurrence vocabulary for each geographical area. To test the multi-keyword instant query performance, all prefixes of the last query keyword in a set were individually combined with the rest keywords as well as the location to generate the prefix queries. Figure 14 presents the performance of the 2-keyword range and kNN instant queries for different prefix lengths. Similar improvements were observed. We observed similar trends for 3+-keyword queries, due to space limitation we do not show the details.

Scalability. Figure 15 shows the scalability on the Businesses data set. The baseline technique on prefixes was not able to index more than 10 million records due its large index size. Similarly, FEH-16 could not index more than 15 million records. We scaled FEHc-16 all the way up to 20 million records, with the index size around 1.8GB. We also plotted the size of the FEH-16 index built on complete keywords (denoted as "FEH" in the figure) for a comparison with the prefix index sizes. It is clear that the index size increased a lot by building FEH on prefixes instead of on complete keywords, as many prefixes need to be added to the filters. Our compression technique greatly alleviated this problem.

We also studied the average kNN query performance for all prefix lengths from 1 to 7. Results in Figure 15(b) showed that the performance of FEHc-16 scaled well when we increased the data size from 5 million to 20 million.

6   Conclusion

In this paper we studied location-based instant search. We first proposed an filtering-effective hybrid index (FEH) that judiciously uses two types of keyword filters based on their selectiveness to do powerful pruning. We then developed indexing and search techniques that utilize the FEH index and store prefix information to efficiently answer instant queries. Our experiments demonstrated the high efficiency and scalability of our techniques.

Acknowledgements. We thank Sattam Alsubaiee for his discussions in this work. This work was partially supported by NSF award IIS-1030002.

References

1. Alsubaiee, S.M., Behm, A., Li, C.: Supporting location-based approximate-keyword queries. In: ACM SIGSPATIAL GIS (2010)
2. Bast, H., Mortensen, C.W., Weber, I.: Output-sensitive autocompletion search. In: Crestani, F., Ferragina, P., Sanderson, M. (eds.) SPIRE 2006. LNCS, vol. 4209, pp. 150–162. Springer, Heidelberg (2006)
3. Bast, H., Weber, I.: Type less, find more: fast autocompletion search with a succinct index. In: SIGIR, pp. 364–371 (2006)
4. Beckmann, N., Begel, H.P., Schneider, R., Seeger, B.: The R*-tree: an efficient and robust access method for points and rectangles. In: SIGMOD (1990)
5. Bolettieri, P., Esuli, A., Falchi, F., Lucchese, C., Perego, R., Piccioli, T., Rabitti, F.: CoPhIR: a test collection for content-based image retrieval. In: CoRR (2009)
6. Chaudhuri, S., Kaushik, R.: Extending autocompletion to tolerate errors. In: SIG-MOD Conference, pp. 707–718 (2009)
7. Chen, Y.-Y., Suel, T., Markowetz, A.: Efficient query processing in geographic web search engines. In: SIGMOD (2006)
8. Cong, G., Jensen, C.S., Wu, D.: Efficient retrieval of the top-k most relevant spatial web objects. PVLDB 2(1) (2009)
9. Felipe, I.D., Hristidis, V., Rishe, N.: Keyword search on spatial databases. In: ICDE (2008)
10. Guttman, A.: R-trees: A dynamic index structure for spatial searching. In: SIG-MOD (1984)
11. Hariharan, R., Hore, B., Li, C., Mehrotra, S.: Processing spatial-keyword (SK) queries in geographic information retrieval (GIR) systems. In: SSDBM (2007)
12. Hjaltason, G.R., Samet, H.: Ranking in spatial databases. In: SSD, pp. 83–95 (1995)
13. Ji, S., Li, G., Li, C., Feng, J.: Efficient interactive fuzzy keyword search. In: WWW, pp. 371–380 (2009)

14. Khodaei, A., Shahabi, C., Li, C.: Hybrid indexing and seamless ranking of spatial and textual features of web documents. In: Bringas, P.G., Hameurlain, A., Quirch-mayr, G. (eds.) DEXA 2010. LNCS, vol. 6261, pp. 450–466. Springer, Heidelberg (2010)
15. Li, G., Ji, S., Li, C., Feng, J.: Efficient type-ahead search on relational data: a TASTIER approach. In: SIGMOD (2009)
16. Morrison, D.R.: Patricia - practical algorithm to retrieve information coded in alphanumeric. J. ACM 15(4), 514–534 (1968)
17. Roussopoulos, N., Kelley, S., Vincent, F.: Nearest neighbor queries. In: SIGMOD Conference, pp. 71–79 (1995)
18. Vaid, S., Jones, C.B., Joho, H., Sanderson, M.: Spatio-textual indexing for ge-ographical search on the web. In: Anshelevich, E., Egenhofer, M.J., Hwang, J. (eds.) SSTD 2005. LNCS, vol. 3633, pp. 218–235. Springer, Heidelberg (2005)
19. Yao, B., Li, F., Hadjieleftheriou, M., Hou, K.: Approximate string search in spatial databases. In: ICDE (2010)
20. Zhang, D., Chee, Y.M., Mondal, A., Tung, A.K.H., Kitsuregawa, M.: Keyword search in spatial databases: Towards searching by document. In: ICDE (2009)
21. Zhang, D., Ooi, B.C., Tung, A.K.H.: Locating mapped resources in web 2.0. In: ICDE (2010)
22. Zhou, Y., Xie, X., Wang, C., Gong, Y., Ma, W.-Y.: Hybrid index structures for location-based web search. In: CIKM (2005)

Continuous Inverse Ranking Queries in Uncertain Streams

Thomas Bernecker[1], Hans-Peter Kriegel[1], Nikos Mamoulis[2], Matthias Renz[1],
and Andreas Zuefle[1]

[1] Department of Computer Science, Ludwig-Maximilians-Universität München
{bernecker,kriegel,renz,zuefle}@dbs.ifi.lmu.de
[2] Department of Computer Science, University of Hong Kong
nikos@cs.hku.hk

Abstract. This paper introduces a scalable approach for continuous inverse ranking on uncertain streams. An uncertain stream is a stream of object instances with confidences, e.g. observed positions of moving objects derived from a sensor. The confidence value assigned to each instance reflects the likelihood that the instance conforms with the current true object state. The inverse ranking query retrieves the rank of a given query object according to a given score function. In this paper we present a framework that is able to update the query result very efficiently, as the stream provides new observations of the objects. We will theoretically and experimentally show that the query update can be performed in linear time complexity. We conduct an experimental evaluation on synthetic data, which demonstrates the efficiency of our approach.

1 Introduction

Recently, it has been recognized that many applications dealing with spatial, temporal, multimedia, and sensor data have to cope with uncertain or imprecise data. For instance, in the spatial domain, the locations of objects usually change continuously, thus the positions tracked by GPS devices are often imprecise. Similarly, vectors of values collected in sensor networks (e.g., temperature, humidity, etc.) are usually inaccurate, due to errors in the sensing devices or time delays in the transmission. Finally, images collected by cameras may have errors, due to low resolution or noise. As a consequence, there is a need to adapt storage models and indexing/search techniques to deal with uncertainty.

Special formulations of queries are required in order to take the uncertainty of the data into account. In this paper, we focus on the *probabilistic inverse ranking* (PIR) query on uncertain streaming data, i.e. the data change with elapsing time. While PIR queries have been studied for static data [1], to the best of our knowledge, there is no previous work in the context of dynamic data or data streams. Given a stream of uncertain objects, a user-defined score function S that ranks the objects and a user-defined (uncertain) query object q, a PIR query computes all the possible ranks of q associated with a probability. The PIR query is important for many real applications including financial data analysis, sensor data monitoring and multi-criteria decision making where one might be interested in the identification of the rank (significance) of a particular

J.B. Cushing, J. French, and S. Bowers (Eds.): SSDBM 2011, LNCS 6809, pp. 37–54, 2011.
© Springer-Verlag Berlin Heidelberg 2011

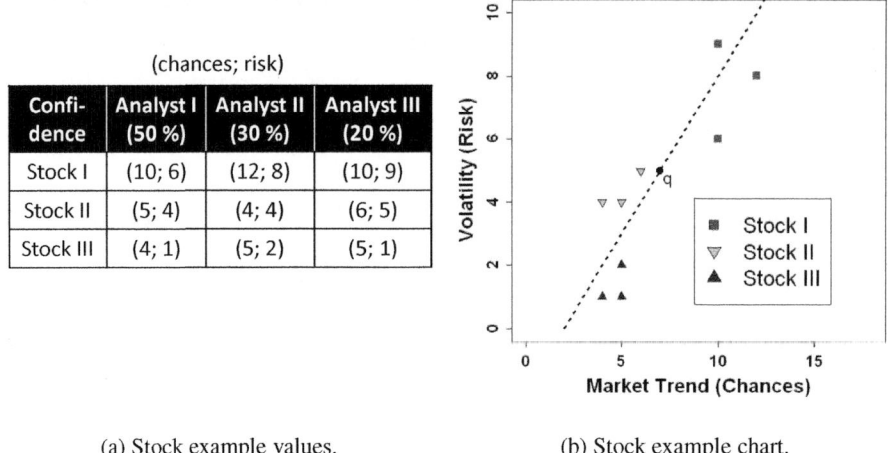

| | (chances; risk) | | |
Confi-dence	Analyst I (50 %)	Analyst II (30 %)	Analyst III (20 %)
Stock I	(10; 6)	(12; 8)	(10; 9)
Stock II	(5; 4)	(4; 4)	(6; 5)
Stock III	(4; 1)	(5; 2)	(5; 1)

(a) Stock example values.   (b) Stock example chart.

Fig. 1. Chances and risk predictions by three analysts for three stocks

object among peers. Consider the exemplary application illustrated in Figure 1(a): A financial decision support system monitors diverse prognostic attributes of a set of three stocks, e.g. predicted market trend (chances) and volatility (risk), which are used to rate the profitability of the stocks according to a given score function. As it can be observed, the chance and risk estimations are not unique among different analysts and each analyst is given a different confidence level. Figure 1(b) shows graphically the three stocks with their respective analyst predictions and the query object q. Here we assume that we are given a score function defined as $S = (Chances - Risk)$. The dotted line in Figure 1(b) denotes all points x where $S(x) = S(q)$, i.e. all points with the same score as q. Any instance located to the right of this line has a higher score than q, while any instance to the left has a lower score. Therefore, we can safely assume that Stock II has a lower score than q while Stock III certainly has a higher score than q. However, the relative ranking of Stock I with respect to q is uncertain. While two of three analysts (with a total confidence of 80%) would rank Stock I higher than q, the third analyst would rank it lower. Thus, the PIR query for q returns that q is on rank two with a probability of 20%, on rank three with a probability of 80% and definitely not on rank one or four. This result can be used to answer questions like "Given a score function, what is the likelihood that a query stock q is one of the global top-3 best stocks?". The problem we study in this paper is how to efficiently update these likelihoods when the analysts release new estimations on a ticker stream.

As another example (taken from [2]), for a newborn, we may be interested in his/her health compared with other babies, in terms of height, weight, and so on. In this case, we can infer the newborn's health from his/her rank among others. Note that data of newborn babies in a hospital are confidential. Thus, for the sake of privacy preservation, such information is usually perturbed by adding synthetic noise or generalized by replacing exact values with uncertain intervals, before releasing them for research purposes. Thus, in these situations, we can conduct a PIR query over uncertain data (perturbed or generalized) in order to obtain all possible ranks that a newborn may have with high confidence. In addition, we may want the distribution of possible ranks of the baby to be dynamically updated, as new data arrive, in order to be confident that the baby's status remains good compared to new cases. Therefore, rank updates for the query (baby) have to be applied, as new measurements arrive from a stream.

The rest of the paper is organized as follows: In the next section, we survey existing work in the field of managing and querying uncertain data streams. In Section 3, we formally define the problem of probabilistic inverse ranking on data streams. Our approach for solving the problem efficiently is described in Section 4. In Section 5, we generalize the problem by additionally considering uncertain queries. We experimentally evaluate the efficiency of our approach in Section 6 and conclude the paper in Section 7.

2 Related Work

In this paper, we focus on inverse similarity ranking of uncertain vector data. A lot of work was performed in the direction of ranking among uncertain data [3,4,5,6,7], but there is limited research on the inverse variant of ranking uncertain data [1]. In a nutshell, there are two models for capturing uncertainty of objects in a multi-dimensional space. In the *continuous* uncertainty model, the uncertain values of an object are represented by a continuous probability density function (pdf) within the vector space. This type of representation is often used in applications where the uncertain values are assumed to follow a specific probability density function (pdf), e.g. a Gaussian distribution [5]. Similarity search methods based on this model involve expensive integrations of the pdf's, thus special approximation techniques for efficient query processing are typically employed [8]. In the *discrete* uncertainty model, each object is represented by a discrete set of alternative values, and each value is associated with a probability. The main motivation of this representation is that, in most real applications, data are collected in a discrete form (e.g., information derived from sensor devices). The uncertain stream data, as assumed in this paper, correspond to the discrete uncertainty model which also complies with the x-relations model used in the *Trio* system [9].

In order to deal with massive datasets that arrive online and have to be monitored, managed and mined in real time, the data stream model has become popular. Surveys of systems and algorithms for data stream management are given in [10] and [11]. A generalized stream model, the probabilistic stream model, was introduced in [12]. In this model, each item of a stream represents a discrete probability distribution together with a probability that the element is not actually present in the stream. There has been interesting work on clustering uncertain streams [13], as well as on processing more complex event queries over streams of uncertain data [14]. [15] presents algorithms that capture essential features of the stream, such as quantiles, heavy-hitters, and frequency moments. To the best of our knowledge, this paper is the first addressing the processing of inverse ranking queries on uncertain streams.

The *inverse ranking query* on static data was first introduced by Li [2]. Chen et al. [1] apply inverse ranking to probabilistic databases by introducing the *probabilistic inverse ranking query (PIR)*. Apart from considering only static data, their PIR query definition varies from ours, since its output consists of all possible ranks for a query object q, for which q has a probability higher than a given threshold. Another approach for answering PIR queries has been proposed by [16] which computes the expected inverse rank of an object. The expected inverse rank can be computed very efficiently, however, it lacks from a semantic point of view. In particular, an object that has a very high chance to be on rank one, may indeed have a expected rank far from one, and may not be in the result using expected ranks. Thus, no conclusion can be made about the actual rank probabilities if the expected rank is used, since the expected rank is an aggregation that drops important information.

3   Problem Definition

In this work, we adopt the discrete *x-relation model* proposed in the TRIO system [9], in which an uncertain database $\mathcal{D}$ consists of n uncertain objects which are each modelled by exactly one *x-tuple*. Each x-tuple T includes a number of tuples, which we call (possible) instances, as its alternatives. Each tuple $t \in T$ is associated with a probability $p(t)$, representing a discrete probability distribution of T. Thus, an x-tuple T is a set of a bounded number of instances, subject to the constraint that $\sum_{t \in T} p(t) \leq 1$. Independence is assumed among the x-tuples. For simplicity, we also assume that $\sum_{t \in T} p(t) = 1$.[1] Following the popular possible worlds semantics, $\mathcal{D}$ is instantiated into a possible world with mutual independence of the x-tuples. An uncertain database $\mathcal{D}$ is instantiated into a possible world as follows:

Definition 1 (Possible Worlds). *Let $\mathcal{D} = \{T_1, ..., T_n\}$ and let $W = \{t_1, ..., t_n\}$ be any subset of tuples appearing in $\mathcal{D}$ such that $t_i \in T_i$. The probability of this world W occurring is $P(W) = \prod_{j=1}^{n} p(t_j)$. If $P(W) > 0$, we say that W is a possible world, and we denote by $\mathcal{W}$ the set of all possible worlds.*

Without loss of generality, we consider uncertain vector objects L in a d-dimensional vector space. That is, each object is assigned to m alternative locations l associated with a probability value. For example, the m alternative positions of an uncertain object are associated with observations derived from m sources of information (sensors). In our stock example the sources correspond to the assessments of the analysts and in the baby ranking example, the sources correspond to m uncertain values uniformly sampled from the corresponding uncertain measurement range.

Definition 2 (Probabilistic Stream). *We define an uncertain data stream, analogously to [12]. A probabilistic stream is a data stream $A = [x_0, ..., x_t, ...]$ in which each item x_t encodes a random variable reported at time t from the stream, corresponding to an object update. In particular, each x_t has the form $\langle O, L \rangle$, where O is an object ID and L is a location vector of length $|L|$. Each element $l \in L$ contains a location $l.loc \in \mathbb{R}^d$ and a probability l.p. In addition, we assume that $\sum_{l \in L} l.p = 1$, i.e. we assume that the object have no existential uncertainty. i.e. that object O is existentially certain.*

---

[1] For the problem of inverse ranking, this assumption means no loss of generality, since existential uncertainty can be modelled by simply adding to T an additional instance with a probability $1 - \sum_{t \in T} p(t)$ and a score of $-\infty$ (that is a distance of ∞ to the query).

Definition 3 (Probabilistic Stream Database). *A probabilistic stream database is an uncertain database connected to at least one probabilistic stream. Each stream item $x_t = \langle O, L \rangle$ at time t denotes an update of the uncertain object $O \in DB$.[2] Therefore, at time t, the x-relation describing object O is replaced by the new location distribution L coming from the stream.*

This probabilistic stream database model is very general and can be easily adapted to simulate popular stream models:

The *sliding window model* of size m can be simulated by imposing the following constraint to the probabilistic stream: For any two stream items $x_t = \langle O, L_t \rangle$, $x_s = \langle O, L_s \rangle$, $t < s$, of the same object O, it holds that if there no other stream items between time t and s concerning the same object, it holds that L_{t+1} is derived from L_t by

- adding exactly one new instance to L_t, and
- removing the oldest instance of L_t if $|L_t| > m$

The probabilities $p(l), l \in L_t$ are often set to $p(l) = \frac{1}{|L_t|}$, but other distributions can be used. In particular, more recently observed instances can be given a higher probability to obtain the *weighted sliding window model*. Additionally, the infinite sliding window model is obtained by setting $k = \infty$. In this work, the stream model is left abstract, as the proposed solutions are applicable for any such model.

Next we define the problem to be solved in this work.

Definition 4 (Probabilistic Inverse Ranking Query). *Given an uncertain database $\mathcal{D}$ of size n, a query object q and a score function*

$$S : \mathcal{D} \to \mathbb{R}_o^+.$$

Assuming that only the top-k ranks are of interest, a probabilistic inverse ranking query $PIR(q)$ returns for each $i \in [1, ..., k]$ the probability $P_q^t(i)$ that q is on rank i w.r.t. S, i.e. the probability that there exist exactly $i - 1$ objects $o \in \mathcal{D}$ such that $S(o) > S(q)$ at time t.

Given a set of n uncertain objects and a probabilistic stream A as defined above, our problem is to compute and update, for a given query object q and a given score function S the result of $PIR(q)$ at each time t, i.e. after each object update. The challenge is to ensure that this can be done correctly in terms of the possible world semantics, and highly efficiently to allow online processing of the probabilistic stream A. Since the number of possible worlds at a time t is exponential in the number n of uncertain stream objects at time t, these two challenges are conflicting. In the following we will propose an approach to compute $PIR(q)$ in $O(n^2)$ from scratch, and to update it in $O(n)$ when a new update is fetched from the stream. In addition, we will show how the result of $PIR(q)$ can be efficiently updated, if the query object q is itself a stream object that changes frequently.

---

[2] O may also be a new object.

Table 1. Table of notations used in this work

	Table of Notations
$\mathcal{D}$	An uncertain stream database.
n	The cardinality of $\mathcal{D}$.
q	A query vector in respect to which a probabilistic inverse ranking is computed.
k	The ranking depth that determines the number of ranking positions of the inverse ranking query result.
o_x	An uncertain stream object corresponding to a finite set of alternative vector point instances.
p_o^t	The probability that object o has a higher score than q at time t.
$P^t(i)$	The result of the inverse ranking at time t: The probability that q is at rank i at time t.
$P_{i,j}^t$	The probability that, out of j processed objects, exactly i objects have a higher score than q at time t.
$P_{PBR}^t(i)$	The result of the Poisson binomial recurrence at time t: The probability that i objects have a higher score than q at time t, if all objects o for which $p_o^t = 1$ are ignored.
$\hat{P}_{PBR}^t(i)$	The adjusted result of the Poisson binomial recurrence at time t: Identical to $P_{PBR}^t(i)$ except that the effect of the object that changes its position at time $t+1$ is removed from the calculation.
C^t	The number of objects o at time t for which $p_o^t = 1$.

4   Probabilistic Inverse Ranking

Consider an uncertain stream database $\mathcal{D}$ of size n, a query object q, a score function S and a positive integer k. Our algorithm basically consists of two modules:

- The *initial computation* of the probabilistic inverse ranking that computes for each rank $i \in [1, ..., k]$ the probability $P^t(i)$ that q is ranked on position i at the initial time t, when the query is issued. We show how this can be performed in $O(k \cdot n)$ time.
- The *incremental stream processing* that updates PIR(q) at time $t+1$, given the probabilistic inverse ranking at time t. Therefore, the probabilities $P^{t+1}(i)$ that Q is ranked on position i at time $t+1$ have to be computed given the $P^t(i)$. We show how this update can be done in $O(k)$ time.

4.1   Initial Computation

For each object $o_j \in \mathcal{D}$ let $p_{o_j}^t$ be the probability that o_j has a higher rank than q at time t, i.e. $p_{o_j}^t = P(S(o_j) > S(q))$. These probabilities can be computed in a single database scan. We can process the $p_{o_j}^t$ successively by means of the *Poisson binomial recurrence* [17], as proposed in [18]. Therefore, let $P_{i,j}^t$ be the probability that, out of the j objects processed so far, exactly i objects have a higher score than q. This probability depends only on the two following events:

- $i - 1$ out of the first $j - 1$ processed objects have a higher score than q **and** o_j has a higher score than q.
- i out of the first $j - 1$ processed objects have a higher score than q **and** o_j does not have a higher score than q.

This observation and the assumption of independence between stream objects can be used to formulate the following Poisson binomial recurrence:

$$P_{i,j}^t = P_{i-1,j-1}^t \cdot p_{o_j}^t + P_{i,j-1}^t \cdot (1 - p_{o_j}^t) \tag{1}$$

with $P_{0,0}^t = 1$ and $P_{i,j}^t = 0$ for $i < 0$ or $i > j$.

When the last object of the database is processed, i.e. $j = n$, then $P_{i,j}^t = P_{i,n}^t \overset{Definition}{=} P^t(i+1)$.[3] Computing the $P_q^t(i+1)$ for $0 \leq i \leq k-1$ yields the probabilistic inverse ranking. In each iteration, we can omit the computation of any $P_{i,j}^t$ where $i \geq k$, since we are not interested in any ranks greater than k, and thus, are not interested in the cases where at least k objects have a higher score than q. In total, for each $0 \leq i < k$ and each $1 \leq j \leq n$, $P_{i,j}^t$ has to be computed resulting in an $O(k \cdot n)$ time complexity.

Equation 1 is only required for objects o_j for which $0 < p_{o_j}^t < 1$. Objects o_j for which $p_{o_j}^t = 0$ can safely be ignored in the initial computation, since they have no effect on the $P^t(i)$. For objects o_j for which $p_{o_j}^t = 1$, we use a counter C^t that denotes the number of such objects. Thus, when o_j is encountered in the initial computation, the Poisson binomial recurrence is avoided and C^t is incremented. The probabilities obtained from the Poisson binomial recurrence by ignoring objects for which $p_{o_j}^t = 1$ are denoted as $P_{PBR}^t(i), 0 \leq i \leq k$.

The probabilistic inverse ranking can be obtained from the $P_{PBR}^t(i), 0 \leq i \leq k$ and C^t as follows:

$$P^t(i) = P_{PBR}^t(i - 1 - C^t), \text{ for } C^t + 1 \leq i \leq C^t + 1 + k \tag{2}$$

$$P^t(i) = 0 \text{ otherwise}$$

Example 1. Assume that a database containing objects $o_1, ..., o_4$ and an inverse ranking query with query object q and $k = 2$. Assume that $p_{o_1}^t = 0.1$, $p_{o_2}^t = 0$, $p_{o_3}^t = 0.6$ and $p_{o_4}^t = 1$. To compute the initial inverse ranking, we first process o_1 using Equation 1:

$$P_{0,1}^t = P_{-1,0}^t \cdot p_{o_1}^t + P_{0,0}^t \cdot (1 - p_{o_1}^t) = 0 \cdot 0.1 + 1 \cdot 0.9 = 0.9$$

$$P_{1,1}^t = P_{0,0}^t \cdot p_{o_1}^t + P_{1,0}^t \cdot (1 - p_{o_1}^t) = 1 \cdot 0.1 + 0 \cdot 0.9 = 0.1$$

Next we process o_2, but notice that $p_2^t = 0$, so o_2 can be skipped. Then, object o_3 requires an additional iteration of Equation 1:

$$P_{0,2}^t = P_{-1,1}^t \cdot p_{o_3}^t + P_{0,1}^t \cdot (1 - p_{o_3}^t) = 0 \cdot 0.6 + 0.9 \cdot 0.4 = 0.36$$

$$P_{1,2}^t = P_{0,1}^t \cdot p_{o_3}^t + P_{1,1}^t \cdot (1 - p_{o_3}^t) = 0.9 \cdot 0.6 + 0.1 \cdot 0.4 = 0.58$$

$P_{2,2}^t$ does not need to be computed since $2 = i \geq k = 2$.

Next we process o_4. Since $p_{o_4}^t = 1$, only C^t has to be incremented to 1. At this point, we are done. We have obtained:

$$P_{PBR}^t(0) = 0.36 \text{ and } P_{PBR}^t(1) = 0.58$$

---

[3] The event that i objects have a higher score than q corresponds to the event that q is ranked on rank $i + 1$.

To get the final inverse ranking at time t, we use Equation 2 to obtain

$$P^t(1) = P^t_{PBR}(1 - 1 - 1) = P^t_{PBR}(-1) = 0$$
$$P^t(2) = P^t_{PBR}(2 - 1 - 1) = P^t_{PBR}(0) = 0.36$$

4.2   Incremental Stream Processing

A naive solution would apply the Poisson binomial recurrence (cf. Equation 1) whenever a new object location o_x is fetched from the stream. However, the expensive update which is linear in the size of the database would make online stream processing impractical for large databases. In the following, we show how we can update $P^{t+1}(i)$ for $1 \le i \le k$ in constant time using the results of the previous update iteration.

Without loss of generality, let o_x be the object for which a new position information is returned by the stream at time $t + 1$. $p^t_{o_x}$ ($p^{t+1}_{o_x}$) denotes the old (new) probability that o_x has a higher score than q.

Our update algorithm uses two phases:

- **Phase 1:** Removal of the effect of the old value distribution of the uncertain object o_x. That is, removal of the effect of the probability $p^t_{o_x}$ from the result $P^t_{PBR}(i), 0 \le i < k$. This yields an intermediate result $\hat{P}^{t+1}_{PBR}(i), 0 \le i < k$.
- **Phase 2** Incorporation of the new value distribution of the uncertain object o_x. That is including the probability $p^{t+1}_{o_x}$ in the intermediate result $\hat{P}^{t+1}(i), 0 \le i < k$ obtained in Phase 1.

Phase 1. The following cases w.r.t. $p^t_{o_x}$ have to be considered:

Case I: $p^t_{o_x} = 0$. This case occurs if o_x is a new object or if it is certain that o_x has a lower score than q at time t. Thus nothing has to be done to remove the effect of $p^t_{o_x} = 0$: $\hat{P}^t_{PBR}(i) = P^t_{PBR}(i)$.

Case II: $p^t_{o_x} = 1$, i.e. if it is certain that o_x has a higher score than q at time t. In this case we just have to decrement C^t by one to remove the effect of $p^t_{o_x}$. Thus $\hat{P}^t_{PBR}(i) = P^t_{PBR}(i)$ and $C^{t+1} = C^t - 1$.

Case III: $0 < p^t_{o_x} < 1$, i.e. it is uncertain whether o_x has a higher score than q at time t. To remove the effect of $p^t_{o_x}$ on all $P^t_{PBR}(i)$ $(1 \le i \le k)$ we look at the last iteration of Equation 1, that was used to obtain $P^t_{PBR}(i), 0 \le i \le k$. Let o_l be the object that was incorporated in this iteration:

$$P^t_{PBR}(i) = P^{t'}_{PBR}(i - 1) \cdot p^t_{o_l} + P^{t'}_{PBR}(i) \cdot (1 - p^t_{o_l}),$$

where $P^{t'}_{PBR}(i)$ describes the probability that i objects have a score higher than q, if (in addition to all objects o_i for which $p^t_{o_i} = 1$) o_l is ignored. Now we observe that the $P^t_{PBR}(i)$'s $(1 \le i \le k)$ are not affected by the order in which the objects are processed within the recursion. In particular, the $P^t_{PBR}(i)$'s do not change, if the objects are processed in an order that processes o_x last, thus we obtain:

$$P^t_{PBR}(i) = \hat{P}^t_{PBR}(i - 1) \cdot p^t_{o_x} + \hat{P}^t_{PBR}(i) \cdot (1 - p^t_{o_x}),$$

This can be resolved to

$$\hat{P}^t_{PBR}(i) = \frac{P^t_{PBR}(i) - \hat{P}^t_{PBR}(i-1) \cdot p^t_{o_x}}{1 - p^t_{o_x}}. \tag{3}$$

With $i = 0$ we obtain

$$\hat{P}^t_{PBR}(0) = \frac{P^t_{PBR}(0)}{1 - p^t_{o_x}}, \tag{4}$$

because the probability $\hat{P}^t_{PBR}(-1)$ that exactly -1 objects have a higher score than q is zero. Since the $P^t_{PBR}(i)$'s for $0 \le i \le k-1$ are known from the previous stream processing iteration, $\hat{P}^t_{PBR}(0)$ can be easily computed using Equation 4. Now we can inductively compute $\hat{P}^t_{PBR}(i+1)$ by using $\hat{P}^t_{PBR}(i)$ for any i and exploiting Equation 3.

Phase 2. In Phase 2, the same cases have to be considered:

Case I: $p^{t+1}_{o_x} = 0$: Object o_x has no influence on the result at time $t+1$. Nothing has to be done. Thus $P^{t+1}_{PBR}(i) = \hat{P}^t_{PBR}(i)$.

Case II: $p^{t+1}_{o_x} = 1$: Object o_x certainly has a higher score than q. Thus $C^{t+1} = C^t + 1$ and $P^{t+1}_{PBR}(i) = \hat{P}^t_{PBR}(i)$.

Case III: $0 < p^{t+1}_{o_x} < 1$: We can incorporate the new probability for o_x to be ranked higher than q, i.e. p^{t+1}_x, to compute the new probabilistic inverse ranking by an additional iteration of the Poisson binomial recurrence:

$$P^{t+1}_{PBR}(i) = \hat{P}^t_{PBR}(i-1) \cdot p^{t+1}_{o_x} + \hat{P}^t_{PBR}(i) \cdot (1 - p^{t+1}_{o_x}).$$

Example 2. Reconsider Example 1 where, at time t, we obtained $C^t = 1$, $P^t_{PBR}(0) = 0.36$ and $P^t_{PBR}(1) = 0.58$. Now, assume that at time $t+1$ object o_3 changes its probability from 0.6 to 0.2, i.e. $p^t_{o_3} = 0.6$ and $p^{t+1}_{o_3} = 0.2$. Phase 1 starts using Case III. Using Equation 4 we get:

$$\hat{P}^t_{PBR}(0) = \frac{P^t_{PBR}(0)}{1 - p^t_{o_3}} = \frac{0.36}{0.4} = 0.9$$

Using Equation 3 we also get:

$$\hat{P}^t_{PBR}(1) = \frac{P^t_{PBR}(1) - \hat{P}^t_{PBR}(0) \cdot p^t_{o_3}}{1 - p^t_{o_3}} = \frac{0.58 - 0.9 \cdot 0.6}{0.4} = 0.1$$

This completes Phase 1. In Phase 2, Case III is chosen and we get:

$$P^{t+1}_{PBR}(0) = \hat{P}^t_{PBR}(-1) \cdot p^{t+1}_{o_3} + \hat{P}^t_{PBR}(0) \cdot (1 - p^{t+1}_{o_3}) = 0 \cdot 0.2 + 0.9 \cdot 0.8 = 0.72$$

$$P^{t+1}_{PBR}(1) = \hat{P}^t_{PBR}(0) \cdot p^{t+1}_{o_3} + \hat{P}^t_{PBR}(1) \cdot (1 - p^{t+1}_{o_3}) = 0.9 \cdot 0.2 + 0.1 \cdot 0.8 = 0.26$$

This completes the update step (C^t remains unchanged, i.e. $C^{t+1} = C^t$). The result is obtained analogously to Example 1 using Equation 2:

$$P^{t+1}(1) = P^{t+1}_{PBR}(1 - 1 - 1) = P^{t+1}_{PBR}(-1) = 0$$

$$P^{t+1}(2) = P^{t+1}_{PBR}(2 - 1 - 1) = P^{t+1}_{PBR}(0) = 0.72$$

Now assume, that at time $t + 2$ object o_4 changes its probability from 1 to 0: In Phase 1, Case II is used and C_t is decremented from 1 to 0 to obtain $C^{t+1} = 0$. In Phase 2, Case I is used and nothing is done. We get:

$$P^{t+2}_{PBR}(0) = \hat{P}^{t+1}_{PBR}(0) = P^{t+1}_{PBR}(0) = 0.72$$

$$P^{t+2}_{PBR}(1) = \hat{P}^{t+1}_{PBR}(1) = P^{t+1}_{PBR}(1) = 0.26$$

We obtain the result using Equation 2:

$$P^{t+2}(1) = P^{t+2}_{PBR}(1 - 1 - 0) = P^{t+2}_{PBR}(0) = 0.72$$

$$P^{t+2}(2) = P^{t+2}_{PBR}(2 - 1 - 0) = P^{t+2}_{PBR}(0) = 0.36$$

The latter example shows why we need to maintain k probability values at each point of time: Even though some of the k probabilities may not be required to obtain the result, they may be required to obtain the result at a later time.

Regarding the computational complexity, the following holds for both Phase 1 and Phase 2: Case I and II have a cost of $O(1)$ since either nothing has to be done, or only C^t has to be incremented or decremented. Case III has a total cost of $O(k)$ leading to a total runtime of $O(k)$ in the update step.

5   Uncertain Query

In the previous section we have assumed that the query object q is fixed, i.e. has a certain position in $\mathbb{R}^d$. We now consider the case in which the query is also given as an uncertain stream object. Similar to the database objects, we now assume that the query object Q^t is represented by a set of m alternative instances $Q = \{q^t_1, ..., q^t_m\}$ at time t. The probabilistic inverse ranking query PIR(Q) w.r.t. an uncertain query object Q can be computed by aggregating the probabilistic inverse ranking query results w.r.t. each instance q_j of Q. Formally,

$$P^t_Q(i) = \sum_{j=1..m} P^t_{q_j}(i) \cdot p(q_j)$$

for all $j \in \{1, \ldots, m\}$, where $p(q_j)$ denotes the probability that the query object is located at q_j and $P^t_{q_j}(i)$ is the probability that instance q_j us located at rank i. $P^t_{q_j}(i)$ can be computed and updated as proposed in Section 4.

In this scenario, the stream may return new position information of the query object as well. Generally, when the stream returns new position information of the query q, the probabilities of all objects being ranked before q may change. Consequently, the inverse ranking result usually needs to be recomputed from scratch, using the technique shown in Section 4.1. However, in most applications, the position of an object only changes slightly. Therefore, the probability of other objects to have a higher score than q normally does not change for most objects. We exploit this property as follows.

Let Q be the query object with alternative instances $q_1^t, ..., q_m^t \in Q$ at time t and let $S_{min}^t(Q)$ and $S_{max}^t(Q)$ denote the minimum and maximum among all possible scores derived from the instances of Q at time t. In the following we assume that new query object instances are reported from the stream at time $t + 1$:

Lemma 1. *If* $S_{min}^t(Q) \leq S_{min}^{t+1}(Q)$, *then for any object* o_x *with* $p_{o_x}^t = 0$ *it holds that* $p_{o_x}^{t+1} = 0$ *assuming* x *has not been updated at time* $t + 1$.

Proof

$$\text{Assumption: } S_{min}^t(Q) \leq S_{min}^{t+1}(Q) \tag{5}$$

$$\text{Assumption: } \forall i : S^t(x_i) = S^{t+1}(x_i) \tag{6}$$

$$\text{Assumption: } p_{o_x}^t = 0 \tag{7}$$

$$(7) \Leftrightarrow \forall q \in Q, \forall x_i \in x : S^t(q) > S^t(x_i) \Leftrightarrow \forall x_i \in o_x : S_{min}^t(Q) > S^t(x_i) \tag{8}$$

$$Def : \forall q \in Q, \forall x_i \in o_x : S^{t+1}(q) \geq S_{min}^{t+1}(Q)$$

$$\overset{5}{\geq} S_{min}^t(Q)$$

$$\overset{8}{\geq} S^t(x_i)$$

$$\overset{6}{=} S^{t+1}(x_1)$$

$$\Rightarrow \forall q \in Q, \forall x_i \in o_x : S^{t+1}(q) \geq S^{t+1}(x_1)$$

$$\Leftrightarrow p_{o_x}^{t+1} = 0$$

Lemma 2. *If* $S_{max}^t(Q) \geq S_{max}^{t+1}(Q)$, *then for any object* o_x *with* $p_{o_x}^t = 1$ *it holds that* $p_{o_x}^{t+1} = 1$.

Proof. Proof analogous to Lemma 1.

With the above Lemmata we can reduce the number of objects that have to be considered for re-computation of the inverse ranking at time $t + 1$. Especially, if $S_{min}^t(Q) \leq S_{min}^{t+1}(Q) \wedge S_{max}^{t+1}(Q) \geq S_{max}^{t+1}(Q)$, then we have to compute $p_{o_x}^{t+1}$ for those objects $o_x \in \mathcal{D}$ for which $p_{o_x}^t \notin \{0, 1\}$. For the remaining objects o we have to update p_o^t and the inverse ranking probabilities considering the cases outlined in Section 4.2. Let us note, that the effectiveness of this pruning scheme highly depends on the grade of uncertainty of the objects. In our experiments, we show that the number of objects pruned from the computation of the inverse ranking can be very large.

A very drastic change of the position of the query object may, in the worst case, cause all probabilities $p_{o_x}^t, o_x \in \mathcal{D}$ to change. The incremental computation of Section 4 requires two computations: The removal of the effect of $p_{o_x}^t$ and the incorporation of $p_{o_x}^{t+1}$ for any object $o_x \in \mathcal{D}$ that changed its probability of having a higher score than q. In contrast, a computation from scratch requires only one computation for each $o_x \in \mathcal{D}$: the incorporation of $p_{o_x}^{t+1}$. Therefore, it is wise to switch to a full re-computation of the PIR if more than $\frac{n}{2}$ objects change their probability.

6   Experiments

In the bigger part of the experimental evaluation, we use a synthetic dataset modelling a data stream with observations of 2-dimensional objects. The location of an object o_x at time t is modelled by m alternatives of a Gaussian distributed random variable X_{o_x} maintained in an array called *sample buffer*. For each $o_x \in \mathcal{D}$, the mean $E(X_{o_x})$ follows a uniform $[-10, 10]$-distribution in each dimension. The probabilistic stream A contains, for each $o_x \in \mathcal{D}$, exactly 10 alternative positions, that are randomly shuffled into the stream. Once a new alternative position of an object o_x is reported by the stream, it is stored in the sample buffer of o_x by replacing the least recently inserted one. We tune three parameters to evaluate the performance of the incremental PIR method described in Section 4: the database size n (default $n = 10,000$), the standard deviation σ of uncertain object instances (default $\sigma = 5$), and the sample buffer size m. For the scalability experiments, we chose $m = 3$. The evaluation of σ was performed with $m = 10$. In addition, we experimentally evaluate the influence of the degree of uncertainty

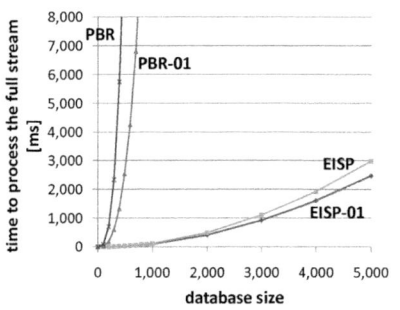

(a) **PBR** vs. **EISP** (full processing)

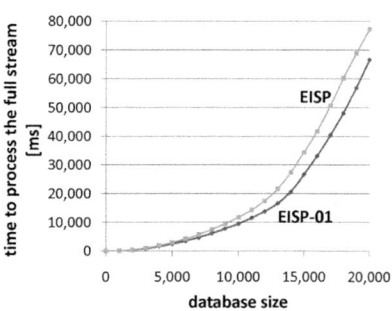

(b) **EISP** vs. **EISP-01** (full processing)

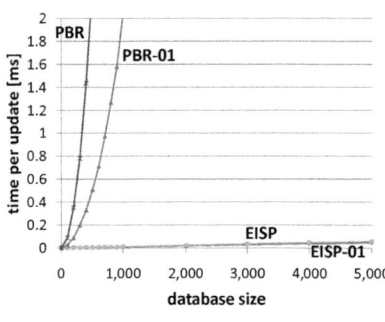

(c) **PBR** vs. **EISP** (single update)

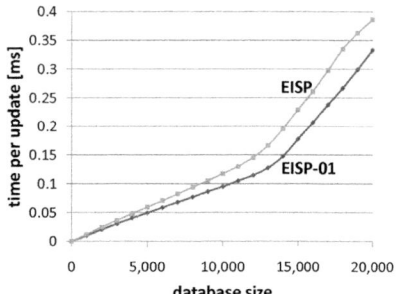

(d) **EISP** vs. **EISP-01** (single update)

Fig. 2. Scalability of the PIR approaches

on the performance of our incremental PIR method (cf. Section 5). Finally, in Section 6.5, we examine the scalability issues on a real-world dataset.

We denote our approach by **EISP** (Efficient Inverse Stream Processing). For comparison, we implemented the Poisson binomial recurrence based algorithm (abbreviated by **PBR**) as proposed by [1] that uses Equation 1, at each point of time where the stream provides a new observation. In addition, we evaluate the effect of the strategy proposed in Section 4 to avoid computation of objects o_x with a probability $p_{o_x}^t \in \{0, 1\}$ of having a higher score than q. This strategy will be denoted as *01-Pruning*. **EISP-01** and **PBR-01** denote the versions of **EISP** and **PBR**, respectively, that use *01-Pruning*.

6.1 Scalability

In the first experiment, we evaluate the scalability of **EISP**, **PBR**, **EISP-01** and **PBR-01** w.r.t. the database size n. We choose $k = n$ because if k is chosen constant and n is scaled up, the number of objects that certainly have a higher score than q will eventually reach k. In this case, *01-Pruning* will immediately notice that q cannot possibly be at one of the first k positions and will prune the computation. Then **EISP-01** and **PBR-01** have no further update costs. The results of these experiments are shown in Figure 2.

Figures 2(a) and 2(b) evaluate the total time required to process the whole stream, i.e. all $10 \cdot n$ object updates. It can be observed that all four algorithms show a superlinear time complexity to process the whole stream (cf. Figure 2(a)). In addition, the utilization of *01-Pruning* leads to an improvement in the runtime. As the number of uncertain objects (i.e. the objects in the database for which it is uncertain whether they have a higher score than q and thus cannot be removed by *01-Pruning*) increases as well as the number of certain objects, we obtain a linear speed-up gain using *01-Pruning*.

For a more detailed evaluation of the update cost in each iteration, consider Figures 2(c) and 2(d): Here, the average time required for an update is shown. Note that the update cost of both **PBR** and **PBR-01** grows fast with n. This is explained by the quadratic cost of $O(k \cdot n)$ (recall that we chose $k = n$) of the Poisson binomial recurrence at each update step. On the other hand, the update cost of $O(k)$ of **EISP** is linear to the number of database objects in this experiments (due to $k = n$). Here, *01-Pruning* has high influence on **PBR** but smaller effect on **EISP** especially for $n \leq 5,000$. The effect of *01-Pruning* may seem low for **EISP**, but note that in our experiments we measured the total time required for an update: This includes the time required to fetch a new location from the stream, compute its score, and recompute the total probability that the respective object has a higher score than q. This overhead is naturally required for any approach.

6.2 Standard Deviation σ

In the next experiment, we test the effect of the standard deviation σ on the distribution of location instances. Here, the total time required to process the whole stream was examined. The results are depicted in Figure 3. As **PBR** has to process all objects in each iteration of the inverse ranking, there is no influence of σ when this method is used (cf. Figure 3(a)). *01-Pruning* is able to reduce the runtime complexity having low values for σ, as many uncertain objects do not overlap with the score function and can

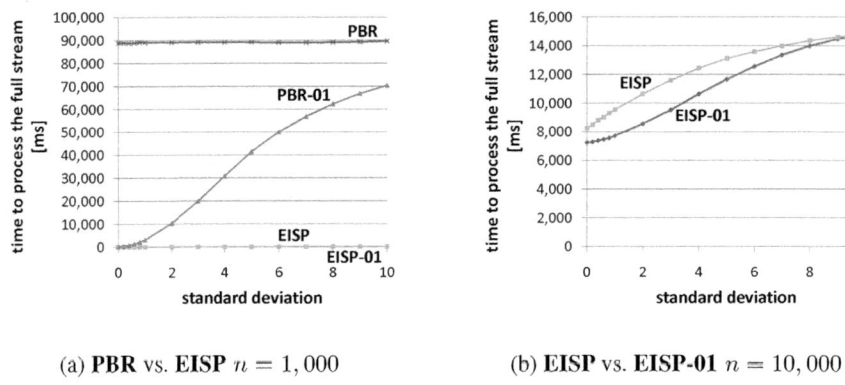

(a) **PBR** vs. **EISP** $n = 1,000$ (b) **EISP** vs. **EISP-01** $n = 10,000$

Fig. 3. Runtime evolution w.r.t. the standard deviation σ

therefore be neglected in each iteration. However, with an increasing value of σ, the cost of **PBR-01** approaches that of **PBR**, as the uncertainty ranges are spread over a greater range of the data space. **EISP** and **EISP-01** outperform the other methods by several orders of magnitude. Figure 3(b) shows that, for a small value of σ, there is a significant effect of *01-Pruning*. This becomes evident considering that the time overhead required to process the stream is about 7000 ms in this experiment. The reason is that for $\sigma = 0$ *01-Pruning* there exists no uncertainty, and thus all objects always have a probability of either 0 or 1 of having a higher score than q. Thus, Case I and Case II (cf. Section 4) are used in each update step and the Poisson binomial recurrence is never required. For $\sigma > 10$ most objects o_x have a probability $0 < p_{o_x}^t < 1$ of having a higher score than q. Thus, Case III is used in each iteration and C^t approaches zero.

6.3 Sample Buffer Size m

Next, the total stream processing time was evaluated w.r.t. the sample buffer size m. Figure 4 shows that m has an impact on all inverse ranking methods. Again, using **PBR**, the number of considered alternatives only influences the required runtime if we apply *01-Pruning* (cf. Figure 4(a)). If m increases, the probability that an object o has both instances with a higher and smaller score than q increases, i.e. it is uncertain whether $S(q) > S(o)$. Figure 4(b) shows that even for $m = 10$, we obtain a relatively high performance gain using *01-Pruning*, since the alternatives remain in the extent of their probabilistic distribution. Thus, for many objects o, $S(q) > S(o)$ can be decided even for a large m.

6.4 Uncertain Query

Finally, we evaluate the case that the query q is given as an uncertain stream object, now denoted by Q. As described in Section 5, the whole inverse ranking has to be recomputed by the **PBR** method if a position update of Q occurs. We test the performance of our adapted **EISP** method for this case.

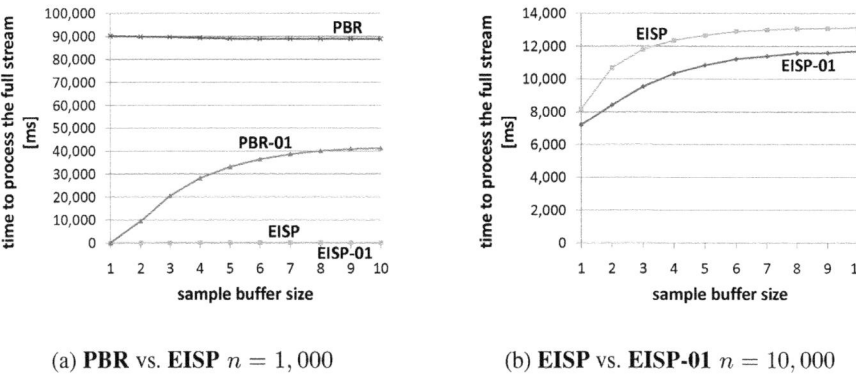

(a) **PBR** vs. **EISP** $n = 1,000$ (b) **EISP** vs. **EISP-01** $n = 10,000$

Fig. 4. Runtime evolution w.r.t. the sample buffer size m

For each time stamp t, we vary a probability value for Q of being updated and compare the versions of **PBR** with **EISP** that use *01-Pruning* in Figure 5(a). A value of 0 corresponds to the case that Q is certain, whereas a value of 1 assumes an update of Q in each iteration and thus forces **EISP-01** to always recompute the actual inverse ranking. It can be observed that the runtime required for processing the whole stream when using **EISP-01** increases linearly with a growing probability of the query object of being uncertain. This effect is due to the fact that the number of updates of Q and thus the number of complete re-computations have to be done according to the chosen probability value. As **PBR-01** does not depend on the uncertainty of Q because it recomputes the inverse ranking in each iteration anyway, its curve defines an upper asymptote to the curve of **EISP-01**.

6.5   Scalability Evaluation on Real-World Data

For an experimental evaluation of the scalability on real-world data, we first utilize the International Ice Patrol (IIP) Iceberg Sightings Dataset[4]. This dataset contains information about iceberg activity in the North Atlantic from 2001 to 2009. The latitude and longitude values of sighted icebergs serve as 2-dimensional values positions up to 6216 probabilistic objects, where each iceberg has been sighted at different positions. The stream consists of up to 10 positions of each iceberg which are ordered chronologically. Here again, we chose $m = 3$. Figure 5(b) indicates that the observations made for synthetic data can be transferred to real-world data. Note that for this dataset, *01-Pruning* is very effective, since the position of an iceberg has a very small variance. Many icebergs even appear to hold their position over time.

The next set of experiments uses the NBA Dataset[5], containing information about North American basketball players. Each of the 3738 records in this dataset corresponds

---

[4] The IIP dataset is available at the National Snow and Ice Data Center (NSIDC) web site (*http://nsidc.org/data/g00807.html*).

[5] The NBA dataset was derived from *www.databasebasketball.com*.

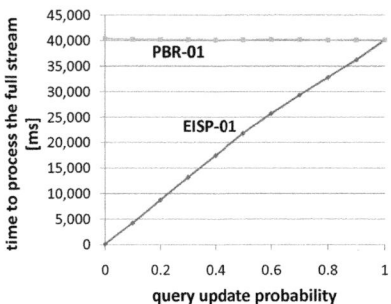

(a) Runtime evolution w.r.t. the probability of updating the query object ($n = 1,000$).

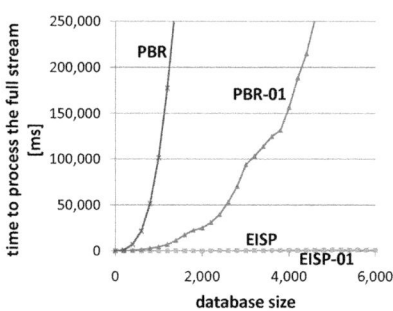

(b) Scalability of the PIR approaches regarding full processing on the IIP dataset.

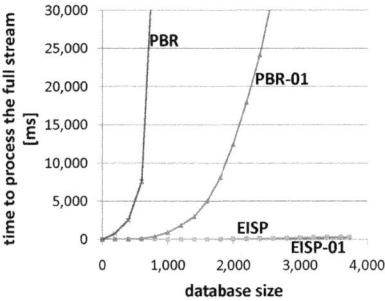

(c) Scalability of the PIR approaches regarding full processing on the NBA dataset.

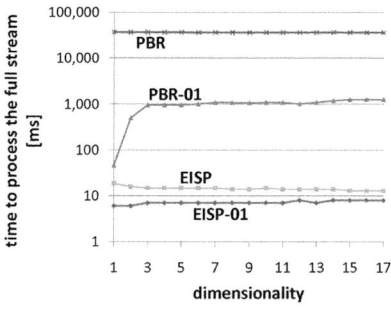

(d) Scalability of the PIR approaches w.r.t. the data dimensionality regarding full processing on the NBA dataset.

Fig. 5. Additional experiments

to the performance of one player in one season. In particular, each record contains a total of 17 dimensions representing the number of games played, the number of points scored, etc. in one given season between the years 1946 and 2006. For our experiments, we model players by uncertain stream objects, using a sliding window model of size $m = 3$, that is, a player is described by his performance in the last three years. The probabilistic stream contains all records of the dataset. For simplicity, the score function $s(x)$ we use is simply the sum of all (normalized) attributes. In this scenario, the semantic of a PIR query is to compute, for any given time, the rank of player Q with respect to all NBA players. First, we evaluated the scalability of our PIR algorithm in Figure 5(c) using all 17 dimensions. It can be observed that the scalability is very similar to the IIP dataset, despite of the increased dimensionality. This is further evaluated in Figure 5(d) where we scale the number of dimensions. For the approach that do not utilize *01-Pruning*, the runtime appears to be constant in the number of dimensions.

This can be explained by the fact that the dimensionality only affects the computation of the score of an object. Since we use the sum of all dimensions, we theoretically expect the algorithm to scale linearly in the number of dimensions, but the impact of this linear computation can be neglected. It can also be observed that, using *01-Pruning*, the runtime increases for low dimensions, and then becomes constant for higher dimensions. This can be explained by the uncertainty of the individual dimensions: The first dimension represents the number of games played by a player, which is a variable with a rather low deviation for each player. Even if a player has a very volatile performance, the number of games he played may be about the same. Therefore, the one dimensional dataset has a rather low uncertainty, and thus, a lower runtime (cf. Section 6.2). However, a bad player may be replaced, and thus not play the full time, which is covered by the second dimension, that aggregates the number of minutes played in a year and has a higher deviation. The third dimension has the highest uncertainty, as it describes the number of points scored by a player in a year. After the third dimension, adding further dimensions does not significantly increase the total deviation of the sum (i.e. the score) of a player. In summary, increasing the dimensionality has no significant effect on the runtime, but may increase the uncertainty of the object, thus indirectly increasing the runtime.

7 Conclusions

In this paper, we proposed a general solution to efficiently answering probabilistic inverse ranking queries on streams. State-of-the-art approaches solving the PIR query problem for static data are not applicable for stream data due to the $O(k \cdot n)$ complexity of the Poisson binomial recurrence. We have shown theoretically and experimentally that the update cost of our approach is $O(k)$ and thus applicable for stream databases. Let us note that our framework can be easily adapted to tackle further variants of inverse ranking/top-k queries on streams: the threshold probabilistic inverse ranking query, that returns exactly those ranking positions i for which $P_q^t(i)$ is greater than a user-specified parameter $\tau \in [0,1]$, as proposed in [1], and the (threshold) probabilistic top-k query, that returns the probability that q is one of the best k objects in the database. The latter has many applications in decision-making environments.

One aspect of future work is to develop an approximate approach, which is able to efficiently cope with continuous data models. The idea is to derive for each database object O, a lower and an upper bound of the probability that O has a higher score than Q. Using these approximations, we can apply the concept of uncertain generating functions [19] in order to obtain an (initial) approximated result of a PIR query, which guarantees that the true result is bounded correctly. The problem at hand is to update these uncertain generating functions efficiently when an update is fetched from the stream.

References

1. Lian, X., Chen, L.: Probabilistic inverse ranking queries over uncertain data. In: Zhou, X., Yokota, H., Deng, K., Liu, Q. (eds.) DASFAA 2009. LNCS, vol. 5463, pp. 35–50. Springer, Heidelberg (2009)

2. Li, C.: Enabling data retrieval: By ranking and beyond. In: Ph.D. Dissertation, University of Illinois at Urbana-Champaign (2007)
3. Yi, K., Li, F., Kollios, G., Srivastava, D.: Efficient processing of top-k queries in uncertain databases with x-relations. IEEE Trans. Knowl. Data Eng. 20(12), 1669–1682 (2008)
4. Cheng, R., Kalashnikov, D., Prabhakar, S.: Evaluating probabilistic queries over imprecise data. In: Proceedings of the ACM International Conference on Management of Data (SIGMOD), San Diego, CA (2003)
5. Böhm, C., Pryakhin, A., Schubert, M.: Probabilistic ranking queries on gaussians. In: SSDBM, pp. 169–178 (2006)
6. Cormode, G., Li, F., Yi, K.: Semantics of ranking queries for probabilistic data and expected results. In: Proceedings of the 25th International Conference on Data Engineering, ICDE 2009, Shanghai, China, March 29-April 2 (2009)
7. Soliman, M., Ilyas, I.: Ranking with uncertain scores. In: Proceedings of the 25th International Conference on Data Engineering, ICDE 2009, Shanghai, China, March 29-April 2, pp. 317–328 (2009)
8. Tao, Y., Cheng, R., Xiao, X., Ngai, W., Kao, B., Prabhakar, S.: Indexing multi-dimensional uncertain data with arbitrary probability density functions. In: Proceedings of the 31st International Conference on Very Large Data Bases (VLDB), Trondheim, Norway, pp. 922–933 (2005)
9. Agrawal, P., Benjelloun, O., Das Sarma, A., Hayworth, C., Nabar, S., Sugihara, T., Widom, J.: Trio: A system for data, uncertainty, and lineage. In: Proceedings of the 32nd International Conference on Very Large Data Bases (VLDB), Seoul, Korea (2006)
10. Babcock, B., Babu, S., Datar, M., Motwani, R., Widom, J.: Models and issues in data stream systems. In: PODS, pp. 1–16. ACM, New York (2002)
11. Muthukrishnan, S.: Data streams: algorithms and applications. Found. Trends Theor. Comput. Sci. 1(2), 117–236 (2005)
12. Jayram, T.S., Kale, S., Vee, E.: Efficient aggregation algorithms for probabilistic data. In: SODA, pp. 346–355. Society for Industrial and Applied Mathematics, Philadelphia (2007)
13. Aggarwal, C.C., Yu, P.S.: A framework for clustering uncertain data streams. In: ICDE, pp. 150–159. IEEE Computer Society, Washington, DC, USA (2008)
14. Ré, C., Letchner, J., Balazinksa, M., Suciu, D.: Event queries on correlated probabilistic streams. In: SIGMOD, pp. 715–728. ACM, New York (2008)
15. Cormode, G., Garofalakis, M.: Sketching probabilistic data streams. In: SIGMOD. ACM, New York (2007)
16. Lee, M.C.K., Ye, M., Lee, W.C.: Reverse ranking query over imprecise spatial data. In: COM. GEO. (2010)
17. Lange, K.: Numerical analysis for statisticians. In: Statistics and Computing (1999)
18. Bernecker, T., Kriegel, H.P., Mamoulis, N., Renz, M., Züfle, A.: Scalable probabilistic similarity ranking in uncertain databases. IEEE Trans. Knowl. Data Eng. 22(9), 1234–1246 (2010)
19. Bernecker, T., Emrich, T., Kriegel, H.P., Mamoulis, N., Renz, M., Züfle, A.: A novel probabilistic pruning approach to speed up similarity queries in uncertain databases. In: Proceedings of the 27th International Conference on Data Engineering (ICDE), Hannover, Germany (2011)

Finding Haystacks with Needles: Ranked Search for Data Using Geospatial and Temporal Characteristics

V.M. Megler and David Maier

Computer Science Department, Portland State University
vmegler@cs.pdx.edu, maier@cs.pdx.edu

Abstract. The past decade has seen an explosion in the number and types of environmental sensors deployed, many of which provide a continuous stream of observations. Each individual observation consists of one or more sensor measurements, a geographic location, and a time. With billions of historical observations stored in diverse databases and in thousands of datasets, scientists have difficulty finding relevant observations. We present an approach that creates consistent geospatial-temporal metadata from large repositories of diverse data by blending curated and automated extracts. We describe a novel query method over this metadata that returns ranked search results to a query with geospatial and temporal search criteria. Lastly, we present a prototype that demonstrates the utility of these ideas in the context of an ocean and coastal margin observatory.

Keywords: spatio-temporal queries, querying scientific data, metadata.

1 Introduction

In the past decade, the number and types of deployed environmental sensors have exploded, with each sensor providing a sequence of observations. Each individual observation has one or more sensor measurements and is associated with a geographic location and a time. Almost a decade ago, this explosion was described as "the Data Deluge" [14], and continued exponential growth in data volumes was predicted [19]. For example, an oceanography observatory and research center with which we collaborate (CMOP, http://www.stccmop.org) now has terabytes of observations spanning more than a decade, reported by a changing set of fixed and mobile sensors. This collection of data provides a rich resource for oceanographic research.

Scientists now research ecosystem-scale and global problems. Marine biologists wish to position their samples within a broader physical context; oceanographers look for comparative times or locations similar to (or dissimilar from) their research target. They want to search these collections of sensor observations for data that matches their research criteria. However, it is getting harder to find the relevant data in the burgeoning volumes of datasets and observations, and the time involved in searching constrains scientist productivity and acts as a limit on discovery. For example, a microbiologist may be looking for "any observations near the Astoria Bridge in June 2009" in order to place a water sample taken there into physical context. Within the

J.B. Cushing, J. French, and S. Bowers (Eds.): SSDBM 2011, LNCS 6809, pp. 55–72, 2011.
© Springer-Verlag Berlin Heidelberg 2011

observatory, there are many observation types that the microbiologist needs to search. Observations range from a point in space at a point in time, such as a group of water samples, through fixed stations, which have a single point in space but may have a million observations spanning a decade, to mobile sensors. The mobile sensors may collect millions of observations over widely varying geographic and temporal scales: science cruises may cover hundreds of miles in the ocean over several weeks, while gliders and autonomic unmanned vehicles (AUVs) are often deployed for shorter time periods – hours or days – and a few miles, often in a river or estuary. Locating and scanning each potentially relevant dataset of observations is time-consuming and requires understanding each dataset's storage location, access methods and format; the scientist may not even be aware of what relevant datasets exist. Once geospatially located, fixed sensors can easily be filtered based on location but must still be searched on time; identifying whether mobile sensors were close by at the appropriate time may require time-consuming individual analyses of each sensor's observations.

The scientists have powerful analysis and visualization tools available to them (e.g., [16, 25, 27]); however, these tools must be told the dataset and data ranges to analyze or visualize. While these tools allow the scientist to find needles in a haystack, they do not address the problem of which haystacks are most likely to contain the needles they want. Visualizing a dataset of observations for the desired location in June may confirm there is no match. However, potentially relevant substitutes "close by" in either time or space (say, from late May in the desired place, or from June but a little further away) are not found using current methods, much less ranked by their relevance. Even with a search tool that can find data in a temporal or spatial range, the scientist may not know how far to set those bounds in order to encompass possible substitutes.

We can meet this need by applying concepts from Information Retrieval. The scientists' problem can be cast as a compound geospatial-temporal query across a collection of datasets containing geospatial and temporal data; the search results should consist of datasets ranked by relevance. The relevance score for each dataset should be an estimate of the dataset content's geographic and temporal relevance to the query. The desire for real-time response implies that the query be evaluated without scanning each dataset's contents.

This paper describes a method for performing such a ranked search. Our contributions are:

1. An approach, described in Section 2, to scoring and ranking such datasets in response to a geospatial-temporal query. We calculate a single rank across both geospatial and temporal distances from the query terms by formalizing an intuitive distance concept. The approach is scalable and light-weight.

2. An approach, described in Section 3, for creating metadata describing the relevant geospatial and temporal characteristics of a collection of scientific datasets to support the ranking method. The metadata supports hierarchical nesting of datasets, providing scalability and flexibility across multiple collection sizes and spatial and temporal scales.

3. A loosely-coupled, componentized architecture that can be used to implement these approaches (Section 4).

4. A tool that implements these ideas and demonstrates their utility in the setting of an ocean observatory, in Section 5. Figure 5 shows the user interface.

We provide additional notes and implications of our approach in Section 6, describe related work in Section 7 and conclude with future research (Section 8).

In devising the details of our approach, we are biased towards identifying computationally light-weight approaches in order to achieve speed and scalability; as noted in considering the success of Google Maps, "Richness and depth are trumped by speed and ease, just as cheap trumps expensive: not always, but often." [22] We are also biased towards exploiting well-studied and optimized underlying functions and techniques wherever possible. We assume that after a successful search the scientist (who we also call the user) will access some subset of the identified datasets; we generically refer to a "download", although it may take other forms.

2   Ranking Space and Time

The scientist identifies a physical area and a time period he or she wishes to explore, which we will refer to as the *query*; we define the query as consisting of both geospatial and temporal *query terms*. The scientists have a qualitative intuition about which observations they consider a complete geospatial or temporal match, a relatively close match, or a non-match for their queries.

The top of Figure 1 shows a temporal query term, denoted T, with a line representing the query time span of "June". We consider the temporal query to have a center and a radius; here, the center is June 15 and the radius 15 days. Lines A(t), B(t), ..., E(t) represent the time spans of observations stored in datasets A, B, ..., E. Span A(t) represents a complete match; all observations in this dataset are from June. Span C(t)'s observations span the month of May and so is "very close"; similarly, Span B(t) is "closer" than Span C(t) but is not a complete match. Span D(t) is further away and Span E(t), with observations in February, is "far" from the June query.

The bottom section of Figure 1 shows a two-dimensional geospatial query term G as drawn on a map, represented by a central point *P* (in our running example, geocoordinate 46.23,-123.88, near the Astoria bridge), and a radius *r* (½ km) within which the desired observations should fall. The marker labeled A(g) represents the geospatial extent of observations in dataset A; here, they are at a single location, for example a fixed station or a set of observations made while anchored during a cruise. Extents B(g), E(g) and F(g) represent single-location datasets further away from the query center. Linear Extents C(g) and D(g) represent transects traveled by a mobile observation station such as a cruise ship, AUV or glider. Polygonal Extents J(g) and K(g) represent the bounding box of a longer, complex cruise track. Point Extent A(g) falls within the radius of the query and so is a complete match to the geographic query term. The qualitative comparison remains consistent across geometry types, with marker B(g) and line C(g) both being considered "very close" and polygon K(g) and marker F(g) being "too far" from the query to be interesting.

Intuitively, these qualitative comparisons can be scaled using a multiple of the search radius. For example, if the scientist searches for "within ½ km of P", then perhaps a point 5 km away from P is "too far". However, if the scientist searches for "within 5 km of P", then 5 km away from P is a match but 50 km is too far. In fact, the scientist is applying an implicit scaling model that is specific to his task [24].

The same intuitive scaling can be applied across both the temporal and spatial query terms; temporal observations at F(t) and spatial observations at marker B(g) could be considered equidistant from their search centers. Further, when considering both the temporal and spatial distances simultaneously, the dataset F, with temporal observations F(t) (quite close) at location F(g) (too far), is further from the query than datasets A ("here" in both time and space), B and C ("quite close" in both time and space). These examples illustrate the situation of one dataset *dominating* another: being closer in both time and space. The more interesting case arises in ranking two datasets where neither dominates the other, such as D and F: F is temporally closer, but D is closer in space. To simplify such comparisons, we propose a numeric distance representation that uses the query radii as the weighting method between the temporal and geospatial query terms. For example, had the spatial portion of the query been "within 5 km of P", D(g) and F(g) would both be considered "here" spatially, but D would now be dominated by F since it is temporally dominated by F.

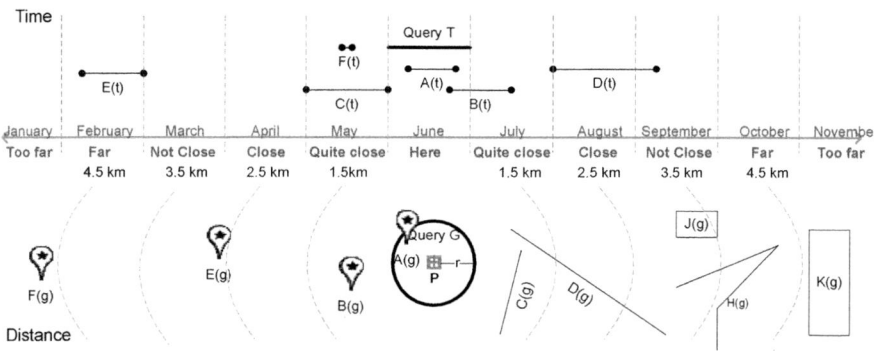

Fig. 1. Example of qualitative geospatial and temporal ranking: the top section shows a temporal query T and the time spans of various observation datasets. Dataset A(t) is a complete match, while datasets B(t), C(t), D(t), E(t) and F(t) are at increasing times from the query. The bottom section shows a geospatial query G, with the geospatial locations and extents of the same observation datasets represented by points (shown by markers), polygons and lines at various distances. In the middle is a qualitative scale that applies to both time and space.

In essence, the observations within a dataset represent a distribution of both temporal and geospatial distances from the query center, with a single point in time or space being the most constrained distribution. Each query term itself represents a distribution of times and locations. In order to rank the datasets, we need a single distance measure to characterize the similarity between the dataset and the query terms. There are many options for representing the proximity of two such entities, with varying computational complexities [23]. A commonly used surrogate for distance between two geographic entities is centroid-to-centroid distance. While it is a poor approximation when the entities are large and close together, it is relatively simple to calculate, at least for simple geometries. However, this measure ignores the radii of the query terms, and does not directly identify overlaps between the geometries. Another well-studied distance measure is minimum (and maximum) distance between two entities. This distance can be estimated by knowing only the bounds of the entities. This latter measure more closely matches our criteria; it can be calculated quickly using information (the bounds) that can be statically extracted from a dataset. This measure can be used to identify key characteristics that will drive our ranking: whether a dataset is within our query bounds and so is a complete match; whether the query and dataset overlap or whether they are disjoint, and if so by how much. This discussion applies equally to the one-dimensional "space" of time. In combining the space and time metrics, we will need to "scale" them by the radii of the respective query terms.

To compute these comparisons across a potentially large number of datasets, we have formulated a numerical similarity value that takes into account query-term radius and dataset distribution and can be cheaply estimated with summary information about temporal or spatial distributions, such as the bounds.

For the temporal term, let Q_{Tmin} and Q_{Tmax} represent the lower and upper bounds of the query time range. Further let d_{Tmin} and d_{Tmax} represent the minimum and maximum times of observations in dataset d. For calculation purposes, all times are translated into a monotonically increasing real number, for example "Unix time". Equation 1 below calculates d_{Rmin}, the distance of dataset d's minimum time from the temporal query "center", i.e., the mean of Q_{Tmin} and Q_{Tmax}, then scales the result by the size of the query "radius", i.e., half its range. Similarly Equation 2 calculates d_{Rmax}, the "scaled time-range distance" of the dataset's maximum time. Equation 3 calculates an overall temporal distance d_{Tdist} for the dataset from the query: the first subcase accounts for a dataset completely within the query range, the second through fourth account for a dataset overlapping the query range above, below, and on both sides, and the last subcase accounts for a dataset completely outside of the query range.

Then, we let s represent a scaling function that converts the calculated distance from the query center into a relevance score, while allowing a weighting factor to be applied to the distance result; per Montello [24], the implicit scaling factor may change for different users or different tasks. Finally, Equation 4 calculates our overall time score d_{Ts} for this dataset by applying the scaling function to d_{Tdist}. In our current implementation, s is $(100 - f * d_{Tdist})$; that is, when the dataset is a complete match it is given a score of 100, whereas if it is f "radii" (currently $f = 10$) from the query center it is considered "too far" and given a score of 0 or less.

Similarly, let C represent the center location of the geospatial query and r the radius. Let the locations of all the observations within a single dataset d be represented by a single geometry g. By convention this geometry can be a point, line (or polyline) or polygon [12]. Let d_{Gmin} and d_{Gmax} represent the minimum and maximum distances of the geometry from C, using some distance measure such as Euclidean distance. Equation 5 calculates the overall distance measure for three subcases: the dataset is completely within the query radius; the dataset overlaps the query circle, or the dataset is completely outside the query circle. Equation 6 gives a geospatial-relevance score d_{Gs} for dataset d by again applying the scaling function s to the calculated overall distance measure.

In Equation 7, the geospatial score d_{Gs} and the temporal score d_{Ts} are composed to give an overall score d_{score}. Combining these two distance measures results in a multi-component ranking, which are the norm in web search systems today [7, 17, 18, 20].

We take a simple average of the two distance scores. Note, however, that each of these rankings has been scaled by the radii of the query terms; thus, the user describes the relative importance of time and distance by adjusting the query terms.

$$d_{R\,min} = \frac{d_{T\,min} - ((Q_{T\,max} - Q_{T\,min})/2 + Q_{T\,min})}{(Q_{T\,max} - Q_{T\,min})/2} \tag{1}$$

$$= (2d_{T\,min} - Q_{T\,max} - Q_{T\,min})/(Q_{T\,max} - Q_{T\,min})$$

$$d_{R\,max} = (2d_{T\,max} - Q_{T\,max} - Q_{T\,min})/(Q_{T\,max} - Q_{T\,min}) \tag{2}$$

$$d_{Tdist} = \begin{cases} 0 & d_{T\,min} \geq Q_{T\,min}, d_{T\,max} \leq Q_{T\,max} \\[2mm] \dfrac{(|d_{R\,max}|-1)^2}{2|d_{R\,max} - d_{R\,min}|} & d_{T\,min} \geq Q_{T\,min}, d_{T\,max} > Q_{T\,max} \\[2mm] \dfrac{(|d_{R\,min}|-1)^2}{2|d_{R\,max} - d_{R\,min}|} & d_{T\,min} < Q_{T\,min}, d_{T\,max} \leq Q_{T\,max} \\[2mm] \dfrac{(|d_{R\,max}|-1)^2 + (|d_{R\,min}|-1)^2}{2|d_{R\,max} - d_{R\,min}|} & d_{T\,min} < Q_{T\,min}, d_{T\,max} > Q_{T\,max} \\[2mm] (|d_{R\,min} + d_{R\,max}|/2)-1 & d_{T\,min} > Q_{T\,max} \vee d_{T\,max} < Q_{T\,min} \end{cases} \tag{3}$$

$$d_{Ts} = s(d_{Tdist}) \tag{4}$$

$$d_{Gdist} = \begin{cases} 0 & d_{Gmax} \leq r \\[2mm] \dfrac{(d_{G\,max}/r - 1)^2}{2(d_{G\,max} - d_{G\,min})/r} & d_{Gmin} \leq r, d_{G\,max} \geq r \\[2mm] (d_{G\,min} + d_{G\,max})/r - 1 & d_{Gmin} > r \end{cases} \tag{5}$$

$$d_{Gs} = s(d_{Gdist}) \tag{6}$$

$$d_{score} = (d_{Gs} + d_{Ts})/2 \tag{7}$$

Given a collection of candidate datasets, each dataset's d_{score} can be calculated. Optionally, datasets with $d_{score} \leq 0$ can be discarded. Remaining datasets are sorted in decreasing order of d_{score} into a ranked list and become the results of the query.

We performed a 40-person user study, asking respondents to rank pairs of datasets in response to spatial, temporal, and spatial-temporal queries. The questions included comparisons with different geometries (e.g., polyline to point or polygon). Except for a small number of outlier cases, across all categories, when agreement amongst respondents is greater than 50% our distance measure agrees with the majority opinion. When there is a large disagreement with our distance measure, there is generally large disagreement amongst the respondents. Not surprisingly, these cases are correlated with small differences in distance between the two options.

3   Metadata Representing a Dataset Collection

The scoring and ranking approach described here assumes availability of suitable metadata against which to apply these equations. This section describes creating this metadata from datasets with geospatial and temporal contents, using the collection of observation datasets at our oceanography center as examples. We focus here only on *inherent metadata* [15], that is, information derived from the datasets themselves.

The base metadata requirements of our ranking and scoring approach are simple: the temporal bounds of each dataset, represented as a minimum and maximum time; the spatial footprint of each dataset, represented by a basic geometry type such as a point, line or polygon; and a dataset identifier. The temporal bounds can easily be extracted by scanning the dataset. Similarly, every dataset's observations fall within a geographic footprint. For a single point location such as a fixed sensor, the dataset's metadata record is created by combining the time range information with the fixed geographic location of the sensor.

Mobile sensors store a series of observations along with the geographic location and time for each observation. The overall dataset can be represented by the time range and the maximum geospatial bounds of the points within the dataset, that is, a rectangle (polygon) within which all points occur. The geospatial bounds can be extracted during the scan of the dataset, identifying the lowest and highest x and y coordinates found. For mobile sensors that follow a path or a series of transects during which the observations are collected (as in our case), a more informative alternative is available; the series of points can be translated into a polyline with each pair of successive points representing a line segment. If appropriate, the polyline can be approximated by a smaller number of line segments. The simplified polyline can be compactly stored as a single geometry and quickly assessed during ranking.

To provide for additional expressiveness across the range of possible dataset sizes and scales, we incorporate the idea of hierarchical, nested metadata. Across our collection of observations, we have locations where a single water sample was collected, locations with millions of sensor observations made over many years, and multi-week ocean cruises where millions of observations were collected across several weeks with tracks that crossed hundreds of miles. The hierarchical metadata allows us to capture a simple bounding box for a complex cruise, but also drill down to the individual cruise segments to identify the subset closest to the query terms.

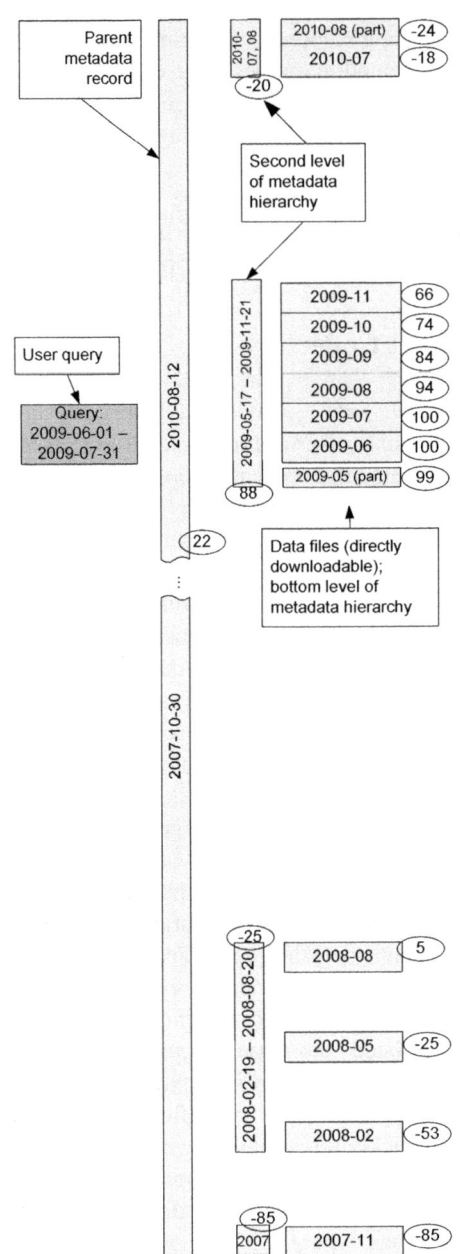

Fig. 2. Scoring example for intermittent data: the right-hand blocks represent downloadable datasets; the left-hand blocks represent the metadata hierarchy and curation choices (one record per year, plus one for the lifetime). Ovals show the scores given each dataset relative to the query.

Metadata records are classified recursively into parents and children. A record with no parent is a *root* record. A parent record's bounds (both temporal and geospatial) must include the union of the bounds of its children. The children's regions might not cover all of the parent, for example, if there are gaps in a time series. A record with no children is a *leaf* record. A metadata collection is made up a set of root records and their children (recursively). The number of levels within the hierarchy is not limited. For instance, we might decompose a cruise temporally by weeks and days within weeks, then segment each spatially.

The scoring method is applied recursively to the collection of metadata records. We initially retrieve and score root metadata records only. If an entry is deemed interesting, it is added to a list of records whose children will be retrieved on the next pass. An entry is deemed interesting if the minimum geographic and time range distance is not "too far", and the minimum and maximum scaled time or geographic range distances are different from each other. The second criterion implies that if subdivisions of this dataset are available, some of these subdivisions may be more highly relevant than the parent dataset as a whole. We repeat until either the list of records to retrieve is empty or no interesting records have children.

Figure 2 demonstrates these concepts. It shows a fixed sensor station that reports data only during some months. Each light-gray block in the diagram represents a metadata record, showing time duration. In this case, three levels of metadata exist: an overall lifetime record, a medium level for the portion in each year that the station reports data, and a detailed level consisting of a record for each month. Next to each metadata record is shown its score for the given query. It can be seen that there are two individual months that score 100; datasets on either side score in the 90s. The year in which those months occur scores 88, whereas years that do not overlap the query range receive negative relevance scores. The overall lifetime record, which overlaps the query at both ends, receives a score of 22. Parent and child records are returned in the query result, allowing the scientist to choose between accessing only the months of interest or the entire year.

4   Architecture

As shown in Figure 3, our architecture extends existing observatory repositories. In general, observatories contain several major components: a network of sensors; a set of processes that collect observations and normalize them (adjust record formats, apply calibrations, etc.); a repository to store the normalized observations; and a set of analysis programs that access the stored observations. There may also be a web interface that allows the user to view the catalog and download specific subsets of the data. To these existing system components, we add four loosely coupled components: a metadata-creation component, a metadata repository, a scoring-and-ranking component and a user interface.

The *metadata-creation component* extracts a minimal set of metadata from the contents of the observation repository to represent the source observations, and stores

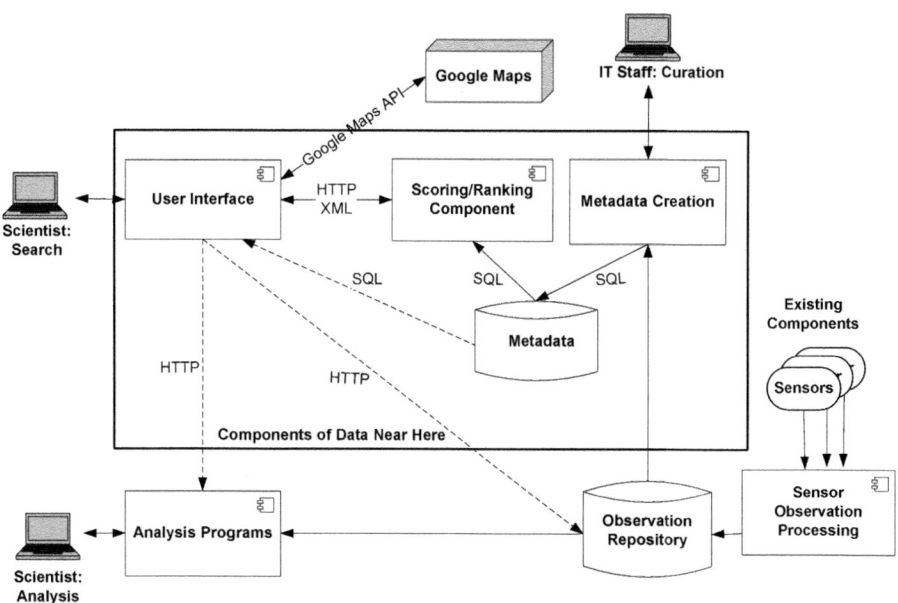

Fig. 3. The combined system and deployment diagram shows existing components and the new components added as part of Data Near Here

the extract into its own mini-repository. The goal is to support fast query access by creating a simple abstraction over a far more complex data repository. The IT staff can add new categories of observations (e.g., new types of mobile devices), change the number or grouping of hierarchical levels used to represent data, or change the representation of a category of observations (e.g., treating cruises solely as lines rather than as lines and bounding boxes at different levels of the hierarchy); this activity is a *data curation* process [13]. At present, these changes involve writing or modifying scripts; an informal set of patterns is emerging and could be formalized if desired.

The *scoring-and-ranking component* receives query terms from the user interface and interacts with the metadata. It scores each candidate metadata record, and returns to the user interface a set of ranked records. The scoring and ranking algorithm is loosely coupled with the metadata and is independent of the user interface, allowing different algorithms to be easily tested without modifying the other components.

The *user interface* is responsible for collecting the geospatial and temporal query terms from the user and presenting the search results; it also provides the user with some control over the presentation (e.g., the number of search results to return). The user interface exploits Google Maps [3] for geospatial representation of the query and results. The sole direct interaction between the user interface and the metadata is when the user interface requests metadata information to populate the query interface's selections (for example, the 'Category' entry field in Figure 5). The search results link to the datasets within the repository and optionally to analysis programs.

The loosely coupled nature of the components allows maximum flexibility in altering the internal design or methods used by any component without altering the remaining components; the additive nature of the architecture minimizes changes to the existing infrastructure necessary to add this capability.

5 "Data Near Here": An Implementation

The approaches described in this paper have been implemented in an internal prototype at the Center for Coastal Margin Observation and Prediction (CMOP). This center's rich inventory of over 250 million observations is available for public download or direct analysis; additional data can be accessed internally via a variety of tools. The observations and associated metadata are stored in a relational database: most datasets are also stored in NetCDF-formatted downloadable files.

The observational sensors can be loosely grouped by their deployment on fixed or mobile platforms. Mobile sensors are deployed in a series of *missions*, each of which may span hours or days or weeks. Observations may be captured many times a second, either continuously or according to some schedule; there may be a half million or more observations per mission. Hierarchically nested metadata is created at multiple scales; for the Astoria Bridge query, a fixed station that is far distant can be recognized and ignored by looking at a single lifetime entry for the station.

A fixed sensor has a single geographic location over time; its dataset can be geospatially characterized as a single point. Its continuous observations are, for convenience, stored in multiple datasets, each containing a single time range such as a month or (for sparser observations) a year. In addition to dataset leaf records, for each year's worth of observations we create a parent record that summarizes that year's data, plus a lifetime record for the overall time duration of the station.

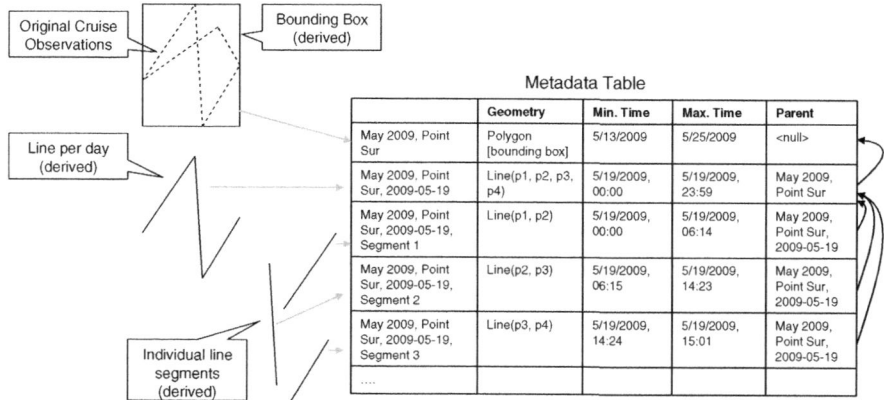

Fig. 4. Space metadata records for mobile stations (here, a multi-week cruise) are created by creating a line from point observations and simplifying it (middle hierarchy level, on line 2 of the table), then splitting the line into detailed line segments for the leaf records and extracting a bounding box for the parent record

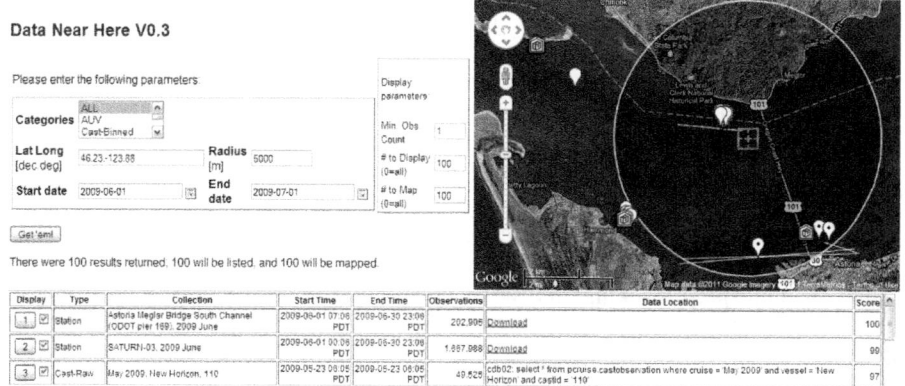

Fig. 5. Map display of Data Near Here search results for the example query in this paper. The map shows a section of the Columbia River near its mouth that includes Highway 101 crossing the Astoria Bridge between Oregon and Washington. The search center and radius are shown along with a set of markers and lines locating the highest-ranked datasets found for the search. The list below the map shows the four highest-ranked results, the first of which is a complete match; the next three are close either in time or space, but are not complete matches.

As is shown in Figure 4, the track for a mobile-sensor mission can be a represented by a polyline. In order to extract the polyline from the observations, we use the PostGIS *makeline* function to convert each day's worth of observations into a polyline, then apply the PostGIS implementation of the Douglas-Peucker algorithm, *simplify*, to create a simplified polyline. The simplified polyline, along with the day's start and end time, is stored as a metadata record. We create an additional metadata record for the lifetime of the mission; this record is simply the bounding box of the polylines with the begin and end times of the overall mission. We then programmatically extract each line segment from the simplified polyline, match the vertices to the time the mission was at that location, and store each line segment with its time range as a leaf metadata record. This three-level hierarchy for mobile sensors can be created quickly, and provides multiple scales of metadata.

At the end of these processes, we have a consistent metadata format for both fixed and mobile sensor observations. We also have the option of storing multiple sets of metadata representing the same (or similar) underlying data, if, for example, alternative groupings of the data are more appropriate for specific user groups (for example, partitioned by day or by tide). A varying number of levels can be used for a subset of the collection or even a subset of sensors within a specific category; we may wish to, for example, add a daily metadata record for specific fixed sensors. In other cases, such as water-sample data, we chose to only have one level in the hierarchy.

Keeping the metadata up-to-date involves adding new metadata records as new missions occur or new datasets are created. For each category of data, this update can occur automatically via a set of scripts and triggers that check for new datasets and execute the predefined steps. The moment a new metadata record is created, it is available to be searched. Setting up a new category of data requires deciding the number of hierarchical levels to be defined and the download granularities to support, and then setting up the appropriate scripts.

Figure 5 shows the tool's user interface. The user interface combines three interacting elements: a set of text query entry fields, a Google map that can be used to locate the geospatial query and on which the geospatial locations of highly ranked results are drawn, and the query results: a table of highly scoring datasets ordered by score. All available categories of observational data can be searched, or the scientist can limit the search to a subset. Scientists can provide both time and location parameters; they can also search for all times in which observations were taken at a specific location by leaving the time fields blank, and vice versa. The top-ranked results will be displayed on the map – the scientist can select how many results to return and to display. Clicking on a displayed dataset pops up a summary.

A "data location" field provides access to the data. Where the data can be directly downloaded, this field contains a download link. This link is built when the metadata is created and can contain parameters that subset the complete dataset to the relevant portion if the download mechanism allows. In cases where direct download is not currently possible, this field provides the scientist with the dataset location and an extract command for the dataset's access tool; for example, where the data is held only in a relational database, this field can contain a SQL Select statement to extract the relevant data. A future version will allow scientists to directly open a selected dataset in a visualization and analysis tool.

The technologies used to implement the shown architecture were selected based on existing technologies in use in the infrastructure, to allow for easy integration, extension and support. Metadata creation is performed in a combination of SQL and scripts. The repository is a PostGIS/Postgres database and is accessed via dynamic SQL; the footprint data is stored in a PostGIS geometry column. The scoring and ranking component is written in PHP. Geometric functions are performed by PostGIS during data retrieval from the repository, with final scoring and ranking performed in the PHP module. The user interface is implemented using Javascript, JQuery and the Google Maps API. Current experience leads us to believe these technologies will scale to support the observatory's repository for some time. For a much larger repository, other technology choices would provide greater speed. The architecture allows us to easily make these choices per component as needed.

6 Discussion

Here we discuss the tradeoff between user performance and the design of the metadata hierarchy. The response time seen by the user is driven by several main factors: data transfer times between the components (scoring component to user interface, metadata repository to scoring component); the number of hierarchical levels of metadata; the total number of metadata records to be scored; and the complexity of the scoring algorithm.

The intent of the metadata hierarchy is to bridge the gap between the dataset granularity and the footprint of the dataset's content, within the context of efficient real-time user search. The more hierarchical levels, the more queries must be issued to process the children of interesting metadata records; however, the hierarchical design should allow fewer metadata records to be scored overall. An alternative is to score all metadata records in a single query; however, as many of the roots will have an increasing number of descendents over time (e.g., stations that continue to collect data month after month), we expect that ruling out descendents by examining only the parent record will balance the overhead of multiple queries and allow for greater scalability. We expect the user, after a successful search, to download or analyze selected datasets from the results presented. Thus, there is an assumed alignment between a single metadata record and a single accessible or downloadable unit (such as a single dataset). However, in many cases the capability exists to group multiple datasets into a single accessible unit (e.g., by appending them), or alternatively to access subsets of a dataset (e.g., by encoding parameters to limit the sections of the dataset to access). The data curation process should consider the typical footprint and the likely utility to the scientist of different aggregations of that data.

From a query-performance perspective, the number of leaf metadata records is optimal when each dataset is described by a single metadata record and thus there is only one record per dataset to score and rank. Where a single dataset is geospatially and temporally relatively homogenous, this arrangement may be a practical choice. Where a dataset is geospatially or temporally very diverse or is too large to conveniently download, users are best served if a leaf metadata record exists for each subcomponent or segment they may wish to download. The hierarchy provides a mechanism for mediating this mismatch; a single metadata record can be created for a larger dataset with children for the subcomponents. The scoring component may be able to eliminate the dataset and its children from further consideration based on the parent, and only score the children when the parent appears interesting.

To provide a tangible example of this tradeoff, Table 1 shows summary counts for our currently existing metadata records, representing a subset of CMOP's repository. The breakdown by category in Table 2 highlights the different curation choices made for different observation categories. At one extreme, the 22 fixed stations have an average of 8.2 million observations each, and here a three-level hierarchy has been created. At the other extreme is the water-sample collection, with two observations taken per location and time. The same "cast" data is represented in two forms: one is the unprocessed, or "raw", collection of observations; the same data has also been binned to specific depths and averaged into a much smaller collection of measurements. Variation in geometric representation is also shown; in cruises, for example, the most detailed level is most commonly represented by line segments representing specific cruise transects, but is sometimes represented by points when the cruise vessel was anchored in a single location for a longer period of time. These different representations are easily discerned programmatically from the data but are difficult for a user to identify from the source data without significant effort.

Table 1. Characterization of Data Near Here Metadata. This table summarizes characteristics of the metadata records representing the 225 million observations currently searchable.

Metadata records	15,516
Number of observation categories	7
Records at each hierarchy level	
Roots without children	6,564
Roots with children	60
Children with children	800
Children with no children	8,092
Observations represented	225,627,211
Average observations per metadata record	14,541

Table 2. Characterization of Existing Metadata Records by Category

Category	Hierarchy Level	Geometry	Number of Records	Number with Children	Total Observations Represented	Average Observations per Record
AUV	1	Polygon, Line	22	11	225,757	10,261
	2	Line	29	0	134,841	4,649
Cast-Binned	1	Point	3,066	0	370,967	120
Cast-Raw	1	Point	2,908	0	33,908,614	11,660
Cruise	1	Polygon	20	20	8,064,259	403,212
	2	Line	607	607	8,064,259	13,285
	3	Line, Point	7,125	0	7,615,222	1,068
Glider	1	Polygon	7	7	2,237,628	319,661
	2	Line	128	128	2,237,628	17,481
	3	Line	357	0	1,670,470	4,679
Fixed Stations	1	Point	22	22	180,818,279	8,219,012
	2	Point	65	65	171,903,806	2,644,673
	3	Point	581	0	180,818,239	311,219
Water Samples	1	Point	579	0	1,707	2

The spatial scoring equations were designed to provide a reasonable approximation of distance for the three primary cases – polygon, polyline and point – while minimizing the number and complexity of spatial calculations needed; the current approach uses a total of two spatial calculations (maximum distance and minimum distance between two geometries) for each metadata record scored. Spatial functions can be slow, so minimizing the number and complexity of geometries handled is beneficial. A more complex spatial scoring system can easily be devised; what is less clear is whether, given the uncertainties in people's views of distance [24], the additional complexity provides a better distance score as perceived by the user. What is clear is that the additional complexity will add to the computation time.

7 Related Work

Adapting a definition from Information Retrieval (IR) [20], a dataset is relevant if the scientist perceives that it contains data relevant to his or her information need. In IR systems, the user provides query terms, usually a list of words, to be searched for against an index representing a library of items (where each item may be, for example, a web page). Each item is summarized as index entries of the words found in the document, created prior to receiving the user's query. In almost all cases, the searches are performed against metadata, which itself varies in source and form. In ranked retrieval, each item is given a score representing an estimate of the item's relevance to the query. The list of items is then ranked by ordering items from highest to lowest score. There is much research (e.g., [4, 20, 21]) into ranked relevance of unstructured text documents against text queries. We adapt these ideas to searching contents of scientific datasets with a query consisting of geospatial-temporal search terms which are themselves ranges. The metadata we extract from the datasets performs the role of the index.

Hill et al. [15] present a system for describing and searching a library's digital collection of geographic items. They apply widely accepted collection concepts from paper-based archives that are based on a textual description of a map series (publisher, title, number in series, etc.) to digital map collections. A single collection may contain a set of maps where each map has a different geographic coverage; however, the specific map's geographic coverage is an access or index key to that map. The challenge is how to represent these collections by searchable metadata. They differentiate *contextual metadata*, which is externally provided (e.g., publisher), from *inherent metadata*, derived from automated analysis of the data (e.g., count of items included in a collection). This automatic data analysis adds to the metadata but does not allow the content itself to be searched. They do not provide hierarchical metadata, nor do they discuss methods for ranked search results.

Grossner et al. [11] provide a summary of progress in the last decade in developing a "Digital Earth", and identify gaps in efforts so far. They note that current geographic and temporal search responses provide matches only on one level of metadata; the contents of cataloged digital objects are not exposed and are not searchable. Goodchild [8] notes that most geographic search systems score items based on word matches against metadata without considering the temporal span or geographic content of the items returned, and recognizes [9] the issue of *containment* as an open research question. That is, a map may be cataloged by the extent of its coverage (e.g., "Alaska") but the search mechanism has no method with which to recognize that this map is a match for an item contained within it, (e.g., a search for "Fairbanks"). Goodchild et al. [10] expand on these concerns in the 2007 review of Geospatial One-Stop (GOS) [1], a state-of-the-art government portal to geographic information. GOS and similar portals such as the Global Change Master Directory's Map/Date Search [2] now allow searches using both geographic and temporal criteria; three spatial tests are supported (the map view intersects, mostly contains, or completely contains the dataset), and temporal search appears binary – if items do not match the criteria they are not returned. Only one level of metadata is considered; if a relevant item is embedded within a larger item (Fairbanks within Alaska), the relevant item is not returned. In contrast, we explicitly rank returned items based on both the temporal and geographic "distance" of the dataset contents from the query, and address the containment issue with multiple levels of metadata.

One widely-used geospatial search system is Google Maps [22], which searches for a place name or a specified latitude and longitude, and provides nearby points of interest ("restaurants near here"). They do not currently expose a temporal search capability. It is possible for a site to explicitly link a dataset to a specific location using KML, but it is not currently possible to search ranges within linked datasets. Egenhofer [6] describes some desired geographic request semantics but does not propose an implementation.

Addressing a different kind of geographic search problem, Sharifzadeh and Shahabi [26] compare a set of data points with a set of query points where both sets potentially contain geographic attributes, and identify a set of points that are not *dominated* by any other points. They do not specifically address time, but could presumably treat it as another attribute. Their approach develops the database query and algorithm to return the best points, but, unlike our approach, they do not return ranked results nor place the queries within the context of a larger application.

Several researchers [16, 25, 27] have addressed the difficulty scientists have in finding "interesting" data — data relevant to the scientist's research question — within the exploding quantity of data now being recorded by sensors by focusing on visualization techniques for a specified set of data. The scientist specifies the dataset and range of data within the dataset. The system then presents a visualization of the specified numeric data. The question of how the scientist finds interesting datasets and ranges to visualize is not addressed; that question is the subject of this research.

8 Conclusion

The rapid expansion of deployed observational sensors has led to collection of more observational data than ever before available. The sheer volume of data is creating new problems for scientists trying to identify subsets of data relevant to their research. Techniques to help scientists navigate this sudden plethora of data are a fruitful area for research. This work is one such contribution, focusing on the problem of finding sets of observations "near" an existing location in both time and geospace.

This paper presents a novel approach to providing compound geospatial-temporal queries across a collection of datasets containing geospatial and temporal data; search results consist of datasets ranked by relevance and presented in real time. The approach combines hierarchical metadata extracted from the datasets with a method for comparing distances from a query across geospatial and temporal extents. This approach complements existing visualization techniques by allowing scientists to quickly identify which subset of a large collection of datasets they should review or analyze. The combination of data represented by its geospatial and temporal footprint, using the metadata for search, the metadata hierarchical design and overall loosely-coupled architecture allows for scalability and growth across large, complex data repositories. The prototype described already supports over quarter of a billion observations and more are being added. User response has been very positive.

We plan to extend this work in several directions, including characterizing the observed environmental variables and supporting more expressive queries. The third geospatial dimension, depth, is currently being added. *Contextual metadata* [15] – ownership, terms and conditions, etc. – will be added as the tool gains wider use. The eventual goal is to combine geospatial-temporal search terms with terms such as "with oxygen below 3 mg/liter, where Myrionecta Rubra are present".

Finding relevant data is key to scientific discovery. Helping scientists identify the "haystacks most likely to contain needles" out of the vast quantities of data being collected today is a key component of reducing their time to discovery.

Acknowledgments. This work is supported by NSF award OCE-0424602. We would like to thank the staff of CMOP for their support.

References

1. Geospatial One Stop (GOS), `http://gos2.geodata.gov/wps/portal/gos`
2. Global Change Master Directory Web Site, `http://gcmd.nasa.gov/`
3. The Google Maps Javascript API V3,
   `http://code.google.com/apis/maps/documentation/javascript/`
4. Baeza-Yates, R., Ribeiro-Neto, B.: Modern Information Retrieval. ACM Press, New York (1999)
5. Douglas, D.H., Peucker, T.K.: Algorithms for the reduction of the number of points required to represent a digitized line or its caricature. Cartographica 10(2), 112–122 (1973)
6. Egenhofer, M.J.: Toward the semantic geospatial web. In: Proceedings of the 10th ACM International Symposium on Advances in Geographic Information Systems, pp. 1–4 (2002)
7. Evans, M.P.: Analysing Google rankings through search engine optimization data. Internet Research 17(1), 21–37 (2007)
8. Goodchild, M.F., Zhou, J.: Finding geographic information: Collection-level metadata. GeoInformatica 7(2), 95–112 (2003)
9. Goodchild, M.F.: The Alexandria Digital Library Project: Review, Assessment, and Prospects (2004),
   `http://www.dlib.org/dlib/may04/goodchild/05goodchild.html`
10. Goodchild, M.F., et al.: Sharing Geographic Information: An Assessment of the Geospatial One-Stop. Annals of the AAG 97(2), 250–266 (2007)

11. Grossner, K.E., et al.: Defining a digital earth system. Transactions in GIS 12(1), 145–160 (2008)
12. Herring, J.R. (ed.): OpenGIS® Implementation Standard for Geographic information - Simple feature access - Part 1: Common architecture (2010)
13. Hey, T., Trefethen, A.: e-Science and its implications. Philosophical Transactions of the Royal Society of London. Series A: Mathematical, Physical and Engineering Sciences 361(1809), 1809 (2003)
14. Hey, T., Trefethen, A.E.: The Data Deluge: An e-Science Perspective. In: Berman, F., Fox, G., Hey, T. (eds.) Grid Computing: Making the Global Infrastructure a Reality, pp. 809–824. John Wiley & Sons, Ltd., Chichester (2003)
15. Hill, L.L., et al.: Collection metadata solutions for digital library applications. J. of the American Soc. for Information Science 50(13), 1169–1181 (1999)
16. Howe, B., et al.: Scientific Mashups: Runtime-Configurable Data Product Ensembles. Scientific and Statistical Database Management, 19–36 (2009)
17. Kobayashi, M., Takeda, K.: Information retrieval on the web. ACM Comput. Surv. 32, 144–173 (2000)
18. Lewandowski, D.: Web searching, search engines and Information Retrieval. Information Services and Use 25(3), 137–147 (2005)
19. Lord, P., Macdonald, A.: e-Science Curation Report (2003), http://www.jisc.ac.uk/uploaded_documents/e-ScienceReportFinal.pdf
20. Manning, C.D., et al.: An introduction to information retrieval. Cambridge University Press, Cambridge (2008)
21. Maron, M.E., Kuhns, J.L.: On relevance, probabilistic indexing and information retrieval. Journal of the ACM (JACM) 7(3), 216–244 (1960)
22. Miller, C.C.: A Beast in the Field: The Google Maps mashup as GIS/2. Cartographica 41(3), 187–199 (2006)
23. Miller, H.J., Wentz, E.A.: Representation and Spatial Analysis in Geographic Information Systems. Annals of the AAG 93(3), 574–594 (2003)
24. Montello, D.: The geometry of environmental knowledge. Theories and Methods of Spatio-Temporal Reasoning in Geographic Space, 136–152 (1992)
25. Perlman, E., et al.: Data Exploration of Turbulence Simulations Using a Database Cluster. In: Proceedings of the 2007 ACM/IEEE Conference on Supercomputing, pp. 1–11 (2007)
26. Sharifzadeh, M., Shahabi, C.: The spatial skyline queries. In: Proc. of VLDB, p. 762 (2006)
27. Stolte, E., Alonso, G.: Efficient exploration of large scientific databases. In: Proc. of VLDB, p. 633 (2002)

Using Medians to Generate Consensus Rankings for Biological Data

Sarah Cohen-Boulakia[1,2], Alain Denise[1,2,3], and Sylvie Hamel[4,*]

[1] LRI (Laboratoire de Recherche en Informatique), CNRS UMR 8623
Université Paris-Sud - France
{cohen,denise}@lri.fr
[2] AMIB Group, INRIA Saclay Ile-de-France - France
[3] IGM (Institut de Génétique et de Microbiologie), CNRS UMR 8621
Université Paris-Sud - France
[4] DIRO (Département d'Informatique et de Recherche Opérationnelle)
Université de Montréal - QC - Canada
sylvie.hamel@umontreal.ca

Abstract. Faced with the deluge of data available in biological databases, it becomes increasingly difficult for scientists to obtain reasonable sets of answers to their biological queries. A critical example appears in medicine, where physicians frequently need to get information about genes associated with a given disease. When they pose such queries to Web portals (e.g., Entrez NCBI) they usually get huge amounts of answers which are not ranked, making them very difficult to be exploited. In the last years, while several ranking approaches have been proposed, none of them is considered as the most promising.

Instead of considering ranking methods as alternative approaches, we propose to generate a *consensus ranking* to highlight the common points of a set of rankings while minimizing their disagreements. Our work is based on the concept of median, originally defined on permutations: Given m permutations and a distance function, the **median** problem is to find a permutation that is the *closest* of the m given permutations. We have investigated the problem of computing a median of a set of m rankings considering different elements and ties, under a generalized Kendall-τ distance. This problem is known to be NP-hard. In this paper, we present a new heuristic for the problem and we demonstrate the benefit of our approach on real queries using four different ranking methods.

Availability: http://bioguide-project.net/bioconsert

1 Introduction

With the increasing development of high throughput technologies, very high amounts of data are produced and stored in public databases to make them

---

* Supported by NSERC through an Individual Discovery Grant and a Digiteo Grant (2-months sabbatical leave at LRI).

J.B. Cushing, J. French, and S. Bowers (Eds.): SSDBM 2011, LNCS 6809, pp. 73–90, 2011.

available to the scientific community. Analysing and interpreting any new experimental result necessitates to compare it to public data. This task thus includes querying public databases using portals such as Entrez NCBI[1] or SRS[2]. However, even a simple query such as the search for *genes possibly associated with a given disease* may return thousands of answers. The need for ranking solutions, able to order answers is crucial for helping scientists organize their time and prioritize the new experiments to be possibly conducted. However, ranking biological data is a difficult task for various reasons: biological data are usually annotation files (e.g., a SwissProt, EntrezGene, or OMIM entry) which reflect expertise, they thus may be associated with various degrees of *confidence* [7]; data are not independent of each other but they are linked by cross-references, the network formed by these links plays a role in the *popularity* of the data; the need expressed by scientists may also be taken into consideration whether the most well-known data should be ranked first, or the freshest, or the most surprising [10]... As a consequence, although several ranking methods have been proposed in the last years [5,16,17,14], none of them has been deployed on systems currently used by the scientific community.

In this paper, our goal is to make the most of the results obtained using several ranking methods applied to biological data by generating a *consensus ranking* reflecting their common points while not putting too much importance on elements classified as "good" by only one or a few rankings. Most importantly, we want to provide a new method to take into account two specific features of our framework. First, ranking methods may apply filters on the results provided to the user to reduce the set of information provided. Ranking methods thus do not necessarily consider the exact same set of elements. Second, they may rank several elements at the same position (each ranking thus provides a list of sets); ties should thus be considered.

Interestingly, while the problem of finding a consensus of a set of rankings has not been addressed so far in the context of real biological data [5,16,17], it has been of great and increasing interest to the database community (e.g., [9,8]) in particular within the domain of Web querying. When the same set of elements is considered among rankings, the general problem of finding a median of rankings with ties is known to be NP-hard. As an answer, approximation algorithms have been proposed (in particular [9] and [1]). However, this kind of approaches has never been tested on biological data and is not able to deal with rankings considering different data sets.

In this paper, we first present a new heuristic for the problem, called *BioConsert* (for generating Biological Consensus ranking with ties) in order to enable the generation of a consensus from very large sets and speed-up the generation of "smaller" consensus. Second, we demonstrate the benefit of our approach on concrete biological queries using four ranking methods, and we compare our results with the approximation algorithms by Ailon [1] and Fagin *et al.* [9].

---

[1] http://www.ncbi.nlm.nih.gov/Entrez
[2] http://srs.ebi.ac.uk

The remainder of this paper is organized as follows. After a description of our application domain, Section 2 presents the mathematical framework of our work and in particular introduces the definition of distance and median we have chosen to follow while motivating our choices by the constraints given by our framework. Section 3 introduces the heuristic we propose to make it possible to deal with important data sets while reducing the time necessary to the generation of consensus. Section 4 presents the methodology we have followed to validate our approach by providing results obtained on random and biological data ; we compare our results to results obtained by the approximation algorithms [1,9]. Section 5 discusses related work and draws conclusions.

2   Context and Preliminaries

2.1   Context of the Study

This work has been done in close collaboration with physicians. We have collected queries daily performed and lists of results obtained by four ranking methods. More information about the ranking methods chosen and the kind of queries considered will be provided in Section 5. In this section, our aim is to present the 3 main requirements we want to answer.

Requirement 1: Comparing different sets. The distance should take into account the fact that ranking methods may play the role of filters. Two lists obtained by different ranking methods may not contain the same sets of data: it may be the case that $R1$ mentions data #X which is not present in $R2$. As a consequence, computing a distance between $R1$ and $R2$ implies taking into consideration those missing elements in rankings.

Requirement 2: Considering ties. The distance should consider the fact that each ranking method output has the form of a list of sets (or ties) of answers. For instance, the following line $R1_{Q1} := [\{12, 21, 35, 36\}, \{41\}, \{3, 22\}]$ indicates that the ranking method R1 for the query Q1 returns 7 answers and proposes a ranking in which 4 data objects are ranked equally first (#12,#21,#35,#36), strictly followed by the data object #41, itself strictly followed by the two data objects #3 and #22 (equally at the last position). We have associated with each data objects (answer of a query) arbitrary Ids from 1 to x, x being the total number of results that can be obtained for a given query.

Requirement 3: Minimizing disagreements. The consensus ranking we want to design should minimize the disagreements between rankings with ties. The distance we want to consider should thus be able to penalize two kinds of disagreements: Besides considering as disagreements the classical cases where element i is ranked before element j in one ranking and after element j in the other, we need also to consider as disagreements cases where elements i, j are ties in one ranking (i.e. they are part of the same set such as #3 and #22 in the example above) but not ties in the other one.

The remainder of this section presents the framework we have chosen to consider based on the requirements exposed above.

2.2   Unifying Sets of Data

We first present the *Unifying* preprocess we propose to follow to deal with different sets of data (Requirement 1) while minimizing disagreements between rankings (Requirement 3). More precisely, our aim is to penalize the fact that one element is considered in a ranking but not in another one. The following steps describe the **Unifying preprocess** which will be applied to all our ranking method outputs to unify the results:

1. Compute the union U of the elements (integers) appearing in each ranking.
2. For each ranking R_i, compute the set of elements contained in U but not in R_i, denoted here $U \setminus R_i$,
3. Augment each ranking the following way: add to R_i one new tie at the latest position with elements from $U \setminus R_i$.

All the rankings obtained using the *Unifying* preprocess are thus over the same sets of elements. Additionally, if any ranking had elements that were not in the other rankings before these changes it will be penalized by the fact that this element will be ranked in the last tie in all the other rankings.

Example 1. *For instance, let us consider three different ranking methods which outputs are the following:*

$$R_1 = [\{1, 7, 8, 15\}, \{2\}, \{3, 9\}]$$
$$R_2 = [\{2, 4, 5\}, \{7, 8, 15\}, \{10, 13\}, \{3\}]$$
$$R_3 = [\{1, 2, 3\}, \{4, 5\}, \{6, 8, 10, 12, 13\}, \{7, 9, 14\}, \{15\}]$$

Here we have that $U = \{1, 2, 3, \ldots, 15\}$, $U \setminus R_1 = \{4, 5, 6, 10, 11, 12, 13, 14\}$, $U \setminus R_2 = \{1, 6, 9, 11, 12, 14\}$ and $U \setminus R_3 = \emptyset$. The rankings processed using the Unifying preprocess are then the followings:

$$R_1 = [\{1, 7, 8, 15\}, \{2\}, \{3, 9\}, \{4, 5, 6, 10, 11, 12, 13, 14\}]$$
$$R_2 = [\{2, 4, 5\}, \{7, 8, 15\}, \{10, 13\}, \{3\}, \{1, 6, 9, 11, 12, 14\}]$$
$$R_3 = [\{1, 2, 3\}, \{4, 5\}, \{6, 8, 10, 12, 13\}, \{7, 9, 14\}, \{15\}]$$

In the remainder of this paper, the unifying preprocess is applied to our rankings before running the generation of any consensus ranking. The next section is now dedicated to the description of how to obtain the consensus of a set of rankings with ties over the same set of integers.

2.3   Kendall-τ and Generalized Kendall-τ Distance

A good dissimilarity measure for comparing two rankings without ties is the Kendall-τ distance [11] which counts the number of pairwise disagreements between positions of elements in those rankings. One way to generate a consensus permutation for a given set of permutations is to find a **median** for this set i.e a permutation that minimizes the sum of Kendall-τ distances between this permutation and all permutations in the given set. The problem of finding the median of a set of m permutations of $\{1, 2, 3, \ldots, n\}$ under the Kendall-τ distance is a NP-hard problem (when $m \geq 4$) that has been well-studied over the past years and for which good heuristics exist [2,3,6,8,12,19].

We introduce here more formally the Kendall-τ distance, defined for permutations (rankings without ties). Then we show how it can be generalized to a distance between rankings with ties [9].

The classical formulation: A **permutation** π is a bijection of $[n] = \{1, 2 \ldots, n\}$ onto itself. It represents a total order of the elements of $[n]$. The set of all permutations of $[n]$ is denoted $\mathcal{S}_n$. As usual we denote a permutation π of $[n]$ as $\pi = \pi_1 \pi_2 \ldots \pi_n$. The **identity permutation** corresponds to the identity bijection of $[n]$ and is denoted $\imath = 12 \ldots n$.

The **Kendall-τ distance**, denoted d_{KT}, counts the number of pairwise disagreements between two permutations and can be defined formally as follows: For permutations π and σ of $[n]$, we have that

$$d_{KT}(\pi^1, \pi^2) = \#\{(i, j) : i < j \text{ and } [(\pi^1[i] < \pi^1[j] \text{ and } \pi^2[i] > \pi^2[j]) \text{ or}$$

$$(\pi^1[i] > \pi^1[j] \text{ and } \pi^2[i] < \pi^2[j])]\}$$

where $\pi[i]$ denotes the position of integer i in permutation π and $\#\mathcal{S}$ the cardinality of set $\mathcal{S}$.

Example 2. *Let* $\pi = [5, 3, 2, 1, 4]$ *and* $\sigma = [1, 4, 5, 3, 2]$ *be two permutations of* $\{1, 2, 3, 4, 5\}$. *We have that* $d_{KT}(\pi, \sigma) = 12$ *since we have the following disagreements between pairs of elements of* π *and* σ:
1 appears **after** *2,3 and 5 in* π *and* **before** *2, 3 and 5 in* σ
2 appears **before** *1 and 4 in* π *and* **after** *1 and 4 in* σ
3 appears **before** *1 and 4 in* π *and* **after** *1 and 4 in* σ
4 appears **after** *2,3 and 5 in* π *and* **before** *2, 3 and 5 in* σ
5 appears **before** *1 and 4 in* π *and* **after** *1 and 4 in* σ

Given any set of permutations $A \subseteq \mathcal{S}_n$ and a permutation π, we have

$$d_{KT}(\pi, A) = \sum_{\pi_a \in A} d_{KT}(\pi, \pi_a).$$

The problem of **finding a median of a set of permutations under the Kendall-τ distance** can be stated formally as follows:

Given $A \subseteq \mathcal{S}_n$, we want to find a permutation π^* of $[n]$ such that

$$d_{KT}(\pi^*, A) \leq d_{KT}(\pi, A), \text{ for all } \pi \in \mathcal{S}_n.$$

Generalization to rankings with ties. Following [9], a **bucket order** on $[n]$ is a transitive binary relation $\lhd$ for which there are non empty sets $\mathcal{B}_1, \ldots, \mathcal{B}_k$ (the **buckets**) that form a partition of $[n]$ such that $x \lhd y$ if and only if there are i, j with $i < j$ such that $x \in \mathcal{B}_i$ and $y \in \mathcal{B}_j$. Now, a **ranking with ties** is defined on $[n]$ as $R = [\mathcal{B}_1, \ldots \mathcal{B}_k]$, where $R[x] = i$ if $x \in \mathcal{B}_i$. That means that a ranking with ties is simply a surjective function $R : [n] \longrightarrow [k]$, with $\sigma^{-1}(i) = \mathcal{B}_i$. The **generalized Kendall-τ distance**, denoted $K^{(p)}$, is defined according to a parameter p such that $0 < p \leq 1$. It writes:

$$
\begin{aligned}
K^{(p)}(R_1, R_2) = \#\{(i,j) : i < j \quad &\text{and } [(R_1[i] < R_1[j] \text{ and } R_2[i] > R_2[j]), \text{ or}\\
&\quad (R_1[i] > R_1[j] \text{ and } R_2[i] < R_2[j]))\}\\
+ p \times \#\{(i,j) : i < j &\text{ and } (R_1[i] = R_1[j] \text{ and } R_2[i] \neq R_2[j]), \text{ or}\\
&\quad (R_1[i] \neq R_1[j] \text{ and } R_2[i] = R_2[j])]\},
\end{aligned}
$$

In other words, the generalized Kendall-τ distance considers the number of disagreements between two rankings with ties (answering *Requirement 2*): a disagreement can be either two elements that are in different buckets in each ranking, where the order of the buckets disagree, and each such disagreement counts for 1 in the distance; or two elements that are in the same bucket in one ranking and in different buckets in the other, and each such disagreement counts for p (answering *Requirement 3*).

The problem of **finding a median of a set of rankings with ties under the generalized Kendall-τ distance** can be stated formally as follows, where $Rank_n$ represents the set of all possible rankings with ties over $[n]$:

Given a set of rankings with ties $\mathcal{R} = \{R_1, \ldots R_t\}$, find a consensus ranking R^* of $[n]$ such that

$$
K^{(p)}(R^*, \mathcal{R}) \leq K^{(p)}(R, \mathcal{R}), \text{ for all } R \in Rank_n.
$$

Complexity of the Problem. The problem of finding a median of a set of rankings with ties under the generalized Kendall-τ distance is NP-hard since it contains as a particular case the problem of finding the median of a set of permutations under the Kendall-τ distance. From a pragmatical point of view, if $\mathcal{S}_{n,k}$ represents the number of subjective functions from $[n]$ to $[k]$, then $Rank_n$ contains

$$
\sum_{k=1}^{n} \mathcal{S}_{n,k} = \sum_{k=1}^{n} k! \left\{ {n \atop k} \right\} \sim \frac{n! \ log_2(e)^{n-1}}{2}
$$

elements, where $\left\{ {n \atop k} \right\}$ represent the Stirling numbers of the second kind. (The asymptotic formula is due to Wilf [18].) To give an idea of how big the set $Rank_n$ may be, it already contains 102 247 563 elements when $n = 10$. Since we want to find the median of a sets of rankings of $[n]$, $30 \leq n \leq 800$, we clearly cannot use a brute-force algorithm that goes through the whole set $Rank_n$ to find a median ranking.

Approximation algorithms. Very interestingly, approximation algorithms have been introduced to this problem.

First, Fagin *et al.* [9] have introduced a *constant ratio approximation algorithm* for finding a consensus for a set of ranking with ties $\mathcal{R}$ on the same set of elements. This means that there is a constant c (depending on p only) such that the consensus ranking R_0 found by the algorithm satisfies

$$K^{(p)}(R_0, \mathcal{R}) \leq c \times K^{(p)}(R^*, \mathcal{R})$$

where R^* is a median of $\mathcal{R}$. The approximation ratio, c, depends on the value of p; for example it is 2 for $p = 1/2$. The algorithm is based on dynamic programming and runs in time $O(nt + n^2)$ where t is the number of rankings and n the number of elements.

Second, Ailon [1] have very recently provided solutions for aggregation of partial rankings. They have introduced two approximation algorithms that we will call 2-Ailon and 3/2-Ailon respectively, based on their approximation ratio. The consensus (called *aggregate ranking* by Ailon) provided by the two algorithms are fully ranked, that is, each tie must contain only one element. In our application domain, this very strong requirement is not mandatory (it may be the case that the consensus ranking places several elements in the same tie without forcing them to be ordered). We have thus considered the 2-Ailon algorithm and a variation of it without considering the step responsible for breaking ties into sets of size one, which we call 2var-Ailon in the following. As for the 3/2-Ailon algorithm, it involves solving a large linear programming problem, which is intrinsically highly memory-consuming and may not fit easily with huge data sets.

In the next section, we will present a new heuristic for the same problem, namely the BioConsert heuristic, and a way to reduce the data (so to speed up computations), in the case when we are dealing with huge data sets (several hundreds of elements in rankings). As we will see in Section 4.1, our heuristic outperforms the approximation algorithms for all data sets, synthetic or biological, that we have considered.

3   Heuristic and Data Reduction

As stated above, generating a median is intrinsically costly. In this section we present two solutions we have adopted to reduce the time necessary to generate consensus, that will hopefully be close to the real median, and even make it possible to generate them in case of huge data sets.

The first one is the most important: the BioConsert heuristic which consists in starting with a given ranking and making elements move around in this ranking so that it gets closer to a median. The ideas behind the BioConsert heuristic will be presented more formally in subsection 3.1.

Another solution we propose is to minimize the sets of data taken as input. Since we will sometimes deal with huge rankings, we want a way to be able to consider only a portion of those rankings, without loosing too much important information. This idea is presented in subsection 3.2.

3.1   The BioConsert Heuristic

Given a set $\mathcal{R} = \{R_1, \ldots R_t\}$ of rankings with ties, the idea of the BioConsert heuristic is to apply a series of *"good"* operations on a starting ranking R_{start}, in order to make it closer to a median. We can then choose different rankings as possible start rankings and apply our BioConsert heuristic on each of them, keeping as **best consensus** the result of the best run.

Here, we present some basic definitions and introduce the BioConsert heuristic designed to find a consensus of a set $\mathcal{R}$ of rankings with ties under a generalized Kendall-τ distance.

Definition 1. *Given $R = [\mathcal{B}_1, \ldots, \mathcal{B}_k]$, a ranking with ties of $[n]$ elements and initially k buckets. Consider $\mathcal{B}_\ell = R[i]$, for $1 \leq \ell \leq k$ (i.e. $i \in \mathcal{B}_\ell$). We define two operations on the integers $i, 1 \leq i \leq n$:*

- changeBucket(i,j) *is an operation that removes integer i from its bucket $\mathcal{B}_\ell$ and puts it in bucket $\mathcal{B}_j$, for $1 \leq j \leq k$. If bucket $\mathcal{B}_\ell$ becomes empty after the removal of i, then it is erased from the result.*
- addBucket(i,j) *is an operation that removes i from its bucket $\mathcal{B}_\ell$, creates a new bucket $\{i\}$ with only i in it and puts it in position j, for $1 \leq j \leq k+1$. In the case where $\mathcal{B}_\ell$ was already the singleton $\{i\}$, addBucket(i,j) corresponds to moving $\mathcal{B}_\ell$ to position j.*

It is easy to see that, with these two operations, it is possible to find a sequence of moves that changes any ranking with ties into any other one. Additionally, each of these operations can be performed in $O(n)$ time requirement.

Example 3. *Given $R = [\{1,3,4\}, \{5\}, \{2,7\}, \{6,8,9\}]$ a ranking with ties on $\{1, \ldots, 9\}$, with buckets $\mathcal{B}_1 = \{1,3,4\}, \mathcal{B}_2 = \{5\}, \mathcal{B}_3 = \{2,7\}$ and $\mathcal{B}_4 = \{6,8,9\}$, we have*

$$\text{changeBucket}(1,2)(R) = [\{3,4\}, \{1,5\}, \{2,7\}, \{6,8,9\}]$$
$$\text{changeBucket}(5,4)(R) = [\{1,3,4\}, \{2,7\}, \{5,6,8,9\}]$$
$$\text{addBucket}(1,3)(R) = [\{3,4\}, \{5\}, \{1\}, \{2,7\}, \{6,8,9\}]$$
$$\text{addBucket}(6,5)(R) = [\{1,3,4\}, \{5\}, \{2,7\}, \{8,9\}, \{6\}]$$
$$\text{addBucket}(5,5)(R) = [\{1,3,4\}, \{2,7\}, \{6,8,9\}, \{5\}]$$

Definition 2. *Given a set $\mathcal{R} = \{R_1, \ldots R_t\}$ of rankings with ties, we say that an operation $\mathcal{O}$ on R_i is **good** if it makes R_i closer to the median of $\mathcal{R}$, i.e if*

$$K^{(p)}(\mathcal{O}(R_i), \mathcal{R}) < K^{(p)}(R_i, \mathcal{R})$$

It is now time to present the BioConsert heuristic whose pseudo-code is depicted in Figure 1. The idea is to begin our search for a best consensus from a ranking R_{start}, and to apply *good* change bucket or add bucket operations to this starting point till there is no more possible *good* movement. Each run of the heuristic will thus provide one consensus (obtained with one given starting ranking). The heuristic will be run as many times as they are starting rankings. Eventually, the user will be provided with the best consensus among all the consensus obtained.

```
Algorithm BioConsert (R_start, R)
    n ← domain(R_start)
    k ← number of buckets of R_start
    bool ← 0 (will be changed to 1 if there is no more possible "good" operation)
    chang ← 0 (will tell us if some operations were made)
    WHILE bool <> 1 DO
        FOR i from 1 to n DO
            FOR j from 1 to k DO
                IF changeBucket(i,j) is a good operation THEN
                    R_start ← changeBucket(i, j)(R_start)
                    chang ← chang +1
                END IF
            END FOR
            FOR j from 1 to k + 1 DO
                IF addBucket(i,j) is a good operation THEN
                    R_start ← addBucket(i, j)(R_start)
                    chang ← chang +1
                END IF
            END FOR
        END FOR
        IF chang = 0 THEN
            bool ← 1
        END IF
    END WHILE
    RETURN R_start
```

Fig. 1. Pseudo-code of the BioConsert heuristic for generating consensus rankings

3.2   Reductions of Rankings to Important Data

To reduce the cost of generating consensus rankings, we have chosen to follow an alternative but possibly complementary approach which consists in reducing the data sets to be considered. Instead of considering all the data obtained by the various ranking methods as input, we propose to consider shortened sets. The rankings considered here are supposed to have been unified by our Unifying preprocess described in Section 2.2, so we are dealing with rankings on the same set of data. Say that we have a small set of elements, denoted Imp, known to be the most important ones for all rankings. Note that considering Imp has been done based on the searching process made by scientists who usually have a very precise idea of part of the results they expect to get and are willing to find complementary information

For each ranking output, the only elements considered in the reduced sets are the elements from Imp or appearing in a bucket together with elements from Imp. More precisely, we follow the three-steps *Input Reducing* procedure described below where Imp represents the set of important elements and R_i, $1 \leq i \leq t$, represent the rankings for which we want to find a consensus:

1. First we compute for each ranking R_i, $1 \le i \le t$ the set of elements from Imp or appearing in a bucket together with elements from Imp, denoted $Imp(R_i)$: Let ℓ_i be the smallest index such that $Imp \subset \cup_{k=1}^{\ell_i} B_k(R_i)$. Then $Imp(R_i) = \cup_{k=1}^{\ell} B_k(R_i)$.
2. We then compute the reduced domain $\mathcal{D}$ of the rankings by taking the union of the sets obtained in step 1. $\mathcal{D} = \cup_{i=1}^{t} Imp(R_i)$.
3. We finally consider in each ranking R_i only the elements from $\mathcal{D}$ *i.e.* we remove from R_i all elements not in $\mathcal{D}$.

Example 4. *Say that we have the two following unified rankings R_1 and R_2, for which the most important elements are $Imp = \{1, 2, 3\}$:*

$$R_1 = [\{1, 7, 8, 15\}, \{2\}, \{3, 9\}, \{4, 5, 6, 10, 11, 12, 13, 14\}]$$
$$R_2 = [\{1, 2\}, \{4, 5\}, \{3\}, \{6, 8, 10, 12, 13\}, \{7, 9, 14\}, \{15\}]$$

So, here $Imp(R_1) = \{1, 2, 3, 7, 8, 9, 15\}$, $Imp(R_2) = \{1, 2, 3, 4, 5\}$ and $\mathcal{D} = \{1, 2, 3, 4, 5, 7, 8, 9, 15\}$. We obtain the following reduced rankings by reducing R_1 and R_2 to domain $\mathcal{D}$:

$$reduced(R_1) = [\{1, 7, 8, 15\}, \{2\}, \{3, 9\}, \{4, 5\}]$$
$$reduced(R_2) = [\{1, 2\}, \{4, 5\}, \{3\}, \{8\}, \{7, 9\}, \{15\}]$$

4   Application to Medical Queries

The results presented in this section have been obtained in close collaboration with oncologists and pediatricians from the Children's Hospital of Philadelphia (Pennsylvania, USA) and the Institut Curie (Paris, France). The complete data sets are available at http://bioguide-project.net/bioconsert and consists in the output obtained using each ranking methods on each query, and the consensus and best consensus obtained using all the techniques described in this paper. Note that this paper presents a generic method able to provide a consensus ranking given a set of input rankings: while this section will very briefly describe the ranking methods chosen to give a concrete use case of our approach, more details on those points are beyond the scope of this paper.

Collected input data. We have collected queries daily performed by scientists, and sorted lists of expected answers. We will call *Gold standard* (GS) the list of ranked answers expected for each input query. In this paper, we consider only one kind of query, namely, looking for *the genes known to be possibly associated with the disease X* where X can take seven alternative values: Breast cancer, Prostate cancer, Neuroblastoma, Bladder cancer, Retinoblastoma, ADHD (Attention Deficit Hyperactivity Disorder), and Long QT syndrome. For each disease, the gold standard indicates the list of genes currently known by the team to be associated with the disease. Very interestingly, physicians have associated each disease with 5 to 30 genes, no more. However, when querying public portals, queries about each of the first three diseases (breast and prostate cancers and Neuroblastoma) several hundreds of answers are returned while queries about the last four queries provide more reasonable sets of answers.

Ranking methods and accessing data. We have based our choice of ranking methods on the current approaches available which have been clearly tested on biological data. Each ranking method described here after exploits the fact that biological data form a graph in which nodes are data objects (pieces of data like a database entry) and arcs are cross-references between objects; the target data objects mentioned below denotes the data objects obtained in the result.

First, PageRank (PR) [15], which has been used for the first time in the context of biological data in [5,16], computes for each node a score representing its authority (popularity), based on the probability that a random walk in the graph stops on the target data after an infinite number of steps. Second and third, InEdge (IE) and PathCount (PC), which have been introduced in [17,13], base their ordering on the number of incoming edges pointing to each target data object and the number of existing paths to reach each target data object, respectively. Fourth, BIoggle (BI), is a ranking method introduced in [14], and takes into account the confidence users may have on databases providing data objects.

Among these methods, the keywords used as input of a query (e.g., the name of a disease) can be used by the ranking(-filtering) methods to reduce the search space. Some methods may consider that all the data crossed in a path should contain the user keyword while others may only consider that some particular nodes of the path should contain the keyword, resulting in several sets of answers.

Queries have been run against 22 biological databases accessible through the EBI SRS platform using the BioGuideSRS system [7]. Ranking modules have been added to BioGuideSRS, each of them made it possible to obtain a list of (sets of) genes associated with each disease.

Computing consensus rankings and medians. We have implemented a module in Maple which computes a consensus ranking given a set of input rankings. More precisely, our module is able to compute the exact median (with the brute-force approach, only for very small values of n, *i.e.* for $2 \leq n \leq 8$), and a best consensus using the BioConsert heuristic while considering or not reduced inputs. As the result of our heuristic is obtained using the ranking chosen as the starting point, we have considered as starting points the rankings given by all the five methods described above (GS, BI, IE, PR and PC) plus one ranking, called the tie ranking, denoted R_{tie}, which is the ranking where all elements are in the same tie, i.e., $R_{tie} = [\{1, 2, \ldots, n\}]$. Last, we have implemented the approximation algorithms of Fagin *et al.* [9] and Ailon [1] and tested them on our data sets to compare their results with our results.

The next subsection provides quantitative results on the generation of the best consensus and the use of the heuristic while subsection 4.2 provides qualitative results on the benefit of using our approach.

4.1 Quantitative experiments

In this subsection, we present our results concerning the use of the BioConsert heuristic and methods for reducing data inputs while comparing them to results obtained by approximation algorithms. By default, the value of p (which determines how strongly elements which are ties in one ranking but not in the other should be penalized) is 1 in this subsection, while a discussion about this point is made in the last part of this subsection.

Benefit of using the heuristic. In this first experiment, we have considered pure random sets of data (the exact median could not be computed for a number of elements higher than 8). We have generated 500 different sets of 4 random rankings of $[n]$ elements, for each n, $4 \leq n \leq 8$.

For each of these sets, we have computed (1) the exact median, using a brute-force algorithm that enumerates all candidate consensus and keeps the best one, (2) the best consensus obtained by the BioConsert heuristic, (3) the approximation of the median obtained by the approach of Fagin *et al*, (4) the approximation given by the 2-Ailon procedure, (5) the approximation given by the 2var-Ailon procedure, and (6) the approximation given by the 3/2-Ailon procedure.

We have then compared independently the consensus obtained by the non exact solutions (2) to (6) with the exact median provided by method (1) by computing the percentage of cases where the consensus provided was equal to the exact median.

The results obtained (see Table 1) using BioConsert are particularly good since in almost all configurations, the consensus obtained is the exact median. In the very rare cases where the percentage is lower than 100% (meaning that the best consensus found is not the exact median), the distance between the exact median and the best consensus obtained using the BioConsert heuristic was 1. Interestingly, results are not as good following Fagin's approach: for very small data sets ($n = 4$), less than half of the consensus obtained are equal to the exact median and the percentage of cases where the consensus is equal the exact median decreases even more when n increases. As for the Ailon's approaches (all Ailon's procedures have been considered here), the exact median is never found.

Considering complete and reduced data sets. In all the following experiments, we have considered both complete and reduced data (following the procedure

Table 1. Percentage of cases where the exact median is found

n	$BioConsert$	$Fagin's\ approach$	$Ailon's\ approaches$
4	100%	48.6%	0%
5	100%	43.2%	0%
6	100%	37.6%	0%
7	99,8%	24.2%	0%
8	99,6%	18.4%	0%

described in Section 3.2). Reduced data sets will systematically appear in the lower part of the tables and will be postfixed by *-red.*

For the queries concerning diseases for which very various forms exist and/or which can be studied in very different contexts (Breast cancer, Prostate cancer, Neuroblastoma), we have considered reduced data sets only to focus on the genes having the same importance as genes already known by the physicians. For information, the size of the reduced data sets are provided on Table 2.

Table 2. Length of unified and reduced rankings for all medical queries

Query	Length of unified rankings	Length of reduced rankings
Long QT Syndrome	35	35
ADHD	45	15
Breast Cancer	930	386
Prostate Cancer	710	218
Bladder Cancer	308	115
Retinoblastoma	402	37
Neuroblastoma	661	431

On the stability of our approach. As seen in section 3.1 our heuristic BioConsert starts from one ranking R_{start} that can either be one of the five rankings described above (GS, BI, IE, PR or PC), or the tie ranking R_{tie} and performs *good* moves until obtaining a consensus ranking (when no more good moves are possible). So, for each query, we can possibly get a set of 6 different consensus, one for each R_{start} rankings considered. As defined in Section 3.1 the **best consensus** rankings (provided to the user) will then be the consensus rankings from this set that minimize the generalized Kendall-τ distance to the set $\{BI, IE, PR, PC\}$. In this part, our aim is to analyze the stability of our approach: how frequent it is to obtain several best consensus. In this case, we want to know how far from each other the best consensus may be.

Stability results obtained for BioConsert are provided on Table 3 which indicates the number of different consensus C_i obtained, the number of different best consensus obtained and the minimum/average and maximum number of moves performed by the heuristic to find the consensus rankings. The number of moves is particularly interesting to consider since it is related to the running time.

As for the results obtained by BioConsert with complete data sets, even when different starting points (rankings) are considered the same consensus can be obtained (it is the case for each query where #consensus is lower than 6). More interestingly, the number of best consensus for each query is equal to 1 meaning that the user will be provided with only one result.

For the reduced data, there is one case where several best consensus are provided: In the Retinoblastoma query, three different best consensus have been found by our heuristics. Let us call them $BC1$, $BC2$, and $BC3$. To test the stability of our heuristic, we have thus computed the pairwise generalized Kendall-τ distance between these three best consensus to see how far from each other they

Table 3. Stability of BioConsert, with complete and reduced data sets

			# of moves		
Query	# Consensus	# Best	Min	Max	Average
LQT Syndrome	5	1	4	82	33.5
ADHD	5	1	9	124	54.2
Bladder Cancer	4	1	229	3813	1827
Retinoblastoma	5	1	6	1723	696.5
LQT Syndrome-red	5	1	4	82	33.5
ADHD-red	2	1	2	31	12.8
Breast Cancer-red	6	1	322	1890	945.33
Prostate Cancer-red	5	1	148	992	507.83
Bladder Cancer-red	4	1	46	805	388.8
Retinoblastoma-red	3	3	3	12	8.4
Neuroblastoma-red	5	1	224	2980	1349.5

might be. Note that the maximal $K^{(1)}$ between two rankings of length n is given by $\frac{n(n+1)}{2}$. In the reduced data on Retinoblastoma, $n = 37$, the maximal distance is thus 703. The results of the pairwise comparisons obtained are particularly encouraging since we got $K^{(1)}(BC1, BC3) = 33$, $K^{(1)}(BC1, BC2) = 31$ and $K^{(1)}(BC3, BC2) = 64$. The mean of these distances is thus 42.6 (< 703).

A last conclusion that we can draw from this set of experiments is on the number of moves performed by the heuristic, which could be in the worst case equal to $\frac{n(n+1)}{2}$ and which is systematically lower, showing that our heuristic is able to provide results pretty fast.

Comparison with approximation algorithms. Table 4 shows the distance between the best consensus and the set of rankings $\mathcal{R}$ given as input, with our heuristic

Table 4. Results of BioConsert and approximation algorithms ($p = 1$)

	$K^{(1)}$ distance to $\mathcal{R} = \{BI, IE, PR, PC\}$			
Query	BioConsert	Fagin *et al.*	2-Ailon	2var-Ailon
Long QT Syndrome	352	434	468	422
ADHD	682	1072	1159	998
Bladder Cancer	23 379	38 867	40 908	29 511
Retinoblastoma	75 183	113 242	117 158	103 439
LQT Syndrome-red	352	434	468	422
ADHD-red	48	93	93	56
Breast Cancer-red	79 027	153 617	175 206	120 794
Prostate Cancer-red	26 392	48 734	53 137	39 201
Bladder Cancer-red	3174	7083	7456	3869
Retinoblastoma-red	653	881	1771	1103
Neuroblastoma-red	56 843	126 401	147 059	93 475.5

and with approximation algorithms. We have considered using Fagin's approach, 2-Ailon and 2var-Ailon. It was possible to run the 3/2-Ailon procedure only on our smallest data set, ADHD reduced, due to the too high memory requirement of this procedure (as introduced in 2.3). For ADHD-red, 3/2-Ailon gave a best consensus with a distance of 92 to $\mathcal{R}$, which is much higher than the distance 48 of the best consensus obtained with BioConsert.

More generally, the BioConsert heuristic provides best consensus which are systematically much closer to the set of rankings $\mathcal{R}$ than the consensus obtained by the Fagin's and Ailon's approaches. The variation of 2-Ailon algorithm we have implemented (removing the step that forces ties to be of size one) provides better results than Fagin's but it is still farther from the real median than the results obtained by BioConsert.

On the role of the parameter p. Recalling that the generalized Kendall-τ distance depends on a parameter p (see Section 2.3) which determines how important we want to consider the case where two elements are ties in one ranking but not ties in the other one. In all the previous experiments, we have chosen to consider to perform our runs with the value of p equals to 1, in accordance with wishes expressed by the physicians: two elements ordered in one ranking but tied in the other should be penalized as strongly as two elements not in the same order.

In this last experiment, we have chosen to study the case where p=1/2 (in which it is twice more important to penalize two elements which order is not the same in two rankings than the case where these elements are ties in one ranking and ordered in the other one). This value was considered in some papers (in particular Fagin's *et al.*) as being appropriate for Web users.

Results are provided in Table 5. Although the difference between results obtained using BioConsert and the approximation algorithms are, as expected, lower when p=1/2 than when p=1, our heuristic still systematically outperforms all the approximation algorithms.

4.2   Qualitative Study

In this section, we have studied very precisely the results obtained in close collaboration with physicians and using the GeneValorization tool [4] to obtain information related to the publications associated with the genes involved in a disease. We have considered reduced data sets.

Using our approach offers three kinds of advantages.

First, providing a consensus ranking clearly helps state the common points of several rankings. For instance, in the ADHD query, while the gold standard did not mention them initially, two genes namely HTR1B (#31) and SLC6A2 (#41) appear to be ranked in the top-10 elements. A study of those two genes allowed us to discover that they were respectively associated with 54 and 392 scientific publications stating the fact that the genes were known to be related to the ADHD disease. The same kind of conclusions has been drawn in the context of other queries. For instance, while the gene #495 (PRKCA) was not originally in the gold standard, it is ranked at position 6 by our consensus ranking which

Table 5. Results of BioConsert and approximation algorithms ($p = 1/2$)

Query	$K^{(1/2)}$ distance to $\mathcal{R} = \{BI, IE, PR, PC\}$			
	BioConsert	Fagin *et al.*	2-Ailon	2var-Ailon
LQT Syndrome	205	244	268	244
ADHD	447.5	691.5	686.5	659.5
Bladder Cancer	17013	24281	25364	19665.5
Retinoblastoma	39155.5	65435.5	67492.5	62807.5
LQT Syndrome-red	205	244	268	244
ADHD-red	32.5	54.5	55.5	38.5
Breast Cancer-red	54 060	76 830	103 816	78 810
Prostate Cancer-red	18 139	26 963	33 198	26 176
Bladder Cancer-red	2307.5	4292	4591	2742.5
Retinoblastoma-red	355	442	919	585
Neuroblastoma-red	39 945.5	74 921.5	93 475.5	47 679.5

is an excellent point since it appeared to be mentioned in 40 publications as associated with Neuroblastoma.

Second, while physicians may consider several genes as equally importantly related to the disease, the consensus ranking may help to provide a finer ranking of their implication. Continuing with the ADHD query, the 3 genes TPH2 (#7), DRD4 (#2), and DRD5 (#3) appeared to be more important than TPH1 (#6) while they are all at the same level in the gold standard. Interestingly, the number of publications associated with each of the three genes has been proved to be clearly higher (each gene is mentioned in more than a hundred of publications as associated with ADHD) than TPH1 (5 associated publications). We have been able to draw the same kind of conclusion in the context of all other queries.

Third, providing a consensus ranking avoids users depending on a result obtained by one ranking only. This can be illustrated by results obtained in the Prostate cancer query, where the gene #219 (GLS) is given in the top-4 answers by some rankings (here IE) while it is absolutely not the case in other rankings. The consensus ranking proposes to minimize the impact of this result by placing this gene in position 36. A quick look at the annotation file associated with this gene showed that its implication to the disease is not anymore a reliable information. We can find similar situations for example in the Neuroblastoma query, for which the PC ranking has returned gene #641 in the top-20 results while this information actually relies on no publication at all. Interestingly, the consensus ranking obtained has not even placed this object in the top-100 answers.

5   Discussion

In this paper, we have designed and implemented a new method providing a consensus from rankings possibly composed of different elements, with lists of sets (ties) while minimizing the disagreements between rankings.

We have proposed a preprocess able to deal with different data sets to then be able to work on rankings with ties on the same data sets. Then, while the general problem of finding a median of a set of rankings with ties is NP hard, we have proposed to follow two complementary approaches. First, we have introduced a method to reduce the input data by focusing on data of interest. Second and more importantly, we have introduced an heuristic and have compared the results it provides to the approximation algorithms currently available in the literature [1,9]. We have demonstrated the benefit of using our approach on real biological data sets.

We now discuss related work and provide hints for on going work.

As seen in Section 4.1, our heuristic performs very well compared to approximation algorithms [9,1]. However, it is worth noticing that the aim of approximation approaches is to provide an upper bound of the distance between the consensus generated by a given approximation algorithm and the exact median. This kind of work is particularly interesting on a general and theoretical point of view. In a sense, our approach is more practical, since we introduce an heuristic able to quickly provide a consensus which should be as close as possible to the exact median. However, although our approach has provided better results than the approximation algorithms available on our data sets, in the general case, we are not able to guarantee what could be the maximal distance between the consensus we generate and the exact median.

We are investigating several directions in ongoing work. We are currently considering randomized versions of our heuristic. More precisely, our current heuristic is greedy: at each step it chooses the first possible operation that lowers the distance to the set of rankings given as input. This way, it might be the case that sometimes a local minimum is reached and the heuristic stops at that point. Considering randomized versions of the heuristic, for example by using a Monte Carlo algorithm would consist in choosing each possible operation randomly with a suitable probability at each step. The probability of choosing an operation that increases the distance to the set of rankings given as input can be nonzero, giving a possibility to exit from a local minimum. More generally, we plan to compare our approach with other heuristic search strategies, such as simulated annealing or genetic algorithm.

Other on going work includes testing our approach on new data sets (which involves large amounts of work with physicians), possibly considering new ranking methods (as the ones we are currently designing in the context of the BioGuide project [7]) or studying more precisely the impact of modifying the value of the parameter p (which helps penalize the fact that two elements are in the same tie or not). We actually plan to test learning functions to find the value of p which would fit the best with the physician's expectations.

References

1. Ailon, N.: Aggregation of Partial Rankings, p-Ratings and Top-m Lists. Algorithmica 57(2), 284–300 (2010)
2. Ailon, N., Charikar, M., Newman, N.: Aggregating inconsistent information: Ranking and clustering. In: Proceedings of the 37th STOC, pp. 684–693 (2005)

3. Betzler, N., Fellows, M.R., Guo, J., Niedermeier, R., Rosamond, F.A.: Fixed-Parameter Algorithms for Kemeny Scores. In: Fleischer, R., Xu, J. (eds.) AAIM 2008. LNCS, vol. 5034, pp. 60–71. Springer, Heidelberg (2008)
4. Brancotte, B., Biton, A., Bernard-Pierrot, I., Radvanyi, F., Reyal, F., Cohen-Boulakia, S.: Gene List significance at-a-glance with GeneValorization. To Appear in Bioinformatics (Application Note) (February 2011)
5. Birkland, A., Yona, G.: Biozon: a system for unification, management and analysis of heterogeneous biological data. BMC Bioinformatics 7, 70 (2006)
6. Blin, G., Crochemore, M., Hamel, S., Vialette, S.: Medians of an odd number of permutations. To Appear in Pure Mathematics and Applications (2011)
7. Cohen-Boulakia, S., Biton, O., Davidson, S., Froidevaux, C.: BioGuideSRS: Querying Multiple Sources with a user-centric perspective. Bioinformatics 23(10), 1301–1303 (2006)
8. Dwork, C., Kumar, R., Naor, M., Sivakumar, D.: Rank Aggregation Methods for the Web. In: Proceedings of the 10th WWW, pp. 613–622 (2001)
9. Fagin, R., Kumar, R., Mahdian, M., Sivakumar, D., Vee, E.: Comparing and Aggregating Rankings with Ties. In: Proceedings of PODS 2004, pp. 47–55 (2004)
10. Hussels, P., Trissl, S., Leser, U.: What's new? What's certain? – scoring search results in the presence of overlapping data sources. In: Cohen-Boulakia, S., Tannen, V.(eds.) DILS 2007. LNCS (LNBI), vol. 4544, pp. 231–246. Springer, Heidelberg (2007)
11. Kendall, M.: A New Measure of Rank Correlation. Biometrika 30, 81–89 (1938)
12. Kenyon-Mathieu, C., Schudy, W.: How to rank with few errors. In: Proceedings of the 39th STOC, pp. 95–103 (2007)
13. Lacroix, Z., Raschid, L., Vidal, M.E.: Efficient techniques to explore and rank paths in life science data sources. In: Rahm, E. (ed.) DILS 2004. LNCS (LNBI), vol. 2994, pp. 187–202. Springer, Heidelberg (2004)
14. Laignel, N.: Ranking biological data taking into account user's preferences, Master thesis report (co-supervised by S. Cohen-Boulakia, C. Froidevaux, and U. Leser), University of Paris-Sud XI (2010)
15. Page, L., Brin, S., Motwani, R., Winograd, T.: The PageRank Citation Ranking: Bringing Order to the Web. Stanford University, Stanford (1998)
16. Shafer, P., Isganitis, T., Yona, G.: Hubs of knowledge: using the functional link structure in Biozon to mine for biologically significant entities. BMC Bioinformatics 7, 71 (2006)
17. Varadarajan, R., Hritidis, V., Raschid, L., Vidal, M., Ibanez, L., Rodriguez-Drumond, H.: Flexible and Efficient Querying and Ranking on Hyperlinked Data Sources. In: Proceedings of the 12th International Conference on Extending Database Technology: Advances in Database Technology, pp. 553–564 (2009)
18. Wilf, H.S.: Generatingfunctionology, p. 147. Academic Press, NY (1990)
19. van Zuylen, A., Williamson, D.P.: Deterministic algorithms for rank aggregation and other ranking and clustering problems. In: Kaklamanis, C., Skutella, M. (eds.) WAOA 2007. LNCS, vol. 4927, pp. 260–273. Springer, Heidelberg (2008)

A Truly Dynamic Data Structure
for Top-k Queries on Uncertain Data*

Manish Patil, Rahul Shah, and Sharma V. Thankachan

Computer Science Department, Louisiana State University,
Baton Rouge, LA, USA
{mpatil,rahul,thanks}@csc.lsu.edu

Abstract. Top-k queries allow end-users to focus on the most important (top-k) answers amongst those which satisfy the query. In traditional databases, a user defined score function assigns a score value to each tuple and a top-k query returns k tuples with the highest score. In uncertain database, top-k answer depends not only on the scores but also on the membership probabilities of tuples. Several top-k definitions covering different aspects of score-probability interplay have been proposed in recent past [20, 13, 6, 18]. Most of the existing work in this research field is focused on developing efficient algorithms for answering top-k queries on static uncertain data. Any change (insertion, deletion of a tuple or change in membership probability, score of a tuple) in underlying data forces re-computation of query answers. Such re-computations are not practical considering the dynamic nature of data in many applications. In this paper, we propose a truly dynamic data structure that uses ranking function $PRF^e(\alpha)$ proposed by Li et al. [18] under the generally adopted model of x-relations [21]. PRF^e can effectively approximate various other top-k definitions on uncertain data based on the value of parameter α. An x-relation consists of a number of x-tuples, where x-tuple is a set of mutually exclusive tuples (up to a constant number) called alternatives. Each x-tuple in a relation randomly instantiates into one tuple from its alternatives. For an uncertain relation with N tuples, our structure can answer top-k queries in $O(k \log N)$ time, handles an update in $O(\log N)$ time and takes $O(N)$ space. Finally, we evaluate practical efficiency of our structure on both synthetic and real data.

1   Introduction

The efficient processing of uncertain data is an important issue in many application domains because of the imprecise nature of data they generate. The nature of uncertainty in data is quite varied, and often depends on the application domain. In response to this need, much efforts have been devoted to modeling uncertain data [21, 7, 5, 17, 19]. Most models have been adopted to possible world semantics, where an uncertain relation is viewed as a set of possible instances (worlds) and correlation among the tuples governs generation of these worlds.

Consider traffic monitoring application data [20] (with modified probabilities) as shown in Table 1, where radar is used to detect car speeds. In this application, data

---

* This work is supported in part by US NSF Grant CCF–1017623 (R. Shah).

J.B. Cushing, J. French, and S. Bowers (Eds.): SSDBM 2011, LNCS 6809, pp. 91–108, 2011.

Table 1. Traffic monitoring data: t_1 ,$\{t_2, t_4\}$, $\{t_3, t_6\}$, t_5

Time	Car Location	Plate Number	Speed	Probability	Tuple Id
11:55	L1	Y-245	130	0.30	t_1
11:40	L2	X-123	120	0.40	t_2
12:05	L3	Z-541	110	0.20	t_3
11:50	L4	X-123	105	0.50	t_4
12:10	L5	L-110	95	0.30	t_5
12:15	L6	Z-541	80	0.45	t_6

is inherently uncertain because of errors in reading introduced by nearby high voltage lines, interference from near by car, human operator error etc. If two radars at different locations detect the presence of the same car within a short time interval, such as tuples t_2 and t_4 as well as t_3 and t_6, then at most one radar reading can be correct. We use x-relation model to capture such correlations. An x-tuple τ specifies a set of exclusive tuples, subject to the constraint $\sum_{t_i \in \tau} Pr(t_i) \leq 1$. The fact that t_2 and t_4 cannot be true at the same time, is captured by the x-tuple $\tau_1 = \{t_2, t_4\}$ and similarly $\tau_2 = \{t_3, t_6\}$. Probability of a possible world is computed based on the existence probabilities of tuples present in a world and absence probabilities of tuples in the database that are not part of a possible world. For example, consider the possible world $pw = \{t_1, t_2, t_3\}$. Its probability is computed by assuming the existence of t_1, t_2, t_3, and the absence of t_4, t_5, and t_6. However since t_2 and t_4 are mutually exclusive, presence of tuple t_2 implies absence of t_4 and same is applicable for tuples t_3 and t_6. Therefore, $Pr(pw) = 0.3 \times 0.4 \times 0.2 \times (1 - 0.3) = 0.0168$. Top-$k$ queries on a traditional certain database have been well studied. For such cases, each tuple is associated with a single score value assigned to it by a scoring function. There is a clear total ordering among tuples based on score, from which the top-k tuples can be retrieved. However, for answering a top-k query on uncertain data, we have to take into account both, ordering based on scores and ordering based on existence probabilities of tuples. Depending on how these two orderings are combined, various top-k definitions with different semantics have been proposed in recent times. Most of the existing work is focused only on the problem of answering a top-k query on a static uncertain data. Though the query time of an algorithm depends on the choice of a top-k definition, linear scan of tuples achieves the best bound so far. Therefore, recomputing top-k answers in an application with frequent insertions and deletions can be extremely inefficient. In this paper, we present a truly dynamic structure of size $O(N)$ that always maintains the correct answer to the top-k query for an uncertain database of N tuples. The structure is based on a decomposition of the problem so that updates can be handled efficiently. Our structure can answer the top-k query in $O(k \log N)$ time, handle update in $O(\log N)$ time.

Outline: In Section 2 we review different top-k definitions proposed so far and try to compare them against a parameterized ranking function $PRF^e(\alpha)$ proposed by Li et al. [18]. We choose $PRF^e(\alpha)$ over other definitions as it can approximate many of the other top-k definitions and can handle data updates efficiently. After formally defining the problem (Section 3), we explain how $PRF^e(\alpha)$ can be computed using divide and conquer approach (Section 4), which forms the basis of our data structure explained in Section 5. We present experimental study with real and synthetic data sets in Section 6. Finally we review the related work in Section 7 before concluding the paper.

2   Top-k Queries on Uncertain Data

Soliman et al. [20] first considered the problem of ranking tuples when there is a score and probability for each tuple. Several other definitions of ranking have been proposed since then for probabilistic data.

- Uncertain top-k (U-Topk) [20]: It returns a k-tuple set that appears as top-k answer in possible worlds with maximum probability.
- Uncertain Rank-k (U-kRanks) [20]: It returns a tuple for each i, such that it has maximum probability of appearing at rank i across all possible worlds.
- Probabilistic Threshold Query (PT-k) [13]: It returns all the tuples with probability of appearing in top-k greater than a user specified threshold.
- Expected Rank (E-Rank) [6]: k tuples with highest value of expected rank ($er(t_i)$) are returned.

$$er(t_i) = \sum Pr(pw)rank_{pw}(t_i)$$

where $rank_{pw}(t_i)$ denotes rank of t_i in a possible world pw. In case t_i does not appear in possible world, $rank_{pw}(t_i)$ is defined as $|pw|$.
- Quantile Rank (Q-Rank) [15]: k tuples with lowest value of quantile rank ($qr_\phi(t_i)$) are returned. The ϕ-quantile rank of t_i is the value in the cumulative distributive function (cdf) of rank(t_i), denoted as $cdf(rank(t_i))$ that has a cumulative probability of ϕ. Median rank is a special case of ϕ-quantile rank where $\phi = 0.5$.
- Expected Score (E-Score) [6]: k tuples with highest value of expected score ($es(t_i)$) are returned.

$$es(t_i) = Pr(t_i)score(t_i)$$

- Parameterized Ranking Function (PRF) [18]: PRF in its most general form is defined as,

$$\Upsilon(t_i) = \sum_r w(t_i, r) \times Pr(t_i, r) \tag{1}$$

where w is the weight function that maps a given tuple-rank pair to a complex number and $Pr(t_i, r)$ denotes the probability of a tuple t_i being ranked at position r across all possible worlds. A top-k query returns those k tuples with the highest Υ values. Different weight functions can be plugged in to the above definition to get a range of ranking functions, subsuming most of top-k definitions listed above. A special ranking function $PRF^e(\alpha)$ is obtained by choosing $w(t_i, r) = \alpha^{r-1}$, where α is a constant. Experimental study in [18] reveals that for some value of α with the constraint $\alpha < 1$, PRF^e can approximate many existing top-k

definitions. These experiments use Kendall distance [9] between two top-k answers as a measure to compare the ranking functions. The "uni-valley" nature of the graphs obtained by plotting Kendall distance versus varying values of α for various ranking functions in [18] suggests there exists a value of α for which the distance of a particular ranking function to PRF^e is very small i.e. $PRF^e(\alpha)$ can approximate that function quite well.

Algorithms for computing top-k answers using the above ranking functions have been studied for static data. Any changes in the underlying data forces re-computation of query answers. To understand the impact of a change on top-k answers, we analyze relative ordering of the tuples before and after a change, based on these ranking functions. Let $T - t_1, t_2, .., t_N$ denote independent tuples sorted in non-increasing order of their score. We choose insertion of a tuple as a representative case for changes in T, and monitor its impact on relative ordering of a pair of tuples (t_i, t_j). For ranking function U-kRanks ordering of tuples (t_i, t_j) may or may not be preserved by insertion and cannot be guaranteed when the score of a new tuple is higher than that of t_i and t_j. Consider a database $T = t_1, t_2, t_3$ with existence probability values 0.1, 0.5, and 0.2 respectively. When all tuples are independent, probability that tuple t_i appears at rank 2 across all possible worlds is given by $Pr(t_i, 2) = p_i \sum_{x=1}^{i-1}(p_x \prod_{y=1, y \neq x}^{i-1}(1-p_y))$ [20]. Hence $Pr(t_2, 2) = 0.05 < Pr(t_3, 2) = 0.1$ and tuple t_3 would be returned as an answer for U-2Ranks query. Insertion of a new tuple t_0 with existence probability 0.25 and score higher than that of t_1, causes relative ordering of tuples t_2, t_3 to be reversed as after insertion $Pr(t_2, 2) = 0.15 > Pr(t_3, 2) = 0.0975$. Thus, existing top-$k$ answers do not provide any useful information for re-computation of query answers making it necessary to go through all the tuples again for re-computation in the worst case. Ranking functions PT-k, E-Rank, Q-Rank may also result in such relative ordering reversal. However, when tuples are ranked using $PRF^e(\alpha)$, the scope of disturbance in the relative ordering of tuples is limited as explained in later sections. This enables efficient handling of updates in the database. Therefore, this ranking function is well suited for answering top-k queries on a dynamic collection of tuples.

3   Problem Statement

Given an uncertain relation T of a dynamic collection of tuples, such that each tuple $t_i \in T$ is associated with a membership probability value $Pr(t_i) > 0$ and a score $score(t_i)$ computed based on a scoring function, the goal is to retrieve the top-k tuples. Without loss of generality, we assume all scores to be unique and let $t_1, t_2, ..., t_N$ denotes ordering of the tuples in T when sorted in descending order of the score ($score(t_i) > score(t_{i+1})$).

We use the parameterized ranking function $PRF^e(\alpha)$ proposed by [18] in this paper. $PRF^e(\alpha)$ is defined as,

$$\Upsilon(t_i) = \sum_r \alpha^{r-1} \times Pr(t_i, r) \tag{2}$$

where α is a constant and $Pr(t_i, r)$ denotes the probability of a tuple t_i being ranked at position r across all possible worlds[1]. A top-k query returns the k tuples with highest Υ values. We refer to $\Upsilon(t_i)$ as the rank-score of tuple t_i. In this paper, we adopt the x-relation model to capture correlations. An x-tuple τ specifies a set of exclusive tuples, subject to the constraint $Pr(\tau) = \sum_{t_i \in \tau} Pr(t_i) \leq 1$. In a randomly instantiated world τ takes t_i with probability $Pr(t_i)$, for $i = 1, 2, ..., |\tau|$ or does not appear at all with probability $1 - \sum_{t_i \in \tau} Pr(t_i)$. Here $|\tau|$ represents the number of tuples belonging to set τ. Let $\tau(t_i)$ represents an x-tuple to which tuple t_i belongs to. In x-relation model, T can be thought of as a collection of pairwise-disjoint x-tuples. As there are total N tuples in an uncertain relation T, $\sum_{\tau \in T} |\tau| = N$. From now onwards we represent $Pr(t_i)$ by short notation p_i for simplicity.

4   Computing $PRF^e(\alpha)$

In this section, we derive a closed form expression for the rank-score $\Upsilon(t_i)$, followed by an algorithm for retrieving the top-1 tuple from a collection of tuples. In the next section we show that this approach can be easily extended to a data structure for efficiently retrieving top-k tuples from a dynamic collection of tuples. We begin by assuming tuple independence and then consider correlated tuples, where correlations are represented using x-tuples.

4.1   Assuming Tuple Independence

When all tuples are independent, tuple t_i appears at position r in a possible world pw if and only if exactly $(r - 1)$ tuples with a higher score value appear in pw. Let $S_{i,r}$ be the probability that a randomly generated world from $\{t_1, t_2, ..., t_i\}$ has exactly r tuples [22]. Then, probability of a tuple t_i being ranked at r is given as

$$Pr(t_i, r) = p_i S_{i-1, r-1} \qquad (3)$$

where,

$$S_{i,r} = \begin{cases} p_i S_{i-1,r-1} + (1 - p_i) S_{i-1,r} & \text{if } i \geq r > 0 \\ 1 & \text{if } i = r = 0 \\ 0 & \text{otherwise.} \end{cases}$$

Using above recursion for $S_{i,r}$ and equation 2, 3,

$$\Upsilon(t_i) = \sum_r \alpha^{r-1} Pr(t_i, r) = \sum_r \alpha^{r-1} p_i S_{i-1, r-1}$$

$$\frac{\Upsilon(t_i)}{p_i} = \sum_r \alpha^{r-1} S_{i-1, r-1} = \sum_r \alpha^r S_{i-1, r}$$

---

[1] $Pr(t_i, r) = 0$, for $r > i$.

Similarly,

$$\frac{\Upsilon(t_{i+1})}{p_{i+1}} = \sum_r \alpha^r S_{i,r}$$

$$= \sum_r \alpha^r (p_i S_{i-1,r-1} + (1 - p_i) S_{i-1,r})$$

$$= \alpha p_i \sum_r \alpha^{r-1} S_{i-1,r-1} + (1 - p_i) \sum_r \alpha^r S_{i-1,r}$$

$$= (1 - (1 - \alpha)p_i)\Upsilon(t_i)/p_i$$

We have the base case, $\Upsilon(t_1) = p_1$. Therefore,

$$\Upsilon(t_i) = p_i \prod_{j<i}(1 - (1 - \alpha)p_j) \tag{4}$$

Now, we analyze the contribution of a tuple t_i towards global ranking over T using the above formula as follows.

- Tuple t_i contributes $m_i = p_i$ for the computation of its own rank-score.
- Tuple t_i contributes $c_i = 1 - (1 - \alpha)p_i$ of computing rank-score for all tuples having score less than that of t_i.

Theorem 1. *When all tuples in T are independent,* rank-score *of a tuple t_i can be computed as follows,*

$$\Upsilon(t_i) = m_i \prod_{j<i} c_j$$

where $m_i = p_i$ and $c_j = 1 - (1 - \alpha)p_j$

□

Answering top-1 query:

We use a divide and conquer approach for answering top-1 query on T, which forms the basis for our data structure in later section. Let the given relation $T = \{t_1, t_2, ..., t_N\}$ be partitioned into sub-reltations $T_l = \{t_1, t_2, ..., t_{\lceil N/2 \rceil}\}$ and $T_r = \{t_{\lceil N/2 \rceil+1}, t_{\lceil N/2 \rceil+2}, ..., t_N\}$. Also let t^l and t^r represent the top-1 answer for T_l and T_r with rank-scores $\Upsilon_{T_l}(t^l)$ and $\Upsilon_{T_r}(t^r)$ respectively, where $\Upsilon_{T_l}(t^l)$ is computed by considering only those tuples $t_j \in T_l$ and $\Upsilon_{T_r}(t^r)$ is is computed by considering only those tuples $t_j \in T_r$. Therefore, for $t_i \in T_l$, $\Upsilon_{T_l}(t_i) = m_i \prod_{j<i,t_j \in T_l} c_j$ and similarly for $t_i \in T_r$, $\Upsilon_{T_r}(t_i) = m_i \prod_{j<i,t_j \in T_r} c_j$.

Now when both the relations T_l and T_r are merged to form T, we make the following observations using the above analysis:

- The contribution of each tuple towards its own rank-score remains unchanged.
- Since all the tuples in T_r have a lower score value than any tuple $t_i \in T_l$ they do not contribute towards the rank-score value of t_i computed over entire relation T. Thus $\Upsilon(t_i) = \Upsilon_{T_l}(t_i)$. Hence t^l still has the highest rank-score value $\Upsilon(t^l)$ among the tuples in T_l.

- Since all the tuples in T_l have higher score value than any tuple $t_i \in T_r$, each $t_j \in T_l$ contributes $1 - (1 - \alpha)p_j$ towards rank-score value of t_i computed over entire relation T. Let $C_l = \prod_{t_j \in T_l} c_j = \prod_{t_j \in T_l} 1 - (1 - \alpha)p_j$ represents overall contribution of sub-relation T_l. Then $\Upsilon(t_i) = C_l \Upsilon_{T_r}(t_i)$. Since rank-score value of every tuple $t_i \in T_r$ gets scaled by the same factor C_l, t^r still has the highest rank-score value $\Upsilon(t^r)$ among the tuples in T_r.

Therefore the top-1 answer over uncertain relation T can be chosen from t^l and t^r based on the their rank-score values computed over the entire relation.

4.2   Supporting Correlations

If tuple t_i has some preceding alternatives, then equation 4 cannot be used to compute its rank-score since the event that t_i appears at a position r in a possible world, is no longer independent of the event that exactly $r - 1$ tuples appear in $\{t_1, t_2, ..., t_{i-1}\}$, as in equation 3. To overcome this difficulty, we convert the relation T to $\bar{T}^i$ where all the tuples are independent [22]. For any tuple t_i, let τ^i be the pruned version of τ such that it consists of all tuples from τ that have higher score value than that of t_i i.e. $\tau^i = \{t_j | t_j \in \tau, j < i\}$. For example, let $T = \{\tau_1, \tau_2, \tau_3\}$ where, $\tau_1 = \{t_1, t_3, t_6\}, \tau_2 = \{t_2, t_7\}$ and $\tau_3 = \{t_4, t_5\}$ then $\tau_1^5 = \{t_1, t_3\}, \tau_2^5 = \{t_2\}$ and $\tau_3^5 = \{t_4\}$. Now for each x-tuple $\tau \in T$, we create an x-tuple $\bar{\tau} = \{\bar{t}\}$ in $\bar{T}^i$ such that:

$$Pr(\bar{\tau}) = Pr(\bar{t}) = \begin{cases} Pr(\tau^i) \text{ if } \tau \neq \tau(t_i) \\ Pr(t_i) \text{ otherwise.} \end{cases}$$

This conversion takes into account the fact that only tuples with a score higher than that of t_i contribute to $Pr(t_i, r)$ as well as to $\Upsilon(t_i)$, and the presence of t_i implies absence of all its related tuples. Combining related tuples into a representative tuple $\bar{t}$ does not affect $\Upsilon(t_i)$ here, since the probability that $\bar{t}$ appears is the same as the probability that any one tuple in $\tau \in T$ with score higher than $score(t_i)$ appears. In other words, $\Upsilon(t_i)$ computed using transformed relation $\bar{T}^i$ is same as $\Upsilon(t_i)$ computed using original relation T. However as all the tuples in $\bar{T}^i$ are independent among themselves, we can now use equation 4 on $\bar{T}^i$ to compute the rank-score of tuple t_i. Therefore,

$$\begin{aligned} \Upsilon(t_i) &= p_i \prod_{\substack{\bar{t} \in \bar{T}^i \\ \bar{\tau}(\bar{t}) \neq \tau(t_i)}} (1 - (1 - \alpha)Pr(\bar{t})) \\ &= p_i \prod_{\substack{\tau \in T \\ \tau \neq \tau(t_i)}} (1 - (1 - \alpha)Pr(\tau^i)) \end{aligned} \qquad (5)$$

Now, we analyze the contribution of an x-tuple towards global ranking over T using the above formula as follows.

- x-tuple τ contributes $m_i = p_i$ for computing rank-score of a tuple $t_i \in \tau$.
- x-tuple τ contributes $c_i = 1 - (1 - \alpha)Pr(\tau^i)$ for computing rank-score of a tuple $t_i \notin \tau$.

Answering top-1 query:

Again, we attempt to use a divide and conquer algorithm for answering top-1 query on T by partitioning relation $T = \{t_1, t_2, ..., t_N\}$ into sub-relations $T_l = \{t_1, t_2, ..., t_{\lceil N/2 \rceil}\}$ and $T_r = \{t_{\lceil N/2 \rceil + 1}, t_{\lceil N/2 \rceil + 2}, ..., t_N\}$ and assuming t^l, t^r represent the top-1 answers for T_l, T_r respectively. If property that t^l and t^r remains highest rank-score tuples in their respective sub-relations even after merging of T_l and T_r, holds true then reporting top-1 for relation T can be done by simply comparing rank-score values of t^l and t^r over entire relation T. Unfortunately, this property may not hold true for t^r.

To illustrate the problem, consider an uncertain relation $T = \{t_1, t_2, t_3, t_4\}$ with $p_1 = 0.35, p_2 = 0.3, p_3 = 0.4, p_4 = 0.45$ and tuples t_2 and t_3 are mutually exclusive. Using equation 5, rank-scores can be computed as follows ($\alpha = 0.8$):

$$\Upsilon(t_1) = 0.35$$
$$\Upsilon(t_2) = 0.3(1 - 0.2 \times 0.35) = 0.28$$
$$\Upsilon(t_3) = 0.4(1 - 0.2 \times 0.35) = 0.37$$
$$\Upsilon(t_4) = 0.45(1 - 0.2 \times 0.35)(1 - 0.2 \times (0.3 + 0.4)) = 0.36$$

Top-1 query on T should return tuple t_3 with highest rank-score value 0.37. By adopting the divide and conquer approach to tackle the problem, we partition the given relation into $T_l = \{t_1, t_2\}$ and $T_r = \{t_3, t_4\}$. Top-1 query is applied to these sub-relations as follows.

$$\Upsilon_{T_l}(t_1) = 0.35 \qquad\qquad \Upsilon_{T_l}(t_2) = 0.3(1 - 0.2 \times 0.35) = 0.28$$
$$\Upsilon_{T_r}(t_3) = 0.4 \qquad\qquad\qquad \Upsilon_{T_r}(t_4) = 0.45(1 - 0.2 \times 0.4) = 0.41$$

Thus t_1 and t_4 will be reported from T_l and T_r as top-1 answers respectively. By simple merge operation, which computes rank-score values for t_1, t_4 over relation T and comparing them, t_1 will be reported as top-1 answer for T. However actual top-1 answer is tuple t_3. The fact that dependance of t_2 and t_3 was ignored while answering top-1 over sub-relation T_r is the root cause behind the disturbance in relative ordering of t_3 and t_4.

Therefore in order to maintain the relative ordering of tuples based on their rank-score over entire relation during merge, we redefine the expressions for contributions as follows. Here we use the notation $\hat{p}_i$ for sum of probabilities of all tuples t_j which are related to t_i and have score greater than the score of t_i (i.e. $j < i$). In the above example $\hat{p}_3 = p_2 = 0.3$.

$$\hat{p}_i = Pr([\tau(t_i)]^i) = \sum_{\substack{\tau(t_i) = \tau(t_j) \\ j < i}} p_j$$

Now equation 5 can be re arranged as follows,

$$\Upsilon(t_i) = \frac{p_i}{(1 - (1 - \alpha)\hat{p}_i)} \prod_{\tau \in T} (1 - (1 - \alpha)Pr(\tau^i))$$

$$\frac{\Upsilon(t_i)}{m_i} = \prod_{\tau \in T} (1 - (1-\alpha)Pr(\tau^i))$$

where $m_i = \frac{p_i}{(1-(1-\alpha)\hat{p}_i)}$

Similarly,

$$\frac{\Upsilon(t_{i+1})}{m_{i+1}} = \prod_{\tau \in T} (1 - (1-\alpha)Pr(\tau^{i+1}))$$

Here note that $Pr(\tau^i) = Pr(\tau^{i+1})$ for all $\tau \neq \tau(t_i)$. From the above two equations,

$$\left(\frac{\Upsilon(t_{i+1})}{m_{i+1}}\right) \Big/ \left(\frac{\Upsilon(t_i)}{m_i}\right) = \frac{1 - (1-\alpha)Pr([\tau(t_i)]^{i+1})}{1 - (1-\alpha)Pr([\tau(t_i)]^i)}$$

$$= \frac{1 - (1-\alpha)(\hat{p}_i + p_i)}{1 - (1-\alpha)\hat{p}_i}$$

$$= c_i$$

The base case is $\Upsilon(t_1) = p_1$. Therefore we can rewrite equation 5 as follows,

$$\frac{\Upsilon(t_{i+1})}{m_{i+1}} = c_i \frac{\Upsilon(t_i)}{m_i} = c_i c_{i-1} \frac{\Upsilon(t_{i-1})}{m_{i-1}} = \ldots = \prod_{j \leq i} c_j \qquad (6)$$

The result is summarized in following theorem.

Theorem 2. *For an uncertain relation T, rank-score of a tuple t_i can be computed as,*

$$\Upsilon(t_i) = m_i \prod_{j < i} c_j$$

where $m_i = \frac{p_i}{(1-(1-\alpha)\hat{p}_i)}$, $c_i = \frac{1-(1-\alpha)(\hat{p}_i+p_i)}{1-(1-\alpha)\hat{p}_i}$ and $\hat{p}_i = \sum t_r$, where t_i and t_r are mutually exclusive and $r < i$. □

This equation is applicable for dependent as well as independent tuples. Note that here m_i and c_i are dependent only on the tuples which are related to t_i, hence can be computed/updated efficiently. Moreover, the contribution c_i of a tuple t_i to the rank-score of a tuple t_j is the same for all $j > i$. Hence, the relative ordering will not change even if we use our divide and conquer approach.

Consider the same example as before. We begin by computing values of m_i and c_i for each tuple.

$m_1 = 0.35$ $m_2 = 0.3$ $m_3 = \frac{0.4}{(1-0.2\times0.3)} = 0.43$ $m_4 = 0.45$

$c_1 = (1 - 0.2 \times 0.35) = 0.93$ $c_2 = (1 - 0.2 \times 0.3) = 0.94$

$c_3 = \frac{(1-0.2\times(0.3+0.4))}{(1-0.2\times0.3)} = 0.91$ $c_4 = (1-0.2\times0.45) = 0.91$

Now, we partition T into $T_l = \{t_1, t_2\}$ and $T_r = \{t_3, t_4\}$ and apply top-1 query to these sub-relations.

$\Upsilon_{T_l}(t_1) = m_1 = 0.35$ $\Upsilon_{T_l}(t_2) = m_2 \times c_1 = 0.3 \times 0.94 = 0.28$
$\Upsilon_{T_r}(t_3) = m_3 = 0.43$ $\Upsilon_{T_r}(t_4) = m_4 \times c_3 = 0.45 \times 0.91 = 0.41$

It can be seen that t_1 and t_3 are chosen as top-1 from T_l and T_r respectively. During next comparison, t_3 ($\Upsilon(t_3) = m_3 \times c_1 \times c_2 = 0.37$) will be reported as the top-1 tuple, which is correct.

Table 2. Calculation of rank-scores of tuples in Table 1 ($\alpha = 0.9$) : t_1 ,$\{t_2, t_4\}$, $\{t_3, t_6\}$, t_5

Tuple	Probability	m	c	Υ
t_1	0.30	0.300	0.970	0.300
t_2	0.40	0.400	0.960	0.388
t_3	0.20	0.200	0.980	0.186
t_4	0.50	0.521	0.948	0.475
t_5	0.30	0.300	0.970	0.260
t_6	0.45	0.459	0.954	0.385

5   Our Data Structure

In the earlier sections, we derived the simple closed form expression for calculating $\Upsilon(t_i)$ for a tuple t_i. Now our task is to maintain a dynamic collection of tuples, such that for a given query k, we retrieve top-k rank-scored tuples efficiently. We use data structural approach for this problem. Our structure is a balanced binary search tree Δ (e.g. Red black tree, AVL tree) such that each leaf corresponds to a tuple in an uncertain relation T. Moreover, leaves in the tree are sorted in decreasing order of the score i.e. leaves $\ell_1, \ell_2, ..., \ell_N$ of the tree represent tuples $t_1, t_2, ..., t_N$ in the same order from left to right, such that $score(t_i) > score(t_{i+1})$. Let T_u represents the sub-relation containing tuples associated with leaves of a subtree rooted at node u. i.e. $T_u = \{t_{u'}, t_{u'+1}, ..., t_{u''}\}$ and $\ell_{u'}$ represents the left-most and $\ell_{u''}$ represents the right-most leaf of node u. At each node u, we store a triplet (top_u, M_u, C_u) such that:

- top_u is the tuple (represented by ℓ_{u^*}) with highest rank-score among tuples in sub-relation T_u. Here $u' \leq u^* \leq u''$.
- M_u is the contribution of all tuples in T_u towards rank-score of tuple top_u.

$$M_u = m_{u^*} \prod_{u' \leq i < u^*} c_i$$

- C_u is the contribution of all tuples in T_u towards rank-score of tuple t_i such that $i > u''$, where $\ell_{u''}$ is the right-most leaf of the subtree rooted at node u.

$$C_u = \prod_{u' \leq i \leq u''} c_i$$

Since our data structure stores only a constant number of information at each node, and the number of nodes are bounded by $O(N)$, the total space requirement of our data structure is $O(N)$.

If node u is a leaf node representing the tuple t_i, then $M_u = m_i, top_u = t_i$ and $C_u = c_i$. If u is an internal node, this information can be computed using the MERGE operation given below. Figure 1 shows an example for the uncertain data in table 2.

MERGE(u):
$v = left - child(u), w = right - child(u)$
if $M_v > C_v \times M_w$ then $top_u = top_v$ else $top_u = top_w$
$M_u = \max(M_v, C_v \times M_w)$
$C_u = C_v \times C_w$

Theorem 3. *The data structure Δ maintains a dynamic collection of tuples such that top-1 tuple, $t^1 = top_{root}$ and $\Upsilon(t^1) = M_{root}$.*

Proof by contradiction: Let t_a be the actual top-1 and $top_{root} \neq t_a$. Let u be the closest node from root, such that $top_u = t_a$, that means $top_{parent(u)} = t_b \neq t_a$. This is because during the merge operation at $parent(u)$, $m_a \prod_{x \leq i < a} c_i < m_b \prod_{x \leq i < b} c_i$, where ℓ_x is the leftmost leaf of $parent(u)$. Multiplying both the sides of the equation with $\prod_{i < x} c_i$, we get $\Upsilon(t_a) < \Upsilon(t_b)$, which is a contradiction to the statement that t_a is the highest rank-scored tuple. Therefore $t^1(= t_a)$ will always be at the root and $M_{root} = m_a \prod_{1 \leq i < a} c_i = \Upsilon(t_a) = \Upsilon(t^1)$. $\square$

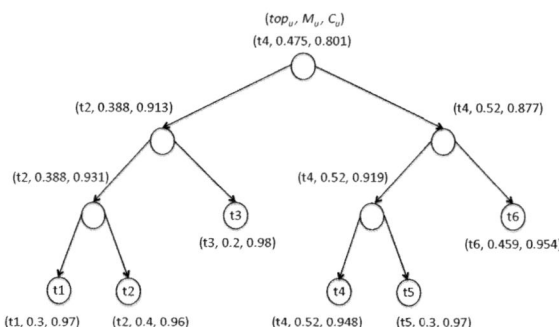

Fig. 1. The data structure for uncertain database in Table 2

In the following subsections, we show how to perform different operations such as `update-leaf`, `insert-leaf` and `delete-leaf` on this tree. Later, we use these operations for retrieving top-k tuples, insertion and deletion of tuples.

5.1 Update-Leaf

The values m_i and c_i within a leaf node ℓ_i can be changed in constant time. But this will change the m and c values at all nodes which are in the path from ℓ_i to root. Therefore we need to perform MERGE operation on all nodes in the path from ℓ_i to root, starting from $parent(\ell_i)$. Since the height of a balanced binary tree is bounded by $O(\log N)$, the total time for update-leaf can also be bounded by $O(\log N)$.

Theorem 4. *The m_i and c_i values of a leaf can be updated in $O(\log N)$ time.*

5.2   Insert-Leaf and Delete-Leaf

We first explain, how a one-to-one correspondence between tree leaves and tuples in relation T can be maintained during insertion or deletion of a leaf.

- *Insert:* To insert a new leaf, we begin by carrying out standard insert procedure of a binary search tree, which would create a new leaf node v. Let w be the parent of this newly created node. Node w being the leaf prior to insertion of v, represents a single tuple from T and should remain as a leaf after insertion of v as well. This can be achieved by creating a new internal node u, which becomes the parent of v and w.
- *Delete:* If deletion of a node results in an internal node with only one child, we perform recursive delete on that internal node.

After insert or delete of a leaf node ℓ_i, we need to update the M and C values at each node along the path of insertion or deletion. This can be achieved by performing MERGE operation in bottom-up fashion beginning with $parent(\ell_i)$. If tree goes out of balance after insert or delete, necessary rebalancing may force further re-computation at nodes whose left or right subtree is changed. However, such nodes are bounded by the height $O(\log N)$ of the tree. Hence Insert-leaf and Delete-leaf operations can be done $O(\log N)$ time.

5.3   Retrieving Top-k Tuples

In theorem 3, we proved that, by MERGE operation the top-1 tuple t^1 will be propagated to root node as top_{root}. Therefore t^1 can be retrieved in constant time. In order to retrieve the top-2 tuple t^2, we use the following strategy. After retrieving t^1, we set $\Upsilon(t^1) = 0$. As a result, the next highest rank-scored tuple t^2 will be propagated as top_{root} instead of t^1. This can be achieved by performing Update-leaf operation on leaf ℓ_j (leaf representing the current $top_{root} = t_j$), with it m_j value set to zero. As c_j remains unchanged, update operation affects only the computation of rank-score of t_j leaving rank-score of all other tuples unchanged. Repeating the same process, we can retrieve top-k tuples with highest rank-score values. We can revert back the changes done in data structure for answering top-k query by restoring the m values for k retrieved tuples using Update-leaf operation. Figure 2 shows an example for retrieving top-2 tuple from the uncertain data in table 1.

Retrieving Top-k:
for $i = 1$ to k
   $t_j = top_{root}$
   report top_{root} as top-i tuple
   Update-leaf(t_j) with $m_j = 0$

Theorem 5. *Top-k* `rank-scored` *tuples can be retrieved in* $O(k \log N)$ *time.*

Proof: For every tuple t_j retrieved for answering top-k query, we perform `Update-leaf` operation twice: once for setting $m_j = 0$ so that tuple with next highest `rank-score` can be retrieved and next after reporting top-k answers so as to restore the tree changes. Since `Update-leaf` is a $O(\log N)$ time operation, total time for top-k retrieval can be bounded by $O(k \log N)$.

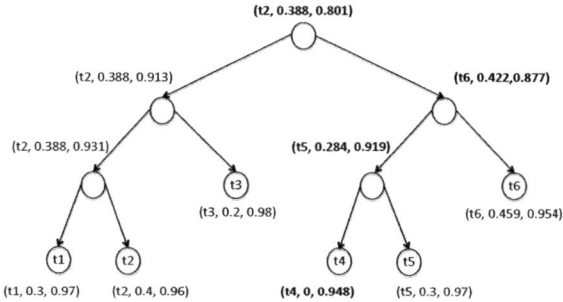

Fig. 2. The data structure in Figure 1 after setting $m_4 = 0$ for retrieving top-2

5.4   Insert-Tuple and Delete-Tuple

Whenever a tuple t_i gets inserted(deleted) from relation T, we modify our data structure as follows:

- We begin by carrying out `Insert-leaf` or `Delete-leaf` operation as necessary. If t_i is an independent tuple then at this point all nodes in the tree Δ have correct values for C and M. Hence no further action is necessary.
- If t_i is not independent, then its insertion(deletion) will change m_j and c_j values for all leaf nodes corresponding to tuples t_j such that $j > i$ and $\tau(t_i) = \tau(t_j)$. These changes can be accommodated by performing `Update-leaf` operation on each ℓ_j.

Figure 3 shows an example of inserting a new tuple t^*(with $score(t_2) > score(t^*) > score(t_3)$) and is mutually exclusive with t_5 in the uncertain data in table 2 and figure 4 shows an example for deletion of a tuple. Thus insertion(deletion) of a tuple can result in one `Insert-leaf` or `Delete-leaf` operation and at max $|\tau(t_i)|$ `Update-leaf` operations. Since any x-tuple can have only constant number of tuples, tuple insertion and deletion can be handled in $O(\log N)$ time. We note that updating of tuples can be simulated by first deleting and then reinserting it with updated values.

   We summarize the space requirement and performance of the proposed data structure in the following theorem.

Theorem 6. *A collection of uncertain data (N tuples) can be maintained using a linear size dynamic data structure, which can retrieve top-k* `rank-scored` *tuples in $O(k \log N)$ time, and can support insertion or deletion of a tuple t in $O(d \log N)$ time, where d is the number of tuples which are related to t.* □

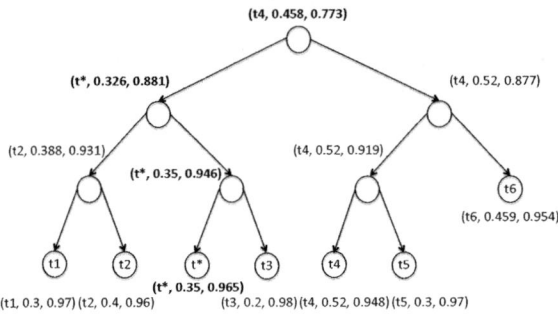

Fig. 3. The data structure in Figure 1 after inserting t*

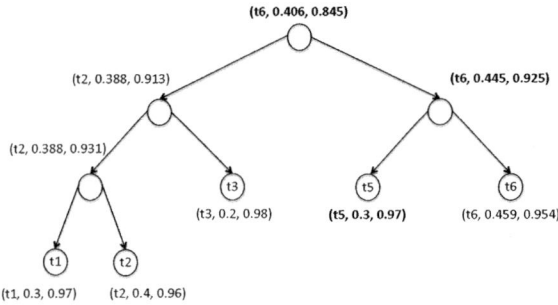

Fig. 4. The data structure in Figure 1 after deleting t_4

6 Experimental Study

In this section, we present an experimental study with both synthetic and real data evaluating effectiveness of the data structure in handling changes in underlying database and answering top-k queries. All experiments were conducted on 2.4 GHz Intel Core 2 Duo machine with 2GB memory running MAC OS 10.6.4.

Datasets: We created a synthetic dataset containing 100,000 tuples. Score of a each tuple is chosen uniformly at random from [0,100000] and it's probability is uniformly distributed in $(0.5 \times 10^{-5}, 1.5 \times 10^{-5})$. The number of tuples involved in each x-tuple follows the uniform distribution (2,10).

Along with synthetic datasets, we also use International Ice Patrol (IIP) Iceberg Sighting Database [1]. Each sighting record in the database contains date, location, number of days the iceberg has drifted, etc. As it is crucial to detect the icebergs drifting for long periods, we use the *number of days drifted* as ranking score. The sighting record also contains a confidence-level attribute according to the source of sighting: R/V (radar and visual), VIS (visual only), RAD (radar only), SAT-LOW (low earth orbit satellite), SAT-MED (medium earth orbit satellite), SAT-HIGH (high earth orbit satellite), and

---
[1] http://nsidc.org/data/g00807.html

EST (estimated). We converted these seven confidence levels into probabilities 0.8, 0.7, 0.6, 0.5, 0.4, 0.3, and 0.4 respectively. We gathered all records from 1981 to 1991 and 1998 to 2004. Based on it then we created 100,000 tuples dataset by repeatedly selecting records randomly.

Results: Experiments in [18] illustrate the effectiveness of ranking function $PRF^e(\alpha)$ at approximating other ranking functions for varying values of α ($\alpha = 1 - 0.9^i, 0 \leq i \leq 200$), where normalized Kendall distance [9] is used to evaluate closeness between the top-100 answers computed using a specific ranking function and $PRF^e(\alpha)$. As revealed by these experiments, ranking functions U-kRanks, PT-k are best approximated by $PRF^e(\alpha)$ for $i \approx 50$, hence we choose $\alpha = 1 - 0.9^{50}$ for all of our experiments. Choice of α only determines the quality of approximation and does not affect the query performance of our data structure.

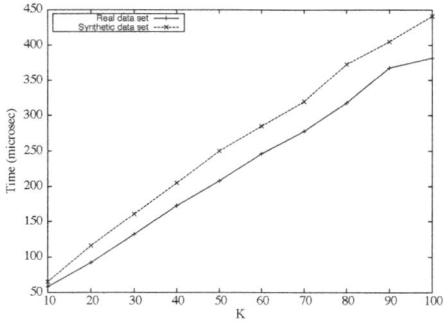

Fig. 5. Top-k query performance on real and synthetic data

We begin by evaluating the query performance of the data structure. We retrieve top-k tuples from both the datasets for k ranging from 10 to 100. Linear dependance of query time as obtained in the time bounds is evident from the results show in Figure 5. Also we can note that, correlations among tuples does not affect the query time of our data structure.

Next set of experiments conducted shows efficiency of our data structure in handling tuple insertions and deletions. Time required for inserting and deleting 100 tuples is measured for datasets of varying sizes. Figure 6 (a) and (b) shows that processing time per tuple increases slowly with data size. Whenever a tuple is inserted or deleted, to maintain the correctness of data structure, we also need to update information for leaves corresponding to its related tuples. As all tuples in real data set are assumed to be independent, average insertion/deletion time of a tuple is less than in case of synthetic data having correlations. For synthetic dataset, an x-tuple is selected at random to which a new tuple is added or from which a existing tuple is deleted. We ensure the x-tuple probability to be less than 1 to which a new tuple is being inserted. Position of a new tuple to be inserted in score-sorted ordering of tuples is selected at random whereas tuple to be deleted is always the highest scored tuple in the victim x-tuple. This results

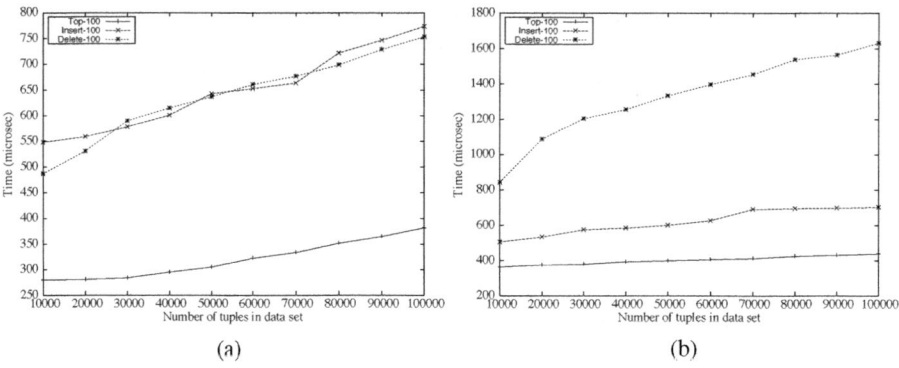

Fig. 6. Processing (insert, delete, top-k) cost on (a) real dataset (b) synthetic dataset

in more number of Update-leaf operations per tuple deleted than for tuple inserted and its effect on tuple insertion/deletion can be seen from figure 6 (b).

The proposed data structure can also be used when data arrives in streaming fashion. Jin et al. [16] have studied the problem of answering top-k queries on sliding windows. Our data structure achieves performance comparable to synopses proposed by them in terms of handling tuple insertion and deletions. Even though our data structure takes linear size as compared to these space efficient synopses, it can be noted that they rely on random order stream model used in streams algorithm community [1, 2, 11] and in worst case would take linear size as well.

7   Related Work

Uncertain data management has attracted a lot of attention in recent years due to an increase in the number of application domains that naturally generate uncertain data. These include sensor networks [8], data cleaning [12] and data integration [10, 3]. Several probabilistic data models have been proposed to capture data uncertainty (e.g TRIO [21], MYSTIQ [7], MayBMS [14], ORION [5], PrDB [19]). Virtually all models have adopted possible worlds semantics. Each data model captures tuple uncertainty (existence probabilities are attached to the tuples of the database), or attribute uncertainty (probability distributions are attached to the attributes) or both. Further distinction can be made among these models based on support for correlations. Most of the work in probabilistic databases has either assumed independence or supports restricted correlations, mutual exclusion being the most common. Recently proposed approaches [19, 17] extend the support for any arbitrary correlations.

Efforts have been made in recent times to extend the semantics of "top-k" to uncertain databases. Soliman et al. [20] defined the problem of ranking over uncertain databases. They proposed two ranking functions, namely U-Top$k$ and U-kRanks, and proposed algorithms for each of them. Improved algorithms for the same ranking functions were presented later by Yi et al. [22]. Hua et al. [13] proposed another top-k definition PT-k (*probabilistic threshold queries*) and proposed efficient solutions. Cormode et al. [6] defined number of key properties satisfied by "top-k" over deterministic data including exact-k, containment, unique-rank, value-invariance, and stability. With each of the existing top-k definition lacking one or more of these properties, Cormode at al. [6] proposed yet another ranking function `expected-rank`. As the list of top-k definitions continued to grow, Li et al. [18] argued that a single specific ranking function may not be appropriate to rank different uncertain databases and empirically illustrated the diverse, conflicting nature of parameterized ranking functions that generalize or can approximate many know ranking functions.

With most of the work for top-k query processing being focused on "one-shot" top-k query for static uncertain data, Chen and Yi [4] were the first to address the dynamic aspect of uncertain data. They proposed a dynamic data structure to support arbitrary insertions and deletions. For an uncertain relation with N tuples, the structure of [4] answers top-k queries in $O(k + \log N)$ time, handles an update in $O(k \log k \log N)$ time and takes $O(N)$ space. However, this structure is tied to a single ranking function i.e. U-Top$k$ and works only for independent tuples. Moreover, it can be built for some fixed k value and cannot answer a top-j for $j > k$. Dependance of time, required for handling update, on k is also not desirable. Recently, Jin et al. [16] proposed a framework for sliding window top-k queries on uncertain streams supporting several ranking functions. This framework assumes random-order stream model which significantly reduces the space requirement as compared to the worst-case scenario in which any data structure will have to remember every tuple in the current window.

8   Conclusions

In this paper we present a dynamic data structure, which can retrieve top-k tuples in $O(k \log N)$ time and has update cost of $O(\log N)$. We also evaluate efficiency of proposed data structure with experiments using synthetic and real data. It is an open question if, we can improve the top-k retrieval time to $O(k + \log N)$ without sacrificing update time or is there any lower bound for this problem?

References

[1] Chakrabarti, A., Cormode, G., McGregor, A.: Robust lower bounds for communication and stream computation. In: STOC, pp. 641–650 (2008)
[2] Chakrabarti, A., Jayram, T.S., Patrascu, M.: Tight lower bounds for selection in randomly ordered streams. In: SODA, pp. 720–729 (2008)
[3] Chaudhuri, S., Ganjam, K., Ganti, V., Motwani, R.: Robust and efficient fuzzy match for online data cleaning. In: SIGMOD Conference, pp. 313–324 (2003)
[4] Chen, J., Yi, K.: Dynamic structures for top-k queries on uncertain data. In: Tokuyama, T. (ed.) ISAAC 2007. LNCS, vol. 4835, pp. 427–438. Springer, Heidelberg (2007)
[5] Cheng, R., Kalashnikov, D.V., Prabhakar, S.: Evaluating probabilistic queries over imprecise data. In: SIGMOD Conference, pp. 551–562 (2003)
[6] Cormode, G., Li, F., Yi, K.: Semantics of ranking queries for probabilistic data and expected ranks. In: ICDE, pp. 305–316 (2009)
[7] Dalvi, N.N., Suciu, D.: Efficient query evaluation on probabilistic databases. In: VLDB, pp. 864–875 (2004)

 [8] Deshpande, A., Guestrin, C., Madden, S., Hellerstein, J.M., Hong, W.: Model-driven data acquisition in sensor networks. In: VLDB, pp. 588–599 (2004)
 [9] Fagin, R., Kumar, R., Sivakumar, D.: Comparing top k lists. In: SODA, pp. 28–36 (2003)
[10] Galhardas, H., Florescu, D., Shasha, D., Simon, E., Saita, C.A.: Declarative data cleaning: Language, model, and algorithms. In: VLDB, pp. 371–380 (2001)
[11] Guha, S., McGregor, A.: Approximate quantiles and the order of the stream. In: PODS, pp. 273–279 (2006)
[12] Halevy, A.Y., Rajaraman, A., Ordille, J.J.: Data integration: The teenage years. In: VLDB, pp. 9–16 (2006)
[13] Hua, M., Pei, J., Zhang, W., Lin, X.: Ranking queries on uncertain data: a probabilistic threshold approach. In: SIGMOD Conference, pp. 673–686 (2008)
[14] Huang, J., Antova, L., Koch, C., Olteanu, D.: MayBMS: a probabilistic database management system. In: SIGMOD Conference, pp. 1071–1074 (2009)
[15] Jestes, J., Cormode, G., Li, F., Yi, K.: Semantics of ranking queries for probabilistic data. IEEE Transactions on Knowledge and Data Engineering PP(99), 1 (2010)
[16] Jin, C., Yi, K., Chen, L., Yu, J.X., Lin, X.: Sliding-window top-k queries on uncertain streams. PVLDB 1(1), 301–312 (2008)
[17] Koch, C., Olteanu, D.: Conditioning probabilistic databases. In: Proc. VLDB Endow., vol. 1, pp. 313–325 (2008)
[18] Li, J., Saha, B., Deshpande, A.: A unified approach to ranking in probabilistic databases. PVLDB 2(1) (2009)
[19] Sen, P., Deshpande, A., Getoor, L.: Prdb: managing and exploiting rich correlations in probabilistic databases. VLDB J. 18(5), 1065–1090 (2009)
[20] Soliman, M.A., Ilyas, I.F., Chang, K.C.C.: Top-k query processing in uncertain databases. In: ICDE, pp. 896–905 (2007)
[21] Widom, J.: Trio: A system for integrated management of data, accuracy, and lineage. In: CIDR, pp. 262–276 (2005)
[22] Yi, K., Li, F., Kollios, G., Srivastava, D.: Efficient processing of top-k queries in uncertain databases. In: ICDE, pp. 1406–1408 (2008)

Efficient Storage and Temporal Query Evaluation in Hierarchical Data Archiving Systems

Hui (Wendy) Wang[1], Ruilin Liu[1], Dimitri Theodoratos[2], and Xiaoying Wu[2]

[1] Department of Computer Science, Stevens Institute of Technology
Hoboken, NJ, USA
{hui.wang,rliu3}@stevens.edu
[2] Department of Computer Science, New Jersey Institute of Technology
Newark, NJ, USA
{dth,xw43}@cs.njit.edu

Abstract. Data archiving has been commonly used in many fields for data backup and analysis purposes. Although comprehensive application software, new computing and storage technologies, and the Internet have made it easier to create, collect and store all types of data, the meaningful storing, accessing, and managing of database archives in a cost-effective way remains extremely challenging. In this paper, we focus on hierarchical data archiving that has been popularly used in the scientific field and web data management. First, we propose a novel compaction scheme for archiving hierarchical data. By compacting both data and timestamps, our scheme substantially reduces not only the amount of needed storage, but also the incremental archiving time. Second, we design a temporal query language to support data retrieval from the compact data archives. Third, as compaction on data and timestamps may bring significant overhead to query evaluation, we investigate how to optimize such overhead by exploiting the characteristics of the queries and of the archived hierarchical data. Finally, we conduct an extensive experimentation to demonstrate the effectiveness and efficiency of both our efficient storage and query optimization techniques.

1 Introduction

Recent years have witnessed an increasing number of enterprises and organizations archive their databases. The reasons of such activity include the mandate to comply with legal and governmental regulations, serving the need for data backup and recovery, and enabling analytical processing to discover data evolutional patterns. For instance, from biology to astronomy, it is necessary to keep track of all previous versions of the scientific data for later verification purposes [8]. However, as increasing volumes of these data being accumulated, their archives can reach a critical mass. Where once megabytes and gigabytes of data needed to be managed, now terabytes and petabytes are a common ground. Although comprehensive application software, new computing and storage technologies, and the Internet have made it easier to create, collect and store all types of data, the meaningful storing, accessing, and managing of the archiving databases in a cost-effective way keeps to be extremely challenging.

In this paper, we focus on archiving hierarchical data in the form of Extensible Markup Language (XML). XML has been largely used in scientific domains [8] and

J.B. Cushing, J. French, and S. Bowers (Eds.): SSDBM 2011, LNCS 6809, pp. 109–128, 2011.
© Springer-Verlag Berlin Heidelberg 2011

many other domains, for example, Web data management. Unfortunately, DBMS vendors and standard groups have not moved aggressively to extend their systems with support for hierarchical data archiving databases. The need for an efficient hierarchical data archiving system is not met and awaits for the development of new techniques.

The management of XML archiving databases involves two fundamental issues: (1) how to store successive versions of XML databases in an archiving database in a cost-effective way, and (2) how to efficiently evaluate queries with temporal expressions over the archiving database. A naive approach for storing successive versions of data consists in storing each version separately. This is undesirable because: (1) the storage can easily grow to be prohibitively expensive as it can become many times larger than the base data [6,15,22], and (2) evaluating temporal queries over databases stored in this way is highly inefficient. On the other hand, careless design of techniques that reduce the storage space may bring significant overhead to the evaluation of temporal queries over the archiving database.

While there has been previous research on the compact storage of archived hierarchical data [8,18,21], little effort has been devoted to the efficient evaluation of temporal queries over the compact storage. To address this problem, we develop an archiving system for hierarchical databases that combines the compact storage with optimization techniques for the evaluation of the temporal constraints of the queries.

To the best of our knowledge, this is the first paper to address the problem of optimizing the evaluation of temporal constraints of queries on compacted archiving XML databases. The main contributions of this paper are:

- We propose a novel compact and updateable storage scheme for XML archiving databases. We show how our storage scheme supports efficient incremental updates to the archiving database upon addition of new database instances (Section 3). Compared to the archiving database scheme of [8], we share the same goal of reducing the storage size. However, in contrast to the timestamp compaction scheme of [8] which recovers the timestamps through a top-down propagation in the archiving database, our approach recovers the timestamps through a bottom-up propagation. Our experiments show that our approach outperforms the top-down approach in terms of update cost (Section 6).
- We design a simple yet expressive language for temporal tree pattern queries in archiving database systems. In order to support evaluation of temporal constraints on the compact storage, we define three temporal evaluation annotations, namely DC, LC, and NC. These annotations cover three situations where the corresponding constraint: (1) must be validated by recovering timestamps in the archive, (2) can be validated locally without navigation, and (3) does not need validation at all (Section 4).
- We formulate the problem of optimizing the evaluation of the temporal constraints. First, we provide a cost model for these annotations, and formally define the optimization problem as finding a minimal annotation for the constraints of a query by replacing as many annotations as possible with cheaper ones (Section 5.1). Second, in order to address this problem, we design inference rules that derive timestamp set containment relationships between query nodes (Section 5.2). The inference rules exploit structural information of the queries and also of the database when a

database schema is available. Third, we design an optimization algorithm for temporal queries, and show that its cost is polynomial on the size of the input query and the database schema (Section 5.3).

- We use extensive experimentation to prove both the efficiency and effectiveness of our techniques on compacting XML databases and incrementally updating XML archives and on optimizing the evaluation of temporal constraints in queries (Section 6).

We present preliminary material in Section 2. We discuss related work in Section 7, and summarize the paper and suggest future work in Section 8.

2   Updating XML databases

Database instances and Archiving databases. An XML database can be modeled as a tree. The nodes in the tree are labeled by element tags, attributes, or values. The edges in the tree represent element-subelement, element-attribute, and element-value relationships. We assume that XML tree nodes are assigned identifiers. The identifier of a node remains unchanged if the node is not deleted. Every newly inserted node gets a new identifier. For simplicity, we assume that the XML instance tree contains only element and value nodes.

The updates on the database create different *database instances* (versions). Each instance is identified by a timestamp which is represented by a version number. Multiple temporal database instances can be merged into a (temporal) XML database, which is called *archiving database*.

Note that here we do not deal with the identification of unchanged nodes across multiple instances [8], or with the detection of changes and the computation of diffs in hierarchical data [11,13,20,27]. These issues are orthogonal to those dealt with in this paper. Algorithms for these tasks can be employed prior to applying the techniques presented in this paper, in order to detect unchanged and new nodes and to assign node identifiers.

Update operations. We consider any type of update operations on XML trees: insertion of a tree below a node, deletion of a subtree rooted at a node, and replacement of a subtree rooted at a node by another tree. An inserted/deleted tree can trivially consist of a single node. For node identifier assignment reasons, a modification of a tree is viewed as a deletion of a subtree followed by an insertion of a tree. Therefore, in the following, an update operation refers to an insertions or a deletion. These operations are denoted as follows (the XML database D is implicit):

- $ins(q, T)$: insert a tree T into the database D. All the nodes in T have new identifiers. The root of T becomes a child of node q in the resulting database.
- $del(p)$: delete from D the subtree rooted at node p.

If a DTD is present, we assume that the initial database and any database resulting by the application of an update operation complies with the DTD.

One can see that the cascading application of a sequence of update operations can be modeled by the application of a set of deletions followed by the application of a set of

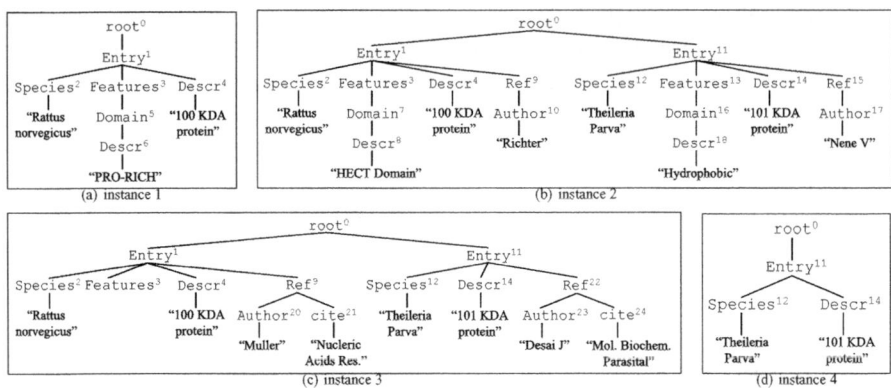

Fig. 1. Four consecutive instances of an extract of the Swiss-Prot Dataset

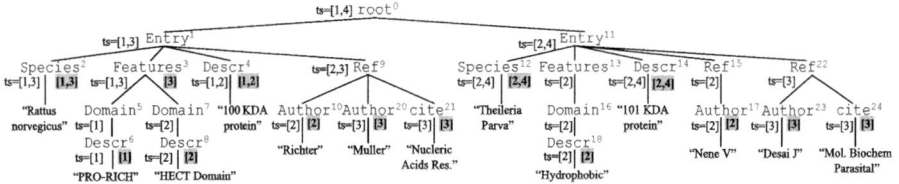

Fig. 2. The compact storage of the four Swiss-Prot database instances of Figure 1

insertions. Specifically, given a database D and a sequence of update operations whose cascading application results in a database D', there is

(a) a set ds of deletion operations $del(p_1), \ldots, del(p_n)$ such that no p_i is an ancestor of a p_j, $i, j \in [1, n]$, (that is, the subtrees rooted at the p_i's are disjoint), and
(b) a set is of insertion operations $ins(q_1, T_1), \ldots, ins(q_m, T_m)$ such that no q_i, $i \in [1, m]$, is a descendant or self of a p_j, $j \in [1, n]$.

that satisfy the following property: database D' can be obtained from D by applying first the operations in ds and then those of is in any order.

In the following we assume that for any two consecutive instances D_i and D_{i+1} of the database that are added to the archiving database, the two sets of update operations ds_i and is_i used to produce D_i from D_{i+1} are available.

3   Compact Storage of Archiving Databases

A naive approach for storing multiple instances of a database D consists in storing these instances separately. This is undesirable because: (i) the storage space increases significantly in the long run especially for databases with frequent updates, and (ii) answering queries across multiple instances may be complex. For example, finding when a particular data piece first appeared in history or when it was last changed may require traversing a very large number of instances. For this reason, XML archiving databases are designed with compact storage schemes which merge multiple instances and compact timestamps.

Merging instances. Successive instances of the XML database D to be archived share a significant portion of identical nodes. This suggests merging these versions by storing the multiple occurrences of the same node only once in the archiving database A. In this direction, with every node in A, we associate a *timestamp set* which contains the timestamps of the instances of D in which this node is valid. The timestamp set of a node p is denoted $ts(p)$. When the timestamps in a timestamp set are successive, $ts(p)$ is denoted by a timestamp interval $[t_l, t_u]$, where (a) t_l is the timestamp of the instance at which node n was first created, and (b) t_u is either the timestamp of the latest instance merged into A or the timestamp of the last instance in which this node is valid (in the latter case, t_u is smaller than the timestamp of the latest merged instance). If $t_l = t_u$, the timestamp set is denoted $[t_l]$. When a new instance of D is inserted into A, the nodes with the same ID are merged and the corresponding timestamps are updated accordingly. Note that once a node is inserted into A it is never removed. Its timestamp set though will be modified. Figure 1 shows four instances of a database. Node IDs are shown as superscripts of the node labels. Figure 2 shows the archiving database resulting by merging the four database instances of Figure 1. The timestamp sets are shown by the nodes preceded by "ts=".

Timestamp set compaction. An important observation is that in hierarchical databases, a node is not inserted before or removed after its ancestors. Because of this, a *monotonicity property* holds between the timestamps of ancestor and descendant nodes in the archiving database which states that given an XML archiving database A and two nodes $p, q \in A$ such that p is an ancestor of q, it is always true that $ts(q) \subseteq ts(p)$. This can lead to a large number of repeated timestamps in the archiving database. To avoid this storage redundancy, we suggest a timestamp set compaction scheme which instead of storing with every node in A its timestamp set, it assigns to some nodes and stores only *timestamp labels* from which the timestamp sets of all the nodes can be computed: given a node p and its children nodes $c_1, \ldots, c_n$ in A, the timestamp label L_p of p is $L_p = ts(p) - \cup_{i=1}^n ts(c_i)$. Intuitively, the timestamp label of a node in A only preserves the timestamps that are not present within the timestamp sets of any of its children. If $L_p = \emptyset$, no timestamp label is assigned to node p in A. Figure 2 shows an example of the new timestamp set compaction scheme. Only some nodes have timestamp labels. They are shown in a gray background by the owning nodes. This is the only timestamp information stored in the archive.

The new compaction scheme contrasts with the compaction scheme of [8] where a timestamp is stored at a child node in A only when it is different from the timestamp of its parent node. In this sense, [8] recovers the timestamp sets of the nodes in A by propagating timestamps top down as opposed to our approach where timestamp sets are computed by propagating timestamps bottom up. In the following, we refer to the compaction scheme in [8] as top-down (TD) and to ours as bottom-up (BU).

Incremental timestamp label computation. When adding a new instance of D into A, a naive method for updating A merges the new database instance into A first, then applies a timestamp set compaction procedure as described above to the merged result. In this process, the new database instance comes with its timestamp labeling all its nodes and these timestamps are added to the timestamp sets of the nodes with the same ID in the resulting archive before applying the compaction process. Since both A and D may be large, this method may incur a significant amount of unnecessary overhead. Therefore, we propose below a procedure for adding a new database instance into A, with the timestamp labels being computed *incrementally*. Let $k - 1$ be the timestamp of the current database instance D_{k-1}, and D_k be the next database instance obtained from D_{k-1} by applying the set of deletions $del(p_1), \ldots, del(p_n)$ and the set of insertions $ins(q_1, T_1), \ldots, ins(q_m, T_m)$ satisfying the properties set forth in Section 2. Let also A_{k-1} be the database archive containing the database instances up to D_{k-1}, we construct A_k, the database archive resulting by adding D_k to A_{k-1}, incrementally through the process described in Listing 1.

Listing 1. Incremental timestamp label computation

Input: k, $del(p_1), \ldots, del(p_n)$, $ins(q_1, T_1), \ldots, ins(q_m, T_m)$, D_{k-1}, A_{k-1}
Output: A_k
1 $A_k := A_{k-1}$
2 **for** every node p of A_k such that $(\forall i \in [1, n], p \neq p_i)$ and $((p$ is a leaf node in D_{k-1} which is not a descendant of a $p_i)$ or (all of p's children in D_{k-1} are p_is)) **do**
3    $L_p := L_p \cup \{k\}$
4 **for** every tree T_i, $i \in [1, m]$ **do**
5   **for** every leaf node q of T_i **do**
6      $L_q := \{k\}$
7   insert the resulting tree into A_k so that its root becomes a child of q_i

---

The incremental process of Listing 1 essentially adds the newly inserted trees $T_1, \ldots, T_m$ to the archive, and also the timestamp k of the latest database instance to all the archive nodes that are leaf nodes in that instance. Note that the newly inserted trees come with the timestamp label $[k]$ in all their leaf nodes and therefore the archive does not need to be accessed for the addition of these timestamps. Only nodes existing in the previous version A_{k-1} of the archive are accessed for the addition of timestamp k.

As we will show later in Section 6, the BU approach exhibits interesting characteristics because it generates a comparable number of timestamp labels in the database archive as the TD approach [8] (which implies that it consumes approximately the same space), while incurring substantially less update costs.

4 Temporal Query Language

Temporal queries on archiving databases comprise three types of constraints: *structural* constraints that are evaluated over the structural relationships of the nodes, *value-based* constraints that restrict the data values of the elements, and *temporal* constraints that are evaluated against the timestamp sets of the nodes. The efficient evaluation of queries with structural and value-based constraints on XML databases has been studied extensively in recent years [7,14,17]. Temporal constraints can be considered as a type of value-based constraint, but in this paper, we consider them separately in order to focus on the optimization of their evaluation, specifically on the compact storage. In this section, we present our temporal query language and discuss implementation issues.

Besides constraints, temporal queries have temporal evaluation annotations that support the evaluation of the temporal constraints. We present these concepts below.

Temporal constraints. We identify a set of temporal constraints that are commonly used in practice.

- $includes(t)$ is satisfied if the relevant timestamp set includes the timestamp t.
- $overlaps(t_a, t_b)$ is satisfied if the relevant timestamp set overlaps with the temporal interval $[t_a, t_b]$.
- $before(t)$ / $after(t)$ is satisfied if the relevant timestamp set precedes / succeeds the timestamp t.
- $contains(t_a, t_b)$ / $is\_contained(t_a, t_b)$ is satisfied if the relevant timestamp set contains / is contained in the interval $[t_a, t_b]$.
- $meets(t_a, t_b)$ is satisfied if either the first timestamp or the last timestamp of the relevant timestamp set touches the boundary of the interval $[t_a, t_b]$.

Other temporal constraints e.g., $start(t_a)$, $end(t_b)$ etc. can be expressed in terms of the temporal constraints introduced above.

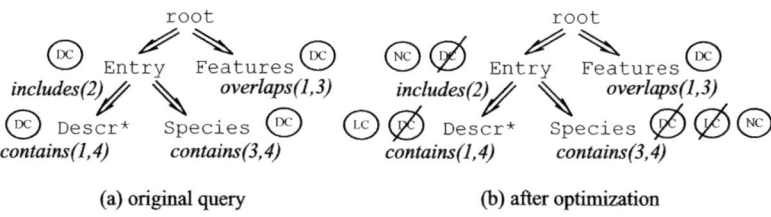

(a) original query   (b) after optimization

Fig. 3. A TTPQ with temporal evaluation annotations

Temporal queries. We support two types of temporal queries, *snapshot* queries that contain point-type temporal constraints and return subtrees of database instances at specific timestamps, and *history trace* queries that contain range-type temporal constraints and return subtrees spanning different instances whose timestamps fall within the query range. These two types of queries are built by adding temporal expressions to the nodes of the query structural pattern. These temporal expressions are Boolean expressions of temporal constraints like those we presented above. Here we assume that the temporal expressions attached to the query nodes are conjunctions of temporal constraints. We focus on tree structural patterns with child (/) and descendant (//) relationships. The resulting queries are called *temporal tree pattern queries* (TTPQs). Similarly to TPQs, TTPQs contain a distinguished node corresponding to the answer node. Figure 3 shows an example of a TTPQ involving descendant relationships. A '*' indicates the answer node. The answer of a query Q on an archiving database A is defined through embeddings of Q to A that satisfy the temporal constraints.

Temporal evaluation annotations. To evaluate a TTPQ, its temporal constraints must be examined against the timestamps of the nodes in the XML archiving database. However, due to the timestamp set compaction scheme, it is possible that in order to compute the timestamp sets of some nodes, the timestamp labels of descendant nodes need to be collected (e.g., the Features nodes in Figure 2). On the other hand, for other nodes their timestamp sets need not be computed as they are equal to their timestamp labels (e.g., the Species nodes in Figure 2). We also observe that it is also possible that for some query nodes, the evaluation of their temporal constraints is not needed (details in Section 5). Based on these observations, we define three temporal evaluation annotations for the query nodes that have temporal constraints. A temporal evaluation annotation determines whether the temporal constraints of the node need to be checked for satisfaction, and if they need how the timestamp set of the image node under an embedding can be computed. Specifically, for a node p in a query Q, the possible annotations and their meaning are the following (M is an embedding of Q to the archiving database):

- DC (Descendant Check): compute the timestamp set of $M(p)$ by unionning all the timestamp labels of those descendant nodes that have one, and check whether the union satisfies the temporal constraints of p.
- LC (Local Check): use the timestamp label of $M(p)$ to check locally whether the temporal constraints of p are satisfied.
- NC (No Check): no operation for validating the temporal constraints of p is needed.

The query in Figure 3 shows examples of temporal evaluation annotations.

5   Optimization of the Evaluation of Temporal Constraints

5.1   Problem Definition

When evaluating a query, for every query node that has a temporal constraint, the timestamp set of its image nodes in the archiving database needs to be computed. In many cases (when the node annotation is DC), such timestamp computation needs a recursive traversal of descendant nodes which is costly, especially for databases with a deep tree structure. However, we observe that sometimes the expensive DC annotation on a node is not necessary and can be replaced by cheaper ones. The following example provides some intuition. Consider the query of Figure 3(a) to be evaluated on an instance of the Swiss-Prot dataset. First observe that as the DTD of Figure 4(a) suggests, all images of the Descr node under an embedding to an archiving database whose instances comply with the DTD (e.g. the archiving database of Figure 2) are leaf nodes. Thus, they always have all their timestamp sets locally specified as timestamp labels. Therefore, the DC annotation on the Descr node is not necessary and can be replaced by LC. Second, due to the monotonicity property of timestamp sets on ancestor and descendant nodes (mentioned in Section 3), the timestamp set of the image node of Entry under an embedding should always contain that of Descr. If the image of Descr satisfies the temporal constraint of Descr (i.e., the timestamp set of the image contains [1,4]), then the image of Entry also satisfies the temporal constraint $includes(2)$ of Entry. Thus, the DC annotation on Entry is not needed and can be replaced by NC.

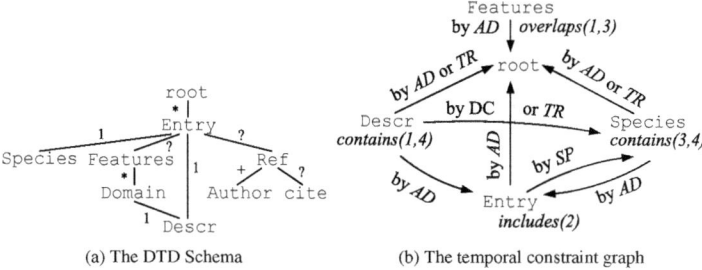

(a) The DTD Schema   (b) The temporal constraint graph

Fig. 4. The temporal constraint graph of the TTPQ of Figure 3 (a)

Cost model. Based on the temporal evaluation annotation DC, LC or NC of a node, we define the cost for evaluating its temporal constraints. Given a query node q, let ℓ be its temporal evaluation annotation, and q' be its image node in the archiving database A under an embedding of the query to A. Let also $|T_{q'}|$ be the number of nodes of the tree rooted at q' in A. Then, the cost of evaluating the temporal constraint of q on A is:

$$C_\ell(q) = \begin{cases} \max_{q'} |T_{q'}| & \ell = \text{DC}; \\ 1 & \ell = \text{LC}; \\ 0 & \ell = \text{NC}. \end{cases}$$

In this definition, we consider for DC the worst case scenario and thus define the cost as the number of nodes of the maximum size tree rooted at an image node of q.

Given a TTPQ Q, let annotation of Q, denoted $\mathcal{A}$, be the vector of annotations in Q for the the nodes $q_1, \ldots, q_n$ which have temporal constraints. The cost of evaluating the temporal constraints of Q on the archiving database A is $C_{\mathcal{A}}(Q) = \sum_{i=1}^{n} C_{l_i}(q_i)$, where $l_i \in \{\text{DC}, \text{LC}, \text{NC}\}$ is the temporal evaluation annotation on the query node q_i.

Minimal temporal evaluation annotations. Based on the cost metric above, we define *minimal temporal evaluation annotations*. We restrict our attention to the set of annotations that correctly return the answers of Q on any archiving database whose database instances comply with the given DTD. Given two different annotations $\mathcal{A}_1, \mathcal{A}_2$ of a TTPQ Q, we say that $\mathcal{A}_1$ *dominates* $\mathcal{A}_2$, denoted $\mathcal{A}_1 \geq \mathcal{A}_2$, if $C_{\mathcal{A}_1}(Q) \geq C_{\mathcal{A}_2}(Q)$ on any archiving database. An annotation $\mathcal{A}$ in a set of annotations S is *minimal* if there is no annotation $\mathcal{B} \in S$ such that $\mathcal{A} > \mathcal{B}$.

The problem. Given a TTPQ Q, initially every query node with a temporal constraint is annotated with the expensive annotation DC. Our optimization goal is to find a minimal temporal evaluation annotation for Q, i.e., we replace as many DC as possible with the cheaper LC or even better with the NC annotation.

5.2   Inference Rules and the Temporal Constraint Graph

For the optimization of the temporal constraints of a query, we use the concept of temporal constraint graph which is first defined in this section. A temporal constraint graph is constructed using inference rules described below.

Inference Rules. The inference rules have the form $P_1, \ldots, P_k \rightarrow R$ meaning that "if the premises $P_1, \ldots, P_k$ are true, then the conclusion R is also true". They infer containment relationships between (the timestamps of the image nodes of) query nodes. We formally define next the concept of containment relationship between query nodes.

Definition 1 (Temporal containment). *Given a query Q and two nodes $p, q \in Q$, we say that node p is* contained *in node q, denoted $p \subseteq_t q$, iff for every embedding of Q to an archiving database, the timestamp set of the image of p is a subset of the timestamp set of the image of q. We say that node p is* equivalent *to node q, denoted $p =_t q$, if $p \subseteq_t q$ and $q \subseteq_t p$. In the presence of a DTD, only archiving databases whose instances comply with the DTD are considered for this definition.*

There are inference rules that depend on the presence of DTD and others not.

Inference rules independent of DTDs. The monotonicity property of the timestamps in the archiving database between ancestor and descendant nodes naturally allows the inference of a containment relationship. Let p, q, r be nodes in a TTPQ Q. Let also $Q \models p//q$ denote the fact that $p//q$ occurs in or can be inferred from Q (e.g., trivially from p/q, or transitively from $p//r$ and $r//q$ in Q). The following inference rule is based on the ancestor-descendant (AD) relationship between query nodes:

$$\text{AD Rule}: \ Q \models p//q \rightarrow q \subseteq_t p$$

Intuitively, the structural constraint $p//q$ in the query forces the images of q to be descendants of the images of p in the archiving database. Thus the monotonicity property holds.

We can also apply a transitivity inference rule to infer containment relationships between nodes:

$$\text{TR Rule}: \ p \subseteq_t q \text{ and } q \subseteq_t r \rightarrow p \subseteq_t r$$

Inference rules dependent on DTDs. In the presence of a DTD for the database instances, additional temporal containment relationships between query nodes can be derived. Given the DTD Δ of a database and a node p of a TTPQ Q, let p_Δ denote the node in Δ which has the same label as p. Let also $SinglePath(p, q)$, denote the fact that there is a single path from p to q in Δ, and all the edges of this path are labeled by '1' (indicating a single mandatory occurrence of a child node). The following inference rule is based on the $SinglePath$ (SP) property:

$$\text{SP Rule}: \ Q \models p//q, \ SinglePath(p_\Delta, q_\Delta) \rightarrow p \subseteq_t q.$$

The reasoning behind the SP rule is that, since the result of any update operation (insertion or deletion) to a database instance complies with Δ: (a) any node having the label of p in a database instance D_i added to the archiving database A, has exactly one descendant node labeled by the label of q in D_i, and (b) if (p_i, q_i) and (p_j, q_j) are two such pairs of nodes in two database instances D_i and D_j, respectively, added to the archiving database, if $p_i = p_j$, $q_i = q_j$. Therefore, any timestamp of the image $M(p)$ of p under an embedding M to A, is also a timestamp of $M(q)$. That is, $p \subseteq_t q$. Since, because of rule AD, $q \subseteq_t p$ also holds, we can derive $q =_t p$.

By combining the SP and AD rules we can derive a temporal containment relationship between two nodes that are both descendants of a node in Q even when no descendant relationship can be derived between them in Q:

DC Rule : $Q \models p//q,\ Q \models p//r,\ SinglePath(p_\Delta, r_\Delta) \rightarrow q \subseteq_t r$

A containment relationship between two nodes that are both descendants of another node in a query can be derived also when the corresponding nodes are appropriately restricted by single path constraints in the DTD:

DE Rule : $Q \models p//q,\ Q \models p//r,\ SinglePath(p_\Delta, r_\Delta), SinglePath(q_\Delta, r_\Delta) \rightarrow r \subseteq_t q$

Because of DC rule we can also infer $q \subseteq_t r$, that is, $r =_t q$. Note that none of rules DC and DE require that nodes q and r lie on the same path in query Q.

Temporal Constraint Graph. Given a temporal query Q, we construct a *temporal constraint graph* G_Q as follows: (a) for every node in Q there is a distinct node in G_Q associated with the label and temporal constraints of the corresponding node in Q, and (b) there is an edge in G_Q from p to q iff the temporal containment relationship $p \subseteq_t q$ can be inferred from Q and the DTD (in case there is one) using the inference rules presented above. Figure 4(b) shows the temporal constraint graph of the query of Figure 3(a) in the presence of the DTD of Figure 4(a). The edges of the graph are annotated by the inference rule(s) which derived the relevant containment relationships.

5.3 Optimization Actions

We distinguish two types of optimization actions: *replacing* DC *with* LC (DC → LC), and *replacing* DC *or* LC *with* NC (DC/LC → NC). These optimization actions exploit also schema information if a DTD is present.

Optimization action DC → LC. This action is based on our observation that if a node p_Δ in a DTD Δ is a sink node (that is, it has no outgoing edges), all nodes labeled by the label of p_Δ in an archiving database A whose instances comply with Δ are leaf nodes. Therefore, any node p in the TTPQ Q having the same label as p_Δ is mapped by any embedding M of Q to A to a leaf node whose timestamp set is equal to its timestamp label. As a consequence, the satisfaction of the temporal constraint of p can be checked locally at every image node of p under M, and the temporal evaluation annotation DC of p can be turned into LC.

In summary, if $sink(p_\Delta)$ denotes the fact that p_Δ is a sink node in DTD Δ, the DC → LC action can be described as follows:

$$\forall p \in Q, sink(p_\Delta) \rightarrow replace\ \text{DC}\ of\ p\ with\ \text{LC}$$

Coming back to the example TTPQ of Figure 3(a) and the DTD of Figure 4(a), the DC → LC action replaces the DC annotation of the Desc and the Species nodes by LC. The resulting TTPQ (after additional optimization actions) is shown in Figure 3(b).

Optimization action DC/LC → NC. Before going into the details, we introduce the notions of *consumed temporal constraint* and *witness node*.

Definition 2 (Consumed Temporal Constraint). *We say that a temporal constraint* $f_1(p)$ *on query node* p *consumes a temporal constraint* $f_2(q)$ *on query node* q *if for every embedding* M *of the query to an archiving database* A *such that the image of* p *under* M *satisfies* f_1, *the image of* p *under* M *satisfies* f_2.

Table 1. Temporal constraint consumption

Consuming temporal constraint on p	Consumed temporal constraint on q
$p \subseteq_t q$	
$includes(t)$	$includes(t)$ $overlaps(t_1, t_2), t \in [t_1, t_2]$
$contains(t_1, t_2)$	$includes(t_3), t_3 \in [t_1, t_2]$ $contains(t_3, t_4), t_3 \geq t_1, t_4 \leq t_2$ $overlaps(t_3, t_4), t_1 < t_3 < t_2$ or $t_1 < t_4 < t_2$
$q \subseteq_t p$	
$is\_contained(t_1, t_2)$	$is\_contained(t_3, t_4), t_3 \leq l_1, t_4 \geq t_2$
$before(t)$	$before(t_1), t_1 \geq t$
$after(t)$	$after(t_1), t_1 \leq t$
$q =_t p$	
$meets(t_1, t_2)$	$meets(t_1, t_2)$

Table 1 shows various cases of temporal constraint consumption for different types of temporal containment relationships between nodes p and q. The reasoning behind these statements is straightforward. For example, with the assumption that $p \subseteq_t q$, when the timestamp set of the image of query node p includes t, the timestamp set of the image of query node q also includes t. Clearly, those statements that are valid for $p \subseteq_t q$ or $q \subseteq_t p$, are also valid for $p =_t q$. Given a query node p, if there exists another query node q whose temporal constraints consume all constraints of p, the evaluation of the constraints on p becomes unnecessary as long as the constraints of q are evaluated. In this case, the satisfaction of (the constraints of) p is witnessed by the satisfaction of (those of) q.

Definition 3 (Witness node). *Given two nodes p, q in a query Q, we say q is a* witness node *of p iff all the temporal constraints of p are consumed by the temporal constraints of q.*

Clearly, the evaluation of the temporal constraints of a witness node makes the evaluation of the constraints on the witnessed query node unnecessary. Therefore, the temporal evaluation annotation of a witnessed query node can be changed to NC irrespectively of whether it was DC or LC as long as the temporal evaluation annotation of a witnessing query node is not NC. A node can be witnessed by and be a witness of multiple other nodes, and the witnessing relationship is transitive. The challenge is to partition the query nodes that participate in witnessing relationships into two sets W and O such that: (a) every node in O is witnessed by a node in W, and (b) replacing in the query the annotation of every node in O by NC produces a minimal annotation for the query. It turns out that in this case, W has a minimal number of nodes. In order to partition the nodes we introduce the concept of witness graph. Given a query Q, a *witness graph* for Q, is a graph W_Q such that: (a) the nodes of W_Q correspond to the nodes of Q, and (b) there is an edge from node p to node q in W_Q, iff p is a witness node of q in Q.

A witness graph for a query Q is constructed using the constraint graph of Q and the consumption relationships of Table 1.

As an example, one can see that the witness graph of the query of Figure 3 (a) whose constraint graph is shown in Figure 4 (b), comprises exactly two edges Desc → Species and Desc → Entry. The edge Desc → Species, for instance, exists because according to the constraint graph in Figure 4 (b), Desc $\subseteq_t$ Species, and therefore, as Table 1 indicates, $contains(1, 4)$ on Desc consumes $contains(3, 4)$ on Species.

Optimization Algorithm. Our temporal constraint optimization algorithm takes as input a TTPQ Q whose nodes are all annotated with DC. It returns query Q with a new annotation. Initially, the algorithm applies the DC → LC optimization action to nodes corresponding to sink nodes in the DTD. Then, it applies exhaustively the inference rules to construct the constraint graph G_Q of Q. Subsequently, it uses G_Q and the consumption relationships of Table 1 to construct the witness graph W_Q of Q in order to apply the optimization action DC/LC → NC. A root node (that is, a node without incoming edges) in W_Q cannot be witnessed by another node. Therefore, such nodes are chosen as witness nodes, and the annotations of all the nodes in Q that are reachable in W_Q by these nodes are turned into NC. The remaining nodes in W_Q all have incoming edges. In every cycle in W_Q, a node with the minimum cost is chosen as a witness node and the annotations of all the nodes reachable by this node in W_Q are turned into NC. Clearly, this process characterizes every node in W_Q as a witness node or as node annotated with NC in the returned query, and the new annotation of query Q is minimal.

The performance of our optimization algorithm is more significant when the nodes with the optimized annotations are not close to leaves of the query, and/or the number of matches of these nodes in the XML archive is important.

The complexity of our optimization algorithm is $O(|\Delta| \times |Q|^2)$, where $|\Delta|$ is the size of the DTD Δ. Note that the optimization result may not be unique. For instance, if there exist two query nodes p and q in a cycle with the same cost, then either one can be chosen as a witness node and both resulting annotations have the same minimal cost.

6 Experimental Evaluation

In this section, we report on the empirical evaluation of our approach for compacting archiving databases and for optimizing the evaluation of the temporal constraints of queries.

6.1 Experimental Setup

Our experiments were performed on an Intel Core 2 CPU 2.40 GHz processor with 4.00 GB of RAM running Windows 7. We used the JDOM engine [1] to parse the XML databases, Wutka DTD parser [2] to parse the XML DTD, and Oracle Berkeley DB XML engine for query evaluation. The algorithms were implemented in Java.

We ran our experiments on both synthetic and real datasets. The synthetic datasets were generated using the IBM XML generator [3] on the DTD of the XMark benchmark [4]. Our real dataset was the *Treebank* dataset [5]. The details of these datasets

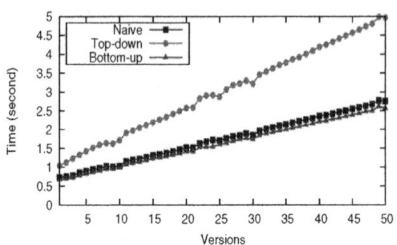

Dataset	Size	# of elements	Max/Avg. depth
Treebank	22.3MB	491108	36/8.5
XMark	14.6MB	160929	21/20.068

Fig. 5. Datasets used in the experiments

Fig. 6. Incremental archiving time

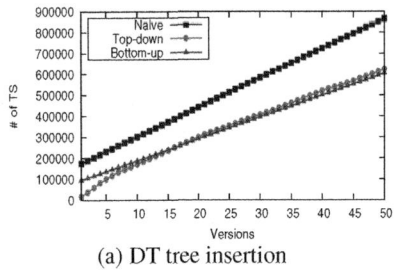

(a) DT tree insertion

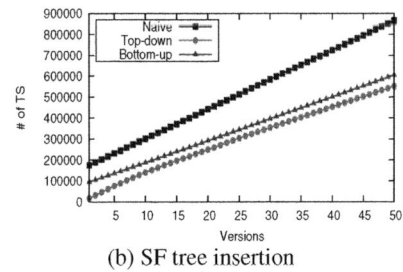

(b) SF tree insertion

Fig. 7. Comparison of the total # of timestamp sets in the archive

are summarized in Figure 5. For the *XMark* and *Treebank* datasets, we created 50 and 20 consecutive database instances respectively. Each instance was generated from the previous one by first deleting and then inserting (sub)trees (see Section 2). The trees represent 10% of the nodes of the database instance. Therefore, the size of the instances remains relatively stable. The size of the archiving database though increases constantly. We describe in Section 6.2 the types of trees that are chosen to be inserted/deleted.

6.2   Archiving Overhead

To measure the efficiency of our compaction scheme, we implemented three storage approaches: (1) the naive approach (NA), which keeps the timestamp sets on every node in the archiving database, (2) the top-down (TD) approach [8] which eliminates the timestamp sets on the children nodes when they are identical to those of their parents, and (3) our bottom-up (BU) approach that eliminates from the timestamp sets of the parent nodes the timestamps that are present in timestamp sets of their children. For these three approaches, we compare the incremental archiving time, the space overhead, the compaction ratio, and the update cost on the XMark archiving database.

Incremental archiving time. We measured the time needed to add a new instance to the archiving database with each one of the three approaches. The results are shown in Figure 6. As expected, the incremental archiving time increases as the size (number of instances) of the archiving database goes up. We observe that the TD approach always achieves the worst time performance. This is due to the fact that when the TD approach updates the timestamp sets in the archiving database, it might need to traverse ancestors of the updated nodes, which is an expensive operation. This operation is not necessary for either the NA or the BU approach. As explained in Section 3, in order to update timestamp sets, the BU approach only accesses nodes in the archiving database that become leaf nodes in the database instance after the deletion of the subtrees (see Listing 1). Second, the performance of the TD approach degrades faster than that of the bottom-up approach; with 50 database versions, the TD approach is almost two times slower than the BU approach.

Space overhead. All three approaches consume the same space for storing the merged data, and it is the different approaches employed for compacting (or not compacting) the timestamp sets that make the difference in the total space consumed by each one of them. To study under what circumstances the BU approach gets an advantage regarding the space overhead on storing the timestamp sets of the archiving database, we consider two database instance update scenarios: (i) insertion of *deep & thin* (DT) trees, in which the inserted nodes form a tree of many levels (4 to 8 levels in our experiments) but with few leaf nodes (1 - 4 nodes in our experiments), and (2) insertion of *shallow & fat* (SF) trees, in which the inserted nodes form a tree of few levels (1 to 4 levels in our experiments) but with many leaf nodes (4 - 8 nodes in our experiments). In both scenarios, the subtrees to be deleted are picked randomly from the database instance. We compare the total number of timestamps of the XMark dataset in the three approaches for both scenarios. The results are shown in Figures 7(a) and (b). First, we observe that due to timestamp set compaction, the total number of timestamp sets of both the TD and BU approaches grows much slower than that of the naive approach. However, which one produces fewer timestamp sets depends on the database instance update scenario: the TD approach loses in the DT tree scenario (Figure 7(a)), but wins in the SF tree scenario (Figure 7(b)). We also observe that for both scenarios, the change on the total number of timestamps by both approaches when the number of instances in the database archive increases is almost the same. Overall, the space consumed by the BU and the TD approaches is comparable.

Compaction ratio. We measure the compaction ratio $R = (S_{NA} - S_X)/S_{NA}$, where S_{NA} is the space consumed by the NA approach and S_X is the space consumed by the BU or the TD approach, of each one of the TD and BU approaches with respect to the NA approach for 50 instances of the XMark dataset.

Table 2. Compaction ratio

Dataset	Bottom-up	Top-down
XMark (shallow&fat)	2.15%	2.72%
XMark (deep&thin)	3.09%	2.75%

Table 2 shows the results. Both approaches achieve compaction ratio of over 2.15%. Furthermore, with both approaches, the SF tree scenario can achieve better compression ratios than the DT trees scenario. Finally, we observe that the differences in the compaction ratios between the two approaches are marginal.

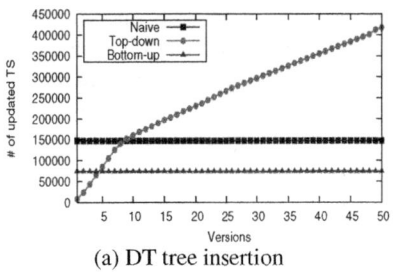 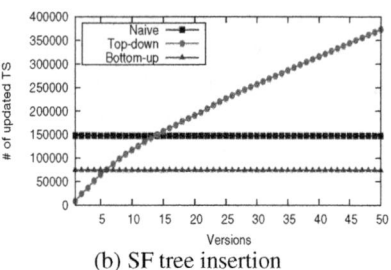

(a) DT tree insertion   (b) SF tree insertion

Fig. 8. Comparison of the number of updated timestamp sets in the archive

Update cost. We also measured the update cost in terms of the number of updated timestamp sets in the archiving database Figure 8 shows the number of updated times-tamps in the two database instance update scenarios when the number of instances in the archiving database increases. The update cost of the TD approach grows significantly with more database instances inserted into the archive. In contrast, the update cost of both the NA and the BU approaches is almost stable with that of the BU approach be-ing smaller. This is so because both the NA and BU approaches do not cause much change on the timestamp sets of the existing nodes in the archiving database, while the TD approach may need to insert new timestamp sets on the existing nodes. With more database instances added to the archive, the update cost may become very high.

Summary. Both the TD and BU approaches can substantially reduce the number of timestamp sets in the archive compared with the NA approach. Depending on the kind of trees inserted to the database instances one or the other approach can yield a smaller number of timestamp sets. In any case, the difference between the two is not signifi-cant. On the other hand, the TD approach always has much worse update cost than the BU approach. This fact, in conjunction with the resulting poor time performance for incremental archiving, makes the BU approach more suitable for archiving databases.

6.3 Temporal Constraint Evaluation Optimization

We now examine the performance of our approach for optimizing the evaluation of the temporal constraints of a query by running a set of TTPQs on archiving databases built on the XMark and the Treebank datasets. The sizes of the XMark dataset and the Treebank dataset are 86.14MB and 80.90MB, respectively. Each reported time is averaged over 5 runs.

We created five TTPQs for both the synthetic XMark and the Treebank dataset. The queries are implemented in XQuery They all contain child (/) and descendant (//) axes. Before optimization, each query contains at least four and at most ten temporal evalua-tion annotations. After optimization, each of them contains only one or two annotations. The other annotations are eliminated by the optimization process. As an example, Query 3 on the XMark dataset in an XPath-like syntax is `/site//person`[*overlap(2,5)*] [name[*include(4)*]] `/address`[*include(4)*]`/city`[*overlap(3, 4)*] before and `/site//person[name]/address`[*include(4)*]`/city` after optimization, while Query 4 on the Treebank dataset is `/FILE/EMPTY`[*include(4)*]`/S`[*include(3)*]`/NP`

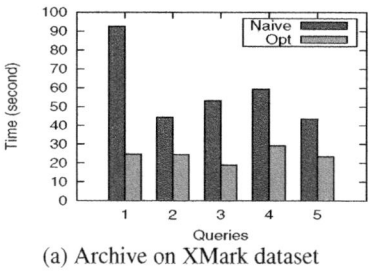

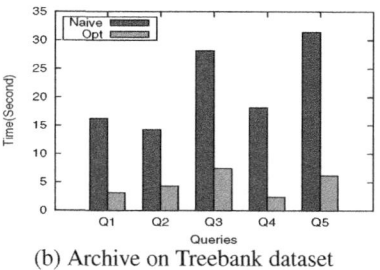

(a) Archive on XMark dataset   (b) Archive on Treebank dataset

Fig. 9. Query evaluation time for different queries

[*include(5)*] [NNP [*include(4)*]] /RB [*contain(3,6)*] before and /FILE/EMPTY/S /NP[NNP [*include(4)*]] /RB [*contain(3,6)*] after optimization. Note that the Treebank dataset does not come with a DTD. Thus, only inference rules independent of DTDs are involved in the optimization of the corresponding TTPQs.

We observe that the optimization process in each query does not exceed 10 millisecond. Compared with the query evaluation time (Figure 9), the optimization overhead is negligible.

The results of the query evaluation time before and after optimization are shown in Figures 9(a) and (b). As we can see, the optimization process can bring significant performance improvement. The benefit varies from 0.45 to 0.75 for the XMark dataset, and from 0.7 to 0.85 for the Treebank dataset.

7   Related Work

There has been considerable work on both the storage and the querying of multiple versions of semi-structured databases

Storage. A few works have proposed to reduce the storage redundancy by storing multiple identical nodes/subtrees as single instance in the archives ([8,21,26]). In particular, Wang et al. [26] consider the nodes that do not change in successive versions as identical. Buneman et al. [8] and Muller et al. [21] treat the nodes with the same key as identical according to some pre-defined notion of key. In the same context, Koltsidas et al. [18] propose an algorithm that sorts hierarchical data in external memory for archiving. Chapman et al. [9] factorize the archiving database by identifying common portions of the database. Out of these works, only [8] considers further compaction on timestamps as in our approach. However, they compact the timestamps in a top-down way, while we do it in bottom-up manner. We experimentally compare these two approaches in terms of storage space and incremental update cost in Section 6.

Along a different line, compact storage of multi-version semi-structured databases is achieved by keeping records of the changes (called *deltas*) between every pair of consecutive versions. For instance, [20,24,28] stores the most current version plus reverse editing scripts that can be used to retrieve previous versions. [10] detects the changes of two XML trees as edit scripts that give the sequence of operations needed to transform one tree into another, and stores the scripts as annotations on nodes and edges of a graph. Chien et al. [12] store the original version in several physical pages; the elements in the new versions whose corresponding pages are not useful (depending on a given usefulness threshold) are copied to new pages, together with new elements. These delta-based schemes suffer from the same problem: retrieving an old version might involve undoing many deltas from the current version. Likewise, the efficiency of finding the evolutionary history of an element is also a problem and may require significant overhead on the inference of deltas.

XML compression [19] shares the goal of reducing the storage of XML database with ours. However, the compressed XML databases do not support direct query evaluation. Although XGRIND [16] does support exact and substring querying of the compressed database, it cannot support the queries that involve temporal and structural constraints, which are common in XML archiving databases. Note though that these compression techniques can be applied to the compacted archiving database, if further reduction of the size is needed.

Query evaluation. There are very few works that address the query evaluation issue on archiving databases. Wang et al. [25,26] list the types of temporal queries that can be evaluated on an (uncompacted) database of versions by specifying temporal queries using XPath and XQuery. Rizzolo et al. [23] store multiple XML databases as graphs, and introduce TXPath, a temporal query language extending XPath 2.0, that can be used on the graphs. Unlike [23], we preserve the tree structure of the database, and use existing XML query engines for the evaluation of temporal queries. Chien et al. [12] assume that some query workload information is available. Based on this assumption, they combine multiple queries in the workload into range version retrieval queries, and propose three index schemes for efficiently evaluating these queries. This technique can be adapted to our optimization framework and further improve the query performance. Wong et al. [28] use XQuery as query language to scan the deltas stored in the system and return the corresponding version(s) that matches the queries. To avoid reconstructing intermediate versions, they design an index structure that stores all the deltas in a single structure indexed on the tags that were involved in an update operation. In any case, the index incurs additional space and update overhead. Finally, Muller et al. [21] implemented XArch, a management system for storing and querying XML archiving databases and designed a declarative query language, XAQL, for querying the archives. None of these approaches addressed the problem of optimizing the evaluation of the temporal constraints as we did in this paper.

8   Conclusion

To address the need for efficiently archiving hierarchical data, we developed an XML archiving system that combines the compact storage of data and timestamps with optimization techniques for the evaluation of queries with temporal constraints. We proposed a novel, updateable scheme that compacts both the successive database instances and the timestamp sets. In order to support the efficient evaluation of temporal queries we introduced three annotations for temporal constraints in queries that can be used to retrieve timestamps from the compact archiving database. We proposed an efficient algorithm which computes an optimal annotation by exploiting structural information of the query and of the database instances when a DTD is available. We experimentally validated our compaction scheme and demonstrated its performance benefits compared to previous ones and the efficiency of our optimization techniques for temporal tree-pattern queries over compact archiving databases.

Future work includes extending the results in this paper by considering additional temporal constraints and unrestricted DTDs that may contain cycles. It is then interesting to further extend the inference rules for temporal containment relationships and the temporal constraint consumption relationships and to design efficient optimization algorithms that can work in this broader framework.

References

1. JDOM XML parser, http://www.jdom.org
2. Wutka DTD parser, http://www.wutka.com/dtdparser.html
3. IBM XML data generator,
   http://www.alphaworks.ibm.com/tech/xmlgenerator
4. XMark XML benchmark project, http://monetdb.cwi.nl/xml/
5. XML Data Repository of University of Washington,
   http://www.cs.washington.edu/research/
   xmldatasets/www/repository.html
6. Annis, J., Zhao, Y., Vockler, J.-S., Wilde, M., Kent, S., Foster, I.T.: Applying chimera virtual data concepts to cluster finding in the sloan sky survey. In: Supercomputing (2002)
7. Bruno, N., Koudas, N., Srivastava, D.: Holistic twig joins: optimal XML pattern matching. In: SIGMOD (2002)
8. Buneman, P., Khanna, S., Tajima, K., Tan, W.-C.: Archiving scientific data. ACM Transactions on Database Systems (2004)
9. Chapman, A.P., Jagadish, H., Ramanan, P.: Efficient provenance storage. In: SIGMOD (2008)
10. Chawathe, S., Garcia-molina, H.: Meaningful change detection in structured data. In: SIGMOD (1997)
11. Chawathe, S.S., Rajaraman, A., Garcia-Molina, H., Widom, J.: Change detection in hierarchically structured information. In: SIGMOD (1996)
12. Chien, S.-Y., Tsotras, V.J., Zaniolo, C., Zhang, D.: Supporting complex queries on multiversion xml documents. ACM Transactions on Internet Technology (2006)
13. Cobena, G., Abiteboul, S., Marian, A.: Detecting changes in xml documents. In: ICDE (2002)
14. Gou, G., Chirkova, R.: Efficiently querying large XML data repositories: A survey. IEEE Trans. Knowl. Data Eng. 19(10), 1381–1403 (2007)
15. Groth, P., Miles, S., Fang, W., Wong, S.C., peter Zauner, K., Moreau., L.: Recording and using provenance in a protein compressibility experiment. In: HPDC (2005)
16. Jayant, P.T., Haritsa, J.R.: Xgrind: A query-friendly xml compressor. In: ICDE (2002)
17. Jiang, H., Wang, W., Lu, H., Yu, J.X.: Holistic twig joins on indexed XML documents. In: VLDB (2003)
18. Koltsidas, I., Muller, H., Viglas, S.D.: Sorting hierarchical data in external memory for archiving. In: PVLDB (2008)
19. Liefke, H., Suciu, D.: XMill: an efficient compressor for XML data. In: SIGMOD (1999)
20. Marian, A., Abiteboul, S., Mignet, L.: Change-centric management of versions in an xml warehouse. In: VLDB (2001)

21. Müller, H., Buneman, P., Koltsidas, I.: Xarch: Archiving scientific and reference data. In: SIGMOD (2008)
22. Pancerella, C., Myers, J.D., Allison, T.C., Amin, K., Bittner, R., Frenklach, M., Green, W.H., ling Ho, Y., Hewson, J., Koegler, W., Yang, C.: Metadata in the collaboratory for multi-scale chemical science. In: Dublin Core Conference (2003)
23. Rizzolo, F., Vaisman, A.A.: Temporal xml: modeling, indexing, and query processing. The VLDB Journal 17, 1179–1212 (2008)
24. Tichy, W.F.: RCS - a system for version control. Software-Practice & Experience (1985)
25. Wang, F., Zaniolo, C.: Temporal queries in XML document archives and web warehouses. In: TIME-ICTL (2003)
26. Wang, F., Zaniolo, C.: Temporal queries and version management in XML-based document archives. Data Knowl. Eng. 65, 304–324 (2008)
27. Wang, Y., DeWitt, D.J., yi Cai, J.: X-Diff: An effective change detection algorithm for XML documents. In: ICDE (2003)
28. Wong, R., Lam, N.: Managing and querying multi-version xml data with update logging. In: DocEng. (2002)

Update Propagation in a Streaming Warehouse

Theodore Johnson and Vladislav Shkapenyuk

AT&T Labs - Research
{johnsont,vshkap}@research.att.com

Abstract. Streaming warehouses are used to monitor complex systems such as data centers, web site complexes, and world-wide networks, gathering and correlating rich collections of events and measurements. Ideally, a streaming warehouse provides both historical data, for deep analysis, and real-time data for rapid response to emerging opportunities or problems. The highly temporal nature of the data and the need to support parallel processing naturally leads to extensive use of horizontal partitioning to manage base tables and layers of materialized views. In this paper, we consider the problem of determining when to propagate updates from base tables to dependent views on a partition-wise basis using autonomous updates. We provide a correctness theory for propagating updates to materialized views, simple algorithms which correctly propagate updates, and examples of algorithms which do not. We extend these results to accommodate needs of production warehouses: repartitioning of tables, mutual consistency, and merge tables. We measure the update propagation delays incurred by two different update propagation algorithms in test and production DataDepot warehouses, and find that only those update propagation algorithms which impose no scheduling restrictions are acceptable for use in a real-time streaming warehouse.

1 Introduction

Data Stream Management Systems (DSMS) have developed in response to the need for on-line monitoring of complex systems such as communication networks [11], financial markets [27], data center management, web transaction logs [46], RFID tracking [47], and so on. A DSMS typically provides a real-time response by processing events in-memory and over a short time window. However, many applications benefit from access to both real-time and historical data in a data warehouse setting, whether to correlate current events with past events, or to provide a seamless warehouse for on-line alerting and analysis as well as deep historical analytics. We term such an information system a *stream warehouse*. Recent papers that describe stream warehouses include application areas such as data center monitoring [4], RFID monitoring [48], web-complex monitoring [46], and very wide scale network monitoring [18]. For example, the Darkstar streaming warehouse at AT&T Labs – Research, built using our stream warehouse system DataDepot [18], runs applications ranging from NICE (deep analysis to find network conditions and events correlated with network problems) to the PathMiner troubleshooting tool (real-time reports on the status of routers along a problematic path), and supports a wide range of networking research activities [30].

J.B. Cushing, J. French, and S. Bowers (Eds.): SSDBM 2011, LNCS 6809, pp. 129–149, 2011.
© Springer-Verlag Berlin Heidelberg 2011

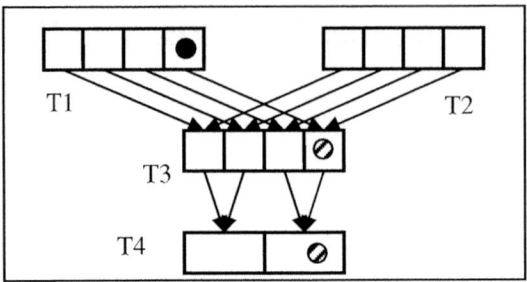

Fig. 1. Partition-wise data flow

An examination of recent literature (especially our system DataDepot [18] at AT&T Labs, Truviso [24] and Everest [46] at Yahoo! Labs but also including Moirae [4]) reveals some common features: temporal partitioning, multi-version concurrency control, and recomputation of partitions to propagate updates, as we discuss further in Section 0. The management of tables in these warehouses has the structure shown in Figure 1. This warehouse has tables T1 through T4, with the data flowing from T1 and T2 to T3 and then T4. However, our concern is with data flowing through partitions. For example, a particular partition in T3 (which is defined as a join between T1 and T2) uses data from particular partitions in T1 and T2.

In this paper, we present a correctness theory for update propagation in a stream warehouse, and algorithms which correctly propagate updates. We assume that data arrives from external sources and is loaded into *base* tables according to some reasonable temporal partitioning of the base data. Updates to base tables are propagated through multiple levels of materialized views. For example, in Figure 1, the partition marked with a dot in base table T1 affects a particular partition of T3, and transitively a partition in T4. Our goal is to propagate all updates without unnecessarily updating a partition of a materialized view.

Our model of incremental view maintenance is that partitions affected by new source data are recomputed entirely, while partitions not affected by new data are left untouched. As discussed in Section 0, partition-wise recomputation is generally more efficient that in-place incremental maintenance, and often is the only option. The problem of incremental table maintenance becomes one of determining which partitions of a derived materialized view are affected by an update to a partition of a source table (whether a base table or another materialized view). Methods for associating a source partition with its derived partitions include query analysis, join dependencies, functional dependencies, or even explicit user specification [14][18].

Stream warehouse updates might be scheduled autonomously for real-time performance (see Section 1.1), in which case view consistency is best captured by either eventual consistency (i.e. *convergence* [50]) or trailing edge consistency [19][18]. As discussed in more depth in Sections 3 and 6, trailing edge consistency – the most recent data that is quiescent - is the best stream warehouse analog of mutual consistency (see [10][51]). The update propagation protocols we present in Section 4 are designed to ensure eventual consistency rather than trailing edge consistency. For one, many real-time applications need the most recent possible data; one doesn't want to wait for a router in Atlanta to deliver its data before diagnosing a network problem in California (see the examples in Section 3). For another, many streaming data feeds are highly disordered [24][29] - in our experience, data can often arrive days to hours late. However, in Section 6, we provide extensions to the basic update protocols which determine the trailing-edge consistency line and which gracefully handle after-the-fact revisions.

1.1   Why Is Update Propagation Interesting?

Given the lengthy literature on materialized view maintenance, one might ask why update propagation is still an open problem. The answer is that new large-scale data analysis systems need more efficient and more flexible view maintenance mechanisms. Much of the existing literature use global orders set by the commit or recovery log [9][10][41][49][51] or a message queue [2][50] to determine what updates need to be propagated. However, a stream warehouse does not need a recovery log because the sources are append-only, and materialized views can be recomputed from their sources. An alternative is to mark the base tables (or base table partitions) that have been updated since the last materialized view refresh, then refresh all views in a batch [14].

The scheduling of updates to base tables and materialized views is best left to a real-time update scheduler [19][26][40]. Therefore, we make no assumptions about the ordering of updates to tables, or whether all updates are applied to a table [19]. Further, the scheduling restrictions inherent in global-order or batch update view maintenance cause excessive update propagation delays for real-time tables, as we show in Section 6.

Autonomous update propagation also simplifies the implementation of a clustered warehouse. The difficulties of applying traditional view maintenance algorithms in modern large-scale analysis systems lead some implementations to provide only minimal consistency guarantees [3].

Our contributions in this paper are:

- We develop a correctness theory for update propagation in a temporally partitioned stream warehouse, which uses a novel form of a vector timestamp.
- We show that the intuitive algorithm similar to the update propagation in make is incorrect under most circumstances.
- We provide simple and low-overhead algorithms which ensure eventual consistency.
- We extend these results to account for complications present in actual stream warehouses, such as limited-size updates and multi-granularity temporal partitioning.
- We develop algorithms which ensure that queries access mutually consistent views under the trailing-edge consistency model.
- We demonstrate the performance benefit of autonomous update propagation with experiments on live stream warehouses.

2 Production Warehouse Examples

The work in this paper was motivated by the Darkstar project [30], which collects a variety of network performance and reliability feeds to support tasks ranging from deep historical research and data mining to real-time problem resolution. The DataDepot stream warehousing tool encouraged the Darkstar developers to push ETL and complex application logic, as well as summary reports, into the warehouse as materialized views.

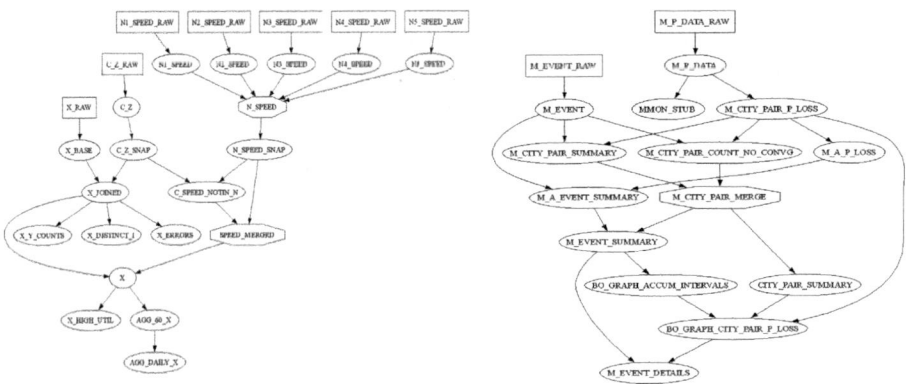

Fig. 2. ETL for table X, and summary tables   **Fig. 3.** Application logic for M events

For the first example, Figure 2 shows the DAG of table dependencies related to loading table X (names have been mangled), and then computing summaries. The edges point from a source to its dependent table(s), and the square boxes are the raw data sources. Approximately 150 million records per day arrive at X_RAW, at five minute intervals (about 350 million raw records/day for the warehouse). The critical real-time path is X_RAW → X_BASE → X_JOINED → X. Each of these tables uses multi-granularity partitioning: the most recent 1-day's worth of data is temporally partitioned into 5-minute buckets, while older data (a 1-year window) is partitioned by 1-hour buckets.

The second example, in Figure 3, shows the complex application logic generating reports related to M data (the names have been mangled), taken from a production warehouse outside of AT&T Labs - Research.

These examples from the Darkstar and other warehouses illustrate the arguments we make in Section 1. First, that a multiple-granularity, temporally partitioned stream warehouse enables the construction of complex warehouses on high-volume data. Second, the mechanism for propagating updates should not constrain the scheduling of updates. The updates to the tables dependent on X_JOINED and X in Figure 2 should not block updates along the critical real-time path from X_RAW to X. Third, a provably correct algorithm for update propagation is critical for the developer to have confidence that complex applications, such as the one shown in Figure 3, will be correctly maintained. Fourth, multiple types of consistency are needed: eventual consistency for table X, trailing edge consistency for the M application.

3 System Model

Our warehouse model (partition-wise dependencies and update propagation, read-isolation at query initiation time, and the lack of a recovery log) are motivated by common features of stream warehouses:

Temporal partitioning: Stream data is inherently temporal, often containing one or more timestamps [11][18] which are monotone increasing (or nearly so [45]) with new data. A method for managing a moving window over temporal data that is commonly used [8][12][18][31], well-known in the folklore [22], and supported by commercial products [7][14] is to partition the table on one of its timestamp fields. New data generally flows into the newest partition; data is expired by dropping the oldest partitions. If the partitions are sized to be the inter-update period, the new data will fit into a new partition allowing for very efficient data ingest and index rebuild.

Multi-version Concurrency Control: The materialized views in a data warehouse are generally maintained by a single update process. The single-updater, multiple-reader access pattern allows for an inexpensive implementation of multi-version concurrency control [38][44]. MVCC is especially attractive for continuously available [12][44][46] real-time warehousing not only because queries are never blocked, but also because updates to a real-time fact table aren't blocked by long-running updates to a summary table. DataDepot [18] performs concurrency control at the partition level, allowing for very low cost concurrency control and recovery.

Recomputation of updated partitions: If a warehouse uses temporal partitioning for both the source and target materialized view, then it's generally more efficient to recompute the partitions of a view that are affected by an update than to propagate incremental updates to them [14] (see [7][34] for an analysis of the optimization problem). Furthermore, the materialized views might be the result of queries that are difficult to maintain efficiently (group-by queries with a Having clause or holistic aggregates [36], non-monotonic queries [20], iceberg queries [42]). Further, many materialized views involve complex analytics not readily expressed in SQL [18][46].

Multi-granularity temporal partitioning: A technique for maintaining a real-time warehouse that is well-known in the folklore [22] is to use multi-granularity temporal partitioning. The most recent data is stored in small time-width partitions to match the update interval; the width can be as narrow as a few seconds [31]. Storing a table in such small partitions makes multi-year windows impractical, so older data is rolled up into longer-duration partitions. DataDepot uses two temporal granularities [18] while SWIFT [31] can use three or more. Re-organizing older data is also desirable because access characteristics change: new data is hot with frequent random access, while old data is cold with sequential access more common. Therefore, an effective storage management strategy is to store recent data on expensive high-RPM disk or SSDs, and reorganize older data onto lower performance but less expensive storage [46].

Deeply nested materialized views: Stream warehouses generally have many levels of materialized views for several reasons. For one, ETL, data cleaning, and data normalization processes can often be best expressed and optimized as a sequence of queries [15]. After the data is loaded, other materialized views correlate, de-normalize, and analyze the data for user convenience as well as to reduce query response times [18][46]. Further materialized views are used to monitor the quality of the data feeds. We present examples of deeply nested views from a production real-time warehouse in Section 3.

In the warehouse, there is a collection of *base* tables which are updated using outside data only. All other tables are d*erived* tables (materialized views), which are updated from new data loaded into base or derived tables (but never updated with outside data). When discussing a particular derived table and its relation to its data sources, the derived table is referred to as the *dependent* table and the tables it depends on are referred to as the *source* tables. Each partition of a dependent table is associated with one or more partitions of its source tables; we refer to these as the dependent partition and the source partitions, respectively (see [14][18] for a discussion of how to compute these associations). The contents of a dependent partition can be computed via its defining query and its source partitions only.

In a stream warehouse, the collection of partitions of a table form a sliding window, extending from the most recent data to some table-dependent time in the past, say two years. Each table in the data warehouse has at least one field that is a timestamp field – its value generally increases over time. Further, each table is partitioned according to a monotonic increasing partitioning function p: if $t_1 < t_2$, then $p(t_1) \leq p(t_2)$ (a table might be partitioned by additional partitioning functions on other fields) The source partitions of a dependent partition generally form a contiguous range for each source table, and their partition identifiers are computed from the dependent partition's identifier using a simple formula. These assumptions are not required for the model, but are useful for extensions discussed in Section 5. Highly parallelized systems such as Everest [46] generally make use of additional dimensions of partitioning. Subdivisions from these additional dimensions can be combined for the purposes of tracking update dependencies, or treated independently.

We make several assumptions about how tables are updated and the system environment.

1) A table is updated by a *table update* program. Only one instance of a table update executes at any time. When the table update starts, the update process computes a collection of partitions to update by applying an *update propagation protocol* to the source and dependent table metadata. The table update process may choose to update some or all of the partitions returned by the update propagation protocol.

2) An update process has read-consistency on its source tables: their contents do not change during the execution of the update process, from the perspective of the update process.

3) If the table update process chooses to update a partition, the partition is recomputed in whole (no special algorithms for incremental partition update). Incremental updates can be implemented through proper support for differential files, but are not the concern of this paper.

4) The system maintains a timestamp which can provide a total order for all events consistent with their actual causal ordering – the *system timestamp*, e.g. the Unix timestamp. A sample of this timestamp is denoted by <time>.

5) There is an *update manager* which invokes updates to tables. The update manager can consult table metadata when determining which table update process to schedule.
6) Each table and each partition is uniquely named, so any set of these names can be uniquely sorted (not a strong assumption).
7) We assume that the collection of (source, destination) dependencies forms a DAG.

In assumption 1, not every stale partition (i.e., selected for update) must necessarily be updated. In practice, the ability to partially update a view provides critical flexibility when managing very large volumes of data. If a view becomes significantly out of date (perhaps because its definition changed) then recomputing the view may take longer than the next scheduled downtime or the mean time to failure of the server [32][33]. Partial view updates ensure that the updates are committed regularly and that the update procedure makes progress. The partial view update may update partitions out-of-order, e.g. update the most recent partitions first. In [19], the authors find that partial updates are a critical technique for minimizing table staleness in a real-time warehouse. Assumption 2 can be satisfied via read/write locks on the source/dependent table partitions. However, multi-version concurrency control is the better option, as noted in Section 1. We do not assume that tables are updated in any particular order, nor that any collection of tables are updated simultaneously – decisions better left to the real-time update scheduler [19]. In a centralized server, accommodating assumption 4) is simple. In a clustered warehouse, coordinated timestamps can be ensured using, e.g. ntp. If coordinated timestamps are problematic, we recommend using protocols which use local timestamps only, for example the source-vector protocol in Section 4.2.

Our goal is to achieve *eventual consistency*: if the base tables remain quiescent then after a sufficient number of updates, every partition of a materialized view converges to the value which reflects the data in the quiescent base tables [50]. While eventual consistency is weaker than mutual consistency [10][51], it is usually appropriate for a stream warehouse. For example, table X in Figure 2 is a critical for real-time diagnostics and needs to be loaded with the most recent possible data regardless of mutual consistency concerns. However, a stream warehouse should provide support for determining the *trailing edge* [18], which marks the boundary between quiescent and non-quiescent partitions, and *leading edge* consistency, which indicates the partitions with the newest possible data. In Section 6, we show that general mutual consistency is difficult to track, but that trailing edge consistency can be determined using a simple protocol.

3.1 Definitions

Let D be a dependent table, and let D_i be its ith partition. At time *ts* we denote them by $D(ts)$ and $D_i(ts)$ respectively. The *generation* of $D_i(ts)$ is the number of times that $D_i(ts)$ has been recomputed (i.e., execution of an update protocol). The generation of $D_i(ts)$ is denoted $g(D_i(ts))$, and the kth generation of D_i is denoted $D_i[k]$. Similarly, the generation of $D(ts)$ is the number of times it (i.e., at least one of its partitions) has been updated, with corresponding notation $g(D(ts))$ and $D[k]$ respectively. The *table-generation* of a partition is the table generation in which the partition was updated.

The table-generation of partition $D_i(ts)$ is denoted $t(D_i(ts))$, and the table-generation of $D_i[k]$ is denoted $t(D_i[k])$.

Associated with each generation of a partition (table) is its *updatestamp*, $u(D_i(ts))$. The updatestamp of a partition (table) must be a strictly increasing function of the partition (table) generation: if $j<k$, then $u(D_i[j]) < u(D_i[k])$ (respectively, $u(D[j]) < u(D[k])$). There is no necessary correlation between updatestamps of any two partitions, or between a partition and a table, other than that imposed by the update propagation protocol.

For a dependent table D, its source tables are $S(D) = (S_1, ..., S_k)$, in sorted order (recall assumption 6). Similarly, for dependent partition D_i, its set of source partitions are

$$S(D_i) = (S_{1,1}, ... S_{1,n1}, ... S_{k,nk}).$$

The vector of generation numbers of the value of $D_i(ts)$ is

$$G(D_i(ts)) = (g(S_{1,1}(ts)), ..., g(S_{1,n1}(ts)), ..., g(S_{k,nk}(ts))),$$

Where $g(S_{i,j}(ts))$ is the generation of $S_{i,j}$ used to compute $D_i(ts)$ and analogously the vector of updatestamps is

$$U(D_i(ts)) = (u(S_{1,1}(ts)), ..., u(S_{1,n1}(ts)), ..., u(S_{k,nk}(ts))).$$

Analogously we define $G(D_i[k])$, and $G(D_i[k])$.

We define $B(D_i)$ as follows: recursively gather $S(S_{i,j})$ and replace the entry of S_{ij} in $S(D_i)$ with $S(S_{ij})$ until all partitions are Base table partitions. For example, suppose that $S(D_5) = (E_3, F_4)$, $S(E_3) = (T_2, R_2, R_3)$, and $S(F_4) = (T_2, T_3, R_3)$ where T and R are Base tables. Then $B(D_5) = ((T_2, R_2, R_3), (T_2, T_3, R_3))$ – note the repetition.

Next we define **GB(D$_i$(ts))** to be the generations of the partitions in $B(D_i)$ used to compute D_i(ts). For example, $GB(D_5(ts)) = ((g(T_2(ts)), g(R_2(ts)), g(R_3(ts))),(g(T_2(ts)), g(T_3(ts)), g(R_3(ts))))$. We define Gb(D$_i$(ts)) as follows : Let $S(D_i) = (S_{1,1}, ..., S_{1,n1}, ..., S_{k,nk})$. Then

$$Gb(D_i(ts)) = (GB(S_{1,1}(ts)), ..., GB(S_{1,n1}(ts)), ..., GB(S_{k,nk}(ts))).$$

We define $UB(D_i(ts))$ and $Ub(D_i(ts))$ similarly, except using the updatestamp instead of the generation number.

At a specific point in time ts, e.g. during an update we will drop the dependence on ts. In this case, $GB(D_i)$ represents the generations of the base table partitions used to compute the current value of D_i, while $Gb(D_i)$ represents the generations of the base table partitions used to compute the current values of the source partitions of D_i, $S(D_i)$. A base table partition may occur multiple times in $B(D_i)$, and its repetitions may have different generation values in $GB(D_i)$. If so, D_i (and perhaps one or more of its direct or transitive source tables) was not computed from mutually consistent views.

In this document we will use the usual notion of vector comparison: Let A, B be two n-dimensional vectors. Then $A > B$ if for each $i=1..n$, $A_i \geq B_i$, and there is some i in $[1..n]$ such that $A_i > B_i$.

Def: An update propagation protocol is *correct* if, when computing partitions of D to update, it selects all those partitions D_i such that $Gb(D_i) > GB(D_i)$.

Def: An update propagation protocol is *minimal* if it selects only those partitions D_i such that $Gb(D_i) > GB(D_i)$.

Lets consider the example in Figure 4. Partition D depends on partitions S_1 and S_2, i.e. $S(D)=(S_1,S_2)$, and in turn $S(S_1)=(B_1, B_2)$ and $S(S_2)=(B_2,B_3)$. When D was last updated at time ts_1, S_1 and S_2 had been computed from the first generation of B_1 through B_3, i.e. $GB(S_1(ts_1))=(1,1)$ and $GB(S_2(ts_1))=(1,1)$, and therefore $GB(D)=((1,1),(1,1))$. At time $ts_2>ts_1$, B_2 was updated, and at time $ts_3>ts_2$, S_1 was updated. Now, at time $ts_4>ts_3$, we consider whether D should be selected for update. $GB(S_1)=(1,2)$, and therefore $Gb(D)=((1,2),(1,1))$. Therefore, a correct protocol must select D for update. Note however that S_1 and S_2 are not mutually consistent. Because updatestamps are strictly increasing functions of generation numbers, we could also express GB, Gb, and the correctness definition in their terms.

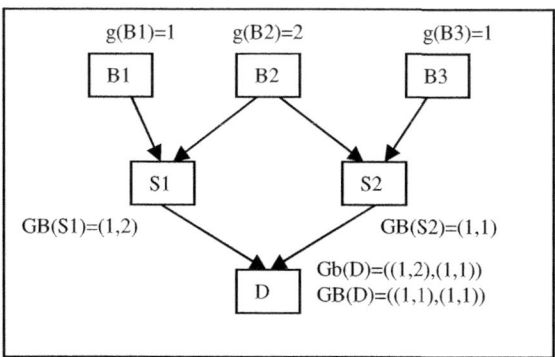

Fig. 4. Generation stamps

We are looking for correct and minimal update propagation protocols. An update protocol should be minimal because otherwise "recompute all partitions" is a correct and acceptable protocol. Note that the update process need not actually perform all of the updates recommended by the update propagation protocol, e.g. for update chopping [19].

A correct update protocol will not by itself ensure the eventual consistency of partitions in the streaming warehouse. The servers might not have sufficient resources to update all tables, or the scheduler and/or table update process might make poor decisions. We assume that the warehouse is maintained by a well-provisioned server (or cluster, etc.), and that the update scheduler and update process make good decisions. However, if the update protocol is not correct, even a well-configured system is liable to fail to propagate all updates.

In a streaming warehouse, new partitions are created as the time window of a table slides forwards; conversely partitions are deleted as they age out of the window. Partitions that have not been created have a generation number and an updatestamp of zero. A trigger for creating a new partition is that a) it is on the advancing side of table's sliding window, and b) it is selected for update. Deleted partitions can retain their generation number / updatestamp, or these can revert to zero. However, a dependent partition that depends on a deleted source partition should not be selected for update, as the information to compute its correct value has been discarded. The warehouse administrator can use the distinction described in [16] of expired vs. deleted partitions (an expired partition is only deleted after all dependent partitions have reached eventual consistency). The closed-partition protocol (Section 6) can help to determine when expired partitions can be safely deleted. Alternatively, the warehouse administrator can generally assume that partitions which have aged out of their table's window have had their final contents propagated and are safe to delete without further checks.

4 Update Protocols

In this section, we present a collection of update protocols which are correct and minimal under various conditions. We define the *level* of a table to be 0 if it is a base table, or else the longest path from the dependent table to a base table source following (source, dependent) table dependencies. Most proofs will use induction on the level of the table.

4.1 Naïve Protocol

Before we present correct update protocols, let us motivate the problem with an example of an incorrect protocol. A first algorithm is motivated by the Unix tool make: update a dependent partition if its last-modified timestamp is less than that of any of its source partitions. Next, consider the example in Figure 5.

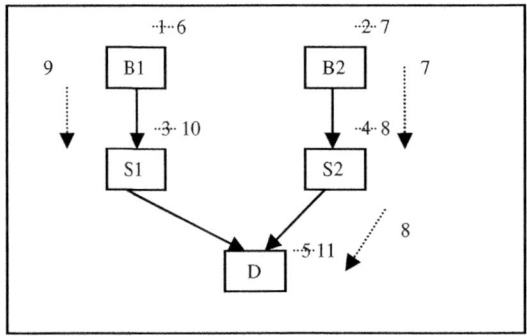

Fig. 5. Propagation by last-modified time fails

Partitions B1, B2, S1, S2, and D initially have last-update times of 1, 2, 3, 4, and 5, respectively. Updates to B1 and B2 occur, ending at times 6 and 7 respectively. At time 7, an update to S2 occurs, ending at time 8. Then, an update to D occurs, starting at time 8 and ending at time 11. Concurrently, an update to S1 starts at time 9 and ends at time 10 (Recall that the update to D has read-consistency on its sources). After these updates occur, D has a larger last-modified time than S1, so the update to B1 will never propagate to D if the base tables remain quiescent.

Update protocols that use only a single timestamp can be made to work, but only with significant scheduling restrictions. In Section 5, we discuss these protocols and present in detail the least restrictive one, the **Starting-timestamp** protocol.

4.2 Source-Vector Protocol

The definition of correctness can be used as an update protocol: Store with each partition D_i its value of $GB(D_i)$. When the update protocol executes, it computes $Gb(D_i)$ from the values of $GB(S_{ij})$, S_{ij} in $S(D_i)$, and recommends D_i for updating if $Gb(D_i)>GB(D_i)$.

The problem with such a protocol is the potentially enormous amount of metadata For example, consider a table that rolls up 5-minute buckets into 1-day buckets (perhaps rolling up 5-minute aggregates into 1-day aggregates). This table requires 288 metadata entries for each table partition[1] – more if the view involves a join. Since large historical tables may contain tens of thousands of partitions, this blowup in partition metadata can become unmanageable.

We can decrease the amount of per-source table metadata required for an update protocol by forcing coordination among the updatestamps of a table's partitions. Recall that the updatestamp of a table is a monotonic increasing function of the generation of a table. Since whenever a table partition's generation increases, the table generation increases, the table updatestamp can be used as the partition updatestamp. The table updatestamp can be a sample of the system timestamp at some point during the execution of the table update.

Source-vector protocol:

1) The table maintains an updatestamp, ut, which is incremented before assigning any new updatestamp to any partition during a table update.

2) Each partition maintains an updatestamp, u, which is the value of ut assigned to the table during the table update.

3) Each partition maintains a *source vector, mu* = $(max(u(S_{1,1}), ..., u(S_{1,n1})). ..., max(u(S_{k,1}), ..., u(S_{k,nk})))$ computed from the source table metadata at the time of the partition's update.

4) when the update protocol runs, it computes for each partition D_i the vector $Mu = (max(u(S_{1,1}), ..., u(S_{1,n1})). ..., max(u(S_{k,1}), ..., u(S_{k,nk})))$. Partition D_i is returned if $Mu(D_i) > mu(D_i)$.

Theorem: The Source-vector Protocol is correct and minimal.

Proof: by induction on the level. Since the protocol assigns valid updatestamps to the partitions in the Base tables, they are correct and minimal.

Correct: Suppose that $Gb(D_i) > GB(D_i)$. Then there is a partition $S_{j,k}$ in source table S_j that has been updated. Therefore table S_j was updated, and obtained a larger updatestamp than any previous; further this updatestamp was assigned to $S_{j,k}$. As a result $Mu[j] > mu[j]$, and therefore $Mu > mu$.

---

[1] There are 288 5-minute intervals in 1 day.

Minimal: Suppose that $Mu > mu$. Then there is at least one j such that $Mu[j] > mu[j]$, meaning that a partition $S_{j,k}$ was updated. By induction, $Gb(Sj) > GB(Sj)$ and there-fore $Gb(D_i) > GB(D_i)$.   □

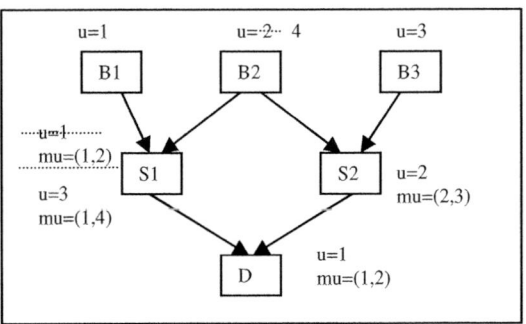

Fig. 6. Source-Vector Prococol

An example of the Source-vector protocol is shown in Figure 6. Initially, B_1, B_2, and B_3 have updatestamps 1, 2, and 3 respectively. S_1 has updatestamp 1 and source vector (1,2), while S_2 has updatestamp 2 and source vector (2,3); D has updatestamp 1 and source vector (1,2). B_2 is updated, which propagates to S_1, changing its updates-tamp to 3, and source vector to (1,4). We can detect that D needs to be updated be-cause S_1's updatestamp is larger than its entry in D's source vector.

While this protocol requires a moderate amount of metadata, the amount is variable depending on the number of source tables. Ideally, we could just use a pair of num-bers, as fixed-size metadata generally enables more efficient implementations.

4.3   Interval-Timestamp Protocol

As we discuss in Section 5, update propagation protocols that use a single timestamp impose significant scheduling restriction. The least restrictive one, the Starting-timestamp protocol, does not allow a dependent table to update while a source is updating. If we allow *two* timestamps, we can nearly eliminate all scheduling restric-tions. We assume that each update has a *read prefix*, which is the period of time from the start of the update to the release of source table metadata, and a *write suffix* during its commit.

Interval-timestamp protocol:

1)   Each partition maintains two updatestamps, *su* and *eu*.
2)   When the update starts, the table samples updatestamp $su(D)$=<time> sometime during the read prefix, and samples updatestamp $eu(D)$=<time> sometime during the write suffix.
3)   A base or derived partition sets $su=su(D)$ and $eu=eu(D)$ when it commits the update (at the end of the write suffix).
4)   For a derived partition, the update protocol selects a partition D_i for update if $su(D_i) < max(eu(S_j) \mid S_j \text{ in } S(D_i))$.

5) The update of a table cannot start during the read prefix or write suffix of its source or dependent tables.

Theorem: the Interval-timestamp protocol is correct and minimal.

Proof: omitted for brevity.

Clause 5) is intended to ensure that all updatestamps which might be compared are unique. Other unique-ifying methods such as a global ticket or read/write locks (on metadata) can be used. In the example in Figure 7, updates of S and D are shown on a timeline, with the read prefix and write suffix indicated by the hashed boxes. The update to D starts at time 7, while an update to S executes concurrently. Because of read consistency, D uses the value of S before its commit – the value of S at time 5. In contrast to the starting-timestamp protocol, D will be able to detect the update to S since $eu(S) = 10 > su(D)=7$.

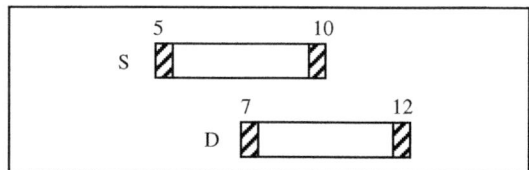

Fig. 7. Read prefix and write suffix

Although the Interval-timestamp protocol uses two updatestamps in comparison to the Starting-timestamp protocol's one, it is a small amount of metadata which is fixed in size – simplifying the task of implementing efficient protocols as compared to the Source-vector protocol.

5 Protocols with a Single Timestamp

In order to reduce the number of metadata variables to one, we will need to make use of a global timestamp: the system timestamp <time>. By comparing these global timestamps, we hope to determine when a partition must be updated. It would seem obvious that a single timestamp should suffice. It can, but only with restrictions on the concurrency of updates. In Section 6, we find that restrictions on concurrency introduce unacceptable delays into the updates of real-time tables; therefore we cannot recommend the use of any single-timestamp protocol. In this Section, we explore single-timestamp protocols and their concurrency restrictions.

The obvious approach is to use an algorithm similar to that used by the Unix tool make. As shown in Section 4.1, a naïve implementation is not correct. The max-source protocol (omitted for brevity) implements this type of algorithm, but it only works in a very restrictive setting. The parallelized version of make, pmake, avoids problems by performing a global analysis of the dependency DAG [6] and rebuilding objects in a depth-first fashion. The algorithm described in [14] also makes a global analysis, but then tries to optimize concurrency.

5.1 Starting-Timestamp Protocol

If table updates are protected by read/write locks, then labeling partitions by their last-updated timestamp is sufficient for a correct update propagation protocol (omitted for brevity). However, this kind of restriction is unacceptable for a real-time or high-availability warehouse, which is why they generally use some kind of versioning [12][18][38][44][46].

The scheduling restrictions of the current-timestamp protocol make it unacceptable for a real-time warehouse. However, the "no concurrent scheduling" restriction is stronger than it needs to be. The Starting-timestamp protocol relies on the read-isolation of updates to loosen the scheduling restrictions.

The Starting timestamp protocol assumes that update processes have read-isolation from a last-lock acquisition point until their commit point. At the commit point, the new metadata for the dependent table becomes visible. The period of time from when the update process starts until it releases the read locks on source table metadata is the *read prefix* of the update process execution.

Starting-timestamp protocol:

1) Each partition maintains an updatestamp u.
2) When the update process starts, the table samples an updatestamp $u(D)$ = <time> sometime during the read prefix.
3) A base or derived partition sets $u=u(D)$ when it updates.
4) For a derived partition, the update protocol selects a partition D_i for update if $u < max(u(S_j) \mid S_j \text{ in } S(D_i))$.
5) The update of a dependent table cannot start if any of its source tables are updating. No update of a source table can start during the read prefix of any of its dependent tables.

Theorem: the Starting-timestamp protocol is correct and minimal.

Proof: by induction on the level.

Correct: Suppose that $Gb(D_i) > GB(D_i)$. Then some source partition $S_{j,k}$ must have been updated since D_i's last update. The updatestamp of $S_{j,k}$ must be larger than $u(D_i)$ because the a) D will not update while S_j is updating, and b) the S_j update process must have started after the read prefix of D's last update process. Therefore D_i will be selected for update.

Minimal: omitted for brevity.

The Starting-timestamp protocol blocks the update of sources during the short metadata processing period of the update of a dependent table. However, the update of a dependent table is blocked during the update of its sources. The problem is that without this scheduling restriction, source partition S might start an update at time 5 and finish at time 10, while dependent partition D might start an update at time 7 and finish at time 12. When comparing timestamps later, no update of D would be triggered even though S has more recent data. However, the experiments in Section 6 show that even these restrictions cause unacceptable update propagation delays for real-time tables.

6 Mutual Consistency

Two tables are *mutually consistent* if they were computed from the same base table snapshots [10][51]. Since we assume that table updates are autonomous, and might not update all stale partitions in a table, we cannot guarantee mutually consistent materialized views. Without any assumptions of the structuring of updates, it is possible to detect when a partition has been computed from mutually consistent sources, and even to determine whether the partitions used for a query are mutually consistent. However, the amount of metadata required for this detection can explode. If we make some mild assumptions about the pattern of updates to base table partitions, then we can provide a simple and efficient protocol for determining mutual consistency.

6.1 The Closed Partition Protocol

If the update patterns in a warehouse are arbitrary, then it seems we cannot ensure mutual consistency without either storing an unmanageable amount of metadata, or requiring an inflexible update procedure. Fortunately, the update patterns in a streaming warehouse are not arbitrary, but rather follow a particular pattern – new data generally has a larger timestamp than older data, and therefore flows into higher numbered partitions.

Previous work in streaming warehouses [18][19] have introduced the notions of *leading-edge* consistency and *trailing-edge* consistency. The leading edge of a table contains the most recent possible data, while the trailing edge contains all data which will not change in the future – meaning that all of its source base partitions are quiescent and the partition has reached eventual consistency. Any two trailing-edge consistent partitions are mutually consistent. Therefore, a simple protocol which ensures mutual consistency is one which detects trailing-edge consistency.

We assume that there is some mechanism for determining when a base table partition has become quiescent. One might receive punctuations [45] in an input stream, or one might assume that a partition sufficiently far from the leading edge has become quiescent.

Closed partition protocol

1. When a new base table partition is created, it is marked *open*.
2. When a base table partition becomes quiescent, it is marked *closed*.
3. Derived partition D_i is marked *open* if any of its source partitions are marked open, and *closed* if all of its source partitions are marked closed, and D_i is up-to-date.

These steps can be performed in conjunction with any of the update propagation protocols discussed in Section 4. Queries which require stable source data can be restricted to querying closed partitions, while queries which require the most recent possible data can use all partitions. The open/closed status of source partitions can also be used to avoid recomputations of dependent partitions for tables which are required to provide mutual consistency: only create and compute a dependent partition if all of its source partitions are closed.

In a stream warehouse, the marking of base partitions as closed is often only an educated guess – late data arrives [29], existing data is found to be corrupted, etc. If a closed base partition is revised with new contents, one can mark all of its transitively dependent partitions open, indicating the loss of mutual consistency until the base partition update has been propagated.

7 Applications and Experiments

The algorithms in this paper were motivated by the DataDepot project [18]. DataDepot originally used a protocol similar to that of make, since this protocol is simple and updates to materialized views were generally done in batch, as in [14]. As DataDepot and its warehouses became more complex and real-time, we realized that the update propagation algorithm suffered from serious problems – the inability to support limited update required for real-time warehousing [19] was a particular motivation. At this point, we initiated a study of update propagation in a stream warehouse.

As the material in Section 5 indicates, the Source-vector protocol most readily supports the needs of industrial stream warehouses, most notably its lack of dependence on a coordinated timestamp – a useful property for a clustered warehouse. An examination of several DataDepot warehouses (in addition to Darkstar) indicates that the additional metadata required to support the Source-vector protocol is small in comparison to the existing metadata – few materialized views involve large joins (see Figure 2 and Figure 3 for examples). Although DataDepot tables typically use thousands to tens of thousands of partitions, the time to process their metadata for update propagation computations is negligible. In the course of extensive monitoring, we found metadata processing to account for less that 1% of the time to perform update processing. The multi-granularity partitioning technique contributes to this efficiency. Some tables use 1-minute partitions, but are stored for one year. Since there are 1440 minutes in a day and 365 days in a year, these tables require more than half a million per-minute partitions. However, per-minute partitions are rolled up into per-day partitions after two days, reducing the number of partitions used to store the table to a manageable level of about 3000.

Although we recommend the Source-vector protocol over the Interval-timestamp protocol, and that over the Starting-timestamp protocol, systems considerations can sway implementation decisions. DataDepot already had a large installed base when research for this paper was performed, and required a backwards-compatible protocol. Therefore, DataDepot switched to the Starting-timestamp protocol, which could most easily integrate the existing table metadata. However, a detailed monitoring of the warehouse showed that the scheduling restrictions of the Starting-timestamp protocol introduced unacceptable delays into the real-time processing chain.

A first example uses the ETL and summary table collection shown in Figure 8. The critical real-time update chain is C_RAW → C_BASE → C_JOINED → C.

We analyzed two days worth of scheduler logs, tracking the time from a first alert that new data had arrived for C_RAW to the data being propagated to C. The logs were collected on a testing warehouse which used the Starting-timestamp protocol and which was tuned to enable updates to propagate as soon as new data had arrived. The results are shown in Figure 9.

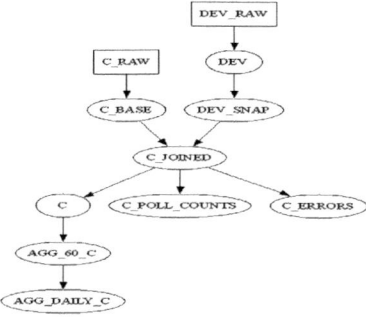

Fig. 8. ETL for table C

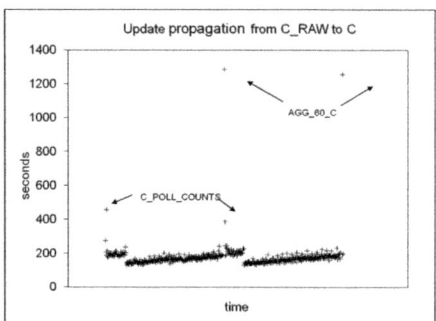

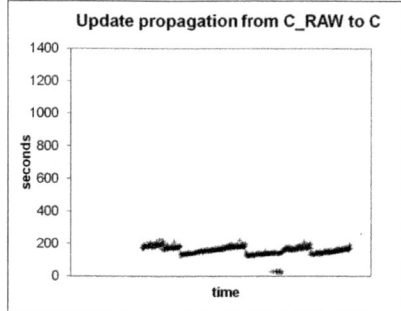

Fig. 9. Update delay for C, Starting-timestamp protocol

Fig. 10. Update delay for C, Interval-timestamp protocol

Most of the time, the update propagation delay is the sum of the update execution times at each table. However, there are four outliers. By tracing through the logs, we confirmed that two of the outliers were due to an update of C_POLL_COUNTS blocking updates to C_JOINED, and two were due to updated to AGG_60_C blocking updates to C. For a near-real time table with a 5-minute (300 second) periods, 1300 second (22 minute) propagation delays are unacceptable.

To eliminate the scheduling delays, we installed a backwards-compatible Interval-timestamp protocol into the test warehouse, and removed the concurrency restrictions from the scheduler. We ran the updated warehouse for approximately three days, and tracked the time to propagate updates from C_RAW to C. The results are shown in Figure 10. The maximum update delay is now much smaller, reduced to 226 seconds – demonstrating the values of asynchronous updates in a real-time warehouse.

Next, we analyzed the update propagation time from the notification that new data has arrived for X_RAW to the data being loaded into X, for the collection of tables shown in Figure 2 (in a production warehouse). We again analyzed about two days worth of update propagation delays using both the Starting-timestamp protocol and the Interval-timestamp protocol. The large number of summary tables derived on the critical real-time path from X_RAW to X caused a delay of nearly 600 seconds, on average, when using the Interval-timestamp protocol, even though the computation time on the critical path is closer to 120 seconds. When we switched to the Interval-timestamp protocol, we reduced the update delay to about 120 seconds. These experiments demonstrate the need for an update propagation protocol that does not restrict the scheduling of updates in a stream warehouse.

In the closest relevant literature [14], the authors describe update propagation in a DAG of temporally partitioned materialized views. While the emphasis of their paper is on optimizing the update queries, they also describe the update propagation procedure as being akin to that used by the make shell tool: partitions of the base tables are marked as "updated" if they receive updates since the last update propagation. When update propagation is initiated, all partitions to be updated in all tables are determined from partition dependencies, and all tables are brought fully up-to-date. While the global update procedure enables global optimizations, the update propagation delays it would incur would exceed those of the Starting-timestamp protocol, which we already find to be unacceptable.

8 Related Work

The primary topic of this paper relates to propagating updates of base tables to materialized views. The bulk of materialized view maintenance has focused on incremental update maintenance using differential update formulas, the earliest reference to which appears in [5]. The idea behind these formulas is to describe the increment to a materialized Select-Project-Join view that is due to an increment to one or more base tables. Formulae extensions for distributive aggregate tables are given in [35]. Non-distributive aggregate tables can also be incrementally maintained, by recomputing only the affected groups [36].

There are different formulas for computing the increment to a view, given different starting assumptions. One significant decision to be made is whether one has the contents of the base table(s) before or after the updates are applied – if the before-image is available the formulas are simpler and more efficient. The attempt to use before-image formulas inappropriately led to the recognition of the "state bug", with a variety of solutions for proper synchronization of update propagation [50][9][10][2].

If all views are not updated together (e.g. during a nightly refresh period), then queries across views can lead to inconsistent results (lack of mutual consistency). Schemes for defining consistency zones and their maintenance are described in [10][51]. By contrast, mutual consistency in a real-time database (not warehouse) refers to database reading all being collected within a short recent time window [23][1]. Trailing-edge consistency was developed as an easy-to-provide consistency model for stream warehouses [18][19], but has precedents in prior work [21]. Some very large systems with autonomous view updates explicitly do not provide any table-level form of mutual consistency [3].

The technique of temporally partitioning a warehouse that stores a sliding window of historical data is commonly used [8][12][18][31][14] and is well-known in the folklore [22]. However, temporal partitioning is little explored in the literature; exceptions include [39][43] and some of the DSMS literature [17][28].

One of the motivations for this work was to be able to provide for autonomous and partial updates of materialized views, as is required for a real-time warehouse

[18][19]. Some previous work has been oriented towards partial or interrupted updates [25][41].

Propagating updates to materialized views generally involves some degree of coordination and ordering. Previous work has generally relied on the global total ordering provided by the commit and recovery log [9][10][41][49][51] or on a message queue [2][50]. Update coordination in this paper ultimately depends on *Gb* and *GB*, which are similar to vector timestamps [13]. However, in *GB* or *Gb*, a partition (e.g. site) is in general represented more than once to account for multiple dependency paths. Previous work which uses vector timestamps for its update propagation modeling includes [21]. Previous work which uses multiple timestamps includes [49]. The terminology and techniques used in this work resemble that of replicated log propagation, e.g. [37], but the goals and the specifics greatly differ.

9 Conclusions

In this paper, we have presented models and algorithms for update propagation in a stream warehouse. To satisfy the needs of a stream warehouse – autonomous, partial, and real-time update propagation – we have developed a novel update propagation model. We use this model to develop simple, efficient, and provably correct update propagation algorithms. In the appendix, we extend these algorithms to handle the needs of real-world warehouses: ensuring mutual consistency, enabling merge tables, and allowing data repartitioning and multiple-granularity temporal partitioning. We implemented these update propagation algorithms in several production warehouses and observed significant improvements in performance.

Stream warehousing is similar in many respects to data stream management. Previous work has argued that efficient DSMS management requires that stream records be organized by temporal partition [17][28]. Therefore, this work is also relevant to DSMS management.

References

[1] Adelberg, B., Garcia-Molina, H., Kao, B.: Applying update streams in a soft real-time database system. In: Proc. ACM SIGMOD Conf. (1995)

[2] Agrawal, D., El Abbadi, A., Singh, A., Yurek, T.: Efficient view maintenance at data warehouses. In: Proc. ACM SIGMOD Conf. (1997)

[3] Agrawakl, P., Silberstein, A., Cooler, B.F., Srivastava, U., Ramakrishnan, R.: Asynchronous view maintenance for VLSD databases. In: Proc. ACM SIGMOD Conf. (2009)

[4] Balazinska, M., Kwon, Y.C., Kuchta, N., Lee, D.: Moirae: History-enhanced Monitoring. In: CIDR (2007)

[5] Blakely, J., Larson, P., Tompa, F.: Efficiently updating materialized views. In: Proc. SIGMOD (1986)

[6] de Boor, A.: Pmake: a tutorial,
     http://www.freebsd.org/doc/en/books/pmake/

[7] Bunger, C.J., et al.: Aggregate maintenance for data warehousing in Informix Red Brick Vista. In: Proc. VLDB Conf. (2001)

[8]   Chen, Q., Hsu, M., Dayal, U.: A data-warehouse/OLAP framework for scalable tele-communication tandem traffic analysis. In: Proc. IEEE Intl. Conf. Data Engineering (2000)

[9]   Colby, L.S., Griffin, T., Libkin, L., Mumick, I.S., Trickey, H.: Algorithms for deferred view maintenance. In: Proc. ACM SIGMOD Conf. (1996)

[10] Colby, L.S., Kawaguchi, K., Lieuwen, F.F., Mumick, I.S., Ross, K.A.: Supporting multiple view maintenance policies. In: Proc. ACM SIGMOD (1997)

[11] Cranor, C., Johnson, T., Spatscheck, O., Shkapenyuk Gigascope, V.: A Stream Database for Network Applications. In: Proc. ACM SIGMOD, pp. 647–651 (2003)

[12] Do, L., Drew, P., Jin, W., Jumani, V., Van Rossum, D.: Issues in developing very large data warehouses. In: Proc. VLDB Conf. (1998)

[13] Fidge, C.J.: Timetamps in message-passing systems that preserve the partial ordering. In: Proc. 11th Australian Computer Science Conference (1988)

[14] Folkert, N., et al.: Optimizing refresh of a set of materialized views. In: Proc. VLDB Conf. (2005)

[15] Galhardas, H., Florescu, D., Shasha, D., Simon, E., Saita, C.-A.: Declarative data cleaning: language, models, and algorithms. In: Proc. VLDB Conf. (2001)

[16] Garcia-Molina, H., Labio, W.J., Yang, J.: Expiring data in a warehouse. In: Proc. VLDB Conf. (1998)

[17] Golab, L., Garg, S., Özsu, M.T.: On indexing sliding windows over online data streams. In: Hwang, J., Christodoulakis, S., Plexousakis, D., Christophides, V., Koubarakis, M., Böhm, K. (eds.) EDBT 2004. LNCS, vol. 2992, pp. 712–729. Springer, Heidelberg (2004)

[18] Golab, L., Johnson, T., Spencer, J.S., Shkapenyuk, V.: Stream Warehousing with Data-Depot. In: Proc. ACM SIGMOD (2009)

[19] Golab, L., Johnson, T., Shkapenyuk, V.: Scheduling updates in a real-time stream warehouse. In: Proc. Intl. Conf. Data Engineering (2009)

[20] Golab, L., Ozsu, M.T.: Update-pattern-aware modeling and processing of continuous queries. In: Proc. ACM SIGMOD (2005)

[21] Hull, R., Zhou, G.: A framework for supporting data integration using the materialized and virtual approaches. In: Proc. ACM SIGMOD Conf. (1996)

[22] Inmon, W.H.: What is a data warehouse? Prism Solutions (1995)

[23] Jha, A.K., Xiong, M., Ramamritham, K.: Mutual consistency in real-time databases. In: Proc. IEEE Real Time Systems Symposium (1996)

[24] Krishnamurthy, S., et al.: Continuous analytics over discontinuous streams. In: Proc. SIGMOD (2010)

[25] Labio, W.J., Wiener, J.L., Garcia-Molina, H., Gorelik, V.: Efficient resumption of inter-rupted database loads. In: Proc. ACM SIGMOD Conf. (2000)

[26] Labrinidis, A., Roussopoulos, N.: Update propagation strategies for improving the quality of data on the web. In: Proc. VLDB Conf. (2001)

[27] Lerner, A., Shasha, D.: The Virtues and Challenges of Ad Hoc Streams Querying in Finance. IEEE Data Engineering Bulletin 26(1), 49–56 (2003)

[28] Li, J., Maier, D., Tufte, K., Papadimos, V., Tucker, P.A.: No pane, no gain: efficient eva-laution of sliding-window aggregates over data streams. SIGMOD Record 34(1), 39–44 (2005)

[29] Li, J., Maier, D., Tufte, K., Papadimos, V., Tucker, P.A.: Semantics and evaluation techniques for window aggregates in data streams. In: Proc. ACM SIGMOD Conf. (2005)

[30] Kalmanek, C.: Exploratory data mining in network and service management. In: IFIP/IEEE Intl. Symp. on Integrated Network Management (2009)

[31] Koutsofios, E., North, S., Truscott, R., Keim, D.: Visualizing Large-Scale Telecommunication Networks and Services. IEEE Visualization, 457–461 (1999)

[32] Labio, W.J., Wiener, J.L., Garcia-Molina, H., Gorelik, V.: Efficient resumption of Interrupted Warehouse Loads. In: Proc. ACM SIGMOD (2000)

[33] Labio, W.J., Yerneni, R., Garcia-Molina, H.: Shrinking the Warehouse Update Window. In: Proc. ACM SIGMOD (1999)

[34] Mistry, H., Roy, P., Sudarshan, S., Ramamritham, K.: Materialized view selection and maintenance using multi-query optimization. In: Proc. ACM SIGMOD (2001)

[35] Mumick, I.S., Quass, D., Mumick, B.S.: Maintenance of Data Cubes and Summary Tables in a Warehouse. In: Proc. ACM SIGMOD (1997)

[36] Palpanas, T., Sidle, R., Cochrane, R., Pirahesh, H.: Incremental maintenance for non-distributive aggregate functions. In: Proc. VLDB (2002)

[37] Petersen, K., et al.: Flexible update propagation for weakly consistent replication. In: Proc. Symp. on Operating System Principles (1997)

[38] Quass, D., Widom, J.: On-line warehouse view maintenance. In: Proc. ACM SIGMOD (1997)

[39] Riederwald, M., Agrawal, D., El Abbadi, A.: Efficient integration and aggregation of historical information. In: Proc. ACM SIGMOD Conf. (2002)

[40] Rohm, U., Bohm, K., Schek, H.-J., Schuldt, H.: FAS – a freshness-sensitive coordination middleware for a cluster of OLAP components. In: Proc. VLDB Conf. (2002)

[41] Salem, K., Beyer, K., Lindsay, B., Cochrane, R.: How to roll a join: asynchronous incremental view maintenance. In: Proc. ACM SIGMOD Conf. (2000)

[42] Fang, M., Shivakumar, N., Garcia-Molina, H., Motwani, R., Ullman, J.: Computing iceberg queries efficiently. In: Proc. VLDB (1998)

[43] Shivakumar, N., Garcia-Molina, H.: Wave-indices : indexing evolving databases. In: SIGMOD (1997)

[44] Taylor, R.: Concurrency in the Data Warehouse. In: Proc. VLDB (2000)

[45] Tucker, P., Maier, D., Sheard, T., Fegaras, L.: Exploiting Punctuation Semantics in Continuous Data Streams. IEEE Trans. Knowledge and Data Engineering 15(3), 555–568 (2003)

[46] Uppsala, K., Johnson, R., Chen, C., Hallmann, J., Hasan, W.: Peta-scale Data Warehousing at Yahoo! In: Proc. ACM SIGMOD Conf. (2009)

[47] Welbourne, E., Koscher, K., Soroush, E., Balazinska, M., Borriello, G.: Longitudinal study of a building-scale RFID ecosystem. In: Proc. Intl. Conf. Mobile Systems, Applications, and Services (2009)

[48] Welbourne, E., et al.: Cascadia: A system for specifying, detecting, and managing RFID events. In: Proc. Intl. Conf. Mobile Systems, Applications, and Services, MobiSys (2008)

[49] Zhou, J., Larson, P.-A., Elmongui, H.G.: Lazy maintenance of materialized views. In: Proc. VLDB Conf. (2007)

[50] Zhuge, Y., Garcia-Molina, H., Hammer, J., Widom, J.: View maintenance in a warehousing environment. In: Proc. ACM SIGMOD (1995)

[51] Zhuge, Y., Wiener, J.L., Garcia-Molina, H.: Multiple view consistency for data warehousing. In: Proc. Intl. Conf. Data Engineering (1997)

Efficient Processing of Multiple DTW Queries in Time Series Databases

Hardy Kremer[1], Stephan Günnemann[1], Anca-Maria Ivanescu[1],
Ira Assent[2], and Thomas Seidl[1]

[1] RWTH Aachen University, Germany
[2] Aarhus University, Denmark
{kremer,guennemann,ivanescu,seidl}@cs.rwth-aachen.de,
ira@cs.au.dk

Abstract. Dynamic Time Warping (DTW) is a widely used distance measure for time series that has been successfully used in science and many other application domains. As DTW is computationally expensive, there is a strong need for efficient query processing algorithms. Such algorithms exist for single queries. In many of today's applications, however, large numbers of queries arise at any given time. Existing DTW techniques do not process multiple DTW queries simultaneously, a serious limitation which slows down overall processing.

In this paper, we propose an efficient processing approach for multiple DTW queries. We base our approach on the observation that algorithms in areas such as data mining and interactive visualization incur many queries that share certain characteristics. Our solution exploits these shared characteristics by pruning database time series with respect to sets of queries, and we prove a lower-bounding property that guarantees no false dismissals. Our technique can be flexibly combined with existing DTW lower bounds or other single DTW query speed-up techniques for further runtime reduction. Our thorough experimental evaluation demonstrates substantial performance gains for multiple DTW queries.

1 Introduction

Time series, i.e., sequences of time-related values, occur in many applications in science and business. The task of finding similar patterns in these sequences is known as Similarity search in time series databases. Dissimilarity measures such as the Euclidean Distance compare time series values at each point in time. In many domains including climate research, speech-processing, and gene expression analysis, however, time series may be "out of sync"; that is, there are local disturbances on the time axis [1,11,17,4]. This is illustrated in Fig. 1: the time series (top blue curve and bottom red curve) contain the same "peak" pattern, only shifted along the time axis. Euclidean Distance (left) compares only values for identical time points, meaning that differences at each point in time (thin vertical black lines) are accumulated, and the similar pattern is not captured. This limitation is overcome by one of the most popular distance measures for time series, Dynamic Time Warping (DTW). It aligns time series by local stretching and shifting along the time axis, i.e. the time is *warped*. This is illustrated in Fig. 1 (right), where DTW (thin black lines) matches more similar values, even if they do not occur at the exact same point in time.

J.B. Cushing, J. French, and S. Bowers (Eds.): SSDBM 2011, LNCS 6809, pp. 150–167, 2011.

   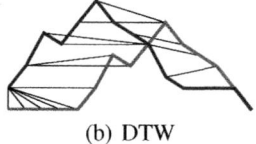

(a) Euclidean Distance   (b) DTW

Fig. 1. The two time series (top blue and bottom red curves) show the same pattern shifted along the time axis; Euclidean Distance (vertical lines) does not capture this, whereas DTW aligns time series and computes the distance between the best match

DTW stems from the speech processing community [8] and has since been successfully used in many other domains [1,11,17,4]. While DTW is an effective measure, it suffers from quadratic complexity in the length of the time series. Therefore, fast query processing approaches have been proposed [18,16,31,24,25,3,5]. All of them target single ad-hoc queries. In many of today's applications, however, massive query amounts need to be processed within limited time. Examples include sensor networks, data mining applications, and interactive visualization [19,9,10]. Fast response times are required for timely reaction to events, scalability, and interactiveness.

We propose a novel method for combined processing of multiple DTW queries that achieves, compared to single-query approaches, substantial speed-up. Our solution is based on a property of many multiple-query scenarios: queries may be similar or even share subsequences. As an example from data mining [9], consider density-based clustering where the transitive closure of the neighborhood (near-by objects) is computed to determine dense regions (clusters), resulting in similar queries for near-by objects. More examples of related queries can, for example, be found in [9].

Our approach exploits similarity among queries to process groups of queries as one. This allows for pruning of time series that are irrelevant to the entire group. To further enhance the pruning power of our concept, we introduce a nested hierarchy of query subgroups. This hierarchy iteratively splits the query group with respect to mutual similarity in order to continue pruning for smaller and more similar query subsets. As shown experimentally, this results in substantial efficiency gains. We prove that this does not incur any false dismissals, i.e. we guarantee completeness of the query results. We show that our approach is orthogonal in nature to existing single DTW query approaches. We demonstrate how DTW filters for single query processing can be extended to multiple query processing and how they can be added to our query hierarchy. We introduce a tree structure that maintains all information necessary for efficient algorithmic processing of query (sub-)groups using any number of filters.

This paper is structured as follows: In Section 2, we discuss related work on multiple query processing and DTW computation. Section 3 introduces our method, defining the multiple DTW query and our general approach in Subsection 3.1. Subsection 3.2 discusses the hierarchy and its representation in our multiple query tree. The combination with extended single query speed-up techniques is presented in Subsection 3.3. An algorithmic solution for processing of k nearest neighbor queries is described in Subsection 3.4, and Subsection 3.5 discusses the special case of range queries with individual thresholds. Section 4 presents experimental results and Section 5 concludes.

2   Related Work

Multiple query processing has been studied in the context of query optimization [27], where the goal is to devise efficient query execution plans. Similar problems have been studied for density-based clustering [30], continuous queries on data streams [7,28], or in sensor networks [14]. While the concept of sharing computations among queries is widely used, these approaches are not designed for time series data.

Dynamic Time Warping (DTW) has been successfully applied in many fields such as music similarity [31], video copy detection [4], biometric signatures [15], or stream monitoring [23]. An extensive empirical study of DTW and other time series distance measures can be found in [12]. Since DTW is an effective measure, but exhibits quadratic complexity in the length of the time series, much research has focused on speeding up the computation. A first group of techniques is based on approximating DTW queries, as e.g. [6]. Here, efficiency gains for single DTW queries come at the cost of accuracy. By contrast, our work for multiple DTW queries guarantees correctness of the result. A second group uses a filter-and-refine architecture with filter functions [13,26]. By showing that the filters indeed lower bound DTW, completeness of the result is ensured [18,16,31,24,17,5]. All of these approaches are for single queries, i.e. they devise a filter-and-refine or approximate search mechanism that processes a single query at a time. For multiple queries, however, these techniques fail to exploit the potential that lies in sharing computations among time series queries.

3   Efficient Multiple Query Processing under DTW

A time series is a sequence of values over time, such as temperature measurements recorded every minute over the period of several years.

Definition 1. *A **time series** t of length n is a temporally ordered sequence $t = [t_1, ..., t_n]$, where t_i denotes the value related to a point in time i.*

Please note that while we define univariate time series here for simplicity of presentation, our proposed method works for multivariate time series as well (where each t_i is a d-dimensional vector $t_i = (t_{i_1}, ..., t_{i_d})$).

Dynamic Time Warping (DTW) is the distance of the best alignment between time series stretched along the time axis. Infinite warping of the time axis is typically not desirable, so warping is usually restricted via *bands* (global constraints). As shown in [12], bands can enhance both accuracy and efficiency of computation. DTW with a band constraint is defined recursively on the length of the sequences.

Definition 2. *k-band DTW. The Dynamic Time Warping distance between two time series s, t of length n, m (w.l.o.g. $n \leq m$) w.r.t. a bandwidth k is defined as:*

$$DTW([s_1, ..., s_n], [t_1, ..., t_m]) =$$

$$dist_{band}(s_n, t_m) + \min \begin{cases} DTW([s_1, ..., s_{n-1}], [t_1, ..., t_{m-1}]) \\ DTW([s_1, ..., s_n], [t_1, ..., t_{m-1}]) \\ DTW([s_1, ..., s_{n-1}], [t_1, ..., t_m]) \end{cases}$$

$$with \quad dist_{band}(s_i, \ t_j) = \begin{cases} |s_i - t_j| & if \ |i - \lceil \frac{j \cdot n}{m} \rceil| \leq k \\ \infty & else \end{cases}$$

$$DTW(\emptyset, \emptyset) = 0, \quad DTW(x, \emptyset) = \infty, \quad DTW(\emptyset, y) = \infty$$

DTW is defined recursively on the minimal cost of possible matches of prefixes shorter by one element. There are three possibilities: match prefixes of both s and t, match s with the prefix of t, or match t with the prefix of s. The difference between overall prefix lengths is restricted to a band of width k in the time dimension by setting the cost of all overstretched matches to infinity. Besides this Sakoe-Chiba band with fixed bandwidth [22], our method works for other types such as R-K bands [21].

DTW can be computed via a dynamic programming algorithm in $O(k * \max\{m, n\})$ time. To reduce the computational overhead, *early stopping* (also called *early abandoning*) checks the minimum value of the current column as the dynamic programming matrix is being filled [24,17]: if this value exceeds the pruning threshold, DTW computation can be stopped immediately. We adopt this approach for any DTW computation in this paper. For the remainder of the discussion we assume that time series are of length n. Time series of different length can be easily interpolated to uniform length without degrading the quality of the final result in a statistically significant way [20].

3.1 Multiple DTW Query

We begin by introducing the problem definition of processing multiple DTW queries. In this subsection, we discuss range queries. The extension to the more complex problem of kNN queries is presented in Sec. 3.4.

Processing multiple range queries corresponds to computing a result set for each query of all time series that are within ε DTW distance from the respective query. Clearly, there might be applications where different thresholds apply per query, and this aspect is discussed in Sec. 3.5.

Definition 3. *Given a set of time series queries $Q = \{q^1, \ldots, q^c\}$ and a time series database DB, a **Multiple DTW ε-query** determines multiple result sets $Res^i = \{t \in DB \mid DTW(q^i, t) \leq \varepsilon\}$ for all $i = 1, \ldots, c$.*

Each query has an individual result set, i.e. the different results are not merged in any way. Thus it is possible to process a multiple DTW query simply by dividing Q into c single, independent queries q^i, such that traditional single query processing algorithms can be applied. This, however, would not exploit the speed-up potential that lies in the knowledge about the overall query set Q. Our novel approach operates directly on the whole set Q, i.e. the queries are examined simultaneously. This technique allows for sharing computations, since pruning of many time series can be done for the entire set at once, thereby speeding up the overall response time considerably.

We introduce the *multiple query distance function*, which uses a single calculation for a set of queries. Using this multiple query distance function, pruning of irrelevant time series is performed for all queries simultaneously, obtaining a substantial reduction of the calculations for the individual result sets. As we will see later on, pruning is done such that false dismissals are avoided, thereby ensuring a complete result set. We then remove false alarms to guarantee correctness of the results.

The general processing scheme is illustrated in Fig. 2: First, an intermediate result
set based on our multiple DTW query distance $multiDTW$ is jointly determined for
all queries. This *shared* set Res^Q is the remainder after pruning the database DB and
is an approximation of the final, individual
result sets Res^i. By design, we guarantee no

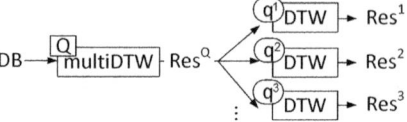

Fig. 2. Multiple DTW query processing
framework

false dismissals of the multiple query distance function: If $t \notin Res^Q$, then $t \notin Res^i$
for any result set. Accordingly, to reject a time series t, just a single calculation of the
multiple query distance function is needed, instead of $|Q|$ individual calculations. Since
the queries share some results, but also differ in some, the set Res^Q needs to be further
processed: Res^Q is split up for each query individually and false alarms are removed,
i.e. further time series are rejected to obtain the results Res^i.

To ensure that the final results are indeed complete, i.e. there are no false dismissals,
the intermediate set has to be a superset of each exact result set. More formally, the
condition $Res^Q \supseteq Res^i$ has to hold for all $i = 1, \ldots, c$. This is the case if the multiple
query distance function fulfills the following property:

Definition 4. *A multiple query distance function multiDist fulfills the **Shared Lower
Bound Property** w.r.t. a distance function dist iff*
$$for\ all\ Q, t \colon \forall q \in Q : multiDist(Q, t) \leq dist(q, t).$$

For each final result $t \in Res^i$ we have $dist(q^i, t) \leq \varepsilon$ and by Def. 4 $multDist(Q, t) \leq
\varepsilon$; thus, if $Res^Q = \{t \in DB \mid multDist(Q, t) \leq \varepsilon\}$, completeness is guaranteed.

The tightest multiple query distance function possible for DTW still fulfilling this
property is $multiDist(Q, t) = \min_{q \in Q} DTW(q, t)$. This, however, entails the calcu-
lation of $|Q|$ many DTW distance values; a performance gain is not realized. To ensure
an efficient processing, we need a compact, single representation that can efficiently be
processed; therefore we aggregate the set Q by a *multiple query bounding box*.

Definition 5. *Given a query set Q, the **Multiple Query Bounding Box** is a sequence
$[(L_1, U_1), \ldots, (L_n, U_n)] = [B_1, \ldots, B_n] = multiBox(Q)$ with $L_i = \min_{q \in Q} q_i$ and
$U_i = \max_{q \in Q} q_i$ (where q_i is the value of time series q at position i).*

By using the minimal and maximal values per time step, this bounding box approxi-
mates the whole query set. This approximation has the benefit of being very compact,
while preserving much information to ensure effective pruning. Based on this bounding
box definition, we introduce our novel *multiple DTW query distance function*.

Definition 6. *The **Multiple DTW Query Distance function** between a multiple query
bounding box $[B_1, .., B_n]$ and a time series t w.r.t. a bandwidth k is defined as:*

$$multiDTW(Q, t) = multiDTW([B_1, .., B_n], [t_1, .., t_m]) =$$

$$dist_{band}(B_n,\ t_m) + \min \begin{cases} multiDTW([B_1, .., B_{n-1}], [t_1, .., t_{m-1}]) \\ multiDTW([B_1, .., B_n],\ [t_1, .., t_{m-1}]) \\ multiDTW([B_1, .., B_{n-1}], [t_1, .., t_m]) \end{cases}$$

$$with\quad dist_{band}(B_i,\ t_j) = \begin{cases} dist(B_i, t_j)\ if\ |i - j| \leq k \\ \infty \qquad\qquad else \end{cases}$$

$$and \quad dist(B_i, t_j) = dist((L_i, U_i), t_j) = \begin{cases} |t_j - U_i| \ if \ t_j > U_i \\ |t_j - L_i| \ if \ t_j < L_i \\ 0 \qquad\quad otherwise \end{cases}$$

$$multiDTW(\emptyset, \emptyset) = 0, \quad multiDTW(x, \emptyset) = multiDTW(\emptyset, y) = \infty$$

Since the distance to the bounding box is used during the DTW calculation, we simultaneously consider all queries and the computational complexity is independent of the actual number of represented queries. We now prove that the shared lower bound property holds, i.e. using $multiDTW$ distance entails no false dismissals.

Theorem 1. $multiDTW$ *fulfills the shared lower bounding property w.r.t. DTW*.

Proof. We need to proof that $\forall p \ \forall q \in Q : multiDTW(Q, p) \le DTW(q, p)$. It suffices to show that $\forall i, j : dist(B_i, p_j) \le dist(q_i, p_j)$ with $q \in Q$ and $B = multiBox(Q)$. Then, no other alignment can be found by $multiDTW$ that would break Theorem 1. The proof is done by cases: We distinguish (a) $L_i \le p_j \wedge p_j \le U_i$, (b) $p_j > U_i$, and (c) $p_j < L_i$. For (a) it holds since $dist(B_i, p_j) = 0$. For (b) the following applies: $dist(B_i, p_j) = |p_j - U_i|$ and $dist(q_i, p_j) = |p_j - q_i|$. Accordingly, we have to show that $|p_j - U_i| \le |p_j - q_i|$. Since $p_j > U_i$ and $U_i \ge q_i$ (Def. 5) this equals to $p_j - U_i \le p_j - q_i$. By subtracting p_j and multiplying with -1 we obtain $U_i \ge q_i$, which is true according to Def. 5. The proof of (c) is analogue to (b).

3.2 Hierarchical Multiple DTW Query

Our multiple DTW query achieves a speed-up compared to single query processing by allowing pruning for a query group. In this section, we explore further pruning through the creation of several subgroups. A single group implies that only a single multi query needs to be processed, which helps reducing the number of DTW computations but still might lead to a relatively large intermediate result set Res^Q. To prevent false alarms, we would need to compare each individual query against this intermediate set. We propose reducing computations at this point by splitting up the query group into smaller subgroups for further pruning. This hierarchy of pruning options is based on the observation that smaller, more similar groups reduce the intermediate result size since the distances for smaller groups are larger, which is caused by tighter bounding boxes. The relation of group size and distance is reflected by the following theorem.

Theorem 2. *Smaller query sets correspond to a more accurate multiDTW, i.e. the distances values are larger:*

$$\forall Q, Q' \subseteq Q, p : \ multiDTW(Q, p) \le multiDTW(Q', p)$$

Proof. Similar to Theorem 1's proof, we need to show that $\forall i, j : dist(B_i, p_j) \le dist(B'_i, p_j)$ with $B = multiBox(Q)$ and $B' = multiBox(Q')$. This ensures that no other alignment can be found by $multiDTW$ that would violate the above theorem, i.e. that the left term in the inequation would be larger than the right term. The proof has three cases: (a) $L_i \le p_j \wedge p_j \le U_i$, (b) $p_j > U_i$, and (c) $p_j < L_i$. For (a) $dist(B_i, p_j) = 0$, $dist(B_i, p_j) \le dist(B'_i, p_j)$ is obviously fulfilled.

For (b) $dist(B_i, p_j) = |p_j - U_i|$. Since $U_i \geq U_i'$ (and thus $p_j > U_i'$) we also have $dist(B_i', p_j) = |p_j - U_i'|$. We have to show that $|p_j - U_i| \leq |p_j - U_i'|$. Since $p_j > U_i$ and $U_i \geq U_i'$ (Def. 5), we obtain $p_j - U_i \leq p_j - U_i'$. By subtracting p_j and multiplying with -1 we get $U_i \geq U_i'$. According to Def. 5 this yields $\max_{q \in Q} q_i \geq \max_{q \in Q'} q_i$. This holds because of $Q \supseteq Q'$.   The proof of (c) is analogue to (b).

Thus, reducing the query set corresponds to increasing the $multiDTW$ distance and thereby the number of time series that can be pruned.

In principle, $multiDTW$ can be used on each partition of an arbitrary partitioning of the query set Q. In the most extreme case of $|Q|$ many partitions, this procedure degenerates to single query processing. With fine grained groups we get small intermediate result sets; however, at high computational costs since many queries need to be processed. On the other hand, just one query group is computationally efficient; however, we get a larger set of candidate objects. We propose to combine different granularities through a hierarchical grouping of the

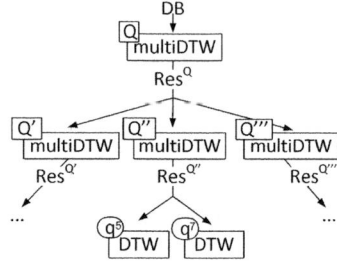

Fig. 3. Multiple Query Tree

query set. We organize queries in our multiple query tree, where the root node represents all queries as a single group (cf. Fig. 3), and by descending into the tree we get more fine grained groups. Eventually, the exact DTW computation for single queries is represented at leaf level.

Definition 7. *Given a query set Q, the **Multiple query tree (MQ-tree)** is defined as*

- *the root node stores a multiple query bounding box for all queries Q*
- *each inner node stores a multiple query bounding box for a query subset $P \subseteq Q$*
- *each leaf node stores the time series of a single query $q \in Q$*
- *all child nodes of a parent represent a complete, disjoint partitioning of their parent, i.e., $P = \bigcup_{P' \in children(P)} P'$*
- *the granularity is refined on each level, i.e., $\forall P' \in children(P) : P' \subset P$*

During query processing, each time series $t \in DB$ is first checked against the root node, i.e. the whole query set. If the $multiDTW$ distance is already larger than the range threshold ε, the time series is pruned. Otherwise the child nodes are analyzed, i.e. subsets of all queries are considered. Again, t is potentially pruned for some of the child nodes or passed on to the next level. Eventually, leaf nodes may be reached. If for the current leaf node/single query the DTW distance is still smaller than ε, t is added to the corresponding result set. The completeness of this method is ensured because a) for each query q^i there exists only one path from the root node to the corresponding leaf, b) all query sets along this path contain the query q^i, and c) the corresponding distance function along this path fulfills the shared lower bounding property.

To find query subgroups that are similar and therefore result in many pruned time series, we use clustering. The hierarchical query groupings of our tree are obtained with OPTICS [2], a hierarchical extension of the density-based clustering algorithm DBSCAN. OPTICS requires no knowledge about the number of clusters a priori, enabling a flexible fan-out in our tree that adjusts automatically to the mutual similarity

Fig. 4. Sequence of traditional filter-and-refine steps for single queries: each query (q^1, q^2, ...) undergoes the same filter steps until pruned or refined completely

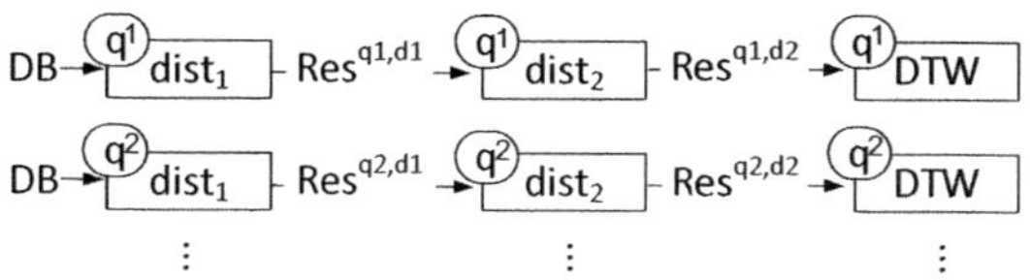

of queries. OPTICS computes a plot of possible clusters at different levels of density. Any horizontal cut in the plot corresponds to a clustering in this hierarchy. OPTICS has two parameters: The maximum density neighborhood range ε', and the minimum number of objects in this neighborhood $minPts$. We set ε' to twice the ε-range of our multiple DTW query and $minPts$ to two for avoiding single queries in inner nodes. Since the number of queries is typically small compared to the number of time series in the database, clustering runtimes are negligible, as our experiments in Sec. 4 confirm.

3.3 Filter-Supported Hierarchical Multiple DTW Query

Our hierarchical multiple DTW technique makes use of similarity among multiple queries, and can exploit a hierarchy of pruning possibilities that greatly reduce DTW computations. Existing work for single queries speed up DTW computations using specific filter functions. In these traditional filter-and-refine algorithms, the idea is to precede the processing of a *single* query by a filter distance computation, that might lead to pruning prior to DTW calculation. If filter functions are lower bounds, i.e. they underestimate the DTW distance, then this filter-and-refine approach is complete [13,26]. Since several lower bounds exist in the literature, one can build a sequence of filters that a query can be subjected to, as illustrated in Fig. 4.

Clearly, this filter concept is orthogonal to our multiple DTW query. Our approach prunes based on the combination of multiple queries into a single representation, whereas the traditional approach uses filters per query. For improved pruning power and thereby most efficient query processing, we propose a combination of the two concepts, which we call *filter-supported hierarchical multiple DTW*.

To realize such a combination, the single query lower bound needs to be extended so that it can serve as a lower bound for a query group, and we go from a traditional filter distance function $dist_f(q, p)$ to a multiple query distance function $multiDist_f(Q, p)$. In this work we exemplarily adapt the well-known and effective LB_{Keogh} lower bounding filter to handle multiple queries [16], which is based on the difference to upper and lower bounds (termed "envelope") within the band constraint.

Traditionally, LB_{Keogh} between a query q and time series t is defined as

$$LB_{Keogh}(q, t) = \sum_{i=1}^{n} \begin{cases} |t_i - \hat{U}_i| & \text{if } t_i > \hat{U}_i \\ |t_i - \hat{L}_i| & \text{if } t_i < \hat{L}_i \\ 0 & \text{otherwise} \end{cases}$$

with $\hat{U}_i = \max\{q_{i-k}, \ldots, q_{i+k}\}$ and $\hat{L}_i = \min\{q_{i-k}, \ldots, q_{i+k}\}$ as lower and upper bounds w.r.t. the DTW bandwidth k.

For a query group, we define the multiple query LB_{Keogh}, which computes the upper and lower bounds for all queries within the band constraint using the previously introduced multiple query bounding box.

Definition 8. *Given a query set Q, its multiple query bounding box $[(L_1, U_1), \ldots, (L_n, U_n)]$ and the DTW bandwidth k, then the **Multiple Query** LB_{Keogh} is defined as*

$$multiLB_{Keogh}(Q, t) = \sum_{i=1}^{n} \begin{cases} |t_i - \widetilde{U}_i| & \text{if } t_i > \widetilde{U}_i \\ |t_i - \widetilde{L}_i| & \text{if } t_i < \widetilde{L}_i \\ 0 & \text{otherwise} \end{cases}$$

with upper bound over all queries $\widetilde{U}_i = \max\{U_{i-k}, \ldots, U_{i+k}\} = \max_{q \in Q}\{\max\{q_{i-k}, \ldots, q_{i+k}\}\}$ and lower bound $\widetilde{L}_i = \min\{L_{i-k}, \ldots, L_{i+k}\} = \min_{q \in Q}\{\min\{q_{i-k}, \ldots, q_{i+k}\}\}$.

To show that this multiple query LB_{Keogh} can be used without incurring false dismissals, we prove the shared lower bounding property.

Theorem 3. *$multiLB_{Keogh}$ fulfills the shared lower bounding property w.r.t. DTW.*

Proof. We need to show that for any query set Q and any time series t, we have $multiLB_{Keogh}(Q, t) \leq DTW(q, t)$.

According to Def. 5, the upper bound $\widetilde{U}_i$ for query set Q is the maximum of the upper bounds for all $q \in Q$. Analogously, the lower bound $\widetilde{L}_i$ is the minimum of the lower bounds for all $q \in Q$. Thus, similar to Theorem 1's proof, we know that for an individual query q in Q the bound can only be higher than for the entire set, i.e. we have that $multiLB_{Keogh}(Q, t) \leq LB_{Keogh}(q, t) \forall q \in Q$. Since LB_{Keogh} is itself a lower bound to DTW [16], i.e. $LB_{Keogh}(q, t) \leq DTW(q, t)$, it holds that $multiLB_{Keogh}(Q, t) \leq DTW(q, t) \forall q \in Q$.

Accordingly, our definition of the $multiLB_{Keogh}$ filter may be safely combined with our hierarchical multiple DTW approach without jeopardizing completeness. Fig. 5 gives an overview: as before, we have different granularities of the query subgroups Q', Q", ... of the entire query set. Additionally, sequential filters, such as $dist_{f1}$ in Fig. 4, extended to multiple queries as $multiDist_1$ in Fig. 5 are added (e.g., LB_{Keogh} to $multiLB_{Keogh}$). In Fig. 5, we illustrate a complete nesting of both paradigms, i.e. for each granularity all possible sequential filters are used. In general, on each level it is possible to select a certain subset of filter distance functions. For example, we could omit the gray highlighted nodes of

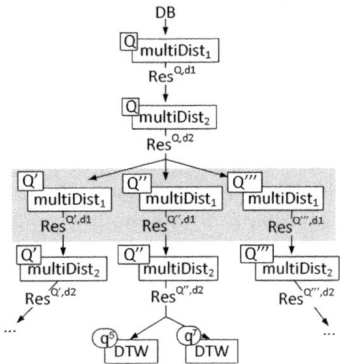

Fig. 5. Filter supported hierarchical multiple DTW

the tree and directly propagate the result to the next level. In this manner, it is possible to flexibly construct a subgroup and filter cascade based on selectivity estimates.

We extend the multiple query tree (Def. 7) to include sequential filters in addition to grouping of queries. is defined as follows:

Definition 9. *Given a query set Q and a series of multiple query distance functions $multiDist_i$ with $i = 1,..,r$ (fulfilling the shared lower bound property w.r.t. DTW), the **Filter-supported Multiple Query Tree (FSMQ-tree)** is a modified MQ-tree:*

- *the root node is a tuple $[Q, 1]$ of all queries Q and the distance function $multiDist_1$*
- *each inner node is a tuple $[P, i]$ representing a query set $P \subseteq Q$ and the distance function $multiDist_i$*
- *each leaf node is a tuple $[\{q\}, r + 1]$ representing a single query $\{q\} \subseteq Q$ and the usual distance function DTW*
- *all child nodes of a parent represent a complete, disjoint partitioning of their parent, i.e., $P = \bigcup_{[P',j] \in children([P,i])} P'$*
- *either the granularity or the distance function is refined, i.e.,*
  $\forall [P', j] \in children([P, i]) : P' \subset P \text{ or } j > i$

Each node of the tree now represents a query subset and a filter function. The hierarchy is used to either go to a new granularity by splitting the query group, or to go to a new lower bounding filter. Since the leafs use DTW itself, and we assume that all filters are lower bounding, the final result is correct and complete.

The query processing is illustrated in Fig. 6. For a query set Q, a database DB and an ε-threshold, we begin by initializing an empty result set for each query. Query processing for database time series starts from the tree root, which represents the entire query set Q and the first filter distance. While we are not at leaf level (line 6), we compare the current multiDist value to the ε-threshold (line 10). We continue processing only for child nodes that

```
1: input: FSMQ-tree with root [Q, 1], ε range, database DB;
2: for each t ∈ DB
3:     process-node([Q, 1],t)
4: return result sets Res^i, i = 1, . . . , |Q|

5: function process-node(node [P, j], time series t)
6:     if( P = {q^i} ∧ j = r + 1 )
7:         if( DTW(q^i, t) ≤ ε )
8:             Res^i = Res^i ∪ {t}
9:     else
10:         if( multiDist_j(P, t) ≤ ε )
11:             for each [P', l] ∈ children([P, j])
12:                 process-node([P', l],t)
13: end function
```

Fig. 6. Processing of multiple DTW ε-queries

cannot be pruned according to this filter (line 12). At leaf level, the final DTW refinement determines whether a time series is included in the respective result set of query q^i (line 8). Thus, the FSMQ-tree maintains all information needed on the current query subset and the respective filter level, making it easy to prune time series that do not contribute to certain result sets.

3.4 Multiple DTW kNN-Query

In the previous sections we discussed the processing of multiple range queries. In range queries, the pruning threshold is known to be exactly the ε-range. We now discuss the case of multiple kNN (k nearest neighbor)-queries, which is more complex in the sense that such a threshold is not known in advance. The result set consists of the k most similar time series.

Definition 10. *Given a set of time series queries $Q = \{q^1, \ldots, q^c\}$ and a time series database DB, a **Multiple DTW kNN-query** determines multiple result sets $Res^i(DB)$, such that $|Res^i(DB)| = k$ and $\forall t \in Res^i(DB) \forall s \in DB \backslash Res^i(DB) :$ $DTW(q^i, t) \le DTW(q^i, s)$ for all $i = 1, \ldots, c$.*

In contrast to the ε-query, where the maximal permitted distance ε is known before-hand, for kNN-queries such a threshold is not given a priori. This threshold, however, plays a crucial role for the early pruning of time series in our hierarchy. To obtain a threshold for kNN-queries, query processing typically retrieves k objects that are iteratively replaced by closer ones as query processing proceeds. The distance of the kth best candidate seen so far acts as a moving ε-threshold. At the end of query processing, it corresponds to the true kth nearest neighbor distance.

For multiple kNN-queries, the question is how to obtain an initial ε-threshold that is as small as possible. If the threshold is too large, only few objects are pruned resulting in larger intermediate result sets and thus higher query times.

Naively, one could process k time series for the group of queries, thus obtaining an initial ε-threshold that corresponds to the maximum kth nearest neighbor distance for all queries. Then, as more time series are processed, the k nearest neighbors of the query set could be adjusted to potentially lower the moving threshold for the query set accordingly. This procedure, however, is not the most efficient one. Since we use a hierarchical grouping of queries, maintaining only a single moving threshold for the entire query set is only useful at root level. Descending into the tree, i.e. using more fine grained partitions, allows calculating tighter thresholds by considering only the queries represented by the current subtree.

Our solution is to maintain a moving threshold per query subgroup for maximizing pruning power: each tree node is enriched by an individual ε-value (i.e., the kth nearest neighbor distance among the time series seen so far).

Definition 11. *Given a query set Q and a series of multiple query distance functions $multiDist_i$ with $i = 1, \ldots, r$ (fulfilling the shared lower bound property w.r.t. DTW), the ε-enriched FSMQ-tree for currently completely processed time series subset $T \subseteq DB$ is defined by:*

- *each node is a 3-tuple $[P, i, \varepsilon]$*
- *a valid FSMQ-tree (cf. Def. 9) is obtained if the 3-tuples are restricted to the first two components*
- *each inner node stores $[P, i, \varepsilon]$ with $\varepsilon = \max_{q^j \in P} \max_{t \in currRes^j(T)} \{DTW(q^j, t)\}$ and $currRes^j(T)$ is q^j's current kNN query result for subset $T \subseteq DB$ (cf. Def. 10).*

The individual moving threshold ε is the maximum among the ε values of its child nodes, which we use for efficient updating during query processing.

Theorem 4. *For each inner node of the ε-enriched FSMQ-tree $[P, i, \varepsilon]$ the following holds:* $\varepsilon = \max_{[P', i', \varepsilon'] \in children([P, i, \varepsilon])} \{\varepsilon'\}$.

Proof. By construction of the FSMQ-tree as a modified MQ-tree (Def. 9), it holds that $P = \bigcup_{P' \in children(P)} P'$ for all child nodes $[P', i', \varepsilon']$ of node $[P, i, \varepsilon]$. Consequently, $\varepsilon = \max_{q^j \in P} \max_{t \in currRes^j(T)} \{DTW(q^j, t)\} =$
$\max_{q^j \in \bigcup_{P' \in children(P)} P'} \max_{t \in currRes^j(T)} \{DTW(q^j, t)\} =$
$\max_{[P', i', \varepsilon'] \in children([P, i, \varepsilon])} \{\varepsilon'\}$ since $\varepsilon' = \max_{q^j \in P'} \max_{t \in currRes^j(T)} \{DTW(q^j, t)\}$. (Note that the maximum over the current result set $currRes^j$ does not change for a given q^j and T.)

Accordingly, as we process time series and include them in the respective result sets of individual queries at leaf level, we can propagate the thresholds up the tree by taking the maximum among child nodes.

For a query set Q, a database DB, and the number of nearest neighbors k, we begin by initializing with k randomly selected time series from the database to fill the result sets. Afterwards the current individual ε thresholds are calculated based on these sets and the ε-enriched FSMQ tree is built. The overall processing algorithm of multiple kNN queries on this tree is shown in Fig. 7. As in the algorithm in Fig. 6, in inner nodes database time series are pruned if they exceed the ε threshold (line 13). There are two major differences in the query processing for multiple DTW kNN queries. First, pruning takes place according to a subgroup-specific threshold, which is only valid for the query subgroup represented by this inner node. Second, if a database time series reaches leaf level (line 6) and the object cannot be pruned by the current individual threshold $currEps^i$, an update of the corresponding result list is performed. The time series with the highest DTW distance to the query is substituted by this more similar time series (lines 8-9). Such an update may lead to improved pruning for the remaining database time series, because the threshold on the kth neighbor distance among time series seen so far for q^i, i.e. $currEps^i$, becomes smaller (line 10). Note that the update of thresholds in line 10 is based on stored values and not on re-computation of the DTW distances. The better threshold is propagated up the tree by recomputing the maxima according to Theorem 4, improving pruning on all levels. Additionally, we improve the tree structure by adjusting the query grouping. As mentioned in the discussion of range queries (cf. Sec. 3.2), OPTICS clustering is used to group queries. If the thresholds change, we refer to the hierarchical plot that OPTICS has generated to re-group the queries. This ensures that we always group the queries based on their similarity with respect to the current pruning options.

In the following, we prove that this multiple kNN query processing can be used without incurring false dismissals, i.e. it ensures completeness of the results.

```
1: input: ε-enriched FSMQ-tree with root [Q, 1, ε₁],
      k, DB;
2: for each t ∈ DB
3:    process-node([Q, 1, ε₁],t)
4: return result sets Resⁱ, i = 1, . . . , |Q|

5: function process-node(node [P, j, ε], time series t)
6:    if( P = {qⁱ} ∧ j = r + 1 )
7:       if( DTW(qⁱ, t) < currEpsⁱ )
8:          currResⁱ =
              currResⁱ\
              {argmax_{x∈currResⁱ}{DTW(qⁱ, x)}}
9:          currResⁱ = currResⁱ ∪ {t}
10:         currEpsⁱ = max_{t∈currResⁱ}{DTW(qⁱ, t)}
11:         propagate currEpsⁱ
12:    else
13:       if( multiDist_j(P, t) ≤ ε )
14:          for each [P', l, ε'] ∈ children([P, j, ε])
15:             process-node([P', l, ε'],t)
16: end function
```

Fig. 7. Processing of multiple DTW kNN-queries

Theorem 5. *Query processing of multiple kNN-queries using the ε-enriched FSMQ-tree is complete, i.e. there are no false dismissals.*

Proof. Time series are only dismissed based on the pruning thresholds in the ε-enriched FSMQ-tree nodes. False dismissals are prevented if the pruning threshold for any query is larger than or at most equal to the final kth nearest neighbor distance [13,26]. This means that completeness is ensured if in all (inner and leaf) nodes of the tree the subgroup-specific thresholds ε are larger than or equal to the exact thresholds of any single query at leaf level in the corresponding subtree. This means that for each node $[P, i, \varepsilon]$ it holds that $\varepsilon \geq \max_{q^i \in P}\{\max_{t \in Res^i(T)}\{DTW(q^i, t)\}\}\forall T \subseteq DB$.

Since in the inner nodes, the corresponding ε can be determined by taking the maximum of the child nodes (cf. Theorem 4), it suffices to prove the above property for the leaf nodes; then it holds for the entire tree.

In the leaf nodes, the query set consists of a single query q^i, i.e. we have $\varepsilon = \max_{t \in currRes^i(T)}\{DTW(q^i, t)\}$. Since the currently processed time series are a subset of the database $T \subseteq DB$, we have that $\varepsilon \geq \varepsilon' := \max_{t \in Res^i(DB)}\{DTW(q^i, t)\}$ which is the final kth nearest neighbor distance.

3.5   Processing of Multiple Range-Queries with Individual Ranges

By augmenting our tree with the concept of individual ε values, we enable the processing of multiple kNN queries. Additionally, this allows processing of multiple DTW range-queries with individual thresholds ε^i. In this case, the ε^i value of each leaf node is fixed, the thresholds in the inner nodes are determined just ones, and there is no dynamic change of the result set $currRes^i$; it is just filled as database time series fall below the individual threshold ε^i (cf. Fig. 6, line 10-11).

4   Experiments

Experiments were run on 3GHz Intel Core 2 CPUs using JAVA implementations. As mentioned in Section 3, all approaches use early stopping of DTW computations [17,24]. Unless stated otherwise, the following defaults were used: Time series length was $n = 512$ and dataset size was 5,000. DTW bandwidth k was 10% of time series length. The number of nearest neighbors retrieved was 5 per multiple query (i.e., 5 per individual query), and for range queries an ε-range was selected that resulted in around 5 nearest neighbors per individual query. As measures we use the wall clock time averaged over 10 multiple queries, the relative number of refinements, i.e. the percentage of database time series that undergo exact DTW calculations, and the relative improvements of our method compared to the baseline method. A multiple query Q is obtained by randomly selecting a set S of seed queries from the database. For each seed $s \in S$, we generate g queries deviating from the seed by a standard deviation of 10%, and a multiple query Q has the cardinality $|Q| = |S| \cdot g$. As default, we use 8 seeds and 5 generated queries per seed, resulting in 40 individual queries per multiple query.

We use synthetic random walk (RW) and real world data. For RW, the t_{i+1}th value is generated by drawing from a normal distribution with parameters $\mu = 0$, $\sigma = 1$ added to the value of t_i : $t_{i+1} = t_i + N(0, 1)$. RW was normalized to an average value of 0. In real data experiments, we used datasets introduced in [16]: From the EEG dataset we used 5,000 time series of length 512; the smaller datasets consist of 500 (burstin), 272 (lightcurb), and 180 (network) time series of length 100. The largest dataset EEG is used throughout the evaluation along with RW data, and an overview over the remaining results is given at the end.

There are no existing solutions for processing multiple DTW queries. As baseline method for comparison, we sequentially process the queries by employing a single-query filter-and-refine method, where the queries are processed independently using

LB_{Keogh} [16]. We chose LB_{Keogh} as the basis of our MultiDTW, but we could also have used other lower bounds as [24,31], since our approach is orthogonal to the concept of lower bounds. We use a simple sequential scan in our framework to process database time series, and our multiple-query technique can also be used with indexing techniques as [3] and dimensionality reduction to further speed up the overall runtime. It has been shown that for very high dimensional data such as time series, linear database scans outperform traditional index structures [29].

The default configuration of our method is FSMQ as introduced in Sec. 3.3. In the FSMQ-tree, we use $multiLB_{Keogh}$ (cf. Def. 8), the adaption of LB_{Keogh} [16], followed by $multiDTW$ (cf. Def. 6).

Query Processing Strategy. We begin our study by comparing the default configuration (FSMQ-tree) to the earlier variants of our method on RW data in Fig. 8. Shown are similar results for both range and kNN queries. The refinement percentages in the right figure validate that both MQ-tree and the FSMQ-tree dramatically reduce the number of DTW computations as opposed to the multiple DTW query. From the runtimes in the left figure, we can infer that even though the differences in the number of refinements are small, it is much more efficient to include filter support, as pruning can be performed much faster. Thus, the FSMQ-tree is an efficient combination of the hierarchical refinement of multiple DTW queries with filter techniques.

Query Set Size and Similarity of the Queries. We now analyze how the number of individual queries per multiple query affects the performance of the FSMQ-tree and the single query processing method. Fig. 9 shows the corresponding experiments; average query times (left y-axis) and relative improvements (right y-axis) are measured. Note that the number of seeds per multiple query remains at the default value of 8. In (a) and (b) range queries and kNN queries on random walk data (RW) are processed. For both query types, the average query times of single query LB_{Keogh} increase much faster than the query times of FSMQ. This is confirmed by the relative improvements: FSMQ outperforms the single query solution significantly: for both range queries and kNN queries we have relative improvements between 10 and 75 percent. It can be concluded that our method performs as intended, i.e. in situations of multiple queries a combined solution clearly surpasses an independent solution. Similar conclusions can be made for the real world in Fig. 9(c), where the relative improvements go up to 87%. The corresponding relative numbers of refinements are shown in Fig. 10(a). As we can see, the percentages are relatively stable for both approaches, but there is a substantial reduction in DTW computations achieved by the multiple query approach FSMQ.

In Fig. 10(b) and 10(c) we show the effect of query similarity. We vary the number of similar queries per group, i.e. how many similar queries are generated for each seed

Fig. 8. Comparison of the three variants of our method: The FSMQ-tree (Sec. 3.3), the MQ-tree (Sec. 3.2), and the simple multiple DTW query (Sec. 3.1); RW

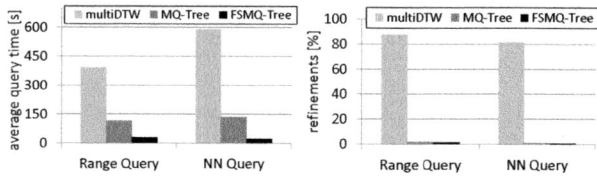

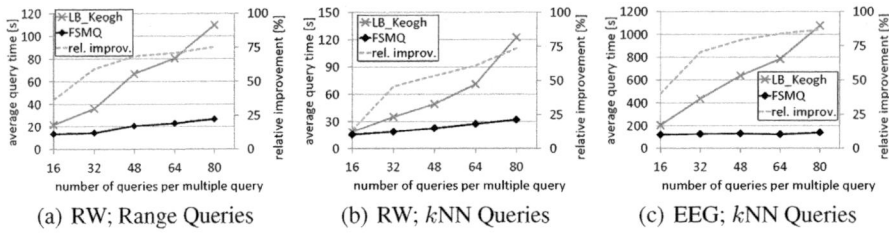

Fig. 9. Varying number of individual queries per multiple query

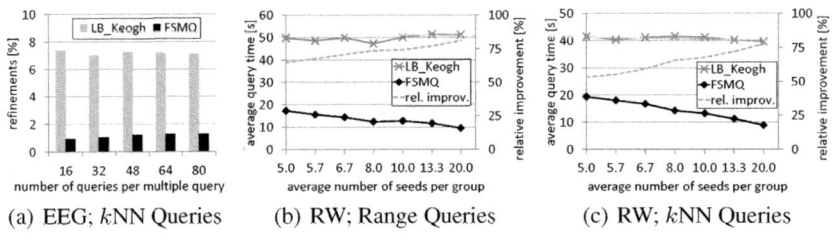

Fig. 10. (a): Number of refinements, i.e. exact DTW computations, for the experiment in Fig. 9(c). (b),(c): Varying number of query seeds for a fixed number individual queries.

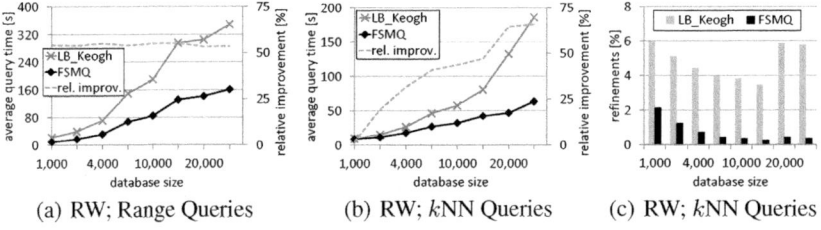

Fig. 11. Database Scalability on random walk data

of a multiple query. The absolute number of queries per multiple query is 40. In both experiments, the query times of our method improve, while the query times of the single query solutions are stable. The increasing relative improvements highlight this aspect.

Database Size. Fig. 11 shows the performance for database sizes between 1,000 and 25,000 times series for range and kNN queries. FSMQ outperforms single query processing for range queries in (a); the relative improvement is stable at around 55%, independently of the database size. With increasing database size, more time series fall within the ε-ranges of the queries. Accordingly, more exact DTW computations are necessary, compensating the positive effects of a larger database size. For the kNN queries, our approach copes with larger database sizes far better than single query processing, i.e. the relative improvement scales with the database size enabling performance gains of up to 70%.

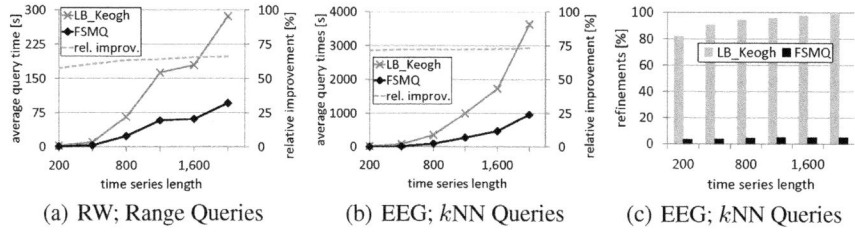

Fig. 12. Varying time series length on random walk and EEG data

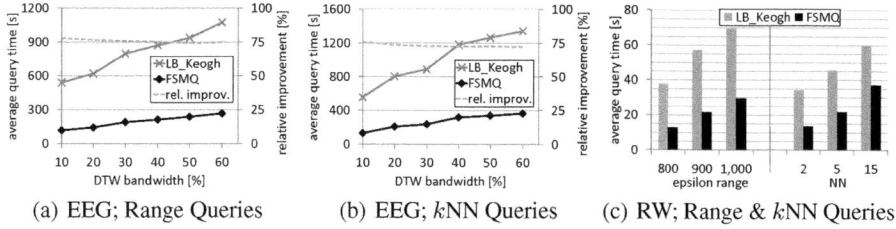

Fig. 13. (a,b): Varying DTW bandwidth k. (c): Varying ε-range and varying number of NN.

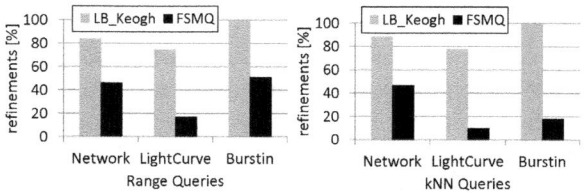

Fig. 14. Performance comparison on real world datasets burstin, lightcurb, and network

Time Series Length. The influence of the time series length is shown in Fig. 12. We used the random walk and EEG data. For both query types stable improvements of about 70% are achieved, independently of the time series length. Fig. 12(c) demonstrates robustness of our approach for scattered values in the time series: the number of refinements required after applying the filter function LB_{Keogh} goes up to nearly 100% of the database, which explains the large query times observed in Fig. 12(b). This can be explained by the large variance in the EEG data: LB_{Keogh} exploits the property that in most time series successive values are very similar by constructing a bounding box (envelope) for all values within the bandwidth constraint. Large scattering of the values leads to large boxes and thereby poor pruning power. Our approach, by contrast is not affected by this scatter, since queries are grouped by their similarity, and the hierarchy opens up pruning possibilities for subgroups as well.

Bandwidth. In Fig. 13(a,b) we study the influence of the DTW bandwidth constraint k on real world data. For both query types, very stable relative improvements of about 70-75% are achieved. While absolute runtimes increase for any approach, since increasing bandwidth means that more points in time are part of the DTW computation, the figures show that our approach reduces runtimes regardless of the bandwidth constraint.

Number of Nearest Neighbors and the ε-range. In Fig. 13(c), we analyze how the parameters epsilon range and number of nearest neighbors of the two query types influence the performance. As for the bandwidth, it is clear that increasing these parameters will lead to higher computational cost, as larger result sets are to be expected. For both query types, FSMQ outperforms the single query processing method. This means that similar to the bandwidth experiment, while absolute runtimes increase, our FSMQ reliably reduces the runtimes by a considerable margin.

Additional Real World Data Experiments. In Fig. 14 we give an overview over the performance on other data sets also used in [16]. For range queries, our FSMQ approach greatly reduces the number of DTW refinements necessary, and for kNN queries, the performance gain of FSMQ over single query processing is even more pronounced.

5   Conclusion

In this work, we address the problem of multiple Dynamic Time Warping (DTW) queries. We group similar DTW queries into groups for joint pruning of irrelevant time series in the database. By introducing a hierarchy of subgroups of multiple DTW queries, further pruning for iteratively smaller groups is achieved. We show that filter functions for single DTW queries can be extended to fit our multiple DTW query processing approach. All information necessary to manage query groups, filter functions, and pruning thresholds is compactly represented in our filter supported multiple query tree (FSMQ-tree). We provide algorithms for processing range queries and k nearest neighbor queries efficiently on the FSMQ-tree. As our experimental evaluation on synthetic and real world data sets demonstrates, we obtain substantial runtime improvements compared to single DTW query processing.

Acknowledgments. This work was supported by the UMIC Research Centre, RWTH Aachen University and the Deutsche Forschungsgemeinschaft (DFG) within the Collaborative Research Center (SFB) 686 "Model-Based Control of Homogenized Low-Temperature Combustion".

References

1. Aach, J., Church, G.M.: Aligning gene expression time series with time warping algorithms. Bioinformatics 17(6), 495–508 (2001)
2. Ankerst, M., Breunig, M.M., Kriegel, H.P., Sander, J.: OPTICS: Ordering points to identify the clustering structure. In: SIGMOD, pp. 49–60 (1999)
3. Assent, I., Krieger, R., Afschari, F., Seidl, T.: The TS-Tree: Efficient time series search and retrieval. In: EDBT, pp. 252–263 (2008)
4. Assent, I., Kremer, H.: Robust adaptable video copy detection. In: Mamoulis, N., Seidl, T., Pedersen, T.B., Torp, K., Assent, I. (eds.) SSTD 2009. LNCS, vol. 5644, pp. 380–385. Springer, Heidelberg (2009)
5. Assent, I., Wichterich, M., Krieger, R., Kremer, H., Seidl, T.: Anticipatory DTW for efficient similarity search in time series databases. PVLDB 2(1), 826–837 (2009)
6. Athitsos, V., Papapetrou, P., Potamias, M., Kollios, G., Gunopulos, D.: Approximate embedding-based subsequence matching of time series. In: SIGMOD, pp. 365–378 (2008)

7. Babu, S., Widom, J.: Continuous queries over data streams. SIGMOD Rec. 30(3), 109–120 (2001)
8. Berndt, D.J., Clifford, J.: Using dynamic time warping to find patterns in time series. In: AAAI Workshop on KDD, pp. 229–248 (1994)
9. Braunmüller, B., Ester, M., Kriegel, H.P., Sander, J.: Efficiently supporting multiple similarity queries for mining in metric databases. In: ICDE, pp. 256–267 (2000)
10. Brochhaus, C., Seidl, T.: Efficient index support for view-dependent queries on CFD data. In: Papadias, D., Zhang, D., Kollios, G. (eds.) SSTD 2007. LNCS, vol. 4605, pp. 57–74. Springer, Heidelberg (2007)
11. Chen, A.P., Lin, S.F., Cheng, Y.C.: Time registration of two image sequences by dynamic time warping. In: Proc. ICNSC, pp. 418–423 (2004)
12. Ding, H., Trajcevski, G., Scheuermann, P., Wang, X., Keogh, E.J.: Querying and mining of time series data: experimental comparison of representations and distance measures. PVLDB 1(2), 1542–1552 (2008)
13. Faloutsos, C.: Searching Multimedia Databases by Content. Kluwer, Dordrecht (1996)
14. Jurca, O., Michel, S., Herrmann, A., Aberer, K.: Continuous query evaluation over distributed sensor networks. In: ICDE, pp. 912–923 (2010)
15. Kar, B., Dutta, P., Basu, T., Viel Hauer, C., Dittmann, J.: DTW based verification scheme of biometric signatures. In: IEEE ICIT, pp. 381–386 (2006)
16. Keogh, E.J.: Exact indexing of dynamic time warping. In: VLDB, pp. 406–417 (2002)
17. Keogh, E.J., Wei, L., Xi, X., Lee, S., Vlachos, M.: LB_Keogh supports exact indexing of shapes under rotation invariance with arbitrary representations and distance measures. In: VLDB, pp. 882–893 (2006)
18. Kim, S.W., Park, S., Chu, W.W.: An index-based approach for similarity search supporting time warping in large sequence databases. In: ICDE, pp. 607–614 (2001)
19. Kusy, B., Lee, H., Wicke, M., Milosavljevic, N., Guibas, L.J.: Predictive QoS routing to mobile sinks in wireless sensor networks. In: IPSN, pp. 109–120 (2009)
20. Ratanamahatana, C.A., Keogh, E.J.: Three myths about dynamic time warping data mining. In: SDM, pp. 506–510 (2005)
21. Ratanamahatana, C.A., Keogh, E.J.: Making time-series classification more accurate using learned constraints. In: SDM, pp. 11–22 (2004)
22. Sakoe, H., Chiba, S.: Dynamic programming algorithm optimization for spoken word recognition. IEEE Trans. Acoust., Speech, Signal Processing 26(1), 43–49 (1978)
23. Sakurai, Y., Faloutsos, C., Yamamuro, M.: Stream monitoring under the time warping distance. In: ICDE, pp. 1046–1055 (2007)
24. Sakurai, Y., Yoshikawa, M., Faloutsos, C.: FTW: fast similarity search under the time warping distance. In: PODS, pp. 326–337 (2005)
25. Salvador, S., Chan, P.: Toward accurate dynamic time warping in linear time and space. Intelligent Data Analysis 11(5), 561–580 (2007)
26. Seidl, T., Kriegel, H.P.: Optimal multi-step k-nearest neighbor search. In: SIGMOD, pp. 154–165 (1998)
27. Sellis, T.K.: Multiple-query optimization. ACM Trans. Database Syst. 13(1), 23–52 (1988)
28. Tok, W.H., Bressan, S.: Efficient and adaptive processing of multiple continuous queries. In: Jensen, C.S., Jeffery, K., Pokorný, J., Šaltenis, S., Hwang, J., Böhm, K., Jarke, M. (eds.) EDBT 2002. LNCS, vol. 2287, pp. 215–232. Springer, Heidelberg (2002)
29. Weber, R., Schek, H.J., Blott, S.: A quantitative analysis and performance study for similarity-search methods in high-dimensional spaces. In: VLDB, pp. 194–205 (1998)
30. Yang, D., Rundensteiner, E.A., Ward, M.O.: A shared execution strategy for multiple pattern mining requests over streaming data. PVLDB 2(1), 874–885 (2009)
31. Zhu, Y., Shasha, D.: Warping indexes with envelope transforms for query by humming. In: SIGMOD, pp. 181–192 (2003)

Probabilistic Time Consistent Queries over Moving Objects

Xiang Lian and Lei Chen

Department of Computer Science and Engineering
The Hong Kong University of Science and Technology
Hong Kong, China
{xlian,leichen}@cse.ust.hk

Abstract. Recently, the wide usage of inexpensive mobile devices, along with broad deployment of wireless and positioning technology, has enabled many important applications such as Delay Tolerant Networks (DTN). In these applications, the positions of mobile nodes are dynamically changing, and are often imprecise due to the inaccuracy of positioning devices. Therefore, it is crucial to efficiently and effectively monitor mobile nodes (modeled as uncertain moving objects). In this paper, we propose a novel query, called *probabilistic time consistent query* (PTCQ). In particular, a PTCQ retrieves uncertain moving objects that *consistently* satisfy query predicates within a future period with high confidence. We present effective pruning methods to reduce the search space of PTCQs, and seamlessly integrate them into an efficient query procedure. Moreover, to facilitate query processing, we specifically design a data structure, namely *UC-Grid*, to index uncertain moving objects. The structure construction is based on a formal cost model to minimize the query cost. Extensive experiments demonstrate the efficiency and effectiveness of our proposed approaches to answer PTCQs.

Keywords: probabilistic time consistent query, uncertain moving object database.

1 Introduction

Recently, the wide usage of mobile devices (e.g., mobile phones and PDAs), along with broad deployment of wireless networks and positioning technology (e.g., GPS), have given rise to many real applications such as *Delay Tolerant Networks* (DTN) [1]. Fig. 1 illustrates an example of DTN, in which each mobile node moves around, transmits replicas of packets to other passing mobile nodes, and eventually relays packets to their destinations. However, since such packet delivery may incur long delay, static base stations are usually deployed to speed up the delivery.

When a base station wants to deliver an important message to a destination, it can distribute the replicas of this message to passing mobile nodes, and then these nodes can spread the message through their surrounding nodes in the network until the destination is reached. Due to the limited bandwidth, the base station usually distributes the message to a limited number of (e.g., k) mobile nodes nearby. In other words, it is desirable for the base station to send the message to its k-nearest neighbors (e.g., 2-nearest neighbors, o_2 and o_3, in Fig. 1(a)). It is important to note that, in the real scenario, it can take

J.B. Cushing, J. French, and S. Bowers (Eds.): SSDBM 2011, LNCS 6809, pp. 168–188, 2011.
© Springer-Verlag Berlin Heidelberg 2011

some time to transfer the message between the base station and mobile nodes. Thus, the base station may only want to send the message to those passing mobile nodes that would *consistently* be the k-nearest neighbors (kNNs) of the base station (query point) within a future period of time. Here, "consistently" means mobile nodes remain to be kNNs of the base station for at least T consecutive timestamps, where T is the length of the time interval needed to deliver a message. Intuitively, these kNN nodes keep to be closest to the base station, and are thus more robust to receive messages.

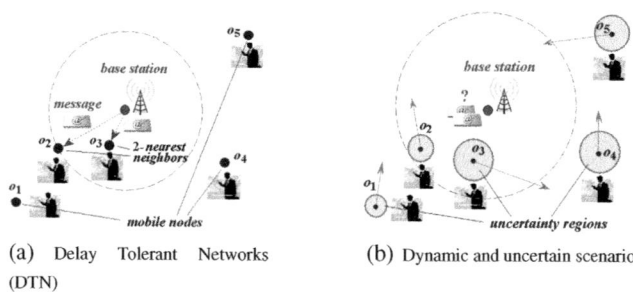

(a) Delay Tolerant Networks (DTN)

(b) Dynamic and uncertain scenario

Fig. 1. Illustration of Applications in Delay Tolerant Networks

There are many challenges to accurately and efficiently answer such a query in a real DTN environment. For example, positions of mobile nodes in DTN can be obtained by positioning devices such as GPS. However, due to the imperfect nature of GPS such as clock errors, ephemeris errors, atmospheric delays, and multipathing and satellite geometry, GPS data are inherently noisy and uncertain [15]. Thus, as shown in Fig. 1(b), the collected (uncertain) data can be modeled by *uncertainty regions* [5] in uncertain databases. Within an uncertainty region, the actual data can appear anywhere following any probabilistic distribution. Thus, in practice, we have to find k consistent nearest mobile nodes with imprecise positional data. It is therefore crucial to guarantee the accuracy of the returned answers. In addition, in Fig. 1(b), due to the movement of mobile nodes, we also need to deal with dynamically moving objects [22] rather than static ones (arrows in the figure indicate moving velocity vectors), which makes efficient monitoring of time consistent kNNs more challenging.

Inspired by the application above, in this paper, we propose an important and useful problem, namely *probabilistic time consistent query* (PTCQ), on uncertain moving objects. Given a query point q, a PTCQ obtains uncertain moving objects that are consistently being kNNs of q with high probabilities for at least T consecutive timestamps in a future period. PTCQ considers a generic data model that captures both uncertain locations and moving velocities[1] of objects in real applications (e.g., imprecise GPS data or changing speeds of GPS users). Moreover, the concept of "time consistent" is not limited to kNN, but can be extended to other query types (e.g., range queries), which we would like to leave as our future work.

---

[1] Our solution can be extended to the case of uncertain moving directions, which will be discussed in Section 5.3.

Different from previous works on continuously monitoring kNNs over moving objects [25,24,14], PTCQ has its own characteristics specific to applications such as DTN. First, PTCQ answers should be kNNs for at least T consecutive timestamps, whereas continuous kNN monitoring does not have this T constraint (i.e., return all possible kNNs within a period, no matter how long they last). Second, PTCQ provides the accuracy guarantee on uncertain data, whereas the existing works of kNN monitoring on certain moving objects cannot be directly applied to the uncertain case.

On the other hand, previous works on uncertain moving objects studied 1-nearest neighbor query (i.e., $k = 1$) either considering a completely different uncertainty model [5], or with a strong assumption of object distributions [21]. In contrast, our work focuses on probabilistic time consistent kNNs for arbitrary integer k (≥ 1), without any symmetric assumption of object distributions. In addition, Cheng et al. [4] studied k-nearest neighbor ($k \geq 1$) in static uncertain databases, and a static R-tree is built for efficient query processing, which is not space- and time- efficient for monitoring dynamically moving objects with high update rates in our PTCQ problem.

In this paper, we aim to tackle the problem of answering PTCQs efficiently and effectively. In particular, since PTCQ has to deal with uncertain data, whose computation involves numerical methods [5] at high cost, we provide effective pruning methods tailored to reducing the PTCQ search space, utilizing the time, velocity, and probabilistic constraints. Further, we propose a data structure, called *UC-Grid*, to index uncertain moving objects, on which we process PTCQ queries. We also give a cost model for the total PTCQ query cost, which can help determine the parameter of the data structure.

In the paper, we make the following contributions.

1. We formalize probabilistic time consistent kNN queries on uncertain moving objects in Section 3.
2. We propose effective pruning methods in Section 4 to filter out those false alarms of PTCQ candidates, considering time, velocity, and probabilistic constraints.
3. We design a data structure, *UC-Grid*, to index uncertain moving objects in Section 5.1, which is based on a formal cost model in Section 5.3 to achieve low query cost.
4. We integrate pruning methods into an efficient PTCQ procedure in Section 5.2.

Section 2 reviews kNN over certain/uncertain moving objects and uncertain static objects. Section 6 illustrates the experimental results. Section 7 concludes this paper.

2   Related Work

In spatial databases (with static and certain data), Roussopoulos et al. [17] proposed a branch-and-bound algorithm to retrieve kNNs of a query point q by traversing the R-tree [7] in a *depth-first* manner. Hjaltason and Samet [8] proposed the *best-first* algorithm to answer kNN queries on R-tree, which can achieve the optimal performance. Some other works [11,18] utilize range queries on R-tree to solve the kNN problem. In contrast, our work considers a different scenario where data uncertainty and object movement are inovlved, and thus previous methods cannot be directly used.

Many works on kNN query over (certain) moving objects often assume that future trajectories of objects are known at query time (e.g., expressed by a linear function

[23,20]). This is the model we use in our PTCQ problem (nonetheless, we also consider uncertain velocities). A TPR-tree [23] was proposed to index both object positions and velocities. A predictive kNN query can be answered by traversing the TPR-tree [19,10]. Other works (e.g., YPK-CNN [25], SEA-CNN [24], and CPM [14]) on continuously monitoring kNNs assume that object velocities are unknown, but object positions can be frequently updated. Grid index is usually used due to its low update cost compared with tree-based indexes [25]. In contrast, our PTCQ retrieves probabilistic time consistent kNNs, rather than kNNs, and we need to handle uncertain data instead of certain ones.

In uncertain databases, *probabilistic nearest neighbor query* (PNN) [5,12,3] was extensively studied, which retrieves uncertain objects that are nearest neighbors of a query point q with probability not smaller than a threshold. Cheng et al. [4] proposed probabilistic kNN queries in static uncertain databases, which obtain sets of k objects with high probabilities of being kNNs of q. These works consider static uncertain objects without movement, thus, static indexes such as R-tree can be constructed. In contrast, our PTCQ is conducted on moving objects, and static tree-based indexes are thus not space- and time- efficient in the dynamic environment with high update rates.

For uncertain moving objects, Cheng et al. [5] modeled the uncertainty introduced by the uncertain movement of objects in a future period, where each object is represented by either a line segment (linear movement) or a circle (free movement). A tree structure, called VCI, is used to index uncertain moving data. In contrast, our work distinguishes the spatial uncertainty from the velocity uncertainty. Thus, under a different uncertain model, their pruning/indexing methods cannot be directly applied to our problem. Trajcevski et al. [22] studied range query on uncertain moving objects, which returns objects being in a region with probabilities either equal to 1 or within $(0, 1)$. Chung et al. [6] worked on range query over 1D uncertain moving objects under the Brownian motion model. To our best knowledge, no previous work studied probabilistic time consistent kNN on 2D data, with the confidence guarantee, and with a generic uncertainty model with any object distributions. Further, Zhang et al. [26] considered range and top-k NN queries over uncertain moving objects, however, this work studied snapshot queries at a future query timestamp t, which is quite different from our PTCQ considering consistent kNNs in a future period. Thus, their proposed techniques for a particular timestamp cannot be directly applied to handing our PTCQ in a period.

3   Problem Definition

3.1   Data Model for Uncertain Moving Objects

Fig. 2 presents a data model for uncertain moving objects. In an uncertain moving object database (UMOD) $\mathcal{D}^U$, each object o at a timestamp t (e.g., $t = 0$) is represented by an *uncertainty region* [5] $UR(o)$. In this paper, we mainly focus on 2-dimensional spatial data, which are related to applications such as DTN [1]. The uncertainty region $UR(o)$ is modeled by a circle centered at C_o with radius $r_o{}^2$, in which object o can reside anywhere following any probabilistic distribution.

---

[2] Note that, here we simply assume uncertainty region has circle shape. Uncertainty regions of other shapes can be transformed to circles that tightly bound the regions.

By adopting the model used for "certain" moving objects [23,16], we assume that each uncertain object o moves linearly with velocity vector $\overline{v_o} = \langle v_o[x], v_o[y] \rangle$, before its new position and velocity are received, where $v_o[x]$ and $v_o[y]$ are the velocities of object o (projected) on x- and y-axes, respectively. Equivalently, the velocity vector $\overline{v_o}$ can be also denoted as $\langle v_o \cdot cos\gamma, v_o \cdot sin\gamma \rangle$. Here, γ is the angle between $\overline{v_o}$ and x-axis, showing the moving direction[3] of object o; and v_o is a random variable within $[v_o^-, v_o^+]$ [9], indicating the speed of object o. Note that, v_o^- and v_o^+ are the minimum and maximum possible velocities of object o, respectively, and the *probability density function* (pdf), $pdf_v(v_o)$, of variable v_o can be obtained by historical speed data of moving object o (reported by positioning devices such as GPS).

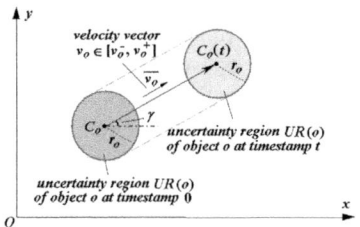

Fig. 2. Data Model for Uncertain Moving Objects

In the example of Fig. 2, with some possible velocity $v_o \in [v_o^-, v_o^+]$, the center of uncertain object o moves from position C_o to position $C_o(t)$ during the period $[0, t]$.

The semantic of our uncertain moving object model above has its root from the uncertainty model in static uncertain databases [12] and moving object model in moving object databases [10,19,23]. We also noticed other models for uncertain moving objects such as [15,5] in which the uncertainty of object locations directly results from the movement of objects within a short period. In contrast, our work distinguishes the uncertainty of spatial locations (introduced by positioning devices like GPS) from that of velocities. Based on this fine uncertainty model, we can predict future positions of objects more accurately, and thus forecast future events (including PTCQs). We would like to leave PTCQs (defined later) with other uncertain data models as our future work.

3.2   Definition of PTCQs

The novelty of PTCQs is in the concept of "time consistent", which, to our best knowledge, has not been studied before. That is, a PTCQ obtains objects that time-consistently satisfy some query predicates in a future time interval. Under different query predicates (e.g., range predicates), PTCQs can cover a broad family of probabilistic time consistent query types, which are useful for many real applications (e.g., mobile networks) that involve uncertain moving objects.

---

[3] Here, we consider a fixed moving direction for the ease of illustration. Our solutions can be easily extended to uncertain moving directions, capturing the case with non-linear moving trajectories, which will be discussed in Section 5.3.

Below, we focus on one typical query type, k-nearest neighbor (kNN) query, and leave other interesting query types as our future work. Specifically, given a UMOD $\mathcal{D}^U$ and a query point q, a probabilistic time consistent kNN query retrieves those uncertain moving objects o that are the kNNs of q with high probabilities, for at least T consecutive timestamps in a future period $[0, t_{ed}]$. We first give the definition of *probabilistic k-nearest neighbors* (PkNN) in a snapshot database at a specific timestamp t_i.

Definition 1. (*Probabilistic k-Nearest Neighbors, PkNN*) *Given a UMOD $\mathcal{D}^U$ at a timestamp t_i, a query point q, an integer k, and a probabilistic threshold $\alpha \in (0, 1]$, a probabilistic k-nearest neighbor (PkNN) of q is an object $o(t_i) \in \mathcal{D}^U$ such that $o(t_i)$ is a kNN of q s with the kNN probability, $Pr_{kNN}(q, o(t_i))$, not smaller than α, that is,*

$$Pr_{kNN}(q, o(t_i)) = \int_{v_o^-}^{v_o^+} pdf_v(v_o) \cdot \int_{o'(t_i) \in UR(o(t_i))} Pr\{dist(q, o'(t_i)) = r\}$$

$$\cdot \sum_{\forall S = \{p_1(t_i), \dots, p_s(t_i)\} \in \mathcal{D}^U \wedge s < k} \left(\left(\prod_{m=1}^{s} Pr\{dist(q, p_m(t_i)) \leq r\} \right) \right.$$

$$\left. \cdot \left(\prod_{p_n \in \mathcal{D}^U \setminus (S \cup \{o\})} Pr\{dist(q, p_n(t_i)) \geq r\} \right) \right) do'(t_i) dv_o \geq \alpha, \qquad (1)$$

where $\overline{o(t_i)} = \overline{o(0)} + \overline{v_o} \cdot t_i$ and velocity v_o ($\in [v_o^-, v_o^+]$) is a variable with pdf $pdf_v(v_o)$.

In Definition 1, an object o is a PkNN answer at timestamp t_i, if and only if its probability of being in the kNN set of q (in Inequality (1)) is not smaller than a given threshold α. In Inequality (1), the outer integral integrates over uncertain velocity, whereas the inner one integrates on possible positions, $o'(t_i)$, of o at timestamp t_i. Within the inner integral, the formula computes the probability that $o'(t_i)$ is one of kNNs of q, i.e., the probability that fewer than k (i.e., s) objects, $p_m(t_i)$, have distances to q smaller than $o'(t_i)$, and meanwhile the rest objects, $p_n(t_i)$, having distances to q never smaller than $o'(t_i)$. If the resulting kNN probability of object o is greater than or equal to threshold α, then o is a PkNN at timestamp t_i. Note that, the PkNN definition generalizes the *probabilistic nearest neighbor query* (PNN) [5] from $k = 1$ to $k \geq 1$.

After giving PkNNs at a timestamp t_i, we are now ready to define the probabilistic time consistent kNN problem.

Definition 2. (*Probabilistic Time Consistent Queries, PTCQ*) *Denote PkNN(t_i) as a set of PkNN objects at a timestamp t_i. Then, given a time constraint T and a future period $[0, t_{ed}](ed + 1 \geq T)$, a probabilistic time consistent query (PTCQ) obtains uncertain moving objects o such that there exists a period $[t_j, t_j + T - 1] \subseteq [0, t_{ed}]$, and $o(t_i)$ is in PkNN(t_i) for all consecutive timestamps $t_i \in [t_j, t_j + T - 1]$, that is,*

$$PTCQ(t_i) = \bigcup_{t_j=0}^{t_{ed}-T+1} \left(\bigcap_{t_i=t_j}^{t_j+T-1} PkNN(t_i) \right). \qquad (2)$$

Note that, due to time and probabilistic constraints, the PTCQ in Definition 2 may not return exactly k answers.

Challenge. From Definition 2, a PTCQ obtains uncertain moving objects that are PkNNs for at least T consecutive timestamps within a future period $[0, t_{ed}]$. Clearly, one straightforward way to answer PTCQs is to directly calculate PkNN answers (i.e., checking Inequality (1)) at each timestamp t_i in period $[0, t_{ed}]$, and then combine the resulting PkNN answers via Eq. (2) in Definition 2. However, the cost of computing the PkNN set is very costly (i.e., $O\left(\sum_{s=1}^{k-1}\binom{N}{s}\right)$ for database size N), involving complex double integral to compute kNN probabilities (in Inequality (1)) via numerical methods [5]. Thus, one major challenge of our PTCQ problem is to improve the efficiency of retrieving probabilistic time consistent kNNs in a dynamic environment with data uncertainty.

4 PTCQ Search over Uncertain Moving Objects

4.1 T-Pruning

Fig. 3 illustrates the heuristics of our T-pruning method by a simple example. Assume that at timestamp 0, we can find k uncertain objects $p_1(0)$, $p_2(0)$, ..., and $p_k(0)$ which are close to a query point q. Since we know the positions and velocities of these k objects (according to the data model mentioned in Section 3.1), we can obtain the maximum possible distance, R_{max}, from these k objects to q in the future period $[0, t_{ed}]$.

As illustrated in Fig. 3, we draw a circle, denoted as $\odot q$, centered at query point q and with radius R_{max}. Clearly, any uncertain object o that is always outside $\odot q$ in period $[0, t_{ed}]$ cannot be kNN of q (due to the existence of k objects $p_1 \sim p_k$). Thus, in turn, object o cannot be the PTCQ answer. Further, if the moving path of any object o passes $\odot q$ for less than T timestamps, then we can prune object o safely as well.

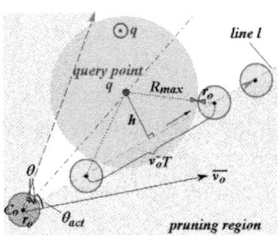

Fig. 3. Heuristics of T-Pruning Method

In the example of Fig. 3, assume object o is completely outside $\odot q$ at timestamp 0. Let θ be the angle between vector $\overline{C_o q}$ and a possible moving path of o which intersects with $\odot q$ for exactly T timestamps (at the lowest speed v_o^-). As a result, if the actual moving direction (i.e., velocity vector $\overline{v_o}$) of object o has the angle, θ_{act}, with vector $\overline{C_o q}$ greater than θ, we can prune object o, which is our basic idea of T-pruning below.

Lemma 1. (T-Pruning) *Let η_q be the angle satisfying $q[x] = ||q|| \cdot cos\eta_q$ and $q[y] = ||q|| \cdot sin\eta_q$, where $||q|| = \sqrt{q[x]^2 + q[y]^2}$. Then, any uncertain object o can be safely pruned, if it holds that:*

$$||q|| \cdot cos(\eta_q - \gamma) < C_o[x] \cdot cos\gamma + C_o[y] \cdot sin\gamma, \text{ or} \tag{3}$$

$$||q|| \cdot sin(\gamma - \eta_q) - (C_o[x] \cdot sin\gamma - C_o[y] \cdot cos\gamma) > (R_{max} + r_o)^2 - (v_o^- \cdot T)^2/4. \tag{4}$$

where $R_{max} + r_o > (v_o^- \cdot T)/2$.

From Lemma 1, we can prune objects o that definitely do not satisfy the T constraint. In the example of Fig. 3, we can safely prune those objects having center trajectories (i.e., C_o) within the shaded region filled with lines (i.e., satisfying $\theta_{act} > \theta$).

The Computation of Radius R_{max}. Up to now, the only issue that remains to be addressed is how to obtain the radius R_{max} of $\odot q$. We observe that the maximum distance from any moving object p_i to query point q within a period $[0, t_{ed}]$ is always achieved at timestamp 0 or t_{ed}. Therefore, we can obtain R_{max} by:

$$R_{max} = max_{j=1}^{k}\{max\{maxdist(q, p_j(0)), maxdist(q, p_j(t_{ed}))\}\} \tag{5}$$

where $maxdist(\cdot, \cdot)$ is the maximum possible Euclidean distance between two objects.

4.2 Period Pruning

After applying T-pruning method, we can obtain a set of PTCQ candidates, we next propose our second pruning method, period pruning, to further reduce the search space.

Recall that, the PTCQ (given in Definition 2) specifies a future period $[0, t_{ed}]$ within which we need to find probabilistic time consistent kNNs. This period can be a time slot to repeatedly send the message to passing mobile nodes, in mobile applications. Different from T-pruning that considers the pruning by only using the moving directions, our period pruning is to filter out those PTCQ false alarms either due to the cut-off timestamp t_{ed} or due to the initial positions of objects at timestamp 0.

As shown in Figs. 4(a) and 4(b), we consider two cases where uncertain object o moves (from outside) into $\odot q$ (case 1) and moves (from inside) out of circle $\odot q$ (case 2), respectively. Due to the period constraint, if object o cannot reside in circle $\odot q$ at the beginning or end of period $[0, t_{ed}]$ for at least T timestamps, o can be safely pruned.

Below, we derive the conditions of period pruning for the two cases above. In the first case where object o is initially outside the circle $\odot q$ at timestamp 0, we consider the extreme situation (such that o can stay in $\odot q$ for T timestamps within period $[0, t_{ed}]$), as

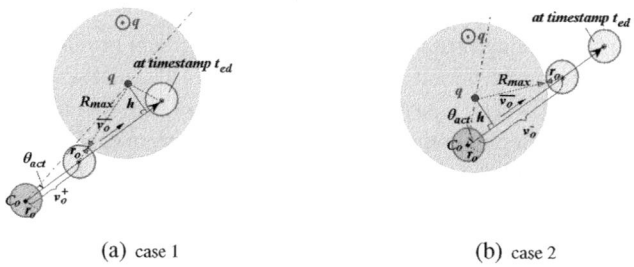

(a) case 1   (b) case 2

Fig. 4. Heuristics of Period Pruning

shown in Fig. 4(a). That is, object o first moves at its highest speed v_o^+ until it intersects with $\odot q$, and then switches to its lowest speed v_o^- to move for T timestamps. After that, if the timestamp is already after t_{ed}, then it indicates that object o cannot stay in $\odot q$ for at least T timestamps, and thus o can be safely pruned; otherwise, o is a candidate.

Similarly, in the second case where object o overlaps with circle $\odot q$ at timestamp 0 (as shown in Fig. 4(b)), we also consider the extreme situation that o moves at its lowest speed v_o^- until it does not intersect with $\odot q$. If this period has length smaller than T, then object o can be also pruned. We summarize the period pruning method below.

Lemma 2. (*Period Pruning*) *Any uncertain object o can be pruned, if it holds that:*

$$
\begin{cases}
||q|| \cdot cos(\eta_q - \gamma) - (C_o[x] \cdot cos\gamma + C_o[y] \cdot sin\gamma) < \frac{v_o^{+2} \cdot (t_{ed} - T)^2 - (R_{max} + r_o)^2 + dist^2(q, C_o)}{2 \cdot v_o^{+2} \cdot (t_{ed} - T)}; \\
\qquad \text{if } dist(q, C_o) > R_{max} + r_o, \\
||q|| \cdot cos(\eta_q - \gamma) - (C_o[x] \cdot cos\gamma + C_o[y] \cdot sin\gamma) < \frac{v_o^{-2} \cdot T^2 - (R_{max} + r_o)^2 + dist^2(q, C_o)}{2 \cdot v_o^{-2} \cdot T}, \\
\qquad \text{otherwise.}
\end{cases}
\tag{6}
$$

Therefore, based on Lemma 2, we can use the period pruning method to prune those false alarms (which cannot be filtered out by T-pruning) that stay in $\odot q$ at the beginning or end of the future period $[0, t_{ed}]$ for less than T timestamps.

4.3   Segment Pruning

Up to now, we have discussed T-pruning and period pruning such that objects staying in circle $\odot q$ (centered at query point q with radius R_{max}) for less than T timestamps. We further consider reducing the PTCQ search space by proposing the segment pruning.

We have an interesting observation that, the radius R_{max} of circle $\odot q$ is only the maximum possible distance from k initial objects p_i to query point q within the *entire* period $[0, t_{ed}]$ (as given in Eq. (5)). Thus, if we consider each timestamp t_i in the period $[0, t_{ed}]$, the maximum possible distance, denoted as $R_{max}(t_i)$, from these k objects to q may change over time. As an example in Fig. 5(a), the size of the circle $\odot q$ may shrink during the period $[0, t_{ed}]$, say radius varying from $R_{max}(0)$ (assuming $R_{max} = R_{max}(0)$ in this example) to $R_{max}(t_{ed})$. In this case, although object o is within the circle $\odot q$ (with radius R_{max}) for at least T timestamps (i.e., cannot be pruned by T- or period pruning), we may still have chance to prune this object, due to the shrinking of circle $\odot q$. For instance, when o starts to intersect with circle $\odot q$ (with radius $R_{max}(0)$) at timestamp t_i, the actual radius is shrinking to $R_{max}(t_i)$ which is smaller than $R_{max}(0)$. Thus, it is possible that object o does not even intersect with the time-varying circle $\odot q$ during the entire period $[0, t_{ed}]$.

Based on this interesting observation, we design a segment pruning method. Specifically, we divide the future period $[0, t_{ed}]$ into segments of equal size $(T - 1)$, that is, $[0, T - 1), [T - 1, 2(T - 1)), ...,$ and $[t_{ed} - T + 1, t_{ed}]^4$. Then, for each segment period $[t_j, t_j + T - 1)$, we can obtain the maximum possible distance, $R_{max}(t_j)$, from the k initial objects p_i to q, and check whether or not a candidate o is completely outside the circle centered at q with radius $R_{max}(t_j)$.

---

[4] Note that, the last segment might have size smaller than $(T - 1)$, which, however, would not affect the correctness of our proposed segment pruning. For simplicity, in this paper, we always assume that $(t_{ed} + 1)$ is a multiple of $(T - 1)$.

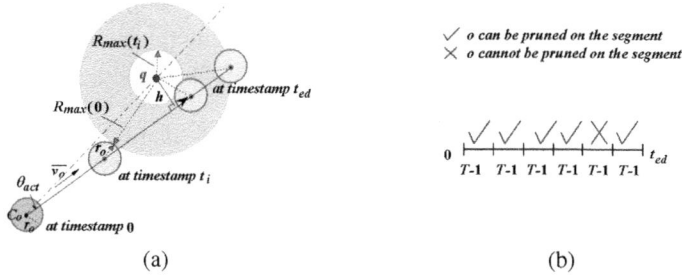

Fig. 5. Heuristics of Segment Pruning

As illustrated in Fig. 5(b), if object o is completely outside circle with in a segment, we say o can be pruned on this segment; otherwise, o cannot be pruned. One interesting observation is that, if object o cannot be pruned on two *consecutive* segments (of size $(T-1)$ each), then o is a possible PTCQ candidate (since o might be probabilistic consistent kNN of q during these $2(T-1)$ timestamps); otherwise, object o can be safely pruned. Here, the intuition of the pruning is that, each segment has length $(T-1)$, whereas the time constraint in PTCQ is at least T. Thus, even if object o cannot be pruned on a segment, as long as o can be pruned on its adjacent segments, o can still be pruned by our segment pruning. This way, we can prune object o in the example of Fig. 5(b), since it cannot be pruned only on a single segment of length $< T$.

We summarize the segment pruning in the following lemma.

Lemma 3. *(Segment Pruning) Assume that we divide $[0, t_{ed}]$ into segments, $[t_j, t_j + T - 1)$. Then, we say any uncertain moving object o can be pruned on segment $[t_j, t_j + T - 1)$, if it holds that, for any $v_o \in [v_o^-, v_o^+], t \in [t_j, t_j + T - 1)$:*

$$v_o^2 t^2 - 2 \cdot v_o \cdot (\|q\| \cdot \cos(\eta_q - \gamma) - (C_o[x] \cdot \cos\gamma + C_o[y] \cdot \sin\gamma)) \cdot t + dist^2(q, C_o)$$
$$- (R_{max}(t_j) + r_o)^2 > 0. \qquad (7)$$

Thus, any object o can be pruned, if o cannot be pruned on any two adjacent segments.

4.4   Filtering with Velocity Distributions

In this and next subsections, we will use velocity and position distributions of uncertain moving objects, respectively, to derive the probability upper bound (i.e., β) and filter out false alarms (if this upper bound is smaller than threshold α). Below, we first discuss how to utilize the velocity parameter to reduce the search space of PTCQs.

As mentioned in Section 3, the velocity v_o of an object o is a random variable within $[v_o^-, v_o^+]$, where v_o^- and v_o^+ are the minimum and maximum possible velocities of o. The distribution of velocity variable v_o can be obtained via historical data (e.g., a histogram containing historical velocity data reported by GPS in mobile applications [1]).

To obtain PTCQ answers, we want to find objects o that are PkNNs (given by Definition 1) with high confidence, for some consecutive timestamps. In other words, at each of consecutive timestamps, t_i, objects o must satisfy Inequality (1), that is, o's kNN probabilities, $Pr_{kNN}(q, o(t_i))$, should be greater than or equal to threshold $\alpha \in (0, 1]$. Due to the high cost of computing $Pr_{kNN}(q, o(t_i))$ directly (involving integral via numerical methods [5]), we propose a filtering method using velocity information. Our basic idea is to compute an upper bound, $UB\_Pr_{kNN}(q, o(t_i))$, of the kNN probability $Pr_{kNN}(q, o(t_i))$ at a low cost. This way, as long as it holds that $UB\_Pr_{kNN}(q, o(t_i)) < \alpha$, we can infer $Pr_{kNN}(q, o(t_i)) < \alpha$, and thus o can be pruned. We summarize our filtering method via velocity distributions below, and prove its correctness.

Lemma 4. (*Filtering with Velocity Distributions*) *Let* $[v_{omin}^{1-\beta}, v_{omax}^{1-\beta}]$ *be a velocity interval in* $[v_o^-, v_o^+]$, *such that:*

$$\int_{v_{omin}^{1-\beta}}^{v_{omax}^{1-\beta}} pdf_v(v_o)dv_o = 1 - \beta. \tag{8}$$

Then, for $\beta < \alpha$, *by using* $v_{omin}^{1-\beta}$ *and* $v_{omax}^{1-\beta}$ *instead of* v_o^- *and* v_o^+, *respectively, in period and segment pruning (in Lemmas 2 and 3, respectively), we can still prune objects.*

Discussions. Lemma 4 shows that we can use the velocity distribution to reduce the search space. In practice, we can maintain a histogram summarizing historical velocity data, for example, the velocity distribution of a mobile node in DTN applications [1]. To enable pruning with velocity, we need to compute velocity interval $[v_{omin}^{1-\beta}, v_{omax}^{1-\beta}]$, for a given α value. Here, we adopt a simple heuristic to set this interval. That is, if object o does not intersect with $\odot q$, then we let $v_{omin}^{1-\beta} = v_o^-$ and continue to find $v_{omax}^{1-\beta}$ such that Eq. (8) holds for $\beta = \alpha$. The intuition is that, by using the velocity interval containing small values, object o is unlikely to reach $\odot q$, and thus has more chance to be pruned by period or segment pruning method. Similarly, if o intersects with $\odot q$, we let $v_{omax}^{1-\beta} = v_o^+$ and find $v_{omin}^{1-\beta}$ satisfying Eq. (8) for $\beta = \alpha$. We use large velocity values such that o can quickly leave $\odot q$ and is more likely to be pruned.

4.5   Filtering with Object Distributions

We next derive another upper bound, β, of the kNN probability from position distributions of objects (orthogonal to velocity distributions in Section 4.4), and prune objects with $\beta < \alpha$. Specifically, we adopt a notion of $(1 - \beta)$ hypersphere proposed in [13].

Definition 3. (*$(1 - \beta)$-Hypersphere [13]*) *Given an uncertain object* o, *a* $(1 - \beta)$-*hypersphere* $o_{1-\beta}$ *is a hypersphere, with center* C_o *and radius* $r_o^{1-\beta}$ *such that* o *resides in* $o_{1-\beta}$ *with probability at least* $(1 - \beta)$.

From Definition 3, we have the lemma below for filtering with object distributions.

Lemma 5. (*Filtering with Object Distributions*) *For* $\beta < \alpha$, *by replacing uncertain object* o *with* $(1 - \beta)$-*hypersphere*, $o_{1-\beta}$, *we can still use* T-, *period, and segment pruning (in Lemmas 1, 2, and 3, respectively) to prune objects safely.*

5   PTCQ Processing

5.1   Data Structure

We next propose a data structure *UC-Grid* for indexing uncertain moving objects, on which our probabilistic time consistent kNN queries can be answered. Specifically, as illustrated in Fig. 6, we divide a 2-dimensional data space $[0, 1]^2$ into $1/\delta^2$ ($\delta < 1$) cells with side length δ, where the setting of the δ value will be discussed in Section 5.3. For each uncertain moving object o, we say that o belongs to a cell if its center location C_o is in that cell. Additional information of objects is also stored in cells, which will be later described. This way, a grid index can be built for uncertain moving objects. Note that, the reason that we choose grid as a basic indexing structure is that it has $O(1)$ time complexity for retrieval and update, which is suitable for dynamic scenario with moving objects. There are also some studies [25] indicating that the grid-based index can result in better query performance on moving objects than tree-based indexes.

Since PTCQ is over moving and uncertain objects, our *UC-Grid*, $\mathcal{I}$, has to store more information. In each cell $c_{i,j}$, we distribute objects into 4 *center lists*, NE, NW, SW, and SE, based on their moving directions with angles $\gamma \in [0, \frac{\pi}{2}), [\frac{\pi}{2}, \pi), [\pi, \frac{3\pi}{2})$, and $[\frac{3\pi}{2}, 2\pi)$, respectively. The 4 center lists naturally represent the 4 quadrants in the grid, which can be easily checked and reduce the search space. Assuming circle $\odot q$ is completely to the top-right direction (NE) of a cell $c_{i,j}$, we only need to access objects in its center list NE, and save the cost of visiting the other 3 lists. In each center list, say NE, we store an object list with objects $(oid, C_o, r_o, \overline{v_o})$, where oid is the object ID, C_o is the center of uncertainty region $UR(o)$, r_o is the radius of $UR(o)$, and $\overline{v_o}$ is the velocity vector of o. Moreover, we also maintain statistics for all objects in list NE, including the minimum/maximum possible velocity v_o (v_{min}^{NE} and v_{max}^{NE}, respectively), angle γ (γ_{min}^{NE} and γ_{max}^{NE}, respectively), radius r_o (r_{omin}^{NE} and r_{omax}^{NE}, respectively), and other statistics (e.g., min_S_1 and max_S_2, discussed later in Section 5.2).

In addition to 4 center lists in cell $c_{i,j}$, we also keep a *cell list* with entries in the form $(cell\_id, count)$, where $cell\_id$ is the id of a cell that contains at least one object intersecting with cell $c_{i,j}$, and $count$ is the number of such objects (when $count=0$, the entry is removed). Since we decide the cell of an object only by its center location, the cell list of $c_{i,j}$ provides an inverted index of cells containing objects that may overlap

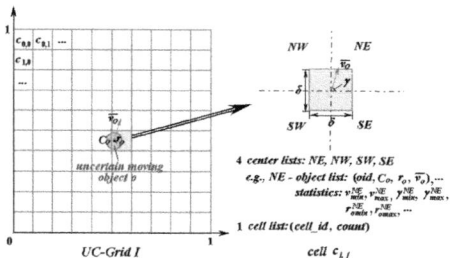

Fig. 6. Illustration of *UC-Grid* Data Structure

with cell $c_{i,j}$, which can be used for PTCQ processing. The dynamic maintenance of *UC-Grid* upon insertions and deletions can be found in Appendix A.

Pruning Heuristics for Cells. Similar to pruning methods in Section 4, we can design pruning rules to prune a group of objects in cells. Please refer to details in Appendix B.

5.2   Query Processing

PTCQ Query Procedure. We next present the procedure of PTCQ processing. We propose a concept of *influence region* (IR), having the property that only those cells intersecting with IR are possible to contain PTCQ answers. In particular, as illustrated in Fig. 7, the influence region is defined as a circle centered at query point q and with radius $(R_{max} + V_{max} \cdot (t_{ed} - T + 1))$, where V_{max} is the maximum speed for all the uncertain moving objects in the data space. In the lemma below, we prove that objects o that are completely outside IR at timestamp 0 cannot be PTCQ answers.

Lemma 6. *(Influence Region) Any uncertain moving object o that does not intersect with influence region at timestamp 0 cannot be PTCQ answer in future period $[0, t_{ed}]$.*

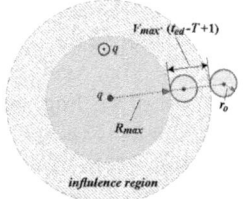

Fig. 7. Illustration of Influence Region

From Lemma 6, PTCQ processing only needs to access a subset of cells in *UC-Grid*. Fig. 8 illustrates the pseudo code of our PTCQ query procedure PTCQ_Processing, which consists of three steps, initialization, pruning, and refinement steps. In the initialization step, we first find k uncertain objects p_i $(1 \leq i \leq k)$ at timestamp 0 which have centers closest to query point q (line 1). The radius R_{max} of $\odot q$ is set to the maximum possible distance from p_i to q in future period $[0, t_{ed}]$ (given by Eq. (5); line 2). Then, in the pruning step, we first retrieve all cells that intersect with the influence region (as defined above), which may contain PTCQ candidates (guaranteed by Lemma 6; line 3). For each retrieved cell, we apply the T-pruning, period, and segment pruning rules for cells given in Section 4 (line 4). Next, we start to check objects within candidate cells by using the pruning methods mentioned in Section 4 (lines 5-7). Finally, in the refinement step, we refine all the remaining candidates by calculating their actual kNN probability (given in Inequality (1) at every timestamp, and obtaining the PTCQ results based on Definition 2. Actual PTCQ answers are reported as the output of PTCQ_Processing.

Refinement Step. In line 8 of procedure PTCQ_Processing, we have to refine candidates by computing the kNN probability $Pr_{kNN}(q, o(t_i))$ in Inequality (1) at each timestamp t_i in the period $[0, t_{ed}]$. This is clearly inefficient due to the double integral

Procedure PTCQ_Processing {
    Input: an uncertain moving object database $\mathcal{D}^U$, a *UC-Grid* $\mathcal{I}$, a query point q, a time constraint T, a future period
           $[0, t_{ed}]$, and a probabilistic threshold $\alpha \in (0, 1]$
    Output: a set of probabilistic time consistent kNN answers
    // *initialization step*
    (1) retrieve k objects p_i that are closest to query point q
    (2) let R_{max} be the maximum distance from p_i to query point q in period $[0, t_{ed}]$
    // *pruning step*
    (3) retrieve all the cells intersecting with the influence region // Lemma 6;
    (4) apply T-, period, and segment pruning rules to prune each retrieved cell $c_{i,j}$ // Section 4;
    (5) for each remaining cell $c_{i,j}$
    (6)   for each object $o \in c_{i,j}$
    (7)   apply pruning methods to prune object o // Lemmas $1 \sim 5$
    // *refinement step*
    (8) refine the remaining candidates based on PTCQ definition
    (9) return the actual PTCQ answers
}

Fig. 8. Procedure of Probabilistic Time Consistent Query

(via numerical methods [5]) in Inequality (1). In addition, we also have to enumerate $(\sum_{s=1}^{k-1} \binom{N}{s})$ object combinations (i.e., $p_m(t_i)$) in Inequality (1), which requires high cost. Therefore, we give a recursive method below to compute kNN probabilities with only linear cost (i.e., $O(N)$).

We denote $G(W, k, r)$ as the probability that among W objects we have seen, there are fewer than k objects with distances to q smaller than r and the rest objects with distances never smaller than r. We then rewrite $Pr_{kNN}(q, o(t_i))$ in Inequality (1) as:

$$Pr_{kNN}(q, o(t_i)) = \int_{v_o^-}^{v_o^+} pdf_v(v_o) \cdot \left(\int_{o'(t_i) \in UR(o(t_i))} Pr\{dist(q, o'(t_i)) = r\} \cdot G(N-1, k, r)do'(t_i) \right) dv_o \quad (9)$$

where $G(W, k, r) = G(W-1, k, r) \cdot Pr\{dist(q, p_W(t_i)) \geq r\} + G(W-1, k-1, r) \cdot Pr\{dist(q, p_W(t_i)) < r\}$ (base case: $G(W, 1, r) = \prod_{j=1}^{W} Pr\{dist(q, p_j(t_i)) \geq r\}$ and $G(k-1, k, r) = 1$). Thus, the time complexity of computing recursive function $G(N-1, k, r)$ in Eq. (9) is $O(N)$.

Optimization. We can further reduce the cost of computing $G(N-1, k, r)$, by *not* considering those objects $p_W(t_i)$ that definitely have distances to q greater than r (since in this case $G(W, k, r) = G(W-1, k, r)$). To achieve this goal, let $maxdist_{cand}$ be the maximum possible distance from all the candidates to q, and let S_{rfn} be a set containing objects having minimum distances to q smaller than $maxdist_{cand}$. Then, our problem of computing $G(N-1, k, r)$ can be reduced to the one of calculating $G(S_{rfn}, k, r)$, which only requires $O(|S_{rfn}|)$ time complexity, where $|S_{rfn}| \ll N$.

5.3 Cost Model

Up to now, we always assume that our PTCQ processing is conducted on *UC-Grid* with cells of size $\delta \times \delta$. However, it is not discussed how to set the parameter δ. Below, we will propose a cost model to formalize PTCQ processing cost, and aim to set appropriate value of parameter δ such that the query cost can be as low as possible.

In particular, the pruning cost in PTCQ procedure consists of two parts, the pruning of the retrieved cells and that of candidates in cells. Since we retrieve those cells intersecting with the influence region (line 3 of procedure PTCQ_Processing), in the worst case, we need to retrieve $\frac{\pi \cdot (R_{max} + V_{max} \cdot (t_{ed} - T + 1) + r_o^{max} + \delta)^2}{\delta^2}$ cells, where r_o^{max} is the maximum possible radius r_o of any uncertain object o. Further, in these cells, the number of object centers in them, $\overline{nb}$, can be estimated by the *power law* [2], considering the *correlation fractal dimension* D_2 of object centers in the 2D data space. In particular, we have $\overline{nb} = (N - 1) \cdot (\pi \cdot (R_{max} + V_{max} \cdot (t_{ed} - T + 1) + r_o^{max} + \delta)^2)^{D_2/2}$. Note that, we use the power law for estimating the number of object centers, since this law is applicable not only to uniform data but also to many other nonuniform data in real applications (e.g., $Zipf$) [2]. As a result, we can obtain the worst-case pruning cost, $cost_{con-kNN}$, of PTCQ processing below.

$$cost_{con-kNN} = \frac{\pi \cdot (C + \delta)^2}{\delta^2} + (N - 1) \cdot (\pi \cdot (C + \delta)^2)^{D_2/2}. \tag{10}$$

where $C = R_{max} + V_{max} \cdot (t_{ed} - T + 1) + r_o^{max}$. Moreover, since R_{max} is the radius of $\odot q$ containing k objects with centers closest to q, by applying the power law again, we have: $(N - 1) \cdot (\pi \cdot (R_{max} - r_o^{max})^{D_2/2} = k$. Thus, R_{max} can be given by $\sqrt{(\frac{k}{N-1})^{2/D_2} \cdot \frac{1}{\pi}} + r_o^{max}$.

We aim to find appropriate δ value for *UC-Grid* which can achieve low $cost_{con-kNN}$ in Eq. (10). Thus, we take the derivative of $cost_{con-kNN}$ with respect to δ, and let it equal to 0, i.e., $\frac{\partial cost_{con-kNN}}{\partial \delta} = 0$, which can be simplified as:

$$(C + \delta)^{D_2-2} \cdot \delta^3 = \frac{2\pi^{1-D_2/2} \cdot C}{N - 1} \tag{11}$$

Thus, we can collect statistics (e.g., C and D_2 in Eq. (11)) from historical data and query logs to estimate appropriate value of δ such that Eq. (11) holds. The resulting δ can achieve low query processing cost, based on our cost model. In the case where we do not have such statistics, we can only assume that data are uniform with $D_2 = 2$, and thus, from Eq. (11), we have $\delta = \sqrt[3]{\frac{2C}{N-1}}$. Then, after statistics are collected, we can reconstruct *UC-Grid* by using appropriate δ according to the available statistics.

Discussions on Uncertain Directions. Our solutions can be easily extended to prune objects with uncertain moving directions. Please refer to details in Appendix C.

6 Experimental Evaluation

In this section, we test PTCQ performance on real and synthetic data sets. Specifically, for synthetic data sets, we generate each uncertain moving object o at timestamp 0 as follows. First, we pick up a point in a data space $[0, 1] \times [0, 1]$ as center, C_o, of object o. Then, we generate the radius, $r_o \in [r_{min}, r_{max}]$, of its uncertainty region, $UR(o)$, for object o. Next, we produce its velocity vector $\overline{v_o}$, by randomly selecting an interval, $[v_o^-, v_o^+]$, of moving velocity v_o within $[V_{min}, V_{max}]$, and generating a moving angle (between moving direction and x-axis), $\gamma \in [0, 2\pi)$, where V_{min} and V_{max} are the minimum and maximum velocity of moving objects, respectively. Here, we consider center *location* C_o following either *Uniform* or *Skew* (with skewness 0.8) distribution (denoted as lU and lS, respectively), and *radius* r_o following either *Uniform* or *Gaussian* (with mean $(r_{min} + r_{max})/2$ and variance $(r_{max} - r_{min})/5$) distribution (denoted as rU and rG, respectively). Thus, we obtain four types of synthetic data sets, $lUrU$, $lUrG$,

$lSrU$, and $lSrG$. For each object o in these data sets, we generate 20 random samples to represent its position distribution in $UR(o)$, and 10 velocity samples within $[v_o^-, v_o^+]$ to represent its velocity distribution. Note that, for data sets with other distributions or parameters (e.g., mean, variance, skewness, or sample size), the query results have similar trend, and thus we do not report all of them here. For the real data, we test a 2D spatial data set, CA, which contains nodes of California Road Network obtained from Digital Chart of the World Server [*http://www.maproom.psu.edu/dcw/*]. We consider each data point in CA as the center C_o of an uncertain moving object o at timestamp 0, and simulate the uncertainty region $UR(o)$ (note: in real DTN applications [1], this uncertainty may come from the inaccuracy of GPS), as well as the moving velocity $\overline{v_o}$, resulting in two data sets, CA_rU and CA_rG, with Uniform and Gaussian distributions of r_o, respectively. For each of the real/synthetic data sets above, we construct a *UC-Grid* structure (mentioned in Section 5.1) over the uncertain moving objects. Then, we also randomly generate 50 query points in the data space to evaluate the PTCQ query performance within a future period $[0, t_{ed}]$.

To our best knowledge, no previous work has studied PTCQ in uncertain moving databases. Thus, we compare the performance of our PTCQ approach, $con\text{-}kNN$, with a straw-man, $Basic$, discussed in Section 3.2, which first computes PkNN answers at each timestamp and then combines results. For fair comparisons, instead of scanning the entire database, $Basic$ computes PkNN answers via *UC-Grid* by retrieving candidate objects intersecting with the influence region. We measure the PTCQ performance, in terms of *filtering time* and *speed-up ratio*. The filtering time is the running time of our pruning methods (mentioned in Section 4), whereas the speed-up ratio is defined as the total time cost of $Basic$ divided by that of our $con\text{-}kNN$ approach.

Table 1 depicts our parameter settings, where numbers in bold font are *default values* of parameters. In the sequel, we only present results with default future period $[0, t_{ed}]=[0, 20]$ and radius range $[r_{min}, r_{max}]=[0, 0.0005]$. We omit similar results with other settings due to the space limitation. Each time we vary one parameter, while setting other parameters to default values. All experiments are conducted on a Pentium IV 3.2GHz PC with 1G memory. The reported results are the average of 50 queries.

Table 1. The Parameter Settings

Parameters	Values
α	$0.1, 0.2, \mathbf{0.5}, 0.8, 0.9$
$[V_{min}, V_{max}]$	$[0, 0.0005], [0, 0.0008], \mathbf{[0, 0.001]}, [0, 0015], [0, 002]$
k	$5, 8, \mathbf{10}, 15, 20$
T	$4, 6, \mathbf{8}, 10, 12$
N	$10K, 20K, \mathbf{30K}, 40K, 50K$

Comparison of the Filtering Effect. We first compare the filtering effect of $Basic$ and $con\text{-}kNN$ in Fig. 9 over 6 real/synthetic data sets (note: here velocity or object distribution is not used to facilitate the pruning, whose effect will be reported in the next set of experiments), in terms of the number of PTCQ candidates to be refined. Recall from Section 5.2 that, $con\text{-}kNN$ applies T-, period, and segment pruning methods on cells and objects, which can greatly reduce the search space; in contrast, $Basic$ has to refine all objects overlapping with the influence region. In the figure, $con\text{-}kNN$

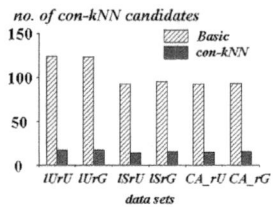

Fig. 9. Filtering Effect vs. Data Sets

has much fewer candidates to be refined, compared with $Basic$, where the difference between the two shows better filtering power of our $con\text{-}kNN$ approach.

Performance vs. Probabilistic Threshold α. Fig. 10 illustrates the effect of threshold α on real and synthetic data, where α varies from 0.1 to 0.9. From experimental results, we find that the filtering time remains low (i.e., about 10^{-4} second). Moreover, the speed-up ratio increases with the increasing α value. This is because PTCQ answers are those objects with kNN probabilities not smaller than α for T consecutive timestamps, and our pruning methods can utilize velocity and object distributions to filter out false alarms via α constraint (i.e., with probabilities $< \alpha$ for at least one out of any T consecutive timestamps). Thus, larger α would result in fewer candidates to be refined, and in turn higher speed-up ratio. In subsequent experiments, since the trend of results on real data is similar to that on synthetic ones, we only report results on synthetic data.

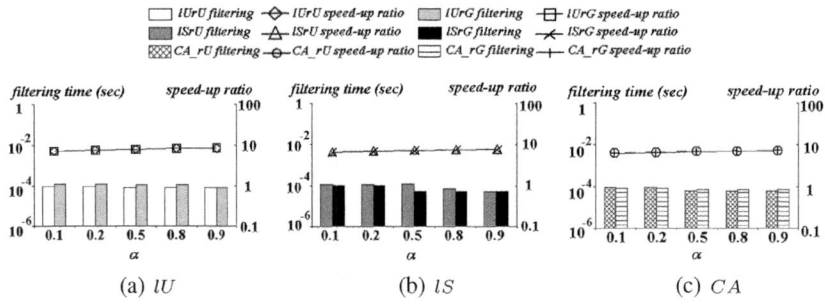

Fig. 10. Performance vs. Probabilistic Threshold α

Performance vs. Velocity Range $[V_{min}, V_{max}]$. Fig. 11 presents the PTCQ performance, where $[V_{min}, V_{max}]$ varies from $[0, 0.0002]$ to $[0, 0.001]$. In figures, the filtering time of $con\text{-}kNN$ slightly increases when the range becomes wider. This is reasonable, since the wider range results in larger influence region and more candidates to be processed. Meanwhile, the speed-up ratio compared with $Basic$ also increases, because $con\text{-}kNN$ uses effective filtering methods and $Basic$ has more candidates to refine. We also did experiments by varying k (from 5 to 20) and the time constraint T (from 4 to 12) with the trends similar to that in Figs. 11 and 10, respectively.

Performance vs. N. Fig. 12 tests the scalability of $con\text{-}kNN$ and $Basic$ against data sizes N from $10K$ to $50K$, where other parameters are set to default values. In fig-

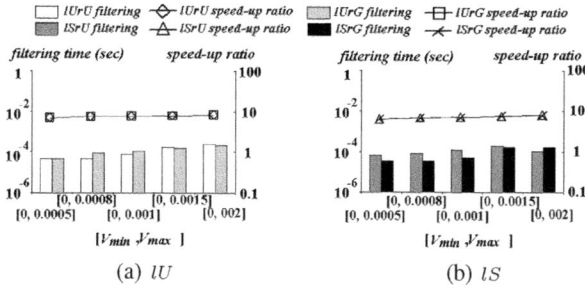

Fig. 11. Performance vs. Velocity Range $[V_{min}, V_{max}]$

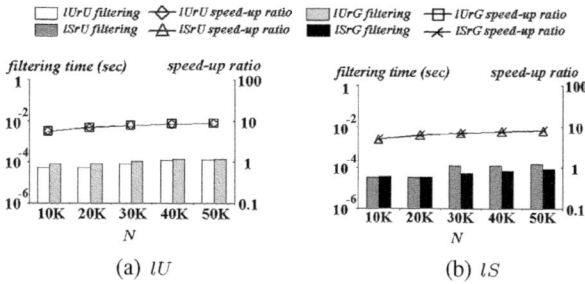

Fig. 12. Performance vs. Data Size N

ures, the filtering time smoothly increases with the increasing data size. Moreover, the speed-up ratio also increases for larger N. This is because for large N, $Basic$ has to refine more candidates, whereas $con\text{-}kNN$ has effective filtering methods to prune false alarms, which confirms good scalability of $con\text{-}kNN$.

7   Conclusions

In this paper, we formulate and tackle a query, namely *probabilistic time consistent query* (PTCQ), on uncertain moving object database. Specifically, we consider one important query type, probabilistic time consistent kNN queries. We provide effective pruning methods utilizing different query constraints, and use them to greatly reduce the search space. To facilitate efficient query processing, we design a data structure, namely *UC-Grid*, to index uncertain moving objects with low update and retrieval costs, whose grid size is determined by our cost model to minimize the query cost. We demonstrate through extensive experiments the efficiency and effectiveness of our approaches.

Acknowledgment. Funding for this work was provided by RGC NSFC JOINT Grant under Project No. N_HKUST61 2/09 and NSFC Grant No. 60736013, 60803105, 60873022, and 60903053.

References

1. Banerjee, N., Corner, M.D., Towsley, D., Levine, B.N.: Relays, base stations, and meshes: enhancing mobile networks with infrastructure. In: MobiCom (2008)
2. Belussi, A., Faloutsos, C.: Self-spacial join selectivity estimation using fractal concepts. Inf. Syst. 16(2) (1998)
3. Beskales, G., Soliman, M., Ilyas, I.F.: Efficient search for the top-k probable nearest neighbors inuncertain databases. In: VLDB (2008)
4. Cheng, R., Chen, L., Chen, J., Xie X.: Evaluating probability threshold k-nearest-neighbor queries over uncertain data. In: EDBT (2009)
5. Cheng, R., Kalashnikov, D., Prabhakar, S.: Querying imprecise data in moving object environments. TKDE 16(9) (2004)
6. Chung, B.S.E., Lee, W.-C., Chen, A.L.P.: Processing probabilistic spatio-temporal range queries over moving objects with uncertainty. In: EDBT (2009)
7. Guttman, A.: R-trees: a dynamic index structure for spatial searching. In: SIGMOD (1984)
8. Hjaltason, G.R., Samet, H.: Distance browsing in spatial databases. TODS 24(2) (1999)
9. Huang, Y.-K., Liao, S.-J., Lee, C.: Evaluating continuous k-nearest neighbor query on moving objects with uncertainty. Inf. Syst. 34(4-5) (2009)
10. Iwerks, G.S., Samet, H., Smith, K.: Continuous k-nearest neighbor queries for continuously moving points with updates. In: VLDB (2003)
11. Korn, F., Sidiropoulos, N., Faloutsos, C., Siegel, E., Protopapas, Z.: Fast nearest neighbor search in medical image databases. In: VLDB (1996)
12. Kriegel, H.-P., Kunath, P., Renz, M.: Probabilistic nearest-neighbor query on uncertain objects. In: Kotagiri, R., Radha Krishna, P., Mohania, M., Nantajeewarawat, E. (eds.) DASFAA 2007. LNCS, vol. 4443, pp. 337–348. Springer, Heidelberg (2007)
13. Lian, X., Chen, L.: Probabilistic group nearest neighbor queries in uncertain databases. TKDE (2008)
14. Mouratidis, K., Papadias, D., Hadjieleftheriou, M.: Conceptual partitioning: an efficient method for continuous nearest neighbor monitoring. In: SIGMOD (2005)
15. Pfoser, D., Jensen, C.S.: Capturing the uncertainty of moving-object representations. In: Güting, R.H., Papadias, D., Lochovsky, F.H. (eds.) SSD 1999. LNCS, vol. 1651, p. 111. Springer, Heidelberg (1999)
16. Raptopoulou, K., Papadopoulos, A.N., Manolopoulos, Y.: Fast nearest-neighbor query processing in moving-object databases. Geoinformatica 7(2) (2003)
17. Roussopoulos, N., Kelley, S., Vincent, F.: Nearest neighbor queries. In: SIGMOD (1995)
18. Seidl, T., Kriegel, H.: Optimal multi-step k-nearest neighbor search. In: SIGMOD (1998)
19. Tao, Y., Papadias, D.: Time-parameterized queries in spatio-temporal databases. In: SIGMOD (2002)
20. Tao, Y., Papadias, D., Sun, J.: The TPR*-tree: an optimized spatio-temporal access method for predictive queries. In: VLDB (2003)
21. Trajcevski, G., Tamassia, R., Ding, H., Scheuermann, P., Cruz, I.F.: Continuous probabilistic nearest-neighbor queries for uncertain trajectories. In: EDBT (2009)
22. Trajcevski, G., Wolfson, O., Hinrichs, K., Chamberlain, S.: Managing uncertainty in moving objects databases. TODS 29(3) (2004)
23. Šaltenis, S., Jensen, C.S., Leutenegger, S., Lopez, M.A.: Indexing the positions of continuously moving objects. SIGMOD Rec. 29(2) (2000)
24. Xiong, X., Mokbel, M.F., Aref, W.G.: SEA-CNN: Scalable processing of continuous k-nearest neighbor queries in spatio-temporal databases. In: ICDE (2005)
25. Yu, X., Pu, K.Q., Koudas, N.: Monitoring k-nearest neighbor queries over moving objects. In: ICDE (2005)
26. Zhang, M., Chen, S., Jensen, C.S., Ooi, B.C., Zhang, Z.: Effectively indexing uncertain moving objects for predictive queries. PVLDB 2(1) (2009)

Appendix

A. Dynamic Maintenance of *UC-Grid* Upon Updates

To insert an uncertain moving object o into *UC-Grid*, we first find the cell $c_{i,j}$ where C_o resides, and then add o to a center list of $c_{i,j}$ based on its direction, which requires $O(1)$ cost. In addition, we also need to find those cells that intersect with object o, and add o to their cell lists (i.e., increasing the counter, $count$, by 1), which requires $O(\frac{\pi(r_o+\delta)^2}{\delta^2})$ cost in the worst case. Similarly, to delete an object o from *UC-Grid*, we first find its location, cell $c_{i,j}$, with $O(1)$ cost, and then remove it from the center list. Finally, for those cells intersecting with o, we decrease their counters (i.e., $count$) in cell lists by 1. In case the counter becomes 0, we remove the entry $(c_{i,j}, 0)$. Thus, the time complexity of deletion is $O(\frac{\pi(r_o+\delta)^2}{\delta^2})$.

B. Pruning Heuristics for Cells

As mentioned in Section 5.1, each cell of *UC-Grid* $\mathcal{I}$ contains 4 center lists, NE, NW, SW, and SE, corresponding to 4 types of moving directions. Thus, each list maintains some statistics, for example, $[\gamma_{min}, \gamma_{max}]$ (note: we omit superscript like NE for brevity), which is the interval of angles γ (between moving directions and x-axis) for all the objects in each list. Similarly, each list also keeps the minimum (maximum) velocities v_{min} (v_{max}), as well as the minimum (maximum) radii r_o, r_{omin} (r_{omax}). Thus, our goal is to derive pruning rules such that: a cell $c_{i,j}$ can be safely pruned, if the pruning conditions (given by pruning methods in Section 4) hold for any uncertain moving object $o \in c_{i,j}$ with $\gamma \in [\gamma_{min}, \gamma_{max}]$, $v_o \in [v_{min}, v_{max}]$, and $r_o \in [r_{omin}, r_{omax}]$.

For T-pruning in Lemma 1, we have its corresponding pruning rule for cells below.

Lemma 7. (*T-Pruning Rule for Cells*) *Any center list in a cell*, $c_{i,j}$, *of UC-Grid* $\mathcal{I}$ *with* $[\gamma_{min}, \gamma_{max}]$, $[v_{min}, v_{max}]$, *and* $[r_{omin}, r_{omax}]$ *can be safely pruned, if it holds that:*

$$||q|| \cdot max_{\gamma \in [\gamma_{min}, \gamma_{max}]}\{cos(\eta_q - \gamma)\} < min\_S_1, \ or \tag{12}$$

$$||q|| \cdot min_{\gamma \in [\gamma_{min}, \gamma_{max}]}\{sin(\gamma - \eta_q)\} - max\_S_2 > (R_{max} + r_{omax})^2 - (v_{min} \cdot T)^2/4. \tag{13}$$

where $min\_S_1 = min_{\forall o \ in \ center \ list}\{C_o[x] \cdot cos\gamma + C_o[y] \cdot sin\gamma\}$ *and* $max\_S_2 = max_{\forall o \ in \ center \ list}\{C_o[x] \cdot sin\gamma - C_o[y] \cdot cos\gamma\}$.

Note that, each object o has constants $(C_o[x] \cdot cos\gamma + C_o[y] \cdot sin\gamma)$ and $(C_o[x] \cdot sin\gamma - C_o[y] \cdot cos\gamma)$. Thus, in Lemma 7, min_S_1 and max_S_2 are the min/max aggregates of these two constants, respectively.

In Inequality (12), to compute $max_{\gamma \in [\gamma_{min}, \gamma_{max}]}\{cos(\eta_q - \gamma)\}$, we consider 2 cases: 1) $\eta_q \notin [\gamma_{min}, \gamma_{max}]$, and 2) $\eta_q \in [\gamma_{min}, \gamma_{max}]$. From the cosine curve, we can obtain:

$$max_{\gamma \in [\gamma_{min}, \gamma_{max}]}\{cos(\eta_q - \gamma)\}$$
$$= \begin{cases} max\{cos(\gamma_{max} - \eta_q), cos(\gamma_{min} - \eta_q)\} & \text{if } \eta_q \notin [\gamma_{min}, \gamma_{max}], \\ 1 & \text{otherwise.} \end{cases} \tag{14}$$

Similarly, for $min_{\gamma \in [\gamma_{min}, \gamma_{max}]}\{sin(\gamma - \eta_q)\}$ in Inequality (13), we have:

$$min_{\gamma \in [\gamma_{min}, \gamma_{max}]}\{sin(\gamma - \eta_q)\}$$

$$= \begin{cases} min\{sin(\gamma_{min} - \eta_q), sin(\gamma_{max} - \eta_q)\} \\ \qquad \text{if} -\frac{\pi}{2}, \frac{3\pi}{2} \notin [\gamma_{min} - \eta_q, \gamma_{max} - \eta_q], \\ -1 \qquad \text{otherwise.} \end{cases} \qquad (15)$$

For the period pruning in Lemma 2, we have the following pruning rule for cells.

Lemma 8. *(Period Pruning Rule for Cells) Any center list in the cell,* $c_{i,j}$*, of UC-Grid* $\mathcal{I}$ *with* $[\gamma_{min}, \gamma_{max}]$*,* $[v_{min}, v_{max}]$*, and* $[r_{omin}, r_{omax}]$ *can be safely pruned, if it holds that:*

$$\begin{cases} ||q|| \cdot max_{\gamma \in [\gamma_{min}, \gamma_{max}]}\{cos(\eta_q - \gamma)\} \\ \quad - min\_S_1 < \frac{v_{max}^2 \cdot (t_{ed} - T)^2 - (R_{max} + r_{omax})^2 + mindist^2(q, c_{i,j})}{2 \cdot v_{max}^2 \cdot (t_{ed} - T)} \\ \quad \text{if } mindist(q, c_{i,j}) > R_{max} + r_{omax}, \\ ||q|| \cdot max_{\gamma \in [\gamma_{min}, \gamma_{max}]}\{cos(\eta_q - \gamma)\} \\ \quad - min\_S_1 < \frac{v_{min}^2 \cdot T^2 - (R_{max} + r_{omax})^2 + mindist^2(q, c_{i,j})}{2 \cdot v_{min}^2 \cdot T}, \\ \quad \text{if } mindist(q, c_{i,j}) \leq R_{max} + r_{omin}. \end{cases}$$

$$\qquad (16)$$

where $max_{\gamma \in [\gamma_{min}, \gamma_{max}]}\{cos(\eta_q - \gamma)\}$ *and* min_S_1 *refer to Eq. (14) and Lemma 7, respectively.*

For the segment pruning, the pruning rule for cell, $c_{i,j}$, considers whether each segment can be pruned (Lemma 3) for any object in $c_{i,j}$. That is, we check whether or not the minimum value of LHS of Eq. (7) for any γ, v, r_o, and t values within their ranges is greater than 0 (RHS of Eq. (7)). If the answer is yes, the cell can be pruned.

C. Discussions on Uncertain Directions.

In Section 3, we assume that objects are moving towards a fixed direction (i.e., with angle γ) in a future period $[0, t_{ed}]$. For uncertain directions of velocities, we can also model them by direction variables $\gamma \in [\gamma^-, \gamma^+]$. Note that, this representation is equivalent to [26] which used a 2D velocity histogram to inherently record uncertain directions. Our solutions can be easily extended by considering the pruning with uncertain directions the same as that for cells (in Lemmas 7 and 8). That is, we replace γ_{min} and γ_{max} in pruning conditions with γ^- and γ^+, respectively. This model with uncertain directions can capture real-world scenarios such as objects with non-linear motions or other motion patterns. Thus, our approach is expected to efficiently process consistent queries for objects with these motion patterns.

Knowledge Annotations in Scientific Workflows: An Implementation in Kepler

Aída Gándara[1], George Chin Jr.[2], Paulo Pinheiro da Silva[1], Signe White[2], Chandrika Sivaramakrishnan[2], and Terence Critchlow[2]

[1] Cyber-ShARE,
The University of Texas at El Paso, El Paso, TX
agandara1@miners.utep.edu, paulo@utep.edu
[2] Pacific Northwest National Laboratory, Richland, WA
{george.chin,signe.white,chandrika.sivaramakrishnan,
terence.critchlow}@pnl.gov

Abstract. Scientific research products are the result of long-term collaborations between teams. Scientific workflows are capable of helping scientists in many ways including collecting information about how research was conducted (e.g., scientific workflow tools often collect and manage information about datasets used and data transformations). However, knowledge about *why* data was collected is rarely documented in scientific workflows. In this paper we describe a prototype system built to support the collection of scientific expertise that influences scientific analysis. Through evaluating a scientific research effort underway at the Pacific Northwest National Laboratory, we identified features that would most benefit PNNL scientists in documenting how and why they conduct their research, making this information available to the entire team. The prototype system was built by enhancing the Kepler Scientific Workflow System to create knowledge-annotated scientific workflows and to publish them as semantic annotations.

Keywords: Scientific Workflows, Knowledge Annotations, Kepler.

1 Introduction

When scientists work collaboratively to conduct scientific research there are many factors that have a direct impact on how research is performed, much of it implicit in the actual process used. For example, when exploring a data set, scientists may use their expertise to select data points and their experience may guide a scientist to impose a certain constraint on the entire dataset. Unfortunately, these decisions are poorly documented and may not be presented to the current research team, much less reflected in any published work. Nonetheless, this knowledge is crucial for conducting research and is the basis for innovation, something that is often needed for scientific research [13].

Scientific workflow tools enable scientists to describe, execute and preserve a research process. In addition, they can be used to annotate data and collect

J.B. Cushing, J. French, and S. Bowers (Eds.): SSDBM 2011, LNCS 6809, pp. 189–206, 2011.
© Springer-Verlag Berlin Heidelberg 2011

provenance about how scientific artifacts were created [8]. However, the knowledge implicit in a workflow reaches beyond its execution, including, for example, the many decisions made to choose algorithms, parameters and datasets that are undocumented in current scientific workflow engines. If that information could be captured in scientific workflows, the associated tools are in a unique position to play an instrumental role in organizing and preserving the implicit and explicit knowledge that is shared among scientists during a research effort.

Similar to many active scientific research projects, the subsurface flow and transport analysis projects at the Pacific Northwest National Laboratory (PNNL), is a collaborative effort where scientists with different knowledge domains, e.g., data collection, model building, and simulation expertise, are working together to perform groundwater modeling. We evaluated the scientific process being used by the groundwater modeling team to understand how collaborative teams conduct scientific research, share knowledge, and produce scientific products. As observed with other teams affiliated with the Cyber-ShARE Center of Excellence[1], we observed that scientists of a highly collaborative team are dependent on the decisions, expertise and results of their colleagues during a research effort, because assumptions made during one scientist's analysis directly affects other scientists on the team. Unfortunately, these decisions and assumptions are often not well documented and their justifications may fade over the course of the project. Providing mechanisms to help scientists manage this knowledge would be a great benefit to assure that the entire team remains aware of relevant research assumptions, constraints and decisions.

This paper explores the process of documenting the implicit side of scientific research using an extension to the Kepler Workflow Management System [10]. Although Kepler has been used to represent groundwater modeling workflows previously, the tool was not able to support documenting the implicit aspects of the research collaboration. In order to represent these decisions and assumptions, Kepler, similar to other executable workflow systems, needed to be extended. To this end, a prototype system was built over Kepler to produce knowledge-annotated scientific workflows. The collected information is published as semantic annotations in RDF [9] with the goal of enabling reuse and integration with related information [14].

In the remainder of this paper, we present a preliminary research prototype designed to address the current limitations of workflow systems and enable documentation of the entire scientific process from initial discussions to executable workflow. Section 2 will provide background information on a subsurface flow and transport project that motivated this prototype as well as technologies that affected our implementation, including Kepler. Section 3 presents the details of our implementation while Section 4 discusses current issues with this prototype and outlines future work for knowledge-annotated scientific workflows, including a comprehensive user study. Finally, Section 5 presents our concluding thoughts.

---

[1] http://cybershare.utep.edu

2   Subsurface Flow and Transport Analysis Case Study

PNNL has extensive research and development capabilities and expertise in the scientific field of subsurface flow and transport, which focuses on the "study of chemical reactions in heterogeneous natural material, with an emphasis on soil and subsurface systems[1]." Some of these capabilities are centered on the construction and application of a PNNL-developed predictive subsurface flow and transport simulator known as STOMP (Subsurface Transport Over Multiple Phases)[16]. Using STOMP along with other subsurface model development software, PNNL groundwater scientists are modeling subsurface flow and transport on a wide variety of internal and external projects.

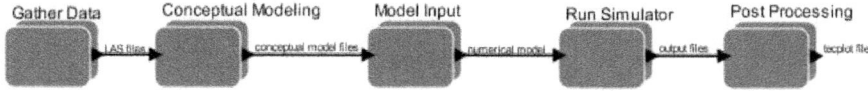

Fig. 1. A high-level workflow describing a groundwater modeling project conducted at PNNL

A groundwater modeling project typically comprises a project manager and several team members. Scientists take on roles within the overall research effort based on their expertise. The groundwater modeling process follows a general data flow pattern of steps that have been summarized as a high-level workflow shown in Figure 1. Each step of the process requires the knowledge and expertise of the scientist performing it, as well as collaboration between steps to understand details about the overall process. The groundwater process starts with the Gather Data step, where data is initially collected by a geoscientist. This geoscientist gathers the data from the field or while running experiments. Next a scientist performs the Conceptual Modeling step where the initial data is analyzed to create a conceptual model (a conceptual understanding of the subsurface geology). The conceptual model is represented in a series of files that are used in the Model Input step. Within the Model Input step a scientist builds a numerical model of the data and annotates it. This numerical model is an implementation of the conceptual model into a discretized, numerical framework. The simulation is executed in the Run Simulator step after which the data is post-processed into target data images and reports.

Throughout the groundwater modeling process, scientists use their expertise to interpret and analyze data, interpolate new data models, run scripts and executables (some of which they have written themselves), visualize data and results, and build and annotate data sets. They must understand details about the overall project and must make simplifying assumptions to account for a lack of data or to take into consideration computational limitations, e.g., available hardware, and project time constraints. Their research is an iterative process, where they might run a series of steps over and over, changing parameters and analysis details, to produce the best results. The collaborative team continuously reviews the results of different steps, makes suggestions, and formulates new assumptions that could alter the overall modeling process: that is, the changes to the process might require that steps be performed again. Sometimes, having to perform steps over could be avoided if each scientist was aware of assumptions or constraints that other team members have made. In many cases, scientists keep journals and notes of what worked and what did not as well as the decisions, assumptions or constraints used to find results. Team discussions normally occur in meetings, via email or by phone. Documentation of these discussions does not always occur.

We found that many of the needs of these scientists are consistent with the needs of other scientists in other domains such as geoinformatics or environmental science. Namely, once an artifact or analysis product is available either during the research process or after, scientists seek to understand not only the "hows" but the "whys" of its creation. Currently, the data and knowledge of running these models are collected in a final report and result data sets. Going back to review and understand the original process requires understanding the final report, visualizing the final results, if they are available, and talking to scientists involved in the initial research and analysis. A scientist's recollection of a specific past project would require that they too have access to their notes and the details of how they conducted scientific analysis and performed the specific steps of a specific process.

2.1   Current Approaches for Collecting the Scientists' Knowledge

One common method that scientists have used to collect their scientific notes is paper-based journals, where scientists can sketch and write the ideas and conclusions of their research. Electronic Laboratory Notebooks (ELNs) provide similar functionality to paper-based journals but with added benefits like electronically organizing scientific notes, relating notes to data, viewing and manipulating related electronic data and multi-user support. More recently, ELNs have been enhanced to function over the Web and to provide interoperable semantic annotations so they can be integrated with other data producing and consuming systems [15]. On a more collaborative front, tools like email, chat tools, and more recently social networking tools like Facebook[2] have been used to elicit discussions and social interactions. These tools support the ability to link data to specific comments, e.g., through attachments, and to talk directly to individuals as well as groups. Although both ELNs and social networking tools promote collaboration and enable the collection of social and scientific information, they are limited in documenting the ongoing research process conducted by scientists. For example, they do not directly support the process definition that is inherently captured in scientific workflow tools.

Executable scientific workflows are beneficial to scientific research and the management of scientific data [8] because their main goal is to collect sufficient information so a process can be executed. As a result, they can capture

---

[2] http://facebook.com

hardware details, user details, operating system details and execution details that can be used to annotate an artifact. Furthermore, the graphical representation that many scientific workflow tools create enables scientists to see the steps in the process and the flow of data. Through this representation, scientists can understand how an artifact was created. As scientific research efforts become more collaborative, scientific workflow environments must consider how they will evolve to support collaborative teams. myExperiment, for example, is a research effort where scientific workflows and other research objects can be published to the myExperiment Portal [7]. The goal of this portal is to support the sharing of scientific research and the reproduction of scientific results. Through myExperiment, users can rate, download and tag workflows as well as maintain discussions about published workflows. The documentation occurs after the workflow is published, thus knowledge annotations occur after the scientific research is completed.

Another benefit of scientific workflows is the ability they provide in reproducing results. The Kepler Scientific Workflow System, for example, publishes workflows and its corresponding data as Kepler Archive (KAR) files. KAR files are encapsulated workflows including all components needed to run the workflow [2]. As a result, KAR files are objects that can be published on myExperiment and then downloaded for other scientists to execute. WDOIT! [12] is a scientific workflow tool that publishes all workflow information as semantic annotations on the Semantic Web. With these annotations, a WDOIT! workflow can classify data with common terminology and link workflows to data and resources on the Web. This open environment enables interoperability and reuse with other semantic-based tools [14].

2.2 Kepler Scientific Workflow System

Kepler is a scientific workflow tool used to build and execute scientific workflows. Kepler provides graphical abstractions to enable scientists to build workflows. For example, scientists describe the steps of a process by adding actors to a canvas (a graphical window); they add ports for data input and output of an actor; and they connect ports between actors to specify a flow of information. Actors can be chosen from a list of available actors; there are menu entries to add ports and actor details, and workflows can further be described by specifying parameters, global values, and execution instructions. Kepler workflows can be described at multiple levels of detail using composite actors. A composite actor is a nested canvas where a scientist can define a subworkflow. The purpose of a composite actor is to hide process details until there is a need to see them, e.g., a scientist chooses to open it. Once all the details for execution have been specified in a workflow canvas, including actors, ports, parameters, and connections, the workflow is ready for execution.

As shown earlier, Figure 1 presents a Kepler workflow representing the groundwater modeling steps identified by groundwater scientists. Building an executable workflow can be quite involved and in some cases distracting, particularly if scientists work in an ad hoc, exploratory framework where they are not certain of the steps they will take to conduct their research. Furthermore, scientists may need to describe a process but not execute it, e.g., if they use their own scripts, programs or manual steps not found in Kepler. Nevertheless, the ability within Kepler for describing processes at different levels of abstraction provides a mechanism for annotating the internal components, e.g., actors and connections, in the workflow with knowledge annotations.

3   Knowledge-Annotated Scientific Workflows

The goal of this research effort was to help the groundwater scientists manage collaborative data that is traditionally generated but not collected during a research effort. Three design principles were incorporated into this prototype. *First design principle*: the prototype needed to enable scientists to primarily describe their research; thus, any workflow construction needed to be a byproduct of the information provided by the scientist, requiring execution details only when necessary. *Second design principle*: the prototype needed to align with the way scientists conducted research and limit the duplication of information and the number of menus and windows needed to document their research. *Third design principle*: the prototype needed to leverage the workflow to manage the annotations, i.e., annotations had to directly relate to steps and connections that the scientist added to their research process, which would in turn enable the prototype to present and publish the result data with a close relation to the process. The prototype enables scientists to build workflows by focusing on scientific analysis and ad hoc scientific exploration. The following sections will work through the prototype system, highlighting the features that were added to support groundwater scientists in documenting the decisions and knowledge behind their research.

3.1   Building Workflows

To enable the groundwater scientists to focus on their research and avoid focusing on executable workflow details, the Kepler interface was modified by adding menu entries, buttons and panels to collect and display the implicit knowledge related to steps scientists take to conduct scientific research. Figure 2 shows the overall prototype interface. The left side of the interface displays the Research Hierarchy panel, which is a tree view of all the process steps defined in a project, and the bottom right displays the Process View panel which shows a graphical representation of the steps in a process. These two panels are always visible, but their content changes based on a scientist's interaction with the tool. The top right displays various tabs that collect annotations related to specific process details, e.g., steps, inputs, outputs. Workflows in the prototype are managed within a project. A project is created by opening up the prototype interface and entering the project details. The Project Description tab is shown at the top of Figure 2. Here a title, description, and the project's team members are specified. When team members are selected for a project, a project manager

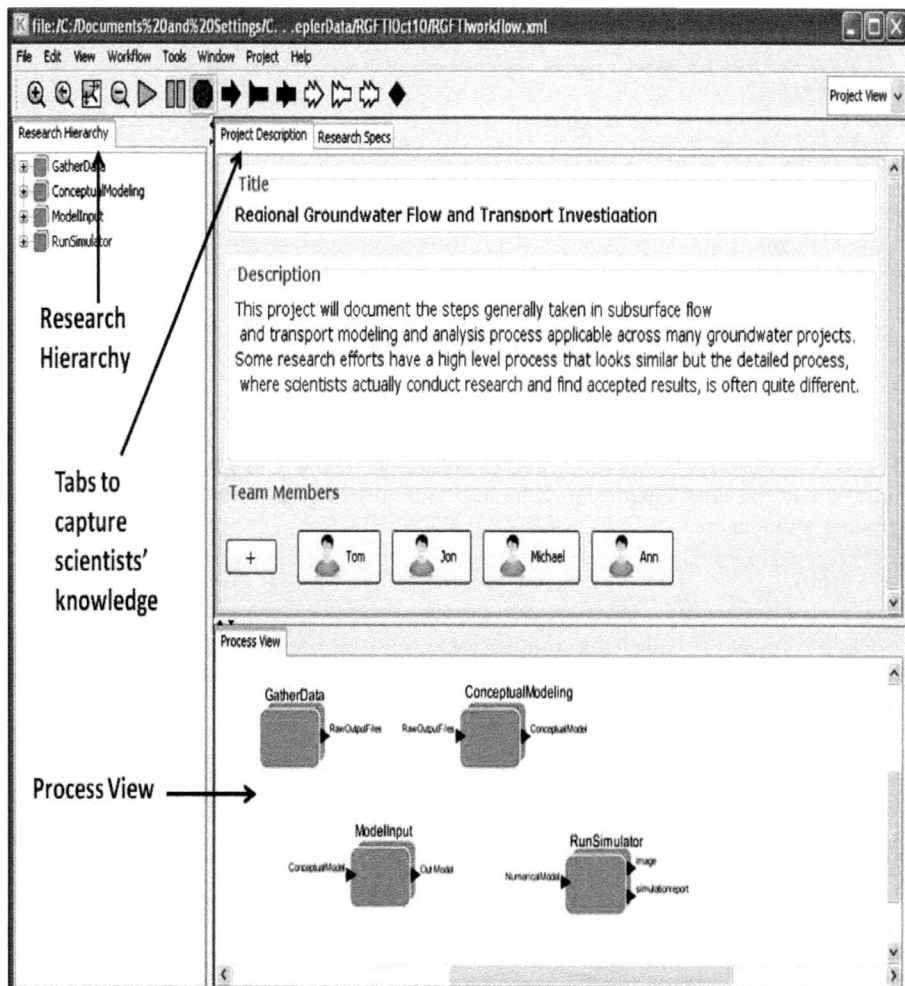

Fig. 2. The prototype interface showing the different panels added to support scientists in annotating their workflows as they describe their research

enters the name and contact information for each team member. In this way, the team has a single place to identify who is actively on the team. Steps can be added to describe the analysis steps that will be performed within a process. As steps are added, a step actor is added to the Process View panel. For example, a subsurface flow and transport project would conduct some variation of the steps defined in the case study found in Section 2. The Process View panel in Figure 2 shows steps for gathering data, building the conceptual model, building the input model, and running the simulation. A scientist can add steps as needed to allow for an ad hoc, exploratory collection of scientific analysis steps. In turn, the

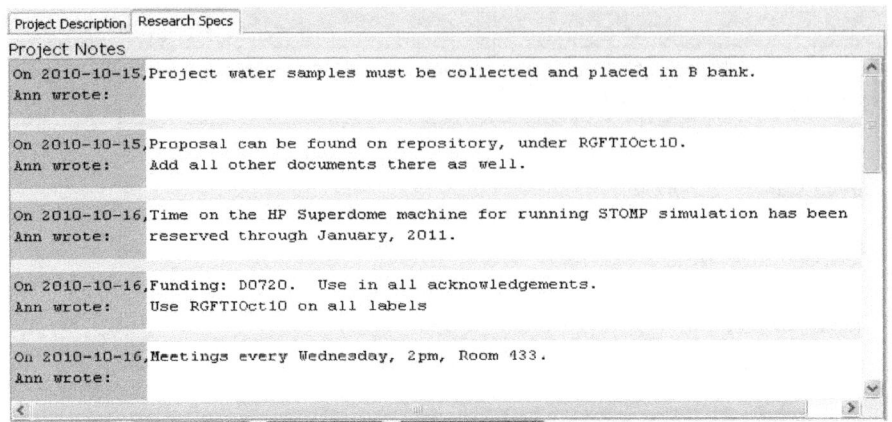

Fig. 3. The Research Specs tab. This tab collects general comments affecting the entire project and provides buttons to help scientists describe their research and annotate process components.

workflow system is collecting annotations, relating them to specific components of the workflow and providing a mechanism for a team of scientists to view the information as it relates to their ongoing research.

Throughout a project, decisions and notes are made that affect the entire project. This information is collected in the Research Specs tab shown in Figure 3. The Research Specs tab has a scrollable entry pane where a project manager or other team member, can enter details about known constraints based on the research proposal, administrative comments as to how the group will function, or ongoing comments about what the group is doing. This tab also has buttons to support the annotation and workflow building process, e.g., Add Step, Add Assumption, and Add Deliverable.

The prototype leverages Kepler's ability to describe processes at multiple levels of abstraction by annotating workflow components, e.g., steps, at the different levels. The Process View panel in Figure 4 describes in more detail the RunSimulator step conducted by a groundwater modeling scientist. At the top of Figure 4 is the Research Notes tab. When a step is created, a Research Notes tab is created with an associated workflow. The Research Notes tab has a scrollable entry pane where scientists can capture their notes about the research, e.g., the decisions they made, what processes worked and why. Scientists can also add more process details that will be reflected as changes to the workflow, e.g., Add Steps, Add Input, Add Output. Scientists can then choose to refine steps to any level of granularity. Figure 5 describes a more detailed level of granularity of the STOMP step, where the groundwater scientist models and annotates the execution of a STOMP actor, an executable actor used at PNNL. This method of defining and refining steps allows scientists to define a hierarchical definition of research where steps are refined if the scientist needs to describe more detail.

More importantly, the process exhibits the integration of annotating a completely abstract process shown in Figure 2 down to a more detailed executable model shown in Figure 5, where scientists can add comments concerning the success or failure of parameters or executed processes.

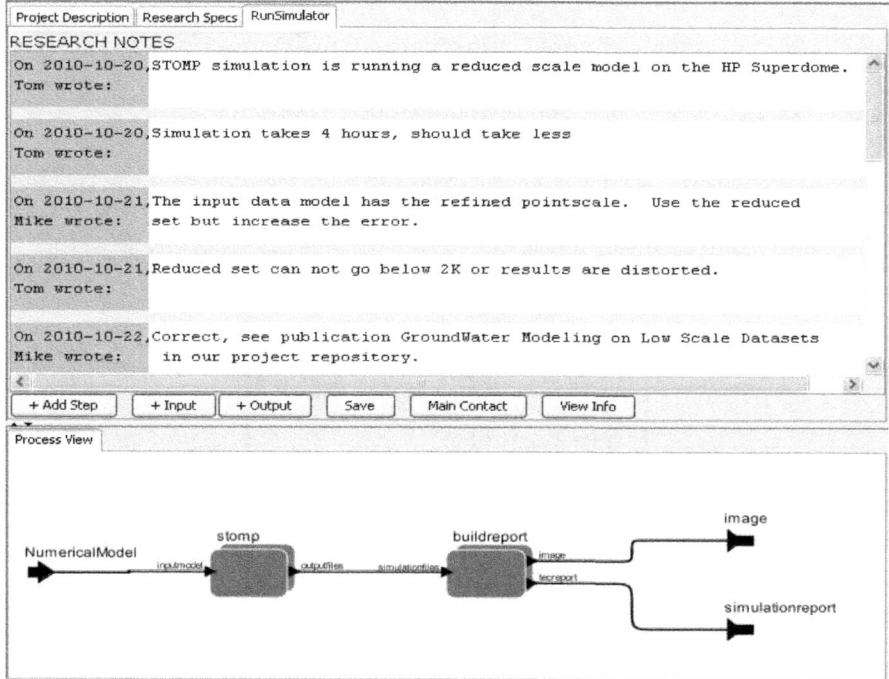

Fig. 4. The knowledge data and workflow of the Run Simulator step. Scientists can refine the details of their research by describing a process, adding research notes and entering details about the inputs and outputs of their research.

As steps are added at any level of the process, the Research Hierarchy panel, shown in Figure 2 is updated with branches added to the appropriate hierarchy tree. When a step is opened (by double-clicking its icon in the Research Hierarchy tree or from the Process View panel), the step's Research Notes panel and Process View panel are displayed. When other scientists need to understand the details about a particular step, they can open the step and see the annotations.

To identify data that will be used or produced in a scientific step, scientists can add inputs or outputs that are displayed as icons on the Process View panel. The Process View panel in Figure 4 shows several input and output icons, e.g., NumericalModel as an input and simulationreport as an output. A scientist can add more knowledge about a specific input or output by opening its Input Details dialog or Output Details dialog, respectively. Figure 6 shows the Input Details dialog for the NumericalModel input from the RunSimulator step. Using this

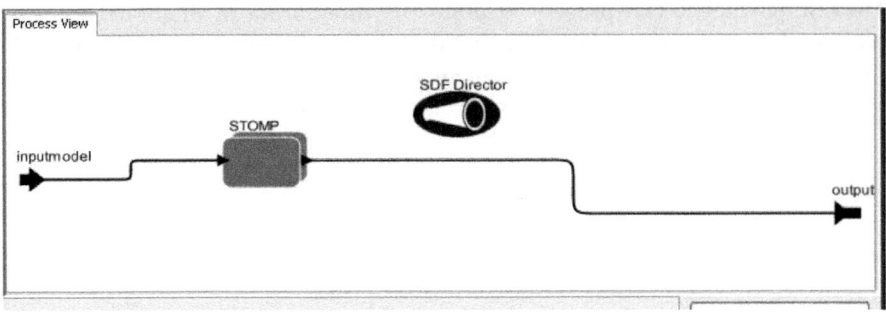

Fig. 5. The Process View panel for the STOMP step. The Process View panel has execution details for running the simulator, e.g. a director, actor, inputs and outputs.

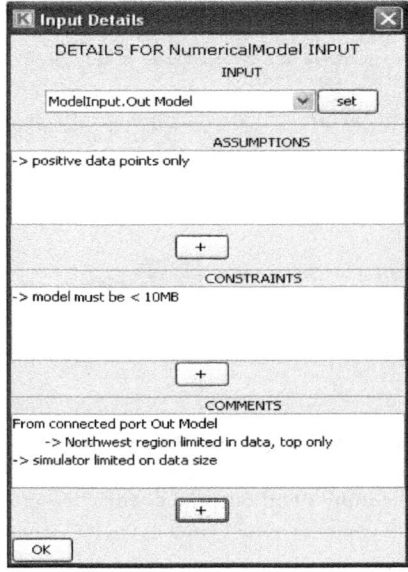

Fig. 6. Input details dialogs for the NumericalModel input of the RunSimulator step. This input is dependent on Outputfiles output of the ModelInput step. Comments from the Outputfiles output can be see here as well.

dialog, scientists can specify what data is needed and from what step, specifying the flow of information and the dependencies between steps. The Output Details dialog (not shown here) allows a scientist to enter similar details related to the outputs produced within a scientific step. These dialogs collect assumptions, constraints, and general comments about the data. Collecting this information as the scientist is conducting research can provide insight to other scientists that depend on it. To make this knowledge more visible, the knowledge data of connected input and output ports are displayed in the respective Input or Output

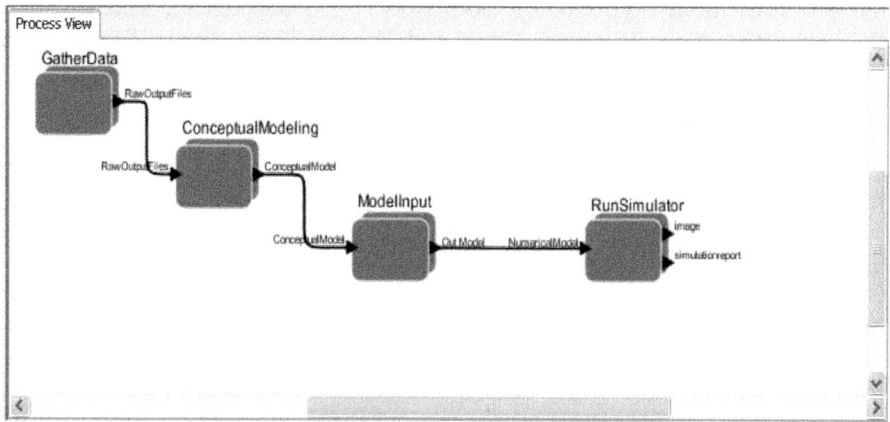

Fig. 7. This workflow was built in the prototype system by collecting knowledge about the research performed by each scientist

Details dialogs. For example, the comment section in Figure 6 also shows the comment from the connected step's (ModelInput) data. Scientists who view a details dialog can review existing details and add their own.

As a result of specifying data dependencies in the details dialogs, connections are made on the Process View panels that contain those actors, building the workflow automatically. Figure 7 shows the complete project workflow after the different steps were opened by the corresponding scientist and research knowledge was collected in them. Notice that the NumericalModel input to the Run-Simulator step is connected to the NumericalModel output from the ModelInput step. This information was specified in the Input Details dialog in Figure 6.

3.2 A View of the Data

Having scientists enter research notes and knowledge details to describe their research is useful for data capture but not sufficient for understanding the data. For scientists to understand a step within the data collection windows provided in the prototype, they would have to open up different tabs and dialogs. By managing this information within the workflow tool, the prototype can help provide this information in a more organized format. The scientists in our case study specifically referenced a need to understand the flow of the data that was coming into their research step and the ability to summarize the details of the step they were viewing. Furthermore, the scientists are often called on to provide the same details that are collected in this prototype for a final project report. To support these needs, we added the ability for scientists to backup or move forward through step connections, build a full summary report and access summary views that can be seen at any level within the research hierarchy.

The backward and forward traversal functionality was added to enable quick jumps between steps. With a single click, a scientist can choose to step back or move forward from a port to the step that it is connected to, which displays the connected step's Research Notes and Process View panels. For example, performing a backward traversal from the NumericalModel input for the Run-Simulator step would open up the Research Notes and display the workflow for the ModelInput step. The advantage to this feature is that it makes it easier for scientists to see the research notes and process view of related steps and then jump back if needed.

To help scientists understand the entire research effort, a full research summary report can be created. The details of scientific research can be quite involved and the goal of this prototype system is to help collaborative groups document and understand the process behind their research. Building a full summary is useful because it summarizes all the details of the scientific process in one document. Another benefit of this feature is that it could facilitate writing documentation, e.g. a research paper or publication, about the scientific process and the accompanying knowledge data. Figure 8 shows the first page of a summary report for the Regional Groundwater Flow and Transport Investigation (RGFTI) project, which is a specific groundwater modeling project currently being performed at PNNL. Understanding what should be in this summary and how it should look is dependent on the group working with the data as well as the needs of each scientist. Further evaluation should help with understanding the appropriate configurations and structure of such a summary.

To help scientists understand the factors that are contributing to the current state of a step, a scientist can view the step's status. This feature will give a summary of steps, connections and connection details from the different levels of refinement of the current step. If the step has several sub-steps and they in turn have sub-steps, the sub-steps are included in the summary.

3.3   The Knowledge-Annotated Data Model

The Kepler source code[3] and the prototype system are Java implementations. The knowledge-annotated data model for the Kepler prototype is stored in a set of five classes, as shown in Figure 9. The CollaborativeProject class is the focal point for a project: it stores project details, MemberInfo objects, JournalEntry objects and StepInfo objects. There is one StepInfo object for each step defined in a project. The StepInfo class stores JournalEntry objects and PortInfo objects. There is one PortInfo object for each input and each output defined in a step. The PortInfo class stores the annotations (assumptions, constraints, and comments) for an input or output defined in a step. The JournalEntry class stores the annotations entered within the project notes or a step's research notes. The MemberInfo class stores information about the team members of a project. Figure 9 also describes the canvasMap variable. When Kepler opens up composite actors, their interface and environment are created in a new process space. For the prototype, this caused difficulty in managing the overall annotations within a project. As a result, the canvasMap variable was added to manage the different contexts that would be annotated within a project.

---

[3] https://code.kepler-project.org/code/kepler/trunk/modules/build-area

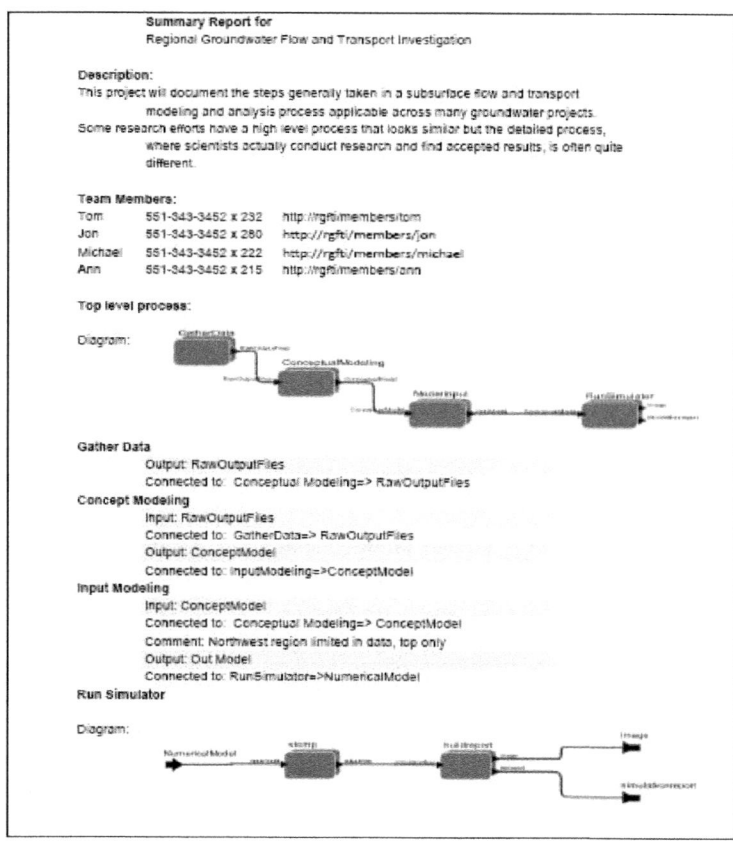

Fig. 8. The first page of a summary report for the RGFTI project. This report gives a summary of the entire project, including project information, e.g., title, team members, process information, steps, and the corresponding knowledge data. This information is provided in a single document.

Most workflow tools store their workflows in tool-specific formats, e.g., Kepler stores all workflow information in KAR files. The issue with tool-specific formats is that they limit interoperability and reuse of workflows; e.g., the encapsulated contents of a KAR file would only be readable by KAR-compatible tools. Although we believe it would be unrealistic that all workflow tools conform to the same representational formats, there is no reason why the annotations collected by the prototype can't be more open to sharing and interoperability. In an effort to enable others to understand what was done and why, the annotations are stored in the Semantically Interlinked Online Communities (SIOC) [5] format. SIOC is an RDF-based OWL model [11]. With this data model, we were able to represent all the annotations collected for a knowledge-annotated scientific workflow project. Our expectation is that this knowledge information, when

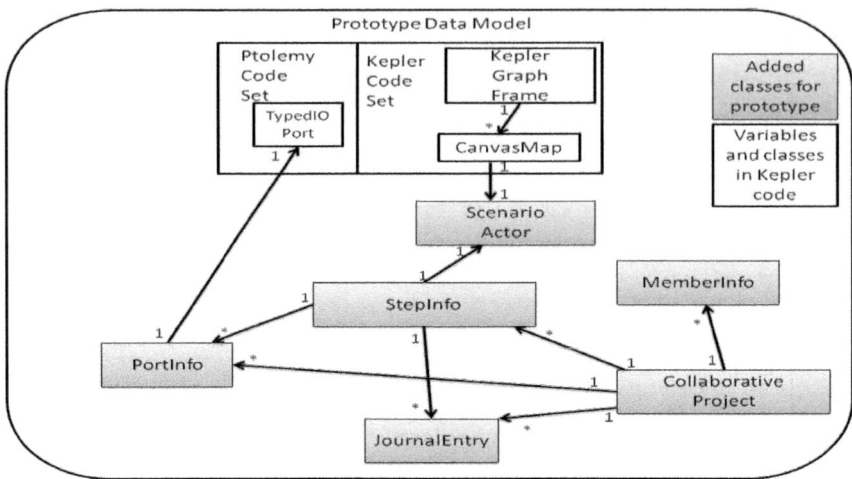

Fig. 9. The knowledge data model for knowledge-annotated scientific workflows in the Kepler prototype. There are five classes used to store all the collected data related to a project.

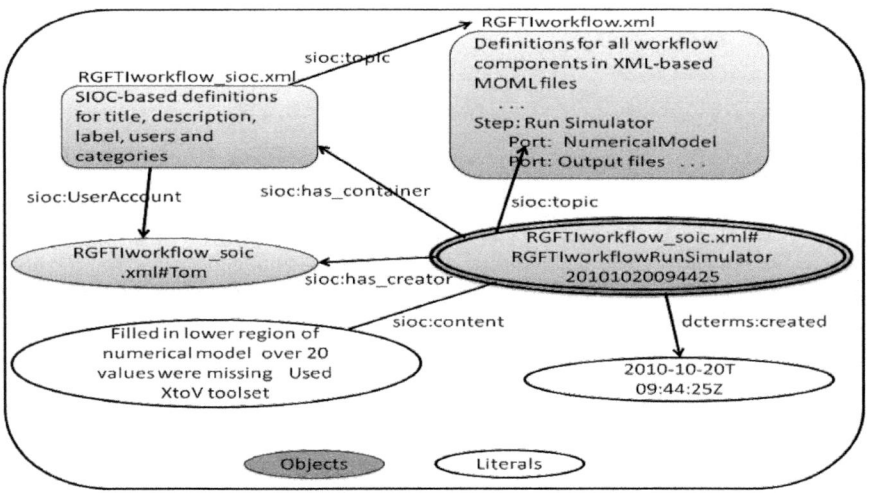

Fig. 10. An RDF graph representation of an SIOC-based description of a comment made in the Run Simulator step. The Run Simulator step is shown in Figure 2

published on the Web, can be linked with different discussions and comments made by other scientists.

Using the unique identifiers for steps, inputs, and outputs in a Kepler workflow as topic resources within the SIOC model (this is how the SIOC model specifies what the annotation is about) the prototype writes a project's annotations to a single SIOC file. Figure 10 shows an RDF graph of the SIOC representation of a comment made about the RunSimulator step's Numerical Model input. The definition of the workflow's SIOC model is contained in the RGFTIworkflow_sioc.xml RDF OWL document. There is an entry in this document for each comment, assumption or constraint related to the workflow, a step or a step's input and output. Following the graph in Figure 10, there is a comment called RGFTI-workflow_soic.xml#RGFTIworkflowRunSimulator20101020094425 that was created by Tom about the NumericalModel input in the RunSimulator step. This comment is found in the RGFTIworkflow_sioc.xml container. The RGFTIwork-flow_sioc.xml container has comments about the workflow in RGFTIworkflow.xml. The SIOC model can be used to describe a variety of details about a comment, including the text in the comment and the creation date.

4   Discussion

Leveraging scientific workflows to document scientific research is enhanced by allowing implicit knowledge annotations to be made during the research process because these annotations reflect why the research was conducted in a specific manner. The prototype presented in this paper has enhanced Kepler by modifying the interface and adding concepts for recording and sharing annotations about the ongoing research. This section discusses issues with the prototype as well as our intentions for future work.

An overriding concern with all information system designs is ensuring that users find the resulting capabilities worth the cost of learning, using, and maintaining information within the system. As with most scientists, the PNNL groundwater scientists are often extremely busy and thus they are hesitant to use technology if it will slow down their work, make them repeat their work, or impose a rigid process that might limit their work. Our main focus while building this prototype was understanding their process and needs. In fact, an executable workflow is not required to capture process and experimental knowledge for our knowledge-annotation system to be useful. Nevertheless, once a scientist focuses on an executable process and specific data sets, significant complexity cannot be avoided. Because Kepler workflows are executable, the prototype extended the framework to include abstract concepts. Through discussing these extensions with the groundwater scientists we were able to confirm that these features and abstractions align with their research methodology. It is our intention to conduct a more formal evaluation of this tool; taking into account different types of scientific scenarios will help us add characteristics to support a more general scientific audience. This evaluation will also help us understand the details in reporting, collaboration and semantic annotations that will make the collection of this data more useful. For example, currently the knowledge annotations are captured as natural language annotations, yet imposing a controlled vocabulary was considered there was a concern with how much additional work would be required by scientists in conforming to a predefined structure. Understanding the most fitting vocabularies requires further evaluation. Moreover, understanding how to capture knowledge annotations and take advantage of their content without distracting scientists is future work for this research.

Mapping the knowledge annotations to a semantic model, i.e. SIOC, has benefits that we must highlight. For one, the knowledge annotations can be leveraged for reasoning within the existing knowledge of a workflow. Bowers and Ludäscher discuss algorithms for propogating semantic annotations through the knowledge of a workflow in order to infer new annotations [4]. Furthermore, integration of knowledge annotations with the process captured by workflows can be integrated with provenance knowledge that has already been used in semantic based workflow tools [6]. Given the content of the knowledge annotations and the descriptions within the SIOC model, software agents can further evaluate the SIOC content for patterns in terminology or collaborative relationships between scientists. By using semantic structures and publishing them on the Semantic Web, the knowledge annotations have the potential for distribution and integration with the Linked Open Data cloud [3]. As a result, software agents from future research efforts could leverage knowledge annotations from one workflow project as supplemental research data.

5   Conclusions

Understanding the needs of the groundwater scientists involved in the RGFTI project highlighted some key facts for this prototype. First, systems that annotate scientific research should also capture the ongoing implicit knowledge associated with this research. Simply collecting comments and discussions after research is conducted or only capturing executable based knowledge, loses important information. Second, by leveraging executable scientific workflow systems to add annotations, scientists can embed their notes within an existing framework already built to describe scientific research. However, the workflow environment must be flexible in its ability to collect these annotations. Many scientists must work in an exploratory mode where their process is ad hoc and the tools must support this. Moreover, scientists do not always know the exact steps they will take to conduct their research beforehand and they do not always have the tools they need instrumented as workflow components. Finally, scientists do not always leverage executable components to describe their process: for example, in some cases they construct data models by hand-picking values. What they do use is expertise, and their expertise is important when they are performing research steps and crucial in understanding *how* and *why* research was conducted to produce scientific results.

Executable scientific workflow tools have an advantage when it comes to documenting research; because they are already documenting the execution of scientific analysis, their definition can be leveraged to manage implicit knowledge collected from scientists conducting the research. Unfortunately, current workflow techniques can be confusing and distracting because scientists are forced into a fixed scientific process before they achieve results. This prototype allows for an ad hoc mode of defining scientific process where scientists focus on documenting research through research notes and workflows through abstract concepts such as steps, inputs and outputs. Furthermore, allowing scientists to see comments about data at strategic points within the research process, gives scientists a process-based and team-driven environment for understanding and describing their work.

Collaborations are characterized not just by a sharing of data but also by the entire process and culture by which scientific research is conducted, data is collected, and knowledge is shared, understood, and reused. One barrier to successful, long-term collaborations is the inability to make specific research artifacts and details available. For example, once data has been collected scientists must decide, amongst other details, how to annotate and publish their data so that it can be used by other scientists. Through knowledge-annotated scientific workflows, scientists are documenting the steps they took to perform their research, the correlation between steps, and why the data was created. Publishing this data to a semantic structure means that the structure of the data is well-defined and enabled for reuse. Knowledge annotated scientific workflows simplify the process of annotating scientific workflows with the scientist's notes - knowledge that is necessary for understanding research but not normally collected.

Acknowledgments

This research was partially funded by the DOE SciDAC Scientific DataManagement Center and by the National Science Foundation under CREST Grant No. HRD-0734825. Any opinions, findings, and conclusions or recommendations expressed in this material are those of the authors and do not necessarily reflect the views of the National Science Foundation.

References

1. EMSL: Capabilities: Subsurface Flow and Transport (January 26, 2011)
2. Kepler KAR: Questions and Answers (January 26, 2011)
3. Bizer, C., Heath, T., Berners-Lee, T.: Linked data - the story so far. International Journal on Semantic Web and Information Systems 5(3), 1–22 (2009)
4. Bowers, S., Ludäscher, B.: A calculus for propagating semantic annotations through scientific workflow queries. In: In Query Languages and Query Processing (QLQP): 11th Intl. Workshop on Foundations of Models and Languages for Data and Objects. LNCS (2006)
5. Brickley, D., Stefan, S., Miles, A., Miller, L., Caoimh, D.O., Neville, C.M.: SIOC Core Ontology Specification. Technical report (March 25, 2003)
6. Da Silva, P.P., Salayandia, L., Del Rio, N., Gates, A.Q.: On the use of abstract workflows to capture scientific process provenance. In: Proceedings of the 2nd Conference on Theory and Practice of Provenance, TAPP 2010, p. 10. USENIX Association, Berkeley (2010)
7. De Roure, D., Goble, C., Stevens, R.: The design and realisation of the virtual research environment for social sharing of workflows. Future Generation Computer Systems 25(5), 561–567 (2009)

8. Deelman, E., Gannon, D., Shields, M., Taylor, I.: Workflows and e-Science: An overview of workflow system features and capabilities. Future Generation Computer Systems 725(5), 528–540 (2008)
9. Lassila, O., Swick, R.: Resource Description Framework (RDF) Model and Syntax Specification. W3C Recommendation (February 22, 1999)
10. Ludäscher, B., Altintas, I., Berkley, C., Higgins, D., Jaeger, E., Jones, M., Lee, E.A., Tao, J., Zhao, Y.: Scientific workflow management and the kepler system: Research articles. Concurr. Comput.: Pract. Exper. 18(10), 1039–1065 (2006)
11. McGuinness, D.L., van Harmelen, F.: OWL Web Ontology Language Overview. Technical report, World Wide Web Consortium (W3C). Proposed Recommendation (December 9, 2003)
12. Salayandia, L., da Silva, P.P., Gates, A.Q., Salcedo, F.: Workflow-driven ontologies: An earth sciences case study. In: Proceedings of the 2nd IEEE International Conference on e-Science and Grid Computing, Amsterdam, Netherlands (December 2006)
13. Senker, J.: The contribution of tacit knowledge to innovation. AI and Society 7, 208–224 (1993)
14. Shadbolt, N., Berners-Lee, T., Hall, W.: The semantic web revisited. IEEE Intelligent Systems 21(3), 96–101 (2006)
15. Talbott, T., Peterson, M., Schwidder, J., Myers, J.D.: Adapting the electronic laboratory notebook for the semantic era. In: Proceedings of the 2005 International Symposium on Collaborative Technologies and Systems (CTS 2005), St. Louis, MO (2005)
16. White, M.D., Oostrom, M.: STOMP Subsurface Transport Over Multiple Phase: User's Guide PNNL-15782(UC 2010). Technical report, Pacific Northwest National Laboratory, Richland, WA (2006)

Improving Workflow Fault Tolerance through Provenance-Based Recovery

Sven Köhler, Sean Riddle, Daniel Zinn,
Timothy McPhillips, and Bertram Ludäscher

University of California, Davis

Abstract. Scientific workflow systems frequently are used to execute a variety of long-running computational pipelines prone to premature termination due to network failures, server outages, and other faults. Researchers have presented approaches for providing fault tolerance for portions of specific workflows, but no solution handles faults that terminate the workflow engine itself when executing a mix of stateless and stateful workflow components. Here we present a general framework for efficiently resuming workflow execution using information commonly captured by workflow systems to record data provenance. Our approach facilitates fast workflow *replay* using only such commonly recorded provenance data. We also propose a *checkpoint* extension to standard provenance models to significantly reduce the computation needed to reset the workflow to a consistent state, thus resulting in much shorter re-execution times. Our work generalizes the rescue-DAG approach used by DAGMan to richer workflow models that may contain stateless and stateful multi-invocation actors as well as workflow loops.

1 Introduction

Scientific workflow systems are increasingly used to perform scientific data analyses [1,2,3]. Often via a graphical user interface, scientists can compose, easily modify, and repeatedly run *workflows* over different input data. Besides automating program execution and data movement, scientific workflow systems strive to provide mechanisms for fault tolerance during workflow execution. There have been approaches that re-execute individual workflow components after a fault [4]. However, little research has been done on how to handle failures at the level of the workflow itself, e.g., when a faulty actor or a power failure takes down the workflow engine itself. Circumstances that lead to (involuntary) workflow failures—for example software errors, power outages or hardware failures—are common in large supercomputer environments. Also, a running workflow might be aborted voluntarily so that it can be migrated to another location, e.g., in case of unexpected system maintenance.

Since typical scientific workflows often contain compute- and data-intensive steps, a simple "restart-from-scratch" strategy to recover a crashed workflow is impractical. In this work, we develop two strategies (namely *replay* and *checkpoint*) that allow workflows to be resumed while mostly avoiding redundant re-execution

J.B. Cushing, J. French, and S. Bowers (Eds.): SSDBM 2011, LNCS 6809, pp. 207–224, 2011.

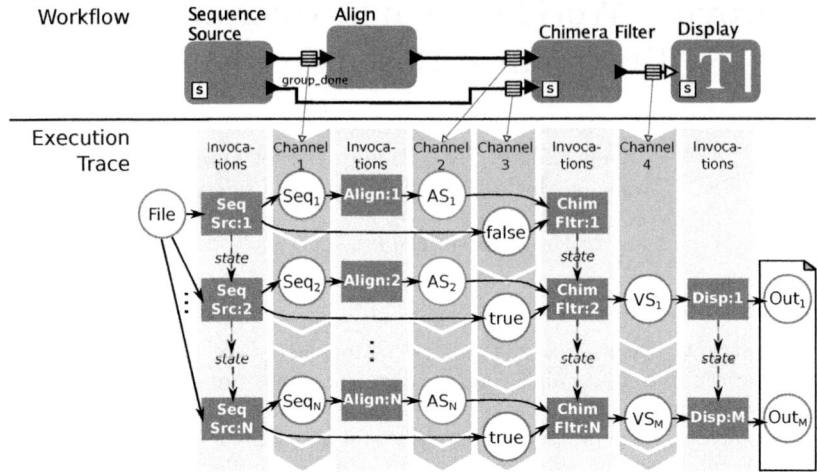

Fig. 1. Example workflow with stateful actors. To recover the workflow execution after a fault, unconsumed tokens inside workflow channels and internal states of all actors except the stateless `Align` have to be restored.

of work performed prior to the fault. The necessary book-keeping information to allow these optimizations is extracted from provenance information that scientific workflow systems often already record for data lineage reasons, allowing our approach to be deployed with minimal additional runtime overhead.

Workflows are typically modeled as dataflow networks. Computational entities (*actors*) perform scientific data analysis steps. These actors consume or produce data items (*tokens*) that are sent between actors over uni-directional FIFO queues (*channels*). In general, output tokens are created in response to input tokens. One round of consuming input tokens and producing output tokens is referred to as an actor *invocation*. For *stateful actors*, the values of tokens output during an invocation may depend on tokens received during previous invocations. The execution and data management semantics are defined by the *model of computation* (MoC).

Commonly used models for provenance are the Read/Write model [5], and the Open Provenance Model (OPM) [6]. In both provenance models, events are recorded when actors consume tokens (`read` or `used_by` events) and produce tokens (`write` or `generated_by` events). Thus, the stored provenance data effectively persists the tokens that have been flowing across workflow channels. We show how this data can be used to efficiently recover faulty workflow executions.

Example. Consider the small scientific pipeline shown in Fig. 1, which carries out two tasks automated by the WATERS workflow described in [1]. As in the full implementation of WATERS, streaming data and stateful multi-invocation actors make an efficient recovery process non-trivial.

The actor `SequenceSource` reads DNA sequences from a text file, emitting one DNA sequence token via the top-right port per invocation. The total number of invocations of `SequenceSource` is determined by the contents of the input file. On the `group_done` port, it outputs a 'true' token when the sequence output is the last of a predefined group of sequences, and 'false' otherwise. `Align` consumes one DNA sequence token per invocation, aligns it to a reference model, and outputs the aligned sequence. The `ChimeraFilter` actor receives the individually aligned sequences from `Align` and the information about grouping from the `SequenceSource`. In contrast to `Align`, `ChimeraFilter` accumulates input sequences, one sequence per invocation, without producing any output tokens until the last sequence of each group arrives. `ChimeraFilter` then checks the entire group for chimeras (spurious sequences often introduced during biochemical amplification of DNA), outputs the acceptable sequences, and clears its accumulated list of sequences.

All actors but `Align` are stateful across invocations: `SequenceSource` and `Display` maintain as state the position within the input file and the output produced thus far, respectively. `ChimeraFilter`'s state is the list of sequences that it has seen so far in the current group. If a fault occurred in the execution of the workflow, the following information will be lost: (1) the content of the queues between actors, i.e., tokens produced by actors but not yet consumed; (2) the point in the workflow execution schedule as observed by the workflow engine; and (3) the internal states of all actors. Correctly resuming workflow execution requires reconstructing all of this information. In many workflow systems, such as Kepler and Taverna it can be challenging to (1) recorde the main-memory actor-actor data transport, i.e. data flowing within the workflow engine without persistent storage[1]; (2) resume workflows that use non-trivial scheduling algorithms for multiple actor invocations based on data availability; and (3) capture the state of stateful actors that are invoked multiple times. In this paper, we show how to do so efficiently with low runtime overhead.
In particular, we make the following contributions:

- We present a general architecture for recovering from workflow crashes, and two concrete strategies (*replay* and *checkpoint*) that provide a balance between recovery speed and required provenance data.
- Our approach is applicable to workflows that can contain both stateful and stateless black-box actors. To the best of our knowledge, this is the first work to consider stateful actors in a fault tolerance context.
- Our approach is applicable to different models of computation commonly used in scientific workflow systems (namely DAG, SDF, PN, and DDF). We achieve this generality by mapping the different models of computations to a common model.
- Our *replay* strategy significantly improves performance over the naïve strategy (77% in our preliminary evaluation). Since this strategy is based on provenance data that is already recorded routinely for data lineage purposes, it adds no runtime overhead.

---

[1] Even if data is persisted to disk due to large data sizes, data handles are usually kept in main memory.

– Finally, we propose an extension to commonly used provenance models, i.e., to record actor states at appropriate points in time. This not only adds information valuable from a provenance point of view, but also enables our *checkpoint* strategy to recover workflows in a very short time span (98% improvement over a naïve re-execution) independent of the amount of work performed prior to the workflow crash.

The rest of the paper is organized as follows. Section 2 presents the fundamentals of our workflow recovery framework. In Section 3, we describe two recovery strategies and how to apply them to different models of computation. Section 4 reports on our prototypical implementation and preliminary evaluation. In Section 5, we provide a brief discussion of related work, and we conclude in Section 6.

2   Fault Tolerance Approach

Our approach generalizes the rescue-DAG method [7,8,9], which is used to recover DAGMan workflows after workflow crashes. DAGMan is a single-invocation model of computation, i.e., all actors are invoked only once with a "read-input—compute—write-output" behavior. The rescue-DAG is a sub-graph of the workflow DAG containing exactly those actors that have not yet finished executing successfully. After a crash, the rescue-DAG is executed by DAGMan, which completes the workflow execution.

To facilitate the execution of workflows on streaming data, several models of computation (e.g., Synchronous DataFlow (SDF) [10], Process Networks (PN) [11], Collection Oriented MOdeling and Design (COMAD) [12] and Taverna [13]) allow actors to have multiple invocations.

If the rescue-DAG approach were applied directly to workflows based on these models of computation, i.e., if all actors that had not completed all of their invocations were restarted, then in many cases a large fraction of the actors in a resumed workflow would be re-executed from the beginning. Instead, our approach aims to resume each actor after its last successful invocation. The difficulties of this approach are the following: (1) The unfolded *trace graph* (which roughly corresponds to the rescue-DAG) is not known a priori but is implicitly determined by the input data. (2) Actors can maintain internal state from invocation to invocation. This state must be restored. (3) The considered models of computation (e.g., SDF, PN, COMAD, Taverna) explicitly model the flow of data across channels, and the corresponding workflow engines perform these data transfers at run time. A successful recovery mechanism in such systems thus needs to re-initialize these internal communication channels to a consistent state. In contrast, data movement in DAGMan workflows is handled by the actors opaquely to the DAGMan scheduler (e.g., via naming conventions) or by a separate system called Stork [14]; materializing on disk all data passing between actors simplifies fault tolerance in these cases.

In the following, we present a simple relational model of workflow definitions and provenance information. We employ this model to define recovery strategies using logic rules. Due to space restrictions, we concentrate here on the SDF and PN models of computation. SDF represents a model that is serially executed according to a statically defined schedule, while PN represents the other extreme of a parallel schedule only synchronized through the flow of data.

2.1   Basic Workflow Model

Scientific workflow systems use different languages to describe workflows and different semantics to execute them. However, since most scientific workflow systems are based on dataflow networks [15,11], a common core that describes the basic workflow structure can be found in every model of computation.

Core Model. Many workflow description languages allow nesting, i.e., embedding a sub-workflow within a workflow. The relation `subworkflow(W,Pa)` supports this nesting in our schema and stores a tuple containing the sub-workflow name `W` and the parent workflow name `Pa`. Each workflow in this hierarchy is associated with a model of computation (MoC) using the relation `moc(W,M)` that assigns the MoC `M` to the workflow `W`.

Actors represent computational functions that are either implemented using the language of the workflow system or performed by calling external programs. The separation of computations in multiple invocations sometimes requires that an actor maintains state across invocations.

Stateless Actor. The values of tokens output during an invocation depend only on tokens input during this invocation.

Stateful Actor. The values of tokens output during an invocation may depend on tokens received during previous invocations.

The predicate `actor(A,W,S)` embeds an actor with unique name `A` into the workflow `W`. The flag `S` specifies whether the actor is stateful or stateless.

Although the data shipping model is implemented differently in various workflow systems, it can be modeled uniformly as follows: Each actor has named ports, which send and receive data tokens. One output port can be connected to many input ports. In this situation, the token is cloned and sent to all receivers. Connecting multiple output ports to one channel is prohibited due to the otherwise resulting write conflicts. Ports are expressed with the predicate `port(A,P,D)` in our schema. The port with name `P` is attached to actor `A`. `D` specifies the direction in which data is sent, i.e., `in` or `out`. Ports are linked through the relation `link(A,P,L)` by sharing the same link identifier `L` (the third parameter of the link relation). A link from the port `p` of actor `a` to the port `q` of actor `b` is encoded as `link(a,p,l)` and `link(b,q,l)`.

Application to Process Networks with Firing. A Process Network (PN), as defined by [15], is a general model of computation for distributed systems. In Kahn PN, actors communicate with each other through unbounded unidirectional FIFO channels. Workflows of the model *PN with firings* [11], a refinement of Kahn PN, can be described with the four core relations `Subworkflow`, `Actor`, `Port`, and `Link`. The PN execution semantics allow a high level of parallelism, i.e., all actors can be invoked at the same time. After an invocation ends, the actor will be invoked again, consuming more data. This procedure stops either when the actor explicitly declares completion or by reaching the end of the workflow execution. A PN workflow ends when all remaining running invocations are deadlocked on reading from an input port.

Application to Synchronous DataFlow (SDF). Besides the data captured by the four core relations (Subworkflow, Actor, Port, and Link), workflow models can provide additional information. As an example, SDF workflow descriptions require annotations on ports. In SDF, output ports are annotated with a fixed *token production rate* and input ports have a fixed *token consumption rate*. Both rates are associated with ports using the predicate token_transfer(A,P,N) in our model. During an invocation, each actor A is required to consume/produce N tokens from the input/output port P.

Another extension is the *firing count* of an actor that specifies the maximum number of actor invocations during a workflow execution. The predicate firing_count(A,N) provides this number (N) for an actor A.

Unlike in PN, where the actors synchronize themselves through channels, the execution of SDF is based on a static schedule that is repeatedly executed in *rounds*. The number of firings of each actor per round is determined by solving balance equations based on token production and consumption rates [10].

2.2   Review of Provenance Model

Another critical part of our approach is the definition of a simple, prototypical provenance model. It defines which observables are recorded during runtime. The Open Provenance Model (OPM) [6] captures the following basic observables: (1) artifact generation, i.e., token production; (2) artifact use, i.e., token consumption; (3) control-flow dependencies, i.e., was_triggered_by relation; and (4) data dependencies, i.e., was_derived_from relation. A more comprehensive provenance schema was defined by Crawl et al. in [16]. It captures the OPM observables in more detail, e.g., it provides timestamps for the beginning and end of invocations. In addition, it records metadata about the workflow execution as well as the evolution of the workflow. All provenance information is recorded by the workflow system transparently without modifications to actors.

Our provenance model uses the basic observables from OPM and adds additional details about events that occurred during an invocation cycle. As soon as an invocation starts, the actor name A and its corresponding invocation number N are stored in the relation invocation(I,A,N,Z) with the status attribute Z set to running. A unique identifier I is assigned to each invocation. Some models of computation allow an actor to indicate that all invocations are completed, for instance if the maximum firing count in SDF is reached. This information is captured in our provenance model as well. When an actor successfully completes an invocation and indicates that it will execute again, the status attribute in the corresponding provenance record is updated to iterating. Otherwise, this attribute status is set to done.

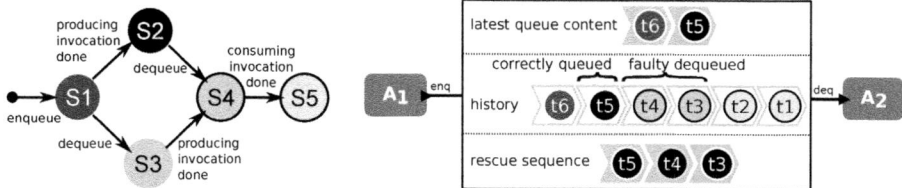

Fig. 2. Input queues with history and token state. Each token produced during workflow execution can be in one of five states. Events on the producing and consuming actors trigger transitions between token states, shown on the left. The right graph shows three views of a channel: (1) the current content of the queue during an execution in the first row, (2) the history of all tokens passed through this channel associated with their state in the middle row, and (3) the *rescue sequence* of tokens that needs to be restored in the third row.

The second observable process in our model is the flow of tokens. Many workflow engines treat channels that define the dataflow as first-class citizens of the model. The dependencies between data tokens are of general interest for provenance. They can be inferred from the core workflow model in combination with the token consumption (read) and production (write) events.

Our model stores read and write events in the `event(Y,T,I,Po,N)` relation. The first entry `Y` determines the event type, i.e., token production events are indicated by the constant `w` while consumption events are encoded with the value `r`. `T` is the data token to be stored. The following two attributes specify which actor invocation `I` triggered this event and on which port `Po` it was observed. The last element `N` in the tuple is an integer value that is used to establish an order of events during the same actor invocation on the same port. Establishing an order using timestamps is not practical because of limited resolution and time synchronization issues.

Based on the `event` relation, the queues of all channels can be reconstructed for any point in time. Figure 2 shows a queue at the time of a workflow failure. Using provenance we can restore the whole history of this queue (shown in the middle right). Based on this history, we can determine the *rescue sequence* of tokens that are independent of failed invocations, i.e., tokens in state S2 and S4.

3   Recovery Strategies

Depending on the model of computation and the available provenance data, different recovery approaches can be used. We will now present our two strategies *replay* and *checkpoint*.

3.1   The Replay Strategy: Fast-Forwarding Actors

Re-running the entire workflow from the beginning is a naïve recovery strategy, which is often impractical, especially when a long-running workflow fails a significant period of time into its execution. The role of provenance in restoring a workflow execution is similar to that of log files used in database recovery.

Stage 1. In the first stage of the *replay* strategy, the point of a failure is determined using provenance information. Invocations of actors that were running when the fault occurred are considered faulty and their effects have to be undone. Query (1) retrieves the invocation identifiers I of faulty invocations.

$$\texttt{faulty\_invoc(I) :- invocation(I,\_,\_,running).} \tag{1}$$

Actors with invocation status done are not recovered, since they are not needed for further execution. All other actors A are retrieved by query (2) and they need to be recovered.

$$\texttt{finished\_actors(A) :- invocation(\_,A,\_,done).}$$
$$\texttt{restart\_actors(A) :- actor(A,\_,\_), not finished\_actors(A).} \tag{2}$$

Stage 2. If an actor is stateless, it is ready to be resumed without further handling. However, if an actor is stateful, its internal state needs to be restored to its *pre-failure state*, i.e., the state after the last successful invocation. Each actor is executed individually by presenting it with all input data the actor received during successful invocations. This input data is retrieved from the provenance log, where it is readily available. The replay(A,I) query (3) extracts the identifiers of all actor invocations that need to be replayed. The tokens needed for those re-invocations are provided by (4). This query retrieves for each port P of actor A the tokens T that are needed to replay invocation I. N is the sequence number of token T at input port (queue) P. The replay does not need to be done in the same order as in the original workflow schedule. All actors can be re-executed in parallel using only the input data recorded as provenance. The actor output can either be discarded or checked against the recorded provenance to verify the workflow execution.

$$\texttt{replay(A,I) :- actor(A,\_,stateful), invocation(I,A,\_,\_),} \tag{3}$$
$$\texttt{   not faulty\_invoc(I).}$$
$$\texttt{replay\_token(A,P,I,T,N) :- replay(A,I), event(r,T,I,P,N).} \tag{4}$$

In order to correctly recover a workflow execution, the problem of side-effects still needs to be addressed. Either stateful actors should be entirely free of side-effects or side-effects should be idempotent. That is, it must not matter whether the side-effect is performed once or multiple times. Examples of side-effects in scientific workflows include the creation or deletion of files, or sending emails. Deleting a file (without faulting if the file does not exist) is an idempotent operation. Further, creating a file is idempotent if an existing file is overwritten. Sending an email is, strictly speaking, not idempotent, since if done multiple times, multiple emails will be sent.

Stage 3. Once all actors are instantiated and in pre-failure state, the queues have to be initialized with the *restore sequence*, i.e., all valid tokens that were present before the execution failed. Tokens created by faulty invocations must be removed, and those consumed by a failed invocation are restored. This information is available in basic workflow provenance and can be queried using (5).

For each port Po of an actor A the query retrieves tokens T with the main order specified by the invocation order N1. However, if multiple tokens are produced in one invocation, the token order N2 is used for further ordering.

The auxiliary view invoc_read(A,P,T) contains all actors A and the corresponding ports P that read token T. The view connect(A1,P1,C,A2,P2) returns all output ports P1 of actor A1 that are connected to actor A2 over input port P2 through channel C. The auxiliary rule (5.1) computes the queue content in state S2 (see Fig. 2), i.e., tokens that were written by another actor but not yet read by actor A2 on port P2. The second rule (5.2) adds back the queue content in state S4, i.e., tokens that were read by a failed invocation of actor A2.

$$\text{current\_queue(A2,P2,T,N1,M1) :- queue\_s2(A2,P2,T,N1,M1).} \tag{5}$$

$$\text{current\_queue(A2,P2,T,N1,M1) :- queue\_s4(A2,P2,T,N1,M1).}$$

$$\begin{aligned} \text{queue\_s2(A2,P2,T,N1,M1) :- connect(A1,P1,C,A2,P2),} \\ \text{invocation(I1,A1,N1,\_), event(w,T,I1,P1,M1),} \\ \text{not invoc\_read(A2,P2,T), not faulty\_invoc(I1).} \end{aligned} \tag{5.1}$$

$$\begin{aligned} \text{queue\_s4(A2,P2,T,N1,M1) :- connect(A1,P1,C,A2,P2),} \\ \text{invocation(I1,A1,N1,\_), event(w,T,I1,P1,M1),} \\ \text{invocation(I2,A2,\_,\_),event(r,T,I2,P2,\_),} \\ \text{faulty\_invoc(I2).} \end{aligned} \tag{5.2}$$

Stage 4. After restoring actors and recreating the queues, faulty invocations of actors that produced tokens which were in state S3 have to be repeated in a "sandbox". This ensures that tokens in state S3 are not sent to the output port after being produced but are discarded instead. If these tokens are sent, then invocation based on them are duplicated. Rule (6) determines tokens T that were in state S3 and it returns the invocation ID I, the port P this token was sent from, and the sequence number in which the token was produced. Query (7) determines which invocations produced tokens in state S3 and therefore have to be repeated in a sandbox environment.

$$\begin{aligned} \text{queue\_s3(I,P,T,N) :- invocation(I,A1,\_,\_),} \\ \text{faulty\_invoc(I), event(w,T,I,P,N),} \\ \text{connect(A1,P,C,A2,P2), invocation(I2,A2,\_,\_),} \\ \text{not faulty\_invoc(I2), event(r,T,I2,P2,\_).} \end{aligned} \tag{6}$$

$$\text{invoc\_sandbox(I) :- faulty\_invoc(I), queue\_s3(I,\_,\_,\_,\_).} \tag{7}$$

After executing the sandbox, the workflow is ready to be resumed. The recovery system provides information about where to begin execution (i.e., the actor at which the failure occurred) to the execution engine (e.g., the SDF scheduler) and then the appropriate model of computation controls execution from that point on.

To summarize, the most expensive operation in the *replay* strategy is the re-execution of stateful actors, which is required to reset the actor to its pre-failure state. Our *checkpoint* strategy provides a solution to avoid this excessive cost.

3.2   The Checkpoint Strategy: Using State Information

Many existing workflow systems are shipped with stateful actors or new actors are developed that maintain state. Because actors in scientific workflows usually have complex and long-running computations to perform, the *replay* strategy can be very time-consuming.

Current provenance models, such as the one used in [16], either do not include the state of actors or record limited information about state as in [17]. The Read-Write-Reset model (as presented in [18]), e.g., records only state reset events, which specify that an actor is in its initial state again. This can be seen as a special case of the *checkpoint* strategy we will present, where states are only recorded when they are equal to the initial state.

To support a faster recovery, we propose to make the actor's state a distinct observation for provenance. Recording state information not only helps to recover workflows, but also makes provenance traces more meaningful: Instead of linking an output token of a stateful actor to all input tokens across its entire history, our model links it to the state input and the current input only.

An actor's state can be recorded by the workflow engine at any arbitrary point in time when the actor is not currently invoked. To integrate checkpointing into the order of events, we store state information immediately after an invocation, using the invocation identifier as a reference for the state. The predicate state(I,S) stores the actor's state S together with the identifier of the preceding invocation I of that actor. The information required to represent an actor state depends on the workflow system implementation.

Given this additional state information the workflow recovery engine can speed up the recovery process. The *checkpoint* strategy is based on the *replay* strategy but extends it with checkpointing.

Stage 1. When normally executing the workflow, state is recorded in provenance. In case of a fault, the recovery system first detects the point of failure. Then the provenance is searched for all checkpoints written for stateful actors. Rule (8) retrieves the state S of each invocation I of a given actor A. If no state was recorded then the invocation will not be contained in this relation:

restored_state(A,I,N,S) :- actor(A,_,stateful),   (8)
    invocation(I,A,N,_),state(I,S), not faulty_invoc(I).

If states were stored for an actor, this actor is updated with the latest available state. Rule (9) will determine the latest recoverable invocation I and the *restorable pre-failure state* S captured after that invocation.

restored_stateGTN(A,I,N2) :- restored_state(A,I,N,_), N > N2.
latest_state(A,I,S) :- restored_state(A,I,N,S),   (9)
    not restored_stateGTN(A,I,N).

Stage 2. Now only those successfully completed invocations that started after the checkpoint have to be replayed. This will use the same methods described above for the *replay* strategy.

Stage 3 and 4. Same as in the *replay* strategy.

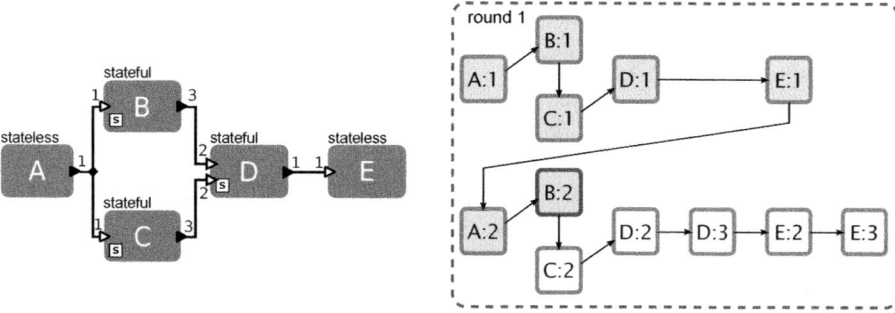

Fig. 3. SDF workflow example with state annotations on the left and a corresponding schedule on the right. The schedule describes the execution order. The red circle indicates the failure during the second invocation of B.

If provenance information is kept only for fault tolerance, not all data needs to be stored persistently. Only data tokens representing the active queues or consumed by stateful actors after the last checkpoint need to be persisted. All other data can be discarded or stored in a compressed representation.

3.3   Recovering SDF and PN Workflows

The SDF example in Figure 3 demonstrates our *checkpoint* strategy. Below, we explain the differences when dealing with PN models.

Synchronous Dataflow (SDF). Figure 3 shows a sample SDF workflow with ports annotated with consumption and production rates for input and output ports, respectively. Actors A and E are stateless while all other actors maintain state. A is a source and will output one data item (token) each time the actor is invoked. A also has a *firing count* of two, which limits the total number of its invocations. Actors B and C consume one token on their input ports and output three tokens per invocation. Outputs are in a fixed, but distinguishable, order, so that an execution can fail between the production of two tokens. Actor D will receive two tokens in each invocation from both B and C and will output one new token.

The schedule in Fig. 3 was computed, as usual, before the workflow execution begins, based on the token production and consumption rates in the model. Actors were then invoked according to the schedule until the workflow crash occurred. All invocations up to the second invocation of A (A:2) completed successfully, invocation B:2 was still running and all other invocations were scheduled for the future. The failed workflow execution, together with checkpointing and data shipping, is summarized in Fig. 4. For recovery, the workflow description as well as the recorded provenance is used by our *checkpoint* strategy. In the following, we will describe the details of the recovery process (shown in Figure 4).

Stateless actors are in the correct state (i.e., pre-failure state) immediately after initialization. That is why actor E is in its proper state after simple initialization

218   S. Köhler et al.

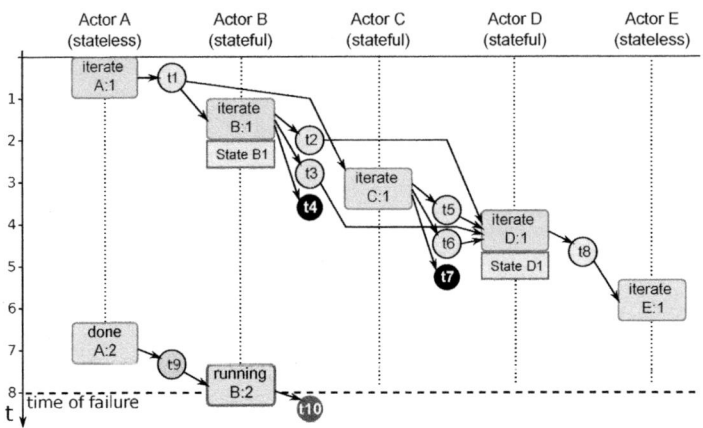

Fig. 4. Workflow execution up to failure in B:2. The states of actors B and D are stored, but no checkpoint exists for C. Token t1 is only send once, but is duplicated by link to both actors B and C. Tokens t4 and t7 are never read. Token t9 is read by a faulty invocation, and t10 is written by a faulty invocation.

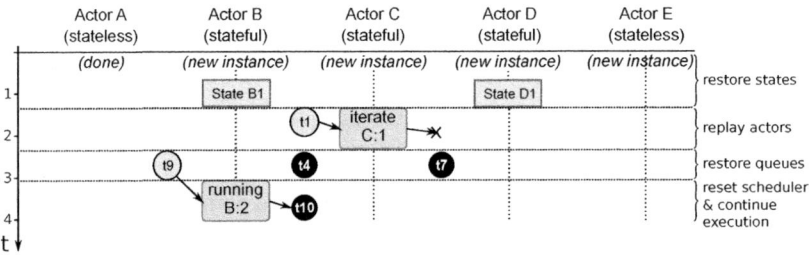

Fig. 5. Individual stages to recover the sample workflow with *checkpoint* strategy. Note how only a small amount of invocations are repeated (compared to Figure 4).

and actor A is identified as done and will be skipped. Both actors B and D are stateful and a checkpoint is found in the provenance. Therefore, the recovery system instantiates these actors and restores them to the latest recorded state. The state at this point in the recovery process is shown in Fig. 5 above the first horizontal line. Second, stateful actors that either have no checkpoints stored (e.g., actor C) or that have successful invocations after the last checkpoint need to have those invocations replayed. To do this for actor C:1, the input token t1 is retrieved from the provenance store. In the next stage, all queues are restored. Since the second invocation of actor B failed, the consumed token t9 is restored back to the queue. Additionally, all tokens that were produced but never consumed (e.g., t4 and t7) are restored to their respective queues. After all queues are rebuilt, the recovery system has to initialize the SDF scheduler to resume execution. This entails setting B as the next active actor, since its invocation was interrupted by the crash. After setting the next active actor, the recovery system can hand over the execution to the workflow execution engine. This final recovery state is shown in Fig. 5. The recovery time is significantly improved compared to the original runtime shown in Fig. 4.

Process Networks (PN). The example SDF workflow shown in Fig. 3 can also be modeled using PN. Since actors under PN semantics have variable token production and consumption rates, these constraints cannot be leveraged to narrow the definition of a faulty invocation. Additionally, repeated invocations are not necessarily required for actors to perform their function. For instance, actor D can be invoked only once, while actor B is invoked multiple times. All invocations in PN run concurrently, and tokens on a port have to be consumed after they are produced and only in the order they were produced. Finally, there are no defined firing limits. Many systems allow an actor to explicitly declare when it is done with all computations, which is recorded in provenance. Actors without that information are invoked until all actors in the workflow are waiting to receive data.

These characteristics have some implications on the recovery process. First, since all actors are executed in parallel, a crash can affect all actors in a workflow. Since actors are invoked in parallel during workflow execution, the recovery engine can safely restore actors in parallel. All actors are instantiated simultaneously at the beginning of the workflow run, in contrast to Fig. 4. Long-running actor invocations reduce the availability of checkpoints and cause longer replay times. Finally, PN uses deadlock detection to define the end of a workflow, which makes it difficult to determine whether a particular actor is actually done (unless it explicitly says so) or just temporary deadlocked. Anything short of all actors being deadlocked by blocking reads (meaning the workflow is done) gives no useful information about which actors will exhibit future activity.

4   Evaluation

To evaluate the performance of the different recovery strategies, we implemented a prototype of our proposed approach in Kepler [19]. The current implementation adds fault tolerance to non-hierarchical SDF workflows.

Implementation. Our fault tolerance framework implements all features necessary for the *checkpoint* recovery strategy (as well as the *replay* strategy) in a separate workflow restore class that is instrumented from the director. We used the provenance system of Crawl et al. [16], which was altered to allow the storage of actor states and tokens. Instead of storing a string representation of a token, which may be lossy, we store the whole serialized token in the provenance database. When using the standard token types, this increases the amount of data stored for each token only slightly. Actors can be explicitly marked as stateless using an annotation on the implementing Java class. Thus, we avoid checkpointing and replay for stateless actors.

During a normal workflow execution, the system records all tokens and actor invocations. Currently, checkpoints for all stateful actors are saved after each execution of the complete SDF schedule. An actor's state is represented by a serialization of selected fields of the Java class that implements the actor. There are two different mechanisms that can be chosen: (1) a blacklist mode that checks fields against a list of certain transient fields that should not be serialized, and (2) a whitelist mode that only saves fields explicitly annotated as state. The serialized state is then stored together with the last invocation id of the actor in the state relation of Kepler's provenance database. The serialization process is based on Java's object serialization and also includes selected fields of super classes.

During a recovery, the latest recorded checkpoint of an actor is restored. All stored actor fields are deserialized and overwrite the actor's fields. This leaves transient member fields intact and ensures that the restored actor is still properly integrated into its parent workflow. Successful invocations completed after a checkpoint or where no checkpoint exists are replayed to restore the correct pre-failure state. For the replay, all corresponding serialized tokens are retrieved from the provenance database. Then, the input queues of an actor are filled with tokens necessary for one invocation and the actor is fired. Subsequently, input and output queues are cleared again before the next invocation of an actor is replayed. The current implementation replays actors serially.

Next, all the queues are restored. For each actor, all tokens are retrieved that were written to an input port of the actor and not read by the actor itself before the fault. These tokens are then placed in the proper queues, preserving the original order. Finally, the scheduling needs modifications to start at the proper point. This process is closely integrated with the normal execution behavior of the SDF director. The schedule is traversed in normal order, but all invocations are skipped until the failed invocation is reached. At this stage, the normal SDF execution of the schedule is resumed.

Preliminary Experimental Evaluation. For an initial evaluation of the practicality of a provenance-based recovery, we created the synthetic workflow shown in Fig. 6. This model simulates a typical scientific workflow with long-running computations and a mix of stateless and stateful actors. We ran this workflow to its completion to measure the running time for a successful execution. We then interrupted the execution during the third invocation of actor C. After this, we loaded the workflow again and resumed its execution using the three different strategies: re-execution from the beginning, *replay* and *checkpoint*.

We ran the experiments ten times for each strategy. In a typical scientific workflow with computation time domination over the data transfer time, provenance recording adds an overhead of 139 milliseconds (with a standard deviation σ of $157.22ms$) to the workflow execution time of 80.7 seconds ($\sigma = 0.16s$). The naïve approach of re-running the whole workflow takes about 80.8 seconds ($\sigma = 0.017s$), repeating 55.8 seconds ($\sigma = 0.037s$) of execution time from before the crash. The *replay* strategy based on standard provenance already achieves a major improvement. The total time for a recovery using this strategy of approximately 12.8 seconds ($\sigma = 0.17s$) is dominated by replaying two invocations of stateful actor C in 10 seconds. The remaining 2.8 seconds are the accumulated overhead for retrieving and deserializing tokens for the replay as well as

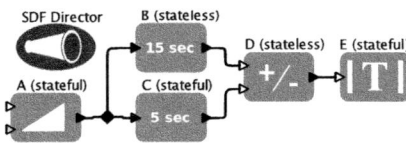

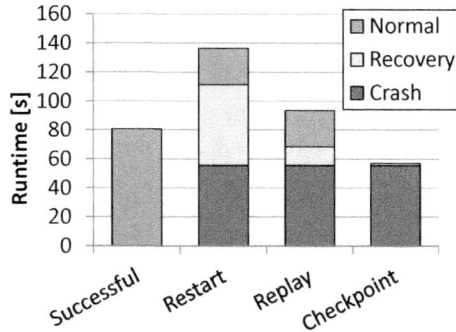

Fig. 6. Synthetic SDF workflow. Actor A is a stateful actor generating a sequence of increasing numbers starting from 0. B is a stateless actor that has a running time of 15 seconds. C is stateful and needs 5 seconds for each invocation. D is fast running stateless actor. E is a stateful "Display" actor.

Fig. 7. Performance evaluation of different recovery strategies.

for restoring the queue content. After the recovery, the workflow execution finishes in 25.8 seconds ($\sigma = 0.02s$). The *replay* strategy reduced the recovery cost from 55.8 seconds to 12.8 seconds, or by 77%, in this workflow. The checkpoint strategy reduced the recovery time to only 1.3 seconds ($\sigma = 0.03s$), including the time for the deserialization process of state as well as the queue restoration. This strategy reduces the recovery time of the synthetic workflow by 97.6% compared to the naïve strategy. Checkpointing is so efficient because it does not scale linearly with the number of tokens sent like the naïve and *replay* strategies. This strategy also benefits from invocation runtimes that are significantly longer than the checkpointing overhead.

5   Related Work

In DAGMan [9], a basic fault tolerance mechanism (rescue-DAG) exists. Jobs are scheduled according to a directed graph that represents dependencies between those jobs. Initially the rescue-DAG contains the whole DAG but as jobs execute successfully, they are removed from the rescue-DAG. If a failure occurs, the workflow execution can thus be resumed using the rescue-DAG, only repeating jobs that were interrupted.

Feng et al. [17] present a mechanism for fault management within a simulation environment under real time conditions. Starting from a "checkpoint" in the execution of an actor, state changes are recorded incrementally and can then be undone in a "rollback". This backtracking approach allows to capture the state of an actor through a preprocessing step that adds special handlers for internal state changes wherever a field of an actor is modified. However, this solution can only be used during runtime of the workflow system. It does not provide checkpoints that cover the full state, and, more importantly, no persistent state storage is available for access after a workflow crash.

Dan Crawl et al. [16] employed provenance records for fault tolerance. Their Kepler framework allows the user to model the reactions upon invocation failures. The user can either specify a different actor that should be executed or that the same actor should be invoked again using input data stored in provenance records. However, they don't provide a fast recovery of the whole workflow system. Neither is the approach applicable for stateful actors.

Fault tolerance in scientific workflows has often been addressed using caching strategies. While still requiring a complete restart of the workflow execution, computation results of previous actor invocations are stored and reused. Swift [20] extends the rescue-DAG approach by adding such caching. During actor execution, a cache is consulted (indexed by the input data), and if an associated output is found, it will be used, avoiding redundant computation. Swift also employs this strategy for optimizing the re-execution of workflow with partially changed inputs. Conceptually, this can be seen as an extension of the rescue-DAG approach. Podhorszki et al. [3] described a checkpoint feature implemented in the ProcessFileRT actor. This actor uses a cache to avoid redundant computations. A very similar approach was implemented by Hartman et al. [1]. Both techniques are used to achieve higher efficiency for computation and allow a faster re-execution of workflows. However, these implementations are highly customized to their respective use cases and integrated in one or several actors rather being a feature of the framework. Also, [3] assumes that only external programs are compute intensive, which is not always the case, as can be seen in [1], where actors perform compute intensive calculations within the workflow system. Furthermore, caching strategies can only be applied to stateless actors, making this approach very limited. In contrast, our approach aims to integrate fault tolerane mechanisms into the workflow engine. Stateless actors are not re-executed during a recovery, since input and corresponding outputs are available in provenance, and the actor state does not need to be restored.

Wang et al. [21] presented a transactional approach for scientific workflows. Here, all effects of arbitrary subworkflows are either completed successfully or in case of a failure undone completely (the dataflow-oriented hierarchical atomicity model is described in [21]). In addition, it provides a dataflow-oriented provenance model for those workflows. The authors assumed that actors are white boxes, where data dependencies between input and output tokens can be observed. They describe a smart re-run approach similar to those presented by Podhorszki et al. and Hartman et al. [1]. Input data of actors is compared to previous inputs, and if an actor is fired with the same data, the output can easily be restored from provenance information rather than re-executing the actor. This white box approach differs from our black box approach that requires setting the internal state of stateful actors. Our system is more generally applicable, as not all actors are available in a white box form that allows for the direct observation of dependencies.

6   Conclusion

We introduced a simple relational representation of workflow descriptions and their provenance information in order to improve fault-tolerance in scientific workflow systems. To the best of our knowledge, our approach is the first to handle not only individual actor failures, but (i) failures of the overall workflow, where workflows (ii) can have a stream-oriented, pipeline-parallel execution model, and (iii) can have loops, and where (iv) actors can be stateful and stateless. Another unique feature of our approach is that the workflow system itself, upon "smart resume" can handle the recovery, i.e., unlike other current approaches, neither actors nor the workflow are burdened with implementing parts of the recovery logic, since the system takes care of everything. To allow for checkpointing of internal state from stateful actors, we have developed an extension to the standard OPM-based provenance models. Information necessary to recover a failed execution of a scientific workflow is extracted from the relational representation via logic rules, allowing our approach to be easily deployed on various provenance stores. We defined and demonstrated a *replay* strategy that speeds up the recovery process by only re-executing stateful actors. Our *checkpoint* strategy improves on *replay* by using the saved checkpoints to significantly reduce actor re-execution. We implemented our approach in the Kepler system. In a preliminary evaluation, we compared our strategies to a naïve re-execution. Here, *replay* and *checkpoint* could reduce recovery times by 77% and 98%, respectively. This highlights the advantage of checkpointing in scientific workflows with compute intensive stateful actors.We plan to add support for other models of computation, e.g. dynamic dataflow (DDF) [22] to our Kepler implementation, in order to add fault tolerance to specific complex workflows [3]. We also plan to port our approach to other systems, e.g., RestFlow [23]. Another enhancement will be to parameterize the time between checkpoint saving as either a number of invocations, or in terms of wall-clock time to balance the overhead of provenance recording and recovery time.

Acknowledgments. Work supported through NSF grant OCI-0722079 and DOE grant DE-FC02-07ER25811.

References

1. Hartman, A., Riddle, S., McPhillips, T., Ludäscher, B., Eisen, J.: Introducing W.A.T.E.R.S.: a Workflow for the Alignment, Taxonomy, and Ecology of Ribosomal Sequences. BMC Bioinformatics 11(1), 317 (2010)
2. Ceyhan, E., Allen, G., White, C., Kosar, T.: A grid-enabled workflow system for reservoir uncertainty analysis. In: Proceedings of the 6th Int'l Workshop on Challenges of Large Applications in Distributed Environments, CLADE 2008 (2008)
3. Podhorszki, N., Ludäscher, B., Klasky, S.A.: Workflow automation for processing plasma fusion simulation data. In: Proceedings of the 2nd Workshop on Workflows in Support of Large-Scale Science, WORKS 2007, New York, NY, USA, pp. 35–44 (2007)
4. Missier, P., Soiland-Reyes, S., Owen, S., Tan, W., Nenadic, A., Dunlop, I., Williams, A., Oinn, T., Goble, C.: Taverna, reloaded. In: Gertz, M., Ludäscher, B. (eds.) SSDBM 2010. LNCS, vol. 6187, pp. 471–481. Springer, Heidelberg (2010)
5. Bowers, S., McPhillips, T., Ludäscher, B., Cohen, S., Davidson, S.: A Model for User-Oriented Data Provenance in Pipelined Scientific Workflows. In: Moreau, L., Foster, I. (eds.) IPAW 2006. LNCS, vol. 4145, pp. 133–147. Springer, Heidelberg (2006)

6. Moreau, L., Freire, J., Futrelle, J., McGrath, R., Myers, J., Paulson, P.: The Open Provenance Model: An Overview. In: Freire, J., Koop, D., Moreau, L. (eds.) IPAW 2008. LNCS, vol. 5272, pp. 323–326. Springer, Heidelberg (2008)
7. Frey, J.: Condor DAGMan: Handling inter-job dependencies. Technical report, University of Wisconsin, Dept. of Computer Science (2002)
8. Deelman, E., Blythe, J., Gil, Y., Kesselman, C., Mehta, G., Patil, S., Su, M.-H., Vahi, K., Livny, M.: Pegasus: Mapping Scientific Workflows onto the Grid. In: Dikaiakos, M.D. (ed.) AxGrids 2004. LNCS, vol. 3165, pp. 11–20. Springer, Heidelberg (2004)
9. Hernandez, I., Cole, M.: Reliable DAG scheduling on grids with rewinding and migration. In: Proceedings of the First Int'l Conference on Networks for Grid Applications, GridNets 2007. pp. 3:1–3:8. ICST (2007)
10. Lee, E.A., Messerschmitt, D.G.: Static scheduling of synchronous data flow programs for digital signal processing. IEEE Trans. Comput. 36, 24–35 (1987)
11. Lee, E., Matsikoudis, E.: The semantics of dataflow with firing. In: From Semantics to Computer Science: Essays in Memory of Gilles Kahn. Cambridge University Press, Cambridge (2008)
12. Dou, L., Zinn, D., McPhillips, T., Köhler, S., Riddle, S., Bowers, S., Ludäscher, B.: Scientific Workflow Design 2.0: Demonstrating Streaming Data Collections in Kepler. In: 27th IEEE Int'l Conference on Data Engineering (2011)
13. Turi, D., Missier, P., Goble, C., De Roure, D., Oinn, T.: Taverna workflows: Syntax and semantics. In: IEEE Int'l Conference on e-Science and Grid Computing, pp. 441–448. IEEE, Los Alamitos (2008)
14. Kosar, T., Livny, M.: Stork: Making data placement a first class citizen in the grid. In: Proceedings of the 24th Int'l Conference on Distributed Computing Systems, 2004, pp. 342–349. IEEE, Los Alamitos (2005)
15. Kahn, G.: The Semantics of a Simple Language for Parallel Programming. In: Information Processing 1974: Proceedings of the IFIP Congress, pp. 471–475. North-Holland, New York (1974)
16. Crawl, D., Altintas, I.: A Provenance-Based Fault Tolerance Mechanism for Scientific Workflows. In: Freire, J., Koop, D., Moreau, L. (eds.) IPAW 2008. LNCS, vol. 5272, pp. 152–159. Springer, Heidelberg (2008)
17. Feng, T., Lee, E.: Real-Time Distributed Discrete-Event Execution with Fault Tolerance. In: Real-Time and Embedded Technology and Applications Symposium, RTAS 2008, pp. 205–214. IEEE, Los Alamitos (2008)
18. Ludäscher, B., Podhorszki, N., Altintas, I., Bowers, S., McPhillips, T.: From computation models to models of provenance: the RWS approach. Concurr. Comput.: Pract. Exper. 20, 507–518 (2008)
19. Ludäscher, B., Altintas, I., Berkley, C., Higgins, D., Jaeger, E., Jones, M., Lee, E.A., Tao, J., Zhao, Y.: Scientific workflow management and the Kepler system: Research Articles. Concurr. Comput.: Pract. Exper. 18, 1039–1065 (2006)
20. Zhao, Y., Hategan, M., Clifford, B., Foster, I., Von Laszewski, G., Nefedova, V., Raicu, I., Stef-Praun, T., Wilde, M.: Swift: Fast, reliable, loosely coupled parallel computation. In: 2007 IEEE Congress on Services, pp. 199–206. IEEE, Los Alamitos (2007)
21. Wang, L., Lu, S., Fei, X., Chebotko, A., Bryant, H.V., Ram, J.L.: Atomicity and provenance support for pipelined scientific workflows. Future Generation Computer Systems 25(5), 568–576 (2009)
22. Zhou, G.: Dynamic dataflow modeling in Ptolemy II. PhD thesis, University of California (2004)
23. McPhillips, T., McPhillips, S.: RestFlow System and Tutorial (April 2011), https://sites.google.com/site/restflowdocs

PROPUB: Towards a Declarative Approach for Publishing Customized, Policy-Aware Provenance

Saumen C. Dey[1], Daniel Zinn[2], and Bertram Ludäscher[1,2]

[1] Dept. of Computer Science, University of California, Davis
[2] Genome Center, University of California, Davis

Abstract. Data provenance, i.e., the lineage and processing history of data, is becoming increasingly important in scientific applications. Provenance information can be used, e.g., to explain, debug, and reproduce the results of computational experiments, or to determine the validity and quality of data products. In collaborative science settings, it may be infeasible or undesirable to publish the complete provenance of a data product. We develop a framework that allows data publishers to "customize" provenance data prior to exporting it. For example, users can specify which parts of the provenance graph are to be included in the result and which parts should be hidden, anonymized, or abstracted. However, such user-defined provenance customization needs to be carefully counterbalanced with the need to faithfully report all relevant data and process dependencies. To this end, we propose PROPUB (Provenance Publisher), a framework and system which allows the user (i) to state provenance publication and customization requests, (ii) to specify provenance policies that should be obeyed, (iii) to check whether the policies are satisfied, and (iv) to repair policy violations and reconcile conflicts between user requests and provenance policies should they occur. In the PROPUB approach, policies as well as customization requests are expressed as logic rules. By using a declarative, logic-based framework, PROPUB can first check and then enforce integrity constraints (ICs), e.g., by rejecting inconsistent user requests, or by repairing violated ICs according to a given conflict resolution strategy.

1 Introduction

A scientific workflow is an executable specification of a computational science experiment. It represents, automates, and streamlines the steps from dataset selection and integration, computation and analysis, to final data product storage, presentation, and visualization [1, 2, 3]. An important advantage of using workflow systems over traditional scripting approaches is that the former provides the ability to automatically capture provenance information [4, 5, 6, 7] about final and intermediary data products. Scientific workflow provenance can be used, e.g., to facilitate reproducibility, result interpretation, and problem diagnosis [8] of computational experiments. However, even in collaborative science settings (e.g., [9, 10, 11]), it may be infeasible or undesirable to publish the complete lineage of a data product. Reasons for a *customizable* provenance publication mechanism include:

- Some provenance information can be far too detailed for the intended audience. For example, workflows often use data from different databases and tools, all requiring

J.B. Cushing, J. French, and S. Bowers (Eds.): SSDBM 2011, LNCS 6809, pp. 225–243, 2011.

inputs in a distinct format. The resulting low-level formatting steps can be useful for someone debugging a faulty workflow, but they may not be relevant for someone trying to understand the "big picture" [12].
- The disclosure of certain information may violate privacy and security policies [13].
- Data publishers might need to protect critical parts of their intellectual property and reveal those parts only later (e.g., after first publishing their findings).

However, simply omitting arbitrary parts of the provenance graph (e.g., by deleting nodes) or changing the graph (e.g., by modifying edges), seems to defeat the purpose of publishing provenance in the first place, i.e., one can no longer trust that the published lineage information is "correct" (e.g., there are no false dependencies) or "complete" (e.g., there are no false independencies). What seems to be needed is a mechanism that allows a data publisher to provide a high-level specification that describes the parts of the provenance graph to be published and the parts to be anonymized, abstracted, or hidden, *while guaranteeing* that certain general provenance policies are still observed.

To this end, we propose PROPUB (Provenance Publisher), a framework and system which allows the user (i) to state provenance publication and customization requests, (ii) to specify provenance policies that should be obeyed, (iii) to check whether the policies are satisfied, and (iv) to repair policy violations and reconcile conflicts between user requests and provenance policies should they occur. A unique feature of PROPUB is that it allows the user to reconcile opposing requests to publish customized lineage information. For example, in our provenance model (which is derived from OPM, the Open Provenance Model [14]), provenance graphs cannot have cycles. Intuitively, edges can be understood as a (weak) form of causal links. Since the effect of an action cannot precede its cause, provenance graphs, like causality graphs, are acyclic. Another structural constraint in OPM and in our model is that the provenance graph is bipartite, i.e., there are no edges between nodes of the same type (no artifact-to-artifact or process-to-process edges).

Provenance vs. Lineage. In the context of scientific workflows and databases, the term provenance usually means the processing history and lineage of data [3,15]. Provenance data may include, e.g., metadata like execution date and time, details of participating users, even information about hardware and software versions used to run the workflow. By *data lineage* we mean provenance information that captures the dependencies between data artifacts and processes. Since our work concentrates on data lineage, we often use the terms data lineage and provenance synonymously in this paper.

Example. Consider the example in Figure 1: The provenance graph[1] in Figure 1(a) shows *data nodes* (circles) and *invocation nodes* (boxes), indicating, e.g., that d_{15} *was generated by* some (actor/process) invocation s_1 and that invocation c_1 *used* d_{15}, denoted by, respectively $s_1 \overset{gby}{\longleftarrow} d_{15}$ and $d_{15} \overset{used}{\longleftarrow} c_1$. Now assume the user wants to *publish the lineage* of d_{18} and d_{19}. A simple recursive query can be used to retrieve all data and invocation nodes upstream from those nodes. Note that this already eliminates all nodes and edges from data node d_{20} to invocation s_3 from further consideration, so these nodes and edges are absent in Figures 1(b) to 1(d). Before publishing the lineage

---

[1] A simplified version of the provenance graph of the First Provenance Challenge [16].

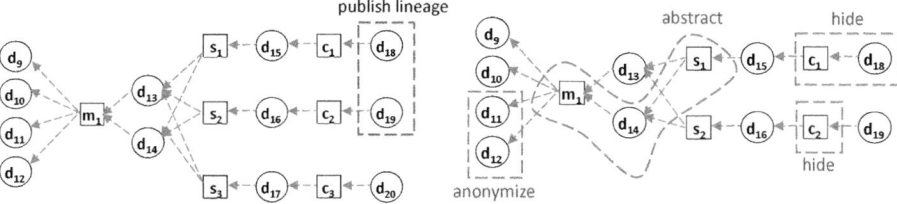

(a) Provenance graph PG and publish request.

(b) User-selected $PG' \subseteq PG$ and customization requests anonymize, abstract, hide.

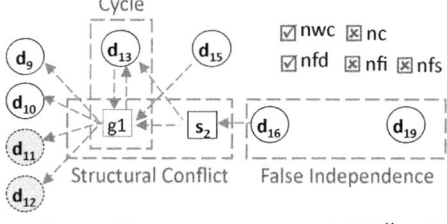

(c) Intermediate provenance graph PG'' with violations of the NC, NFI, and NFS policies.

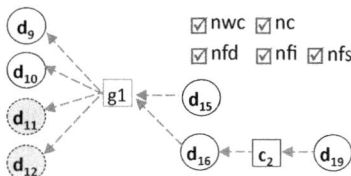

(d) Resulting customized provenance graph CG with policy guarantees.

Fig. 1. In (a) the user wants to publish the lineage of d_{18} and d_{19}, while anonymizing, abstracting, and hiding certain parts in (b). The direct conflict on d_{18} is resolved by the user, stating that hide$\{d_{18}, c_1\}$ should override the publish request in (a). The resulting intermediate graph PG'' in (c) induces several provenance policy violations, which are repaired by conflict resolution strategies, leading to the final provenance graph with provenance guarantees in (d).

of the two selected nodes, the user also requests certain customizations in Figure 1(b): First, data nodes d_{11} and d_{12} should be *anonymized*, i.e., while the existence of those data items can be revealed, the contents of that data should not be accessible. The user also wants to *abstract* a set of nodes $\{m_1, d_{14}, s_1\}$, i.e., "zoom out" (cf. [17]) and create a new abstract node which represents (and thus hides) all the abstracted nodes. Finally, the user indicates a number of nodes to be individually *hidden*, i.e., they should not be in the published provenance graph.

Repairing Conflicts. First, note that the user has created an immediate conflict on d_{18}: that node cannot both be published and hidden at the same time. Such direct conflicts are of course easily detectable from the original user request. However, more subtle constraint violations can be caused when generating the consequences of user requests, and when taking provenance policies (modeled as integrity constraints) into account: Figure 1(c) shows various policy violations when applying the induced changes $\Delta PG'$ (generated from the original user requests) to the provenance graph PG', obtaining another intermediate graph PG'': e.g., by abstracting the three nodes from Figure 1(b) into a single component, a cyclic dependency between d_{13} and the new abstract node g_1 is introduced, violating the acyclicity constraint for provenance graphs. Here, the reason lies in the "non-convex" nature of the user's abstraction request: the node d_{13} is both generated by a node in the abstract group (m_1), and another node (s_1) in that group uses

d_{13}. In other words, there is a dependency path that starts in the to-be-abstracted group at (s_1), then leaves the group (d_{13}) and then returns back into the group (m_1). One way to guarantee that the acyclicity constraint is satisfied is to close the user's abstraction request by computing its *convex hull*, then abstracting all nodes in the now closed set into the new abstract node g_1. Figure 1(d) shows the resulting final provenance graph: the node d_{13} has been "swallowed" by the new abstract node g_1, thus eliminating the problematic cyclic dependency. The edge $g_1 \leftarrow s_2$ creates a type error in Fig. 1(c): similar to OPM graphs, our graphs are bipartite, so edges between nodes of the same type (invocation to invocation or data to data) are not allowed. This conflict can also be resolved by "swallowing" the problematic node. Finally, the user's request to hide c_2 would result in a *false independence* of d_{16} and d_{19}, which is why the user's request to hide c_2 is overridden by the policy in the final customized provenance graph CG.

Dealing with Ramifications. When repairing a constraint violation using a certain strategy (e.g., swallowing additional nodes when abstracting nodes into a new abstract component), new unintended constraint violations can result from the repair action. For example, *false dependencies* can be introduced that way. In general it is difficult to foresee all possible ramifications of complex rules and strategies interacting with one another. The advantage of the rule-based logic approach employed by PROPUB is that well-established declarative semantics can be used to compute the effects and ramifications in complex settings.[2]

Contributions and Outline. We present PROPUB, a framework and system that allows a user to declaratively specify provenance publication and customization requests. In PROPUB, provenance policies are expressed as logic rules. The user specifies high-level provenance publication requests, including "customizations" to anonymize, abstract, hide, or retain certain parts of the provenance graph. By employing a logic-based framework, the system can determine implied actions, e.g., to abstract additional nodes that the user did not directly specify, but that any consistent request has to include. Some implied actions will result in conflicts with user requests or consequences of those requests. In that case, conflict resolution strategies can be applied: e.g., a user request may override a provenance policy, or vice versa, a policy may override a user request. A unique feature of our approach is that it allows the user to publish not only *customized provenance* information, but at the same time, *guarantees* can be given about which policies are satisfied and violated for a published provenance graph (see Fig. 1).

The remainder of this paper is organized as follows. In Section 2, we first present our extensions to the OPM [14] and the overall architecture of our PROPUB framework. Section 3 describes how user requests and provenance policies are modeled as logic queries and constraints. Section 4 then presents a more detailed exposition of the key components of our system, in particular: direct conflict detection and lineage selection; composition of user requests and provenance policies; and handling of conflicts between those. Related work is discussed in Section 5, and Section 6 presents some concluding remarks and suggestions for future work.

---

[2] For example, one can model the conflicts between user requests and provenance policies as a "game" between two players, arguing whether or not a certain action should be executed [18].

2   Overview and Architecture of Provenance Publisher (PROPUB)

At the core of the PROPUB framework are logic rules (Datalog with negation) that act as a declarative specification of user requests (queries), implied actions (Δ-relations that insert, or delete nodes or edges), provenance policies (modeled as integrity constraints), and conflict resolution strategies. These rules are also executable by a Datalog engine, yielding a prototypical implementation of the PROPUB system. Before describing the PROPUB architecture in more detail, we first introduce our provenance model.

2.1   Provenance Model

Our starting point is OPM, the Open Provenance Model [14] and our earlier work [19]: A *provenance* (or *lineage*) *graph* is an acyclic graph $G = (V, E)$, where the nodes $V = D \cup I$ represent either *data* items D or actor *invocations* I. The graph G is bipartite, i.e., the edges $E = E_{\mathrm{use}} \cup E_{\mathrm{gby}}$ are either *used* edges $E_{\mathrm{use}} \subseteq I \times D$ or *generated-by* edges $E_{\mathrm{gby}} \subseteq D \times I$. Here, a *used* edge $(i, d) \in E$ means that invocation i has read d as part of its *input*, while a *generated-by* edge $(d, i) \in E$ means that d was *output* data, written by invocation i. An invocation can use many data items as input, but a data item is written by exactly one invocation.

To facilitate anonymization, we use opaque identifiers for data and invocation nodes, i.e., the IDs provided by D and I can *not* be used to retrieve the actual data or identify the actual function (actor) being used for an invocation. For access to the actual data or actors, we use explicit lookup functions data: $D \rightarrow \mathbb{R}$, and actor: $I \rightarrow \mathbb{A}$, with *data references* $\mathbb{R}$ and *actor references* $\mathbb{A}$, respectively. In this way, e.g., data anonymization can be expressed simply as a deletion from the lookup table $\mathrm{data}(D, \mathbb{R})$. In our logical formulation, we use the following relations to represent provenance information:

```
used(I,D)   % invocation I used data item D
gen_by(D,I)   % data item D was generated by invocation I
actor(I,A)   % invocation I was executed by actor A
data(D,R)   % data item D can be retrieved using reference R
```

We also use auxiliary relations, e.g., a general *dependence* relation dep, which describes that node X depends on node Y, irrespective of the node types of X and Y:

```
dep(X,Y)   :- used(X,Y).
dep(X,Y)   :- gen_by(X,Y).
```

2.2   Framework Architecture

Figure 2 depicts the overall architecture of the PROPUB framework. The user submits a set of publication requests U_0, to publish the lineage of certain nodes, while abstracting or anonymizing parts of the provenance graph as described via customization requests. The first component of PROPUB detects *direct* conflicts within the given user-requests. For example, a data item that was explicitly selected for publication can accidentally be part of an abstraction request, which would cause the data item to not be published—an obvious conflict. The module Direct-Conflict-Detection materializes these conflicts.

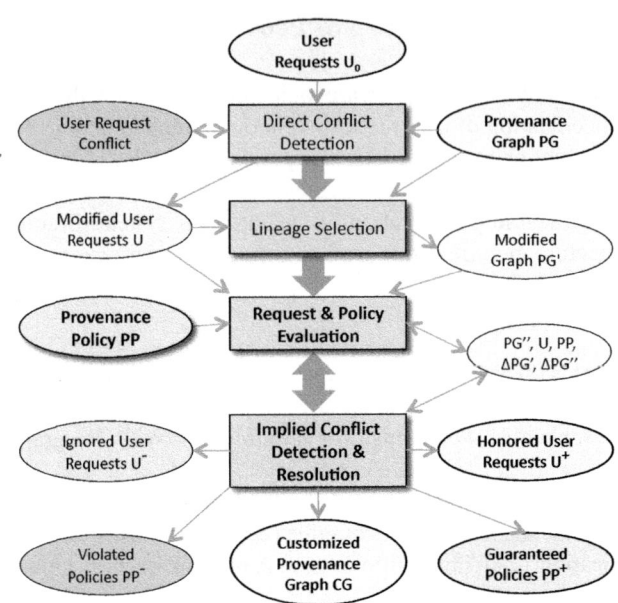

Fig. 2. PROPUB Architecture: First, direct conflicts in the user requests are removed and relevant lineage is computed. Then, user requests and policy constraints are combined, and possible implied conflicts are eliminated. Finally, the customized provenance graph is published together with the satisfied policies.

Based on this data, the user can update her original requests until all direct conflicts are resolved, resulting in a direct conflict-free user request U. In the subsequent Lineage-Selection step, a subgraph PG' is computed, which contains all to-be-published data items together with their complete provenance. The Request & Policy Evaluation module is central to PROPUB and performs important inferences: First, a set of candidate updates $\Delta PG'$ is created that when applied to PG' will create a customized provenance graph CG that satisfies the user requests. However, as already demonstrated in the introductory example in Figure 1, user requests need to be completed or "closed" and take into account policy constraints. As a consequence, graph update actions may be triggered that create new, *implied conflicts*. In a final conflict resolution step using the module Implied-Conflict-Detection-Resolution, the system first detects all such implied conflicts, and then finds user requests and policies that can be satisfied together. Our current prototype uses a greedy search, based on an initial user-defined preference ordering of user requests and policies. Other strategies are also possible, e.g., user-interactive "what-if" scenario management, or automatic approaches that systematically generate multiple "stable solutions" [20].

The result of the PROPUB framework is a customized provenance graph, a list of guaranteed (and also violated) policies, as well as a list of honored and ignored user requests. The user will typically publish the customized graph together with the

```
lineage(D). % publish the provenance for data item D
anonymize(N). % scrub the actor identity or the data reference from the node N
hide(N). % do not show the invocation or data node N
hide_dep(N1,N2). % do not show the dependency edge from N1 to N2
abstract(N,G). % zoom-out all nodes N mapped to the same abstract group G
retain(N). % definitely keep the node N in the customized provenance
retain_dep(N1,N2). % definitely keep the dependency edge (N1,N2)
```

Fig. 3. Schema for user publication requests and user customization requests to remove or to keep nodes and/or edges from a provenance graph

guaranteed policies. The original provenance graph together with the honored user requests can be kept private, to be optionally published at a later point in time when it is safe to do so.

3   Logical Formulation of User Requests and Policies

3.1   User Requests

The user requests supported by the PROPUB framework are summarized in Figure 3. PROPUB expects user requests to be formulated as relations (or facts). We envision that in the future these relations are created by graphical user-interfaces, based on provenance browsing tools (such as [5]), extended with capabilities to mark sets of edges and nodes that are then used to populate the relations associated with the user requests.

In the following, we explain the various user requests via examples.

Publish Lineage. This request allows the user to select a set of data products which should be published along with their provenance information. The selected data items and their provenance yield $PG'(\subseteq PG)$, the relevant sub-graph of the original graph PG. All other requests and policies are then applied to PG'. Note that a provenance-browser that supports provenance selection and navigation can be easily adapted to create input for such lineage requests.

Example. Assume the provenance as shown in Fig. 1(a) to be the complete provenance as recorded by the scientific workflow tool. If the user wants to publish the data product d_{18} along with its lineage, she inserts d_{18} into the lineage relation. This can either be done with the help of a GUI tool, or by stating the fact lineage(d_{18}). The selected provenance graph PG', containing all items that d_{18} depends on, is shown in Fig. 4(a). The user may wish to publish multiple data products together by stating multiple lineage facts. The system then combines the lineage of all individual requests into a single, combined lineage graph. For example, if the user wants to publish the data products d_{18} and d_{19}, she would state the requests shown in Fig. 4(c), resulting in a PG' as shown in Fig. 4(d). □

Anonymize. A publisher may like to share all data products used and generated by an invocation, but may not want to reveal the actor. The anonymize user request can be used to achieve this requirement by anonymizing an invocation i. Applying this

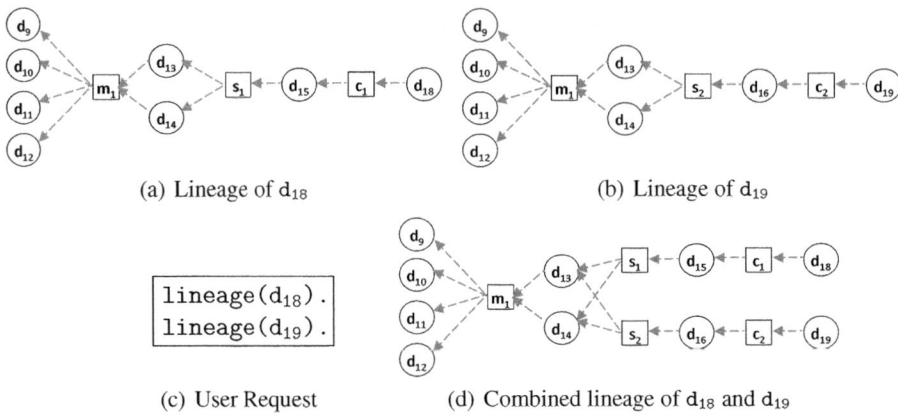

(a) Lineage of d_{18}

(b) Lineage of d_{19}

```
lineage(d18).
lineage(d19).
```

(c) User Request

(d) Combined lineage of d_{18} and d_{19}

Fig. 4. User publication request `lineage` in action: Selecting data items from PG will include them and their lineage in PG', which is then used for further refinement

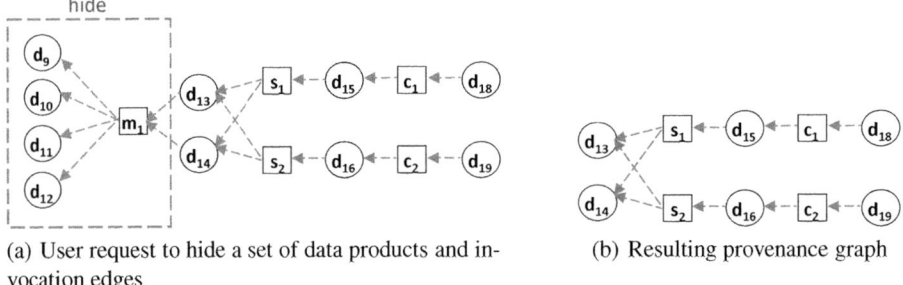

(a) User request to hide a set of data products and invocation edges

(b) Resulting provenance graph

```
hide(d9). hide(d10). hide(d11). hide(d12). hide(m1).
hide_dep(d9,m1). hide_dep(d10,m1).
hide_dep(d11,m1). hide_dep(d12,m1).
```

(c) Issued user request.

Fig. 5. Hiding nodes and edges

user request, the system removes the reference actor(i, name), but does not remove the selected nodes nor adjacent edges in the provenance graph. In a similar way, a publisher may anonymize a data product d by removing the tuple data(d, r) and with it the data value or reference (URL) r, which otherwise could be used to obtain the data value.

Hide. The provenance graph PG' may still have some data or invocation nodes which are sensitive or not relevant for publication. A publisher can remove such nodes and edges using the `hide` user request. The system will remove the selected node and all adjacent edges or the selected edges. An example is shown in Fig. 5.

Abstract. This user request allows the user to replace a set of nodes (either data or invocation or a combination thereof) from PG' and replace them by an abstract

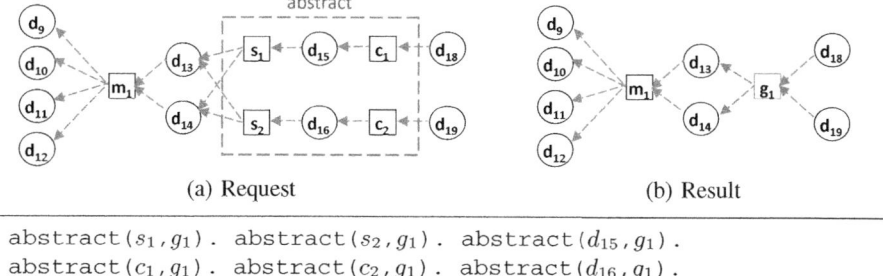

(a) Request   (b) Result

abstract(s_1,g_1). abstract(s_2,g_1). abstract(d_{15},g_1).
abstract(c_1,g_1). abstract(c_2,g_1). abstract(d_{16},g_1).

(c) Issued user request.

Fig. 6. Abstracting Structural Properties

place-holder node. Several abstraction requests can be used to create different abstract nodes. The user requests are given in the binary relation abstract(N, G). All nodes that are associated with the same group id are collapsed into one abstract node, resulting in as many new abstract nodes as there are distinct G values in abstract. An example with only one group is given in Fig. 6.

Retain. This user request can be used to explicitly mark nodes or edges in the provenance graph PG$'$ to be retained in the to-be-published provenance graph. To fix provenance policies, PROPUB may automatically apply deletions to existing nodes and edges in PG$'$. A *retain* user request will create an implied conflict if the marked node or associated edges are to be removed from PG$'$, notifying the user either to relax the provenance policy or to remove the *retain* user request.

3.2 Provenance Policies

Now, we briefly explain the provenance policies (PP) considered by PROPUB. Provenance policies are observed using a set of integrity constraints (IC) via witness relations. The witness relations are defined in Fig. 14.

NWC (No-Write Conflict). A write conflict occurs when there are multiple *generated by* edges for a single data product. This situation can arise if a set of multiple data nodes is selected for abstraction (resulting in a grouped data node) and at least two of the selected data nodes have *generated_by* in-edges (see Fig. 7). We use the *wc(X,Y)* witness relation to observe if invocations X and Y have created the same data artifact.

NC (No-Cycle). Provenance graphs are directed acyclic graphs since they record the causal relationship between invocations and data products. However, when multiple nodes are contracted to a group node by an *abstract* request, cycles can be introduced. An example was already given in Fig. 1(b): the user requests to abstract m_1, d_{14}, and s_1 into an abstraction node g_1. Fig. 1(c) shows the cycle between nodes g_1 and d_{13} after the abstract request has been executed. We use the *cycle(X,Y)* witness relation to observe if there is a cycle between nodes X and Y.

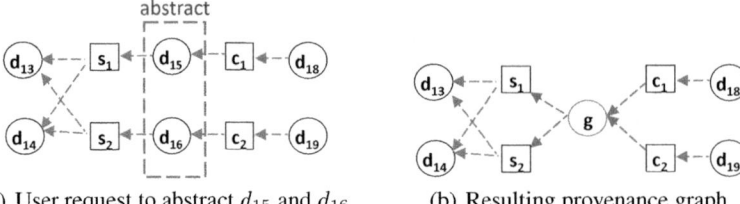

(a) User request to abstract d_{15} and d_{16} (b) Resulting provenance graph

Fig. 7. A Write Conflict. In Fig. 7(b) a new node g is introduced with *generated by* dependencies on invocations s_1 and s_2. This is a write conflict as any data product can only be created by one invocation.

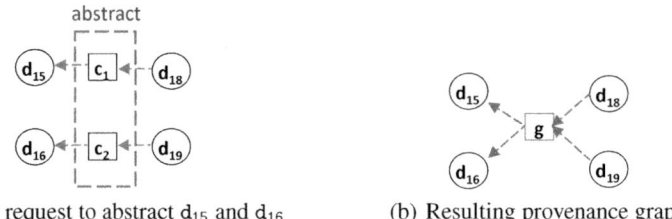

(a) User request to abstract d_{15} and d_{16} (b) Resulting provenance graph

Fig. 8. False Dependency: In (b) d_{18} depends on d_{16}, which does not exist in the original graph as shown in (a)

NFS (No-False Structure). Provenance graphs are bipartite, i.e., data nodes are connected with invocation nodes and vice versa. A false structure (type mismatch) is when there is a node dependent on another of the same type. In Fig. 1(b) the user requests to abstract m_1, d_{14}, and s_1 into an abstraction node g_1. Fig. 1(c) shows that there is a structural conflict between nodes g_1 and s_2 should the abstract request be executed. We use the *fs(X,Y)* witness relation to observe if nodes X and Y are of same type.

While the previous policies (NWC, NC & NFS) describe structural properties of the customized provenance graph CG itself, the following two criteria relate the customized provenance graph CG with the actually recorded provenance PG. Thus, these policies *add* information to the published provenance graph for a collaborator that has access only to CG but not to PG.

NFD (No-False Dependency). A customization from PG to CG exhibits a *false dependency*, if two concrete (i.e., non-abstracted) nodes n_1 and n_2 in CG are transitively dependent on each other but the corresponding nodes in the original provenance graph PG are not. For example, our framework can introduce false dependencies if an abstraction request contains nodes from two independent paths, i.e., paths that do not share common nodes. This case is illustrated in Fig. 8. The user requests to abstract invocation nodes c_1 and c_2. In the resulting provenance graph, d_{18} depends on d_{16}— a dependency that did not exist in the original provenance graph. We use the *fd(X,Y)* witness relation to observe if there is a false dependence between nodes X and Y.

NFI (No-False Independency). A customization from PG to CG exhibits a *false independency* if two concrete nodes n_1 and n_2 are not transitively dependent on each other in CG, but there exists a transitively dependency between d_1 and d_2 in the original graph PG. For example, a honored `hide` request may create false independencies. In Fig. 1(b) the user requests to hide c_2. The resulting graph in Fig. 1(c) shows there would be no dependency between nodes d_{19} and d_{16}, although d_{19} dependents on d_{16} in PG. We use the *fi(X,Y)* witness relation to observe if nodes X and Y are independent.

4   PROPUB Modules

We now describe the PROPUB system, based on its individual modules and their interactions with each other.

4.1   Direct Conflict Detection and Lineage Selection

The first step of PROPUB is to detect obvious conflicts among all user requests. These can occur if some user requests that require nodes or edges (i.e., `lineage`, `retain`) to be carried over to CG from PG but some other user requests (i.e., `hide`, or `abstract`) require the same model elements to be deleted in CG.

We detect direct conflicts via the logic rules given in Fig. 9. If a `conflict(X)`, a `conflict_dep(X, Y)`, or a `conflict_abst(X)` has been derived, we show its derivation tree (including the URs that lead to this conflict) to the user. The user has then the opportunity to prioritize one of the conflicting requests in order to resolve the conflict. Once the user has converged to a set of user requests that do not have direct conflicts, the Lineage-Selection step computes the subgraph PG' based on a reachability query as shown in Fig. 10.

4.2   Evaluating and Reconciling User Requests and Provenance Policies

The core work of the PROPUB framework is to combine user requests and provenance policies that can be satisfied simultaneously. That is, given the selected provenance

```
keep(X)  :- lineage(X).      remove(X)  :- anonymize(X).
keep(X)  :- retain(X).       remove(X)  :- hide(X).
                             remove_dep(X,Y)  :- hide(X), dep(X,Y).
                             remove_dep(X,Y)  :- hide(Y), dep(X,Y).
keep_dep(X,Y) :-             remove_dep(X,Y)  :- hide_dep(X,Y).
    retain_dep(X,Y).         remove(X)  :- abstract(X,_).
                             remove_dep(X,Y)  :-
                                 abstract(X,_), dep(X,Y).
                             remove_dep(X,Y)  :-
                                 abstract(Y,_), dep(X,Y).
conflict(X)  :- remove(X), keep(X).
conflict_dep(X,Y) :- remove_dep(X,Y), keep_dep(X,Y).
conflict_abst(X)  :-  abstract(X,G1), abstract(X,G2), not G1=G2.
```

Fig. 9. Detection of conflicts

```
dep'(X,Y)  :- lineage(X), dep(X,Y).  % initialize with selected set
dep'(Y,Z)  :- dep'(X,Y), dep(Y,Z).   % copy (not a transitive closure) all
                                      % dependent edges
node'(X)   :- dep'(X,_).  % auxiliary relation containing all nodes of PG',
node'(X)   :- dep'(_,X).  % which is used to subset all subsequent relations in PG
used'(X,Y) :- used(X,Y), dep'(X,Y).
gen_by'(X,Y) :- gen_by(X,Y), dep'(X,Y).
actor'(I,A)  :- actor(I,A), node'(I).
data'(D,R)   :- data(D,R), node'(D).
```

Fig. 10. The provenance graph (PG$'$) derived after applying the publication user requests (lineage) on PG

```
  { NWC } ≻ { NFD } ≻ { NC } ≻ { NFI } ≻ { NFS }
≻ { abstract(m1,g1), abstract(d14,g1), abstract(s1,g1) }
≻ { hide(c1) } ≻ { hide(d18) } ≻ { hide(c2) }
```

Fig. 11. Preference order for user requests and provenance policies for introductory example in Fig. 1

graph PG$'$ from the Lineage-Selection step and a set of user requests and provenance policies, PROPUB ideally should return a customized graph CG that reflects the applied user requests and conforms to the provenance policies. Unfortunately, as discussed earlier, applying user requests can invalidate provenance policies. Here, PROPUB tries to apply basic repairs for violated policies. However, these repairs in turn may *undo* certain user requests. To automatically select a subset of user requests and provenance policies that can be satisfied, we propose that the user specifies a ranking among user requests and provenance policies representing a preference order. A greedy algorithm can then be used to find a balance between honored user requests, and satisfied provenance policies. In the following, we detail this process further. In particular, after presenting our encoding of user-preferences and the overarching algorithm, we detail the logic formulation of its sub-routines for applying user requests, checking policy violations, applying basic repairs, and detecting implied conflicts.

User Preferences. In PROPUB, user preferences are declared via an ordered partition of the set of all provenance policies and user requests. An example for such a preference order is given in Fig. 11. Here, satisfying the provenance policies is valued higher than the abstract request, which in turn is preferred over honoring hide requests. Note that all abstract requests with a group identifier (here, g1) are contained in one subset since these rules form a logic unit that should be honored together as a group. The proposed static preference schema is only one possibility to automatically guide user request and provenance policy reconciliation. A more interactive approach where conflicts are visualized based on the participating user requests and policies is also possible. To reduce the specification burden posed to the user, preferences can also be declared based on request classes (relations) rather than on concrete request facts; e.g., the user might specify: policies ≻ abstract ≻ hide ≻ retain.

Algorithm: CUSTOMIZE_PROVENANCE
INPUT: User-selected provenance graph PG′,
           user customization requests U and provenance policies PP.
           Let $X_1 \dot\cup X_2 \dot\cup \ldots \dot\cup X_n$ be a partition (disjoint union) of $U \cup PP$,
           such that $X_1 \succ X_2 \succ \ldots \succ X_n$ is a user-defined order (by importance).
OUTPUT: Customized provenance graph CG,
           honored user requests U^+ and satisfied provenance policies PP^+

BEGIN:
1 **FOR** $k = n, n-1, \ldots, 1$:   // try to satisfy as many requests as possible
2    $\mathbf{X}_k := \bigcup_{i \le k} X_i$ // select the k most important user-requests and policies
3    $U^+ := \mathbf{X}_k \cap U$ // user requests that should be honored
4    $PP^+ := \mathbf{X}_k \cap PP$ // provenance policies that should be satisfied
5    $\Delta PG' := \text{apply}(U^+, PG')$ // inserts and deletes to PG′ induced by U^+ (see Fig. 13)
6    $PG'' := PG' \pm \Delta PG'$ // result after applying the changes (inserts and deletes)
7   ImpliedConflicts := false
8   **DO**
9   **IF** $PG'' \models PP^+$ // policies are satisfied (see Fig. 14)
10       $CG := PG''$; $PP^+ := PP^+ \cup \{p \in PP \setminus PP^+ \mid CG \models p\}$ // (see Fig. 14)
11   **RETURN**
12     $\Delta PG'' := \text{apply}(PP^+, PG'')$ // extend abstract requests to satisfy PP^+ (see Fig. 15)
13   ImpliedConflicts := true **if** $\Delta PG'$ and $\Delta PG''$ have conflicts (see Fig. 9)
14     $PG'' := PG'' \pm \Delta PG''$ // apply the fixes
15   **UNTIL** ImpliedConflicts
16 **ENDFOR**
17 **ABORT** Not even the most important set X_1 could be satisfied!
END

Fig. 12. Greedy algorithm to select a subset of user requests and provenance policies that can be fulfilled without implied conflicts

Customization Algorithm. Figure 12 shows the greedy algorithm to compute the customized provenance graph. Given the selected provenance graph PG′, together with prioritized user requests U and provenance policies PP, the algorithm computes the customized provenance graph CG together with the honored user requests U^+ and satisfied policies PP^+. The main for-loop (lines 1–16) starts with selecting all user requests and policies trying to satisfy them together. Whenever no solution can be found for a specific subset $\mathbf{X}_k$ of requests and policies, the least important request (or policy) is dropped in the next iteration of the for-loop. Within one iteration, the following steps are performed to find a possible customized graph: First, a set of direct consequences $\Delta PG'$ (or changes to PG′) from applying the user requests U^+ to PG′ are computed (line 5). These are then applied to PG′, yielding PG″. Now, the provenance policies are tested and repaired if necessary. If PG″ already satisfies the selected provenance policies PP^+ (line 9), then PG″ is returned as customized graph CG together with the honored user requests U^+ and the satisfied policies PP^+. We check if any further (of the previously removed) provenance policies are also satisfied by CG and add these to the list of satisfied policies. If PG″ does not satisfy the selected policies, then—as a last resort—we determine repairs $\Delta PG''$ that, when applied to PG″ would satisfy the

```
% updates for anonymize user request:
del_node(N)  :- anonymize(N).
ins_actor(N,A) :- anonymize(N), actor'(N,_), A=anonymized.
ins_data(N,R)  :- anonymize(N), data'(N,_), R=anonymized.

% updates for hide user request:
del_node(N)  :- hide(N).
del_dep(X,Y) :- hide(X), dep'(X,Y).
del_dep(X,Y) :- hide(Y), dep'(X,Y).
del_dep(X,Y) :- hide_dep(X,Y).

% updates for retain user request:
ins_actor(I,A) :- retain(I), actor'(I,A).
ins_data(D,R) :- retain(D), data'(D,R).
ins_dep(X,Y) :- retain_dep(X,Y).

% updates for abstract user request:
del_node(N)  :- abstract(N,_).
ins_actor(I,A)  :- abstract(_,I), A=abstracted.
del_dep(X,Y) :- abstract(X,_), dep'(X,Y).
del_dep(X,Y) :- abstract(Y,_), dep'(X,Y).
int_dep(X,Y) :- abstract(X,G), abstract(Y,G), dep'(X,Y).
ins_dep(G,Y) :- abstract(X,G), dep'(X,Y), not int_dep(X,Y).
ins_dep(X,G) :- abstract(Y,G), dep'(X,Y), not int_dep(X,Y).
```

Fig. 13. Direct consequences ΔPG' of user requests U

provenance policies (line 12). However, before we apply these repairs to PG'', we need to check whether these repairs conflict with explicit user requests or their consequences (ΔPG'). If there is a conflict, then the currently chosen set of user requests and policies cannot be satisfied due to implied conflicts, and the do-until-loop will finish to start the next iteration of the outer for-loop. In case there are no implied conflicts, changes ΔPG'' are applied to PG'' (line 14). Since these changes could violate other policies, we go back to line 8 to test this, and possibly repair violations again, etc. The algorithm returns either with a customized graph CG in line 11, or aborts in line 17 if even the partition that is top-priority for the user cannot be satisfied as a whole.

The termination of the do-until-loop is guaranteed since we repair policy violations by *abstracting* more nodes into existing abstraction nodes, a process that necessarily terminates due to the finiteness of PG''.

We now detail the algorithm's major sub-routines:

Handling of User Requests. User requests are expressed as Datalog rules that represent updates to the selected provenance graph PG'. The rules, which are computing the changes ΔPG' for all our user requests, are given in Fig. 13.

Detecting Policy Violations. The provenance policies as described in Section 3.2 can be modeled as integrity constraints over provenance graphs. We use the rules given in Fig. 14 to check if there are nodes that violate any provenance policy.

Policy Repairs. To fix the three structural policy (i.e., NWC, NC, and NFS) violations, we "swallow" the violating nodes into the adjacent abstraction group. To do so, we create an equivalence relation same_group for the nodes in PG″ (see Fig. 15). Nodes that are not taking part in an abstract user request will have their "own" equivalence class. All nodes n that are mapped to the same group id g will be in the same class $[g] = \{n \mid \text{same\_group}(n, g)\}$. When applying the ΔPG″ updates (containing the same_group relation) to PG″, we proceed as follows: A class with only a single member, is replaced by the member itself. Classes that contain more than one member are replaced by the group ID that represents this class.

```
tcdep''(X,Y)  :- dep''(X,Y).  % transitive closure of dependencies in PG''
tcdep''(X,Y)  :- tcdep''(X,Z), tcdep''(Z,Y).
tcdep(X,Y)  :- dep(X,Y).    % transitive closure of dependencies in PG
tcdep(X,Y)  :- tcdep(X,Z), tcdep(Z,Y).

% Provenance policy witness relations:
wc(X,Y)  :- gen_by''(D,X), gen_by''(D,Y), not X=Y.
fs(X,Y)  :- dep''(X,Y), data''(X), data''(Y).
fs(X,Y)  :- dep''(X,Y), actor''(X), actor''(Y).
cycle(X,Y)  :- tcdep''(X,Y), tcdep''(Y,X), not X=Y.
fi(X,Y)  :- tcdep''(X,Y), not tcdep(X,Y).
fd(X,Y)  :- tcdep(X,Y), not tcdep''(X,Y).
```

Fig. 14. Detecting policy violations in the graph PG″ using the witness relations

Implied Conflicts. Implied conflicts are checked like direct conflicts (see Fig. 9); we test whether any retain requests are violated due to the additional abstract requests we introduced via the same_group relation.

5 Related Work

Scientific workflows have become increasingly popular in recent years as a means to specify, automate, and share the computational parts of scientific experiments [2, 1, 21, 22]. One of the advantages of using a workflow approach is the ability to capture, store, and query data lineage and provenance information. Provenance can then be used, e.g., to interpret results, diagnose errors, fix bugs, improve reproducibility, and generally to build trust on the final data products and the underlying processes [23, 24, 8, 25]. In addition, provenance can be used to enhance exploratory processes [6, 26, 27], and techniques have been developed to deal with provenance efficiently [28, 29].

While provenance information is immensely useful, it often carries sensitive information causing privacy concerns, which can be tied to data, actors, and workflow specifications. Without required access privileges the value of a data product, the functionality (being able to guess the output of the actor given a set of inputs) of an actor (module), or the execution flow of the workflow should not be revealed to a user [13].

S. Dey, D. Zinn, and B. Ludäscher

```
same_group(X,Y)  :- cycle(X,Y). % adding NC policy violators
same_group(X,Y)  :- fs(X,Y). % adding NFS policy violators
same_group(X,Y)  :- wc(X,Y). % adding NWC policy violators
same_group(X,X)  :- same_group(X,_). % reflexive
same_group(X,X)  :- same_group(_,X). % reflexive
same_group(X,Y)  :- same_group(Y,X). % symmetric
same_group(X,Y)  :- same_group(X,Z), same_group(Z,Y). % transitive

smaller(X,Y)  :- same_group(X,Y), X < Y. % relation is within one group
minimum(X)  :- node(X), not smaller(_,X). % minima for each group
abstract(X,G)  :- same_group(X,G), minimum(G), % add abstract
    same_group(X,Y),X!=Y. % request when more than one member in same_group
```

Fig. 15. Resolving policy violations by creating (or extending existing) new *abstract* relations to abstract violated nodes. The updates $\Delta PG''$, which will be applied on PG'', is calculated using the rules shown in Fig. 13 for the *abstract* user request.

The security view approach [8] provides a partial view of the workflow through a role-based access control mechanism, and by defining a set of access permissions on actors, channels, and input/output ports as specified by the workflow owner at design time. The provenance information is limited by the structure of the workflow and the security specifications.

To avoid "information overload", the ZOOM*UserViews approach [17] provides a partial, zoomed-out view of a workflow, based on a user-defined distinction between relevant and irrelevant actors. Provenance information is restricted by the definition of that partial view of the workflow. This is somewhat similar to our abstract operation, where the user wishes to abstract away certain details. However, ours appears to be the first work that studies the problem of publishing custom provenance while simultaneously satisfying a set of given publication policies.

6 Conclusions

We have presented PROPUB, a logic-based framework for publishing customized provenance in the presence of publication policies that should be observed when abstracting or otherwise hiding data lineage information. We believe that our ongoing work is but the first step to a new line of provenance research that aims to reconcile the inherently opposing goals of publishing detailed provenance information on one hand, and keeping parts of that information private and secure, on the other. Indeed, there is a nascent but rapidly growing area of secure provenance which focuses on this trade off. For example, the authors of [30] state that:

> "There is thus an inherent tradeoff between the utility of the information provided in response to a search/query and the privacy guarantees that authors/owners desire."

In this paper, we have made first steps to a more comprehensive treatment of "custom provenance" by employing the well-studied machinery of logic rules. Our future work is aimed, e.g., at navigation operators as described in [31] that can handle nested, collection-oriented models of computation and provenance. The navigation along different axes (e.g., horizontally, corresponding to time, and vertically, corresponding to the collection structure) provides an elegant way to view the provenance graph at different granularities and levels of detail. In future work, we also plan to investigate how a logic-based framework like ours could be extended to handle collection-oriented models of data, and how one could employ the workflow structure directly (e.g., for series-parallel graphs) when customizing provenance, and when trying to resolve inherent logical conflicts. We also expect this work to be relevant for large-scale data and tool collaborators such as DataONE[3]. For example, the DataONE provenance workgroup has developed a prototype for publishing interoperable provenance information in a collaborative environment [10]. That prototype currently publishes the complete provenance graph. An improved system can be obtained by incorporating PROPUB, thereby allowing scientists to reveal as much (or as little) information to the wider community as they want. Our framework has the advantage that it is possible to test whether a subsequent, more complete publication of provenance information is consistent with an earlier provenance publication by the same user(s). This is another area of future research.

Acknowledgements. This work was in part supported by NSF grant OCI-0722079 & 0830944, and DOE grant DE-FC02-07ER25811. The second author thanks LogicBlox for supporting his post-doctoral work.

References

1. Gil, Y., Deelman, E., Ellisman, M., Fahringer, T., Fox, G., Gannon, D., Goble, C., Livny, M., Moreau, L., Myers, J.: Examining the challenges of scientific workflows. Computer 40(12), 24–32 (2007)
2. Taylor, I.J., Deelman, E., Gannon, D., Shields, M.S. (eds.): Workflows for eScience. Springer, Heidelberg (2007)
3. Ludäscher, B., Bowers, S., McPhillips, T.M.: Scientific Workflows. In: Encyclopedia of Database Systems, pp. 2507–2511 (2009)
4. Davidson, S.B., Boulakia, S.C., Eyal, A., Ludäscher, B., McPhillips, T.M., Bowers, S., Anand, M.K., Freire, J.: Provenance in Scientific Workflow Systems. IEEE Data Engineering Bulletin 30(4), 44–50 (2007)
5. Anand, M., Bowers, S., Ludascher, B.: Provenance browser: Displaying and querying scientific workflow provenance graphs. In: International Conference, pp. 1201–1204 (2010)
6. Davidson, S., Freire, J.: Provenance and scientific workflows: challenges and opportunities. In: SIGMOD Conference, Citeseer, pp. 1345–1350 (2008)
7. Miles, S., Deelman, E., Groth, P., Vahi, K., Mehta, G., Moreau, L.: Connecting Scientific Data to Scientific Experiments with Provenance. In: Proceedings of the Third IEEE International Conference on e-Science and Grid Computing, pp. 179–186. IEEE Computer Society, Washington, DC, USA (2007)
8. Chebotko, A., Chang, S., Lu, S., Fotouhi, F., Yang, P.: Scientific workflow provenance querying with security views. In: The Ninth International Conference on Web-Age Information Management, WAIM 2008, pp. 349–356. IEEE, Los Alamitos (2008)

---

[3] http://www.dataone.org/

9. Altintas, I., Anand, M.K., Crawl, D., Bowers, S., Belloum, A., Missier, P., Ludäscher, B., Goble, C., Sloot, P.: Understanding Collaborative Studies through Interoperable Workflow Provenance. In: McGuinness, D.L., Michaelis, J.R., Moreau, L. (eds.) IPAW 2010. LNCS, vol. 6378, pp. 42–58. Springer, Heidelberg (2010)

10. Missier, P., Ludäscher, B., Bowers, S., Dey, S., Sarkar, A., Shrestha, B., Altintas, I., Anand, M., Goble, C.: Linking multiple workflow provenance traces for interoperable collaborative science. In: Workflows in Support of Large-Scale Science (WORKS), pp. 1–8. IEEE, Los Alamitos (2010)

11. Altintas, I.: Collaborative Provenance for Workflow-Driven Science and Engineering. PhD thesis, University of Amsterdam (February 2011)

12. Biton, O., Cohen-Boulakia, S., Davidson, S., Hara, C.: Querying and managing provenance through user views in scientific workflows. In: International Conference, pp. 1072–1081 (2008)

13. Davidson, S., Khanna, S., Roy, S., Boulakia, S.: Privacy issues in scientific workflow provenance. In: Proceedings of the 1st International Workshop on Workflow Approaches to New Data-centric Science, pp. 1–6. ACM, New York (2010)

14. Moreau, L., Clifford, B., Freire, J., Gil, Y., Groth, P., Futrelle, J., Kwasnikowska, N., Miles, S., Missier, P., Myers, J., Simmhan, Y., Stephan, E., den Bussche, J.V.: The Open Provenance Model - core specification (v1.1). Future Generation Computer Systems (2010)

15. Cheney, J.: Causality and the Semantics of Provenance. In: CoRR, abs/1006.1429 (2010)

16. Moreau, L., Ludäscher, B., Altintas, I., Barga, R., Bowers, S., Callahan, S., Chin, J., Clifford, B., Cohen, S., Cohen-Boulakia, S., et al.: Special issue: The first provenance challenge. Concurrency and Computation: Practice and Experience 20(5), 409–418 (2008)

17. Biton, O., Cohen-Boulakia, S., Davidson, S.: Zoom* userviews: Querying relevant provenance in workflow systems. In: Proceedings of the 33rd International Conference on Very Large Data Bases, pp. 1366–1369. VLDB Endowment (2007)

18. Ludäscher, B., May, W., Lausen, G.: Referential Actions as Logic Rules. In: PODS, pp. 217–227 (1997)

19. Anand, M., Bowers, S., McPhillips, T., Ludäscher, B.: Exploring scientific workflow provenance using hybrid queries over nested data and lineage graphs. In: Winslett, M. (ed.) SSDBM 2009. LNCS, vol. 5566, pp. 237–254. Springer, Heidelberg (2009)

20. May, W., Ludäscher, B.: Understanding the global semantics of referential actions using logic rules. ACM Transactions on Database Systems (TODS) 27, 343–397 (2002)

21. Ludäscher, B., Altintas, I., Bowers, S., Cummings, J., Critchlow, T., Deelman, E., Roure, D.D., Freire, J., Goble, C., Jones, M., Klasky, S., McPhillips, T., Podhorszki, N., Silva, C., Taylor, I., Vouk, M.: Scientific Process Automation and Workflow Management. In: Shoshani, A., Rotem, D. (eds.) Scientific Data Management: Challenges, Existing Technology, and Deployment. Chapman & Hall/CRC (2009)

22. Goble, C., Bhagat, J., Aleksejevs, S., Cruickshank, D., Michaelides, D., Newman, D., Borkum, M., Bechhofer, S., Roos, M., Li, P., De Roure, D.: MyExperiment: a repository and social network for the sharing of bioinformatics workflows. Nucleic Acids Research (2010)

23. Bose, R., Frew, J.: Lineage retrieval for scientific data processing: a survey. ACM Computing Surveys (CSUR) 37(1), 1–28 (2005)

24. Simmhan, Y., Plale, B., Gannon, D.: A survey of data provenance in e-science. ACM SIGMOD Record 34(3), 31–36 (2005)

25. Freire, J., Koop, D., Santos, E., Silva, C.T.: Provenance for Computational Tasks: A Survey. Computing in Science and Engineering 10(3), 11–21 (2008)

26. Freire, J., Silva, C., Callahan, S., Santos, E., Scheidegger, C., Vo, H.: Managing rapidly-evolving scientific workflows. In: Moreau, L., Foster, I. (eds.) IPAW 2006. LNCS, vol. 4145, pp. 10–18. Springer, Heidelberg (2006)

27. Silva, C., Freire, J., Callahan, S.: Provenance for visualizations: Reproducibility and beyond. Computing in Science & Engineering, 82–89 (2007)
28. Heinis, T., Alonso, G.: Efficient Lineage Tracking For Scientific Workflows. In: Proceedings of the 2008 ACM SIGMOD Conference, pp. 1007–1018 (2008)
29. Anand, M., Bowers, S., Ludäscher, B.: Techniques for efficiently querying scientific workflow provenance graphs. In: Proceedings of the 13th International Conference on Extending Database Technology, pp. 287–298. ACM, New York (2010)
30. Davidson, S., Khanna, S., Roy, S., Stoyanovich, J., Tannen, V., Chen, Y., Milo, T.: Enabling Privacy in Provenance-Aware Workflow Systems. In: Conference on Innovative Data Systems Research, CIDR (2011)
31. Anand, M., Bowers, S., Ludäscher, B.: A navigation model for exploring scientific workflow provenance graphs. In: Proceedings of the 4th Workshop on Workflows in Support of Large-Scale Science, pp. 1–10. ACM, New York (2009)

Provenance-Enabled Automatic Data Publishing

James Frew, Greg Janée, and Peter Slaughter

Earth Research Institute
University of California, Santa Barbara
{frew,gjanee,peter}@eri.ucsb.edu
http://eri.ucsb.edu

Abstract. Scientists are increasingly being called upon to publish their data as well as their conclusions. Yet computational science often necessarily occurs in exploratory, unstructured environments. Scientists are as likely to use one-off scripts, legacy programs, and volatile collections of data and parametric assumptions as they are to frame their investigations using easily reproducible workflows. The ES3 system can capture the provenance of such unstructured computations and make it available so that the results of such computations can be evaluated in the overall context of their inputs, implementation, and assumptions. Additionally, we find that such provenance can serve as an automatic "checklist" whereby the suitability of data (or other computational artifacts) for publication can be evaluated. We describe a system that, given the request to publish a particular computational artifact, traverses that artifact's provenance and applies rule-based tests to each of the artifact's computational antecedents to determine whether the artifact's provenance is robust enough to justify its publication. Generically, such tests check for proper curation of the artifacts, which specifically can mean such things as: source code checked into a source control system; data accessible from a well-known repository; etc. Minimally, publish requests yield a report on an object's fitness for publication, although such reports can easily drive an automated cleanup process that remedies many of the identified shortcomings.

Keywords: provenance, publishing, curation.

1 Introduction and Background

In computing environments there is often tension between freedom of expression and exploration on the one hand, and constraint and assertion on the other. More precisely, the creation of a computational artifact may require creativity and freeform exploration, but at some end point the environment requires that certain assertions be true. The question is, where, when, and how should those assertions be enforced?

1.1 Programming Environments

Programming languages provide a familiar example for many. Creating a program is a fundamentally creative activity, but in the end, the computational

J.B. Cushing, J. French, and S. Bowers (Eds.): SSDBM 2011, LNCS 6809, pp. 244–252, 2011.

environment requires that the program be syntactically and semantically valid[1] to be executed. Different programming environments have take different approaches to enforcing program validity. At the flexibility-maximizing end of the spectrum, many environments allow programs to be created with near-complete freedom using text editors. Assertion checking about programs is then deferred to a compilation step (for compiled languages) and/or to execution time (for interpreted languages). Other environments, particularly "visual" or GUI-based environments, provide syntax-aware and even limited-semantics-aware editors that constrain expressions during creation. These environments attempt to maintain the assertion that the program is at least syntactically correct during its creation. In a kind of *reductio ad absurdum*, we might imagine an environment that requires that a program be valid at *every* step of its construction, thus obviating the need for any later assertion checking. But this is a direction that programming environments have not gravitated toward. Even in the visual programming environments, there is a recognized need that programmers be allowed to create expressions that violate language rules and semantics, at least temporarily. Code may be sketched out initially; there may be references to entities that don't exist yet; test code insertions may create deliberate errors; and so forth.

1.2   Scientific Programming

The creation of science data products is analogous. In the end, we require that a data product be of sufficient quality, and that it have been produced using sufficiently rigorous and repeatable methods, that it is worthy of entering the scientific record. But creating the product requires experimentation and creativity, particularly in the formative stages, and many assertions that we would like to be true at the end will not be true during the entire process of creation. We may, for example, use false or test data, shortcut production steps out of expediency, and many other things of that nature. Moreover, computational science data is produced by software, so the points in the preceding section regarding software creation apply as well.

Scientific workflow environments provide assertion support similar to syntax-aware programming language editors in that desirable end assertions are maintained through the whole process of development. However, we argue that much data production is done outside workflow systems, and even for those scientists working in workflow systems, the flexibility scientists require means that they may drop out of the environment to perform one-off experiments.

1.3   Publishing Science Data Products

By "data publication" we mean the packaging and distribution of scientific datasets. As in traditional publication of scientific literature, data publication involves selecting and formatting content; naming and attributing the resulting artifact, and then making it available via well-known distribution channels.

---

[1] Semantically valid only in the sense that all language requirements are satisfied; we consider no deeper semantics here.

Significantly, only the first step–selection–is an intrinsic result of scientific computation, namely as the computation's result set, and thus happens automatically. The remaining steps are often an added burden at publish time. Publication standards often dictate a specific format (e.g., GeoTIFF[2] or KML[3] for Earth science array and vector data, respectively) that differs from that used internally in a scientific computation environment. Similarly, published data objects must often adhere to naming conventions (e.g., DOIs for datasets or persistent URIs for data granules or services) that differ from more convenient internal names (e.g., filenames or database queries). These formats in turn often dictate or enable specific distribution mechanisms (e.g., geospatial web services[4].)

In addition to requiring specific formats and naming conventions, data publication requires the availability of additional descriptive information. Much of this *metadata* is available as an automatic consequence of the computational environment (e.g., array dimensions, creation datetimes, etc.), but much traditionally is not (e.g., descriptive text, names of responsible parties, etc.) and must be discovered or supplied as part of the publication process. These metadata collectively serve the attribution role analogous to authorship in traditional publishing. Historically, a key missing piece of this attribution has been the connections between a data object and its antecedents, such as can now be supplied by provenance.

2 Provenance-Enabled Automatic Data Publishing

We believe that provenance can greatly simplify the transition between experimental and published products by simplifying the assembly and validation of the metadata necessary to publish a science data object. Key to our approach is exploiting the ES3 system, which captures the necessary provenance without imposing restrictions on relatively unstructured experimental environments. In this section we define provenance as ES3 implements it, describe the ES3 system, and then describe how ES3-collected provenance can help automate the publication process.

2.1 Computational Provenance

Computational provenance refers to knowledge of the origins and processing history of a computational artifact such as a data product or an implementation of an algorithm [1]. Provenance is an essential part of metadata for Earth science data products, where both the source data and the processing algorithms change over time. These changes can result from errors (e.g., sensor malfunctions or incorrect algorithms) or from an evolving understanding of the underlying systems and processes (e.g., sensor recalibration or algorithm improvement). Occasionally such changes are memorialized as product or algorithm "versions," but more

---

[2] http://trac.osgeo.org/geotiff
[3] http://opengeospatial.org/standards/kml
[4] http://ogcnetwork.net/services

often they manifest only as mysterious differences between data products that one would otherwise expect to be similar. Provenance allows us to better understand the impacts of changes in a processing chain, and to have higher confidence in the reliability of any specific data product.

2.2 Transparent Provenance Collection with ES3

ES3 is a software system for automatically and transparently capturing, managing, and reconstructing the provenance of arbitrary, unmodified computational sequences [3]. *Automatic* acquisition avoids the inaccuracies and incompleteness of human-specified provenance (i.e., annotation). *Transparent* acquisition avoids the computational scientist having to learn, and be constrained by, a specific language or schema in which their problem must be expressed or structured in order for provenance to be captured.

Unlike most other provenance management systems [1,6,9], ES3 captures provenance from running processes, as opposed to extracting or inferring it from static specifications such as scripts or workflows. ES3 provenance management can thus be added to any existing scientific computations without modifying or re-specifying them.

ES3 models provenance in terms of processes and their input and output files. We use "process" in the classic sense of a specific execution of a program. In other words, each execution of a program or workflow, or access to a file, yields new provenance events.

Relationships between files and processes are observed by monitoring read and write accesses. This monitoring can take place at multiple levels: system calls (using `strace`), library calls (using instrumented versions of application libraries), or arbitrary checkpoints within source code (using automatically invoked source-to-source preprocessors for specific environments such as IDL[5]). Any combination of monitoring levels may be active simultaneously, and all are transparent to the scientist-programmer using the system. In particular, system call tracing allows ES3 to function completely independently of the scientist's choice of programming tools or execution environments.

An ES3 provenance document is the directed graph of files and processes resulting from a specific invocation event (e.g., a "job"). Nested processes (processes that spawn other processes) are correctly represented. In addition to retrieving the entire provenance of a job, ES3 supports arbitrary forward (descendant) and/or reverse (ancestor) provenance retrieval, starting at any specified file or process.

ES3 is implemented as a provenance-gathering client and a provenance-managing server (Figure 1). The client runs in the same environment as the processes whose provenance is being tracked.

The client is a set of *logger* processes that intercept raw messages from the various monitoring modes (*plugins*) and write them to log files. A common *transmitter* client asynchronously scans the log files, assembles the provenance events

---

[5] `http://www.ittvis.com/idl`

Collector / Data Submission

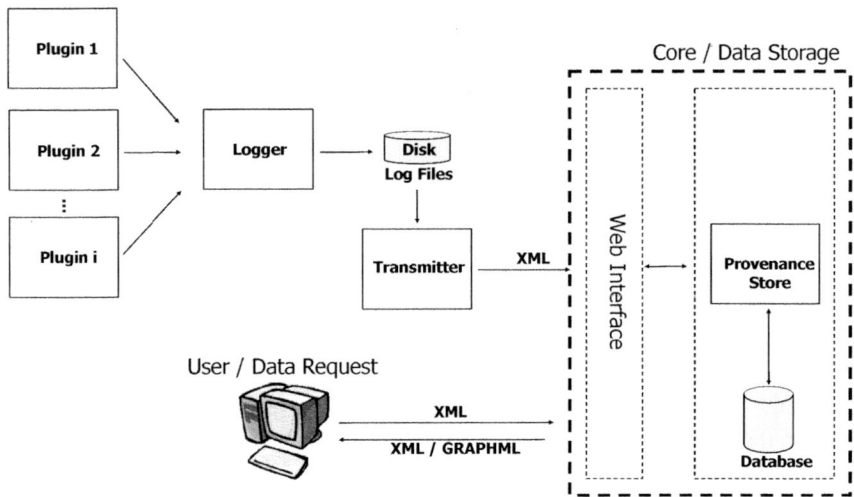

Fig. 1. ES3 architecture

into a time-ordered stream, assigns UUIDs to each file and process being tracked, and submits a raw provenance report to the *ES3 core* (server).

The ES3 core is an XML database with a web service middleware layer that supports insertion of file and provenance metadata, and retrieval of provenance graphs. File metadata allows ES3 to track the one-to-many correspondence between external file identifiers (e.g., pathnames) and internal (UUID) references to those files in provenance reports. Provenance queries cause the ES3 core to assemble a provenance graph (by linking UUIDs) starting at a specified process or file and proceeding in either the ancestor or descendant direction. The graphs are returned serialized in various XML formats (ES3 native, GraphML [2], etc.), which can be rendered by graph visualization clients (e.g., Graphviz[6]; yEd[7]).

ES3's native provenance model is a proper subset of the Open Provenance Model (OPM) [7]. ES3 *processes* and *files* correspond to OPM *processes* and *artifacts*, respectively. ES3 does not support OPM's notion of *agent*, since this entity role cannot be transparently determined from the events ES3 monitors.

2.3   Using Provenance to Drive Publication Decisions

The provenance collected by ES3 represents the complete processing history of a digital object and its antecedents (assuming those antecedents were likewise generated under ES3's observation, or have provenance provided by a similarly capable environment). While extremely valuable as metadata [6] in its own right,

---

[6] http://graphviz.org
[7] http://yworks.com/yed

this provenance can also be exploited to help determine an object's fitness for publication.

Our basic assumption is that a large part of what determines an object's fitness for publication is the *process* by which that object was produced, and the *components* which were combined or transformed in order to produce it. An object produced by scientific codes with known significant bugs, or from source datasets with known errors, may not be suitable for publication, regardless of how well formatted, documented, and presented it is. Or, even if the codes and data used to produce an object are demonstrably acceptable, they may not be suitably *curated*, and this lack of a guaranteed level of future availability to users of the object may be sufficient to disqualify it for publication.

A provenance-driven publication process thus involves traversing a candidate object's provenance to some suitable depth and evaluating whether the object's antecedents justify a decision to publish the object. We envision, and are now prototyping, this decision process as *assertion-based*, with different sets of assertions applied to different categories of antecedents (e.g., programs vs. data files). Managing the assertions separately from the provenance allows us to tailor the assertions to differing levels of rigor—for example, publishing data to a small community of experts may impose fewer constraints on the data's antecedents than publishing to a public repository.

2.4   Comparable Work

One of the primary rationales for collecting provenance information is to enable reproducibility of a process [4]. Our approach to data publication implies that objects that satisfy our publication criteria are more likely (albeit not guaranteed) to be reproducible. An alternative would be to attempt to recreate the entire environment in which a process executed, in order to enable its precise reproduction.

We are aware of two unpublished systems which take this approach. CDE [5] uses system call tracing to package the executable code, input data, and Linux environment (system libraries, language interpreters, etc.) necessary to run a process into a single distributable object, runnable on any Linux system. lbsh [8] uses command-line scraping to determine a process' inputs and outputs, and optionally prepare a (less comprehensive than CDE) package intended to facilitate the process' re-execution.

Our approach is less concerned with *immediate* reproducibility, and more with *long term* reproducibility. Instead of immediately preparing a package that will guarantee an object's reproducibility, we strive to ensure, through our publishability assertions, that such a package *could be* prepared as needed at some arbitrary future time.

3   Worked Example: Publishing Global Ocean Color Data

To illustrate our proposed approach, let us consider a sample computational process, in this case a process that derives an ocean color product from an antecedent product. Figure 2 is a greatly simplified (i.e., many inputs and outputs have been elided) version of an automatically gathered ES3 provenance trace. It depicts a shell script (the outer box) invoking an IDL interpreter (the inner

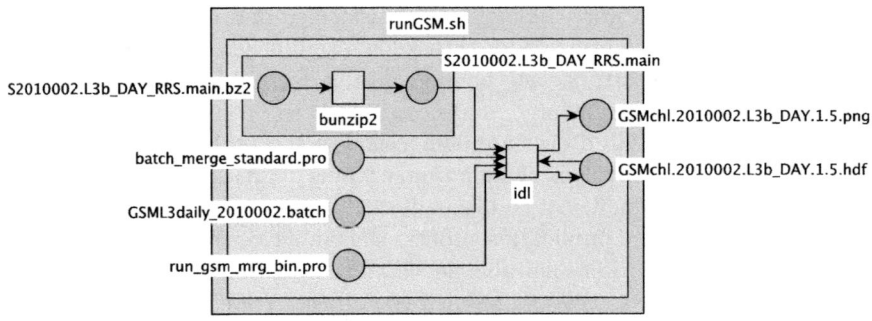

Fig. 2. provenance trace

box), which in turn executes an IDL program, which in turn does the real work: reads an input data file, writes an HDF output file, reads the HDF file back in, and finally produces a PNG preview file.

This type of provenance trace is entirely typical and there is nothing in its structure to indicate its purpose: it could be producing a product to be published, or it could be any manner of test or experiment. Our assertion checking process is initiated only when an output is identified by the scientist as a candidate for publishing. That is, many such traces may be automatically captured by ES3, and most at some point culled, and it is only the act of publishing that triggers any deeper examination of the traces.

There are no fixed assertions to be checked, as the assertions are configurable and will depend greatly on the context. Generally, however, assertions fall into two categories:

- Provenance, or, What did we do? Do we have sufficient information describing the processing that would allow a reader to unambiguously re-create the processing, in principle if not in actuality?
- Confirmation, or, Did we really do what we think we did? It is all too easy to mistakenly operate on the "wrong" file, because the file has the same name as the correct file or because the file's contents have changed without our knowledge. We believe that a publishing system should be able to catch such mistakes.

Returning to our sample process in Fig. 2, the provenance assertions include:

- Does the input data file have a provenance statement? Does it have, for example, ES3 or OPM metadata or another, less structured statement of its source?

- Is the IDL code held in a source code repository, and do the versions of the code used in the processing correspond to committed (i.e., registered) versions in that repository?
- Has information about the computational environment been recorded, e.g., the version of IDL?

The confirmation assertions include:

- If a checksum of the input data file's contents was recorded at the time of its download from an external source, does the checksum match the checksum at the time of the file's use?
- If a correspondence has been made between source code versions and the overall data product version, were the "correct" versions of the source code used?

Notice how the preceding assertions are relevant only when publishing a data product. During development and experimentation there are any number of reasonable reasons why they may be violated.

Implementing such publish-time assertion checking in ES3 requires that we be able to distinguish and characterize different kinds of files participating in the computational process. In a raw ES3 trace, all files referenced during the process are equivalent in ES3's view, being distinguished only as inputs or outputs. But the types of publish-time assertions we want to make depend on the different roles files play in the computational process. What distinguishes an input data file from a source code file in Fig. 2 is that the former was obtained from an external source and is expected to have a provenance statement, while the latter was locally developed and is managed in a source code system, and thus the assertions to be checked are correspondingly different. In our proposed framework, such distinctions can be made ad hoc and on a file-by-file basis, but we expect that general configuration rules will obviate the need for most such fine-grained specification. For example, input data files may be defined to be any files residing in a designated directory. Residency in source code systems can be determined opportunistically, by interrogating source code repositories (e.g., the local CVS server) and by looking for repository artifacts (e.g., RCS and Mercurial repository directories).

A generic restriction on the publication process is the *accessibility* of the published object's antecedents. We assume that any files we wish to check against our publication rules are either directly accessible to the publication process, or have sufficiently complete and trustworthy provenance that assertion checking can be based on metadata alone.

4   Conclusion

Our provenance-driven data publication scheme is a work-in-progress. We are implementing it as a stand-alone `publish` service that, given a digital object, will proceed as follows:

1. Request the object's provenance from ES3.
2. For each antecedent object, test that object against the appropriate publication assertions.
3. List which assertions were violated.
4. If additionally directed, and where possible, take automatic actions (e.g., check code into a repository) to remedy assertion violations.

Of course there will be situations where step 4 fails to automatically render an object fit for publication—for example, publication may require the availability of antecedent files or programs that have been deleted since the object was created. In such cases, we believe the "fitness report" generated by publish will be invaluable documentation, especially if, lacking suitable alternatives, the objects must be published anyway.

References

1. Bose, R., Frew, J.: Lineage Retrieval for Scientific Data Processing: A Survey. ACM Computing Surveys 37(1), 1–28 (2005), doi:10.1145/1057977.1057978
2. Brandes, U., Eiglsperger, M., Herman, I., Himsolt, M., Marshall, M.S.: GraphML progress report: structural layer proposal. In: Mutzel, P., Jünger, M., Leipert, S. (eds.) GD 2001. LNCS, vol. 2265, pp. 109–112. Springer, Heidelberg (2002), doi:10.1007/3-540-45848-4
3. Frew, J., Metzger, D., Slaughter, P.: Automatic capture and reconstruction of computational provenance. Concurrency and Computation: Practice and Experience 20, 485–496 (2008), doi:10.1002/cpe.1247
4. Gil, Y., Cheney, J., Groth, P., Hartig, O., Miles, S., Moreau, L., da Silva, P.P.: Provenance XG Final Report. W3C Provenance Incubator Group (2010), http://www.w3.org/2005/Incubator/prov/XGR-prov-20101214/
5. Guo, P.: CDE: Automatically create portable Linux applications, http://www.stanford.edu/~pgbovine/cde.html
6. Moreau, L.: The Foundations for Provenance on the Web. Foundations and Trends in Web Science 2(2-3), 99–241 (2010), doi:10.1561/1800000010
7. Moreau, L., Clifford, B., Freire, J., Futrelle, J., Gil, Y., Groth, P., Kwasnikowska, N., Miles, S., Missier, P., Myers, J., Plale, B., Simmhan, Y., Stephan, E., Van den Bussche, J.: The Open Provenance Model core specification (v1.1). Future Generation Computer Systems (2010) (in press), doi:10.1016/j.future, 07.005
8. Osterweil, E., Zhang, L.: lbsh: Pounding Science into the Command-Line, http://www.cs.ucla.edu/~eoster/doc/lbsh.pdf
9. Simmhan, Y.L., Plale, B., Gannon, D.: A survey of data provenance in e-science. ACM SIGMOD Record 34, 31–36 (2005), doi:10.1145/1084805.1084812

A Panel Discussion on Data Intensive Science: Moving towards Solutions

Terence Critchlow

Pacific Northwest National Laboratory
902 Battelle Blvd, Richland WA 99352
Terence.Critchlow@pnnl.gov

Over the past several years, a number of groups, including the National Academy of Engineering, have identified grand challenge problems facing scientists from around the world [1]. While addressing these problems will have global impact, solutions are years away at best – and the next set of challenges are likely to be even harder to solve. Because of the complexity of questions being asked, meeting these challenges requires large, multi-disciplinary teams working closely together for extended periods of time. Enabling this new type of science, involving distributed teams that need to collaborate despite vastly different backgrounds and interests, is the cornerstone of Data Intensive Science.

As noted by books such as The Fourth Paradigm [2] and Scientific Data Management: Challenges, Technology, and Deployment [3], our ability to collect and analyze data is providing new and exciting opportunities to extract new knowledge from, and in fact perform novel scientific experiments on, these large data sets. Simply collecting this data, however, is only the beginning. We envision a future where scientists will be able to easily interact with colleagues around the world to identify, search for, acquire and manipulate useful scientific data as easily as they currently work with collaborators down the hall on data generated in their own lab. Only in an environment that supports this type of highly collaborative science will solutions to the grand challenge problems be possible.

Unfortunately, while there has been a lot of excitement in the area of Data Intensive Science over the past couple of years, including some impressive scientific results, much of this potential has remained untapped. In part, this is because the community remains fractured along traditional discipline lines, without a coherent vision of what needs to be done or how it can best be accomplished.

The purpose of this panel is for several leaders in Data Intensive Science to briefly present their views of this future and the key technological breakthroughs that will allow it to be realized. To that end, the panel brings together Malcolm Atkinson (Director of the e-Science Institute), Stefan Heinzel (Rechenzentrum Garching der Max-Plank-Gesellschaft), Tony Hey (Microsoft Research), and Kerstin Kleese Van Dam (Pacific Northwest National Laboratory) and asks them to envision what scientific collaboration could look like in 20 years, as well as identify the technology breakthroughs that need to happen for that vision to be realized.

In particular, the panelists have been asked to answer 3 questions:

> What will scientific collaborations look like in 20 years?
> What will the role of technology be in facilitating this vision?
> What are the biggest breakthroughs required to realize this vision?

J.B. Cushing, J. French, and S. Bowers (Eds.): SSDBM 2011, LNCS 6809, pp. 253–254, 2011.
© Springer-Verlag Berlin Heidelberg 2011

Given the diverse backgrounds, expertise, and experiences represented on this panel, it is expected that both the individual presentations and the subsequent discussions should provide an insightful look into how the technology behind Data Intensive Science may transition from its current state into a form that will truly enable a new age of science.

References

[1] National Academy of Engineering, "Grand Challenges for Engineering",
    http://www.engineeringchallenges.org/cms/challenges.aspx
[2] Hey, A.J.G., Tansley, S., Tolle, K.M.: The Fourth Paradigm: Data-Intensive Scientific Discovery. Microsoft Research (2009)
[3] Shoshani, Rotem, D. (eds.): Scientific Data Management: Challenges, Technology, and Deployment. Chapman & Hall/CRC Computational Science Series (December 2009)

Querying Shortest Path Distance with Bounded Errors in Large Graphs

Miao Qiao, Hong Cheng, and Jeffrey Xu Yu

Department of Systems Engineering and Engineering Management
The Chinese University of Hong Kong
Hong Kong, China
{mqiao, hcheng, yu}@se.cuhk.edu.hk

Abstract. Shortest paths and shortest path distances are important primary queries for users to query in a large graph. In this paper, we propose a new approach to answer shortest path and shortest path distance queries efficiently with an error bound. The error bound is controlled by a user-specified parameter, and the online query efficiency is achieved with prepossessing offline. In the offline preprocessing, we take a reference node embedding approach which computes the single-source shortest paths from each reference node to all the other nodes. To guarantee the user-specified error bound, we design a novel coverage-based reference node selection strategy, and show that selecting the optimal set of reference nodes is NP-hard. We propose a greedy selection algorithm which exploits the submodular property of the formulated objective function, and use a graph partitioning-based heuristic to further reduce the offline computational complexity of reference node embedding.

In the online query answering, we use the precomputed distances to provide a lower bound and an upper bound of the true shortest path distance based on the triangle inequality. In addition, we propose a linear algorithm which computes the approximate shortest path between two nodes within the error bound. We perform extensive experimental evaluation on a large-scale road network and a social network and demonstrate the effectiveness and efficiency of our proposed methods.

1 Introduction

Querying shortest paths or shortest path distances between vertices in a large graph has important applications in many domains including road networks, social networks, biological networks, the Internet, and so on. For example, in road networks, the goal is to find shortest routes between locations or find nearest objects such as restaurants or hospitals; in social networks, the goal is to find the closest social relationships such as common interests, collaborations, citations, *etc.*, between users; while in the Internet, the goal is to find the nearest server in order to reduce access latency for clients. Although classical algorithms like breadth-first search (BFS), Dijkstra's algorithm [1], and A^* search algorithm [2] can compute the exact shortest paths in a network, the massive size of the modern information networks and the online nature of such queries make it infeasible to apply the classical algorithms online. On the other hand, it is space

J.B. Cushing, J. French, and S. Bowers (Eds.): SSDBM 2011, LNCS 6809, pp. 255–273, 2011.
© Springer-Verlag Berlin Heidelberg 2011

inefficient to precompute the shortest paths between all pairs of vertices and store them on disk, as it requires $O(n^3)$ space to store the shortest paths and $O(n^2)$ space to store the distances for a graph with n vertices.

Recently, there have been many different methods [3,4,5,6,7,8,9,10,11,12] for estimating the shortest path distance between two vertices in a graph based on graph embedding techniques. A commonly used embedding technique is reference node embedding, where a set of graph vertices is selected as reference nodes (also called landmarks) and the shortest path distances from a reference node to all the other nodes in a graph are precomputed. Such precomputed distances can be used online to provide an estimated distance between two graph vertices. Although most of the above mentioned methods follow the same general framework of reference node embedding, they differ in the algorithmic details in the following aspects: (1) reference node selection – some (*e.g.*, [6,7,10,11,12]) select reference nodes randomly, while others (*e.g.*, [3,4,8,9]) propose heuristics to select reference nodes; (2) reference node organization – [8,10,11,12] proposed a hierarchical embedding where reference nodes are organized in multiple levels, while most of the other methods use a flat reference node embedding; and (3) an error bound on the estimated shortest path distances – [6,10] analyzed the error bound of the estimated distances with random reference node selection, while most of the other methods have no error bounds or guarantees of the estimated distances.

A theoretical error bound can guarantee the precision of the estimated distance, but the derivation of an error bound is closely related to the reference node selection strategy. Random selection [6,7] or heuristic selection strategies (*e.g.*, based on degree or centrality) [9] cannot derive an error bound to control the precision of the estimated distance. In this paper, we propose a reference node embedding method which provides a distance estimation within a user-specified error bound ϵ. Specifically, we formulate a coverage-based reference node selection strategy, *i.e.*, every node in a graph should be "covered" by some reference node within a radius $c = \epsilon/2$. The coverage property will lead to a theoretical error bound of ϵ. Importantly, allowing a user-specified error bound increases the flexibility of our method in processing queries at different error tolerance levels – when a user specifies an error bound ϵ he can tolerate, we can compute the corresponding radius c for coverage and then the number of reference nodes that are necessary to ensure the error bound. On the other hand, if a user specifies the number of reference nodes he selects, we can find the corresponding value of c and the error bound. We will also show through experimental study that by adjusting the radius c, we can achieve a tradeoff between the theoretical error bound and the offline computational time of the reference node embedding process.

Our main contributions are summarized as follows.

- We take the reference node embedding approach and formulate the optimal reference node set selection problem in a coverage-based scheme. The coverage-based strategy leads to a theoretical error bound of the estimated distance. We show that selecting the minimum set of reference nodes is an NP-hard problem and then propose a greedy solution based on the submodular property of the proposed objective function.
- The reference node embedding can be used to compute an upper bound and a lower bound of the true shortest path distance between any two vertices based on the

triangle inequality. We show that the estimated distance is within a user-specified error bound of the true distance. To further reduce the offline computational complexity of the embedding approach, we propose a graph partitioning-based heuristic for reference node embedding with a relaxed error bound.
- Based on the estimated distances, we propose a linear algorithm to compute the approximate shortest path between two vertices. This algorithm improves the shortest path query efficiency of A^* search by three to five orders of magnitude, while the distance of the approximate shortest path is very close to the exact shortest distance.
- We performed extensive experiments on two different types of networks including a road network and a social network. Although these two types of networks exhibit quite different properties on vertex degree distribution, network diameter, *etc.*, our methods can achieve high accuracy and efficiency on both types of networks.

The rest of the paper is organized as follows. Section 2 introduces preliminary concepts and formulates the distance estimation problem. Section 3 presents our proposed algorithms for reference node selection, shortest path distance estimation and approximate shortest path search. A graph partitioning-based heuristic technique for reference node embedding with a lower offline complexity is proposed in Section 4. Section 5 presents extensive experimental results. We survey related work in Section 6 and conclude in Section 7.

2   Preliminaries and Problem Statement

The input is an edge weighted graph $G = (V, E, w)$, where V is a set of vertices, E is a set of edges, and $w : E \rightarrow R^+$ is a weighting function mapping an edge $(u, v) \in E$ to a positive real number $w(u, v) > 0$, which measures the length of (u, v). We denote $n = |V|$ and $m = |E|$. For a pair of vertices $s, t \in V$, we use $D(s, t)$ to denote the true shortest path distance between s and t. If $(s, t) \in E$, $D(s, t) = w(s, t)$. In this work, we focus on undirected graphs. Our problem can be formulated as follows.

Problem 1 (Distance Estimation with a Bounded Error). Given a graph G and a user-specified error bound ϵ as input, for any pair of query vertices (s, t), we study how to efficiently provide an accurate estimation of the shortest path distance $\widehat{D}(s, t)$, so that the estimation error $|\widehat{D}(s, t) - D(s, t)| \leq \epsilon$.

To efficiently provide a distance estimation, the basic idea is to use a *reference node embedding* approach. Consider a set of vertices $\mathcal{R} = \{r_1, \ldots, r_l\}$ ($\mathcal{R} \subseteq V$), which are called *reference nodes* (also called *landmarks*). For each $r_i \in \mathcal{R}$, we compute the single-source shortest paths to all vertices in V. Then for every node $v \in V$, we can use a l-dimensional vector representation as

$$\overrightarrow{D}(v) = \langle D(r_1, v), D(r_2, v), \ldots, D(r_l, v) \rangle$$

This approach is called reference node embedding. This embedding can be used to compute an upper bound and a lower bound of the true shortest path distance between two vertices (s, t). In the rest of the paper, we will discuss the following questions:

1. Given a graph G and an error bound ϵ, how to select the minimum number of reference nodes to ensure the error bound ϵ in the distance estimation?
2. How to estimate the shortest distance with an error bound given a query (s, t)?
3. How to efficiently compute an approximate shortest path P given a query (s, t)?

3   Proposed Algorithm

The quality of the estimated shortest path distance is closely related to the reference node selection strategy. Given a graph G and an error bound ϵ, we will first formulate a coverage-based reference node selection approach to satisfy the error bound constraint. We will then define an objective function over a set of reference nodes and discuss how to select the minimum set of reference nodes according to the objective function.

3.1   Reference Node Selection

Definition 1 (Coverage). *Given a graph $G = (V, E, w)$ and a radius c, a vertex $v \in V$ is covered by a reference node r if $D(r, v) \leq c$.*

The set of vertices covered by a reference node r is denoted as C_r, *i.e.*, $C_r = \{v | v \in V, D(r, v) \leq c\}$. In particular, we consider a reference node r is covered by itself, *i.e.*, $r \in C_r$, since $D(r, r) = 0 \leq c$. Here we formulate the problem of optimal reference node selection.

Problem 2 (Coverage-based Reference Node Selection). Given a graph $G = (V, E, w)$ and a radius c, our goal is to select a minimum set of reference nodes $\mathcal{R}^* \subseteq V$, *i.e.*, $\mathcal{R}^* = \arg\min_{\mathcal{R} \subseteq V} |\mathcal{R}|$, so that $\forall v \in V - \mathcal{R}^*$, v is covered by at least one reference node from $\mathcal{R}^*$.

Given a user-specified error bound ϵ, we will show in Section 3.3, when we set $c = \epsilon/2$, the coverage-based reference nodes selection method can guarantee that the error of the estimated shortest path distance is bounded by ϵ.

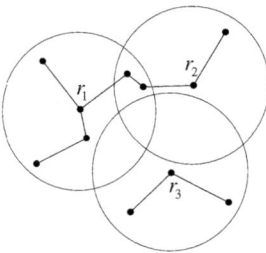

Fig. 1. Coverage-based Reference Node Selection

Example 1. Figure 1 shows a graph with three reference nodes r_1, r_2 and r_3. The three circles represent the area covered by the three reference nodes with a radius c. If a vertex lies within a circle, it means the shortest path distance between the vertex and the corresponding reference node is bounded by c. As shown in the figure, all vertices can be covered by selecting the three reference nodes.

Besides the coverage requirement, a reference node set should be as compact as possible. To evaluate the quality of a set of reference nodes $\mathcal{R}$, we define a gain function over $\mathcal{R}$.

Definition 2 (Gain Function). *The gain function over a set of reference nodes $\mathcal{R}$ is defined as*

$$g(\mathcal{R}) = | \bigcup_{r \in \mathcal{R}} C_r| - |\mathcal{R}| \tag{1}$$

In Figure 1, $g(\{r_1\}) = 5$, $g(\{r_2\}) = 3$, $g(\{r_3\}) = 2$ and $g(\{r_1, r_2, r_3\}) = 8$. The gain function g is a submodular function, as stated in Theorem 1.

Definition 3 (Submodular Function). *Given a finite set N, a set function $f : 2^N \to R$ is submodular if and only if for all sets $A \subseteq B \subseteq N$, and $d \in N \setminus B$, we have $f(A \cup \{d\}) - f(A) \geq f(B \cup \{d\}) - f(B)$.*

Theorem 1. *For two reference node sets $A \subseteq B \subseteq V$ and $r \in V \setminus B$, the gain function g satisfies the submodular property:*

$$g(A \cup \{r\}) - g(A) \geq g(B \cup \{r\}) - g(B)$$

Proof. According to Definition 2, we have

$$\begin{aligned} g(A \cup \{r\}) - g(A) &= |C_A \cup C_r| - (|A| + 1) - |C_A| + |A| \\ &= |C_A \cup C_r| - |C_A| - 1 \\ &= |C_r - C_A| - 1 \end{aligned}$$

where $C_r - C_A$ represents the set of vertices covered by r, but not by A.

Since $A \subseteq B$, we have $C_r - C_B \subseteq C_r - C_A$, hence $|C_r - C_B| \leq |C_r - C_A|$. Therefore, the submodular property holds. $\qquad\square$

As our goal is to find a minimum set of reference nodes $\mathcal{R}^*$ to cover all vertices in V, it is equivalent to maximizing the gain function g:

$$\max_{\mathcal{R}} g(\mathcal{R}) = \max_{\mathcal{R}} (| \bigcup_{r \in \mathcal{R}} C_r| - |\mathcal{R}|) = |V| - \min_{\mathcal{R}} |\mathcal{R}| = g(\mathcal{R}^*)$$

In general, maximizing a submodular function is NP-hard [13]. So we resort to a greedy algorithm. It starts with an empty set of reference nodes $\mathcal{R}_0 = \emptyset$ with $g(\mathcal{R}_0) = 0$.

Then it iteratively selects a new reference node which maximizes an additional gain, as specified in Eq.(2). In particular, in the k-th iteration, it selects

$$r_k = \arg \max_{r \in V \setminus \mathcal{R}_{k-1}} g(\mathcal{R}_{k-1} \cup \{r\}) - g(\mathcal{R}_{k-1}) \tag{2}$$

The algorithm stops when all vertices in V are covered by the reference nodes. The greedy algorithm returns the reference node set $\mathcal{R}$.

Continue with our example. According to the greedy selection algorithm, in the first step, we will select r_1 as it has the highest gain. Given $\mathcal{R}_1 = \{r_1\}$, we have $g(\{r_1, r_2\}) - g(\{r_1\}) = 1$ and $g(\{r_1, r_3\}) - g(\{r_1\}) = 2$. So we will select r_3 in the second step. Finally we will select r_2 to cover the remaining vertices. Note that to simplify the illustration, we only consider selecting reference nodes from r_1, r_2, r_3 in this example. Our algorithm actually considers every graph vertex as a candidate for reference nodes.

To effectively control the size of $\mathcal{R}$, we can further relax the requirement to cover all vertices in V. We observe that such a requirement may cause $|\mathcal{R}|$ unnecessarily large, in order to cover the very sparse part of a graph or the isolated vertices. So we set a parameter *Cover Ratio* (CR), which represents the percentage of vertices to be covered. The above greedy algorithm terminates when a fraction of CR vertices in V are covered by $\mathcal{R}$.

3.2   Shortest Path Distance Estimation

Given $\mathcal{R}$, we will compute the shortest path distances for the node pairs $\{(r, v) | r \in \mathcal{R}, v \in V\}$. This is realized by computing the single-source shortest paths for every $r \in \mathcal{R}$. Given a query node pair (s, t), we have

$$|D(s, r) - D(r, t)| \le D(s, t) \le D(s, r) + D(r, t)$$

for any $r \in \mathcal{R}$, according to the triangle inequality. Figure 2 shows an illustration of the shortest path distance estimation between (s, t), where the circle represents the area covered by a reference node r with a radius c. In this example, s is covered by r, while t is not.

By considering all reference nodes, we have tighter bounds

$$D(s, t) \le \min_{r \in \mathcal{R}} (D(s, r) + D(r, t)) \tag{3}$$

and

$$D(s, t) \ge \max_{r \in \mathcal{R}} |D(s, r) - D(r, t)| \tag{4}$$

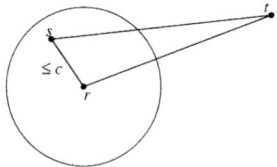

Fig. 2. Distance Estimation

Both the upper bound $\min_{r \in \mathcal{R}}(D(s,r)+D(r,t))$ and the lower bound $\max_{r \in \mathcal{R}}|D(s,r)$ $- D(r,t)|$ can serve as an approximate estimation for $D(s,t)$. We denote them as $\widehat{D}_U$ and $\widehat{D}_L$, respectively. However, [4] reported that the upper bound achieves very good accuracy and performs far better than the lower bound in the internet network. We confirmed this observation on both a social network and a road network in our experiments. Thus we adopt the shortest path distance estimation as

$$\widehat{D}_U(s,t) = \min_{r \in \mathcal{R}}(D(s,r) + D(r,t))$$

3.3   Error Bound Analysis

In this section, we will show that, given a query (s,t), when s or t is covered within a radius c by some reference node from $\mathcal{R}$, the estimated distance $\widehat{D}_U(s,t)$ is within a bounded error of the true distance $D(s,t)$.

Theorem 2. *Given any query (s,t), the error of the estimated shortest path distance $\widehat{D}_U(s,t)$ can be bounded by $2c$ with a probability no smaller than $1-(1-CR)^2$, where c is the coverage radius and CR is the cover ratio.*

Proof. Given a query (s,t) and a reference node set $\mathcal{R}$, assume s is covered by a reference node, denoted as r^*, i.e., $D(s,r^*) \leq c$. Without loss of generality, we assume $D(s,r^*) \leq D(r^*,t)$. Note that the following error bound still holds if $D(s,r^*) > D(r^*,t)$. The error of the estimated shortest path distance between (s,t) is bounded by

$$\begin{aligned}
err(s,t) &= \widehat{D}_U(s,t) - D(s,t) \\
&= \min_{r \in \mathcal{R}}(D(s,r) + D(r,t)) - D(s,t) \\
&\leq D(s,r^*) + D(r^*,t) - D(s,t) \\
&\leq D(s,r^*) + D(r^*,t) - |D(s,r^*) - D(r^*,t)| \\
&= 2D(s,r^*) \\
&\leq 2c
\end{aligned}$$

The first inequality holds because $\min_{r \in \mathcal{R}}(D(s,r) + D(r,t)) \leq D(s,r^*) + D(r^*,t)$; and the second inequality holds because we have the lower bound property $D(s,t) \geq |D(s,r^*) - D(r^*,t)|$.

The error bound holds when either s or t, or both are covered by some reference nodes. When neither s nor t is covered by some reference nodes within a radius c, $err(s,t)$ is unbounded. The probability for this case is $(1 - CR)^2$. However, in this case, if a reference node r happens to lie on the shortest path from s to t, we have the estimated distance $\widehat{D}_U(s,t) = D(s,t)$, i.e., the error is still bounded by $2c$. Therefore, the probability that the error of an estimated distance is unbounded is at most $(1-CR)^2$. Thus we have $P(err(s,t) \leq 2c) \geq 1 - (1 - CR)^2$. □

Given a user-specified error bound ϵ, we will have $P(err(s,t) \leq \epsilon) \geq 1 - (1 - CR)^2$, when we set $c = \epsilon/2$.

3.4 Approximate Shortest Path Computation

With the shortest distance estimation, we propose a heuristic algorithm SPC to compute an approximate shortest path P for a query (s, t). The SPC algorithm works as follows: let $r = \arg\min_{v \in \mathcal{R}}(D(s, v) + D(v, t))$. We use such r to break down the path into two segments as $P(s, t) = SP(s, r) + SP(r, t)$. Here, $SP(s, r)$ represents the exact shortest path from s to r. To compute $SP(s, r)$ in linear time, we can follow the criterion

$$next(s) = \arg\min_{v \in N(s)}(D(s, v) + D(v, r))$$

where $N(s)$ denotes the the neighbor set of s and $next(s)$ denotes the successive neighbor of s that lies on the shortest path from s to r. Here we determine $next(s)$ based on the exact shortest distances $D(s, v)$ and $D(v, r)$. We iteratively apply the above criterion to find every vertex on the shortest path from s to r. Similarly, to compute $SP(r, t)$, we can follow the criterion

$$prev(t) = \arg\min_{v \in N(t)}(D(r, v) + D(v, t))$$

where $prev(t)$ is the preceding neighbor of t that lies on the shortest path from r to t.

The SPC algorithm computes an approximate shortest path whose distance equals $\widehat{D}_U(s, t)$. The time complexity is $O(|\mathcal{R}| + deg \cdot |P|)$, where $O(|\mathcal{R}|)$ is the time for finding the reference node r to break down the path, and deg is the largest vertex degree in the graph.

4 Graph Partitioning-Based Heuristic

For the reference node embedding method we propose above, the offline complexity is $O(|E| + |V|\log|V|)$ to compute the single-source shortest paths for a reference node $v \in \mathcal{R}$. It can be simplified as $O(n\log n)$ $(n = |V|)$ when the graph is sparse. Therefore, the total embedding time is $O(|\mathcal{R}|n\log n)$, which could be very expensive when $|\mathcal{R}|$ is large. In this section, we propose a graph partitioning-based heuristic for the reference node embedding to reduce the offline time complexity with a relaxed error bound. To distinguish the two methods we propose, we name the first method *RN-basic* and the partitioning-based method *RN-partition*.

4.1 Partitioning-Based Reference Node Embedding

The first step of RN-partition is reference node selection, which is the same as described in Section 3.1. In the second step, we use KMETIS [14] to partition the graph into K clusters $C_1, \ldots, C_K$. As a result, the reference node set $\mathcal{R}$ is partitioned into these K clusters. We use $\mathcal{R}_i$ to denote the set of reference nodes assigned to C_i, i.e., $\mathcal{R}_i = \{r | r \in \mathcal{R} \ and \ r \in C_i\}$. It is possible that $\mathcal{R}_i = \emptyset$ for some i. For a cluster C_i with $\mathcal{R}_i = \emptyset$, we can select the vertex from C_i with the largest degree as a within-cluster reference node, to improve the local coverage within C_i. Note that the number of such within-cluster reference nodes is bounded by the number of clusters K, which is a small number compared with $|\mathcal{R}|$.

The idea of the partitioning-based reference node embedding is as follows. For the cluster C_i, we compress all reference nodes in $\mathcal{R}_i$ as a supernode SN_i and then compute the single-source shortest paths from SN_i to every vertex $v \in V$. The reference node compression operation is defined as follows.

Definition 4 (Reference Node Compression). *The reference node compression operation compresses all reference nodes in $\mathcal{R}_i$ into a supernode SN_i. After compression, for a vertex $v \in V \setminus \mathcal{R}_i$, $(SN_i, v) \in E$ iff $\exists r \in \mathcal{R}_i$, s.t. $(r, v) \in E$, and the edge weight is defined as $w(SN_i, v) = \min_{r \in \mathcal{R}_i} w(r, v)$.*

Then the shortest path between SN_i and v is actually the shortest path between a reference node $r \in \mathcal{R}_i$ and v with the smallest shortest path distance, *i.e.*,

$$D(SN_i, v) = \min_{r \in \mathcal{R}_i} D(r, v)$$

and we denote the closest reference node $r \in \mathcal{R}_i$ to v as $r_{v,i}$, which is defined as

$$r_{v,i} = \arg \min_{r \in \mathcal{R}_i} D(r, v)$$

Note $D(SN_i, v) = D(r_{v,i}, v) = \min_{r \in \mathcal{R}_i} D(r, v)$. In the following, we will use $D(SN_i, v)$ and $D(r_{v,i}, v)$ interchangeably.

The time complexity for computing shortest paths from the supernodes in each of the K clusters to all the other vertices in V is $O(Kn \log n)$. In addition, we compute the shortest path distances between every pair of reference nodes within the same cluster. The time complexity of this operation is $O(|\mathcal{R}|n/K \log n/K)$, if we assume the nodes are evenly partitioned into K clusters. We further define the diameter d for a cluster as follows.

Definition 5 (Cluster Diameter). *Given a cluster C, the diameter d is defined as the maximum shortest distance between two reference nodes in C, i.e.,*

$$d = \max_{r_i, r_j \in C} D(r_i, r_j)$$

where $D(r_i, r_j)$ is the shortest distance between r_i and r_j.

Then the diameter of the partitioning $C_1, \ldots, C_K$ is defined as the maximum of the K cluster diameters, *i.e.*,

$$d_{max} = \max_{i \in [1, K]} d_i$$

4.2 Partitioning-Based Shortest Path Distance Estimation

Given a query (s, t), for the supernode SN_i representing a cluster C_i, based on the triangle inequality we have

$$D(s, t) \leq D(s, SN_i) + D(r_{s,i}, r_{t,i}) + D(t, SN_i)$$

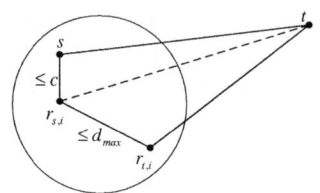

Fig. 3. Distance Estimation in RN-partition

Figure 3 shows an illustration of the shortest path distance estimation between (s, t) in RN-partition, where the circle represents a cluster C_i. Note that in general s and $r_{s,i}$ may not necessarily belong to the same cluster, and the shortest path distance $D(s, r_{s,i})$ may not necessarily be bounded by the radius c. But these factors will not affect the distance estimation strategy.

By considering all K clusters, we have a tighter upper bound

$$D(s,t) \leq \min_{i \in [1,K]} (D(s, SN_i) + D(r_{s,i}, r_{t,i}) + D(t, SN_i))$$

We denote this estimated distance upper bound as $\widehat{D}_U^P(s, t)$.

4.3 Error Bound Analysis

In the following theorem, we will show that, when s or t is covered within a radius c by some reference node from a cluster C_i for some i, the estimated distance $\widehat{D}_U^P(s, t)$ is within a bounded error of the true distance $D(s, t)$.

Theorem 3. *Given any query (s, t), the error of the estimated shortest path distance $\widehat{D}_U^P(s, t)$ by RN-partition can be bounded by $2(c + d_{max})$ with a probability no smaller than $1 - (1 - CR)^2$, where c is the coverage radius, CR is the cover ratio and d_{max} is the maximum cluster diameter.*

Proof. Given a query (s, t), assume s is covered by at least one reference node from $\mathcal{R}$ within a radius c. Without loss of generality, assume such a reference node is from the cluster C_i for some i and denote it as $r_{s,i}$. According to the triangle inequality, we have

$$D(r_{s,i}, t) - D(s, t) \leq D(s, r_{s,i}) \leq c$$

By adding $D(s, r_{s,i})$ on both sides, we have

$$D(s, r_{s,i}) + D(r_{s,i}, t) - D(s, t) \leq 2D(s, r_{s,i}) \leq 2c \tag{5}$$

Denote the closest reference node in C_i to t as $r_{t,i}$. Then we have

$$D(r_{t,i}, t) - D(r_{s,i}, t) \leq D(r_{s,i}, r_{t,i}) \leq d_{max}$$

Since $r_{s,i}, r_{t,i}$ belong to the same cluster, their distance is bounded by d_{max}. By adding $D(r_{s,i}, r_{t,i})$ on both sides, we have

$$D(r_{s,i}, r_{t,i}) + D(r_{t,i}, t) - D(r_{s,i}, t) \leq 2D(r_{s,i}, r_{t,i}) \leq 2d_{max} \tag{6}$$

By adding Eq.(5) and Eq.(6), we have

$$D(s, r_{s,i}) + D(r_{s,i}, r_{t,i}) + D(r_{t,i}, t) - D(s, t) \leq 2(c + d_{max})$$

As we have defined $\widehat{D}_U^P(s, t) = \min_{i \in [1,K]}(D(s, SN_i) + D(r_{s,i}, r_{t,i}) + D(t, SN_i))$, the error of the estimated shortest path distance between (s, t) is bounded by

$$
\begin{aligned}
err^P(s, t) &= \widehat{D}_U^P(s, t) - D(s, t) \\
&\leq D(s, r_{s,i}) + D(r_{s,i}, r_{t,i}) + D(r_{t,i}, t) - D(s, t) \\
&\leq 2(c + d_{max})
\end{aligned}
$$

The error bound holds when either s or t, or both are covered by some reference nodes with a radius c. When it happens that neither s nor t is covered by some reference nodes, Eq.(5) does not hold in general, thus $err^P(s, t)$ is unbounded. The probability for this case is $(1 - CR)^2$. For a similar reason as explained in Theorem 2, i.e., even when neither s nor t is covered, if there are reference nodes $r_{s,i}$, $r_{t,i}$, for some i, on the shortest path from s to t, we can still have an accurate estimation which satisfies the error bound. Therefore the probability that the error of an estimated distance is unbounded is at most $(1 - CR)^2$. Thus, we have $P(err^P(s, t) \leq 2(c + d_{max})) \geq 1 - (1 - CR)^2$ □

Compared with RN-basic, RN-partition reduces the offline computational complexity to $O(Kn \log n + |\mathcal{R}|n/K \log n/K)$. As long as we choose a reasonably large K such that $|\mathcal{R}|/K \leq K$, the complexity of RN-partition is dominated by $O(Kn \log n)$. As a tradeoff, the error bound is relaxed from $2c$ to $2(c + d_{max})$. The cluster diameter d_{max} is determined by the size of the graph and the number of clusters K. Table 1 compares RN-basic and RN-partition on time/space complexity and the error bound. In experimental study, we will study the relationship between K, the offline computation time and the accuracy of the estimated distances.

Table 1. Comparison between RN-basic and RN-partition

	RN-basic	RN-partition
Offline Time Complexity	$O(|\mathcal{R}|n \log n)$	$O(Kn \log n + |\mathcal{R}|n/K \log n/K)$
Offline Space Complexity	$O(|\mathcal{R}|n)$	$O(Kn + |\mathcal{R}|^2/K)$
Distance Query Complexity	$O(|\mathcal{R}|)$	$O(K)$
Error Bound	$2c$	$2(c + d_{max})$

5 Experiments

We performed extensive experiments to evaluate our algorithms on two types of networks – a road network and a social network. The road network and the social network exhibit quite different properties on: (1) degree distribution, i.e., the former roughly follows a uniform distribution while the latter follows a power law distribution; and (2) network diameter, i.e., the social network has the shrinking diameter property [15] and the small world phenomenon, which, however, do not hold in the road network. All experiments were performed on a Dell PowerEdge R900 server with four 2.67GHz six-core CPUs and 128GB main memory running Windows Server 2008. All algorithms were implemented in Java.

5.1   Comparison Methods and Evaluation

We compare our methods RN-basic and RN-partition with two existing methods:

- **2RNE** [8] by Kriegel et al. uses a two level reference node embedding which examines K nearest reference nodes for both nodes in a query to provide a distance estimation. We select reference nodes uniformly and set $K = 3$.
- **Centrality** [9] by Potamias et al. selects reference nodes with low closeness centrality. According to [9], the approximate centrality measure is computed by selecting a sample of S random seeds, where we set $S = 10,000$ in our implementation.

For a node pair (s, t), we use the relative error to evaluate the quality of the estimated distance

$$rel\_err(s, t) = \frac{|\widehat{D}_U(s, t) - D(s, t)|}{D(s, t)}$$

As it is expensive to exhaustively evaluate all node pairs in a large network, we randomly sample a set of $10,000$ node pairs in the graph as queries and evaluate the average relative error on the sample set.

5.2   Case Study 1: Road Network

We use the New York City road network, which is an undirected planar graph with $264,346$ nodes and $733,846$ edges. A node here represents an intersection or a road endpoint while the weight of an edge represents the length of the corresponding road segment. The data set can be downloaded from http://www.dis.uniroma1.it/∼challenge9/.

The degrees of most nodes in the road network fall into the range of $[1, 4]$ and the network has no small world phenomenon. For the $10,000$ random queries we generate, we plot the histogram of the shortest path distance distribution in Figure 4. The average distance over the $10,000$ queries is $d_{avg} = 26.68$KM. So if we set the radius $c = 0.8$KM, the average relative error can be roughly bounded by $2c/d_{avg} = 0.06$.

Parameter Sensitivity Test on CR. In this experiment, we vary the cover ratio CR and compare the average error, the reference node set size and offline index time by RN-basic and RN-partition with $K = 100, 250, 500$, respectively. We fix the radius $c = 0.8$KM.

Figure 6 shows the average error of RN-basic and RN-partition with different K values. The average error of RN-basic is below 0.01 and slightly decreases as CR increases. The average error of RN-partition decreases very sharply when the number of partitions K increases and becomes very close to that of RN-basic when $K = 500$.

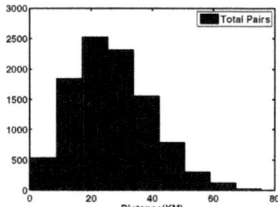
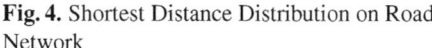

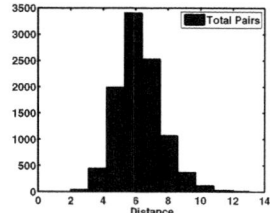

Fig. 4. Shortest Distance Distribution on Road Network

Fig. 5. Shortest Distance Distribution on Social Network

Figure 7 shows that the number of reference nodes $|\mathcal{R}|$ increases linearly with CR. As RN-basic and RN-partition have the same reference node selection process, the number is the same for both methods. When $CR = 1.0$, we need $9,000$ reference nodes to cover the road network with $264,346$ nodes.

Figure 8 shows the offline index time in logarithmic scale for RN-basic and RN-partition to compute the single-source shortest paths from every reference node. RN-partition reduces the index time of RN-basic by one order of magnitude. In addition, as the number of reference nodes $|\mathcal{R}|$ increases linearly with CR, the index time of RN-basic also increases linearly with CR, because the time complexity is $O(|\mathcal{R}|n \log n)$. On the other hand, the index time of RN-partition remains quite stable as CR increases, because RN-partition only computes the shortest paths from each of the K clusters as the source.

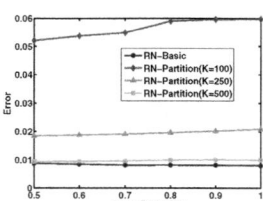

Fig. 6. Average Error vs. Cover Ratio (CR) on Road Network

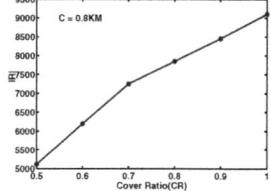

Fig. 7. Reference Node Set Size vs. Cover Ratio (CR) on Road Network

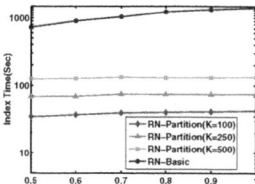

Fig. 8. Index Time vs. Cover Ratio (CR) on Road Network

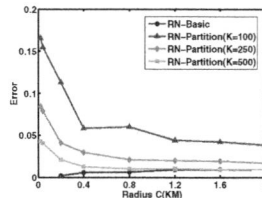

Fig. 9. Average Error vs. Radius (c) on Road Network

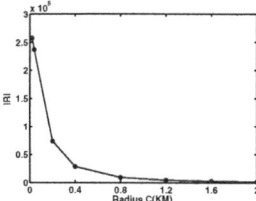

Fig. 10. Reference Node Set Size vs. Radius (c) on Road Network

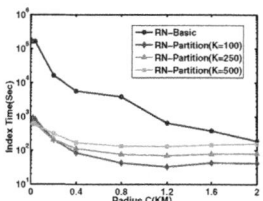

Fig. 11. Index Time vs. Radius (c) on Road Network

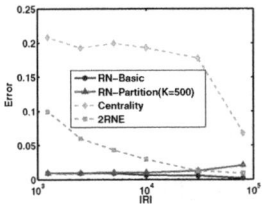
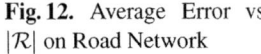
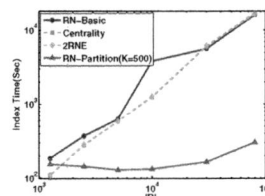
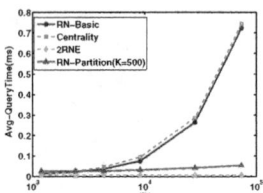

Fig. 12. Average Error vs. $|\mathcal{R}|$ on Road Network

Fig. 13. Index Time vs. $|\mathcal{R}|$ on Road Network

Fig. 14. Average Query Time vs. $|\mathcal{R}|$ on Road Network

Parameter Sensitivity Test on c. In this experiment, we vary the radius c and compare the average error, the reference node set size and offline index time by RN-basic and RN-partition with $K = 100, 250, 500$, respectively. We fix the cover ratio $CR = 1.0$.

Figure 9 shows the average error of RN-basic and RN-partition with different K values. We can make the following observations from the figure: (1) RN-partition ($K = 500$) achieves an average error very close to that of RN-basic when $c \geq 0.8$KM; (2) The average error of RN-basic monotonically increases with c, which is consistent with the theoretical error bound of $2c$; and (3) Different from RN-basic, the average error of RN-partition shows a decreasing trend with c. When c is very small, the number of reference nodes is very large. So RN-partition may choose suboptimal reference nodes for distance estimation, which leads to a larger error.

Figure 10 shows that the number of reference nodes $|\mathcal{R}|$ decreases with c. When $c < 0.4$KM, $|\mathcal{R}|$ decreases sharply with c. Figure 11 shows the offline index time of RN-basic and RN-partition in logarithmic scale. As $|\mathcal{R}|$ decreases with c, the index time of RN-basic also decreases with c. RN-partition reduces the index time of RN-basic by two orders of magnitude or more when $c < 0.2$KM but the difference becomes smaller as c increases. RN-basic cannot finish within 10 hours when $c \leq 0.08$KM. On the other hand, the index time of RN-partition increases moderately when c decreases to 0.2KM or below.

Comparison with 2RNE and Centrality. We compare our approaches with 2RNE [8] and Centrality [9] in terms of average error, index time and average query time, as we vary the number of reference nodes. For our methods, we set $CR = 1.0$. From Figure 12 we can see that both RN-basic and RN-partition (for most cases) outperform 2RNE and Centrality by a large margin in terms of average error. Figure 13 shows that RN-partition reduces the index time of the other three methods by up to two orders of magnitude. The index time of RN-basic, 2RNE and Centrality increases linearly with $|\mathcal{R}|$, as they all have the same time complexity of $O(|\mathcal{R}|n \log n)$, while RN-partition slightly increases the index time. Figure 14 shows that the query time of RN-partition and 2RNE remain almost constant, while that of RN-basic and Centrality increase linearly with the number of reference nodes.

Shortest Path Query Processing. In this experiment, we evaluate the efficiency and quality of the SPC procedure for computing the shortest paths. For comparison, we implemented A^* algorithm using $\widehat{D}_L$ as the h function, since it provides a lower bound

Table 2. Comparison between SPC and A^* on Road Network

	Average Error	Average Query Time (millisec)
SPC	0.012	0.19
A^*	0	141.79

distance estimation. We evaluate both methods on the $10,000$ random queries. We set $|\mathcal{R}| = 20,000$. Table 2 shows that SPC finds approximate shortest paths with an average error of 0.012 while A^* computes the exact shortest paths. But SPC is about 750 times faster than A^*, since it is a linear algorithm.

5.3   Case Study 2: Social Network

We download the DBLP dataset from http://dblp.uni-trier.de/xml/ and construct an undirected coauthor network, where a node represents an author, an edge represents a coauthorship relation between two authors, and all edge weights are set to 1. This graph has several disconnected components and we choose the largest connected one which has $629,143$ nodes and $4,763,500$ edges. The vertex degree distribution follows the power law distribution.

We randomly generate $10,000$ queries and plot the histogram of the shortest path distance distribution in Figure 5. The average distance between two nodes over the $10,000$ queries is $d_{avg} = 6.34$, which conforms with the famous social networking rule "six degrees of separation". Given $2c/d_{avg}$ as a rough estimation of the relative error bound, if we set $c = 3$, the relative error bound is $2 \times 3/6.34 = 94.64\%$. Therefore, we only test our methods given $c \in \{1,2\}$, to control the relative error bound in a reasonably small range. Note that $c = 1$ defines the coverage of a node based on the number of its neighbors, *i.e.*, degree; while $c = 2$ measures the coverage based on the number of neighbors within two hops.

Parameter Sensitivity Test on CR. We vary the cover ratio CR and compare the average error, the reference node set size and offline index time by RN-basic and RN-partition with $K = 100, 200, 300$, respectively. We fix the radius $c = 1$.

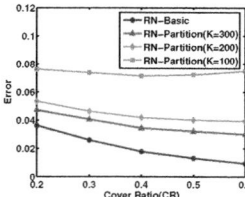
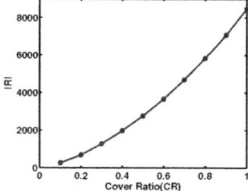
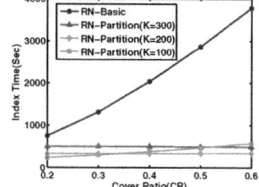

Fig. 15. Average Error vs. Cover Ratio (CR) on Social Network

Fig. 16. Reference Node Set Size vs. Cover Ratio (CR) on Social Network

Fig. 17. Index Time vs. Cover Ratio (CR) on Social Network

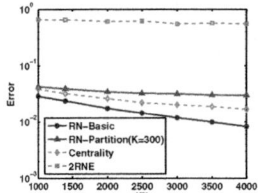
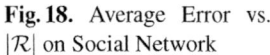
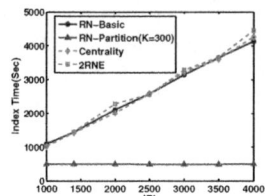
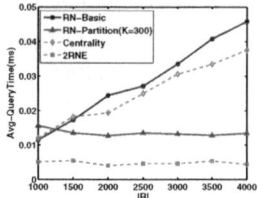

Fig. 18. Average Error vs. $|\mathcal{R}|$ on Social Network

Fig. 19. Index Time vs. $|\mathcal{R}|$ on Social Network

Fig. 20. Average Query Time vs. $|\mathcal{R}|$ on Social Network

Figure 15 shows that the average error of RN-basic is in the range of $[0.009, 0.04]$ and it decreases quickly as CR increases. The average error of RN-partition is slightly higher than that of RN-basic and it decreases as K increases.

Figure 16 shows the number of reference nodes $|\mathcal{R}|$ as we vary CR in the range of $[0, 1.0]$. Different from the road network which shows a linear relationship between $|\mathcal{R}|$ and CR, we observe that $|\mathcal{R}|$ increases slowly when CR is small, but much faster when CR is large. This is due to the power law degree distribution in the social network – we first select the authors with the largest number of collaborators as reference nodes; but in the later stage, with the decrease of node degrees, we need to use more reference nodes to achieve the same amount of coverage.

Figure 17 shows the offline index time for RN-basic and RN-partition. We observe that the index time of RN-basic increases quickly when CR increases. When $CR = 0.6$, RN-basic is about 10 times slower than RN-partition. We also observe that the index time of RN-partition slightly increases with CR when $K = 100$. This is because a large portion of time is spent on computing the shortest path distances between all pairs of reference nodes within the same partition. When CR increases, the number of reference nodes falling into the same partition is larger, which causes the time increase.

Parameter Sensitivity Test on c. In this experiment, we vary the radius $c \in \{1, 2\}$ and compare the average error, the reference node set size and offline index time by RN-basic and RN-partition ($K = 300$). We fix $CR = 0.6$. Table 3 shows that the number of reference nodes is reduced by 100 times when c is increased to 2. As a result, the offline index time for RN-basic is also reduced by 100 times with the increase of c because the time complexity is $O(|\mathcal{R}|n \log n)$. The index time for RN-partition is seven times smaller than RN-basic when $c = 1$, but slightly higher when $c = 2$ due to the within partition computational overhead. The average error of RN-partition is slightly higher than that of RN-basic, and the error of both methods increases with c, which is consistent with the theoretical error bound.

Table 3. Parameter Sensitivity Test on Radius c on Social Network

		RN-basic		RN-partition	
Radius c	$|\mathcal{R}|$	Average Error	Index Time (sec)	Average Error	Index Time (sec)
1	3653	0.009	3778.17	0.030	485.88
2	31	0.138	30.70	0.144	65.71

Comparison with 2RNE and Centrality. We compare our approaches with 2RNE and Centrality in terms of average error, index time and average query time, as we vary the number of reference nodes. For our methods, we set $c = 1$. Figure 18 shows that RN-basic achieves the smallest error, followed by Centrality and RN-partition. 2RNE performs the worst, because it selects reference nodes uniformly, rather than selecting reference nodes with large degrees. Figure 19 shows that the index time of RN-partition remains stable when $|\mathcal{R}|$ increases, while the time of the other three methods increases linearly with $|\mathcal{R}|$. Figure 20 shows that the query time of RN-partition and 2RNE remain almost constant, while that of RN-basic and Centrality increase linearly with the number of reference nodes.

Shortest Path Query Processing. We compare SPC with A^* on shortest path query on the $10,000$ random queries on the DBLP network. We set $|\mathcal{R}| = 4,000$. Table 4 shows that SPC finds approximate shortest paths with an average error of 0.008 while A^* computes the exact shortest paths. But SPC is about $100,000$ times faster than A^*.

Table 4. Comparison between SPC and A^* on Social Network

	Average Error	Average Query Time (millisec)
SPC	0.008	0.046
A^*	0	4469.28

6   Related Work

Dijkstra's algorithm [1] computes the single-source shortest path in a graph with non-negative edge weights with a time complexity of $O(|E| + |V| \log |V|)$. The Floyd-Warshall algorithm [16] computes the shortest paths between all pairs of vertices with a dynamic programming approach. Its time complexity is $O(|V|^3)$. The A^* search algorithm [2,17] uses some heuristics to direct the search direction.

In the literature, graph embedding techniques have been widely used to estimate the distance between two nodes in a graph in many applications including road networks [5,8], social networks and web graphs [7,9,11,12] and the Internet [3,4]. Kriegel et al. [8] proposes a hierarchical reference node embedding algorithm for shortest distance estimation. Potamias et al. [9] formulates the reference node selection problem to selecting vertices with high betweenness centrality. [3] proposes an architecture, called IDMaps which estimates the distance in the Internet and a related work [4] proposes a Euclidean embedding approach to model the Internet. [6] defines a notion of slack – a certain fraction of all distances that may be arbitrarily distorted as a performance guarantee based on randomly selected reference nodes. [10] and its follow up studies [11,12] provide a relative $(2k - 1)$-approximate distance estimation with $O(kn^{1+1/k})$ memory for any integer $k \geq 1$. A limitation of many existing methods is that, the estimated shortest path distance has no error bound, thus it is hard to guarantee the estimation quality. In contrast, our approach provides an absolute error bound of the distance estimation by $2c$ in RN-basic or by $2(c + d_{max})$ in RN-partition.

Computing shortest paths and processing k-nearest neighbor queries in spatial networks have also received a lot of attention. Papadias et al. [18] propose to use the Euclidean distance as a lower bound to prune the search space and guide the network expansion for refinement. [19] uses first order Voronoi diagram to answer KNN queries in spatial networks. Hu et al. [20] propose an index, called distance signature, which associates approximate distances from one object to all the other objects in the network, for distance computation and query processing. Samet et al. [21] build a shortest path quad tree to support k-nearest neighbor queries in spatial networks. [22] proposes TEDI, an indexing and query processing scheme for the shortest path query based on tree decomposition.

7 Conclusions

In this paper, we propose a novel coverage-based reference node embedding approach to answer shortest path and shortest path distance queries with a theoretical error bound. Our methods achieve very accurate distance estimation on both a road network and a social network. The RN-basic method provides very accurate distance estimation, while the RN-partition method reduces the offline embedding time of RN-basic by up to two orders of magnitude or more with a slightly higher estimation error. In addition, our methods outperform two state-of-the-art reference node embedding methods in processing shortest path distance queries.

Acknowledgment

The work was supported in part by grants of the Research Grants Council of the Hong Kong SAR, China No. 419109, No. 419008 and No. 411310, and the Chinese University of Hong Kong Direct Grants No. 2050446.

References

1. Dijkstra, E.W.: A note on two problems in connexion with graphs. Numerische Mathematik 1(1), 269–271 (1959)
2. Hart, P.E., Nilsson, N.J., Raphael, B.: A formal basis for the heuristic determination of minimum cost paths. IEEE Transactions on Systems Science and Cybernetics SSC4 4(2), 100–107 (1968)
3. Francis, P., Jamin, S., Jin, C., Jin, Y., Raz, D., Shavitt, Y., Zhang, L.: IDMaps: A global internet host distance estimation service. IEEE/ACM Trans. Networking 9(5), 525–540 (2001)
4. Ng, E., Zhang, H.: Predicting internet network distance with coordinates-based approaches. In: INFOCOM, pp. 170–179 (2001)
5. Shahabi, C., Kolahdouzan, M., Sharifzadeh, M.: A road network embedding technique for k-nearest neighbor search in moving object databases. In: GIS, pp. 94–100 (2002)
6. Kleinberg, J., Slivkins, A., Wexler, T.: Triangulation and embedding using small sets of beacons. In: FOCS, pp. 444–453 (2004)
7. Rattigan, M.J., Maier, M., Jensen, D.: Using structure indices for efficient approximation of network properties. In: KDD, pp. 357–366 (2006)

8. Kriegel, H.-P., Kröger, P., Renz, M., Schmidt, T.: Hierarchical graph embedding for efficient query processing in very large traffic networks. In: Ludäscher, B., Mamoulis, N. (eds.) SS-DBM 2008. LNCS, vol. 5069, pp. 150–167. Springer, Heidelberg (2008)
9. Potamias, M., Bonchi, F., Castillo, C., Gionis, A.: Fast shortest path distance estimation in large networks. In: CIKM, pp. 867–876 (2009)
10. Thorup, M., Zwick, U.: Approximate distance oracles. Journal of the ACM 52(1), 1–24 (2005)
11. Gubichev, A., Bedathur, S., Seufert, S., Weikum, G.: Fast and accurate estimation of shortest paths in large graphs. In: CIKM, pp. 499–508 (2010)
12. Sarma, A.D., Gollapudi, S., Najork, M., Panigrahy, R.: A sketch-based distance oracle for web-scale graphs. In: WSDM, pp. 401–410 (2010)
13. Khuller, S., Moss, A., Naor, J.: The budgeted maximum coverage problem. Information Processing Letters 70, 39–45 (1999)
14. Karypis, G., Kumar, V.: A fast and high quality multilevel scheme for partitioning irregular graphs. SIAM Journal on Scientific Computing 20(1), 359–392 (1999)
15. Leskovec, J., Kleinberg, J., Faloutsos, C.: Graph evolution: Densification and shrinking diameters. ACM Trans. Knowledge Discovery from Data 1(1) (2007)
16. Floyd, R.W.: Algorithm 97: Shortest path. Communications of the ACM 5(6), 345 (1962)
17. Goldberg, A.V., Harrelson, C.: Computing the shortest path: A* search meets graph theory. In: SODA, pp. 156–165 (2005)
18. Papadias, D., Zhang, J., Mamoulis, N., Tao, Y.: Query processing in spatial network database. In: VLDB, pp. 802–813 (2003)
19. Kolahdouzan, M., Shahabi, C.: Voronoi-based k nearest neighbor search for spatial network databases. In: VLDB, pp. 840–851 (2004)
20. Hu, H., Lee, D.L., Lee, V.C.S.: Distance indexing on road networks. In: VLDB, pp. 894–905 (2006)
21. Samet, H., Sankaranarayanan, J., Alborzi, H.: Scalable network distance browsing in spatial databases. In: SIGMOD, pp. 43–54 (2008)
22. Wei, F.: TEDI: Efficient shortest path query answering on graphs. In: SIGMOD, pp. 99–110 (2010)

PG-Join: Proximity Graph Based String Similarity Joins

Michail Kazimianec and Nikolaus Augsten

Faculty of Computer Science, Free University of Bozen-Bolzano,
Dominikanerplatz 3, 39100 Bozen, Italy
kazimianec@inf.unibz.it, augsten@inf.unibz.it

Abstract. In many applications, for example, in data integration scenarios, strings must be matched if they are similar. String similarity joins,
which match all pairs of similar strings from two datasets, are of particular interest and have recently received much attention in the database
research community. Most approaches, however, assume a global similarity threshold; all string pairs that exceed the threshold form a match in
the join result. The global threshold approach has two major problems:
(a) the threshold depends on the (mostly unknown) data distribution,
(b) often there is no single threshold that is good for all string pairs.

In this paper we propose the *PG-Join* algorithm, a novel string similarity join that requires no configuration and uses an *adaptive threshold*.
PG-Join computes a so-called proximity graph to derive an individual
threshold for each string. Computing the proximity graph efficiently is
essential for the scalability of PG-Join. To this end we develop a new and
fast algorithm, *PG-I*, that computes the proximity graph in two steps:
First an efficient approximation is computed, then the approximation
error is fixed incrementally until the adaptive threshold is stable. Our
extensive experiments on real-world and synthetic data show that PG-I
is up to five times faster than the state-of-the-art algorithm and suggest
that PG-Join is a useful and effective join paradigm.

1 Introduction

String data is ubiquitous and represents textual information in relational
databases, Web resources, data archives, etc. Finding similar matches for a given
string has received much attention from different research communities and is
widely applied in data cleaning, approximate query answering, and information
retrieval. String similarity joins, which match all pairs of similar strings from
two datasets, are of particular interest. They are used, for example, to join tables on string attributes that suffer from misspellings, typographical errors, or
inconsistent coding conventions [1,2,3,4]. In data cleaning scenarios, a similarity
join matches misspellings to their correct counterparts, and the misspelling is
replaced.

An example are tables with a street name attribute. Street names are often
spelled differently in different databases [5]. One database may call a street 'via

J.B. Cushing, J. French, and S. Bowers (Eds.): SSDBM 2011, LNCS 6809, pp. 274–292, 2011.

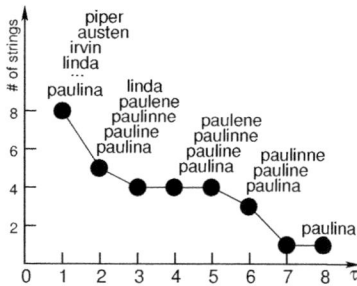

Fig. 1. Proximity Graph for the String 'paulina'

Maso d. Pieve', an other database calls the same street 'via M. della Pieve'. An exact join fails to match the two streets. The similarity join matches two strings if they are similar enough.

A key challenge of similarity joins is the choice of a good similarity threshold, i.e., fixing the degree of similarity that allows a pair of strings to match. Most approaches require the user to fix a global threshold, and the same threshold is used for all string pairs. This approach has two major problems: (a) The threshold depends on the underlying data distribution, which is often unknown to the user. Further, the numeric value of the threshold depends on the similarity function in use and may be hard to interpret, e.g., does 0.8 match the user's intuition of "similar"? (b) In many cases there is no single threshold that matches all string pairs that should match. A high similarity threshold may prevent many good pairs from matching; but if the threshold is decreased to include these pairs, also many bad pairs are included.

In this paper we introduce a new join paradigm and develop *PG-Join*, a string similarity join algorithm that does not require configuration and uses adaptive thresholds. PG-Join computes proximity graphs to derive an individual threshold for each string. Proximity graphs where developed by Mazeika and Böhlen [6] to automatically detect the cluster borders in their GPC clustering algorithm.

The proximity graph is computed for a center string and shows, for each similarity threshold τ, the number of dataset strings within the threshold from the center, i.e., the number of strings sharing τ q-grams (substrings of length q) with the center. Figure 1 shows the proximity graph for the center string *paulina* in a dataset with eight strings. The x-axis shows similarity threshold τ, the y-axis the number of strings within the respective neighborhood. The neighborhoods are computed from right to left, i.e. from the greatest similarity threshold to the smallest one. After each neighborhood computation the cluster center is adjusted.

The adaptive join threshold for a string is derived from its proximity graph. Intuitively, the threshold is increased until further increasing it does not increase the number of strings in the neighborhood. More precisely, the threshold is defined by the rightmost endpoint of the longest horizontal line in the proximity graph. In Figure 1, the threshold is $\tau = 5$ and the center string *paulina* is matched to *pauline*, *paulinne*, and *paulene*, but not to the other strings.

PG-Join computes a proximity graph for each string in both datasets and its performance critically depends on an efficient proximity graph algorithm. The original algorithm by Mazeika and Böhlen [6] has been considerably improved in recent works [7,8]. All these works compute the neighborhoods in the proximity graph in isolation and do not make use of previously computed results. The reason is that the center is adjusted after each neighborhood computation such that the neighborhoods are not directly comparable.

We take a radically different approach. In a first step we compute a fast approximation of the proximity graph that assumes a constant center. In a second step we fix the error introduced by our assumption. We show that this can done incrementally from right to left and we can stop early when the join threshold is stable. Together with an incremental update of the center this leads to, *PG-I*, a highly efficient algorithm for the proximity graph computation. Our experiments show that our algorithm is up to five times faster then the fastest known algorithms.

Problem definition. Our goal is a new string similarity join that

- is based on q-gram similarity,
- does not require a predefined similarity threshold,
- adapts the threshold dynamically to the underlying data distribution,
- and is computed efficiently.

Contribution Summarizing, we make the following contributions:

- We propose PG-Join, a new string similarity join algorithm. PG-Join is different from other algorithms in that it uses an adaptive similarity threshold instead of a global threshold, leading to better join results.
- We develop PG-I, an efficient algorithm for computing the proximity graph, which is used in PG-Join. PG-I also improves the efficiency of other proximity graph based algorithms, for example, the GPC clustering algorithm [6].
- In our experimental evaluation on synthetic and real-world datasets we evaluate the effectiveness of PG-Join with respect to the global threshold join and show that PG-I is up to five times faster than the fastest known algorithm for proximity graphs.

Overview. The remaining paper is organized as follows. We present background material in Section 2, introduce PG-Join in Section 3, and present the PG-I algorithm for efficient computing of the proximity graph in Section 4. We discuss related work in Section 5. After the experimental evaluation of our solutions in Section 6 we conclude in Section 7.

2   Background

In this section we give a short introduction to proximity graphs and their use for detecting cluster borders. In our join algorithm, proximity graphs are used to compute an adaptive threshold for each string.

2.1 Proximity Graph

Let s be a string, and the extended string $\bar{s}$ be s prefixed and suffixed with $q-1$ characters dummy '#'. The **profile** $P(s,q)$ of s is the bag of all substrings of $\bar{s}$ of length q, called q-grams. The **overlap** of two profiles $P(s',q)$ and $P(s'',q)$ is the cardinality of their intersection, i.e., $o(P(s',q), P(s'',q)) = |P(s',q) \cap P(s'',q)|$. The q-gram overlap measures the similarity between two strings; the higher the overlap value, the more similar the strings are.

Let D be a set of strings, and P be a profile. The **neighborhood** of P in D for similarity threshold τ (τ-neighborhood) is the subset of all strings of D that have an overlap of at least τ with P, $N(D,P,\tau) = \{s \in D : o(P, P(s,q)) \geq \tau\}$; we denote the τ-neighborhood as N_τ if P and D are clear from the context. The **center** $P_c(N_\tau, q)$ of the neighborhood N_τ is the profile that consists of the K most frequent q-grams in N_τ, i.e., in $\biguplus_{s \in N_\tau} P(s,q)$, where $K = \sum_{s \in N_\tau} |P(s,q)|/|N_\tau|$ is the average profile size in N_τ.

The **proximity graph** of string s is defined as $PG(s,D,q) = ((1,|N_1|), (2,|N_2|), \ldots, (m,|N_m|))$, $m = |P(s,q)|$, where N_τ is recursively defined as follows:

$$N_\tau = \begin{cases} \{s\} & \text{if } \tau = |P(s,q)|, \\ N(D, P_c(N_{\tau+1}, q), \tau) \cup N_{\tau+1} & \text{otherwise.} \end{cases} \quad (1)$$

The proximity graph maps similarity thresholds τ, $1 \leq \tau \leq |P(s,q)|$, to the size of the respective neighborhood N_τ (Figure 1) and is computed from the largest to the smallest threshold. The neighborhood corresponding to the greatest threshold is defined to be $\{s\}$. For the remaining thresholds the neighborhood is computed around the center of the previous neighborhood.

2.2 Automatic Border Detection

Let $PG = \{(1,|N_1|),(2,|N_2|),\ldots,(m,|N_m|)\}$ be a proximity graph. We define the horizontal lines in the proximity graph by their endpoints. The set of all **horizontal lines** in the proximity graph PG is defined as $H(PG) = \{(i,j)|\ (i,|N_i|),(j,|N_j|) \in PG, |N_i| = |N_j|, i \leq j\}$. The length of a horizontal line (i,j) is $j - i$.

The **border**, $border(PG) = \{j|\ (i,j) \in H(PG), \forall(x,y) \in H(PG) : y - x \leq j - i\}$, is the right endpoint of the rightmost horizontal line of maximal length. The **GPC cluster** of s is the neighborhood N_b for the similarity threshold $b = border(PG)$.

Example 1. Consider the proximity graph presented in Figure 1. For the center string $s =$ *paulina*, the cluster border is the rightmost endpoint $b = 5$ of the horizontal line $(3,5)$, and the GPC cluster is $C = N_5 = \{$*paulina, pauline, paulinne, paulene*$\}$.

3 Proximity Graph Join

In this section we present our string similarity join algorithm, PG-Join, that relies on a new join paradigm. Instead of joining string pairs based on a user defined threshold, an individual threshold is computed for each string. The algorithm is based on proximity graphs that have successfully been used by the GPC clustering algorithm to detect cluster borders automatically.

Intuitively, for each string s in one dataset we compute a GPC cluster in the other dataset with the cluster center s. The strings in the cluster are the matching partners of s in the similarity join. Thus the cluster border of the GPC cluster is the individual threshold for string s. This is a major improvement over a global, user defined threshold, since the threshold adapts to the string distribution around each string. In the following we formally define the matches of a string s in a dataset D.

Definition 1 (*Similarity Matches of a String*). *Let s be a string, D a set of strings, $PG(s, D \cup \{s\})$ the proximity graph of s in $D \cup \{s\}$, and N_b the respective GPC cluster. The similarity matches of string s in the dataset D are defined as $M(s, D) = \{(s, m) | m \in N_b \cap D\}$.*

For the strings of two datasets D and D', the similarity matches are not necessarily symmetric, e.g., a pair $(s, s') \in D \times D'$ might be in $M(s, D')$ but not in $M(s', D)$. The reason is that the proximity graphs are not symmetric. For GPC clusters this means that the choice of the cluster centers influences the clustering result. This is, however, not the case for our PG-Join. We compute the proximity graph for *each* string in both sets D and D', leaving no room for randomness. The join result is the union of all matches in both directions, leading to a symmetric join operator. Next we formally define the PG-Join.

Definition 2 (*PG-Join*). *The PG-Join $\mathbb{M}(D, D')$ between two sets of strings D and D' is defined as the union of the matches of all strings in both directions,*

$$\mathbb{M}(D, D') = \bigcup_{s \in D} M(s, D') \cup \bigcup_{s' \in D'} M^{-1}(s', D).$$

The PG-Join between two string sets D and D' is symmetric modulo the order of the output pairs, i.e., $\mathbb{M}(D, D') = \mathbb{M}^{-1}(D', D)$. The symmetry of the join is important to guarantee a unique join result and to allow optimizations when the join appears in a larger query.

An alternative definition of the PG-Join, that favors precision over recall, could use the intersection between left and right matches instead of the union. However, our empirical analysis showed that an intersection based join typically misses too many correct matches and the small increase in precision is punished with a large drop in recall, leading to overall worse results.

Algorithm 1 shows the pseudo-code for PG-Join. The input are two sets of strings D and D', and the q-gram size q. The output is the join result $\mathbb{M}(D, D') \subseteq D \times D'$. PG-Join computes an inverted list index for each string set. In the two loops over the string sets D and D', the matches in the respective other set are computed (Lines 4-9). For each string $s \in D$, a proximity graph in the other dataset, D', is computed such that the resulting GPC cluster consists of $s \in D$ and strings from D'. The center string s is matched to all strings

---

Algorithm 1. PG-Join(D, D′, q)

Data: D, D': sets of strings; q: size of q-grams;
Result: PG-join result $\mathbb{M}(D, D') \subseteq D \times D'$

1 **begin**
2      $IL \leftarrow getInvertedIndex(D, q)$; $IL' \leftarrow getInvertedIndex(D', q)$;
3      $\mathbb{M}$: empty list of string pairs from $D \times D'$;
4   **foreach** $s \in D$ **do**
5          $C \leftarrow PG\text{-}I(s, IL', q)$; // compute the GPC cluster of $s \in D$ in D'
6   **if** $|C| > 1$ **then** $\mathbb{M} \leftarrow \mathbb{M} \cup (\{s\} \times (C \setminus \{s\}))$;

7   **foreach** $s \in D'$ **do**
8          $C \leftarrow PG\text{-}I(s, IL, q)$; // compute the GPC cluster of $s \in D'$ in D
9   **if** $|C| > 1$ **then** $\mathbb{M} \leftarrow \mathbb{M} \cup ((C \setminus \{s\}) \times \{s\})$;

10   **return** $\mathbb{M}$;

Fig. 2. The PG-Join Algorithm

in its cluster (except itself). The same procedure is symmetrically applied to each string $s' \in D'$. The algorithm for the computation of the proximity graphs and the respective GPC clusters uses the inverted lists and will be discussed in Section 4.

4   Fast and Incremental Proximity Graph Computation

PG-Join computes a proximity graph for each string in both datasets to derive the individual thresholds. Fast proximity graph algorithms are the key to an efficient evaluation of PG-Join. The fastest algorithm in literature is PG-Skip [8]. This algorithm substantially improves over previous approaches by avoiding neighborhood computations that are not relevant for detecting the threshold.

A key problem of the proximity graph computation is the center that is updated for each neighborhood. This imposes an order on the neighborhood computations (right to left), and a neighborhood can be computed only after all the neighborhoods to its right are known. This forces proximity graph algorithms to compute each neighborhood from scratch and limits the optimization options.

Our approach is different from previous attempts. Instead of recomputing each neighborhood from scratch we proceed in two steps. In the first step we assume that the center is never updated and compute an approximation for the proximity graph (Section 4.1). Our assumption allows us to compute the whole proximity graph by a single scan of the inverted lists that represent the constant center. The effort of this step is comparable to a single neighborhood computation. In the second step we fix the error introduced by the approximation (Section 4.2). We proceed from right to left and carefully update the neighborhoods with respect to the moving center. The updates are much smaller than the neighborhoods and can be performed fast. Following the pruning criterion of PG-Skip, we stop the updates as soon as the cluster border is detected.

The cluster center of a neighborhood N_τ is a function of neighborhood $N_{\tau+1}$. All state-of-the-art algorithms compute this function from scratch for each neighborhood. We propose a new algorithm that leverages the deltas that fix the approximation error to incrementally update the cluster center (Section 4.3). This leads to a major improvement in the overall runtime. In Section 4.4 we combine all components into our new and incremental proximity graph algorithm, PG-I.

4.1   Quick and Dirty: A Fast Proximity Graph Approximation

In this section we present a fast algorithm that computes an approximation of the proximity graph and assumes a constant center. The center for all neighborhood computations is the profile of the string, for which the proximity graph is computed, i.e., for a proximity graph $PG(s, D, q)$ the center is $P(s, q)$.

The approximation algorithm is based on an inverted q-gram index, $IL(D, q)$, that indexes all strings $s \in D$. The inverted index is an array of inverted q-gram lists, $L(\kappa, D) = (s \mid s \in D, \kappa \in P(s, q))$, where the list for q-gram κ lists all string IDs that contain κ. In a preprocessing step we assign consecutive integer IDs to both q-grams and strings and use them as array indexes.

Example 2. Let $D = \{s_1, s_2, \ldots, s_8\} = \{paulina, pauline, linda, paulene, irvin,$
$austen, paulinne, piper\}$, and $q=2$. The inverted list index of D consists of 29 inverted lists. The inverted lists of $P = \{\#p, pa, au, ul, li, in, na, a\#\}$ are

$$
\begin{array}{ll}
L(\#p, D) = (s_1, s_2, s_4, s_7, s_8) & L(li, D) = (s_1, s_2, s_3, s_7) \\
L(pa, D) = (s_1, s_2, s_4, s_7) & L(in, D) = (s_1, s_2, s_3, s_5, s_7) \\
L(au, D) = (s_1, s_2, s_4, s_6, s_7) & L(na, D) = (s_1) \\
L(ul, D) = (s_1, s_2, s_4, s_7) & L(a\#, D) = (s_1, s_3)
\end{array}
$$

The approximation algorithm further uses the following global data structures to compute the proximity graph of a string s with profile P:

- *Proximity Graph* $PG[1..|P|]$: $PG[\tau]$ stores the size of the τ-neighborhood, i.e., the number of strings that have overlap at least τ with the center P, $PG[\tau] = |\{r \mid o(P, P(r, q)) \geq \tau\}|$. The array is initialized with zeros.
- *Counter* $AC[1..|D|]$: $AC[r]$ stores the overlap between the profile of string r and the center P. AC is initialized with zeros.
- *Dirty List DL*: List of all strings that ever changed their count in the counter AC. Initialized with an empty list.

We compute the proximity graph for a string s with a single scan of the inverted lists $L(\kappa, q)$, $\kappa \in P$. For each string r on a scanned list, (a) the counter $AC[r]$ is incremented to $c = AC[r] + 1$, (b) r is appended to the dirty list DL, and (c) $PG[c]$ is incremented. When all inverted lists of P are processed, PG stores the (approximate) proximity graph and N the respective neighborhoods.

The algorithm is fast and computes the approximate proximity graph in a single scan over the inverted q-gram lists of the center profile. Since the center is assumed to be constant, the resulting proximity graph is an approximation and may differ from the exact one.

Example 3. We continue Example 2 and compute the approximate proximity graph for the center $P = \{\#p, pa, au, ul, li, in, na, a\#\}$. We scan the inverted

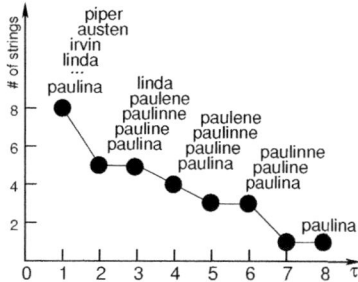

Fig. 3. Proximity Graph for the Constant Center 'paulina'

lists of P (see Example 2) one by one. For each string on the inverted list, we increment its count in AC by 1. Then we use the updated count of this string as the index value in the array PG and also increment by 1 the respective count in PG. After processing all inverted lists of P, the arrays AC and PG are:

Array of Counts, AC

value:	8	6	3	4	1	1	6	1
id:	s_1	s_2	s_3	s_4	s_5	s_6	s_7	s_8

Proximity Graph, PG

value:	8	5	5	4	3	3	1	1
id:	1	2	3	4	5	6	7	8

The neighborhoods are the following: $N_8 = N_7 = \{s_1\}$, $N_6 = N_5 = \{s_1, s_2, s_7\}$, $N_4 = \{s_1, s_2, s_4, s_7\}$, $N_3 = N_2 = \{s_1, s_2, s_3, s_4, s_7\}$, and $N_1 = \{s_1, s_2, s_3, s_4, s_5, s_6, s_7, s_8\} = D$. The respective proximity graph is illustrated in Figure 3. It is different from the exact one in Figure 1.

4.2 Cleaning Up: Fixing the Approximation Error

In this section we present the algorithm *Update-N* that corrects the error that the approximation algorithm introduces. The approximation algorithm uses the center of the rightmost neighborhood, e.g., *paulina* in Figure 3, to compute all neighborhoods in the proximity graph, whereas the correct center to be used is the center of the previous neighborhood, e.g., the center of N_5 for the computation of N_4 (see Eq. 1).

The approximation algorithm computes the neighborhoods for a (possibly) wrong center. *Update-N* fixes this error based on the differences (Δ^+ and Δ^-) between the wrong and the correct center, thus avoiding to compute the neighborhood from scratch. This approach is based on the observation that the deltas are typically much smaller than the center. Often they are empty or contain only very few q-grams.

Update-N receives the deltas and the threshold τ, for which the neighborhood must be updated. It traverses the inverted lists of only those q-grams that are in the deltas. For each string r on a scanned list, we decrease/increase the overlap $AC[r]$. Since the overlap of the strings changes, we also update the sizes of the respective neighborhoods in PG.

The proximity graph is a cumulative function, i.e., the strings that appear in a neighborhoods of overlap $o > \tau$ must also appear in the neighborhood

N_τ. We prevent the change of the overlap value of a string r that appears in a neighborhood N_o by setting its overlap $AC[r]$ to the negative value $-o$.
In addition to the global data structures defined in Section 4.1 we define:

- *Pseudo-Neighborhoods* $S[1..m]$, $m = |P(s,q)|$. For given τ, the list $S[\tau]$ contains those strings of D that satisfy the condition $\forall r \in S[\tau] : o(P, P(r,q)) = \tau$. The neighborhood N_τ can be computed as $\bigcup_{i=\tau}^m S[i]$.
- *Difference Counter* $AD[1..|D|]$. $AD[r]$ stores the change of the overlap between the string r and the new center P, and is initialized with zeros.

The pseudo code for *Update-N* is shown in Algorithm 2. For each q-gram κ of Δ^+ and Δ^-, *Update-N* traverses its inverted list and for each string r of the list either increments (if $\kappa \in \Delta^+$) or decrements (if $\kappa \in \Delta^+$) the counter $AD[r]$. The dirty list DL remembers all strings for which the count has changed. Whenever the counter AD is incremented or decremented, also the proximity graph PG is updated respectively. The overlap of r with the center (computed so far) is $o = AC[r]+AD[r]$. If $AD[r]$ is incremented, then $PG[o]$ is incremented, otherwise $PG[o + 1]$ is decremented (Lines 5-11).

After AD and PG are updated, *Update-N* scans the dirty list DL, i.e., the strings r for which the overlap has changed, and updates the respective overlap $AC[r]$ (Lines 12-18). If the overlap $AC[r]$ is greater/equal to τ, the string is added to the pseudo-neighborhood $S[\tau]$. The string r is marked as appearing in the neighborhood N_τ by setting $AC[r] = -\tau$. String r is not added to any $S[o]$, $o > \tau$, since these neighborhoods have already been corrected in an earlier step.

Finally, *Update-N* checks the overlap of strings, for which the counts were not updated but which have τ common q-grams with the correct center. For that it scans the pseudo-neighborhood $S[\tau]$ and marks the strings with overlap τ by assigning $AC[r] = -\tau$ (Lines 19-20).

Note that it is possible that a string is added more than once to $S[\tau]$. Further, since we only add but never delete, $S[\tau]$ may contain strings that no longer share τ q-grams with the center. We call these strings *false-positives*. The exact τ-neighborhood is the union of all strings in the pseudo-neighborhoods $S[x]$, $x \geq \tau$, for which $AC[x] = -x$.

Example 4. Let us update the neighborhood for $\tau = 5$ of the proximity graph shown in Figure 1 and the dataset D given in Example 2. The current arrays AC and AD are as follows:

Array of Differences, AD

value:	0	0	0	0	0	0	0	0
id:	s_1	s_2	s_3	s_4	s_5	s_6	s_7	s_8

Array of Counts, AC

value:	-8	-6	3	4	1	1	-6	1
id:	s_1	s_2	s_3	s_4	s_5	s_6	s_7	s_8

The lists of S are the following: $S[8] = \{s_1\}$, $S[7] = \emptyset$, $S[6] = \{s_2, s_7\}$, $S[5] = \emptyset$, $S[4] = \{s_4\}$, $S[3] = \{s_3\}$, $S[2] = \emptyset$, $S[1] = \{s_5, s_6, s_8\}$.

Let $P = \{\#p, pa, au, ul, li, in, ne, e\#\}$. Let $\Delta^+ = \{ne, e\#\}$, and $\Delta^- = \{na, a\#\}$. The according inverted lists are: $L(ne) = \{s_2, s_4, s_7\}$, $L(e\#) = \{s_2, s_4, s_7\}$, $L(na) = \{s_1\}$, and $L(a\#) = \{s_1, s_3\}$. We traverse the lists and update the arrays:

Algorithm 2. Update-N$(\tau, \Delta^+, \Delta^-)$

Data: τ: similarity threshold; Δ^+, Δ^-: diffs between wrong and correct center (q-gram sets);
Result: fixes neighborhood N_τ

```
 1 begin
 2   │ if Δ⁺ ≠ ∅ ∨ Δ⁻ ≠ ∅ then
 3   │ │ DL: empty list of string IDs;
 4   │ │ AD[1..|D|]: difference counter initialized with 0's;
 5   │ │ foreach r ∈ IL[κ], κ ∈ Δ⁺ do
 6   │ │ │ if AD[r] = 0 then DL.add(r);
 7   │ │ │ AD[r] ← AD[r] + 1; o = AC[r] + AD[r];
 8   │ │ │ if AC[r] ≥ 0 ∧ o ≤ τ then PG[o] = PG[o] + 1;
 9   │ │ foreach r ∈ IL[κ], κ ∈ Δ⁻ do
10   │ │ │ AD[s] ← AD[r] − 1; o = AC[r] + AD[r];
11   │ │ │ if AC[r] ≥ 0 ∧ o < τ then PG[o + 1] = PG[o + 1] − 1;
12   │ │ foreach r ∈ DL do
13   │ │ │ if AC[r] ≥ 0 then
14   │ │ │ │ AC[r] ← AC[r] + AD[r];
15   │ │ │ │ if AC[r] ≥ τ then
16   │ │ │ │ │ S[τ].add(r); AC[r] ← −τ;
17   │ │ │ │ else
18   │ │ │ │ │ if AC[r] ≠ 0 then S[AC[r]].add(r);
19   │ foreach r ∈ S[τ] do
20   │ │ if AC[r] = τ then AC[r] ← −τ;
```

Fig. 4. Neighborhood Update

Array of Differences, AD

value:	-2	2	-1	2	0	0	2	0
id:	s_1	s_2	s_3	s_4	s_5	s_6	s_7	s_8

Array of Counts, AC

value:	-8	-6	2	-5	1	1	-6	1
id:	s_1	s_2	s_3	s_4	s_5	s_6	s_7	s_8

The array AD records the change of the overlap of strings after the lists are processed. We compute new overlap values for AC by summing up the respective values in AC and AD. The strings s_1, s_2, and s_7 are marked as strings appearing in the resulting neighborhood (are assigned with a negative number in AC), therefore we do not update their counts. The lists of S are the following after the update: $S[8] = \{s_1\}$, $S[7] = \emptyset$, $S[6] = \{2,7\}$, $S[5] = \{s_4\}$, $S[4] = \{s_4\}$, $S[3] = \{s_3\}$, $S[2] = \{s_3\}$, $S[1] = \{s_5, s_6, s_8\}$. Note that the same string may appear in different lists. However, this does not affect the final neighborhood since we can always retrieve the real list ID of the string from the array AC .

4.3 Optimizing for Speed: Incremental Cluster Center Updates

In this section we present *Adjust-C*, an algorithm that incrementally updates the cluster centers based on the differences between adjacent neighborhoods.

Remember that the center of a neighborhood N_τ consists of its k most frequent q-grams, where k is the average size of the string profiles, and the neighborhood N_τ is computed around the center of $N_{\tau+1}$ (see Section 2.1). The state-of-the-art algorithms use a straight forward approach and compute the center of each

Algorithm 3. Adjust-C(τ, num)

Data: τ : compute center for neighborhood N_τ; *num*: number of q-grams in $N_{\tau+1}$;
Result: *heap*: center of N_τ; *num*: number of q-grams in N_τ

```
1  begin
2  |   N+ ← ∅;
3  |   foreach r ∈ S[τ] do
4  |   |   if AC[r] = −τ then
5  |   |   |   N+ ← N+ ∪ {r};
6  |   |   |   num ← num + |r| + q − 1;
7  |   k ← num/PG[τ];
8  |   k_d ← k − |heap|;
9  |   if k_d > 0 then
10 |   |   K ← set of k_d most frequent q-grams in hist that are not in heap;
11 |   |   foreach κ ∈ K do push(heap, (hist[κ], κ), addr);
12 |   else for 1 to k_d do pop(heap, addr);
13 |   foreach κ ∈ P(s, q), r ∈ N+ do
14 |   |   hist[κ] ← hist[κ] + 1; // increment the frequency of κ in the histogram
15 |   |   if |heap| < k or  hist[κ] > top(heap) then
16 |   |   |   if addr[κ] = −1 then // κ not in the heap
17 |   |   |   |   push(heap, (hist[κ], κ), addr);
18 |   |   |   else // update frequency of κ in heap
19 |   |   |   |   update(heap, addr[κ], (hist[κ], κ), addr);
20 |   |   if |heap| > k then pop(heap, addr);
21 |   return (heap, num);
```

Fig. 5. Incremental Computation of the Neighborhood Center

neighborhood from scratch. Our *Adjust-C* algorithm uses the *new strings* $N^+ = N_\tau \setminus N_{\tau+1}$ in N_τ to incrementally update the center of $N_{\tau+1}$ to N_τ.

The pseudo code of *Adjust-C* is shown in Algorithm 3. The algorithm proceeds in two steps. In the first step, the new strings N^+ are computed, in the second step the center is updated with respect to the q-grams in N^+.

(1) Computation of the new strings N^+. Computing the set difference $N^+ = N_\tau \setminus N_{\tau+1}$ is almost as expensive as computing the neighborhood from scratch. Fortunately we can do much better. We use the global array S maintained by *Update-N* (Section 4.2) to compute N^+ efficiently. $S[\tau]$ points to a list of strings that contains N^+ and some false positives. We remove the false positives with a single scan of $S[\tau]$. A string is in N^+ iff $AC[\tau] = -\tau$ (Lines 3-5).

(2) Center Update. We compute the center of N_τ from the center of $N_{\tau+1}$. The difference between the two centers results from the q-grams of the strings in N^+. The center update maintains two global data structures, the *center heap* and the *q-gram histogram*. The center heap for neighborhood N is a min-heap of (count, q-gram)-pairs that stores the k most frequent q-grams, i.e., the q-grams that belong to the center. The top of the heap is the least frequent q-gram. The histogram stores the frequency of each q-gram in the neighborhood. Initially, center heap and histogram store the values for neighborhood $N_{\tau+1}$.

The size of the center of N_τ may differ from $N_{\tau+1}$. The number of q-grams in N_τ is computed by adding the profile sizes of the strings in N^+ to the number of q-grams in $N_{\tau+1}$. The center size k is the average profile size (Lines 6-7).

Algorithm 4. PG-I(s, IL, q)

Data: s: center string; IL: inverted index list of q-grams; q: gram size
Result: GPC cluster around the center string s

1 **begin**
2   initialize global data structures (see Sections 4.1 and 4.2);
3   compute approximate proximity graph (see Section 4.1),
4      $l_{max} \leftarrow 0; l \leftarrow 0;$ // lengths of the longest and current horizontal lines
5      $m \leftarrow |P(s,q)|;$ // rightmost similarity threshold
6      $b \leftarrow m;$ // cluster border
7   **for** $\tau = m - 2$ **downto** 1 **do**
8          $P[\tau + 1] \leftarrow Adjust\text{-}C(\tau, num);$
9          $\Delta^+ \leftarrow P[\tau + 1] \setminus P[\tau + 2]; \Delta^- \leftarrow P[\tau + 2] \setminus P[\tau + 1];$
10   **if** $|\Delta^+| + |\Delta^-| \leq |P[\tau + 1]|$ **then**
11             $Update\text{-}N(\tau, \Delta^+, \Delta^-);$
12   **else** compute N_τ from scratch for the center $P[\tau + 1];$
        // compute cluster border
13   **if** $PG[\tau] = PG[\tau + 1]$ **then** $l \leftarrow l + 1;$ // increase current horizontal line
14   **else** $l \leftarrow 0;$ // start new horizontal line
15   **if** $l > l_{max}$ **then** $l_{max} \leftarrow l; b \leftarrow \tau + l;$
        // stop proximity graph computation if possible
16   **if** $(b = \tau + l \wedge \tau - 2 \leq l_{max}) \vee (\tau - 1 + l \leq l_{max})$ **then break**;
    // retrieve cluster for border b
17     $C \leftarrow \emptyset;$
18   **foreach** $r \in S[b] \cup S[b + 1] \cup \cdots \cup S[m - 1]$ **do**
19   **if** $-b \geq AC[r]$ **then** $C \leftarrow C \cup \{r\};$
20   **return** $C;$

Fig. 6. Proximity Graph Computation Algorithm

If k is larger than the current heap size, the $k - |heap|$ most frequent q-grams that are not yet in the center heap are pushed; otherwise $|heap| - k$ elements are popped. We maintain a auxiliary heap with the most frequent q-grams that are not in the center heap to avoid a scan of the histogram. The auxiliary heap is of size $max - |heap|$, where max is the size of the largest profile in D (Lines 8-12).

Finally we produce the q-grams of all strings in N^+ and update the frequencies in the histogram. If the frequency of a q-gram κ grows larger than the key on top of the center heap, two cases must be distinguished: (a) if κ is already in the heap, its frequency is updated, (b) otherwise the top of the heap is replaced (Lines 13-20). In order to update the frequency of an element in the heap it must be located. To avoid scanning the heap we maintain an array $addr$ that stores the position of the q-grams in the center heap and is -1 for all other q-grams.

After all q-grams of N^+ are processed, the histogram stores the frequencies of the q-grams in N_τ and the q-grams of the heap are the center of N_τ.

Example 5. We continue Example 4 and compute the center of the neighborhood $N_\tau = 5$. Before the update the heap elements are: $(3, \#p)$, $(3, pa)$, $(3, au)$, $(3, ul)$, $(3, li)$, $(3, in)$, $(2, ne)$, $(2, e\#)$. $S[5] = \{s_4\} = \{paulene\}$, and the center size is 8. After processing the q-grams of s_4, the heap elements are: $(4, \#p)$, $(4, pa)$, $(4, au)$, $(4, ul)$, $(3, li)$, $(3, in)$, $(3, ne)$, $(3, e\#)$; the center did not change.

4.4   Putting It All Together: The Incremental Proximity Graph Algorithm

In this section we combine the algorithms of the previous subsections into our new proximity graph algorithm *PG-I*. The input is a string s, the inverted list index IL of a string set D, and the q-gram size q; the output is the GPC cluster around s in D (which is used by *PG-Join* to find matches for the string s).

The approximation algorithm computes all neighborhoods around the center $P(s, q)$ of the rightmost neighborhood $m = |P(s, q)|$. This is correct for $\tau = m-1$, but the other neighborhoods need to be updated to reflect the center changes.

The neighborhoods are updated in the loop in Lines 7-11 (Algorithm 4). In each iteration with similarity threshold τ the following holds: the neighborhoods $N_1, .., N_{\tau+1}$ are correct with respect to the center of neighborhood $N_{\tau+2}$. Since a neighborhood must be computed around the center of the neighborhood to its right, $N_{\tau+1}$ is correct, but N_τ needs to be updated. This is done in three steps: (1) Compute the center of $N_{\tau+1}$ using the incremental center update in Section 4.3, (2) compute the deltas Δ^+ and Δ^- for the new center, (3) use the deltas to update the neighborhood N_τ with *Update-N* of Section 4.2. The delta Δ^+ stores the q-grams that must be added to the center, Δ^- the q-grams that must be removed.

If the added sizes of Δ^+ and Δ^- are greater than the size of the center of $N_{\tau+1}$, the incremental update is not beneficial and we compute the neighborhood N_τ from scratch (Line 12). This amounts to computing the approximate proximity graph for the remaining of thresholds ($\leq \tau$) around the center of $N_{\tau+1}$.

We keep track of the longest horizontal line l_{max} in the proximity graph and maintain the current cluster border b. We stop the computation when further computations can not change the border (Lines 13-16). Intuitively, the algorithm stops when the remaining neighborhoods cannot result in a horizontal line longer than l_{max}. This approach has been shown to be correct and optimal [8].

The GPC cluster for s is the neighborhood N_b in the proximity graph. PG-I retrieves $C = N_b$ by taking the union of all pseudo-neighborhoods for the threshold $\tau \geq b$ and filtering false-positives from each list $S[\tau]$ (Lines 18-19).

5   Related Work

Similarity joins have been intensively studied by the database research community and are closely related to fuzzy string matching [9], record linkage [10,11], and near-duplicate detection [12,13]. For strings, a number of similarity measures based on q-grams [14,15] or edit distances [16] have been proposed that may be used in the predicate of a similarity join. In our approach, the overlap between the q-grams of two strings is used as a similarity measure.

Gravano *et al.* [1] propose a q-gram based string similarity join for an off-the-shelf databases system that makes use of the capabilities of the underlying relational engine. In addition, in a filter-and-verify approach the result for the q-gram based join may be used as candidates for a join based on the edit distance. The Ed-Join algorithm proposed by Xiao *et al.* [2] employs additional filters for the similarity join with edit distance constraints and improves the performance. These methods use q-grams as an efficient lower bound for the edit distance, while our q-gram based PG-Join uses the overlap of q-grams as a similarity measure. In a recent work, Ribeiro and Härder [17] improve the efficiency of the candidate generation phase at the cost of the verification phase and show that they overall outperform approaches that focus on reducing the runtime for the verification phase. Chaudhuri *et al.* [3] propose an operator for set similarity joins, which can also be used for string joins based on q-grams. All these approaches assume a global, user specified threshold. Our PG-Join algorithm uses an adaptive threshold which is automatically computed for each string.

Jestes *et al.* [4] develop a probabilistic string similarity join. They introduce the probabilistic string-level and character-level models to capture string probabilities and defines the expected edit distance as a similarity measure. In our work we do not assume that probabilities for strings are given.

PG-Join is based on proximity graphs, which where developed by Mazeika and Böhlen [6] for the Proximity Graph Cleansing (GPC) algorithm. GPC is a self-tuning clustering algorithm and uses proximity graphs to automatically detect cluster borders. Our join algorithm uses the same technique to define an individual threshold for each string.

The efficient computation of the proximity graph is crucial for both GPC and our PG-Join algorithm. Mazeika and Böhlen [6] proposed an approximation algorithm based on sampling. Kazimianec and Augsten [7] presented the first efficient algorithm, PG-DS, for computing the exact proximity graph. PG-DS is based on a solution for the τ-occurrence problem developed by Li *et al.* [18] and outperforms the sampling technique. A recent development, PG-Skip [8], prunes unnecessary neighborhood computations to improve the efficiency. All these algorithms compute the neighborhoods of the proximity graph in isolation. Our PG-I algorithm uses the results for smaller neighborhoods to compute larger ones incrementally. The efficiency of PG-I is shown in our experiments.

6   Experiments

In this section we evaluate the effectiveness of our PG-Join on real-world datasets and compare it to the standard string similarity join that uses a global threshold. We further evaluate the efficiency of our incremental proximity graph algorithm, PG-I, and compare it to the fastest state-of-the-art algorithm, PG-Skip [8].

6.1   Experimental Setup

In our experiments we use six real-world datasets with different characteristics:

-   **Bozen**: 1313 street names of Bozen in 403 clusters (4–35 characters; average: 12 characters);[1]

---

[1] http://www.inf.unibz.it/~augsten/publ/tods10

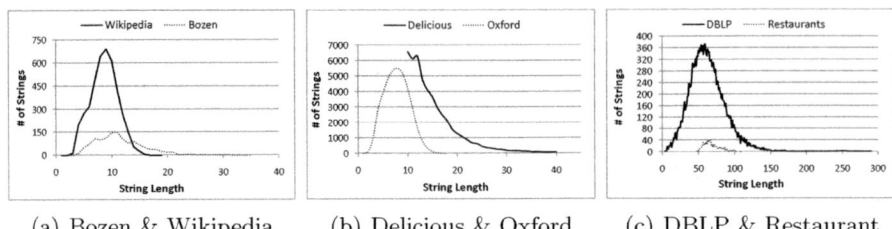

(a) Bozen & Wikipedia   (b) Delicious & Oxford   (c) DBLP & Restaurant

Fig. 7. Distribution of String Length for Real-World Datasets

- **Oxford**: a natural language collection of unique misspellings of the Oxford text archives with 39030 strings in 6003 clusters (1–18 characters; average: 8 characters);[2]
- **Wikipedia**: 4153 unique misspellings found on Wikipedia in 1917 clusters (1–19 charaters; average: 9 characters);[2]
- **Restaurants**: a joint collection of 864 restaurant records from the Fodor's and Zagat's restaurant guides in 752 clusters (48–119 characters; average: 69 characters);[3]
- **DBLP**: 10000 article titles (up to 280 characters; average: 61 characters);[4]
- **Delicious**: 48397 bookmark tags (10–40 characters; average: 15).[5]

The string length distribution for all datasets is shown in Figure 7. The first four datasets are clustered, and the ground truth is known. We produce from each of these datasets two new datasets, A and B, that we join. The join datasets are produced by splitting clusters, and the correct join result, $\mathbb{M}^{gt} \subseteq A \times B$, follows from the clusters in the original dataset, $A \cup B$. We use the well-established F-measure [19] to assess the quality of a join result, $\mathbb{M}$, with respect to the ground truth, $\mathbb{M}^{gt}$. The F-measure is the harmonic mean of precision and recall, $F = 2 \cdot \frac{precision \cdot recall}{precision + recall}$, where $precision = |\mathbb{M} \cap \mathbb{M}^{gt}|/|\mathbb{M}|$, and $recall = |\mathbb{M} \cap \mathbb{M}^{gt}|/|\mathbb{M}^{gt}|$. If the clustering of a dataset is unknown, we randomly partition the dataset into two subsets of similar size for the join experiments.

6.2 Evaluation of Effectiveness

We compare the quality of PG-Join with the standard string similarity join that uses a global threshold. Both algorithms run on real-world datasets with known ground truth and different string length distributions. The standard similarity join uses the *normalized q-gram distance* [5] between two strings s and r,

$$d(s,r) = \frac{|P(s,q)| + |P(r,q)| - 2 \cdot o(P(s,q), P(r,q))}{|P(s,q)| + |P(r,q)| - o(P(s,q), P(r,q))},$$

---

[2] http://www.dcs.bbk.ac.uk/~roger/corpora.html
[3] http://www.cs.utexas.edu/users/ml/riddle/data.html
[4] http://www.informatik.uni-trier.de/~ley/db
[5] http://www.delicious.com/

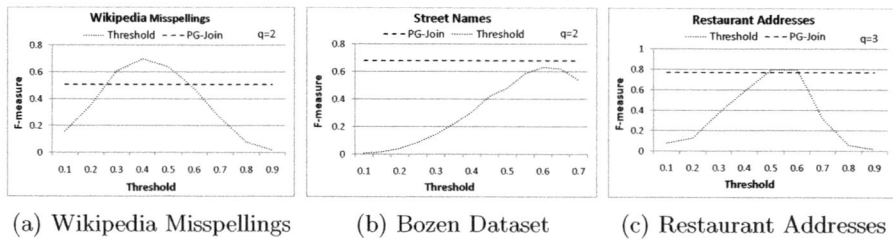

(a) Wikipedia Misspellings   (b) Bozen Dataset   (c) Restaurant Addresses

Fig. 8. Effectiveness: PG-Join vs. Threshold Matching

where $o(P(s,q), P(r,q))$ is the *overlap* between the profiles of s and r (see Section 2.1). The normalization accounts for the difference in the string length and results in values between zero and one. For the standard approach, we observed better results on our test datasets for the normalized distance rather than for the non-normalized distance or the q-gram similarity based on overlap.

The standard approach uses a global threshold τ that depends on the dataset, in particular on the distances between pairs of objects that should match vs. the distance between objects that should not match. We expect a poor choice of the threshold to significantly reduce the quality of the join result. PG-Join needs no configuration parameters and uses an adaptive threshold.

We test the standard join for a large range of thresholds. Figure 8 shows the results for the short, middle, and long strings of Wikipedia, Bozen, and Restaurant, respectively. We use $q = 2$ for all datasets except Restaurant, for which $q = 3$ gives better results since the strings are longer. We compare the standard join to PG-Join, which does not require a global threshold and is based on the overlap similarity (see Section 2). The results confirm that the threshold join performs well only for well chosen thresholds. The values of the good thresholds vary substantially between the datasets, i.e., it is not straightforward to choose a good threshold. For Bozen and Restaurant, PG-Join is as good as the threshold matching with the best threshold. For the short strings of Wikipedia, the threshold approach wins for some thresholds and looses for others.

6.3 Evaluation of Efficiency

We evaluate the efficiency of our proximity graph algorithm, PG-I, and compare it to the fastest state-of-the-art algorithm, PG-Skip [8].

In our first experiment we plug both algorithms into our PG-Join. We run the experiment on three real-world datasets with different characteristics: Oxford, Delicious, and DBLP (see Section 6.1). Figure 9 shows the runtime for subsets of different size. PG-I outperforms PG-Skip for all datasets. The advantage increases with the average string length in the dataset: On short and medium length strings (Oxford and Delicious), PG-I is approximately 1.5-2 times faster than PG-Skip; on the long DBLP paper titles, PG-I is 2.5 times faster.

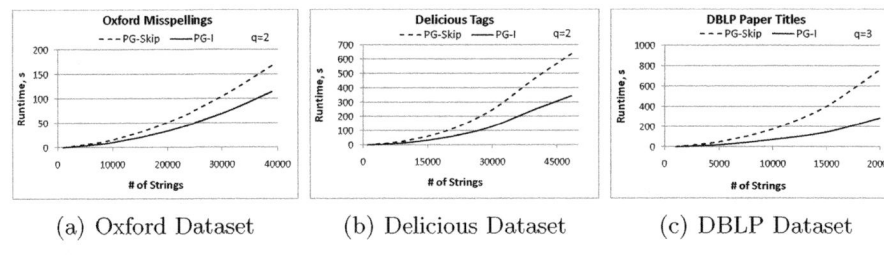

(a) Oxford Dataset   (b) Delicious Dataset   (c) DBLP Dataset

Fig. 9. PG-Join Runtime on Real-World Data: PG-I vs. PG-Skip

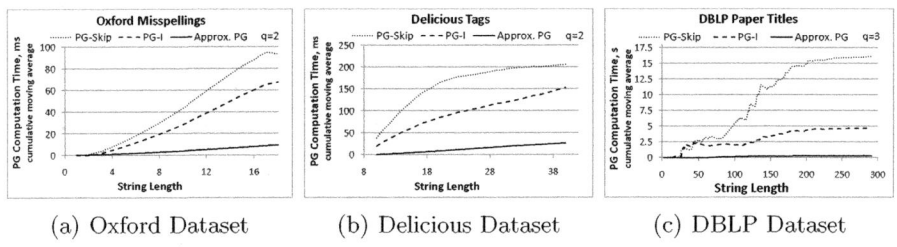

(a) Oxford Dataset   (b) Delicious Dataset   (c) DBLP Dataset

Fig. 10. Proximity Graph Computation

In Figure 10 we analyze the impact of the string length on the runtime of the proximity graph algorithms. We group the strings by their length and measure the runtime for each string length separately. The graphs show the cumulative average of the runtime. The cumulative average is more informative than a per-length average since for many length values no strings or very few strings exist.

PG-I is faster than PG-Skip for most string lengths. The runtime difference typically grows with the string length, i.e., the advantage of PG-I is greater for longer strings. For some strings PG-Skip is as fast or slightly faster. The proximity graphs of these strings have a long horizontal line and the clusters are very small, i.e., the proximity graph computation is fast for both algorithms. In this case, the overhead of PG-I for computing an approximation in the first step does not pay off (cf. the computational time of the approximate proximity graph, Figure 10(b)-10(c)). On average, PG-I is substantially faster (see Figure 10).

6.4   Scalability on Synthetic Data

We perform a controlled experiment on synthetic data and generate a dataset of random strings. From each string we produce noisy strings by randomly renaming, inserting, and deleting characters in order to get clusters of similar strings. We further produce two join datasets, A and B, by splitting each cluster of the original dataset into two subclusters of equal size.

In Figure 11(a) we vary the length of the strings, while the number of strings and the size of the clusters remain constant. PG-I outperforms PG-Skip, and the advantage increases with the string length. For strings of length 20, PG-I is

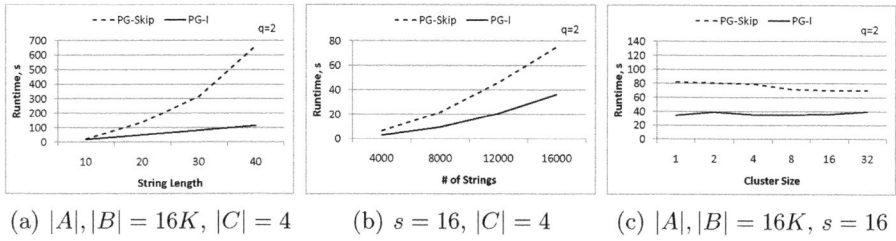

(a) $|A|, |B| = 16K, |C| = 4$ (b) $s = 16, |C| = 4$ (c) $|A|, |B| = 16K, s = 16$

Fig. 11. PG-Join Runtime on Synthetic Data: PG-I vs. PG-Skip

two times faster, and more than five times for strings of length 40. PG-I scales better with the number of strings (Figure 11(b)). The runtime of both methods changes slightly with the cluster size (Figure 11(c)) since the PG-Join algorithm computes one proximity graph for each string, and the cluster size has no serious impact on the runtime. PG-I is faster for all cluster sizes.

6.5   Summary

The quality of the PG-Join is encouraging, in particular on middle-length and long strings. PG-Join uses an adaptive threshold and needs no data dependent configuration. This opens a new perspective on similarity joins of datasets, for which little is known about the data distribution. Our incremental proximity graph algorithm, PG-I, outperforms the fastest known algorithm, PG-Skip, in almost all settings. PG-I substantially improves the runtime of our PG-Join.

7   Conclusions

In this work we proposed the PG-Join algorithm, a new string similarity join that uses so-called proximity graphs to compute an adaptive similarity threshold. We developed the PG-I algorithm to compute the proximity graph efficiently. PG-I is up to five times faster than the best proximity graph algorithms in literature. We evaluated effectiveness and efficiency of our solution on a variety of datasets. Our experiments suggest that our approach is both useful and scalable.

References

1. Gravano, L., Ipeirotis, P.G., Jagadish, H.V., Koudas, N., Muthukrishnan, S., Srivastava, D.: Approximate string joins in a database (almost) for free. In: Proceedings of the 27th Int. Conf. on Very Large Data Bases, VLDB 2001, pp. 491–500. Morgan Kaufmann Publishers Inc., San Francisco (2001)
2. Xiao, C., Wang, W., Lin, X.: Ed-join: an efficient algorithm for similarity joins with edit distance constraints. In: Proc. VLDB Endow., vol. 1, pp. 933–944 (2008)
3. Chaudhuri, S., Ganti, V., Kaushik, R.: A primitive operator for similarity joins in data cleaning. In: Proceedings of the 22nd Int. Conf. on Data Engineering, ICDE 2006, p. 5. IEEE Computer Society, Los Alamitos (2006)

4. Jestes, J., Li, F., Yan, Z., Yi, K.: Probabilistic string similarity joins. In: Proceedings of the 2010 Int. Conf. on Management of Data, SIGMOD 2010, pp. 327–338. ACM, New York (2010)
5. Augsten, N., Böhlen, M., Gamper, J.: The pq-gram distance between ordered labeled trees. ACM Trans. Database Syst. 35, 4:1–4:36 (2008)
6. Mazeika, A., Böhlen, M.H.: Cleansing databases of misspelled proper nouns. In: CleanDB (2006)
7. Kazimianec, M., Augsten, N.: Exact and efficient proximity graph computation. In: Catania, B., Ivanović, M., Thalheim, B. (eds.) ADBIS 2010. LNCS, vol. 6295, pp. 289–304. Springer, Heidelberg (2010)
8. Kazimianec, M., Augsten, N.: PG-skip: Proximity graph based clustering of long strings. In: Yu, J.X., Kim, M.H., Unland, R. (eds.) DASFAA 2011, Part II. LNCS, vol. 6588, pp. 31–46. Springer, Heidelberg (2011)
9. Bilenko, M., Mooney, R., Cohen, W., Ravikumar, P., Fienberg, S.: Adaptive name matching in information integration. IEEE Intelligent Systems 18, 16–23 (2003)
10. Jin, L., Li, C., Mehrotra, S.: Efficient record linkage in large data sets. In: Proceedings of the 8th Int. Conf. on Database Systems for Advanced Applications, DASFAA 2003, p. 137. IEEE Computer Society, Los Alamitos (2003)
11. Hjaltason, G.R., Samet, H.: Incremental distance join algorithms for spatial databases. In: Proceedings of the 1998 Int. Conf. on Management of Data, SIGMOD 1998, pp. 237–248. ACM, New York (1998)
12. Elmagarmid, A.K., Ipeirotis, P.G., Verykios, V.S.: Duplicate record detection: A survey. IEEE Trans. on Knowl. and Data Eng. 19, 1–16 (2007)
13. Henzinger, M.: Finding near-duplicate web pages: a large-scale evaluation of algorithms. In: Proceedings of the 29th Int. Conf. on Research and Development in Information Retrieval, SIGIR 2006, pp. 284–291. ACM, New York (2006)
14. Ukkonen, E.: Approximate string-matching with q-grams and maximal matches. Theoretical Computer Science 92, 191–211 (1992)
15. Li, C., Wang, B., Yang, X.: Vgram: improving performance of approximate queries on string collections using variable-length grams. In: Proc. of the 33rd Int. Conf. on Very Large Data Bases, VLDB 2007, pp. 303–314. VLDB Endow. (2007)
16. Navarro, G.: A guided tour to approximate string matching. ACM Comput. Surv. 33, 31–88 (2001)
17. Ribeiro, L.A., Härder, T.: Generalizing prefix filtering to improve set similarity joins. Information Systems 36(1), 62–78 (2011)
18. Li, C., Lu, J., Lu, Y.: Efficient merging and filtering algorithms for approximate string searches. In: Proceedings of the 24th Int. Conf. on Data Engineering, ICDE 2008, pp. 257–266. IEEE Computer Society, Los Alamitos (2008)
19. Rijsbergen, C.J.V.: Information Retrieval, 2nd edn. Butterworth-Heinemann, Butterworths (1979)

A Flexible Graph Pattern Matching Framework via Indexing

Wei Jin and Jiong Yang

Department of Electrical Engineering and Computer Science
Case Western Reserve University
{wei.jin,jiong.yang}@case.edu

Abstract. In recent years, pattern matching has been an important graph analysis tool in various applications. In previous existing models, each edge in the query pattern represents the same relationship, e.g., the two endpoint vertices have to be connected or the distance between them should be within a certain uniform threshold. However, various real world applications may require edges representing different relationships or distances, some may be longer while others may be shorter. Therefore, we introduce the flexible pattern matching model where a range $[min_e, max_e]$ is associated with an edge e in the query pattern, which means that the minimum distance between the matched endpoints of e is in the range of $[min_e, max_e]$. A novel pattern matching algorithm utilizing two types of indices is devised. In addition to the traditional pattern matching scheme, a top-k matches generation model is also proposed. Extensive empirical studies have been conducted to show the effectiveness and efficiency of our indices and methods.

1 Introduction

Graphs are natural representations of relationships between objects in many domains, and they are amenable to computational analysis. Accordingly, graph data is increasingly important in a variety of scientific and engineering applications. For example, graphs are used to represent molecular and biological networks of various kinds, social networks, computer networks, power grids, and computer software. Many types of computational tools, e.g., subgraph search, frequent subgraph mining are used to analyze the graph data. The *pattern matching query* [19,2,13] is one of the most important and popular analysis tools. In the general format, a pattern Q is a graph with m vertices. An edge (v_1, v_2) in Q represents a relationship between v_1 and v_2. The relationship could be in various forms, e.g., connected, shortest distance within a parameter δ, etc. Given a database G and a query pattern Q with m vertices, a match of Q is a subset of m vertices in G satisfying the following conditions: (1) these m vertices in G have the same labels as the corresponding vertices in Q, and (2) for any two adjacent vertices v_i and v_j in Q, the corresponding vertices u_i and u_j in G satisfy the relationship specified on the edge (v_i, v_j). In previous work, the relationship is uniform across all edges in Q. For example, in [2], an edge represents that the two endpoint vertices are connected in G while in [19], an edge indicates that the corresponding endpoint vertices in G are within a distance of δ and δ is uniform for all edges.

J.B. Cushing, J. French, and S. Bowers (Eds.): SSDBM 2011, LNCS 6809, pp. 293–311, 2011.
© Springer-Verlag Berlin Heidelberg 2011

These models have been proven useful in many applications. However, in some applications, a more general model is needed. The following are two example applications.

In a protein interaction network (PIN), a vertex represents a protein while an edge between two proteins indicates that two proteins interact with each other. A signaling traduction pathway may pass a signal from one protein to another via a set of intermediate proteins. The number of intermediate proteins may vary in different species. Therefore, biologists may be interested in a pattern involving several of important proteins. However, since in different species the intermediate proteins may be different, it is impossible to specify all proteins. Thus, the query can only be expressed as a graph where each vertex is a protein and the edge between two proteins indicate the possible distance between these corresponding proteins in the PIN.

In object-oriented programming, developers and testers handle multiple objects of the same or different classes. The object dependency graph of a program run, where each vertex is an object and each edge is an interaction between two objects through a method call or a field access, helps developers and testers understand the flow of the program and identify bugs. A software bug may involve several objects. However, since multiple objects may have similar functions, object invocations may not go through the same chain of objects. As a result, a bug can be represented as a graph g where each vertex is an object and an edge between objects v and u represents that there exists a chain of invocations from v to u with various length. [11].

Based on these examples, we propose a flexible pattern matching framework. In [19], an edge (v_1, v_2) in a query pattern represents that the matches of v_1 and v_2 should be within the δ distance and this δ is the same for all edges. While in our framework, there is a range $[min_e, max_e]$ associated for each edge e. Different edges may have different ranges. The range specifies the shortest distance between the matches of the endpoints (vertices) should be at least min_e and at most max_e. To the best of our knowledge, our flexible patterns is the first attempt to model different relationships in a single pattern, which previous models cannot represent.

Although existing methods can be modified to find the matches of our flexible patterns, the resulting algorithms are not efficient. The work most related to ours is the Distance Join (D-Join) [19]. With the D-Join method, k sets of vertices $(S_1, S_2, \ldots, S_k)$ in G are randomly chosen as anchors. Each vertex u' in G is mapped into a k dimensional space as a vector. The ith value of the vector is the minimum shortest distance from u' to any vertex in S_i. By using the triangle inequality, the lower bound of the shortest distance between any two vertices in G can be estimated based on the values in the k dimensional space, which is much cheaper than directly computing the shortest distance in G. Thus, for an edge (v_1, v_2) in Q with a range $[min_{1,2}, max_{1,2}]$, we can know whether the shortest distance of u_1 (v_1's match) and u_2 (v_2's match) is larger than $max_{1,2}$. If so, this pair of vertices can be pruned. However, this method suffers from the following shortcomings. (1) The number of sets of vertices has to be large for an accurate estimate, which would require a large amount of space and time. (2) It can only be used for the estimation of the upper bound, but not for the lower bound. (3) The estimated results may not be tight.

We design a new distance (global) index method, which would give an estimate on the shortest distance between every pair of vertices. Since the number of vertices in the database graph G could be very large, it is practically infeasible to store all pair-wise distances. In the global index, G is partitioned into m clusters. The shortest distance between any two clusters and the diameter of each cluster are stored. The shortest distance between two vertices can be estimated by using the cluster-wise distance and diameters of clusters.

The global index can give an estimate on the shortest distance between two vertices. The estimate tends to be more accurate when the two vertices are far away. However, in real applications, it is much more common for a user to specify a close relationship than a far-away relationship. For instance, in a biological motif, a user is more interested in patterns with clustered proteins. Therefore, it is beneficial to design a more accurate index for close neighbors of a vertex. In this paper, a local index is designed to capture the surrounding vertices for each vertex in the database graph. For each vertex u, the vertices and labels within a certain distance are considered as in the neighborhood of u. The neighborhood of u is divided into ring structures so that it is easy to have an accurate estimation on the distance of two vertices.

Another issue is that in real applications, people may not be interested to analyze thousands of matches returned during a query. A top-k matches generation scheme is designed. The matches are evaluated by some scoring function and during a query, only k matches with the highest(lowest) scores are returned.

The following is a list of contributions of this work. (1) A flexible pattern matching model is proposed. (2) An index based pattern matching algorithm is devised, which includes the following innovations: (i) The query graph Q is preprocessed to obtain more implicit and tighter relationships on edges. (ii) Two types of indices, the global and local indices, are constructed for fast match pruning. (3)A top-k matches generation scheme is proposed to quickly retrieve the top ranked matches. (4) A large number of real and synthetic data sets are used to show the effectiveness and efficiency of our model comparing to alternative methods.

The remainder of this paper is organized as follows. The related work is briefly discussed in Section 2. Section 3 presents some preliminaries and the problem statement. The global and local indexing structures are presented in Section 4 and 5, respectively. We present the query matching algorithm and the top-k pattern discovery algorithm in Section 6 and 7. The empirical studies are presented in Section 8. Final conclusions are drawn in Section 9.

2   Related Work

Graph database research has attracted a great amount of attention recently. Related work on graph matching can be divided into the following categories: the subgraph isomorphism test, graph and subgraph indexing, approximate graph matching, and graph pattern matching.

The subgraph isomorphism test is an NP-Complete problem and several algorithms have been proposed over the past few decades. Ullmann's method [14], probably the earliest one, uses backtracking to reduce the state search space, which is very expensive for querying on a large graph. Christmas's algorithm [3] is efficient but non-deterministic, thus cannot guarantee to find the correct solution. Cordella [5] proposed a new subgraph isomorphism algorithm for large graphs with the help of a set of feasibility rules during the matching process. All these algorithms perform the query directly on the database graph without any indexing structure.

Graph indexing is used in many applications, e.g., biological networks, social networks, etc. Many index-based graph matching and searching frameworks have been proposed. Under such frameworks, the database consists of many small or medium sized graphs, and the goal is to find the set of database graphs that contain or are contained by the query graph. In some works, substructures are extracted to index the graphs, e.g., tree-pi [16], tree-Delta [17], g-Index [15]. The authors of Closure-tree [9] use the closure to represent a set of graphs and then a data structure similar to R-tree is used to organize these closures. Some other work deals with subgraph indexing on a very large database graph, e.g. GADDI [16], Graphgrep [8], SPath [18]. The subgraph indexing problem is to locate all matches of a query graph. The index usually captures the information of the database graph in terms of paths, neighborhoods etc., and accelerates the query process.

Overall, the above approaches are designed for subgraph search and cannot be directly applied on the graph pattern query problem. Recently, two schemes were proposed for the graph pattern matching. One is R-Join [2], which handles the connectivity query on each edge of the graph pattern. The matches to two vertices of an edge in the query pattern graph must be connected in the database graph. A more general version, D-Join [19], allows setting a uniform upper bound on every edge of the graph pattern. The shortest distance of the matches in the database graph must be no larger than the uniform upper bound. We generalize D-Join further so that users can specify various degree of relationships of two vertices in the query pattern graph.

3   Preliminaries

In this section, the formal definitions and the problem statement are presented. We investigate the flexible pattern matching problem on a large undirected and connected database graph with weighted edges and labeled vertices. Without the loss of generality, it is easy to extend the model and method developed in this work to directed and unweighted-edge graphs.

Definition 1. *A **labeled database graph** G is a five element tuple $G = (V_G, E_G, \Sigma_{V_G}, L_V, W_E)$ where V_G is a set of vertices and $E_G \subseteq V_G \times V_G$ is a set of edges. Σ_{V_G} is the set of vertex labels. L_V is a labeling function that defines the mapping from $V_G \rightarrow \Sigma_{V_G}$ and W_E maps each edge in E_G to a positive weight. For an unweighted graph, $W_E(e) \equiv 1$ for every edge e.*

Definition 2. *The **query pattern graph** $Q = (V_Q, E_Q, L_V, \Delta)$, where V_Q and E_Q are the vertex and edge sets, L_V maps each vertex in V_Q to a label, and Δ maps each edge e into a range $[min_e, max_e]$ satisfying $1 \leq min_e \leq max_e$.*

Definition 3. *Given a database graph G and a query graph Q with vertices $v_1, v_2, ..., v_y$, a **match** of Q in G is a set of y distinctive vertices $u_1, u_2, ..., u_y$ in G such that:(1)*

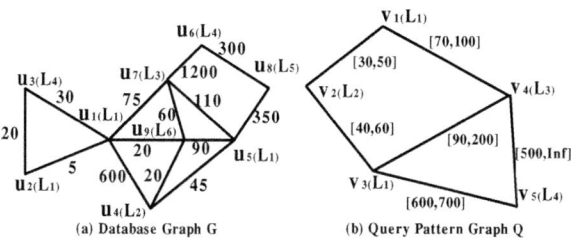

Fig. 1. The Problem and Matches

Each vertex v_i in Q has the same label as its matched counterpart vertex u_i in G, i.e. $L_V(v_i) = L_V(u_i)$;(2) For each edge (v_i, v_j) in Q, the shortest distance between u_i and u_j in G is within the range of $\Delta(v_i, v_j)$ in Q.

For query Q with vertices $(v_1, v_2, v_3, v_4, v_5)$ in Figure 1(b), there are two matches of Q: $(u_1, u_4, u_5, u_7, u_6)$ and $(u_2, u_4, u_5, u_7, u_6)$ in Figure 1(a).

Problem Statement: Given a large database graph G, and a query graph Q, we want to efficiently find all the matches of Q in G with the help of the indexed information.

4   Global Index

The global indexing structure is similar to the highway system. The database graph G is first partitioned into K overlapping clusters. Vertices in a cluster are close to each other while vertices in different clusters are farther apart. On average, a vertex belongs to A clusters. The global index consists of two parts. The first part is the highway distance between any two clusters. There is a center c_i in each cluster C_i. The highway distance between clusters C_x and C_y (D_{xy}) is the shortest distance between c_x and c_y. The second part includes the following information. Let's assume that a vertex v belongs to clusters $C_1, C_2, \ldots, C_A$. The shortest distance from v to the centers of these A clusters are also recorded. The overall space of the global index is $O(K^2 + A|V|)$. Since K is usually in thousands while A is often less than 10, the overall space complexity is well under control.

With this global index, the shortest distance between any two vertices can be estimated in the following manner. Suppose that vertices v_i and v_j belong to clusters C_x and C_y, respectively, then the shortest distance between v_i and v_j is within interval $[D_{xy} - d(x, v_i) - d(y, v_j), D_{xy} + d(x, v_i) + d(y, v_j)]$ where $d(x, v_i)$ is the shortest distance from v_i to the center of cluster C_x. The interval can be further refined (tightened) if v_i or v_j belongs to multiple clusters. Thus, the lower bound is $max_{\forall x, \forall y} (D_{xy} - d(x, v_i) - d(y, v_j))$ and the upper bound is $min_{\forall x, \forall y} (D_{xy} + d(x, v_i), d(y, v_j))$.

For example, in Figure 2(a), we have four clusters. Four highways connect the clusters and their lengths are shown in the figure. Vertex v_1 belongs to C_1 and C_2, and v_2 belongs to C_3 and C_4. Suppose $d(c_1, v_1) = 2$, $d(c_2, v_1) = 2$, $d(c_4, v_2) = 3$ and $d(c_3, v_2) = 2$. To estimate the shortest distance between v_1 and v_2, clusters C_1 and C_4 would yield an interval of $[4, 14]$. We can also obtain three other intervals $[3, 11]$,

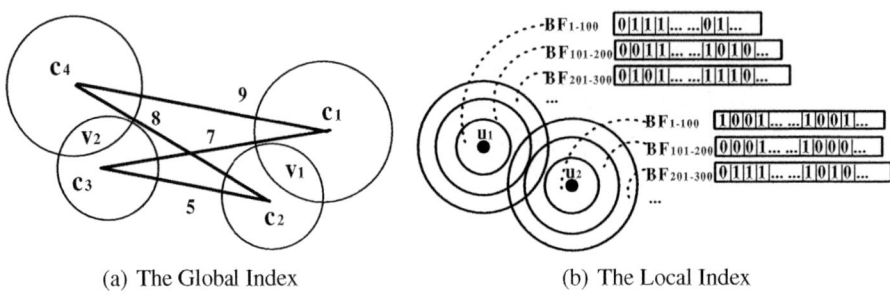

(a) The Global Index   (b) The Local Index

Fig. 2. Global and Local Indices

$[3, 13]$, and $[1, 9]$ from cluster pairs C_1 and C_3, C_2 and C_4, and C_2 and C_3. The final estimated interval would be $[4, 9]$.

4.1   Graph Partition

Two important parameters control the clustering process: the total number of clusters K and the average number of A clusters that one vertex belongs to. A variation of the k-medoids algorithm is devised to partition the database graph. To avoid the expensive pairwise shortest distance calculation, a heuristic algorithm is employed. K vertices are randomly generated as seeds and each one is expanded to form a cluster by the best-first search [6]. Since we know that the expected number of vertices ev in a cluster is $V \times A/K$, the search terminates until ev vertices are visited. After K clusters are generated, there may exist some vertices that are not included in any cluster. These vertices are assigned to A closest seeds. Thus, the real average number of vertices in a cluster is slightly higher than ev. The seed of a cluster can be considered as the center of the cluster. The highway distance can be discovered via Dijkstra Algorithm. The distance from a vertex to a center has also been obtained during the cluster formation step, thus no more computation is needed for this. Moreover, a table, which maps a vertex to a list of clusters that the vertex belongs to, is also generated. The formal description of this index building process is shown in Algorithm 1.

Algorithm 1. *Building the Global Index*

Input: Database Graph G, K, A
Output: highway distance for each pair of clusters C_x and C_y; shortest distances between a vertex to its centers; a table that maps a vertex to a list of clusters

1: randomly generate K vertices as seeds;
2: grow each seed to a cluster by the best-first search until $V \times A/K$ vertices are reached;
3: assign the unvisited vertices to its closest seed;
4: calculate D_{xy}, and the mapping table

4.2 Analysis of Global Index

Our objective is to produce an accurate estimation, i.e., the width of the estimation interval is small. For any fixed value A, larger K will generate a larger number of clusters and lead to smaller clusters. In this case, the distance from a vertex to its center will be small. As a result, the estimation interval will be tight. However, large K means large index size and building time. On the other hand, for a given fixed value of K, it is uncertain whether large A or small A would yield a better estimation. With large A, a vertex resides in more clusters and there exist multiple highways connecting vertices, in turn, the distance estimation interval would be tighter. However, in this case, each cluster is large and the distance from a vertex to its center is also large. Therefore, it is very difficult to determine whether large or small A will yield a better estimation interval. This depends on many factors: distance distribution function, connectivity, etc. Therefore, we empirically analyze the effects of A and K.

A 100k-vertex graph with average degree of 5 is randomly generated by the gen-graph_win [21]. The weight of each edge is randomly assigned to a value between 1 and 1000. After building the global index, we randomly draw 500 pairs of vertices to estimate their shortest distance. Figure 3 (a) shows the average width of the estimation interval with various of K and A. It is evident that there is a turning point for the interval width with respect to A, i.e., when $A = 3$, the estimation interval has a smallest width. This is mainly due to the fact that when A is too large, each vertex belongs to too many clusters. On the other hand, if A belongs to too few clusters, the intersection of interval effects would not take place.

In general, larger K will produce tighter interval width. However, this effect diminishes when K is larger than a few thousand. On the other hand, the global index building time is linear with respect to K, as shown in Figure 3 (b). Therefore, in real applications, A should be set to 3 while and K is set to a few thousand based on available computation resources.

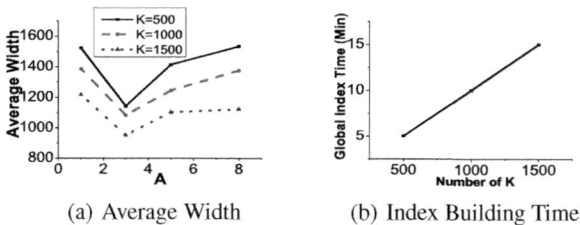

(a) Average Width   (b) Index Building Time

Fig. 3. Effects of parameters on the Global Index

In the global index, the width of the estimation interval on the shortest distance between u_1 and u_2 can be considered as the degree of inaccuracy of the estimation, which is depended on the distance between u_1 or u_2 to their respective centers and is independent on the true shortest distance between u_1 and u_2 ($dist$). We use a ratio ($width/dist$) to represent the estimation accuracy. If the shortest distance $dist$ is large, the inaccuracy ratio is small. On the other hand, the inaccuracy ratio tends to be larger when u_1 and u_2 is closer. In many applications, users are likely to specify closer relationship. Thus, it is necessary to create a local index structure for the neighborhood close to a vertex. The local index structure is described in the next section.

5   Local Index

In this paper, a local index is used to capture the information of the neighborhood around a vertex u, e.g., which set of vertices and labels are close to u. To match a query vertex v to a database vertex u, the set of vertex labels in u's neighborhood should be a superset of those in v's neighborhood. Thus, it is beneficial to index the labels first for fast pruning. After this pruning, a large amount of false positives still remain since multiple vertices may share the same label. Next, the vertices in the each neighborhood are also indexed.

When building the local index, two thresholds are employed: Max_Dist and Num_Interv. Max_Dist is the maximum radius of the neighborhood of u. To accommodate the lower bound of a range, the neighborhood is partitioned into a set of rings. The parameter Num_Interv controls the number of rings in the neighborhood. A ring is defined by the interval $[min\_dist, max\_dist]$, which means that vertices whose shortest distance to u is within the interval are associated with the ring. For example, if $Max\_Dist = 1000$, and $Num\_Interv = 10$, then the ten rings are $[1, 100], [101, 200]$, $\dots, [901, 1000]$.

Given a database graph G and two thresholds, two types of local indices, vertex and label indices, are constructed. Specifically, the vertex index on a vertex u is constructed as the following. u's neighborhood is searched up to Max_Dist by a shortest path discovery algorithm, e.g., Dijkstra algorithm [6], and Num_Interv buckets are constructed as the vertex index, each of which is for one of the rings. Vertices whose shortest distance to u between min_dis and max_dis are inserted into the bucket associated with interval $[min\_dis, max\_dis]$. Similarly, for the label index, instead of vertices, the labels of these vertices are added into the buckets. Clearly, the label index is smaller than the vertex index since multiple vertices may share the same label. Figure 2(b) shows the general structure of the indices.

The local index can naturally support the lower bound distance edge query, which could not be modeled by existing methods. For example, for an edge (v_i, v_j) in Q with a range $[150, 300]$, we want to determine whether (u_i, u_j) in G is a candidate match for (v_i, v_j). If $Max\_Dist = 1000$ and $Num\_Interv = 10$, we can simply search whether u_j is in one of u_i's bucket associated with rings $[101, 200]$ and $[201, 300]$ since these two rings fully cover the range $[150, 300]$. If not, then the pair (u_i, u_j) could not be a candidate for (v_i, v_j).

Searching whether or not a vertex or a label exists in a bucket is an essential operation and could be performed many times during a search. Thus, it is very important that the operation can be performed efficiently. In this paper, the bloom filter data structure is used to implement the buckets in the local index. A **bloom filter** B contains an m-bit vector and a set of k independent hash functions [1]. It is used to determine whether an element x is a member of a set X. Initially, all bits in B are 0s. Each of the k hash functions maps an element into an integer between 1 and m and then the corresponding k bits in B are set to 1. To build a bloom filter for the set X, every member of X is mapped and the corresponding bits are set. Later, to determine whether an element x is in X, x is mapped and we check whether all the corresponding bits are 1. If at least one of these bits is 0, then x is not a member of X. Otherwise x is a member of X with a high confidence. There is no false negative in the bloom filter. However, there could be false positives, i.e., if all mapped bits of x are 1 in B, then there is still a chance that x is not a member of X. The false positive rate α depends on m, n (number of elements in X), and k [12], which is

$$\alpha \approx (1 - e^{-kn/m})^k \tag{1}$$

Obviously, larger m or smaller n would decrease the false positive rate. Given m and n, the value of k that minimizes the probability of false positive is around $0.7m/n$. According to equation (1), to maintain a false positive rate of 1%, m should be set to approximately 9.6n, which means that it takes 9.6 bits to encode an element. The number of the hash functions is 7.

The Max_Dist and Num_Interv are two important parameters that determine the total size of the indexing structures. Let d and w be the average degree of G and the average weight on each edge of G. The average radius (in number of edges) of the indexed region of a vertex is $\frac{Max\_Dist}{w}$ and the total number of vertices in the region is $d^{\frac{Max\_Dist}{w}}$. Assume that l bits are used to encode a vertex. l can be determined by the false positive rate allowed for the bloom filters. Therefore, the total size of vertex index for one vertex is $l \times d^{\frac{Max\_Dist}{w}}$, which means that the total size of vertex index is $|V| \times l \times d^{\frac{Max\_Dist}{w}}$ where $|V|$ is the number of vertices in G. The size of the label index is less than that of the vertex index since multiple vertices may have the same label. Thus, with the available space for indices, we can estimate the value of Max_Dist. With the larger value Num_Interv, the neighborhood will be partitioned into more rings and in turn more accurate results will be found. When the ring width is set to w, it is equivalent to the width of each ring being 1.

The space complexity of the local index is $O(|V(G)| \times N(Max\_Dist))$, where $N(Max\_Dist)$ is the average number of vertices in a neighborhood of radius Max_Dist.

6   Query Algorithm

In this section, we present the query algorithm. Given a database graph G with the global and local indices, and a query graph pattern Q, our goal is to retrieve all matches from G that satisfy the conditions specified in Q according to Definition 3. The query algorithm have the following major steps: 1. Processing the query graph to obtain implicit relationships among vertices; 2. Generating candidate vertex matches by the local label index; 3. Generating candidate pattern matches via the local vertex index and global index; 4. Verifying the candidate matches to eliminate false positives and processing edges that are not captured by indices.

6.1 Query Preprocessing

An edge in Q represents a restriction on the shortest distance relationship between two vertices. Some relationships are explicit, i.e., specified by users. Others are implicit which can be inferred from the explicit relationships. For each vertex v in Q, the matches for v can be pruned if more restrictions on the relationships between v and other vertices in Q are present. Therefore, we infer the implicit relationships based on the triangle inequality.

Q can be considered as a complete graph. Some of edges have user specified ranges while others are *virtual edges* which are not specified by the user. Initially, the range on these virtual edges (v_1, v_2) are $[0, \infty]$ which represents that any distance between v_1 and v_2 will satisfy the requirement. Let the range on edges (v_1, v_2) and (v_2, v_3) be $r_{1,2} = [min_1, max_1]$ and $r_{2,3} = [min_2, max_2]$. The upper bound of the range $r_{1,3}$ on edge (v_1, v_3) will be $max_1 + max_2$. The lower bound of the range on edge (u, w) is 0 if r_1 and r_2 overlap. Otherwise, the lower bound is $min_2 - max_1$ assuming $min_2 \geq max_1$. For example, if $r_1 = [30, 50], r_2 = [70, 100]$, then $r_3 = [20, 150]$.

The upper bounds on all virtual edges in Q can be obtained by the all-pair shortest path discovery algorithm, e.g., Floyd-Warshall algorithm [7]. In this algorithm, only the upper bound of a range is concerned. The initial upper bound on a virtual edge is set to infinity. After obtaining the upper bounds, we will update the lower bounds in the following manner. Initially, the lower bound of a virtual edges is 0. The lower bound updating algorithm is a dynamic programming method similar to that of Floyd-Warshall algorithm. When updating the lower bound of a range on an edge, we compare the existing lower bound with all possible lower bounds computed with the above method, the highest lower bound is retained. The ranges on the real edges may be updated also if a tighter range can be obtained. The complexity of this step is $O(|V(Q)|^3)$ where $|V(Q)|$ is the number of vertices in Q. Since Q usually consists of a smaller number of vertices, the query preprocessing step can be performed efficiently. Figure 4 shows an example of the pre-processing.

It is possible that on some virtual edge, the lower bound of the range is larger than the upper bound. For example, ranges on (v_1, v_2), (v_2, v_3) and (v_1, v_3) are $[10, 20]$, $[35, 40]$, and $[10, 14]$, respectively. After applying the range updating algorithm, the lower bound in the range of the third edge is 15 which is larger than its upper bound. In this case, the query pattern is invalid and there could not be any match for this pattern. Therefore, we will not process the query further and report that this query is *invalid*.

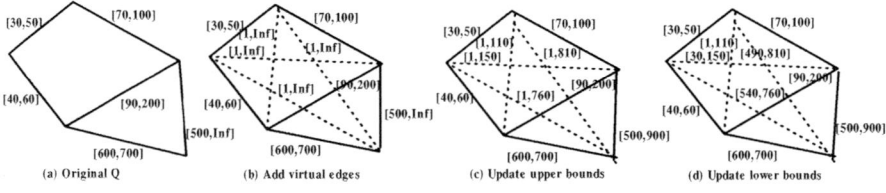

Fig. 4. Pre-Processing Query Pattern Graph Q

6.2 Candidate Vertex Match Generation Based on Local Label Index

For each vertex v_i in Q, a candidate set CL_i is constructed. CL_i contains all vertices in G that share the same label as v_i. These vertices are potential matches of v_i. Assume there is a table that maps vertex labels to vertices. CL_i can be obtained by searching the table.

After obtaining CL for each vertex in Q, these candidate sets can be pruned by the label index. The neighbors of v_i in Q are used to prune CL_i. Since Q is a complete graph after the pre-processing, neighbors of v_i contain all other vertices in Q. If the range r_{ij} on edge (v_i, v_j) is within $[0, Max\_Dist]$, then the label of v_j is used to prune CL_i. For each vertex $u \in CL_i$, we find the bloom filters of u that covers the r_{ij}. If the label of v_j does not appear in any of these bloom filters, then u cannot be a candidate for v_i and thus it can be removed from CL_i. On the other hand, if r_{ij} is not within $[0, Max\_Dist]$, then there does not exist any set of Bloom filters that can cover r_{ij} and thus the label of v_j could not be used to prune CL_i. After all vertices in Q are applied to prune CL_i based on the label indices (some vertices may not be useful), the candidate pattern matches are generated based on the vertex index.

6.3 Candidate Pattern Match Generation via Local Vertex Index and Global Index

A match M of Q consists of m vertices in G (assuming Q has m vertices). Each vertex in M is from one of the CL sets. Thus, there are potentially $\prod_1^m |CL_i|$ matches if not considering false positives. Each pair of (u_i, u_j) where $u_i \in CL_i$ and $u_j \in CL_j$ matches an edge in Q. The match discovery process can be considered as a repeated join process.

Each edge in Q is processed as a join. For example, let (v_1, v_2) be an edge in Q. For each vertex u_1 in CL_1, we need to find all vertices u_2 in CL_2 such that the shortest distance of u_1 and u_2 is within the specified range r_{12} on the edge. This join could be very expensive if it has been done on G directly. In this work, the global index and local vertex index will be utilized to reduce the search time. To find the matches of (v_1, v_2), for each vertex u_1 in CL_1, the local vertex index is employed if r_{12} intersects with $[0, Max\_Dist]$. Let rings $B_1, B_2, \ldots, B_y$ of u_1 be these covering the range r_{12}. We want to find which vertices in CL_2 are in these bloom filters. If there does not exist any vertex in CL_2 that is in one of these filters, then u_1 would not be a candidate match of v_1 and can be removed from CL_1. Otherwise, let $u_2^1 \in CL_2$ and $u_2^2 \in CL_2$ be the vertices in one of the filters in $B_1, B_2, \ldots, B_y$. Then (u_1, u_2^1) and (u_1, u_2^2) are two potential matches of (v_1, v_2).

The local vertex index can only be used for these edges whose ranges intersect with $[1, Max\_Dist]$. For other edges whose range exceeds Max_Dist, the global index is used. When the global index is used, we obtain all clusters contain v_1 or v_2 by the mapping table. Then the distance interval between v_1 and v_2 is estimated through the method described in the previous section. There are three possible relationships between the interval reported by the index and the range of an edge. If the interval is contained within the range, it is definitely a match. If they intersect, there is a potential match. Otherwise, it is not a match.

The range of a virtual edge is derived from the ranges on real edges, and thus it is redundant. However, the virtual edges are still used in this step due to the following reason. The bloom filters of a local vertex index have false positives. By performing the joins on the virtual edges, some false positives can be pruned. Since directly calculating the exact distance between two vertices on a database graph is much more expensive than checking the bloom filters, the additional computation of performing joins on virtual edges would be beneficial. For example, for a triangle (v_1, v_2, v_3), where (v_1, v_2), (v_2, v_3) are real edges and (v_2, v_3) is an inferred virtual edge, the false positives produced by the bloom filters of a match of v_2 when checking whether the matches of v_3 are qualified may be removed by the join processing of the virtual edge (v_1, v_3). The goal of this step is to prune the candidates with low computation costs.

The cost of processing an edge (v_1, v_2) in Q is $O(|CL_1| \times |CL_2|)$ where CL_1 and CL_2 are the set of candidate matches of v_1, v_2. Thus, the joins are performed in the order of the processing cost. The edge with the lowest cost is performed first. After processing edge (v_1, v_2), a "link" is used to represent a match. For instance, there is a match (u_1, u_2) for (v_1, v_2), a link is set between u_1 and u_2. It is possible that an element in CL_1 is linked to multiple elements in CL_2, and vice versa. There could be some unlinked elements u' in CL_1. This means that there does not exist any match of v_2 within r_{12} of u' and thus u' could not be a match of v_1. Therefore, all unlinked elements of CL_1 and CL_2 are pruned. Then the processing costs of edges adjacent to v_1 and v_2 are updated. Next the edge with the new lowest processing cost is chosen and processed. This procedure terminates when all edges in Q are processed.

For example, suppose $Max\_Dist = 800$ and the pre-processed Q is shown in Figure 5(a). Figure 5(b) presents a partial matching process. A column under v_i represents the candidate set CL_i for v_i and u_{ik} represents the kth vertex in the column under v_i. Suppose edge (v_3, v_4) has the lowest processing cost and is chosen to be matched first. u_{33} (marked as X) will be pruned since no link is attached to it. Next edge (v_2, v_3) is processed and links are generated for the matches of this edge. Then the edge (v_2, v_4) is processed. u_{42} cannot be matched to any vertex in CL_2 and thus it is removed. This is due to a false positive reported by the bloom filters associated with u_{32}, and u_{42} cannot be matched to any vertex in CL_2 during the double check when processing the virtual edge (v_2, v_4). This ensures a low false positive rate, which is smaller than the one of the bloom filter. Figure 5(c) shows the result, a graph R, after processing all the edges in Figure 5(a).

The candidate pattern matches can be obtained based on the graph R guided by Q in the following manner. First a candidate set CL_i is chosen as the starting point. For each vertex u_{ij} in CL_i, a breadth-first traversal is performed via the "links". For each edge adjacent to v_i, (v_i, v_k), in Q, if there does not exist any link between u_{ij} to any vertex in CL_k, then u_{ij} could not be in any match of Q and the search will start from a new vertex in CL_i. On the other hand, assume there are x links to vertices in CL_k. Each of these links (u_{ij}, u_{kl}) corresponds to a match of Q. The search continues on each matched vertex u_{kl} (through the links) in CL_k. This process continues until all matches of Q are generated.

For instance, assume that we start from CL_1. There are two vertices u_{11} and u_{12} in CL_1. u_{11} is chosen to start first. Three edges are adjacent to v_1 in Q, which connect v_1

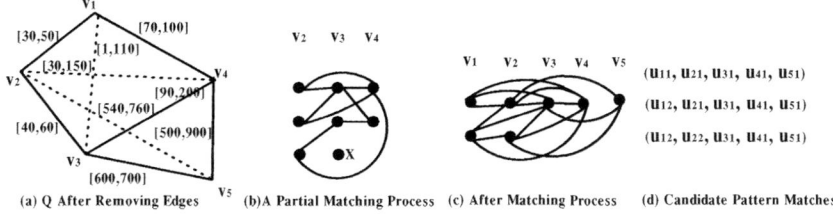

Fig. 5. Candidate Pattern Match Generation via Vertex Index

Algorithm 2. *Candidate Pattern Match Generation via Vertex Index*

Input: the PreProcessed Query Pattern Graph Q
Output: Candidate Pattern Matches

1: Generate candidate sets, each of which corresponds to a vertex in Q;
2: Prune the vertices in each candidate set by label vertex;
3: **repeat**
4: Choose an edge e from Q with lowest processing cost;
5: Perform a join for e over the two candidate sets, and place a link between any two vertices that satisfy the range of e according to one vertex's bloom filters or the highway-like index, which results a new graph R;
6: Remove any vertex without any link;
7: Update the edge cost;
8: **until** all edges in Q are processed
9: Traverse the graph R in a breadth first manner to generate the candidate pattern matches
10: return a set of candidate pattern matches

---

to v_2, v_3, and v_4. Thus, we will check whether u_{11} has links to CL_2, CL_3, and CL_4. In this example, u_{11} has links connecting u_{11} to u_{21}, u_{31}, and u_{41}, respectively. Then we check u_{21}, it has links to u_{11}, u_{31}, u_{41}, and u_{51}. Next vertex u_{31} is checked and so on. The candidate pattern matches generated from Figure 5(c) are shown in Figure 5(d). The formal description of the candidate pattern generation algorithm is presented in Algorithm 2.

6.4 Match Verification

After obtaining the set of candidate pattern matches in the previous step, it is necessary to further verify these candidates due to the following reasons. (1) If the interval reported by the global index intersects with the range of an edge but is not contained, it is not guaranteed that the shortest distance falls into the range. (2) There are false positives in the local label and vertex indices.

Due to the above reasons, a final verification step has to be performed on the database graph G. Since this step could be expensive, it is only applied on the original real edges of Q, but not the virtual edges since these virtual edges are inferred from real edges. To verify a pair of vertices in the candidate patterns from G are indeed a match of an edge, we compute the shortest distance of the two vertices from a candidate pattern by the

2-hop labeling technique [4]. Each vertex u in G is assigned with two sets $L_{in}(u)$ and $L_{out}(u)$ that are subsets of $V(G)$. Members of the two sets are called centers and the shortest distance between the vertex and any center are pre-computed and stored. Then the shortest distance of u_1 and u_2 can be obtained by

$$D(u_1, u_2) = min_{w \in L_{out}(u_1) \cap L_{in}(u_2)} \{D(u_1, w) + D(w, u_2)\}$$

The space complexity is $O(|V(G)| |E(G)|^{1/2})$ for storing the pre-computed results and the time complexity is $O(|E(G)|^{1/2})$ for computing the shortest distance between any two vertices.

7   Top-k Matches Generation

In many real-world applications, it may not be useful to analyze tens of thousands of matches. Thus, a top-k matches generation scheme is designed such that matches are evaluated by a scoring function, and during a query, only k matches with the highest(lowest) scores are returned. In this paper, the score of a match is defined as the sum of the shortest distances on all edges in the match. Without a loss of generality, lower score matches are preferred since they represent more close relationships. We made the

---

Algorithm 3. *A Top-k Matches Generation Scheme*

Input: Database Graph G, k, and the indices
Output: The top-k matches with k lowest scores

1: Traverse the candidate list graph and place qualified links by a depth first search until k complete matches are reached;
2: insert partial matches to Q_h;
3: insert complete matches to Q_k;
4: $th \leftarrow Score(End(Q_k))$;
5: **repeat**
6:    $WorkingMatch \leftarrow Dequeue(Q_h)$
7:    $M \leftarrow$ expand $WorkingMatch$ one more edge;
8:    **if** $Score(M) > th$ **then**
9:        prune M;
10:   **end if**
11:   **if** M is a complete match **then**
12:       $Enqueue(Q_k, M)$;
13:       $Remove(End(Q_k))$ and update th;
14:   **else**
15:       $Enqueue(Q_h, M)$;
16:   **end if**
17: **until** $IsEmpty(Q_h)$ or $Score(WorkingMatch) > th$
18: return the matches in Q_k;

---

following observation that once k matches are generated, the score of the kth match can be used as a threshold to prune other partial matches. For example, if the score of a partial match already exceeds the threshold, then we can simply discard it and do not need match it further. Based on the observation, a hybrid searching algorithm to generate the top-k matches is presented. Two priority queues are maintained. One queue Q_k is for the current top-k matches with size k, the other queue Q_h stores the current partial matches. At the first phase, we traverse the candidate list graph and place links in a depth first fashion until top-k complete matches are generated and stored in Q_k. The shortest distance of a link can be calculated by the 2-hop labeling technique. The highest overall distance of any match in Q_k is obtained as the threshold th. Next, we extend partial matches in Q_h. The match with the smallest score is extended first and links are inserted in a breadth first style. New resulting partial matches with a higher score than th are removed, and others are inserted back to Q_h. If a new complete match M is generated with a score lower than th, then the highest score match in Q_k is replaced by M and the threshold th is updated. The algorithm terminates when Q_h is empty. The formal description is shown in Algorithm 3.

8   Experimental Results

To empirically evaluate the performance of our flexible graph pattern framework (FGP), FGP is compared with D-Join, the most related work in this field, on a set of real and synthetic data. The original D-Join only allows a uniform upper bound assigned on each edge of the query graph. Thus, we extend the D-Join method to handel both bounds. There are four innovations in FGP: global index, local index, query preprocessing, and Top-K query. To analyze each innovation of FGP, four versions of FGP are designed: FGP-global (only consisting of the global index), FGP-both (consisting of global and local vertex indices), FGP (the full version containing both indices and the query preprocessing), and FGP-topK. All the methods are implemented with C++ and running on a PC with 2.6 GHz dual-core CPUs and 2 GB main memory.

8.1   Experiments on a Real Data Set

A protein interaction network for homo sapiens is used here. Each vertex is a protein and the label of the vertex is its gene ontology term from [20]. An edge in the graph

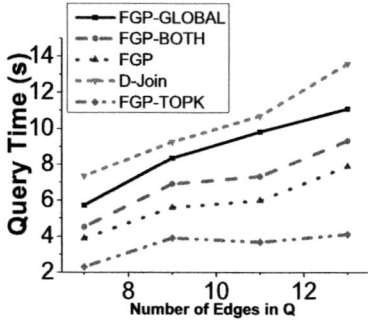

Fig. 6. Query Performances on a Protein Interaction Network

represents an interaction between the two proteins. The graph contains 6410 vertices, 27005 edges, and 632 distinct labels. The average degree of a vertex is 8.4. The weight on each edge is 1.

The local index of a vertex v includes the neighborhood with a radius of two, which covers 20% of the entire graph on average. The parameters for the global index are: $K = 600$, $A = 3$. FGP spends 174 seconds to build 1.4MB indices while it takes 238 seconds for D-Join to build a 3.9MB LLR embedding based index. Four known signal transduction pathways from the KEGG database [10] are used as the query pattern Q. The pathways are from species other than homo sapiens, e.g., yeast, fly. To obtain a more flexible query pattern Q, we remove some vertices in Q to represent missing information and set the range on each edge to [1,4]. The number of edges on the four pathways are 7, 9, 11, 13. When using top-k matching, k is set to 100. The query times with respect to the number of edges are shown in Figure 6.

FGP takes about half time of that of D-Join on average, and the acceleration is more evident when the number of edges of the query pattern graph increases because more edges means more joins need to be done. The bloom filter based local index can achieve 99% accuracy, when the shortest distance of a pair vertices is within Max_Dist, they can be quickly identified. Otherwise the global index is used to further prune the candidate vertices. While D-Join's pruning power is much lower than that of the FGP method. Moreover, by using the preprocessing and the label index of the local index, initially the candidate sets can be shrunk 15% to 30%. The top-k query time grows mildly.

8.2 Experiments on Synthetic Data Sets

The performances of the FGP and D-Join methods are analyzed on a set of randomly generated connected graphs. The graphs are obtained from a tool called gengraph_win [21]. The labels are randomly generated from 1 to 1000. We use four parameters, the number of vertices in G, the average degree of G, Max_Dist, and Num_Interv to analyze the performance of FGP. The default values for the four parameters are 50K, 7.5, 1000, and 10. Each time one parameter is varied while others remain as the default values.

Results of the indexing (including index size and building time) are shown in Figure 7. The space complexity of the local index is $O(|V(G)| \times N(Max\_Dist))$, where $N(Max\_Dist)$ is the average number of vertices in a neighborhood of radius Max_Dist, depending on the average degree of the graphs. The space for the global index is $O(A \times |V| + K^2)$. In D-Join, each vertex u is mapped into a k-dimensional space, where k is equal to $log^2 |V(G)|$. Thus, the space complexity of D-Join is $O(|V(G)| * log^2 |V(G)|)$.

When the average degree of the database graph is fixed, the size of the FGP indices mainly depends on the number of vertices and K. In our configuration, A is set to 3 and K is set to 1k, 2.5k, 5k, 10k, and 10k on the 5 graphs. As shown in Figure 7(a), because for each vertex u in G, the number of vertices in Max_Dist neighborhood does not vary significantly, the size and building time of the FGP index (both global and local) grows approximately linearly with respect to $|V(G)|$. The size of D-Join's index grows much faster. Next the average degree of G is varied from 5 to 20. Figure 7(c) shows the size of the FGP's local index increases with the average degree since many more vertices are within a neighborhood with denser graphs. When the average degree is no more than 20, FGP index is smaller than D-Join. Many real world application graphs fall into this range, e.g., biological networks and system dependence graphs. The index size of D-Join is not affected by the average degree. Figure 7(e) gives the index size growing trend with respect to Max_Dist. It can be seen that the index size of the FGP is smaller than that of the D-Join until the local index captures vertices three hops away on average. Num_Interv does not affect the size because when the number of elements in the bloom filters and the false positive rate are fixed, the total length of the bit vector of the bloom filters are fixed. Large Num_Interv means more bloom filters, each of which is shorter. Overall, the index size of FGP is more sensitive to the density of the graph while the one of the D-Join is more affected by the number of vertices in the graph. Considering the graphs in real applications are usually sparse with small average degree, the FGP index is more scalable.

For the index construction time, FGP needs to explore the neighborhood of a vertex and calculate the pairwise shortest distances between each pair of K centers, while D-Join needs to perform the shortest distance computation for every vertex in G to $log^2(|V(G)|)$ anchors. Thus, the FGP method has a much smaller index construction time especially when the graph is large and dense as shown in Figure 7 (b), (d). When we increase Max_Dist, more vertices need to be visited, the time grows moderately as shown in Figure 7 (g). The time is slightly affected by Num_Interv since more bloom filters are created and maintained during the index construction process.

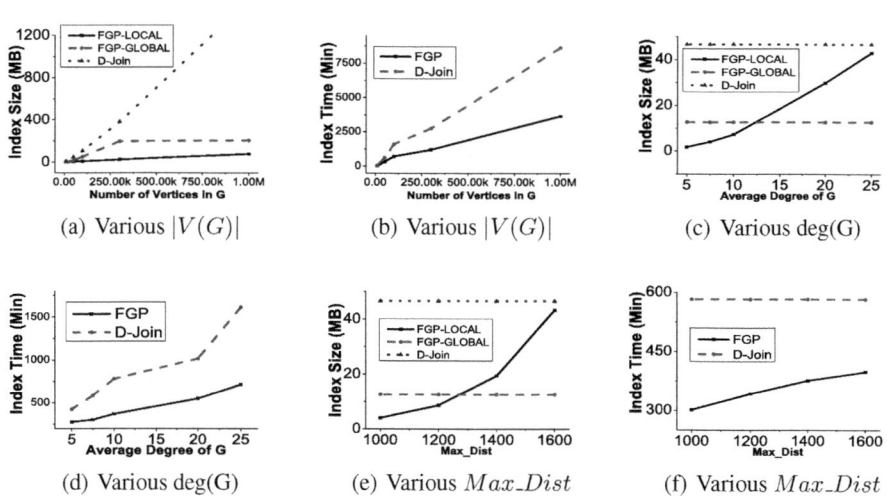

Fig. 7. Index Construction Comparison on Synthetic Data Sets

To analyze the query time of the FGP methods and the D-Join method, we use a default query chain graph with five vertices and four edges. The label of a vertex in the query graph is randomly generated. The default range on each edge is randomly generated as a subset of $[1, 2000]$.

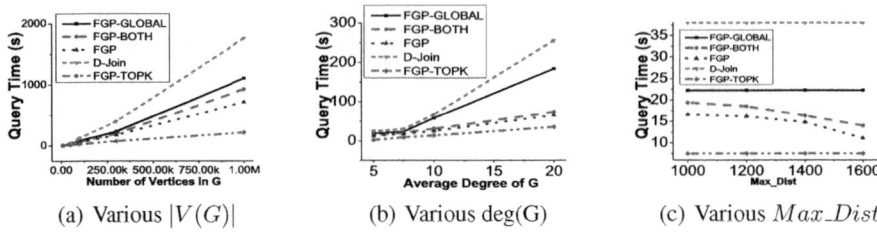

(a) Various $|V(G)|$ (b) Various deg(G)   (c) Various Max_Dist

Fig. 8. Query Time Comparison on Synthetic Data Sets

The scalability is measured with respect to the number of vertices and average degree of the database graphs and the results are shown in Figure 8 (a) and (b). FGP-Global outperforms D-Join, especially in large and dense graphs due to the following reasons. Large and dense graphs lead to a large number of possible matches and also more time is needed for the verification step. The global index reduces the number of candidate matches based on the intersection results to avoid the unnecessary and costly verification computation while D-Join's embedding index is not as powerful as FGP's global index. The difference between FGP-GLOBAL and FGP shows the effects of the local index. The local index of FGP ensures an efficient and accurate pruning of unqualified matches within Max_Dist. In addition, compared with other curves, FGP-TOPK grows much more slowly with respect to the size of the graph, which does not discover all matches.

Next the effects of Max_Dist is analyzed and shown in Figure 8 (c). The local index depends on this parameter. With the growth of Max_Dist from 1000 to 1600, more vertices can be covered by the local index. The acceleration effect of the local index for FGP is shown as the decrease of the execution time. However, this comes at the price of the increase of the index size and construction time.

Next we analyze the effect of Num_Interv on the query time. With the larger value of Num_Interv, each bloom filter covers shorter intervals and the estimates of the shortest distance within Max_Dist could be more accurate. However, more bloom filters also need to be checked. Thus, it has a mix effect and our experiments reflect this fact. Considering the cost of longer index construction time with large Num_Interv, a moderate value of Num_Interv would be more appropriate.

Overall, in this set of experiments, the efficiency of our FGP method is demonstrated with various number of database graphs and query graphs. FGP outperforms the D-Join method with a wide margin especially with larger and denser graphs due to the pruning power of the local and global indices.

9   Conclusions

In this work, we presented a flexible pattern matching model. The main differences between our model and exiting models are that each edge of the query pattern is associated with a range which specifies the lower and upper bounds of distance on the endpoint vertices. This gives the flexibility on specifying various relationships on the vertices in the pattern. To facilitate these pattern matching queries, two types of indices: local and global index, are constructed. A novel matching algorithm is devised to efficiently retrieve the matches for a given pattern with the help of these indices. In addition, we also provide a top-k matching scheme for the pattern discovery process. The efficiency of our method has been demonstrated on both real and synthetic data sets.

References

1. Bloom, B.H.: Space/time trade-offs in hash coding with allowable errors. Communications of the ACM 13(7) (1970)
2. Cheng, J., Yu, J.X., Ding, B., Yu, P.S., Wang, H.: Fast graph pattern matching. In: Proc. of ICDE (2008)
3. Christmas, W.J., Kittler, J., Petrou, M.: Structural matching in computer vision using probabilistic relaxation. IEEE Trans. on PAMI 17(8) (1995)
4. Cohen, E., Halperin, E., Kaplan, H., Zwick, U.: Reachability and distance queries via 2-hop labels. SIAM Journal of Computing 32(5) (2003)
5. Cordella, L., Foggia, P., Sansone, C., Vento, M.: A (sub)graph isomorphism algorithm for matching large graphs. IEEE Trans. on PAMI 26(10) (2004)
6. Dijkstra, E.W.: A note on two problems in connection with graphs. Numerische Mathematik 1, 269–271 (1959)
7. Floyd, R.W.: Algorithm 97: shortest path. Communications of the ACM 5(6) (1961)
8. Giugno, R., Shasha, D.: GraphGrep: A fast and universal method for querying graphs. In: Proc. of ICPR (2002)
9. He, H., Singh, A.K.: Closure-Tree: An index structure for graph queries. In: Proc. of ICDE (2006)
10. Kanehisa, M., Goto, S.: KEGG: Kyoto encyclopedia of genes and genomes. Nuc. Ac. Res. 28, 27–30 (2000)
11. Nguyen, T., Nguyen, H., Pham, N., AI-Kofahi, J., Nguyen, T.: Graph-based mining of multiple object usage patterns. In: Proc. of the Joint Meeting of ESEC and ACM SIGSOFT (2009)
12. Starobinski, D., Trachtenberg, A., Agarwaln, S.: Efficient PDA synchronization. IEEE Trans. on Mobile Computing 2(1) (2003)
13. Tong, H., Faloutsos, C., Gallagher, B., Eliassi-Rad, T.: Fast best-effort pattern matching in large attributed graphs. In: Proc. of the KDD (2007)
14. Ullmann, J.: An algorithm for subgraph isomorphism. Journal of the ACM (1976)
15. Yan, X., Yu, P., Han, J.: Graph Indexing, a Frequent Structure-based Approach. In: Proc. of Sigmod (2004)
16. Zhang, S., Li, S., Yang, J.: Gaddi: distance index based subgraph matching in biological networks. In: Proc. of EDBT (2009)
17. Zhang, S., Hu, M., Yang, J.: Treepi: a novel graph indexing method. In: Proc. of ICDE (2007)
18. Zhao, P., Han, J.: On graph query optimization in large networks. In: Proc. of VLDB (2010)
19. Zou, L., Chen, L., Ozsu, M.T.: Distance-Join: pattern match query in a large graph database. In: Proc. of Int. Conf. on Very Large Data Bases (2009)
20. Gene Ontology, http://www.geneontology.org/
21. gengraph_win,
    http://www.cs.sunysb.edu/ãlgorith/implement/viger/distrib/

Subgraph Search over Massive Disk Resident Graphs

Peng Peng[1], Lei Zou[1,*], Lei Chen[2], Xuemin Lin[3], and Dongyan Zhao[1,4]

[1] Peking University, China,
{pengpeng,zoulei,zdy}@icst.pku.edu.cn
[2] Hong Kong of Science and Technology, Hong Kong, China
leichen@cse.ust.hk
[3] University of New South Wales, Australia
lxue@cse.unsw.edu.au
[4] Key Laboratory of Computational Linguistics (PKU), Ministry of Education, China
zoulei@icst.pku.edu.cn

Abstract. Due to the wide applications, *subgraph queries* have attracted lots of attentions in database community. In this paper, we focus on subgraph queries over a large graph G. Different from existing feature-based approaches, we propose a bitmap structure based on edges to index the graph. At run time, we decompose Q into a set of *a*djacent *e*dge *p*airs (AEP). We develop *e*dge *j*oin (EJ) algorithms to address AEP subqueries. The bitmap index can reduce both I/O and CPU cost. More importantly, the bitmap index has the linear space complexity instead of the exponential complexity in feature-based approaches, which confirms its good scalability. Extensive experiments show that our method outperforms existing ones in both online and offline performances significantly.

1 Introduction

As an alterative to relational database, graph database utilizes graph as the underlying model, which represents and stores information by nodes and connecting edges. The key feature in graph databases is that query processing is optimized for structural queries, such as shortest-path queries [4,1,7,3], reachability queries [4,14,12,2], and subgraph queries [9,15]. In this paper, we focus on *subgraph queries*. Due to the wide applications, subgraph queries have attracted lots of attentions in database community [9,15,6,10,11,19,17]. Generally speaking, there are two scenarios of graph database models in these literatures. The first scenario is that graph database has a large number of small-size connected data graphs. Given a query Q, subgraph query retrieves all data graphs containing Q. In the second scenario, there is a single large graph (may not be connected) G, such as biological networks. Given a query Q, subgraph query needs to locate all embeddings of Q in G. For example, given a biological network G and a structural motif Q, we want to locate all embeddings of Q.

In this paper, we focus on the second scenario, i.e., finding all embeddings of Q over a single large graph G. The hardness of this problem lives in its exponential search space. Obviously, it is impossible to employ some subgraph isomorphism algorithm, such as ULLMANN [13] and VF2 [5] algorithms, to find all embeddings on the fly. In order to speed up query processing, we need to create indexes for large graphs and rely

---

* Corresponding author.

J.B. Cushing, J. French, and S. Bowers (Eds.): SSDBM 2011, LNCS 6809, pp. 312–321, 2011.
© Springer-Verlag Berlin Heidelberg 2011

these indexes to reduce the search space. To be an efficient index, its space cost should be as small as possible.

In this paper, we propose an index that can meet the above requirement. Firstly, we classify edges according to their endpoint labels. Then, for each cluster, we assign two bit strings to summarize all edges. Given a large graph G, the space complexity of the indexing structure is $O(|E(G)|)$, where $|E(G)|$ is the number of edges in G.

At run time, given a query Q, based on the cost model, we first decompose it into a set of adjacent edge pairs (AEP) subqueries. Then, we develop a kind of edge join algorithms to address AEP subqueries. In this algorithms, the search space can be reduced significantly with the help of the bitmap index. To summarize, in this work, we have made the following contributions:

1) We propose a novel bitmap index for a large graph G, which has both linear space cost and light maintenance overhead.

2) We decompose a subgraph query into a set of AEP subqueries. Based on our proposed bitmap index, we develop an efficient *Edge Join* (EJ) algorithm to answer AEP subqueries. We propose a cost model and a histogram-based cost estimation method to guide query decomposition.

3) Finally, we conduct extensive experiments over both real and synthetic datasets to evaluate our proposed approaches.

2   Related Work

Recently, subgraph search over a single large graph has began to attract researchers' attentions, such as GADDI [16], Nova [18] and SPath [17]. All of these methods construct some indexes to prune the search space of each vertex to reduce the whole search space. GADDI proposes an index based on *n*eighboring *d*iscriminating *s*ubstructure (NDS) distances, which needs to count the number of some small discriminating substructures in the intersecting subgraph of each two vertices. Nova proposes an index named *nIndex*, which is based on the label distribution and is integrated into a vector domination model. Both GADDI and Nova are memory-based algorithm, meaning that they cannot scale to very large graphs. SPath constructs an index by neighborhood signature, which utilizes the shortest paths within the k-neighborhood subgraph of each vertex. Because the index is built based on the k-neighborhood subgraph of each vertex, the index building time is very expensive, especially for a large graph. Distance-join [19] is our earlier work, in which, we propose a distance join algorithm for pattern match query over a single large graph. The match definition in [19] is not subgraph isomorphism. Thus, the method in [19] cannot be used to answer subgraph query.

3   Background

In this section, we review the terminology that we will use in this paper, and formally define our problem. In this work, we study subgraph search over a large *directed vertex-labeled* graph (Definition 1). In the following, unless otherwise specified, the term "graph" refers to a directed vertex-labeled graph. Note that, it is easy to extend this method to large "undirected" graph by specifying the edge an unique direction using a fixed rule.

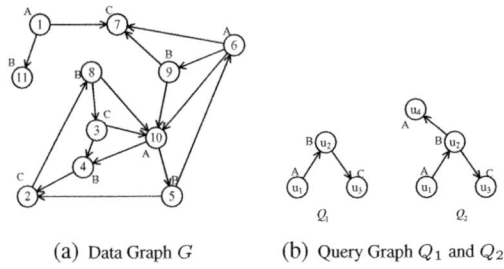

(a) Data Graph G (b) Query Graph Q_1 and Q_2

Fig. 1. Graph G and Query Q

Definition 1. *A directed vertex-labeled graph G is denoted as $G=\{V(G), E(G), L, F\}$, where (1) $V(G)$ and $E(G)$ are set of vertices and directed edges, respectively; (2) L is a set of vertex labels, and (3) the labeling function F defines the mapping $F : V(G) \rightarrow L$. Furthermore, according to the alphabetical order, we can define the total order for all distinct vertex labels in L.*

Figure 1(a) shows a running example of a directed vertex-labeled graph. Note that, the numbers inside the vertices are *vertex IDs* that we introduce to simplify description of the graph; and the letters beside the vertices are *vertex labels*. An directed edge from v_1 to v_2 is denoted as $\overrightarrow{v_1v_2}$.

Definition 2. *A labeled graph $G = \{V(G), E(G), L_V, F\}$ is isomorphic to another graph $G' = \{V'(G'), E'(G'), L'_V, F'\}$, denoted by $G \approx G'$, if and only if there exists a bijection function $g : V(G) \rightarrow V'(G')$ s.t.*

    $1) \forall v \in V(G), F(v) = F'(g(v)); 2) \forall v_1, v_2 \in V(G), \overrightarrow{v_1v_2} \in E \Leftrightarrow \overrightarrow{g(v_1)g(v_2)} \in E'$
    Given two graphs Q and G, Q is subgraph isomorphic to G, denoted as $Q \subseteq G$, if Q is isomorphic to at least one subgraph G' of G, and G' is a match of Q in G.

Definition 3. *(Problem Statement) Given a large data graph G and a query graph Q, where $|V(Q)| \ll |V(G)|$, the problem that we conduct in this paper is defined as to find all matches of Q in G, where matches are defined in Definition 2.*

4 Index

Definition 4. *Given vertex labels l_1 and l_2 in graph G,*
    $F^{-1}(l_1)$ *is defined as* $F^{-1}(l_1) = \{v | F(v) = l_1 \wedge v \in V(G)\}$. *Furthermore, we order all vertices in $F^{-1}(l)$ in the ascending order of the vertex IDs.*
    $F^{-1}(\langle l_1, l_2 \rangle) = \{(v_1, v_2) | F(v_1) = l_1 \wedge F(v_2) = l_2 \wedge \overrightarrow{v_1v_2} \in E(G)\}$ *and $F^{-1}(\langle l_1, l_2 \rangle) |_{l_1} = \{v_1 | (v_1, v_2) \in F^{-1}(\langle l_1, l_2 \rangle)\}$, where $F^{-1}(\langle l_1, l_2 \rangle)$ denotes all directed edges with two ending points are l_1 and l_2, respectively.*

Take graph G in Figure 1 for example. $F^{-1}(A) = \{1, 6, 10\}$ denotes all vertices whose labels are 'A', and $F^{-1}(\langle A, B \rangle) = \{(6, 9), (10, 5), (1, 11), (10, 4)\}$ denotes all edges whose labels pair is $\langle A, B \rangle$. Obviously, $F^{-1}(\langle A, B \rangle)|_A = \{1, 6, 10\}$ and $F^{-1}(\langle A, B \rangle)|_B = \{4, 5, 9, 11\}$.

In order to index edges of each labels pair, we propose the following indexing structures.

Definition 5. *Given a labels pair* $\langle l_1, l_2 \rangle$, $F^{-1}(l_1) = \{v_1, ..., v_m\}$ *and* $F^{-1}(l_2) = \{v'_1, ..., v'_n\}$, *its* start signature *and* end signature *(denoted as* $SB(\langle l_1, l_2 \rangle)$ *and* $EB(\langle l_1, l_2 \rangle)$*) are defined as follows:*

$SB(\langle l_1, l_2 \rangle)$ *is a length-m bit-string, denoted as* $SB(\langle l_1, l_2 \rangle) = [a_1, ..., a_m]$, *where each bit* a_i *(*$i = 1, .., m$*) corresponds to one vertex* $v_i \in F^{-1}(l_1)$, *and* $\forall i \in [1, m]$ $a_i = 1 \Leftrightarrow v_i \in F^{-1}(\langle l_1, l_2 \rangle)|_{l_1}$.

$EB(\langle l_1, l_2 \rangle)$ *is a length-n bit-string, denoted as* $EB(\langle l_1, l_2 \rangle) = [b_1, ..., b_n]$, *where each bit* b_i *(*$i = 1, .., n$*) corresponds to one vertex* $v'_i \in F^{-1}(l_2)$, *and* $\forall i \in [1, n]$ $b_i = 1 \Leftrightarrow v'_i \in F^{-1}(\langle l_1, l_2 \rangle)|_{l_2}$.

Let us recall the shaded area corresponding to labels pair $\langle A, B \rangle$. Since $F^{-1}(A) = \{1, 6, 10\}$ and $F^{-1}(B) = \{4, 5, 8, 9, 11\}$, thus, $|SB(\langle A, B \rangle)| = 3$ and $|EB(\langle A, B \rangle)| = 5$. Since $F^{-1}(\langle A, B \rangle)|_A = \{1, 6, 10\}$, thus, $SB(\langle A, B \rangle) = [111]$. Since $F^{-1}(\langle A, B \rangle)|_B = \{4, 5, 9, 11\}$, thus, $EB(\langle A, B \rangle) = [11011]$.

Besides start and end signatures, for each labels pair, we also define *start* and *end lists* as follows.

Definition 6. *Given a labels pair* $\langle l_1, l_2 \rangle$, $F^{-1}(l_1) = v_1, ..., v_m$ *and* $F^{-1}(l_2) = v'_1, ..., v'_n$,

its start list *is defined as follows:*

$$SL(\langle l_1, l_2 \rangle) = \{[i, (\overrightarrow{v_i v_j})] | \overrightarrow{v_i v_j} \in F^{-1}(\langle l_1, l_2 \rangle)\}$$

where $[i, (\overrightarrow{v_i v_j})]$ *denotes one edge* $\overrightarrow{v_i v_j} \in F^{-1}(\langle l_1, l_2 \rangle)$ *and* i *is called* start index, *which denotes the i-th bit that corresponds to* v_i *in* $SB(\langle l_1, l_2 \rangle)$.

its end list *is defined as follows:*

$$EL(\langle l_1, l_2 \rangle) = \{[j, (\overrightarrow{v_i v_j})] | \overrightarrow{v_i v_j} \in F^{-1}(\langle l_1, l_2 \rangle)\}$$

where $[j, (\overrightarrow{v_i v_j})]$ *denotes one edge* $\overrightarrow{v_i v_j} \in F^{-1}(\langle l_1, l_2 \rangle)$ *and* j *is called* end index, *which denotes the j-th bit that corresponds to* v_j *in* $EB(\langle l_1, l_2 \rangle)$.

There are four edges $\overrightarrow{6, 9}$, $\overrightarrow{10, 5}$, $\overrightarrow{1, 11}$, $\overrightarrow{10, 4}$ with the label labels pair $\langle A, B \rangle$. Since 1, 6 and 10 correspond to the 1-st, 2-nd and 3-rd bit in $SB(\langle A, B \rangle)$, respectively, thus, $SL(\langle A, B \rangle) = \{[1, (\overrightarrow{1, 11})], [2, (\overrightarrow{6, 9})], [3, (\overrightarrow{10, 5})], [3, (\overrightarrow{10, 4})]\}$. Analogously, $EL(\langle A, B \rangle) = \{[1, (\overrightarrow{10, 4})], [2, (\overrightarrow{10, 5})], [4, (\overrightarrow{6, 9})], [5, (\overrightarrow{1, 11})]\}$.

Besides, we use $SL(\langle l_1, l_2 \rangle)|_i = \{(v_i, v_j) | [i, \overrightarrow{v_i v_j}] \in SL(\langle l_1, l_2 \rangle)\}$ and $EL(\langle l_1, l_2 \rangle)|_j = \{(v_i, v_j) | [j, \overrightarrow{v_i v_j}] \in EL(\langle l_1, l_2 \rangle)\}$ to denote the projection on the list, where i (j) denotes the bit position that corresponds to v_i (v_j) in $SB(\langle l_1, l_2 \rangle)$ ($EB(\langle l_1, l_2 \rangle)$). For example, $SL(\langle A, B \rangle)|_3 = \{\overrightarrow{10, 4}, \overrightarrow{10, 5}\}$, since vertex 10 corresponds to the 3-rd bit in $SB(\langle A, B \rangle)$.

As discussed above, for each labels pair $\langle l_1, l_2 \rangle$, we assign it four associated data structures to it, that are start signature, end signature, start list and end list.

In order to access the start and end lists efficiently, we can build clustered B$^+$-trees over these lists to save I/O cost by avoiding the sequential scan. Besides, in order to save the space cost of bitmap index, we propose to use the *compression version* of bitmap index, i.e., only recording the non-zero bit positions in start and end signatures. It is straightforward to know there are $2 \times |E(G)|$ non-zero bits in all start and end signatures in total, and there are $2 \times |E(G)|$ edges in all start and end lists in total.

5 AEP-Based Query Evaluation

In this section, we discuss how to evaluate a subgraph query over a large graph G. Firstly, we propose an edge join algorithm to answer an adjacent edge pair query (AEP query, Definition 7) in Section 5.1. Given a general query Q having more than 2 edges, we decompose Q into a set of AEP queries, and find matches of Q by joining all AEP query results. In order to optimize query Q, we propose a cost model to guide the decomposition over Q. The cost model and the general subgraph query algorithm will be discussed in Section 5.2 and 5.3, respectively.

5.1 Edge Join

Definition 7. *Given a query Q having two edges e_1 and e_2, if e_1 is adjacent to e_2, Q is called an adjacent edge pair (AEP for short) query.*

Let us recall an AEP query Q_1 in Figure 1, which has two adjacent edges $e_1 = \overrightarrow{u_1u_2}$ and $e_2 = \overrightarrow{u_2u_3}$ with one common vertex u_2. The labels pairs of e_1 and e_2 are $\langle A, B \rangle$ and $\langle B, C \rangle$, respectively. Considering labels pair $\langle A, B \rangle$, all edges in $F^{-1}(\langle A, B \rangle)$ are matches of e_1, i.e., $M(e_1) = F^{-1}(\langle A, B \rangle)$ (defined in Definition 4). Due to the same reason, $M(e_2) = F^{-1}(\langle B, C \rangle)$. The baseline algorithm is to perform a natural join between $M(e_1)$ and $M(e_2)$ based on the common column u_2. Therefore, the join cost can be modeled as follows:

$$
\begin{aligned}
Cost &= C_{IO} \times (|M(e_1)| + |M(e_2)|)/P_{disk} + C_{cpu} \times (|M(e_1)| * |M(e_2)|) \\
&= 8 \times C_{IO}/P_{disk} + 16 \times C_{cpu}
\end{aligned}
\tag{1}
$$

where C_{IO} is the average I/O cost for one disk page access, and $(|M(e_1)| + |M(e_2)|)$ $/P_{disk}$ is the number of disk pages for storing $M(e_1)$ and $M(e_2)$, and C_{cpu} is the average CPU cost.

Algorithm 1. Edge Join (EJ) Algorithm

Require: Input: Given a AEP query p with two adjacent edges $e_1 = \overrightarrow{u_1u_2}$ and $e_2 = \overrightarrow{u_2u_3}$.
   Their labels pairs are $\langle l_1, l_2 \rangle$ and $\langle l_2, l_3 \rangle$, respectively. Assume that e_1 and e_2 are ES joined.
   Output: $M(e_1 \cup e_2)$.
1: According to HT index, we load $EB(\langle l_1, l_2 \rangle)$ and $SB(\langle l_2, l_3 \rangle)$ into memory.
2: Let $b = EB(\langle l_1, l_2 \rangle) \wedge SB(\langle l_2, l_3 \rangle)$.
3: **if** b is a bit-string in which all bits are 0 **then**
4:      $M(e_1 \cup e_2) = \phi$.
5: **else**
6:      **for** $i=1,...,|b|$ **do**
7:          **if** $b[i] = 1$ **then**
8:              $M(e_1 \cup e_2) = M(e_1 \cup e_2) \cup (EL(\langle l_1, l_2 \rangle)|_i \bowtie SL(\langle l_2, l_3 \rangle)|_i)$
9: Return $M(e_1 \cup e_2)$.

In order to speed up query processing, we need to reduce the cost in Equation 1. Actually, it is not necessary to join the whole $M(e_1)$ and $M(e_2)$. Instead, we can reduce

$M(e_1)$ and $M(e_2)$ to $M'(e_1)$ and $M'(e_2)$, respectively. Then, the matches of $(e_1 \cup e_2)$ (denoted as $M(e_1 \cup e_2)$) can be obtained by performing $M'(e_1) \bowtie M'(e_2)$. The following lemma shows the pruning strategy. We can prove that Lemma 1 satisfies *no-false-negative* requirement.

Lemma 1. *Given two edges* $e_1 = \overrightarrow{u_1 u_2}$ *and* $e_2 = \overrightarrow{u_2 u_3}$ *in query* Q, *where* $F(u_1) = l_1$, $F(u_2) = l_2$ *and* $F(u_3) = l_3$, *the matches of* $(e_1 \cup e_2)$ *can be found as follows:*
1) if $EB(\langle l_1, l_2 \rangle) \wedge SB(\langle l_2, l_3 \rangle)$ *is a bit-string in which each bit is 0, then* $M(e_1 \cup e_2)$ =NULL;
2) if $EB(\langle l_1, l_2 \rangle) \wedge SB(\langle l_2, l_3 \rangle)$ *is a bit-string in which the* I_i-*th bit is 1,* $i = 1, ..., n$, *then* $M(e_1 \cup e_2)$ *can be evaluated by the following equation:*

$$M(e_1 \cup e_2) = \bigcup_{i=1}^{i=n} \left(EL(\langle l_1, l_2 \rangle)|_{I_i} \bowtie SL(\langle l_2, l_3 \rangle)|_{I_i} \right)$$

Based on Lemma 1, we propose *Edge Join* algorithm in Algorithm 1. For example, the labels pairs of two adjacent edges in Q_1 are $\langle A, B \rangle$ and $\langle B, C \rangle$. $EB(\langle A, B \rangle) \wedge SB(\langle B, C \rangle)$= [11010]. The non-zero bit positions are $I_1 = 1, I_2 = 2, I_3 = 4$. Actually, these non-zero bit positions correspond to vertices 4, 5, 9 in G. According to Lemma 1, we can find $M(e_1 \cup e_2)$ as follows:

$$M(e_1 \cup e_2) = (EL(\langle A, B \rangle)|_1 \bowtie SL(\langle B, C \rangle)|_1) \cup (EL(\langle A, B \rangle)|_2 \bowtie SL(\langle B, C \rangle)|_2)$$
$$\cup (EL(\langle A, B \rangle)|_4 \bowtie SL(\langle B, C \rangle)|_4) = (10, 4) \bowtie (4, 2) \cup (10, 5) \bowtie (5, 2) \cup (6, 9) \bowtie$$
$$(9, 7) = \{(10, 4, 2), (10, 5, 2), (6, 9, 7)\}$$

In this case, the join cost can be evaluated as follows:

$$\begin{aligned}
Cost(e_1, e_2) &= \delta + C_{IO} \times \left(\sum_{i=1}^{i=n} \left(|EL(e_1)|_{I_i}| + |SL(e_2)|_{I_i}| \right) / P_{disk} \right. \\
&+ C_{cpu} \times \sum_{i=1}^{i=n} \left(|EL(e_1)|_{I_i}| \times |SL(e_2)|_{I_i}| \right) \\
&= \delta + 6 \times C_{IO} + 3 \times C_{cpu}
\end{aligned} \quad (2)$$

where δ is the average cost for bitwise AND operation, which is small enough to be neglected. Obviously, the cost in Equation 2 is less than that in Equation 1.

As discussed early, due to the clustered B$^+$-tree over the start and end lists, we can save I/O cost in the selections over these lists and employ the merge join instead of the nested loop join.

5.2 Cost Estimation

In this subsection, we propose a method to estimate the join cost in EJ algorithm, which will be used in Section 5.3 to answer subgraph query. Let us recall the cost model in Equation 2. It is easy to estimate δ, P_{disk} and C_{IO} and C_{CPU} from the collected statistics of query data. The key issue is how to estimate $|EL(e_1)|_{I_i}|$ and $|SL(e_2)|_{I_i}|$. In order to address this problem, we propose a histogram-based approach. For each labels pair $\langle l_1, l_2 \rangle$, we build two histograms, denoted as $SH(\langle l_1, l_2 \rangle)$ and $EH(\langle l_1, l_2 \rangle)$.

Definition 8. *Given a labels pair* $\langle l_1, l_2 \rangle$, *the* start histogram *for* $\langle l_1, l_2 \rangle$ *is a length-n number array, denoted as* $SH(\langle l_1, l_2 \rangle) = [h_1, ..., h_n]$, *and each number* h_i ($i = 1, ..., n$) *corresponds to one vertex* v_i *in* $F^{-1}(l_1)$, *and* $\forall i \in [1, n]$, $h_i = |SL(\langle l_1, l_2 \rangle)|_i|$.

The end histogram *for* $\langle l_1, l_2 \rangle$ *is a length-m number array, denoted as* $EH(\langle l_1, l_2 \rangle) = [h_1, ..., h_m]$, *and each bit* h_i $(i = 1, ..., m)$ *corresponds to one vertex* v_i *in* $F^{-1}(l_2)$, *and* $\forall i \in [1, m], h_i = |EL(\langle l_1, l_2 \rangle)|_i|$.

Given two ES join edges $e_1 = \overrightarrow{u_1 u_2}$ and $e_2 = \overrightarrow{u_2 u_3}$. We first compute $r = EB(e_1) \wedge SB(e_2)$. For i-th bit in r $(i = 1, ..., |r|)$, if it is a non-zero bit, we can estimate $|EL(e_1)|_i|$ and $|SL(e_2)|_i|$ according to the i-th element in $EH(e_1)$ and $SH(e_2)$, respectively. Finally, we estimate the join cost by Equation 2.

5.3   AEP-based Query Algorithm

In order to answer a subgraph query Q $(|E(Q)| > 2)$, we first decompose Q into a set of AEP (Definition 7) queries. Then, for each AEP $(e_i \cup e_j)$, we employ EJ algorithm to find $M(e_{i_1} \cup e_{i_2})$ and estimate $Cost(p_i)$ by Equation 2. For ease of presentation, we use p_i to denote an AEP $(e_{i_1} \cup e_{i_2})$. Finally, we join all of $M(p_i)$ and get the results.

Definition 9. *Given a set of AEP, denoted as* $AS = \{p_i = (e_{i_1} \cup e_{i_2})\}$ *in* Q, *we say that* AS *covers* Q *if and only if* $\bigcup \{p_i\} = Q$.

We say that AS *is a* minimal cover *over* Q, *if and only if* AS *satisfies the following two conditions: (1)*AS *covers* Q; *and (2)Removing any AEP from* AS *will lead that* AS *cannot cover* Q.

Obviously, we only need to decompose Q into a minimal cover and answer the AEP queries of it. Hence, The cost of answering Q can be evaluated by the sum of all edge joins. Thus, we use $Cost(AS) = \sum Cost(p_i)$ to estimate the cost for answering Q, where $p_i = (e_{i_1} \cup e_{i_2}) \in AS$.

Given a query Q, there may exist more than one minimal cover over Q. In order to optimize subgraph query processing, we need to find the *optimal minimal cover*, which is the cover that has minimal cost. Unfortunately, finding the optimal minimal cover over Q is at least NP-hard. We can prove the hardness of finding the optimal decomposition by reducing the *minimal set cover problem* (MSC). Practically, we propose an approximate solution to find the decomposition of query Q.

In our algorithm, we adopt the similar greedy strategy in MSC problem. Initially, we set Q' =NULL. Let S be the set of all possible AEPs in query Q. We select one AEP p with the minimal estimation cost from S. Then, we employ EJ algorithm to find matches $M(p)$, and insert p into Q' and remove it from S. Iteratively, we select one AEP p (in S) with the minimal estimation cost among all AEPs (in S) adjacent to Q'.We also employ EJ algorithm to find $M(p)$.Then, according to the topological relationship between Q' and p, we perform the corresponding natural join, meaning that we update $M(Q') = M(Q') \bowtie M(p)$ and $Q' = Q' \cup p$. The above steps are iterated until $Q' = Q$.

Now, we illustrate AEP query by Q_2 in Figure 1(b). There are three AEP queries in Q_2, i.e., $S = \{p_1, p_2, p_3\}$, where $p_1 = \langle A, B \rangle \wedge \langle B, A \rangle$, $p_2 = \langle A, B \rangle \wedge \langle B, C \rangle$, $p_3 = \langle B, A \rangle \wedge \langle B, C \rangle$. According to cost estimation, we can know $Cost(p_1) = 4 \times C_{IO}/P_{disk} + 2 \times C_{CPU}$, $Cost(p_2) = 6 \times C_{IO}/P_{disk} + 3 \times C_{CPU}$, $Cost(p_3) = 6 \times C_{IO}/P_{disk} + 3 \times C_{CPU}$. Therefore, we first select p_1 and evaluate p_1 by EJ algorithm, since $Cost(p_1)$ is minimal. The result set is $RS(p_1) = \{(6, 9, 10), (10, 5, 6)\}$. Then, $S = S - \{p_1\}$ and $Q' = Q' + p_1$. Since $Cost(p_2) = Cost(p_3)$ and both p_2 and p_3 are adjacent to Q', we randomly select p_2 and insert p_2 into Q'. We also evaluate p_2

by EJ algorithm and obtain $RS(p_2) = \{(6, 9, 7), (10, 5, 2), (10, 4, 2)\}$. Now, $Q' = Q$. We join $RS(p_1)$ and $RS(p_2)$ based on the common vertices u_1 and u_2, i.e., $RS(Q) = RS(p_1) \bowtie_{u_1, u_2} RS(p_2) = \{(6, 9, 10, 7), (10, 5, 6, 2)\}$.

6   Experiments

In this section, we evaluate our method AEP over both synthetic and real data sets, and compare them with some state-of-the-art algorithms, such as GADDI [16], Nova[18] and SPath [17]. Our method has been implemented using standard C++. The experiments are conducted on a P4 2.0GHz machine with 2Gbytes RAM running Linux. Furthermore, GADDI and Nova's softwares are provided by authors. We use best-effort re-implementation according to [17]. Since all competitor are designed for undirected graphs, thus, we extend our method for undirected graphs in the following experiments. Further details about the extension can be found in our technical report [8].

Data Sets. *a) Erdos Renyi Model*: This is a classical random graph model. It defines a random graph as N vertices connected by M edges, chosen randomly from the $N(N - 1)/2$ possible edges. In experiments, we vary N from 10K to 100K. The default average degree is set to be 5 and the default number of vertex labels (denoted as $|L|$) is 250. This dataset is denoted ER data.

b) Yago dataset is a RDF dataset. We build a RDF graph, in which vertices corresponds to subjects and objects, and edges correspond to properties. For each subject or object, we use its corresponding class as its vertex label. We ignore the edge labels in our experiments. There are $368, 587$ vertices, $543, 815$ edges and $45, 450$ vertex labels in Yago graph.

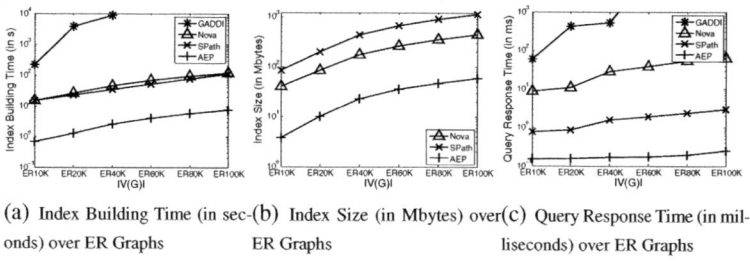

(a) Index Building Time (in sec- (b) Index Size (in Mbytes) over (c) Query Response Time (in mil-
onds) over ER Graphs   ER Graphs   liseconds) over ER Graphs

Fig. 2. Performance VS. $|V(G)|$

Experiment 1.(Performance VS. $|V(G)|$) This experiment is to study the scalability of our methods with increasing of $|V(G)|$. In this experiment, we use ER datasets and fix $|V(Q)|$ (i.e., the number of vertices in query Q) to be 10. Figure 2(a) shows that our method (AEP) have the linear index building time, which outperforms Nova, GADDI and SPath by orders of magnitude. Figures 2(b) shows that our method (AEP) have much smaller index sizes than those in Nova and SPath. Note that, GADDI cannot finish index building in reasonable time (within 24 hours) when $|V(G)| \geq 60K$. Besides, the GADDI software provided by authors cannot report index size, thus, we ignore the

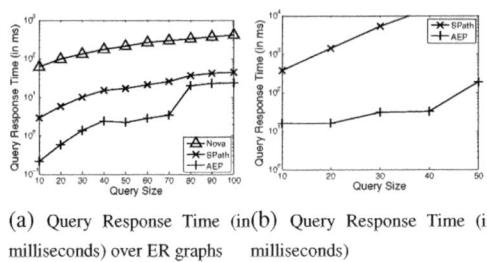

(a) Query Response Time (in (b) Query Response Time (in
milliseconds) over ER graphs   milliseconds)

Fig. 3. Online Performance

comparison with GADDI in index sizes. We also report the average query response times in Figure 2(c), which show that AEP does not increase greatly when varying $|V(G)|$ from 10K to 100K, which confirms the good scalability of our method. Note that, our methods are faster than other methods by at least 2 orders of magnitude in query processing.

Experiment 2.(Performance versus $|V(Q)|$) In this experiment, we evaluate the performance of our methods with the increasing of query size, i.e. $|V(Q)|$. In this experiment, we fix $|V(G)|$ to be 100K. Figure 3(a) shows that query response time are increasing in all methods when varying $|E(Q)|$ from 10 to 100. Our method (AEP) has the best performance.

Experiment 3.(Performance over Real Datasets) We also test our methods in a real dataset Yago. Note that Nova and GADDI cannot work on Yago dataset due to running out of memory. Therefore, we only compare our method with SPath. Obviously, The online processing in our method is faster, as shown in 3(b). In Yago dataset, we can finish index building in less than 3 minutes, which is much faster than that in SPath.

7   Conclusions

In order to address subgraph query over a single large data graph G, in this paper, we propose a novel bitmap structure to index G. At run time, we propose a subgraph query algorithm in this paper. Aimed by the bitmap index, we can reduce the search space and improve the query performance significantly. Extensive experiments over both real and synthetic data sets confirm that our methods outperforms existing ones in both offline and online performances by orders of magnitude.

Acknowledgments. Peng Peng, Lei Zou and Dongyan Zhao were supported by NSFC under Grant No.61003009 and RFDP under Grant No. 20100001120029. Lei Chen was supported in part by RGC NSFC JOINT Grant under Project No. N_HKUST61_2/09, and NSFC Grant No. 60736013 and 60803105. Xuemin Lin was supported by ARC grants (DP110102937, DP0987557, DP0881035).

References

1. Chan, E.P.F., Lim, H.: Optimization and evaluation of shortest path queries. VLDB J. 16(3) (2007)
2. Chen, Y., Chen, Y.: An efficient algorithm for answering graph reachability queries. In: ICDE (2008)
3. Cheng, J., Yu, J.X.: On-line exact shortest distance query processing. In: EDBT (2009)
4. Cohen, E., Halperin, E., Kaplan, H., Zwick, U.: Reachability and distance queries via 2-hop labels. SIAM J. Comput. 32(5) (2003)
5. Cordella, L.P., Foggia, P., Sansone, C., Vento, M.: A (sub)graph isomorphism algorithm for matching large graphs. IEEE Trans. Pattern Anal. Mach. Intell. 26(10), 1367–1372 (2004)
6. Jiang, P.Y.H., Wang, H., Zhou, S.: Gstring: A novel approach for efficient search in graph databases. In: ICDE (2007)
7. Jing, N., Huang, Y.-W., Rundensteiner, E.A.: Hierarchical encoded path views for path query processing: An optimal model and its performance evaluation. IEEE Trans. Knowl. Data Eng. 10(3) (1998)
8. Peng, P., Zou, L., Zhao, D., Chen, L., Lin, X.: Technical report, http://www.icst.pku.edu.cn/intro/leizou/subgraphsearch.pdf
9. Shasha, D., Wang, J.T.-L., Giugno, R.: Algorithmics and applications of tree and graph searching. In: PODS (2002)
10. Tian, Y., McEachin, R.C., Santos, C., States, D.J., Patel, J.M.: Saga: a subgraph matching tool for biological graphs. Bioinformatics 23(2) (2007)
11. Tian, Y., Patel, J.M.: Tale: A tool for approximate large graph matching. In: ICDE, pp. 963–972 (2008)
12. Trißl, S., Leser, U.: Fast and practical indexing and querying of very large graphs. In: SIGMOD (2007)
13. Ullmann, J.R.: An algorithm for subgraph isomorphism. J. ACM 23(1) (1976)
14. Wang, H., He, H., Yang, J., Yu, P.S., Yu, J.X.: Dual labeling: Answering graph reachability queries in constant time. In: ICDE (2006)
15. Yan, X., Yu, P.S., Han, J.: Graph indexing: A frequent structure-based approach. In: SIGMOD (2004)
16. Zhang, S., Li, S., Yang, J.: Gaddi: distance index based subgraph matching in biological networks. In: EDBT, pp. 192–203 (2009)
17. Zhao, P., Han, J.: On graph query optimization in large networks. In: VLDB (2010)
18. Zhu, K., Zhang, Y., Lin, X., Zhu, G., Wang, W.: NOVA: A novel and efficient framework for finding subgraph isomorphism mappings in large graphs. In: Kitagawa, H., Ishikawa, Y., Li, Q., Watanabe, C. (eds.) DASFAA 2010. LNCS, vol. 5981, pp. 140–154. Springer, Heidelberg (2010)
19. Zou, L., Chen, L., Özsu, M.T.: Distancejoin: Pattern match query in a large graph database. PVLDB 2(1)

BR-Index: An Indexing Structure for Subgraph Matching in Very Large Dynamic Graphs

Jiong Yang and Wei Jin

Department of Electrical Engineering and Computer Sciecne
Case Western Reserve University
{jiong.yang,wei.jin}@case.edu

Abstract. Subgraph indexing, i.e., finding all occurrences of a query graph Q in a very large connected database graph G, becomes an important research problem with great practical implications. To the best of our knowledge, most of subgraph indexing methods focus on the static database graphs. However, in many real applications, database graphs change over time. In this paper, we propose an indexing structure, BR-index, for large dynamic graphs. The large database graph is partitioned into a set of overlapping index regions. Features (small subgraphs) are extracted from these regions and used to index them. The updates to G can be localized to a small number of these regions. To further improve the efficiency in updates and query processing, several novel techniques and data structures are invented, which include feature lattice, maximal features, and overlapping regions. Experiments show that the BR-index outperforms alternatives in queries and updates.

1 Introduction

With the emergence of applications in social networks, software engineering, and computational biology, more data are represented as graphs. The size of these graphs can be in excess of millions of vertices and edges. Finding occurrences of a subgraph pattern in these large graphs becomes an important research problem with great practical applications. The problem studied in this paper is the subgraph indexing, which is to find all occurrences of a given query graph in a very large database graph G changing over time.

There is a large body of work in graph indexing and matching areas. Related work includes subgraph search and matching [1,4,6,7,8,14,15,18,19,21], approximate subgraph matching, and similar graph search [3,5,9,12,13,16]. However, the amount of work in subgraph indexing is still relatively small, e.g., Graphgrep[4], TALE[9], GADDI[19], SAPPER[20], etc. In addition, these work only deals with static database graphs, i.e., the database graph does not change over time. To the best of our knowledge, there does not exist any work on subgraph indexing for large evolving graphs. In many applications, the large database graph changes over time. For example, a social network may consist of millions of persons (as vertices). Each edge may represent a relationship between persons. The average degree of a vertex can be in the range of scores. Users may want to find occurrences of some special pattern (e.g., a student who has a sibling working as a doctor, etc.) for targeted marketing, homeland security, etc. The social network could change frequently with new persons and relationships added to the network. Since the graph is very large, it may not be feasible to rebuild the entire indexing structure once the network changed. Therefore, it is necessary to devise a dynamic indexing structure that can be efficiently updated and maintained.

J.B. Cushing, J. French, and S. Bowers (Eds.): SSDBM 2011, LNCS 6809, pp. 322–331, 2011.

Since our targeted database graph is very large, the only feasible dynamic indexing structure should be built on local information. In such a case, during an update only a small portion of the indexing structure needs to be modified. In this paper, we propose a partition-based indexing structure. The database graph G is partitioned into a set of (possibly overlapping) regions. When an edge or vertex is inserted or deleted, only a small number of these regions needs be modified. In previous research, e.g., [19], it has been proven that feature based indexing has been very useful. Thus, in this paper, a set of small features (subgraphs) are extracted from all these regions and are used to represent them. During the query processing, the features in the query graph Q are identified and used to find the index regions containing some of these features. Finally the occurrences of Q are assembled from these index regions.

There exist several challenges on designing such a dynamic indexing structure. First, a feature may occur on the boundary of two regions and may not be fully contained in one of the regions, which would lead to missing occurrences. To avoid this problem, the *bounding region property* is identified. Second, there are potentially a very large number of features. We need to search which of these features are in the query graph and an index region may contain many these features. It is necessary to design an efficient and systematic method to search these features. In this paper, several novel techniques are used here, including the concept of *maximal features* and the *feature lattice* data structure The third challenge relies on the selectivity of features. We identify the **overlapping region property** to perform further pruning by using overlapping index regions for overlapping feature occurrences. Then the selectivity can be improved dramatically. Overall, the **BR-index** data structure is proposed for indexing large dynamic graphs, which utilizes the techniques above.

The remainder of this paper is organized as follows. We present the preliminaries and properties in Section 2. Section 3 shows the BR-index structure. In Section 4, we present the algorithms to construct and maintain the BR-index. Section 5 shows the query/match algorithms with the BR-index. Empirical results are shown in Section 6. Final conclusions are shown in Section 7.

2   Preliminaries

Definition 1. *Given a vertex v in a graph G and an integer k, we define the k-neighborhood of v, denoted as $N_k(G, v)$, as a set of vertices in G such that $N_k(G, v) = \{v' | d(G, v', v) \le k\}$ where $d(G, v', v)$ is the shortest distance between vertices v' and v in graph G.*

Definition 2. *Given a subset of vertices $V_1 \subseteq V$ of graph G, an **induced subgraph** of V_1 is the subgraph composed of V_1 and any edge whose both endpoints belong to V_1. We denote the induce subgraph of V_1 as $S(V_1)$.*

Definition 3. *A **core region** CR is an induced subgraph of a vertex set V' in G. A **partition** of a graph G is a set of core regions $\{CR_1, CR_2, ..., CR_n\}$, which have the vertex set $\{V_{CR1}, V_{CR2}, \ldots, V_{CRn}\}$ respectively such that*

1. $V = \bigcup_{i=1}^{n} V_{CRi}$, and
2. $\forall i, j \leq n, i \neq j, V_{CRi} \cap V_{CRj} = \varnothing$.

Given a core region CR, an k-index region, IR^k, is an extension of CR. Let the vertex set V_{IR}^k be the union of the k-neighborhoods of all vertices in the vertex set, V_{CR}, of the core region CR, i.e., $V_{IR}^k = \bigcup N_k(G, v), \forall v \in V_{CR}$. The k-index region IR^k is equal to $S(V_{IR}^k)$ which is an induced subgraph of vertex set V_{IR}^k in graph G. k is called the **extension** *to the core region in the index region.*

Problem Statement: Given a large database graph G, we want to build an index structure of G such that for any query graph Q, the index structure can be used to find all occurrences of Q in G efficiently. In addition, the indexing structure should be able to efficiently handle the incremental updates of the graphs, e.g., insertion and deletion of edges and vertices.

Note that index regions can overlap with other index regions while core regions do not. To clarify the notation in this paper, the term *vertices* is used to refer to the vertices in the database and query graphs while the term *nodes* is employed to refer to the nodes in the indexing structure.

Next, we introduce two very important properties used in this paper. To clearly present the bounding region property, we first define the center of a graph.

Definition 4. *The* **eccentricity** *of a vertex v, $ecc(v)$, in a connected graph G is the maximum shortest distance from v to any other vertex. The* **radius** *of G, $radius(G)$, is the minimum eccentricity among all vertices in G. The* **center** *of G, $center(G)$, is the set of vertices with the eccentricity equal to the radius.*

The database graph G is partitioned into a set of non-overlapped core regions SCR. The set of k-index regions SIR^k which are obtained by extending from each core region in SCR, has the *bounding region* property. Two subgraphs in a graph g are **overlapping** if these subgraphs share at least one common vertex in g.

Property 1. For any subgraph g of G $(radius(g) \leq k)$, there exists at least one index region IR_i^k in any partition of G, such that g is a subgraph of IR_i^k.

Property 2. Let o_1 and o_2 be the overlapping occurrences of features f_1 and f_2, respectively. For any occurrence o of Q in the database graph G, let IR_1 and IR_2 be the two index regions that contain o_1 and o_2 in o. IR_1 and IR_2 must overlap in G.

The proofs of these two properties are omitted due to the space limitation.

3 BR-Index Data Structure

Armed with the bounding region property, we invent a novel subgraph indexing structure, called BR-index. A BR-index consists of three parts: a vertex lookup table, a set of features $\mathcal{F}$, and a set of index regions $\mathcal{I}$. First, the vertex lookup table is a hash table that maps a vertex v into two elements: (1) the index region whose core region contains v and (2) the set of index regions that contain v. The vertex and index regions are identified by their IDs. The vertex lookup table is used for inserting and removing a vertex.

Second, a feature is a small subgraph with the radius at most r. Let f_1 and f_2 be two distinct features and f_2 is a supergraph of f_1, then every index region that contains f_2 must also contain f_1. Thus, there exists a large amount of redundant information. As a result, in the BR-Index, we use the concept of **maximal features**. A feature $f \in \mathcal{F}$ is a maximal feature of an index region IR if IR contains f and there does not exist another feature $f' \in \mathcal{F}$ such that f' is a proper supergraph of f and IR contains f' also.

For a given query graph Q, we need to find all maximal features in $\mathcal{F}$ contained in Q. Thus, the features are organized in a lattice which would provide an efficient way to find maximal features in Q. Each feature in $\mathcal{F}$ is represented as a node. There exists an edge between two nodes f_1 and f_2 if f_1 is an **immediate supergraph** of f_2. f_1 is an immediate supergraph of f_2 if (1) f_1 is a proper supergraph of f_2 and (2) there does not exist another feature $f_3 \in \mathcal{F}$ such that $f_1 \supset f_3 \supset f_2$. Each feature f is associated with a list of index region IDs (ir) where these index regions contain f as a maximal feature. Therefore, if we want to find all index regions that contain the feature f, then we need to take the union of all ir fields of f and f's descendants.

Third, there are five fields in an index region IR : ID, core vertices (cv), index region subgraph (irs), overlapping index region IDs (oir), and feature IDs (fid). The ID is a unique integer assigned to each index region. The cv are those vertices in the core region of IR. The parameters c_{max} and c_{min} control the maximum and minimum number of vertices in a core region. The irs is the induced graph of vertices in the index region of IR. The adjacency list is used to represent the irs since it is more space compact and time efficient to manipulate. The oir of IR contains the IDs of these index regions that are overlapping with IR. The IDs of the maximal features of IR are stored in the fid field.

4   BR-index Construction and Maintenance

Vertex Insertion. At the beginning, the BR-index BR contains a null graph. Then all vertices and edges in the database graph are inserted into BR sequentially. When a vertex v is inserted into BR, we first check whether v already exists. If so, the insertion is invalid and will be rejected. Otherwise, v is put into a special index region IV which stores all isolated vertices. The vertex lookup table is updated to reflect the existence of v. Assume that the label of the newly inserted vertex is l_v. Let f_v be the feature with only one vertex of label l_v. If f_v does not exist in the feature set, it will be created. IV is added into the set of index regions that contain f_v as a maximal feature.

Edge Insertion. Vertex InsertionWhen an edge (u, v) is inserted into BR-index, we first check whether u and v exist or not. If not, these vertices are inserted first. Next, the core regions containing u and v can be found via the vertex lookup table. There are four cases for vertices u and v. (1) Both u and v are isolated vertices. They will be removed from the IV and put into the core region with the fewest vertices. (2) One vertex is in IV and the other is in some other core region. Without a loss of generality, let's assume that u is in IV while v is in another core region CR. u will be moved to CR and connected to v. (3) u and v are in the same core region. We need update the irs of all index regions containing both u and v to include the newly inserted edge. (4) u and v are in different core regions. Let IR_v and IR_u be the index regions that contain v and u as a core vertex, respectively. u is inserted into IR_v while v is inserted into IR_u. The irs field of all index regions containing both u and v needs to be updated to include the new edge. In addition, IR_v and IR_u become overlapping so that the oir field of IR_v and IR_u needs to be updated. Adding a new edge into an index region IR may lead to the change of features contained in IR. For each maximal feature f contained in IR, we search whether any super feature f' of f is contained in IR. If so, f is replaced by f' in $IR.fid$. In the case that the number of vertices in a core region exceeds c_{max}, a split of the core region will be invoked and this procedure is described later.

Edge Deletion. To remove an edge (u, v), we first determine the index regions that contain both vertices u and v, and then the edge (u, v) is removed from the irs of these index regions. For the index region IR_u, v and all its adjacent edges will be removed from IR_u if v is not connected with any other core vertex. In this case, since IR_u is no longer containing v, we need check whether IR_u still overlaps with these index regions containing v. If not, the oir field of the respective index regions will be updated. The same process will be applied to the index region IR_v. After deleting the edge, we need check whether u or v becomes isolated. If so, they will be moved to the special index region IV. Deleting an edge may also affect the current maximal features contained in the index regions. If a maximal feature does not exist in the index region any longer, we check its ancestors in the feature lattice (sub-features) to determine the new maximal features associated with these index regions. In the case that the number of vertices in a core region falls below c_{min}, a core region merge process will be invoked, which will be explained below.

Vertex Deletion. If the vertex v is in the IV region, we simply delete it. Otherwise, we locate all regions that contain v by the lookup table. First all the edges attached to v are deleted, and next v is deleted. If the number of vertices in the core region CR_v falls below c_{min}, then CR_v will be merged with the the index region that has the highest connectivity with CR_v.

Core Region Merge. The vertices in two core regions are merged into one core region and then the core region is grown to an index region. If the number of vertices in the new core region exceeds c_{max}, we invoke the core region split procedure to produce two new core regions. Otherwise, the maximal features in the new index region are rebuilt since we may have larger maximal features. In other words, we extend previous maximal features in the two original index regions and see if they exist in the new index region.

Core Region Split. The goal is to split a big core region into two smaller balanced core regions. First the min-cut algorithm is applied on the set of vertices to create two partitions. If the partitions are not balanced, then border vertices are moved from the group with more vertices to the group with less vertices. When the two groups have the same number of vertices, then they become two core regions. Finally these two core regions are extended to form two index regions. The maximal feature sets for these two new index regions are recomputed.

Feature Maintenance. Features are organized in a lattice. Initially, after adding all isolated vertices, we have L single-vertex maximal features, where L is the number of distinct labels. During the updates of G, more index regions are associated with each feature. Denote X to be the total number of index regions in G. Assuming we have sufficient working memory to store the feature lattice, if a feature f is associated with $s \times X$ index regions where s is a small fraction, then f needs to be updated, e.g., extending f to several super features. (s can be chosen in various manners. Without a loss of generality, s is set to 0.01 in this paper.)

Since there exists a large number of super features with one more edge for a sub-feature, we require that features be small graphs of radius one. Assume that f has l edges. In theory, an l edge feature could have $O(l^2 + l * |L|)$ super features with $l + 1$ edges, where $|L|$ is the number of distinct labels. It is infeasible to generate all these super features since it could take too much memory. Thus, f is only extended to p super features with $l + 1$ edges. The p super features are generated randomly chosen. After the new features are generated, for each of those new features, we first check whether it is contained by at least one index region. If not, the new feature is discarded. The remaining features are added into the lattice. For each index region IR associated with f, if IR includes one super feature (f') of f, then IR is removed from $f.ir$ and added to $f'.ir$. Also $IR.fid$ is updated to reflect that f' is the maximal feature instead of f.

5 Subgraph Query

During the query, a set of features in the query graph Q are identified and these features are used to find the set of candidate regions that may contain a partial occurrence (match) of Q. These partial occurrences are finally assembled to obtain the matches of Q. The query process consists of three phases: (1) feature extraction, (2) index region pruning, and (3) occurrence discovery.

We assume that there exists a lexicographical order among all vertex labels. The search starts with the highest lexicographical order label in Q. Let's assume that l is such a label. In the feature set $\mathcal{F}$, the feature f of a single vertex with label l is located. Next, a depth-first search process is used to find the maximal features as a subgraph of Q by traversing the feature lattice from f. Suppose that f_1 is a maximal feature of Q. The index regions of G containing occurrences of f_1 is the union of ir fields of f_1 and f_1's descendants in the feature lattice. The process terminates when all maximum features containing f have been identified. Next we start searching the maximal features involving the second highest lexicographical order label l' and so on. The entire feature extraction procedure terminates after we have found all maximal features starting from all vertex labels in Q. By the end of this phase, we have the index regions containing all occurrences of maximal features in Q.

Based on the overlapping region property, the index regions can be pruned in the following manner. Let o_{ij} be the jth occurrence of feature i in Q. First, a graph representing the overlapping occurrences of features is constructed for all maximal features of Q. Each occurrence is a node in the graph and there is an edge between two occurrences if they share at least one vertex in Q. At the beginning, each node o_{ij} in the overlapping graph is associated with a set of index regions containing the maximal feature f_i. Next, the index regions associated on each node in the overlapping graph are pruned. We start from the node o_{i1} with the least number of index regions. A depth-first traversal is invoked to prune the index regions on the two ending points of an edge. When processing an edge (o_{ik}, o_{jl}), the set of index regions on o_{jl} (I_{jl}) is used to prune the index regions associated with o_{ik} (I_{ik}). For each index region IR in I_{ik}, we check whether one of IR's overlapping regions is in I_{jl}. If so, IR can remain in I_{ik}. Otherwise, IR is removed from I_{ik}. The same process is performed on all index regions in I_{ik}.

In the occurrence discovery step, we assemble the occurrences of the maximal features in index regions to form the matches of the query graph. A depth first traversal is performed on the overlapping features graph staring from the node with the least number of index regions. A match of Q is generated as the following. The first step is to locate the occurrence of a maximal feature f in an index region IR, this can be done very efficiently since the feature is a small graph of radius one. It is possible that f may have multiple occurrences in IR, and these occurrences will be processed sequentially. After obtaining one occurrence of f in IR, we find the matches of f_1 (which is adjacent to f in the overlapping feature graph) in IR_1 (where IR is overlapping with IR), and furthermore, verify the overlapping vertex (vertices) in G between the occurrences of f and f_1. This process continues until all the maximal features in the overlapping feature graph are matched and connected. It is possible that some vertices in Q are not covered by the maximal features, and thus not matched. The final step is to extend the current occurrences of all the features to these uncovered vertices.

6   Experimental Results

In this section, we empirically analyze the performance of the BR-index. Since there are several innovations of BR-index, several versions of BR-index are implemented to show the effects of each individual innovation. BR-index is the full version. BR-index1 only uses index regions: G is partitioned into a set of index regions and each region is indexed by a set of features. BR-index2 uses index regions and the overlapping region information for better pruning power, but without the maximal feature technique, i.e. in the feature lattice, each feature is associated with IDs of all index regions that contain this feature, rather than IDs of index regions that contain this feature as a maximal feature. We compare various versions of BR-index with Closure-Tree (C-tree) [5]. Although TALE [9], GADDI [19] and GraphGrep [4] can also be used for subgraph indexing, they are not designed to handle very large graphs, and TALE is an approximate subgraph indexing method which may not find all occurrences of a subgraph. C-tree is a graph indexing technique to find all graphs that contain a given query subgraph. It introduces the concept of graph closure which is a structural union of the underlying database graphs. It organizes graphs hierarchically to build a C-tree where each node summarizes its descendants by a graph closure. In this experiment, C-tree is modified for subgraph indexing. As a result, a leaf node of C-tree is an index region. BR-index and C-tree are implemented with C++. All experiments are conducted on a Dell PowerEdge 2950, with two 3.33GHZ quad-core CPUs and 32GB main memory, using Linux 2.6.18.

We use the query time, index construction time, index update time and index size to empirically analyze these methods. After building the index, we add 500 edges to the graph and then randomly delete 500 edges. The processing time is recorded as the index update time. The query graphs are obtained directly from the database graph.

6.1   Parameter Settings

There are two important parameters for BR-index: the maximum number of vertices in a core region C_{max} and the branch factor of a node p in the feature lattice. We test BR-index with various C_{max} and p against query time and index update time. With higher C_{max}, less core regions will be formed but each index regions will be larger. Less index regions lead to faster traversal of BR-index while larger index regions equate to a longer time to match a feature. Large p results in more features included in the lattice, which will accelerate the query in general. However, it comes at the price of larger space for the lattice and longer index construction time and update time because at each level of the lattice, more features need to be tested.

Figure 1 (a) and (b) show the index update time and query time with various C_{max} and p on the flickr graph. When varying C_{max} ($p = 20$), 30 seems to be a break point since after it, both update and query time grows dramatically. For p ($C_{max} = 30$), the query time decrease slowly since although the query graph may contain more features, essentially, the determining factor is the maximal feature. While the index update time grows quickly after 30. Therefore, in all experiments, C_{max} is set to 30 and p to 25. To improve the disk utilization, C_{min} is set to half of C_{max}, which is 15.

6.2   Experiment on a Real Data Set

The real-world graph is generated from user account information obtained by crawling the flickr website. A user account denotes a vertex, and there is an edge between two user accounts if they have added each other as friends. The flickr graph has 1,286,641 vertices and the average degree is 25.1. The label for each vertex is randomly assigned between 1 to 1000.

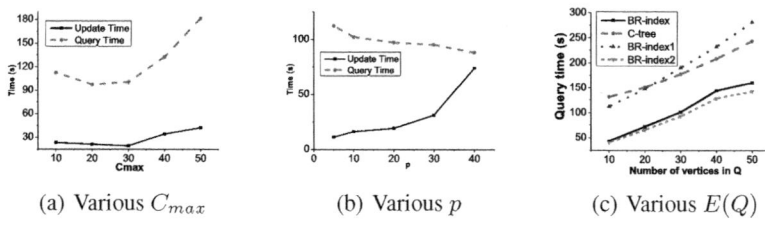

(a) Various C_{max} (b) Various p (c) Various $E(Q)$

Fig. 1. Experiments on a Real Data Set

Table 1. Index Comparisons on flickr

	Construction Time(s)	Update Time(s)	Size(GB)
BR-index	7310	22	2.5
BR-index1	6782	12	2.2
BR-index2	18210	77	2.9
C-tree	8235	112	2.7

Table 1 shows the index construction time, update time, and size of BR-index and C-tree. BR-index has a similar size to C-tree but has a shorter index construction time. The main computation of BR-index are graph partition, feature generation and matching. C-tree needs to compute the graph closures for its internal nodes, which is a very time-consuming process. The index update time of BR-index is much smaller than C-tree. For BR-index, the updates are only applied on a set of index regions, which are small graphs. But C-tree has to recompute the graph closure from leaf nodes along the path to the root. BR-index1 is a simplified version of BR-index without the overlapping region information. Thus, it has the smallest index construction time, update time, and size. BR-index2, without the maximal feature innovation, needs to maintain every feature and its associated index regions, and vice versa. Thus, it has the longest index size, construction time, and update time.

Figure 1 (c) shows the query time of different versions of BR-index against C-Tree on the flickr graph. Q is extracted from the flickr graph. We vary the size of Q in terms of vertices. The average degree of Q is around 3.5. The query time of all methods increases as the size of Q grows since large query graphs usually contain more features, which need to be connected to form final matches of Q. Overall, BR-index outperforms C-tree due to the following reasons. Most of the query time is spent on the final verification by extending matches of the connected features to the occurrences of Q. By using the overlapping region property, BR-index only needs to load the overlapping index regions. C-tree, however, employs a Pseudo Subgraph Isomorphism test to traverse the database graph, which is much more time consuming.

7   Conclusions

In this paper, we addressed the problem of indexing large dynamic graphs. The large database graph is partitioned to a set of overlapping index regions. Features are extracted from these regions. A feature lattice is constructed to maintain and search these features. By utilizing the maximal feature mechanism, the space for storing the features can be reduced. When processing query graph Q, the set of maximal features in Q are extracted, and the index regions containing these features are identified. A feature occurrence/overlapping graph is constructed to capture the overlap of these features in Q. The index regions are pruned based on these overlapping relationship. Finally, the matches of Q are located. According to our knowledge, this is the first attempt on indexing a very large dynamic graph.

References

1. Cheng, J., Ke, Y., Ng, W., Lu, A.: fg-index: Towards verification-free query processing on graph databases. In: Proc. of SIGMOD (2007)
2. Cordella, L., Foggia, P., Sansone, C., Vento, M.: A (Sub)Graph Isomorphism Algorithm for Matching Large Graphs. In: PAMI (2004)
3. Dost, B., Shlomi, T., Gupta, N., Ruppin, E., Bafna, V., Sharan, R.: QNet: A tool for querying protein interaction networks. In: Speed, T., Huang, H. (eds.) RECOMB 2007. LNCS (LNBI), vol. 4453, pp. 1–15. Springer, Heidelberg (2007)
4. Giugno, R., Shasha, D.: GraphGrep: A Fast and Universal Method for Querying Graphs. In: Proc. of ICPR (2002)
5. He, H., Singh, A.K.: Closure-tree: an index structure for graph queries. In: Proc. of ICDE (2006)
6. He, H., Singh, A.K.: Graphs-at-a-time: Query Language and Access Methods for Graph Databases. In: Proc. of SIGMOD (2008)
7. Jiang, H., Wang, H., Yu, P., Zhou, S.: Gstring: A novel approach for efficient search in graph databases. In: Proc. of ICDE (2007)
8. Shasha, D., Wang, J., Giugno, R.: Algorithmic and applications of tree and graph searching. In: PODS (2002)
9. Tian, Y., Patel, J.: TALE: A Tool for Approximate Large Graph Matching. In: Proc. of ICDE (2008)
10. Ullmann, J.: An algorithm for subgraph isomorphism. J. ACM (1976)
11. Viger, F., Latapy, M.: Efficient and simple generation of random simple connected graphs with prescribed degree sequence. In: Wang, L. (ed.) COCOON 2005. LNCS, vol. 3595, pp. 440–449. Springer, Heidelberg (2005)
    www.cs.sunysb.edu/ãlgorith/implement/viger/implement.shtml
12. Wang, X., Smalter, A., Huan, J., Lushington, G.: G-Hash: Towards Fast Kernel-based Similarity Search in Large Graph Databases. In: Proc. of EDBT (2009)
13. Williams, D., Huan, J., Wang, W.: Graph database indexing using structured graph decomposition. In: Proc. of ICDE (2007)
14. Xiao, Y., Wu, W., Pei, J., Wang, W., He, Z.: Efficiently indexing shortest paths by exploiting symmetry in graphs. In: Proc. of EDBT (2009)
15. Yan, X., Yu, P., Han, J.: Graph indexing, a frequent structure-based approach. In: Proc. of SIGMOD (2004)
16. Yan, X., Yu, P., Han, J.: Substructure similarity search in graph databases. In: Proc. of SIGMOD (2005)
17. Zhang, S., Li, J., Gao, H., Zou, Z.: A Novel Approach for Efficient Supergraph Query Processing on Graph Databases. In: Proc. of EDBT (2009)
18. Zhang, S., Hu, M., Yang, J.: Treepi: a new graph indexing method. In: Proc. of ICDE (2007)
19. Zhang, S., Li, S., Yang, J.: GADDI: Distance Index based Subgraph Matching in Biological Networks. In: Proc. of EDBT (2009)
20. Zhang, S., Yang, J., Jin, W.: SAPPER: Subgraph Indexing and Approximate Matching in Large Graphs. In: Proc. of VLDB (2010)
21. Zhao, P., Yu, J.X., Yu, P.S.: Graph indexing: Tree + delta $\geq$ graph. In: Proc. of VLDB (2007)

CloudVista: Visual Cluster Exploration for Extreme Scale Data in the Cloud

Keke Chen, Huiqi Xu, Fengguang Tian, and Shumin Guo

Ohio Center of Excellence in Knowledge Enabled Computing
Department of Computer Science and Engineering
Wright State University, Dayton, OH 45435, USA
{keke.chen,xu.39,tian.6,guo.18}@wright.edu

Abstract. The problem of efficient and high-quality clustering of extreme scale datasets with complex clustering structures continues to be one of the most challenging data analysis problems. An innovate use of data cloud would provide unique opportunity to address this challenge. In this paper, we propose the Cloud-Vista framework to address (1) the problems caused by using sampling in the existing approaches and (2) the problems with the latency caused by cloud-side processing on interactive cluster visualization. The CloudVista framework aims to explore the entire large data stored in the cloud with the help of the data structure *visual frame* and the previously developed VISTA visualization model. The latency of processing large data is addressed by the *RandGen* algorithm that generates a series of related visual frames in the cloud without user's intervention, and a hierarchical exploration model supported by cloud-side subset processing. Experimental study shows this framework is effective and efficient for visually exploring clustering structures for extreme scale datasets stored in the cloud.

1 Introduction

With continued advances in communication network technology and sensing technology, there is an astounding growth in the amount of data produced and made available throughout cyberspace. Cloud computing, the notion of outsourcing hardware and software to Internet service providers through large-scale storage and computing clusters, is emerging as a dominating technology and an economical way to host and analyze massive data sets. Data clouds, consisting of hundreds or thousands of cheap multi-core PCs and disks, are available for rent at low cost (e.g., Amazon EC2 and S3 services). Powered with distributed file systems, e.g., hadoop distributed file system [26], and MapReduce programming model [7], clouds can provide equivalent or better performance than traditional supercomputing environments for data intensive computing.

Meanwhile, with the growth of data volume, large datasets[1] will often be generated, stored, and processed in the cloud. For instance, Facebook stores and processes user activity logs in hadoop clusters [24]; Yahoo! used hadoop clusters to process web documents and generate web graphs. To explore such large datasets, we have to develop

---

[1] The concept of "large data" keeps evolving. with existing scales of data, roughly, we consider $< 10^3$ records to be small, $10^3 - 10^6$ to be medium, $10^6 - 10^9$ to be large, and $> 10^9$ to be extreme scale.

J.B. Cushing, J. French, and S. Bowers (Eds.): SSDBM 2011, LNCS 6809, pp. 332–350, 2011.

novel techniques that utilize the cloud infrastructure and its parallel processing power. In this paper we investigate the problem of large-scale data clustering analysis and visualization through innovative use of the cloud.

1.1 Challenges with Clustering Extreme Scale Data

A clustering algorithm tries to partition the records into groups with certain similarity measure [15]. While a dataset can be large in terms of the number of dimensions (dimensionality), the number of records, or both, a "large" web-scale data usually refer to those having multi-millions, or even billions of records. For example, one-day web search clickthrough log for a major commercial web search engine in US can have tens of millions of records. Due to the large volume of data, typical analysis methods are limited to simple statistics based on linear scans. When high-level analysis methods such as clustering are applied, the traditional approaches have to use data reduction methods.

Problems with Sampling
The three-phase framework, sampling/summarization → data analysis on sample data → postprocessing/validation is often applied to clustering large data in the single workstation environment (as shown in Figure 1).

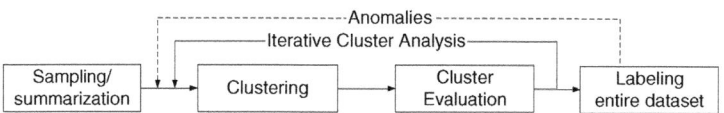

Fig. 1. Three phases for cluster analysis of large datasets

This framework can temporarily address some problems caused by large datasets in *limited scale*. For instance, dealing with complex clustering structures (often the case in many applications) may need clustering algorithms of nonlinear complexity or visual cluster analysis, which cannot be applied to the entire large dataset. With data reduction, the most costly iterative analysis is on the reduced data in the second phase, while we assume the number of iteration involving the three phases is small.

Due to the sampling or summarization phase there is a mismatch between the clustering structure discovered on the sample dataset and that on the entire dataset. To fully preserve the clustering structure, the sampling rate has to be higher than certain lower bound that is determined by the complexity of the clustering structure and the size of the dataset [11]. While the size of entire dataset keeps growing rapidly, the amount of data that the second phase can handle stays limited for a typical workstation, which implies a decreasing sample rate. The previous work in the three-phase visual cluster analysis framework [4] has addressed several problems in extending the clustering structure to the entire dataset under low sample rate, such as missing small clusters, abnormal visual cluster patterns, cluster boundary extension, and unclear secondary clustering structure.

These problems become more severe with lower sample rate. Therefore, new processing strategies are needed to replace the three-phase framework for extreme scale datasets.

Problems with Visual Cluster Exploration

Previous studies have shown that visual cluster exploration can provide unique advantages over automated algorithms [3,4]. It can help user decide the best number of clusters, identify some irregular clustering structures, incorporate domain knowledge into clustering, and detect errors.

However, visual cluster exploration on the data in the cloud brings extra difficulties. First, the visualization algorithm should be parallelizable. Classical visualization methods such as Principal Component Analysis and projection pursuit [13] involve complex computation, not easy to scale to large data in the parallel processing environment. Second, cloud processing is not optimized for low-latency processing [7], such as interactive visualization. It would be inappropriate to respond to each user's interactive operation with a cloud-based processing procedure, because the user cannot tolerate long waiting time after each mouse click. New visualization and data exploration models should be developed to fit the cloud-based data processing.

1.2   Scope and Contributions

We propose the cloud-based interactive cluster visualization framework, **CloudVista**, to address the aforementioned challenges for explorative cluster analysis in the cloud. The CloudVista framework aims to eliminate the limitation brought by the sampling-based approaches and reduce the impact of latency to the interactivity of visual cluster exploration.

Our approach explores the entire large data in the cloud to address the problems caused by sampling. CloudVista promotes a collaborative framework between the data cloud and the visualization workstation. The large dataset is stored, processed in the cloud and reduced to a key structure "visual frame", the size of which is only subject to the resolution of visualization and much smaller than an extreme scale dataset. Visual frames are generated in batch in the cloud, which are sent to the workstation. The workstation renders visual frames locally and supports interactive visual exploration.

The choice of the visualization model is the key to the success of the proposed framework. In the initial study, we choose our previously developed VISTA visualization model [3] for it has linear complexity and can be easily parallelized. The VISTA model has shown effectiveness in validating clustering structures, incorporating domain knowledge in previous studies [3] and handling moderately large scale data with the three-phase framework [4].

We address the latency problem with an automatic batch frame generation algorithm - the RandGen algorithm. The goal is to efficiently generate a series of meaningful visual frames without the user's intervention. With the initial parameter setting determined by the user, the RandGen algorithm will automatically generate the parameters for the subsequent visual frames, so that these frames are also continuously and smoothly changed. We show that the statistical properties of this algorithm can help identify the clustering structure. In addition to this algorithm, we also support a hierarchical exploration model to further reduce the cost and need of cloud-side processing.

We also implement a prototype system based on Hadoop/MapReduce [26] and the VISTA system [3]. Extensive experiments are conducted to study several aspects of the framework, including the advantages of visualizing entire large datasets, the performance of the cloud-side operations, the cost distribution between the cloud and the application server, and the impact of frame resolution to running time and visualization quality. The preliminary study on the prototype has shown that the CloudVista framework works effectively in visualizing the clustering structures for extreme scale datasets.

2   CloudVista: the Framework, Data Structure and Algorithms

CloudVista works differently from existing workstation-based visualization. Workstation-based visualization directly processes each record and renders the visualization after the visual parameters are set. In the CloudVista framework, we clearly divide the responsibilities between the cloud, the application server, and the client (Figure 2). The data and compute intensive tasks on large datasets are now finished in the cloud, which will generate the intermediate visual representations - the visual frames (or user selected subsets). The application server manages the visual frame/subset information, issues cloud processing commands, gets the results from the cloud, compresses data for transmission, and delivers data to the client. The client will render the frames, take care of user interaction, and, if the selected subsets are small, work on these small subsets directly with the local visualization system.

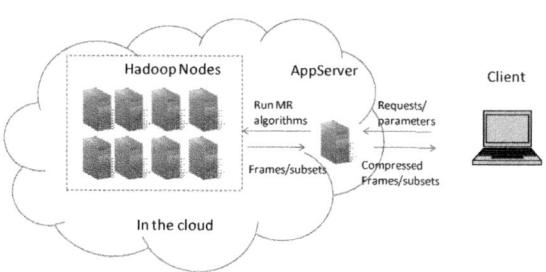

Fig. 2. The CloudVista framework

We describe the framework in three components: the VISTA visualization model, the key data structure "visual frame", and the major data processing and visualization algorithms. We will also include a cost analysis on cloud-side operations at the end of this section.

2.1   The VISTA Visualization Model

The CloudVista framework uses our previously developed VISTA visualization model [3] for it has linear complexity and can be easily parallelized. To make the paper self-contained, we describe the definition of this model and its properties for cluster visualization.

VISTA visualization model is used to map a k-dimensional point to a two dimensional point on the display. Let $\mathbf{s}_i \in \mathbb{R}^2$, $i = 1, \ldots, k$ be unit vectors arranged in a "star shape" around the origin on the display. $\mathbf{s}_i$ can be represented as $\mathbf{s}_i = (\cos(\theta_i), \sin(\theta_i))$, $\theta_i \in [0, 2\pi]$, i.e., uniquely defined by θ_i. Let a k-dimensional normalized data point $\mathbf{x} = (x_1, \ldots x_i, \ldots, x_k)$, $x_i \in [-1, 1]$ in the 2D space and $\mathbf{u} = (u_1, u_2)$ be $\mathbf{x}$'s image on the two dimensional display based on the VISTA mapping function. $\alpha = (\alpha_1, \ldots, \alpha_k)$, $\alpha_i \in [-1, 1]$ are dimensional weights and $c \in \mathbb{R}^+$ (i.e., positive real) is a scaling factor. Formula 1 defines the VISTA model:

$$f(\mathbf{x}, \alpha, \theta, c) = c \sum_{i-1}^{k} \alpha_i x_i \mathbf{s}_i. \tag{1}$$

α_i, θ_i, and c provide the adjustable parameters for this mapping. For simplicity, we leave θ_i to be fixed that equally partitions the circle, i.e., $\theta_i = 2i\pi/k$. Experimental results showed that adjusting α in $[-1, 1]$, combined with the scaling factor c is effective enough for finding satisfactory visualization [3,4].

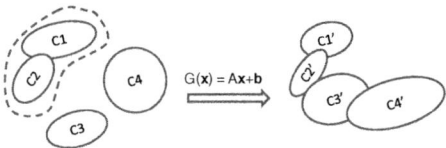

Fig. 3. Use a Gaussian mixture to describe the clusters in the dataset

This model is essentially a simple linear model with dimensional adjustable parameters α_i. The rationale behind the model is

Proposition 1. *If Euclidean distance is used as the similarity measure, an affine mapping does not break clusters but may cause cluster overlapping.*

Proof. Let's model arbitrary shaped clusters with a Gaussian mixture [8]. Let μ be the density center, and Σ be the covariance matrix of the Gaussian cluster. A cluster C_i can be represented with

$$\mathcal{N}_i(\mu_i, \Sigma_i) = \frac{1}{(2\pi)^{k/2}|\Sigma_i|^{1/2}} \exp\{-(\mathbf{x} - \mu_\mathbf{i})'\Sigma^{-1}(\mathbf{x} - \mu_\mathbf{i})/2\}$$

Geometrically, μ describes the position of the cluster and Σ describes the spread of the dense area. After an affine transformation, say $G(\mathbf{x}) = A\mathbf{x} + \mathbf{b}$, the center of the cluster is moved to $A\mu_i + \mathbf{b}$ and the covariance matrix (corresponding to the shape of dense area) is changed to $A\Sigma_i A^T$. And the dense area is modeled with $\mathcal{N}_i(A\mu_i + \mathbf{b}, A\Sigma_i A^T)$. Therefore, affine mapping does not break the dense area, i.e., the cluster. However, due to the changed shapes of the clusters, $A\Sigma_i A^T$, some clusters may overlap each other. As the VISTA model is an affine model, this proposition also applies to the VISTA model.   □

Since there is no "broken cluster" in the visualization, any visual gap between the point clouds reflects the real density gaps between the clusters in the original high-dimensional space. The only challenge is to distinguish the distance distortion and cluster overlapping introduced by the mapping. Uniquely different from other models, by tuning α_i values, we can scrutinize the multidimensional dataset visually from different perspectives, which gives dynamic visual clues for distinguishing the visual overlapping[2].

In addition, since this model is a record-based mapping function, it is naturally parallel and can be implemented with the popular parallel processing models such as MapReduce [7] for large scale cloud-based data processing. Therefore, we use the VISTA model in our framework. Note that our framework does not exclude using any other visualization model if it can efficiently implement the functionalities.

2.2   The Visual Frame Structure

A key structure in CloudVista is the *visual frame* structure. It encodes the visualization and allows the visualization to be generated in parallel in the cloud side. It is also a space-efficient data structure for passing the visualization from the cloud to the client workstation.

Since the visual representation is limited by display size, almost independent of the size of the original dataset, visualizing data is naturally a data reduction process. A rectangle display area for a normal PC display contains a fixed number of pixels, about one thousand by one thousand pixels[3]. Several megabytes will be sufficient to represent the pixel matrix. In contrast, it is normal that a large scale dataset may easily reach terabytes. When we transform the large dataset to a visual representation, a data reduction process happens, where the cloud computation model, e.g., MapReduce, can nicely fit in.

We design the visual representation based on the pixel matrix. The visual data reduction process in our framework is implemented as an aggregation process. Concretely, we use a two dimensional histogram to represent the pixel matrix: each cell is an aggregation bucket representing the corresponding pixel or a number of neighboring pixels (which is defined by the *Resolution*). All points are mapped to the cells and then aggregated. We name such a 2-D bucket structure as "visual frame". A frame can be described as a list of tuples $\langle u_1, u_2, d \rangle$, where (u_1, u_2) is the coordinate of the cell and $d > 0$ records the number of points mapped to the cell. The buckets are often filled sparsely, which makes the actual size of a frame structure is smaller than megabytes. Low resolution frame uses one bucket representing a number of neighboring pixels, which also reduces the size of frame.

Such a visual frame structure is appropriate for density-based cluster visualization, e.g., those based on the VISTA model. The following MapReduce code snippet describes the use of the visual frame based on the VISTA model.

---

[2] A well-known problem is that the VISTA model cannot visually separate some manifold structures such nested spherical surfaces, which can be addressed by using spectral clustering [19] as the preprocessing step.

[3] Note that special displays, such as NASA's hyperwall-2, needs special hardware, which are not available for common users, thus do not fit our research scope.

```
1: map(i, x)
2: i: record id, x: k-d record.
3: (u₁, u₂) ← f(x, α, θ, c);
4: EmitIntermediate((u₁, u₂), 1)

1: reduce((u₁, u₂), v)
2: (u₁, u₂): coordinate, v: list of counts.
3: d ← 0;
4: for each vᵢ in v do
5:     d ← d + vᵢ;
6: end for
7: Emit(⟨u₁, u₂, d⟩);
```

The VISTA visualization model maps the dense areas in the original space to separated or overlapped dense areas on the display. With small datasets, clusters are visualized as dense point clouds, where point-based visualization is sufficient for users to discern clustering structures. With large datasets, all points are crowded together on the display. As a result, point-based visualization does not work. We can use the widely adopted heatmap method to visualize the density information - the cells with high density are visualized with warmer colors. With the heatmap method, we can still easily identify clusters from the visualization. We will see some visualization results based on this design in Section 3.

2.3   Algorithms Improving Interactivity

In this section, we describe two major algorithms addressing the latency caused by cloud-side data processing. The first algorithm, RandGen, randomly generates a batch of related frames based on the first frame. The user can then explore the batch of frames locally with the workstation. To further reduce the effect of latency and the need of cloud-side operations, we also develop the algorithms supporting the hierarchical exploration model.

RandGen: Generating Related Frames in Batch. Visualization and dimension reduction techniques inevitably bring distance distortion and cause overlapped clusters in lower dimensional space. While it is possible to use algorithms to generate a set of "best" candidate visualization results as projection pursuit [5] does, it is often too costly for large data. Another approach is to allow the user to tune the visual parameters and observe the data in different perspectives to find the possible visual overlapping, which was employed by the VISTA system [3].

In the single workstation mode for medium-size data, the workstation can quickly respond to user's interactive operation and re-generate the visualization by applying the VISTA model to the entire dataset or sample data. However, this interactive model is not realistic if the data processing part is in the cloud. In this section, we develop the RandGen algorithm that can automatically generate a batch of related frames in the cloud based on the parameter setting for the first frame. The collection of frames are passed to the client and the user can spend most time to understand them locally in the workstation. We also prove that the batch of frames generated with RandGen can help users identify the clustering structure.

The RandGen algorithm is a random perturbation process that generates a collection of related frames. Starting from the initial α values that are given by the user, RandGen applies the following small stochastic updates to all dimensional weights simultaneously, which are still limited to the range -1 to +1. Let α_i^ϕ represent the α parameter for dimension i in frame ϕ, the new parameter $\alpha_i^{\phi+1}$ is defined randomly as follows.

$$\delta_i = t \times B,$$

$$\alpha_i^{\phi+1} = \begin{cases} 1 & \text{if } \alpha_i^\phi + \delta_i > 1 \\ \alpha_i^\phi + \delta & \text{if } \alpha_i^\phi + \delta_i \in [-1,1] \\ -1 & \text{if } \alpha_i^\phi + \delta_i < -1, \end{cases} \tag{2}$$

where t is a predefined step length, often set to small, e.g., $0.01 \sim 0.05$, and B is a coin-tossing random variable - with probability 0.5 it returns 1 or -1. δ_i is generated independently at random for each dimension. $\alpha_i^{\phi+1}$ is also bounded by the range $[-1,1]$ to minimize the out-of-bound points (those mapped out of the display). This process repeats until the α parameters for a desired number of frames are generated. Since the adjustment at each step is small, the change between the neighboring frames is small and smooth. As a result, sequentially visualized these frames will create continuously changing visualization. The following analysis shows why the RandGen algorithm can help identify visual cluster overlapping.

Identifying Clustering Patterns with RandGen. We formally analyze why this random perturbation process can help us identify the clustering structure. The change of visualization by adjusting α values can be described by the random movement of each visualized point. Let $\mathbf{v}_1$ and $\mathbf{v}_2$ be the images of the original data record $\mathbf{x}$ for the two neighboring frames, respectively. Then, the point movement is represented as

$$\Delta_{\mathbf{u}} = c \sum_{i=1}^{k} \delta_i x_i \mathbf{s}_i.$$

By definition of B, we have $E[\delta_i] = 0$. Since δ_i are independent of each other, we derive the expectation of $\delta_i \delta_j$

$$E[\delta_i \delta_j] = E[\delta_i]E[\delta_j] = 0, \text{ for } i \neq j.$$

Thus, it follows the expectation of point movement is zero: $E[\Delta_{\mathbf{u}}] = 0$. That means the point will randomly move around the initial position. Let the coordinate $\mathbf{s}_i$ be (s_{i1}, s_{i2}). We can derive the variance of the movement $\text{var}(\Delta_{\mathbf{u}}) =$

$$c^2 t^2 \text{var}(B) \begin{pmatrix} \sum_{i=1}^{k} x_i^2 s_{i1}^2 & \sum_{i=1}^{k} x_i^2 s_{i1} s_{i2} \\ \sum_{i=1}^{k} x_i^2 s_{i1} s_{i2} & \sum_{i=1}^{k} x_i^2 s_{i2}^2 \end{pmatrix} \tag{3}$$

There are a number of observations based on the variance. (1) The larger the step length t, the more actively the point moves; (2) As the values s_{ix} and s_{iy} are shared by all points, the points with larger vector length $\sum_{i=1}^{k} x_i^2$ tends to move more actively.

Since we want to identify cluster overlapping by observing point movements, it is more interesting to see how the relative positions change for different points. Let $\mathbf{w}_1$

and $\mathbf{w}_2$ be the images of another original data record $\mathbf{y}$ for the neighboring frames, respectively. With the previous definition of $\mathbf{x}$, the visual squared distance between the pair of points in the initial frame would be

$$\Delta_{\mathbf{w},\mathbf{v}}^{(1)} = ||\mathbf{w}_1 - \mathbf{v}_1||^2 = ||c\sum_{i=1}^{k}\alpha_i(x_i - y_i)\mathbf{s}_i||^2. \tag{4}$$

Then, the change of the squared distance between the two points is

$$\begin{aligned}\Delta_{\mathbf{w},\mathbf{v}} &= 1/c^2(\Delta_{\mathbf{w},\mathbf{v}}^{(2)} - \Delta_{\mathbf{w},\mathbf{v}}^{(1)})\\ &= (\sum_{i-1}\delta_i(x_i - y_i)s_{i1})^2 + (\sum_{i=1}\delta_i(x_i - y_i)s_{i2})^2\\ &+ 2(\sum_{i=1}\delta_i(x_i - y_i)s_{i1})(\sum_{i=1}\alpha_i(x_i - y_i)s_{i1})\\ &+ 2(\sum_{i=1}\delta_i(x_i - y_i)s_{i2})(\sum_{i=1}\alpha_i(x_i - y_i)s_{i2}).\end{aligned}$$

With the independence between δ_i and δ_j for $i \neq j$, $E(\delta_i) = 0$, $s_{i1}^2 + s_{i2}^2 = 1$, and $E^2[\delta_i] = t^2\mathrm{var}(B) = 0.25t^2$, it follows the expectation of the distance change is

$$E[\Delta_{\mathbf{w},\mathbf{v}}] = \sum_{i=1}^{k}E^2[\delta_i](x_i - y_i)^2 = 0.25t^2\sum_{i=1}^{k}(x_i - y_i)^2$$

where , i.e., the average change of distance is proportion to the original distance between the two points. That means, if points are distant in the original space, we will have higher probability to see them distant in the visual frames; if the points are close in the original space, we will more likely observe them move together in the visual frames. This dynamics of random point movement helps us identify possible cluster overlapping in a series of continuously changing visual frames generated with the RandGen method.

Bootstrapping RandGen and Setting the Number of Frames. One may ask how to determine the initial set of α parameters for RandGen. We propose a bootstrapping method based on sampling. In the bootstrapping stage, the cloud is asked to draw a number of samples uniformly at random (μ records, defined by the the user according to the client side's visual computing capacity). The user then locally explores the small subset to determine an interesting visualization, the α parameters of which are sent back for RandGen. Note that this step is used to explore the sketch of the clustering structure. Therefore, the problems with sampling we mentioned in Introduction are not important.

Another question is how many frames are appropriate in a batch for the RandGen algorithm. The goal is to have sufficient number of frames so that one batch is sufficient for finding the important cluster visualization for a selected subset (see the next section for the extended exploration model), but we also do not want to waste computing resources to compute excessive frames. In the initial study, we found this problem is sophisticated because it may involve the proper setting of the step length t, the complexity of the clustering structure, and the selection of the initial frame. In experiments, we will simply use 100 frames per batch. Thorough understanding of this problem would be an important task for our future work.

Supporting Hierarchical Exploration. A hierarchical exploration model allows the user to interactively explore the detail of any part of the dataset based on the current visual frame. Such an exploration model can also exponentially reduce the data to be processed and the number of operations to be performed in the cloud side.

We develop algorithms to support such an exploration model. Figure 4 shows the flowchart how the client interacts with the cloud side in this exploration model. Depending on the size of the selected subset of data (ν records), the cloud side may have different processing strategies. If the selected data is small enough to fit in the client's visualization capacity, i.e., μ records, the cloud will return the subset directly (Case 1). If the rate $\mu/\nu > \xi$, where ξ is an acceptable sampling rate set by the user, e.g., 5%, a uniform sampling is performed on the selected subarea in the cloud to get μ sample records (Case 2). In Case 1 and 2, the subsequent operations on the subset will be handled locally at the client side. Otherwise, if the rate $\mu/\nu < \xi$ that sampling is not an appropriate option, the cloud side will start the RandGen algorithm (Case 3). We will formally analyze the cloud-related cost based on this exploration model in Section 2.4.

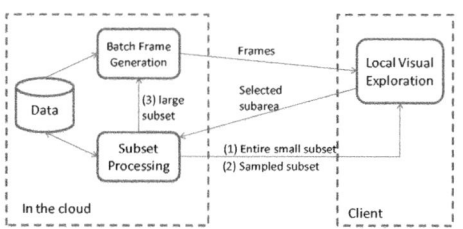

Fig. 4. Interactions between the client and the cloud

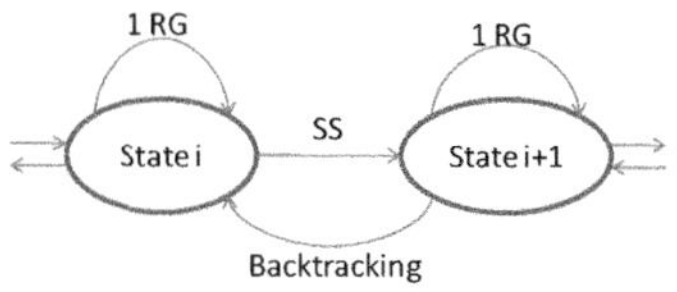

Fig. 5. State transition in terms of operations

The key operation, subset selection and sampling, should be supported in the cloud. The definition of the selected subset is derived based on the user selected subarea on the current visual frame, and then passed to the cloud together with other visualization parameters. We design a MapReduce algorithm to filter out the selected records based on the area definition. The sampling step can also be appropriately integrated into this step. The details of the algorithms are skipped due to the space limitation.

2.4   A Cost Model for CloudVista

In cloud computing, an important problem is resource provisioning [1]. To understand the interactivity of the system, it is also important to estimate how frequently an exploration will be interrupted for getting results from the cloud. In this section, we will model the exploration process with a Markov chain and derive an estimate to the number of cloud-side operations. The average cost of each operation will be studied in experiments.

The cloud-client interaction can be roughly represented with a Markov chain. Figure 5 shows two sample states of the chain; other states are similarly modeled. The user's interactive exploration can be described as a number of drill-downs on the interested visual areas. Thus, the length of the chain is correlated the number of cloud operations. If the user starts with the state i, she/he may require a RandGen (RG) operation for which the size of data keeps unchanged - let's denote it N_i. Or, she/he can perform a subset selection (SS) to drill down, which moves to the state $i+1$ and the size of dataset is changed to N_{i+1}, correspondingly. This chain extends until the subset can be fully handled locally.

We estimate the length of the chain as follows. Assume a visualization covers n cells, i.e., the aggregation buckets, on the display area on average, and thus the average density of the cells is N_i/n for state i. We also assume the area the user may select for subsect exploration is about λ percentage of the n cells. So the size of data at state $i+1$ is $N_{i+1} \approx \lambda N_i$. It follows $N_{i+1} = \lambda^{i+1} N_0$. We have defined the client's visualization capacity μ and the acceptable sampling rate ξ. For N_{i+1} records to be handled fully locally by the client, the boundary condition will be $N_i > \mu/\xi$ and $N_{i+1} \leq \mu/\xi$. Plugging $N_{i+1} = \lambda^{i+1} N_0$ into the inequalities, we get

$$\log_\lambda \frac{\mu}{\xi N_0} - 1 \leq i < \log_\lambda \frac{\mu}{\xi N_0},$$

i.e., $i = \lfloor \log_\lambda \frac{\mu}{\xi N_0} \rfloor$. Let the critical value be $\rho = i + 1$. Assume only one RandGen with sufficient number of frames is needed for each state. Since the number of interesting subareas for each level are quite limited, denoted by κ, the total number of cloud operations is $O(\kappa\rho)$. A concrete example may help us better understand the number ρ. Assume the client's visualization capacity is 50,000 records, there are 500 million records in the entire dataset, the acceptable sampling rate is 5%, and each time we select about 20% visual area, i.e., $\lambda = 0.2$, to drill down. We get $\rho = 4$. Therefore, the number of interrupts caused by cloud operations can be quite acceptable for an extreme scale dataset.

3   Experiments

The CloudVista framework addresses the sampling problem with the method of exploring whole dataset, and the latency problem caused by cloud data processing with the RandGen algorithm and the hierarchical exploration model. We conduct a number of experiments to study the unique features of the framework. First, we show the advantages of visualizing the entire large data, compared to the visualization of sample data. Second, we investigate how the resolution of the visual frame may affect the quality of visualization, and whether the RandGen can generate useful frames. Third, we present the performance study on the cloud operations. The client-side visual exploration system (the VISTA system) has been extensively studied in our previous work [3,4]. Thus, we skip the discussion on the effectiveness of VISTA cluster exploration, although the frame-based exploration will be slightly different.

3.1   Setup

The prototype system is setup in the in-house hadoop cluster. This hadoop cluster has 16 nodes: 15 worker nodes and 1 master node. The master node also serves as the application server. Each node has two quad-core AMD CPUs, 16 GB memory, and two 500GB hard drives. These nodes are connected with a gigabit ethernet switch. Each worker node is configured with eight map slots and six reduce slots, approximately one map slot and one reduce slot per core as recommended in the literature. The client desktop computer can comfortably handle about 50 thousands records within 100 dimensions as we have shown [4].

To evaluate the ability of processing large datasets, we extend two existing large scale datasets to larger scale for experiments. The following data extension method is used to preserve the clustering structure for any extension size. First, we replace the categorical attributes (for KDD Cup data) with a sequence of integers (starting from 0), and then normalize each dimension[4]. For a randomly selected record from the normalized dataset, we add a random noise (e.g., with normal distribution $N(0, 0.01)$) to each dimensional value to generate a new record and this process repeats for sufficient times to get the desired number of records. In this way the basic clustering structure is preserved in the extended datasets. The two original datasets are (1) **Census 1990 data** with 68 attributes and (2)**KDD Cup 1999 data** with 41 attributes. The KDD Cup data also includes an additional label attribute indicating the class of each record. We denote the extended datasets with Census_{ext} and KDD_{ext} respectively.

3.2 Visualizing the Whole Data

In this experiment, we perform a comparative study: analyzing the visualization results generated with the original VISTA system and the CloudVista framework, on sample datasets and on the entire dataset, respectively. The experiment uses two Census datasets: a sample set of 20,000 records for the VISTA system and an extended dataset of 25 million records (5.3 GB in total) for the CloudVista.

Figure 6 shows the clustering structure with the VISTA system[5]. There are three major clusters - the dense areas in the visualization. This result has been validated with the BestK plot method [4]. Since the Census dataset has been discretized, i.e., all continuous domains are partitioned and discretized, categorical clustering analysis is also applicable. We apply the categorical cluster validation method: BestK plot method to find the best clustering structure [4], which confirms the result visualized in Figure 6. The BestK plot on 1,000 samples shows that the optimal clustering structure has three clusters and a secondary structure has two (these two clustering structures is a part of the hierarchical clustering structure, i.e., two of the three clusters are more similar (closer) to each other than to the third cluster).

Correspondingly, the visualization result in Figure 6 also shows a hierarchical structure based on density: there are three clusters C1, C2.1, and C2.2, while C2.1 and C2.2 are close to each other to form the secondary clustering structure. Except for these major clustering structures, on Figure 6 we have questions about other structural features: (1) Can we confirm that C1 consists of many small clusters? (2) Are there possibly small clusters or outliers between C1 and C2.2? (3) How closely are C2.1 and C2.2 related? These questions are unclear under the visualization of the sample data.

---

[4] The commonly used methods include max-min normalization or transforming to standard normal distribution.

[5] The dark circles, lines, and annotations are not a part of the visualization (for both Figure 6 and 7). They are manually added to highlight the major observations.

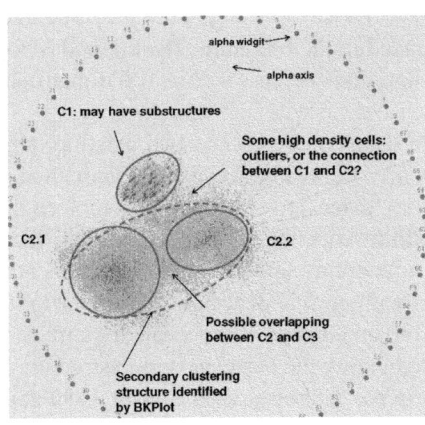

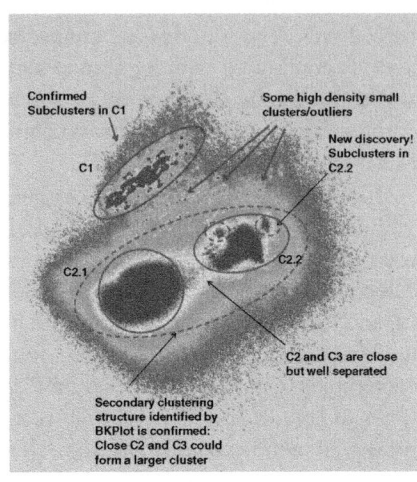

Fig. 6. Visualization and Analysis of Census data with the VISTA system

Fig. 7. Visualization and Analysis of 25 Million Census records (in 1000x1000 resolution)

To compare the results, we use the same set of α parameters as the starting point and generate a series of frames with small step length (0.01) on the 25 million records with the CloudVista framework. Figure 7 shows one of these frames. We can answer the above question more confidently with the entire dataset. (1) C1 indeed consists of many small clusters. To further understand the relationship between them, we may need to drill down C1. (2) Small clusters are clearly observed between C1 and C2.2. (3) C2.1 and C2.2 are closed related, but they are still well separated. It is also confirmed that the margin between C1 and C2.x is much larger and clearer than that between C2.1 and C2.2, which is consistent with the secondary structure identified by BKPlot. In addition, we also find some small sub-clusters inside C2.2, which cannot be observed in Figure 6.

We summarize some of the advantages of visualizing entire large data. First, it can be used to identify the small clusters that are often undetectable with sample dataset; Second, it helps identifying delicate secondary structures that are unclear in sample data. Sample data has its use in determining the major clustering structure.

3.3 Usefulness of Frames Generated by RandGen

We have shown the statistical properties of the RandGen algorithm. In a sufficient number of randomly generated frames by RandGen, the user will find the clustering pattern in the animation created by playing the frames and distinguish potential visual cluster overlaps. We conduct experiments on both the $Census_{ext}$ and KDD_{ext} datasets with the batch size set to 100 frames. Both the random initial frame and the bootstrapping initial frame are used in the experiments. We found in five runs of experiments, with this number of frames, we could always find satisfactory visualization showing the most

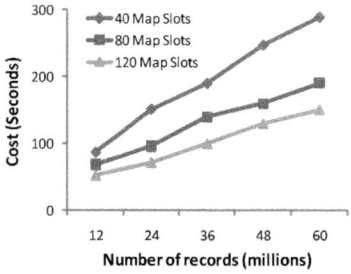

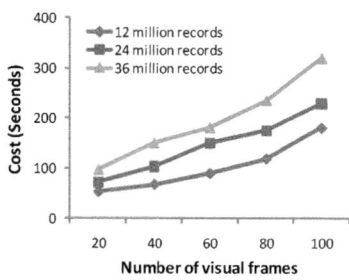

Fig. 8. Running time vs data size for Census$_{ext}$ data (RandGen for 100 frames)

Fig. 9. Running time vs the number of frames for Census$_{ext}$ data

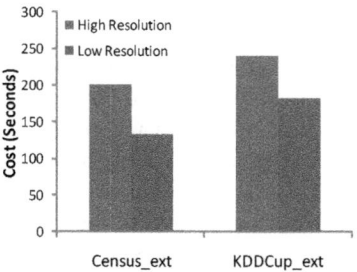

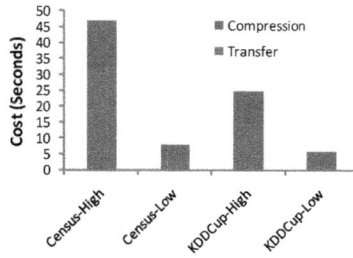

Fig. 10. Cloud processing time vs resolutions for RandGen (100 frames, Census_ext: 25 Million records, KDD-Cup_ext: 40 Million records)

Fig. 11. Cost breakdown (data transfer + compression) in app server processing (100 frames, Census-*: 25 Million records, KDDCup-*: 40 Million records, *-high: 1000x1000 resolution, *-low: 250x250 resolution)

detailed clustering structure. The video at http://tiny.cc/f6d4g shows how the visualization of Census$_{ext}$ (with 25 millions of records) changes by playing the 100 frames continuously.

3.4 Cost Evaluation on Cloud-Side Data Processing

In this set of experiments, we study the cost of the two major cloud operations: the RandGen algorithm and subset processing. We also analyze the cost distribution between the cloud and the app server.

Lower resolution can significantly reduce the size of the frame data, but it may miss some details. Thus, it represents a potential tradeoff between system performance and visual quality. Figure 7 in previous discussion is generated with 1000x1000 resolution, i.e., 1 aggregation cell for 1 pixel. Comparing with the result of 250x250 resolution, we find the visual quality is slightly reduced, but the major clustering features are well preserved for the $Census_{ext}$ data. Reducing resolution could be an acceptable method to achieve better system performance. We will also study the impact of resolution to the performance.

RandGen: Figure 8 demonstrates the running time of MapReduce RandGen algorithm with different settings of map slots for the extended census data. We control the number of map slots with Hadoop's fair scheduler. We set 100 reduces corresponding to 100 frames in a batch for all the testing cases[6]. Note that each number in the figures is the average of 5 test runs. The variance is small compared to the average cost and thus ignored in the figures. The running time shows that the MapReduce RandGen algorithm is about linearly scalable in term of data size. With increasing number of map slots, the cost also decreases proportionally. Figure 9 shows the cost also increases about linearly within the range of 100 frames.

We then study the cost distribution at the server side (cloud + application server). The total cost is split into three parts: cloud processing, transferring data to app server from the cloud, and compressing. The following settings are used in this experiment. For RandGen of 100 frames, we compare two extended datasets: 25 million records of Census ($Census_{ext}$) data and 40 million records of KDD Cup (KDD_{ext}) data on 15 worker nodes. The results are generated in two resolutions: 1000x1000 (aggregation bucket is 1x1 pixel) and 250x250 (aggregation bucket is 4x4 pixels), respectively. Since the cloud processing cost dominates the total cost, we present the costs in two figures. Figure 10 shows the cost of cloud processing. KDD_{ext} takes more time since its data size is much larger. Also, lower resolution saves a significant amount of time. Figure 11 shows the cost breakdown at the app server, where the suffixes of the x-axis names: "-L" and "-H" mean low and high resolutions, respectively. Interestingly, although KDD_{ext} data takes more time in cloud processing, it actually returns less data in frames, which implies a smaller number of cells are covered by the mapped points. By checking the high-resolution frames, we found there are about 320 thousands of covered cells per frame for census data, while only 143 thousands for KDD cup data, which results in the cost difference in app server processing.

Table 1 summarizes the statistics for different resolutions. We use the amount of data generated by the cloud to represent the communication cost between the cloud and the client (the "compressed data" in Table 1). "Frame size" represents the average number of covered aggregation buckets in each frame; "total time" is the sum of times for cloud processing, transferring from the cloud to the app server, and compressing data. It shows low resolution will have significant cost saving. Low resolution visualization will be very appropriate for exploring higher level clustering structure, where details are less important.

Subset Processing: Subset exploration results in three possible operations: subset Rand-Gen, subset fetching and subset sampling. We have analyzed the number of cloud operations based on the hierarchical exploration model. In this experiment, we let a trained user interactively select interested high-density spots in the frames generated

---

[6] We realized this is not an optimal setting, as only 90 reduce slots available in the system, which means 100 reduce processes need to be scheduled in two rounds in the reduce phase.

Table 1. Summary of the RandGen experiment.

	resolution	frame size	compressed frames	total time(sec)
Census$_{ext}$	High	320K	100MB	247
	Low	25K	9.7MB	141
KDD$_{ext}$	High	143K	45MB	265
	Low	12K	4.6MB	188

with RandGen and then evaluate how many each of the three operations may be triggered. In each round, 100 frames are generated in each batch with 15 worker nodes on 5.3GB Census$_{ext}$ data or 13.5GB KDD$_{ext}$ data in high resolution. The user browses the frames and randomly selects the high-density subarea to drill down. Totally, 60 drill-down operations are recorded for each dataset.

We summarize the result in Table 2. "Size of Selected Area" represents the average size of the selected area with $\pm$ representing the standard deviation. "Direct" means the number of subsets that will be fully fetched. "Sampling" means the number of subsets that can be sampled. "SS-RG" means the number of subsets, the sizes of which are too large to be sampled - the system will perform a subset RandGen to preserve the structure. "D&S Time" is the average running time (seconds) for each "Direct" or "Sampling" operation in the cloud side processing, excluding the cost of SS-RG, since we have evaluated the cost of RandGen in Table 1.

Table 2. Summary of the subsect selection experiment

	Size of Selected Area	# of Cloud Operations			D&S Time(sec)
		Direct	Sampling	SS-RG	
Census$_{ext}$	13896 $\pm$ 17282	4	34	22	36
KDD$_{ext}$	6375$\pm$9646	9	33	18	43

Interestingly, the selected areas are normally small: on average about 4% of the entire covered area for both datasets. Most selections, specifically, 63% for Census$_{ext}$ and 70% for KDD$_{ext}$ data, can be handled by "Direct" and "Sampling" and their costs are much less than RandGen.

4   Related Work

Most existing cluster visualization methods cannot scale up to large datasets due to their visual design. Parallel Coordinates [14] uses lines to represent multidimensional points. With large data, the lines are stacked together, cluttering the visual space. Its visual design also does not allow a large number of dimensions to be visualized. Scatter plot matrix and HD-Eye [12] are based on density-plots of pairwise dimensions, which are not convenient for finding the global clustering structure and are not scale to the number of dimensions. Star Coordinates [16] and VISTA [3] models are point-based models and have potential to be extended to handle really large datasets - the work described in the paper is based on the VISTA visualization model. IHD [27] and Hierarchical Clustering Explorer [23] are used to visualize the clustering structures discovered by clustering algorithms, which are different from our purpose of using the visualization system to discover clusters.

Cluster visualization is also a dimensionality reduction problem in the sense that it maps the original data space to the two dimensional visual space. The popularly used dimensionality reduction algorithms such as Principal Component Analysis and Multidimensional Scaling [6] have been applied in visualization. These methods, together with many dimensionality reduction algorithms [21,22], are often costly - nonlinear to the number of records and thus they are not appropriate for large datasets. FastMap [9] addresses the cost problem for large datasets, but the choice of pivot points in the mapping may affect the quality of the result. Random projection [25] only preserves pairwise distances approximately on average and the precision is subject to the number of projected dimensions - the lower projected dimensions the worse precision. Most importantly, all of these dimensionality reduction methods do not address the common problems - how to detect and understand distance distortion and cluster overlapping. The projection-based methods such as Grand Tour and Projection Pursuit [5] allow the user to interactively explore multiple visualizations to discover possible distance distortion and cluster overlapping, but they are too costly to be used for large datasets. The family of star coordinates systems [16,3] address the visual distortion problem with a more efficient way, which is also the basis of our approach. The advantage of stochastic animation in finding patterns, as we do with RandGen, is also explored in graph visualization [2]

The three-phase framework "sampling or summarization – clustering/cluster analysis – disk labeling" is often used to incorporate the algorithms of high time complexity in exploring large datasets. As the size of data grows to very large, the rate between the size of the sampled or summarized dataset to the original size becomes very small, affecting the fidelity of the preserved clustering structure. Some clustering features such as small clusters and the connection between closely related clusters are not easy to be discovered with the sample set [4]. Therefore, there is a need to explore the entire large dataset.

Recently, several data mining algorithms have been developed in the cloud, showing that the hadoop/MapReduce [7] infrastructure is capable to reliably and efficiently handle large-scale data intensive problems. These instances include PLANET [20] for tree ensemble learning, PEGASUS [17] for mining peta-scale graphs, and text mining with MapReduce [18]. There is also an effort on visualizing scientific data (typically, low dimensional) with the support of the cloud [10]. However, none has been reported on visualizing multidimensional extreme scale datasets in the cloud.

5   Conclusion

The existing three-phase framework for cluster analysis on large scale data has reached its limits for extreme scale datasets. The cloud infrastructure provides a unique opportunity to address the problem of scalable data analysis - terabytes or even petabytes of data can be comfortably processed in the cloud. In this paper, we propose the Cloud-Vista framework to utilize the ability of scalable parallel processing power of the cloud, and address the special requirement of low-latency for user-centered visual analysis.

We have implemented the prototype system based on the VISTA visualization model and Hadoop/MapReduce. In experiments, we carefully evaluate the unique advantages of the framework for analyzing the entire large dataset and the performance of cloud-side algorithms. The initial results and the prototype system have shown this framework works effectively for exploring large datasets in the cloud. As a part of the future work, we will continue to study the setting of the batch size for RandGen and experiment with larger hadoop cluster.

References

1. Armbrust, M., Fox, A., Griffith, R., Joseph, A.D., Katz, R., Konwinski, A., Lee, G., Patterson, D., Rabkin, A., Stoica, I., Zaharia, M.: Above the clouds: A berkeley view of cloud computing. Technical Report, University of Berkerley (2009)
2. Bovey, J., Rodgers, P., Benoy, F.: Movement as an aid to understanding graphs. In: IEEE Conference on Information Visualization, pp. 472–478. IEEE, Los Alamitos (2003)
3. Chen, K., Liu, L.: VISTA: Validating and refining clusters via visualization. Information Visualization 3(4), 257–270 (2004)
4. Chen, K., Liu, L.: iVIBRATE: Interactive visualization based framework for clustering large datasets. ACM Transactions on Information Systems 24(2), 245–292 (2006)
5. Cook, D., Buja, A., Cabrera, J., Hurley, C.: Grand tour and projection pursuit. Journal of Computational and Graphical Statistics 23, 155–172 (1995)
6. Cox, T.F., Cox, M.A.A.: Multidimensional Scaling. Chapman&Hall/CRC, Boca Raton, FL (2001)
7. Dean, J., Ghemawat, S.: MapReduce: Simplified data processing on large clusters. In: USENIX Symposium on Operating Systems Design and Implementation (2004)
8. M.J. (ed.) (1998)
9. Faloutsos, C., Lin, K.-I.D.: FastMap: A fast algorithm for indexing, data-mining and visualization of traditional and multimedia datasets. In: Proceedings of ACM SIGMOD Conference, pp. 163–174 (1995)
10. Grochow, K., Howe, B., Barga, R., Lazowska, E.: Client + cloud: Seamless architectures for visual data analytics in the ocean sciences. In: Proceedings of International Conference on Scientific and Statistical Database Management, SSDBM (2010)
11. Guha, S., Rastogi, R., Shim, K.: CURE: An efficient clustering algorithm for large databases. In: Proceedings of ACM SIGMOD Conference, pp. 73–84 (1998)
12. Hinneburg, A., Keim, D.A., Wawryniuk, M.: Visual mining of high-dimensional data. In: IEEE Computer Graphics and Applications, pp. 1–8 (1999)
13. Huber, P.J.: Projection pursuit. Annals of Statistics 13(2), 435–475 (1985)
14. Inselberg, A.: Multidimensional detective. In: IEEE Symposium on Information Visualization, pp. 100–107 (1997)
15. Jain, A., Murty, M., Flynn, P.: Data clustering: A review. ACM Computing Surveys 31, 264–323 (1999)
16. Kandogan, E.: Visualizing multi-dimensional clusters, trends, and outliers using star coordinates. In: Proceedings of ACM SIGKDD Conference, pp. 107–116 (2001)
17. Kang, U., Tsourakakis, C.E., Faloutsos, C.: Pegasus: Mining peta-scale graphs. Knowledge and Information Systems, KAIS (2010)
18. Lin, J., Dyer, C.: Data-intensive text processing with MapReduce. Morgan & Claypool Publishers, San Francisco (2010)
19. Ng, A.Y., Jordan, M.I., Weiss, Y.: On spectral clustering: Analysis and algorithm. In: Proceedings Of Neural Information Processing Systems NIPS (2001)

20. Panda, B., Herbach, J.S., Basu, S., Bayardo, R.J.: Planet: Massively parall learning of tree ensembles with mapreduce. In: Proceedings of Very Large Databases Conference, VLDB (2009)
21. Roweis, S.T., Saul, L.K.: Nonlinear dimensionality reduction by locally linear embedding. Science 290(5500), 2323–2326 (2000)
22. Saul, L.K., Weinberger, K.Q., Sha, F., Ham, J., Lee, D.D.: Spectral methods for dimensionality reduction. In: Semi-Supervised Learning. MIT Press, Cambridge (2006)
23. Seo, J., Shneiderman, B.: Interactively exploring hierarchical clustering results. IEEE Computer 35(7), 80–86 (2002)
24. Thusoo, A., Shao, Z., Anthony, S., Borthakur, D., Jain, N., Sen Sarma, J., Murthy, R., Liu, H.: Data warehousing and analytics infrastructure at facebook. In: Proceedings of ACM SIGMOD Conference, pp. 1013–1020. ACM, New York (2010)
25. Vempala, S.S.: The Random Projection Method. American Mathematical Society (2005)
26. White, T.: Hadoop: The Definitive Guide. O'Reilly Media, Sebastopol (2009)
27. Yang, J., Ward, M.O., Rundensteiner, E.A.: Interactive hierarchical displays: a general framework for visualization and exploration of large multivariate datasets. Computers and Graphics Journal 27, 265–283 (2002)

Efficient Selectivity Estimation by Histogram Construction Based on Subspace Clustering

Andranik Khachatryan, Emmanuel Müller, Klemens Böhm, and Jonida Kopper

Institute for Program Structures and Data Organization (IPD)
Karlsruhe Institute of Technology (KIT), Germany
{khachatryan,emmanuel.mueller,klemens.boehm}@kit.edu

Abstract. Modern databases have to cope with multi-dimensional queries. For efficient processing of these queries, query optimization relies on multi-dimensional selectivity estimation techniques. These techniques in turn typically rely on histograms. A core challenge of histogram construction is the detection of regions with a density higher than the ones of their surroundings. In this paper, we show that subspace clustering algorithms, which detect such regions, can be used to build high quality histograms in multi-dimensional spaces. The clusters are transformed into a memory-efficient histogram representation, while preserving most of the information for the selectivity estimation. We derive a formal criterion for our transformation of clusters into buckets that minimizes the introduced estimation error. In practice, finding optimal buckets is hard, so we propose a heuristic. Our experiments show that our approach is efficient in terms of both runtime and memory usage. Overall, we demonstrate that subspace clustering enables multi-dimensional selectivity estimation with low estimation errors.

1   Introduction

Query optimization is an essential component of every database management system. The optimizer relies on accurate size estimates of sub-queries. To this end, the optimizer estimates the selectivity of query predicates, i.e., the number of tuples that satisfy the predicate. A predicate can refer to several attributes. If these attributes are correlated, statistics on their joint distribution are essential to come up with a good estimate. Multi-dimensional histograms are a prominent class of such multi-attribute statistics [15,23,12,1,7,9,26,11,19]. However, with increasing number of dimensions they fall prey to the so-called "curse of dimensionality" [8]. The construction costs of the histogram increase, and the precision decreases. In particular, memory efficiency is a major challenge, as histograms need to fit in a few disk pages [23,9,26].

Self-tuning histograms are state-of-the-art methods for selectivity estimation. They use the query execution results (feedback) to refine themselves [9,26,11,19]. They focus on the refinement steps during query processing to achieve high precision, arguing that even a good initial configuration provides only a short-term benefit. Thus, a central hypothesis with self-tuning histograms has been that

J.B. Cushing, J. French, and S. Bowers (Eds.): SSDBM 2011, LNCS 6809, pp. 351–368, 2011.

their refinement techniques are enough to ensure high precision, while initialization is a minor tweak. We show that this is only one side of the coin: Doing without initialization techniques has a serious drawback. First, histograms need many queries in order to adapt to the data set. Second, even given a large number of queries to train, the uninitialized histogram still cannot match the precision of our initialized version.

Another problem with multi-dimensional histograms is that they focus on capturing correlated data regions in full-dimensional space. This can be wasteful, because in different regions of the data space only a small set of attributes may be correlated, while additional attributes only add noise. Thus, traditional selectivity estimation methods spend too much memory and achieve only low estimation quality. Similar challenges have been observed for traditional clustering and solved by recent subspace clustering techniques.

In this paper we focus on pre-processing steps to initialize a self-tuning histogram based on subspace clustering. Having detected high density regions (dense subspace clusters) with many objects we build memory-efficient histograms based on this. We make use of subspace clustering as a novel data mining paradigm. As highlighted by a recent study [21], subspace clustering can detect groups of objects in any projection of high-dimensional data. In particular, the resulting subspace clusters represent dense regions in projections of the data and capture the local correlation of attribute sets for each cluster individually. Thus, our hypothesis is that these dense subspace regions can be utilized to initialize the histogram and enable good selectivity estimations. Our experiments confirm this hypothesis. In order to initialize a histogram with dense subspace regions we have to transform subspace clusters into efficient histogram structures, in order to meet the memory constraints of the histogram. There are various ways to do this, but any transformation introduces some estimation error. We need to minimize this error. To this end, we formally derive a criterion which lets us compare different transformations and choose the better one. We also define special classes of transformations with interesting properties. For these classes, we are able to compute the transformation which is best according the aforementioned criterion. We show however that finding the optimal solution is too expensive in the general case. We propose an efficient heuristic with high quality estimation.

As mentioned above, a central hypothesis with self-tuning histograms has been that initialization yields only a short-term benefit [26,9]. We use six subspace clustering algorithms [21] to initialize a histogram, and use the uninitialized version as a baseline. We make the following important observations:

1. *Good* initialization makes a difference. One out of the six methods we tried has shown consistent improvement of estimation precision over the uninitialized version, even after a significant number of training queries.
2. Self-tuning histograms can achieve high precision using refinement. Namely, an uninitialized self-tuning histogram was able to catch up with the remaining five initialized histograms, each based on different subspace clustering paradigm.

A related observation is that initialized histograms need less memory to provide similar or even better estimation quality. Through different evaluation settings, they need about 1/4 to 1/8 of the memory required in uninitialized histograms to produce estimates of the same precision.

2 Related Work

Static Selectivity Estimation
Multi-dimensional histograms capture the joint distribution of attributes for selectivity estimation [23,12,7,27,22]. The MHist histogram [23] recursively partitions the data set, starting with one bucket which represents the whole data set. In each iteration it splits an existing bucket into two, until the storage space is exhausted. The bucket to split and the split dimension are chosen greedily, based on one-dimensional histograms. HiRed histograms [7] split each bucket into half across all dimensions, then compute an error measure for the split. If it does not exceed a certain threshold, the split is discarded, otherwise it becomes permanent. GENHIST [12] partitions the data set into a fine-grained grid structure, identifies the dense parts in the data set, and assigns them to buckets. Then it removes those data points and starts over again. GENHIST provides good selectivity estimates but is expensive to construct and store. [7] shows that given a fixed memory budget, the simpler HiRed histogram can outperform GENHIST. A problem which all static histograms share is that they need to be rebuilt regularly to reflect changes in the data set.

Self-Tuning Selectivity Estimation
Self-tuning histograms [9,1,26] address this problem by using the query workload to adjust themselves. This leads to histograms which are more sensitive to the query workload. If the user accesses parts of the data set frequently, the histogram automatically becomes more detailed for this part. However, information about previously unaccessed regions remains coarse. We introduce the general idea of [9] in Section 3. Some significant improvements of this "base" version have been proposed, but they are orthogonal to our work. For instance, [19] provides a method which enhances the estimation precision near histogram-bucket boundaries, using interpolation methods. [11] shows how the histogram can be compressed significantly by aligning buckets and quantizing coordinates relative to parent bucket coordinates. Finally, [26] introduces a possibly more efficient selectivity estimation procedure which relies less on feedback and uses information-theoretic measures such as entropy to compute tuple densities.

As [9] shows, if enough queries to adapt to the data set are given, the self-tuning histograms will perform as well as the best static histograms. From here, an implicit assumption is made that initializing a self-tuning histogram would merely provide a short-term benefit. However, we show that this is not always the case. Initialized histograms based on subspace clustering provide better estimates even after a large number of queries have been executed. This is because self-tuning histograms construct and compress buckets greedily, and greedy algorithms are often sensitive to initial values. However, it has never be confirmed due to the strong belief in the refinement step in self-tuning approaches. In contrast to this, we will show in the following sections how some subspace clustering algorithms can provide significant quality improvements by initialization of self-tuning histograms.

Overall, all of these traditional selectivity estimation techniques have missed the detection of locally relevant attributes. Histograms should have been adapted to the local correlation of objects in each of the dense regions. In particular, since initialization techniques have been a topic of minor interest, we observe limited quality improvement in today's self-tuning methods. They are unable to capture the intrinsic structure of the data by their histogram construction.

Clustering Algorithms
Clustering is an unsupervised data mining task for grouping of objects based on mutual similarity [14]. As an unsupervised task, it reveals the intrinsic structure of a data set without prior knowledge about the data. Clusters share a core property with histograms, as they represent dense regions covering many objects. However, the detection of meaningful clusters in high-dimensional data spaces is hindered by the "curse of dimensionality" as well [8]. Irrelevant attributes obscure the patterns in the data. Global dimensionality techniques such as Principle Components Analysis (PCA) try to reduce the number of attributes [16]. However, the reduction may yield only a clustering in one reduced space. With locally varying attribute relevance, this means that clusters that do not show up in the reduced space will be missed.

Recent years have seen increasing research in subspace clustering, which aims at identifying locally relevant attribute sets for each cluster. Subspace clustering was introduced in the CLIQUE approach, which detects dense grid cells in subspace projections [3]. In the past decade, several approaches have extended this clustering paradigm [25,24,28]. Furthermore, traditional clustering, such as the k-medoid and DBSCAN algorithm, have been extended to cope with subspace projections [2,17]. For example, density-based subspace clustering detects arbitrarily-shaped clusters in projections of the data space [17,5,6,20]. Overall, various cluster definitions focusing on different objective functions have been proposed.

In the following we will show how these clusterings can be transformed into a memory-efficient histogram and provide high quality initializations for selectivity estimation. As a common property, we use the ability of subspace clustering to detect high density regions in projections of high-dimensional data. We abstract from specific clustering models, focusing on compactness, density-connected and arbitrarily shaped clusterings. This makes our general transformation of subspace clusters into histograms applicable to a wide range of subspace clustering algorithms.

3 Selectivity Estimation Overview

We now briefly describe the state-of-the-art selectivity estimation technique STHoles [9], on which our work is based on. In general, selectivity estimation tries to estimate the number of objects satisfying the query predicate. Following [15,9,26,12,23] we consider conjunctive query predicates which refer to several attributes of the relation. Given a relation $R = (A1, A2, \ldots, Ad)$ in a d-dimensional data space, a *range query* on this relation has the form

```
SELECT <attr-list>
FROM   R
WHERE  l1 <= A1 <= h1 AND ...  ld <= Ad <= hd
```

The WHERE condition of the query is an axis-aligned hyper-rectangle in the attribute value space.

Histogram structure

Buckets in STHoles are rectangular, can be nested into each other, but cannot overlap partially. Thus, the histogram divides the data space hierarchically, similar to an R^+-tree. Figure (1) shows a 2-dimensional histogram with 3 buckets. b_0 is the root bucket, it includes the the whole attribute value space. b_1 and b_2 are children of b_0, and b_3 is a child of b_2.

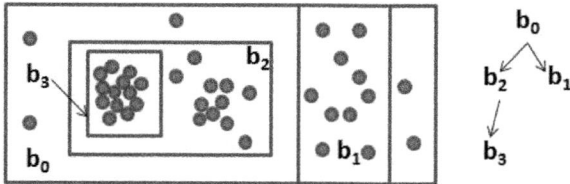

Fig. 1. Histogram with 3 buckets

Each bucket b stores the number of tuples in it, denoted $count(b)$. It does not include the tuples contained in child buckets; so $count(b_2)$ in Figure (1) can be less than $count(b_3)$. Accordingly, the *volume* of bucket b_2, $vol(b_2)$, is the volume of the bounding box of b_2 minus the volume of the box of b_3. An equivalent notion is the tuple density, defined as $dens(b) = count(b)/vol(b)$.

An essential property of STHoles is its ability to store regions of arbitrary dimensionality. The histogram can have buckets which are lower dimensional than the data space. In Figure (1) the bucket b_1 is in fact 1-dimensional, because vertically it spans the whole attribute value range. One of the reasons we chose STHoles as the underlying histogram for our approach is that it natively supports lower-dimensional buckets.

Subspace Clustering and Selectivity Estimation

Subspace clusters provide groups of objects with locally relevant attribute combinations. Similarly to the bucket b_1, irrelevant attributes simply span the complete data range in these dimensions. Thus, a histogram should store only the relevant dimensions. Only these attributes provide the information required for selectivity estimation. In contrast to traditional clustering techniques that focus on all dimensions, subspace clustering can assist histogram construction also in the detection of relevant attribute sets. This approach has a high potential for memory-efficient and high quality selectivity estimation.

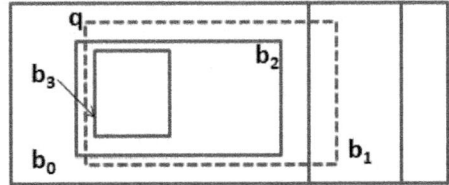

Fig. 2. Histogram with 3 buckets and a query (dashed rectangle)

Selectivity Estimation based on Histograms.
Based on the histogram one can estimate the selectivity of the query using the *uniform spread assumption*. It states that tuples are distributed uniformly inside the bucket, and allows us to limit the stored information to the overall object count instead of a complex data distribution function inside each bucket.

$$est(q) = \sum_b dens(b) \cdot \frac{vol(q \cap b)}{vol(b)} \qquad (1)$$

Here, q is the query, and the sum iterates over all histogram buckets. Thus, according to the uniform spread assumption, each bucket which intersects with the query region contributes tuples proportionally to the volume of the intersection and the density of that bucket. Figure (2) shows the histogram with a query (the dashed rectangle).

Self-Tuning During Query Processing
In addition to the calculation of selectivity estimates, STHoles uses the query processing for self-tuning. The result stream is used to update/create new buckets in the histogram. We present the so-called bucket drilling step [9] in the following.

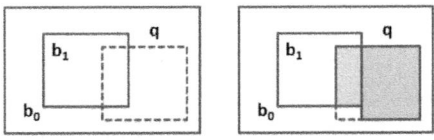

Fig. 3. Drilling buckets using feedback

Algorithm 1. Obtaining rectangular buckets candidates from non-rectangular intersection

---

1: $c \leftarrow q \cap b$
2: $pp \leftarrow \{b_i \in children(b) | c \cap b_i \neq \emptyset \land b_i \not\subseteq c\}$
3: **while** $pp \neq \emptyset$ **do**
4:     Select bucket $b_i \in pp$ and dimension j such that shrinking c along j by excluding b_i results in the smallest reduction of c
5:     Shrink c along dimension j
6:     Update pp
7: **end while**
8: **return** c

Figure (3) (left) shows a histogram with 2 buckets and a query (dashed). After the query is executed, the intersections with b_0 and b_1 are calculated. The intersection with b_0 is not rectangular, so it is shrunk across one of the dimensions, to bring it to rectangular shape (cf. Algorithm 1). After the bucket candidates are created, the actual number of tuples falling into them can be approximated or calculated directly using the query feedback. Thus, buckets can be updated/created. Furthermore, when the number of buckets exceeds the allowed maximum, a compression step takes place (cf. [9]).

Evaluation of Estimation Quality
To measure the quality of selectivity estimates we use the normalized absolute error (NAE) metric, which is a standard metric for assessing selectivity estimates. Let Q be the query workload, and $real(q)$ be the real selectivity of the query, then the NAE is

$$err = \frac{1}{|Q|} \sum_{q \in Q} \frac{|real(q) - est(q)|}{real(q)} \qquad (2)$$

4   Cluster Transformation

The goal of histogram construction is to have a concise summary of data which enables precise selectivity estimates. Clustering algorithms in turn may report clusters in different ways, often by simply listing all elements. Furthermore, clusters can have different shapes, depending on the definition.

In order to make clusters usable for selectivity estimations, we need to transform clusters into memory-efficient histogram structures. The goal of such a transformation is to obtain a histogram which produces estimation error as small as possible. In this section, we address this issue first from a theoretical and then from a practical point of view. Namely, we

- formalize the transformation of a cluster to a bucket
- define classes of transformations with useful properties
- show that strictly optimal transformations are overly expensive
- introduce heuristics which find good representations

Histogram buckets usually have a strict form, e.g., are axis-aligned and rectangular. Our goal is to transform the output of the clustering algorithm to a set of rectangles. We call these rectangles *Representative Rectangles*, or *RRs*.

One seemingly straightforward idea is to use minimal bounding rectangles as *RRs*. However, as our theoretical analysis shows, such *RRs* can be far from optimal or even useless. This emphasizes the importance of choosing *RRs* carefully. We formalize the notion of quality of transformation and optimality in Section 4.1 and present several formal results regarding optimal *RRs*. We use these results in 4.2 when we calculate *RRs*.

4.1   The Optimal RR for Selectivity Estimation

Histogram buckets span (axis-aligned hyper-)rectangles in attribute-value space. A cluster is a set of points: our aim is to transform it to a rectangle while making sure that the transformation "falsifies" the cluster as little as possible.

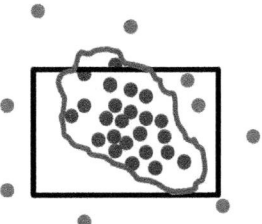

Fig. 4. A cluster and a candidate RR

We denote the set of all rectangles in the data space as $\Re$. $\Re$ can be finite or infinite, depending on the data domain. The transformed rectangles serve as histogram buckets. In the histogram we essentially substitute the cluster C with RR. Figure (4) shows a cluster C and a candidate RR. We now look at clusters not as a discrete set of points but as regions with an extent in space and density, to bring rectangles and clusters into the same domain.

Definition 1. *Given a cluster C, we denote by $|C|$ the volume of its extent. The density of the cluster, $dens(C)$, is the number of objects in the cluster divided by $|C|$.*

Because $C \neq RR$ in general, substituting C with RR introduces an estimation error. Suppose that the density of the cluster is $dens(C)$, and outside of the cluster it is roughly 0, and the density of RR is $dens(RR)$. Then, as a result of substituting C with RR, the following density changes occur:

- $RR - C$ has density 0, but instead we estimate its density to be $dens(RR)$
- $C - RR$ has density $dens(C)$, instead we estimate its density to be 0.
- $C \cap RR$ has density $dens(C)$, instead we estimate its density to be $dens(RR)$.

The overall estimation error resulting from the substitution of a fixed C with RR is given by the function $\epsilon(RR, dens(RR))$. It is the sum of errors of the three regions mentioned above:

$$\epsilon(RR, dens(RR)) = \int_{RR \cup C} |est(u) - real(u)|\, du =$$

$$\int_{RR \cap C} |dens(RR) - dens(C)|\, du + \int_{RR-C} dens(RR) du + \int_{C-RR} dens(C) du =$$

$$(|dens(RR) - dens(C)|)\, |RR \cap C| + dens(RR)\, |RR - C| + dens(C)\, |C - RR| \tag{3}$$

Definition 2. *A rectangle RR with density $dens(RR)$ is called optimal (w.r.t. $\Re$), denoted by $RR = opt(\Re)$ if*

$$\epsilon(RR, dens(RR)) = \min_{r \in \Re} \epsilon(r, dens(r))$$

We first prove that the density of $opt(\Re)$ is upper-bounded by the density of the cluster:

Lemma 1. *For any cluster C with density $dens(C)$, if $RR = opt(\Re)$, then $dens(RR) \leq dens(C)$*

Proof. Let us assume that the opposite is true, for some $\alpha > 0$ $dens(RR) = dens(C) + \alpha$, RR is optimal, which means

$$\epsilon(RR, dens(R)) = \alpha\, |RR \cap C| + dens(C)\, |C - RR| + (dens(C) + \alpha)\, |RR - C|$$

is minimal. Take $dens'(RR) = dens(C) - \alpha$,

$$\epsilon'(RR, dens'(R)) = \alpha\, |RR \cap C| + dens(C)\, |C - RR| + (dens(C) - \alpha)\, |RR - C|$$

$\epsilon'(RR, dens'(R)) < \epsilon(RR, dens(R))$, which contradicts the assumption that ϵ is minimal. □

Figure (4) illustrates why $dens(RR)$ should not exceed $dens(C)$. RR possibly contains regions which are not in C, and does not necessarily cover all of C. So instead of some part of C with high density, RR contains a part which has density 0.

Using Lemma (1), we can simplify Equation (3)

$$\epsilon(RR, dens(RR)) = dens(C) \cdot |C| + dens(RR) \cdot (|RR - C| - |RR \cap C|) \tag{4}$$

We can now find the expression for the optimal value of $dens(RR)$.

Lemma 2. *For a fixed rectangle RR, the value of $dens(RR)$ which minimizes $\epsilon(RR, dens(RR))$ is given by:*

$$dens(RR) = \begin{cases} dens(C) & \text{if } |RR \cap C| > |RR - C| \\ 0 & \text{otherwise} \end{cases}$$

Proof. In Equation (4), the part depending on dens(RR) is

$$dens(RR) \cdot (|RR - C| - |RR \cap C|)$$

In case $|RR - C| > |RR \cap C|$, it is positive. To minimize it, we put $dens(RR) = 0$. In case $|RR \cap C| > |RR - C|$, it is negative, and we put $dens(RR) = dens(C)$, which is the largest value for $dens(RR)$ according to Lemma (1). ☐

The first implication from this lemma is that if $|RR - C| > |RR \cap C|$ then the rectangle RR is not useful and can be omitted. RR is useless when the space contained in RR not belonging to C is larger than the common part of C and RR (Figure (4)). However, when $|RR \cap C| > |RR - C|$, then the best strategy is to minimize the estimation for the region $|RR \cap C|$. This is achieved by putting $dens(RR) = dens(C)$. Below, we always consider RRs with $|RR \cap C| > |RR - C|$, and their density $= dens(C)$. Finding the optimal RR is not straightforward, however. Before turning to optimal RRs, we discuss some "obvious" RRs, such as minimal bounding rectangle.

Definition 3. *We denote the set of all rectangles which enclose C by $\Re_C^+$.*

$$\Re_C^+ = \{R | R \in \Re, C \subseteq R\} \tag{5}$$

Obviously, the minimal bounding rectangle of C is in $\Re_C^+$.

Definition 4. *We denote the set of all rectangles enclosed in C by $\Re_C^-$*

$$\Re_C^- = \{r | r \in \Re, r \subseteq C\} \tag{6}$$

The maximal inbound rectangle of a cluster is in $\Re_C^-$.

Lemma 3. $\Re^+$ *contains a unique optimal RR or is empty.*

Proof. We construct a rectangle R_0 such that $\forall R \in \Re_C^+$, $R_0 \subseteq R$. For dimension j, project all points on j, find the minimum and maximum – those would be the sides of the rectangle parallel to dimension j. Repeating this for all dimensions we will obtain the rectangle. Obviously, any rectangle in $\Re_C^-$ contains R_0. If R_0 satisfies the condition $|R_0 \cap C| > |R_0 - C|$ then $R_0 = opt(\Re_C^+)$, otherwise $\Re_C^+$ does not contain any RRs.

Consider again Figure (4) for an example in 2-dimensional space. To find the minimal rectangle in $\Re^+$, take the up-most point of the cluster and draw a line parallel to the x-axis, do the same with the lowest point. Now, take the rightmost point and draw a line parallel to the y-axis, same with the leftmost point. The rectangle which is bounded by those 4 lines is R_0.

We now proceed as follows: We first present an algorithm which finds $opt(\Re_C^-)$. It constructs a convex hull of the cluster and fits the largest RR into it. In practice, this approach has limitations. In particular, it is too expensive for large clusters. As an alternative, we describe a heuristic which is both fast and effective.

4.2 Cluster-to-Bucket Transformation

Finding opt($\Re_C^-$)

As a first step, we compute the convex hull of the cluster. The complexity of this is $O(n \cdot log(n))$, where n is the number of data points [10]. Given the convex hull of the cluster, we fit the largest axis-aligned rectangle into it. The algorithm in [4] transforms this problem to a convex optimization problem. The complexity is $O(2^d \cdot h)$, where h where h is the number of vertices of the polygon, and d is the dimensionality of the data space. The complexity of the overall procedure is $O(n \cdot log(n) + 2^d \cdot h)$. This does not scale well against the dimensionality or the number of data objects.

Fig. 5. A cluster with rectangles RR (solid) and RR' (dashed)

Heuristic

We propose an alternative heuristic which is computationally affordable. It starts with a rectangle that fulfills the condition $|RR \cap C| > |R - C|$ and expands it iteratively. Figure (5) shows a rectangle RR which is expanded along the y-axis downwards. The expanded rectangle RR' is dashed. RR' is a better bucket than RR if $\epsilon(RR') < \epsilon(RR)$, i.e., it approximates the cluster with less error than RR. Given Equation (4), this is equivalent to computing

$$\lambda(RR, RR') = |RR' \cap C| - |RR \cap C| + |RR - C| - |RR' - C| \qquad (7)$$

and comparing it to 0. In order to compute λ, we have to compute $|R \cap C|$ and $|R - C|$ for a rectangle R. This is not straightforward because C has an arbitrary shape. We compute $R \cap C$ using the following idea: if we generate M data points uniformly distributed inside R, then the expected number of points m that will fall inside $R \cap C$ will be proportional to its area:

$$\frac{E[m]}{M} = \frac{|R \cap C|}{|R|}$$

From here we obtain

$$|R \cap C| = |R| \cdot \frac{E[m]}{M} \qquad (8)$$

Now we can compute $\lambda(RR', RR)$, using Equation (8) and the fact that $|R - C| = |R| - |R \cap C|$.

In Line 1, we initialize *bestRR* with some RR. In our implementation, we do this as follows: The center of the rectangle is the median. To compute the length

Algorithm 2. Greedy algorithm for finding a RR

```
 1: bestRR ← initial()
 2: repeat
 3:     best ← ∞
 4:     RR ← bestRR
 5:     for all possible expansions e do
 6:         RR' ← expand(RR, e)
 7:         if λ(RR, RR') > best then
 8:             best ← λ(RR, RR')
 9:             bestRR ← RR'
10:         end if
11:     end for
12: until best = ∞
```

of the projection of the rectangle on dimension i, we first project all points of the cluster to dimension i, let this be C_i. We define $diam(C_i) = max(C_i) - min(C_i)$. We took 1/10-th of $diam(C_i)$ as the length of dimension i.

5   Experiments

In the experiments we compare how different subspace clustering algorithms (MineClus [28], PROCLUS [2], CLIQUE [3], SCHISM [25], INSCY [6] and DOC [24]) perform as histogram initializers. Implementations were used out of the *OpenSubspace* repository [21]. We first run each clustering algorithm against a data set, obtain the output, transform it into a bucket set and then measure the selectivity estimation error. We look at the following issues:

- *Precision.* We measure the estimation error of various clustering algorithms. We vary the number of tuples, the dimensionality of the dataset and the number of histogram buckets allowed.
- *Memory consumption.* For a fixed estimation error threshold, we measure how much memory can be saved if we use initialized histograms vs. non-initialized STHoles.
- *Scalability.* The runtime cost of initialization depending on dimensionality.

5.1   Setup

We used synthetic multi-dimensional data sets, consisting of multiple overlapping Gaussian bells. Such data sets are used both for selectivity estimation experiments [9] and to assess the quality of clustering algorithms [13,18]. We conducted two sets of experiments – for accuracy and for scalability. We generated 20 to 50 Gaussian bells with standard deviation = 50 in the data domain $[0, \ldots, 1000]^d$. The number of tuples t, the dimensionality d and the number of histogram buckets B are the main parameters to vary. This is because both clustering and selectivity estimation algorithms are sensitive to these parameters. In order to obtain clusters of different size, we generated tuple counts for each cluster according to a Zipfian distribution with $skew = 1$. Thus, we assign different numbers of tuples to each cluster. Table 1 gives an overview of parameter values for the accuracy and scalability experiments.

Table 1. Parameters values of experiments

Experiment	Parameter	Value
Accuracy	d: dimensionality	2 to 10
	t: tuple count	10,000 to 50,000
	B: buckets	25 to 200
Scalability	d: dimensionality	10 to 20
	t: tuple count	500,000 to 1,000,000
	B: buckets	200

We used 2,000 queries in the experiments. The query centers are uniformly distributed, and each query spans 1% of the volume of the data set. We used the first 1,000 queries for the algorithms to learn, this is the number from [9] and is considered enough for STHoles to fully adapt to the data set. Essentially, 1,000 random queries having 1% of the volume of the dataset means the whole data set is being accessed $1000 \cdot 0.01 = 10$ times. This should be more than enough to provide good quality estimations. Thus, we start calculating the estimation error only after 1,000 queries have have run and all algorithms have learned. The error measure is the normalized absolute error, see Equation (2).

5.2   Estimation Precision

We first look at the selectivity estimation error for traditional STHoles compared to our initialized histograms. Figures 6, 7, 8 show the dependency of the average normalized error on the number of histogram buckets, for different algorithms. Each figure consists of two parts, one for 10,000 tuples and one for 50,000. We varied the dimensionality d, so each figure is for a different d. Comparing the underlying clustering algorithms for our transformation, we observe one clear winner in all figures, namely the MineClus algorithm. MineClus clearly provides a high quality initialization particularly robust w.r.t. the dimensionality of the data space. While other clustering approaches show highly varying precision, MineClus is the only algorithm with low estimation errors in all data sets.

More specifically, looking at the plots for $d \geq 4$ (Figures 7, 8), we can see MineClus in the lower end of the graph, with other algorithms (STHoles included) in the upper part. PROCLUS is interesting: It is the best of this worse-performing pack for $d = 4$ and $d = 5, t = 10,000$. Starting with $d = 5, t = 50,000$ it becomes better and for $d = 10$ actually joins MineClus. In all experiments MineClus outperformed STHoles, while SCHISM and INSCY where consistently worse. Furthermore, a general observation from the figures is that adding more buckets increases the quality of histograms. However, the estimation quality

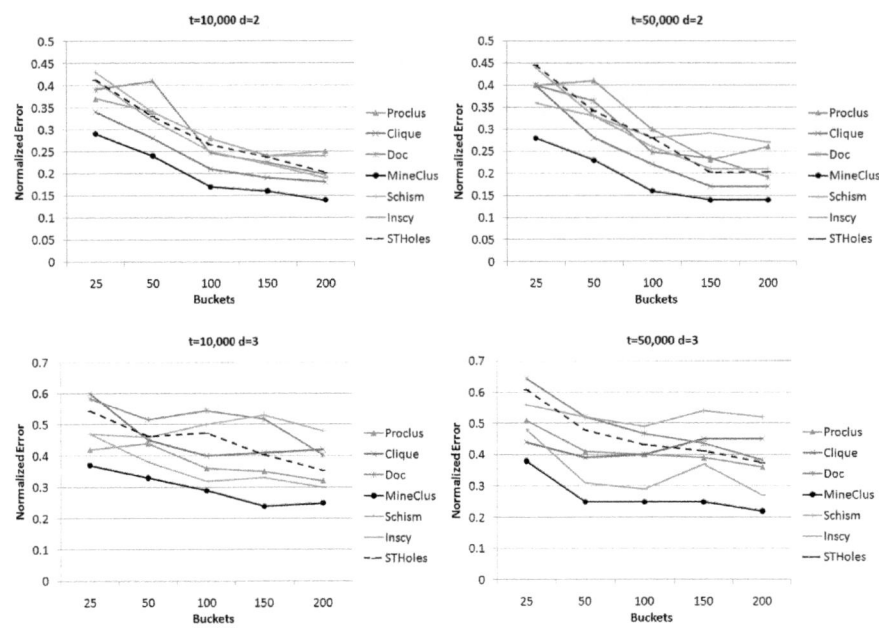

Fig. 6. Error vs bucket count for 2- and 3-dimensional space

highly depends on the type of initialization. MineClus is the best method in all settings we have tried.

Looking at the methods which performed worse, we can see that STHoles is usually in the middle of the pack. This shows that:

- Not every initialization is good or meaningful. After 1,000 queries for learning, STHoles performs about as good as most of the initialized methods. This confirms the statement in the original STHoles paper [9].
- However, the underlying clustering methods make the difference between good and best. MineClus is the winner. It provides a high quality initialization that yields better estimations than the original STHoles method.

With increased dimensionality of the data space, MineClus continues to perform better compared to STHoles, tackling the challenges of high-dimensional data better than the traditional selectivity estimation techniques. Further experiments with various high-dimensional data sets are required to find out how persistent this effect is throughout different application domains. Overall, we can see that initialization based on subspace clustering algorithms shows a clear benefit in terms of estimation precision.

5.3   Memory-Efficiency w.r.t. Different Initializations

Table 2 shows a different perspective on the previous experiments. It highlights the memory-efficiency of our initialization compared to the relatively high mem-

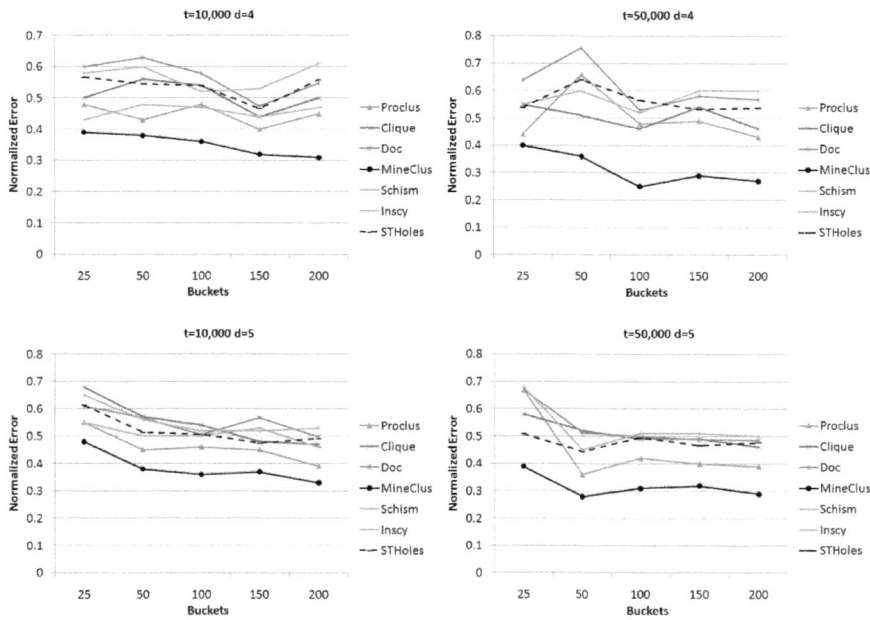

Fig. 7. Error vs bucket count for 4- and 5-dimensional space

ory consumption of STHoles. For each combination of parameter values, it shows how many buckets MineClus needs to be at least as accurate as STHoles. For instance, for 10,000 tuples, 2-dimensional data set and 200 buckets allocated for STHoles, MineClus needs only 100 buckets to produce estimates of equal or better quality. We can obtain this from the upper-left plot on Figure (6), by drawing a horizontal line at about 0.2, which is the error of STHoles for this setting and 200 buckets. This horizontal line intersects the MineClus curve between 50 and 100 buckets. So 100 buckets is a conservative estimate of the buckets needed for MineClus to match the precision of STHoles with 200 buckets. The table shows that when the dimensionality of the data set $d \geq 3$, 50 buckets for MineClus are enough to match the precision of STHoles with 200 buckets. Even more surprisingly, out of 24 rows in the table (rows corresponding to $d \geq 3$) only in two cases MineClus needs 50 buckets, otherwise only 25 suffice to match the precision of STHoles with 100-200 buckets. This means that our initialization reduces the memory consumption by a factor of up to 8 in most cases.

5.4 Scalability w.r.t. Data Dimensionality

In general, the key parameter for subspace clustering is the dimensionality of the data space [21]. It affects the runtime performance of the clustering algorithm and thus is essential for our transformation as well: We fix the number of buckets to 200 and evaluate the influence of the dimensionality. Subspace clus-

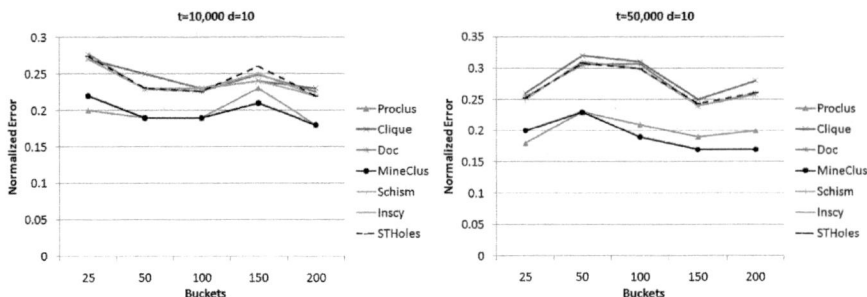

Fig. 8. Error vs bucket count for 10-dimensional space

Table 2. Description of data sets

Tuples	Dim	ST Buckets	MineClus Buckets	Tuples	Dim	MaxBucketNr	Buckets Needed
10,000	2	100	50	50,000	2	100	25
		150	50			150	100
		200	100			200	100
	3	100	25		3	100	25
		150	25			150	25
		200	50			200	25
	4	100	25		4	100	25
		150	25			150	25
		200	25			200	25
	5	100	25		5	100	25
		150	25			150	25
		200	25			200	25
	10	100	25		10	100	25
		150	25			150	25
		200	50			200	25

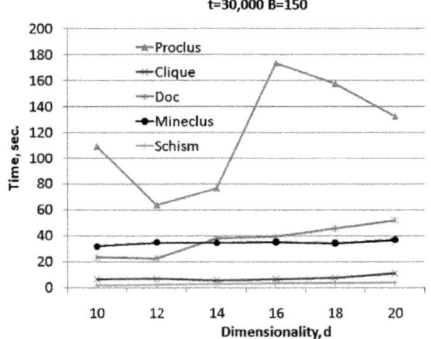

Fig. 9. Execution time against the dimensionality

tering algorithms have been designed for efficient pruning in high-dimensional data. Thus, most of them show efficient and scalable runtime results (cf. Figure (9)). We also ran the algorithms against data sets with 500,000 to 1,000,000 tuples; experiments finished within reasonable time, and we have not observed any serious scalability issues with this number of tuples.

6   Conclusions and Future Work

Uninitialized self-tuning histograms need an excessively large number of queries to *learn* the data set. In this paper we studied initialization of self-tuning histograms using subspace clustering results. With our transformation of subspace clusters to memory-efficient histogram buckets, we could achieve significant improvement over traditional self-tuning selectivity estimators. In contrast to the traditional assumption that self-tuning can compensate the benefits of initialization, we show that our initialization is of clear benefit. Combining initialization with self-tuning results in a high quality histogram with low estimation errors.

In this work, our intention has been to use subspace clustering algorithms as a black box, without modifications. Future work will strive for more customized subspace clustering techniques. Based on the best performance of MineClus, we will work on improved sampling techniques inside the clustering algorithm. Namely, MineClus could be adapted in its choice of subspace clusters to fit the desired objective function in selectivity estimation.

References

1. Aboulnaga, A., Chaudhuri, S.: Self-tuning histograms: building histograms without looking at data. In: SIGMOD 1999 (1999)
2. Aggarwal, C.C., Wolf, J.L., Yu, P.S., Procopiuc, C., Park, J.S.: Fast algorithms for projected clustering. SIGMOD Record 28(2) (1999)
3. Agrawal, R., Gehrke, J., Gunopulos, D., Raghavan, P.: Automatic subspace clustering of high dimensional data for data mining applications. SIGMOD Record 27(2) (1998)
4. Amenta, N.: Bounded boxes, hausdorff distance, and a new proof of an interesting helly-type theorem. In: SCG 1994 (1994)
5. Assent, I., Krieger, R., Müller, E., Seidl, T.: DUSC: Dimensionality unbiased subspace clustering. In: ICDM 2007 (2007)
6. Assent, I., Krieger, R., Müller, E., Seidl, T.: INSCY: Indexing subspace clusters with in-process-removal of redundancy. In: ICDM 2008 (2008)
7. Baltrunas, L., Mazeika, A., Bohlen, M.: Multi-dimensional histograms with tight bounds for the error
8. Beyer, K.S, Goldstein, J., Ramakrishnan, R., Shaft, U.: When is nearest neighbor meaningful? In: Beeri, C., Bruneman, P. (eds.) ICDT 1999. LNCS, vol. 1540, pp. 217–235. Springer, Heidelberg (1998)
9. Bruno, N., Chaudhuri, S., Gravano, L.: STHoles: a multidimensional workload-aware histogram. SIGMOD Record (2001)
10. Cormen, T., Leiserson, C., Rivest, R., Stein, C.: Introduction to Algorithms. MIT Press, Cambridge (2009)

11. Fuchs, D., He. Z., Lee, B.S.: Compressed histograms with arbitrary bucket layouts for selectivity estimation. Inf. Sci. 177, 680–702 (2007)
12. Gunopulos, D., Kollios, G., Tsotras, V.J., Domeniconi, C.: Approximating multi-dimensional aggregate range queries over real attributes. SIGMOD Record (2000)
13. Halkidi, M., Vazirgiannis, M.: Clustering validity assessment: finding the optimal partitioning of a data set. In: ICDM 2001 (2001)
14. Han, J., Kamber, M.: Data Mining: Concepts and Techniques. Morgan Kaufmann, San Francisco (2001)
15. Ioannidis, Y.: The history of histograms (abridged). In: VLDB 2003 (2003)
16. Jolliffe, I.: Principal Component Analysis. Springer, New York (1986)
17. Kröger, P., Kriegel, H.P., Kailing, K.: Density-connected subspace clustering for high-dimensional data. In: SDM 2004 (2004)
18. Liu, Y., Li, Z., Xiong, H., Gao, X., Wu, J.: Understanding of internal clustering validation measures
19. Luo, J., Zhou, X., Zhang, Y., Shen, H.T., Li, J.: Selectivity estimation by batch-query based histogram and parametric method. In: ADC 2007 (2007)
20. Müller, E., Assent, I., Günnemann, S., Krieger, R., Seidl, T.: Relevant Subspace Clustering: mining the most interesting non-redundant concepts in high dimensional data. In: ICDM 2009 (2009)
21. Müller, E., Günnemann, S., Assent, I., Seidl, T.: Evaluating clustering in subspace projections of high dimensional data. In: PVLDB, vol. 2(1) (2009)
22. Muthukrishnan, S., Poosala, V., Suel, T.: On rectangular partitionings in two dimensions: Algorithms, complexity, and applications
23. Poosala, V., Ioannidis, Y.E.: Selectivity estimation without the attribute value independence assumption. In: VLDB 1997 (1997)
24. Procopiuc, C.M., Jones, M., Agarwal, P.K., Murali, T.M.: A monte carlo algorithm for fast projective clustering. In: SIGMOD 2002 (2002)
25. Sequeira, K., Zaki, M.J.: Schism: A new approach for interesting subspace mining. In: ICDM 2004 (2004)
26. Srivastava, U., Haas, P., Markl, V., Kutsch, M., Tran, T.: ISOMER: Consistent histogram construction using query feedback. In: ICDE 2006 (2006)
27. Wang, H., Sevcik, K.C.: A multi-dimensional histogram for selectivity estimation and fast approximate query answering. In: CASCON 2003 (2003)
28. Yiu, M.L., Mamoulis, N.: Frequent-pattern based iterative projected clustering. In: ICDM 2003 (2003)

Finding Closed MEMOs*

Htoo Htet Aung and Kian-Lee Tan

School of Computing, National University of Singapore

Abstract. Current literature lacks a thorough study on the discovery of meeting patterns in moving object datasets. We (a) introduced MEMO, a more precise definition of meeting patterns, (b) proposed three new algorithms based on a novel data-driven approach to extract closed MEMOs from moving object datasets and (c) implemented and evaluated them along with the algorithm previously reported in [6], whose performance has never been evaluated. Experiments using real-world datasets revealed that our filter-and-refinement algorithm outperforms the others in many realistic settings.

1 Introduction

We will start our discussion with a motivating example. Suppose a commander is laying out a battle plan through computer simulation, which is a common practice. If he knows that the enemy possesses a tactical weapon, a single strike of which takes w minutes to (aim and) fire and is capable of destroying all ships within its circular target area of radius r meters, he may want to ensure that his battle plan is not vulnerable to such a strike, which would destroy a significant portion of his fleets, say m ships – i.e. he may want to ensure that his battle plan has no instance, in which m ships gather in a circular area smaller than that of a radius r meters for more than w minutes (so that the enemy can aim and fire). Such instances he is looking for (and trying to eliminate) in his battle plan are the meeting patterns, the subject of this paper.

Informally, a meeting is formed when a group of objects comes to and stays in a fixed area for a while. Information of meeting patterns can be analyzed to discover knowledge on trends of meeting places (and times), in which the meetings are formed. For instance, meetings of mobile users show trends in popular places, those of students show seasonal preference of campus facilities used for group activities and those of wild-animals show changes in their natural habitats. Such information can be used by market-researchers, school administrators and scientists to plan advertisements, facilities and further researches etc.

The information of meeting places can be used in planning the deployment of mobile service centers. For example, in developing countries, the concept of a mobile library becomes very popular as it allows library resources (books, personnel etc) to be shared by patrons residing in different geographic areas. In

---

* This project is partially supported by a research grant TDSI/08-001/1A from the Temasek Defense Systems Institute.

J.B. Cushing, J. French, and S. Bowers (Eds.): SSDBM 2011, LNCS 6809, pp. 369–386, 2011.

order to maximize the utilization of a mobile library, it should be deployed in such a way that it is available (within r meters) to a number of (say m) patrons. To match a stationary library's service as close as possible, a mobile library must be accessible for a sufficient amount of time (say w hours) for each patron so that he can come to the library at his convenience (e.g during lunch hour, after work etc). On the other hand, if the mobile library is not accessible long enough (e.g. the patron can only glimpse it while he is rushing to work), the patron may choose not to visit the mobile library in favor of his tight schedule, reducing the utility of the mobile library. Knowledge of the past meetings formed by patrons can assist in efficient planning of the deployment of mobile libraries and other similar services like mobile medical centers and mobile Internet cafe.

To discover meetings, one may opt to count objects using proximity sensors (RFID readers) at the potential meeting places rather than mining moving object datasets like GPS traces. However, this approach has several drawbacks. Firstly, proximity sensors have limited ranges, thus, it often requires to aggregate data from multiple sensors to discover a single meeting. Since an object can be detected by multiple sensors, resulting in duplicate readings, aggregation is not a straight-forward task. Secondly, as the sensors need to be deployed at potential meeting places, meetings formed in unexpected places will not be discovered. Moreover, tokens (RFID tags) must be attached to the objects in advance, which is highly impractical. On the other hand, many people are accustomed to sharing their GPS traces with friends and businesses (or mobile libraries) through location-sharing services like Google Latitude.

To the best of our knowledge, there has been only a limited amount of studies on discovery of meeting patterns and no implementation (experimental evaluation) exists. Our contributions include a) introducing a more accurate and realistic definition of meeting patterns b) developing three new algorithms to discover meeting patterns from moving object datasets and c) experimenting them along with our adaptation of the algorithm proposed in [6].

2   Problem Definition

Definition 1. *For a given set of objects $O = \{o_1, o_2, ..., o_n\}$, time-stamps $T = \{t_1, t_2, ..., t_\tau\}$ and a two-dimensional spatial-space $\mathbb{R}^2$, a moving object dataset $\mathcal{D}$ is a set of records of the form $\langle o, t, loc \rangle$ where $o \in O, t \in T$ and $loc \in \mathbb{R}^2$.*

In a moving object dataset, a pair of o and t uniquely determines loc. However, a given moving object dataset $\mathcal{D}$ may have missing $\langle o, t, loc \rangle$ for some pairs of o and t due to limitations in real-life settings. Although time is presented as discrete steps, generality of Def. 1 is not undermined since the inter-step interval can be arbitrarily small.

Definition 2. *For given parameters: $m > 1$, $r > 0$ and $w \geq 1$, a set of objects M forms a MEeting of Moving Objects, or a MEMO, during the time-interval $I(M)$ at the circular region $loc(M)$ if (i) M has at least m objects, (ii) $I(M)$ spans for at least w consecutive time-stamps, (iii) $loc(M)$ has a minimal radius $r(M) \leq r$ and (iv) all objects $o \in M$ reside in $loc(M)$ in each $t \in I(M)$.*

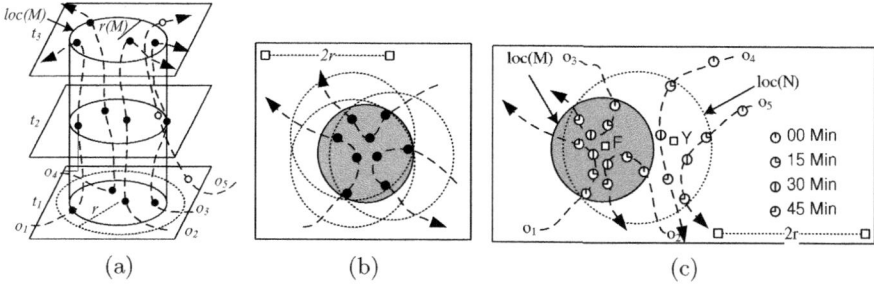

Fig. 1. An example of (a) a MEMO, (b) an accurate meeting place and (c) two overlapping closed MEMOs

Definition 2 defines a MEMO formed by m or more objects. In order to form a MEMO, the participating objects must stay in a circle, whose radius is not larger than r, for at least w time-stamps. In Fig. 1a, which shows movement of five objects for three time-stamps and a circle (drawn in dotted lines) with radius r, $M = \{o_1, o_2, o_3, o_4\}$ forms a MEMO from t_1 to t_3 for parameters : $m = 3$ and $w = 2$. The meeting place of M is the circular region $loc(M)$, whose radius is $r(M) \leq r$ and in which all objects in M stay.

Our definition of the meeting pattern, MEMO, is more precise than the one given in [6] as it explicitly defines the meeting place as the smallest possible circle in contrast to a circle of fixed radius r. As a result, it can report more accurate and non-ambiguous meeting places. For example, for the meeting pattern formed by three animals in Fig. 1b, the most accurate place of their habitat (the shaded circle), rather than a larger, less accurate circular regions of radius r (three shown as dotted circles), can be reported.

Definition 3. *A MEMO M is a closed MEMO if and only if there is no other MEMO $M' \neq M$ such that M' contains all members of M and the life-span of M' completely covers that of M.*

Definition 3 formally defines the concept of a closed MEMO of maximal interval and maximal set of members. For example, in Fig. 1a, for parameters : $m = 3$ and $w = 2$, $M_1' = \{o_1, o_2, o_3\}$ from t_1 to t_3 and $M_2' = \{o_1, o_2, o_3, o_4\}$ from t_1 to t_2 are non-closed MEMOs as there is a closed MEMO $M = \{o_1, o_2, o_3, o_4\}$ from t_1 to t_3 covering them.

Overlaps between closed MEMOs are possible and, in fact, necessary. Consider two docks, dock F serving class-F ships and dock Y serving class-Y ships as shown in Fig. 1c. For parameters $m = 3$ and $w = 30$ minute, both $M = \{o_1, o_2, o_3\}$ from 0 to 45 minute and another MEMO $N = \{o_1, o_2, o_3, o_4, o_5\}$ from 15 to 45 minute are closed MEMOs as one does not cover the other. This result intuitively agrees with the observation that the commander must be informed of both meetings because changes made to eliminate one does not necessarily eliminate another.

Definition 4. *Given a moving object dataset $\mathcal{D}$ and parameters : $m > 1$, $r > 0$ and $w \geq 1$, the task of **Finding closed MEMOs** is to list the complete information of all closed MEMOs formed in $\mathcal{D}$ according to m, r and w.*

3   Related Works

Given a moving object dataset $\mathcal{D}$, a set of objects M containing $k \geq m$ objects and a time-interval $I(M)$ spanning at least w time-stamps, whether M forms a MEMO according to the given parameters: m, w and r can be verified by checking if the minimum circle $loc(M)$ which covers the set of points $P = \{p | \langle o, t, p \rangle \in \mathcal{D}$, $o \in M$ and $t \in I(M)\}$, is smaller than a circle having radius r. The earliest algorithm to calculate the smallest circle enclosing a finite set of points is found in the translated text [13]. It starts with an arbitrarily large circle and shrinks (and move) the circle until no more shrinking is possible. Since it is difficult to implement on computers, Elzinga and Hearn [4] proposed a new algorithm called Euclidean Messenger Boy algorithm (EMB) that monotonously increases the radius until all points are covered. The studies in [3] reveals that, for a set of points P, its smallest enclosing circle $C(P)$ is unique and always exists.

In order to discover all closed MEMOs in a dataset $\mathcal{D}$ containing movement records of a set of objects O, we need to check each subset $O' \subseteq O$ if O' forms a MEMO. *Apriori* algorithm, the first data-driven algorithm to traverse the power set $\mathcal{P}(O)$ of a given set O having the *apriori*-properties – if $M \subseteq O$ is interesting, then its subset $M' \subseteq M$ must be interesting – appears in [1]. Starting with all the interesting sets with exactly one member each, it systematically builds up interesting sets containing $(k + 1)$ members from those containing k members.

Since the *Apriori* algorithm requires a large amount of memory, Zaki [15] proposed Equivalence CLAss Transformation (ECLAT). Using ECLAT, the power set $\mathcal{P}(O)$ of a given set $O = \{o_1, o_2, o_3, ..., o_n\}$ can be divided into n equivalent classes $C_1, C_2, C_3, ..., C_n$. The kth equivalent class C_k is defined as $C_k = \{M | o_k \in M$ and if $o_i \in M$ then $o_k \preceq o_i\}$ for an arbitrary partial-order $\preceq$ on O. Each equivalent class C_k, which is a lattice of 2^k sets following *apriori*-properties, is recursively divided into sub-classes unless it fits entirely into the memory for processing by the *Apriori*-algorithm. It limits the memory requirement of frequent-itemset-mining at the expense of redundant processing. FP-growth, a dialect of ECLAT algorithm, is proposed in [7].

One of the earliest studies on extracting aggregate spatial-temporal patterns formed by multiple objects was RElative MOtion (REMO) framework [10], which worked on speed and direction of objects in each snapshot to uncover complex patterns. REMO framework was later extended in [11] to include location and was capable of discovering more complex patterns like encounter pattern, which is a meeting pattern that last exactly a single time-stamp. Gudmundsson *et al.* [5] proposed a better algorithm to discover encounter patterns. Popular place, which is an extension to another REMO pattern, convergence, is discussed in [2]. Unlike the meeting pattern we present in this paper, entities visiting a popular place need not be present simultaneously.

A fixed flock, a relaxed version of the meeting pattern, is defined as a set F of m or more objects, which are within a (moving) circle of radius $r > 0$ in each of the w or more consecutive time-stamps (r, m and w are given). Gudmundsson and Kreveld [6] reported computing longest-duration fixed flocks is NP-Hard. They also reported that computing the longest-duration meeting patterns from a given dataset is $O(n^4\tau^2 log(n) + n^2\tau^2)$, where n is the number of unique objects and τ is the number of trajectory segments. They did not provide experimental evaluations for their algorithms.

Hwang et al. [8] proposed an Apriori based algorithm and an FP-growth based algorithm to find frequent moving groups, which is related to the flock pattern. These algorithms are extended to find maximal groups in [14].

4   Algorithms for Finding Closed MEMOs

We developed three new algorithms to discover closed meetings of moving objects (closed MEMOs). The first algorithm uses the *apriori*-properties of MEMOs, while the second employs ECLAT partitioning to divide the search space into partitions, achieving practical efficiency. The last algorithm introduces a filtering step for the first and second algorithms. We will present some preliminaries before a detailed discussion on our algorithms.

Definition 5. *A set of objects containing exactly k members will be called a k-object-set.*

Definition 6. *For given parameters: $r > 0$ and $w \geq 1$, a k-object-set O_k forms a k-MEMO M_k during the time-interval $I(M_k)$ at the circular region $loc(M_k)$ if (i) $I(M_k)$ spans for at least w consecutive time-stamps, (ii) $loc(M_k)$ has a minimal radius $r(M_k) \leq r$ and (iii) all objects $o \in O_k$ resides in $loc(M_k)$ in each $t \in I(M_k)$.*

Definition 7. *A k-object-set O_k forms a k-closed MEMO M_k if and only if O_k does not form another k-MEMO $M_k' \neq M_k$ such that the life-span of M_k' covers that of M_k.*

Definition 8. *For a given k-object-set O_k, the set of all its k-closed MEMOs $L(O_k) = \{M_{k1}, M_{k2}, M_{k3}, ..., M_{kj}\}$ is called the MEMO-List (M-List) of O_k.*

Definition 5 defines a k-object-set, which can form zero or more k-MEMOs in a given dataset $\mathcal{D}$. Definition 6 is a relaxed version of Def. 2. A k-MEMO has exactly k participants (in contrast to the fact that a MEMO must have at least m participants). For $k \geq m$, a k-MEMO is a MEMO. Following Def. 7, a k-closed MEMOs cannot be covered by another MEMO having k (or fewer) members. Therefore, all closed MEMO having k participants are k-closed MEMO (the reverse is not always true). Definition 8 defines a MEMO-List, which groups all k-closed MEMO formed by a single k-object-set.

4.1   An Apriori-Based Closed MEMO Miner (A-Miner)

Lemma 1. *If a set of points P is covered by a minimum covering circle C, the minimum covering circle C' of its subset $P' \subseteq P$ is not larger than C.*

Proof. For any set of points P', the minimum covering circle C' always exists and is unique [3]. Therefore, for any circle D, which covers P', $r(C') \le r(D)$, where $r(X)$ is the radius of circle X. Since C covers $P' \subseteq P$, $r(C') \le r(C)$. □

Using Lemma 1, we can derive the *apriori*-properties of MEMOs as follow: for a given time-interval I, if a set of objects M forms a MEMO, there is a circle $loc(M)$ having a radius $r(M) \le r$ and enclosing the set of locations (points) $L = \{loc | \langle o, t, loc \rangle \in \mathcal{D}, o \in M$ and $t \in I\}$. The corresponding set of locations $L' = \{loc | \langle o, t, loc \rangle \in \mathcal{D}, o \in M'$ and $t \in I\}$ of $M' \subseteq M$ is, by Lemma 1, covered by a circle $loc(M')$ not larger than $loc(M)$, i.e. $r(M') \le r(M)$. Thus, $r(M') \le r$ and M' forms a MEMO during the interval I. Therefore, all subsets of a MEMO are MEMOs. In other words, for any time-interval, M does not form a valid MEMO if any of its subset does not.

The *Apriori*-based closed MEMO miner (A-miner), adapted from the *Apriori*-algorithm in [1], exploits the *apriori*-properties of the MEMOs to systematically discover the MEMOs formed by $(k + 1)$-object-sets only when those formed by its subsets, k-object-sets, exist. An outline of the A-miner is given in Algorithm 1. The function Closed-MEMO($\mathcal{D}$, w, r, O_k) returns the sorted list (M-List) of k-closed MEMOs formed by O_k.

Algorithm 1. Apriori-based closed MEMO Miner.

Input: $\mathcal{D}$, r, m and w.
Output: A set of closed MEMO $\mathcal{M}$.
1: The set of 1-object-sets $\mathcal{C}_1 \leftarrow \emptyset$, $\mathcal{M} \leftarrow \emptyset$ and $k \leftarrow 1$
2: **for all** $o \in O$ **do**
3:     Object set $O_1 \leftarrow \{o\}$ and M-List $L(O_1) \leftarrow$ Closed-MEMO($\mathcal{D}$, w, r, O_1)
4:     **if** $L(O_1)$ is not empty **then**
5:         $\mathcal{C}_1 \leftarrow \mathcal{C}_1 \cup \{O_1\}$
6: **while** $\mathcal{C}_k \ne \emptyset$ **do**
7:     **for all** $O_k \in \mathcal{C}_k$ **do**
8:         **if** $k \ge m$ **then**
9:             $\mathcal{M} \leftarrow \mathcal{M} \cup L(O_k)$
10:     The set of $(k + 1)$-object-sets $\mathcal{C}_{k+1} \leftarrow \emptyset$
11:     **for all** $O_k, O'_k \in \mathcal{C}_k$ such that $|O_k \cap O'_k| = k - 1$ **do**
12:         $O_{k+1} \leftarrow O_k \cup O'_k$ and $L(O_{k+1}) \leftarrow$ Closed-MEMO($\mathcal{D}$, w, r, O_{k+1})
13:         **if** $L(O_{k+1})$ is not empty **then**
14:             $\mathcal{C}_{k+1} \leftarrow \mathcal{C}_{k+1} \cup \{O_{k+1}\}$
15:     $k \leftarrow k + 1$
16: $\mathcal{M} \leftarrow \mathcal{M} - \{M | M$ is not a closed-MEMO$\}$.

The A-miner initializes the M-Lists of 1-object-sets, which are likely to form larger MEMOs (lines 2-5). Starting with $k = 1$, the M-List of each $k + 1$-object-set is built only if two of its subset k-object-sets have non-empty M-Lists (lines

Table 1. A trace of A-miner

k	C_k	$\mathcal{M}$
1	$\{\{a\},\{b\},\{c\},\{d\}\}$	$\emptyset$
2	$\{\{a,b\},\{a,c\},\{b,c\},\{b,d\}\}$	$\{\{a,b\},\{a,c\},\{b,c\},\{b,d\}\}$
3	$\{\{a,b,c\}\}$	$\{\{a,b\},\{a,c\},\{b,c\},\{b,d\},\{a,b,c\}\}$
4	$\emptyset$	$\{\{a,b\},\{a,c\},\{b,c\},\{b,d\},\{a,b,c\}\}$

10-15). In doing so, if $k \geq m$, the MEMOs in the M-Lists of the k-object-sets are potential closed MEMOs, thus, they are put into the result set $\mathcal{M}$ (lines 7-9), which is finally filtered (line 16).

Figure 2a shows an example of moving object dataset containing four objects. For parameters: $m = 2$ and $w = 3$, the lattice, which represent the corresponding search space is shown in Fig. 2b, while Table 1 shows the corresponding trace of execution of Algorithm 1. A-miner starts at the bottom of the lattice with 1-object sets $C_1 = \{\{a\},\{b\},\{c\},\{d\}\}$, each of which are attached with its corresponding M-List in the lattice. For example, since, during each of the intervals $[t_1, t_3]$ and $[t_4, t_6]$, the locations of b is covered by a circle, $\{b\}$ is attached with two entries, 1-3 and 4-6. From 1-object-sets, 2-object-sets are extracted in the first iteration $k = 1$. For example, from Fig. 2a, we can see a and b are together in t_1, t_2 and t_3. Therefore, $\{a, b\}$ is put into C_2. However, a and d never met (never were inside the same small circle with) each other, therefore, C_2 does not contain $\{a, d\}$. In the next iteration $k = 2$, the 2-MEMOs in the M-Lists of each 2-object-set in C_2 are put into potential result set $\mathcal{M}$. The 3-object-set $\{a, b, c\}$ is put into $\mathcal{M}$ in the last iteration $k = 3$, in which no 4-object-set has non-empty M-List. A-miner will finally remove $\{b, c\}$ from $\mathcal{M}$ as the k-MEMO it forms is covered by the one formed by $\{a, b, c\}$ and, hence, is not a closed-MEMO.

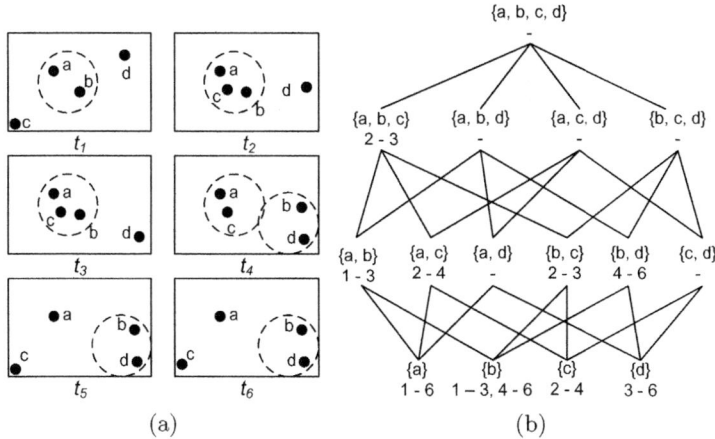

Fig. 2. An example of (a) movement of four objects and (b) its corresponding lattice

Definition 9. *For a set of objects $S \subseteq O$ containing at least k objects and an order $\prec$ defined on O, the list $P_k(S)$ containing the first k elements of S sorted according to $\prec$ is called a k-prefix of S. 0-prefix is always an empty set.*

In A-miner, finding pairs of k-object-sets sharing $k - 1$ objects to build the M-Lists of $(k+1)$-object-sets (line 11) is an expensive operation. Moreover, for each $k + 1$-object-set, O_{k+1}, there are $k^2 + k$ possible pairs of $O_k, O'_k \in C_k$ such that $|O_k \cap O'_k| = k - 1$, $O_k \subset O_{k+1}$ and $O'_k \subset O_{k+1}$ leading to redundant calculations of its M-List $L(O_{k+1})$. Therefore we use an order among the moving objects in order to have a canonical form of each object set. If two k-object-sets, O_k and O'_k, share the same $k - 1$-prefix, then we build the M-List of $O_{k+1} = O_k \cup O'_k$, ignoring other pairs of the subsets of O_{k+1}.

Building M-List of O_{k+1}. A frequent component in A-Miner is calculating the M-List of $(k + 1)$-object-set (line 12). A naive method to do so is to check if O_{k+1} forms a MEMO in each maximal time-interval I spanning w or more time-stamps as outlined in Algorithm 2. For each potential start-time *start*, the minimum covering circle C of all locations of $o \in O_{k+1}$ from *start* to *ts* is calculated (lines 5-6); *ts* is increased until the minimum circle C becomes larger than that of radius r — if the interval spans w or more time-stamps, the algorithm finds a k-closed MEMO in the maximal interval spanning from *start* to $ts - 1$ at location C' and appends it to the result $L(O_{k+1})$ (lines 7-10).

Algorithm 2. Closed-MEMO.

Input: $\mathcal{D}$, w, r and O_{k+1}.
Output: A sorted list of $(k + 1)$-closed MEMO $L(O_{k+1})$.
1: $start \leftarrow min(\{t | \langle o, t, loc \rangle \in \mathcal{D}$ and $o \in O_{k+1}\})$
2: $end \leftarrow max(\{t | \langle o, t, loc \rangle \in \mathcal{D}$ and $o \in O_{k+1}\})$
3: $ts \leftarrow start$, $L \leftarrow \emptyset$, $C' \leftarrow null$
4: **while** $ts \leq end$ **do**
5:     $P \leftarrow \{loc | \langle o, t, loc \rangle \in \mathcal{D}, o \in O_{k+1}$ and $t_{start} \leq t \leq ts\}$
6:     $C' \leftarrow C$, $C \leftarrow$ Min-Covering-Circle(P)
7:     **if** $radius(C) > r$ and $ts - start \geq w$ **then**
8:         $members(M) \leftarrow O_{k+1}$, $loc(M) \leftarrow C'$, $t_{start}(M) \leftarrow start$, $t_{end}(M) \leftarrow ts - 1$
9:         Append M to $L(O_{k+1})$, $start \leftarrow start + 1$
10:     $ts \leftarrow ts + 1$
11: **if** $end - start + 1 \geq w$ **then**
12:     $members(M) \leftarrow O_{k+1}$, $loc(M) \leftarrow C$, $t_{start}(M) \leftarrow start$, $t_{end}(M) \leftarrow end$
13:     Append M to $L(O_{k+1})$

We introduced two optimizations to Algorithm 2. In the first optimization, we further exploited the *apriori*-properties of MEMOs. For a given interval I, if O_{k+1} forms a valid $(k + 1)$-closed MEMO, a k-object-set $O_k \subset O_{k+1}$ and $O_{k'} \subset O_{k+1}$ such that $|O_{k'} - O_k| = 1$ must have valid k-closed MEMO(s) and k'-closed MEMO(s) covering I. Therefore, we utilized the M-Lists of O_k and O'_k, which are readily available in memory, to compute the M-List of O_{k+1}.

Our implementation applies Algorithm 2 only on the intervals covered by the
k-closed MEMOs $M_i \in L(O_k)$ and $M'_j \in L(O'_k)$. Since k-closed MEMOs cannot
cover each other, each M-List can be sorted in the temporal order, enabling us
to utilize a simple sort-merge-join algorithm to efficiently check such intervals.
We only use the naive approach to calculate the (sorted) M-List of 1-object-sets.

In Algorithm 2, calculating the minimum covering circle (line 6) from scratch
using the Euclidean Messenger Boy algorithm (EMB) described in [3] dominates
a substantial amount of runtime during our initial tests. Therefore, as the second
optimization, we developed an incremental version of EMB [1] that can derive the
new circle C from C' (and P') to introduce further improvements to A-Miner.

4.2   An ECLAT-Based Closed MEMO Miner (E-Miner)

In the worst case scenario, the *Apriori*-based closed MEMO miner (A-miner)
needs $\binom{n}{k} = \frac{n!}{k!(n-k)!}$ M-Lists of k-object-sets in memory in order to calculate
those of $(k+1)$-object-sets. For datasets containing records of a large number of
moving objects (large n values), the memory requirements of A-miner is tremendous even for modern workstations equipped with several gigabytes of physical
memory. Therefore, Equivalent CLAss Transformation (ECLAT), proposed in
[15], is used to partition the search space in our ECLAT-based close MEMO
Miner (E-miner).

Definition 10. *A k-equivalent-class, denoted as $C(Q_k, k)$ contains all object-
sets, each of which has at least k objects and has Q_k as their k-prefix, i.e.
$C(Q_k, k) = \{S | P_k(S) = Q_k \text{ and } |S| \geq k\}$.*

Definition 10 defines the equivalent-class $C(Q_k, k)$, which contains all object-sets
(not necessarily of the same size) having the same k-prefix, Q_k. For example, for
the moving objects shown in Fig. 2a, $\{b\}$, $\{b, c\}$, $\{b, d\}$ and $\{b, c, d\}$ belong to the
1-equivalent-class $C(\{b\}, 1)$ as they all have the same 1-prefix, $\{b\}$. All object-sets
having more than k objects in $C(P_k, k)$ can be divided into $(k + 1)$-equivalent-
classes, each of which having one of the $(k + 1)$-object-sets in $C(P_k, k)$ as its
$(k+1)$-prefix. For example, $C(\{b\}, 1)$ has two 2-object-sets $\{b, c\}$ and $\{b, d\}$ and,
thus, the object-sets having more than 1 object, i.e. $\{b, c\}$, $\{b, d\}$ and $\{b, c, d\}$,
can be divided into two 2-equivalent-classes, $C(\{b, c\}, 2) = \{\{b, c\}, \{b, c, d\}\}$
and $C(\{b, d\}, 2) = \{\{b, d\}\}$. Algorithm 3 shows an outline of E-miner, which,
starting with 0-equivalent-class (the whole search space), recursively divides the
k-equivalent-class into $(k + 1)$-equivalent-classes until each can fit in memory.
E-miner maintains a stack of equivalent-classes that needs further partitioning.

E-miner starts by pushing the 0-equivalent-class, represented by $C(\emptyset, k)$, onto
the stack (line 1). For each top-most k-equivalent-class $C(Q_k, k)$ popped out
from stack, E-miner checks if its prefix Q_k forms a k-MEMO first and maintains
the result list $\mathcal{M}$ accordingly (lines 3-6). Then, E-miner checks if the $C(Q_k, k)$
can fit into the memory (line 7). If so, it is processed in the same fashion as

---

[1] Since it is fairly straight-forward to derive incremental EMB from original EMB, we
omit its details to preserve space.

Algorithm 3. ECLAT-based Closed MEMO Miner.

Input: $\mathcal{D}$, r, m, w and an order $\prec$ on O
Output: A set of closed MEMO $\mathcal{M}$.
 1: $\mathcal{M} \leftarrow \emptyset$ and Push 0-equivalent-class $C(\emptyset, 0)$ to stack.
 2: **while** Stack is not empty **do**
 3:     Pop k-equivalent-class $C(Q_k, k)$ from stack
 4:     M-List $L(Q_k) \leftarrow$ Closed-MEMO$(\mathcal{D}, w, r, Q_k)$
 5:     **if** $L(Q_k)$ is not empty **then**
 6:         $\mathcal{M} \leftarrow \mathcal{M} \cup L(Q_k)$
 7:     **if** $C(Q_k, k)$ fits in memory **then**
 8:         Prefix $Q \leftarrow Q_k$, $k \leftarrow k+1$ and $\mathcal{C}_k \leftarrow \emptyset$
 9:         **for all** $O_k \in \{Q \cup \{o_i\}|$ if $o_j \in Q$ then $o_j \prec o_i$ or $o_j = o_i\}$ **do**
10:             $\mathcal{C}_k \leftarrow \mathcal{C}_k \cup \{O_k\}$
11:         **while** $\mathcal{C}_k \neq \emptyset$ **do**
12:             **for all** $O_k \in \mathcal{C}_k$ **do**
13:                 **if** $k \geq m$ **then**
14:                     $\mathcal{M} \leftarrow \mathcal{M} \cup L(O_k)$
15:             The set of $(k+1)$-object-sets $\mathcal{C}_{k+1} \leftarrow \emptyset$
16:             **for all** $O_k, O_k' \in \mathcal{C}_k$ s.t $|O_k \cap O_k'| = k-1$ **do**
17:                 $O_{k+1} \leftarrow O_k \cup O_k'$ and $L(O_{k+1}) \leftarrow$ Closed-MEMO$(\mathcal{D}, w, r, O_{k+1})$
18:                 **if** $L(O_{k+1})$ is not empty **then**
19:                     $\mathcal{C}_{k+1} \leftarrow \mathcal{C}_{k+1} \cup O_{k+1}$
20:             $k \leftarrow k+1$
21:     **else**
22:         **for all** $Q_{k+1} \in \{Q_k \cup \{o_i\}|$ if $o_j \in Q_k$ then $o_j \prec o_i\}$ **do**
23:             Push $C(Q_{k+1}, k+1)$ to stack
24: $\mathcal{M} \leftarrow \mathcal{M} - \{M|M$ is not a closed-MEMO$\}$.

---

in A-miner (lines 8-20). Otherwise, the k-equivalent-class is divided into $k+1$-equivalent-classes (lines 22-23).

Table 2 shows a partial trace of E-miner on the search space shown in Fig. 2b for the same set of parameters: $m = 2$ and $w = 3$. Let us assume the memory can only hold two object-sets (and their M-Lists). In step 1, it checks the whole search space lattice, $C(\emptyset, 0)$, and, since it cannot fit into the memory (it contains 16 object-sets), pushes four 1-equivalent-classes, $C(\{a\}, 1)$, $C(\{b\}, 1)$, $C(\{c\}, 1)$ and $C(\{d\}, 1)$ onto the stack. In steps 2 and 3, $C(\{d\}, 1) = \{\{d\}\}$ and $C(\{c\}, 1) = \{\{c\}, \{c, d\}\}$ are popped and, since these equivalent-classes can fit into the memory, their members are examined (but nothing is put into the result set, $\mathcal{M}$ since $\{c\}$, $\{d\}$ and $\{c, d\}$ does not form any MEMO for the given parameters). In step 4, $C(\{b\}, 1) = \{\{b\}, \{b, c\}, \{b, d\}, \{b, c, d\}\}$ is popped from the stack. Since it has four object-sets and cannot fit in the memory, it is divided into two 2-equivalent-classes $C(\{b, c\}, 2)$ and $C(\{b, d\}, 2)$, which are pushed onto the stack for later processing. In steps 5 and 6, $C(\{b, d\}, 2) = \{\{b, d\}\}$ and $C(\{b, c\}, 2) = \{\{b, c\}, \{b, c, d\}\}$ are popped out, checked and $\{b, d\}$ and $\{b, c\}$ are inserted into the result set $\mathcal{M}$. In the next steps, the equivalent-class $C(\{a\}, 1)$ containing eight object-sets will be divided and processed.

Table 2. A partial trace of E-miner

Step	$C(Q_k, k)$	Stack	$\mathcal{M}$
1	$C(\emptyset, 0)$	$\{C(\{a\}, 1), C(\{b\}, 1), C(\{c\}, 1), C(\{d\}, 1)\}$	$\emptyset$
2	$C(\{d\}, 1)$	$\{C(\{a\}, 1), C(\{b\}, 1), C(\{c\}, 1)\}$	$\emptyset$
3	$C(\{c\}, 1)$	$\{C(\{a\}, 1), C(\{b\}, 1)\}$	$\emptyset$
4	$C(\{b\}, 1)$	$\{C(\{a\}, 1), C(\{b, c\}, 2), C(\{b, d\}, 2)\}$	$\emptyset$
5	$C(\{b, d\}, 2)$	$\{C(\{a\}, 1), C(\{b, c\}, 2)\}$	$\{\{b, d\}\}$
6	$C(\{b, c\}, 2)$	$\{C(\{a\}, 1)\}$	$\{\{b, d\}, \{b, c\}\}$
...	...	...	

In E-Miner, calculating M-List of the k-Prefix Q_k popped out from the stack (line 4) is a frequent component. Since it is costly to use the naive computation described in Algorithm 2, in our implementation, their M-Lists are computed before they are pushed onto the stack. We maintain the M-List of all 2-object-sets and, for any Q_{k+1} about to be pushed onto the stack, its M-List $L(Q_{k+1})$ is computed from the M-List of Q_k and any 2-object-set O_2' such that $|O_2' - Q_k| = 1$.

4.3 A Filter-And-Refinement Closed MEMO Miner (FAR-Miner)

In A-miner and E-miner, the dataset is referred for the locations of moving objects in calculating the minimum covering circles to verify if the objects actually form a MEMO for the given parameters. Those queries (and computation of the circle) are often wasted when the objects do not form a MEMO (when the radius of the circle is larger than the given r). In Filter-And-Refinement-based Closed MEMO Miner (FAR-miner), we introduced a filtering step, which needs less access to the dataset, to avoid computation of minimum covering circles.

Lemma 2. *If a set of points P is covered by a minimum covering circle C, whose radius $r(C) \leq r$, then two points $p, q \in P$ cannot be further apart than $2r$.*

Proof. No two points $p, q \in P$, which are either inside or on the edge of C, can be further apart than the length of its diameter, i.e. $distance(p, q) \leq 2r(C)$, and $2r(C) \leq 2r$. Therefore, $distance(p, q) \leq 2r$. □

Lemma 2 claims that if the distance between two points $p, q \in P$ is more than $2r$, the minimum covering circle C of P must have a radius larger than r. In other words, if the distance between location of object o_i at t_j and that of object o_i' at t_j' is further than $2r$, o_i and o_i' do not form a MEMO at interval I containing t_j and t_j'.

Definition 11. *For given parameters : m, w and r, a subset of the dataset $\mathcal{D}' \subseteq \mathcal{D}$, which contains all movement records of a set of objects O' in an interval $I(\mathcal{D}')$ is termed as a potential-MEMO if (i) O' has at least m objects, (ii) $I(\mathcal{D}')$ spans for at least w consecutive time-stamps and (iii) all locations the objects $o \in O'$ visited during $I(\mathcal{D}')$ are not further than $2r$ from each other.*

Definition 11 defines a potential-MEMO, which is likely to form a MEMO for the given parameters. Closed potential-MEMO, k-potential-MEMO and k-closed potential-MEMO can be defined in ways similar to Def. 3, 6 and 7, respectively. It is also apparent that potential-MEMOs also have *apriori*-properties. FAR-miner consists of two steps (i) the **Filtering step**, which finds the set of all closed potential-MEMOs $\mathcal{M}' = \{\mathcal{D}'|\mathcal{D}'$ is a closed potential-MEMO$\}$ and (ii) the **Verification step**, which, for each potential-MEMO $\mathcal{D}' \in \mathcal{M}'$, verifies if the objects actually form a MEMO.

To perform the filtering step, we use A-miner (or E-miner), using a slightly modified version of Algorithm 2 as its subroutine, since potential-MEMOs also have the *apriori*-properties. Therefore for the filtering step, instead of building minimum covering circles and checking their radii (lines 6-7), the modified algorithm would simply check the distance between all $p, q \in P$. It is easy to show that, if no two points in each of the sets, $A \cup B$, $A \cup C$ and $B \cup C$ are further than $2r$, no two points in the set $A \cup B \cup C$ are. Therefore, when O_k and O'_k (such that $O_k - O'_k = \{o_i\}$ and $O'_k - O_k = \{o_j\}$) are known to form potential-MEMOs in interval I, whether $O_{k+1} = O_k \cup O'_k$ forms a potential-MEMO in I can be easily derived by checking if all the locations of o_i and o_j in I are within distance $2r$ of each other. In other words, by maintaining all 2-closed potential-MEMOs in memory, when $k \geq 2$, the potential-MEMOs formed by O_{k+1} object-sets can be derived without referring the dataset for the actual locations of the objects. To perform the verification step, we directly apply A-miner (or E-miner) discussed in Sect. 4.1 (4.2) on $\mathcal{D}'$ with the given parameters.

Since it is possible to perform the filtering and verification using either A-miner or E-miner dialects, there are four possible flavors of FAR-miner. However, in Sect. 5, we report the performance of filtering and verification steps, both using A-miner dialect, as all flavors show similar performance during our initial tests.

As a hindsight, we noted that, due to the *apriori*-properties of the MEMOs, intermediate mining results (k-MEMOs and potential k-MEMOs) obtained using parameters : m, w and r can be easily reused for subsequent runs on the same dataset using different parameters : m', w' and r', bounded by the criterion: $r' \leq r$ and $w' \geq w$. However, it is not practical to save all intermediate results of each previous run. Therefore, we recommend to save only the intermediate results obtained using a short w and/or a large r as they are more likely to be reused. The intermediate results in question can be further trimmed down to its subsets based on the domain knowledge. For example, in an application where meetings of 3 objects are common, only k-MEMOs for $k \geq 4$ are to be saved for reuse. In our experiments in Sect. 5, we do not reuse any intermediate result.

5   Experiments

5.1   Experiment Setup

We implemented the algorithms in Java and conducted all the experiments on a Red-Hat Enterprise Linux server equipped with Intel Xeon X5365 CPU running at 3.00GHz and a physical memory capped at 8GB. We copied the whole dataset

Table 3. A summary of the datasets

Name	Object Count	Covers	No. of Records
Statefair	19	3hr	17,545
Orlando	41	14hr	133,076
New York	39	22hr	118,584
NCSU	35	21hr	128,417
KAIST	92	23hr	404,981
SF-Cab21	482	8hr	1,156,458
SF-Cab22	477	8hr	1,236,497

into physical memory prior to the experiments although our algorithms (A-miner, E-miner and FAR-miner) can also work on disk-based datasets.

We adapted the column-sweeping algorithm (CS-miner) proposed in [6] for reference because it is the only work in the literature, which, theoretically, can report all closed MEMOs accurately. However, in practice, it is impossible to continuously rotate a 3D-column, whose base (height) represents place (duration) of a meeting, around each location an object visited in each time-stamp (and check the order, in which the objects enter/leave the column) as specified. Therefore, we decided to rotate the column discreetly by a user-defined angular step, θ, which, in our experiments, is set to $1°$ unless otherwise stated. CS-miner reports less accurate results [2] when θ becomes larger (relative to r).

We used five datasets – Statefair, Orlando, New York, NCSU and KAIST – of human movement [9] and larger datasets, SF-Cab21 and SF-Cab22, of taxi movement extracted from [12]. New York consists of traces of the volunteers commuting by subways, by buses and on foot, while NCSU and KAIST consist of traces of students on campuses. SF-Cab21 (SF-Cab22) consists of taxi movement from 8AM to 4PM in San Francisco Bay Area on 21-Apr-08 (22-Apr-08). The time interval between each pair of consecutive time-stamps is 10 seconds in all datasets. We pre-processed all datasets to eliminate missing records in gaps shorter than an hour.

5.2   Results and Analysis

The outcome of the first set of experiments, comparing the performance of the algorithms on human movement datasets, is shown in Fig. 3a (note that the y-axis is in log-scale). We were looking for meetings of at least two people ($m = 2$) lasting for at least 15 minute ($w = 15$ minute), which were reasonable choices of parameters for the corresponding datasets. In NCSU and KAIST, we even discovered meetings of up to 3 and 5 students. Our proposed data-driven algorithms, A-miner and E-miner, run faster than CS-miner to find the closed MEMOs as they ignore the fast-moving objects in building M-List of 1-object-sets while

---

[2] CS-Miner often has missing results (false-negatives) as well as reports several over-lapping (non-closed) MEMOs in place of a single MEMOs.

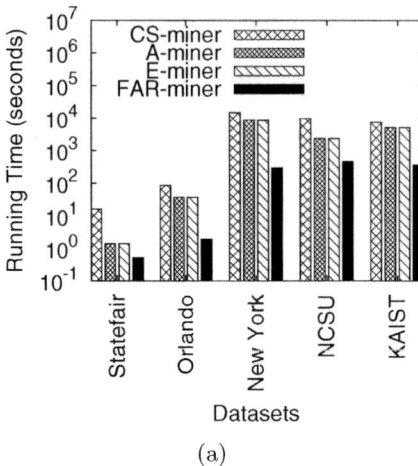

Dataset	A-Miner, E-Miner and FAR-Miner	CS-Miner
Statefair	5	5
Orlando	15	15
New York	16	16
NCSU	48	46
KAIST	126	123

(a)   (b)

Fig. 3. Comparison of (a) the performance of the algorithms and (b) the number of MEMOs they found on human-movement datasets using $m = 2$, $w = 15$ minutes and $r = 10$ meters

CS-miner attempts to build MEMOs containing them in vain. FAR-miner outperforms A-miner and E-miner by a large order of magnitude due to its cheap pruning mechanism. We also noted that, even using a reasonably low θ value of $1°$, CS-miner still missed some results for NCSU and KAIST (see Fig. 3b).

In the next set of experiments, whose outcome is plotted in Fig. 4a and 4b, we assessed the scalability of the algorithms on different sizes of datasets. We randomly picked 30, 60 and 90 moving objects from the largest human-movement dataset, KAIST, and 150, 300, 450 moving objects from SF-Cab21 for these experiments. We set the value of r to 30 meters in order to find MEMOs in smaller subsets of KAIST. For executions on subsets of SF-Cab21, we intuitively chose parameters ($m = 5$, $w = 30$ minute and $r = 25$ meter) to reflect taxis waiting in taxi queues and severe traffic jams. We noticed that the larger value of r we set, the smaller value of θ we should use to maintain accuracy for CS-miner since it reports inaccurate results, when we set its internal parameter $\theta = 1°$ in KAIST. Therefore, for experiments on SF-Cab21, we used a θ value of $0.5°$. Our algorithms (A-Miner, E-Miner and FAR-Miner) scale well on the increasing dataset size but CS-Miner does not – exceeding eight hours to process data of 90 moving objects in KAIST. Among our proposed algorithms, FAR-miner performs better than the others except in the largest subset of KAIST, when the very large value of $r = 30$ meters increased the number of potential MEMOs to nearly three thousand and A-miner outperforms FAR-miner.

Analyzing the run-time statistics of FAR-miner in previous experiments given in Table 4, we found that run-time is dominated by the verification step as the filtering step took only a few seconds regardless of the dataset. The verification time is not dependent on the size of the dataset but on the number of potential results ($\mathcal{D}'$). To verify this claim, we conducted another experiment on

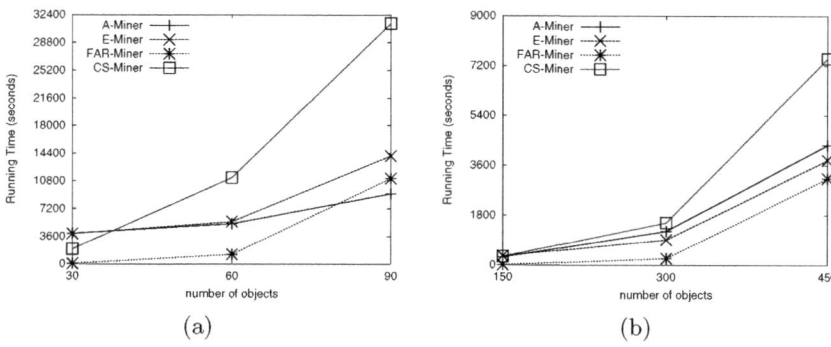

Fig. 4. Impact of the size of the dataset on performance of the algorithms (a) for $m = 2$, $w = 15$ minutes and $r = 30$ meters on subsets of KAIST and (b) for $m = 5$, $w = 30$ minutes and $r = 25$ meters on subsets of SF-Cab21

SF-Cab22, which is comparable in size to SF-Cab21, using the same set of parameters: $m = 5$, $w = 30$ minutes and $r = 25$ meters. It turns out that there are fewer MEMOs for the given parameters in SF-Cab22 as well as fewer potential results. Subsequently, verification time (and total running time) of SF-Cab22 is significantly smaller than that of SF-Cab21. Since, in typical circumstances, the number of meetings formed in a dataset are supposed to be few, we noted that FAR-miner will give reasonable performance regardless of the size of input dataset. Even when there are many meetings formed in a dataset, FAR-miner outperforms A-miner and E-miner as they take longer to complete in SF-Cab21 for the same parameters (see Fig. 4b).

Table 4. Run-time decomposition of FAR-miner for verification and filtering steps

Dataset	Filtering (seconds)	Verification (seconds)	Total (seconds)	No. of MEMOs	No. of $\mathcal{D}'$
Statefair	0.2	0.2	0.4	5	3
Orlando	1.2	0.6	1.8	15	18
New York	17.5	284.1	301.6	16	19
NCSU	6.2	452.1	458.3	48	64
KAIST	9.9	356.6	365.5	126	182
SF-Cab21	18.4	3110.3	3128.7	561	1230
SF-Cab22	19.8	1139.8	1159.6	10	129

In the subsequent sets of experiments, we studied the impact of parameter values on the performance of the algorithm. Figure 5a and 5b show the impact of value of r on the performance of the algorithms. Increasing r relaxes the conditions by allowing MEMOs with larger meeting places and increases the number of closed MEMO in a dataset. Thus, it, in turn, increases the run-time of

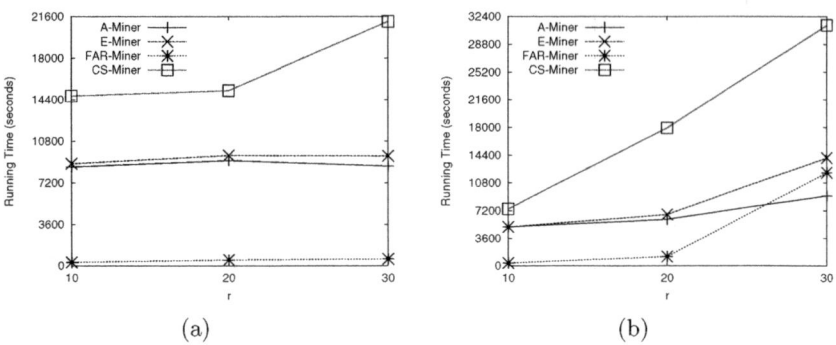

Fig. 5. Impact of r on performance of the algorithms for $m = 2$ and $w = 15$ minutes on (a) New York and (b) KAIST

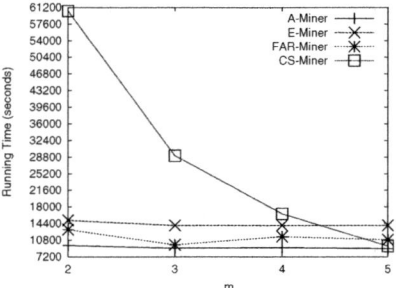

Fig. 6. Impact of m on performance of the algorithms for $w = 15$ minutes and $r = 30$ meters on KAIST

all algorithms. However, performance of CS-miner degraded rapidly (especially in KAIST) as r increases while our algorithms' performance were stable. Most of the time, FAR-miner significantly outperforms the rest except in KAIST at $r = 30$ meters. In this peculiar instance, there was nearly three thousand potential MEMOs to verify and, since verification time dominates the run-time and depends on the number of potential MEMOs as we noted earlier, FAR-miner took a few minutes more than A-miner to complete.

Figure 6 shows the impact of value of m on the performance of algorithms. We used a lower value of $\theta = 0.5°$ to improve CS-miner's accuracy ($r = 30$ meter). Increasing m reduces the number of MEMOs found in a particular dataset. Therefore, CS-miner finishes faster than A-miner and E-miner when $m = 5$ as there are a very few meetings of 5 or more students in KAIST. FAR-miner still performs better than the others due to its powerful filtering step. All our algorithms' performance are stable regardless the value of m given.

Figure 7a and 7b shows the impact of value of w on the performance of the algorithms. Increasing w puts more restriction by demanding participants to stay

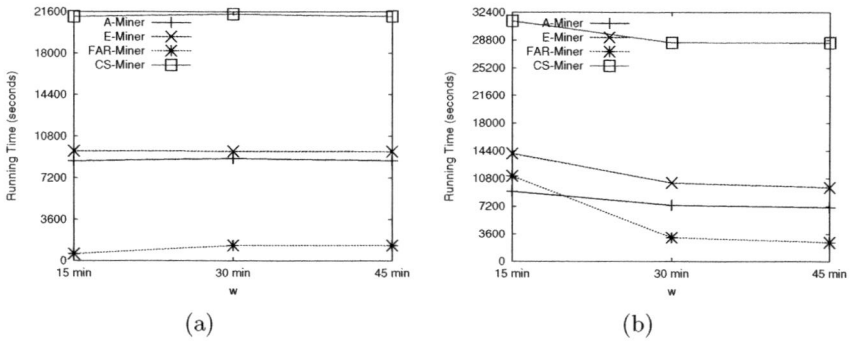

Fig. 7. Impact of w on performance of the algorithms for $m = 2$ and $r = 30$ meters on (a) New York and (b) KAIST

still longer and decreases the number of closed MEMO in a dataset. Thus, it, in turn, decreases the run-time of all algorithms. Our data-driven algorithms still outperform CS-miner in all cases and, most of the time, FAR-miner significantly performs better than the rest.

From our experiments, we concluded that our proposed data-driven algorithms, A-miner and E-miner, performed better than CS-miner in many realistic settings. Although E-miner took slightly longer to complete than A-miner, it can limit its memory needs and is suitable for larger datasets. In real-life scenarios, where few MEMOs are expected, we recommend to use FAR-miner as its fast filtering step would improve performance significantly.

6   Conclusion

In this paper, we introduced a more precise definition of meeting patterns called MEMO, taking the accuracy of reported meeting place into account. We developed three novel algorithms to discover closed MEMOs. Experiments on real-life datasets showed that our proposed algorithms can perform better than the existing one. A general framework to discover all aggregate movement patterns is considered as future research goal.

References

1. Agrawal, R., Srikant, R.: Fast algorithms for mining association rules in large databases. In: Bocca, J.B., Jarke, M., Zaniolo, C. (eds.), Proceedings of 20th International Conference on Very Large Data Bases, VLDB 1994, Santiago de Chile, Chile, September 12-15, pp. 487–499. Morgan Kaufmann, San Francisco (1994)
2. Benkert, M., Djordjevic, B., Gudmundsson, J., Wolle, T.: Finding popular places. In: Proc. 18th International Symposium on Algorithms and Computation (2007)

3. Drager, L.D., Lee, J.M., Martin, C.F.: On the geometry of the smallest circle enclosing a finite set of points. Journal of the Franklin Institute 344(7), 929–940 (2007)
4. Elzinga, J., Hearn, D.W.: Geometrical Solutions for Some Minimax Location Problems. Transportation Science 6(4), 379–394 (1972)
5. Gudmundsson, J., Kreveld, M., Speckmann, B.: Efficient detection of motion patterns in spatio-temporal data sets. In: Proceedings of the 13th International Symposium of ACM Geographic Information Systems, pp. 250–257 (2004)
6. Gudmundsson, J., van Kreveld, M.: Computing longest duration flocks in trajectory data. In: GIS 2006: Proceedings of the 14th Annual ACM International Symposium on Advances in Geographic Information Systems, pp. 35–42. ACM, New York (2006)
7. Han, J., Pei, J., Yin, Y.: Mining frequent patterns without candidate generation. In: Proceedings of the 2000 ACM SIGMOD International Conference on Management of Data, SIGMOD 2000, pp. 1–12. ACM, New York (2000)
8. Hwang, S.-Y., Liu, Y.-H., Chiu, J.-K., Lim, E.: Mining mobile group patterns: A trajectory-based approach. In: Ho, T.-B., Cheung, D.W.-L., Liu, H. (eds.) PAKDD 2005. LNCS (LNAI), vol. 3518, pp. 713–718. Springer, Heidelberg (2005)
9. Jetcheva, J.G., Chen Hu, Y., Palchaudhuri, S., Kumar, A., David, S., Johnson, B.: Design and evaluation of a metropolitan area multitier wireless ad hoc network architecture, pp. 32–43 (2003)
10. Laube, P., Imfeld, S.: Analyzing relative motion within groups of trackable moving point objects. In: Egenhofer, M.J., Mark, D.M. (eds.) GIScience 2002. LNCS, vol. 2478, p. 132. Springer, Heidelberg (2002)
11. Patrick Laube, S.I., van Kreveld, M.: Finding remo - detecting relative motion patterns in geospatial lifelines. In: Proceedings of the 11th International Symposium on Spatial Data Handling, pp. 201–215 (March 2004)
12. Piorkowski, M., Sarafijanovoc-Djukic, N., Grossglauser, M.: A Parsimonious Model of Mobile Partitioned Networks with Clustering. In: The First International Conference on COMmunication Systems and NETworkS (COMSNETS) (January 2009)
13. Rademacher, H., Toeplitz, O.: The spanning circle of a finite set of points. The Enjoyment of Mathematics: Selection from Mathematics for the Amateur, 103–110 (1957)
14. Wang, Y., Lim, E.-P., Hwang, S.-Y.: Efficient algorithms for mining maximal valid groups. The VLDB Journal 17(3), 515–535 (2008)
15. Zaki, M.J.: Scalable algorithms for association mining. IEEE Transactions on Knowledge and Data Engineering 12, 372–390 (2000)

Density Based Subspace Clustering over Dynamic Data

Hans-Peter Kriegel, Peer Kröger, Irene Ntoutsi, and Arthur Zimek

Institute for Informatics, Ludwig-Maximilians-Universität München
http://www.dbs.ifi.lmu.de
{kriegel,kroeger,ntoutsi,zimek}@dbs.ifi.lmu.de

Abstract. Modern data are often high dimensional and dynamic. Subspace clustering aims at finding the clusters and the dimensions of the high dimensional feature space where these clusters exist. So far, the subspace clustering methods are mainly static and cannot address the dynamic nature of modern data. In this paper, we propose a dynamic subspace clustering method, which extends the density based projected clustering algorithm PreDeCon for dynamic data. The proposed method efficiently examines only those clusters that might be affected due to the population update. Both single and batch updates are considered.

1 Introduction

Clustering is the unsupervised classification of data into natural groups (called clusters) so that data points within a cluster are more similar to each other than to data points in other clusters. Due to its broad application areas, the clustering problem has been studied extensively in many contexts and disciplines, including Data Mining. As a result, a large number of clustering algorithms exists in the literature (see [16] for a thorough survey). However, modern data impose new challenges and requirements for the clustering algorithms due to their special characteristics. First of all, a *huge amount* of data is collected nowadays as a result of the wide spread usage of computer devices. This possibility of cheaply recording massive data sets may also be the reason for another new characteristic of modern data; the *high dimensionality* of objects. While years ago, data recording was more expensive and, thus, the relevance of features was carefully evaluated before recording, nowadays, people tend to measure as much as they can. As a consequence, an object might be described by a large number of attributes. Many of these attributes may be irrelevant for a given application like cluster analysis and there might be correlations or overlaps between these attributes. In addition to their quantity and high dimensionality, today's data is often highly *dynamic*, i.e., new data records might be inserted and existing data records might be deleted, as time goes by.

As an example, consider the data derived from the Bavarian newborn screening program [20]. For each newborn in Bavaria, Germany, the blood concentrations of 43 metabolites are measured in the first 48 hours after birth producing

J.B. Cushing, J. French, and S. Bowers (Eds.): SSDBM 2011, LNCS 6809, pp. 387–404, 2011.

a vast amount of high dimensional data that is highly dynamic (new individuals are added usually in a batch on a daily or weekly basis). The analysis of these data shall help doctors in the diagnosis the exploration of known and new metabolic diseases. Clustering the data is a crucial step in this process. However, for different diseases, it is very likely that different metabolites are relevant. Thus, clusters representing groups of newborns with a homogeneous phenotype, e.g. suffering from a similar disease, can usually only be found in subspaces of the data. As batches of new individuals are coming in every day or week, the detected clustering structure needs to be updated as well. Due to the huge amount of data, the update of the clustering structure should be done incrementally only for the changing part of the structure, rather than re-computing the complete structure from scratch. Let us note that such screening projects are implemented in a large number of states/countries so that there are many data sets having similar characteristics that need to be analyzed.

The scenario described above represents a general *data warehouse environment*. With this term, we do not associate a certain architecture, but describe an environment in which changes in the transactional database are collected over some period (e.g. daily) and the data warehouse is updated using batch operations. Beside data originating from scientific experiments, also many companies store terabytes of corporate data in such an environment. Applications like scientific data analysis or industrial decision support systems in such environments require not only high accuracy from data analysis methods but also fast availability of up-to-date knowledge — a prohibitive demand for many data mining algorithms which are able to gain knowledge only from scratch using highly complex operations. Rather, to cope with the problem of updating mined patterns in a data warehouse environment, algorithms preferably should permanently store the acquired knowledge in suitable data structures and facilitate an efficient adaptation of this stored knowledge whenever the raw data changes.

Lately, a lot of work has been carried out on adapting traditional clustering algorithms in order to meet the requirements of modern systems or on proposing new algorithms that are specialized on handling data with the above features. In particular, several methods have been proposed for each of the aforementioned problems separately, like for clustering of large amounts of data, e.g. [22,11,5], for clustering over data streams, e.g. [15,2], for change detection and monitoring over evolving data, e.g. [1,21], as well as for clustering high dimensional data (see [19] for a survey). Less work though has been done to tackle the complete list of challenges in a single, unified approach.

We propose a new algorithm, based on the density based subspace clustering algorithm PreDeCon [6] providing a solution to the problem of high dimensionality by finding both clusters and subspaces of the original feature space where these clusters exist. The original PreDeCon works upon static datasets. In this work, we propose an incremental version of PreDeCon, which also deals with the issue of dynamic data.[1] The new algorithm can also serve as a framework for

---

[1] A preliminary version of this paper has been discussed at the StreamKDD 2010 workshop [18].

monitoring clusters in a dynamic environment. We choose the algorithm PRE-DECON [6] because it already addresses the problem of high dimensional data (for static scenarios) and it relies on a density-based clustering model such that updates usually do not affect the entire clustering structure but rather cause only limited local changes. This is important to explore update procedures for dynamic data.

The rest of the paper is organized as follows. In Section 2, we discuss the related work and our contribution. In Section 3, we present the basic notions of PREDECON which are necessary for the understanding of the incremental method. In Section 4, we present the incremental algorithm, INCPREDECON. We distinguish between a single update scenario and a batch update scenario (Section 5 and 6, respectively). Experimental results are reported in Section 7. Section 8 concludes our work.

2   Related Work and Contributions

2.1   Subspace Clustering

The area of subspace clustering has lately emerged as a solution to the problem of the high dimensionality of the data. Its goal is to simultaneously detect both clusters (i.e.,, sets of objects) and subspaces of the original feature space where these clusters exist. This is in contrast to the traditional clustering that searches for groups of objects in the full dimensional space [16]. Also this is in contrast to global dimensionality reduction techniques like Principal Component Analysis (PCA) that search for clusters in the reduced (though full) dimensional space. In subspace clustering different features might be relevant for different clusters and the goal is to find both the clusters and the features that form these clusters.

Recent work on subspace clustering (see e.g. [19] for a review) so far focus on finding clusters in different subspaces of the original feature space in static data. None of these methods are suitable to efficiently keep track of changes of the clustering structure over time. Rather, the clustering structure can only be updated by computing the entire clustering from scratch.

2.2   Incremental Clustering

Traditional incremental clustering methods rely on the old clustering at time point $t-1$ (based on dataset $\mathcal{D}_{t-1}$) and on the update operations at time point t in order to derive the new clustering at t. In this category belong methods like incDBSCAN [10] which is the incremental version of the density based algorithm DBSCAN [11] and incOPTICS [17] which is the incremental version of the density based hierarchical clustering algorithm OPTICS [5]. Both incDBSCAN and incOPTICS methods exploit the fact that, due to the density based nature of the corresponding static algorithms, an update operation affects only some part of the old clustering instead of the whole clustering. The update process works directly upon raw data. Both methods produce the same results with the corresponding static methods when the latest are applied over the accumulative dataset $\mathcal{D}_t$. Charikar et al.[8] present an incremental K–Means method which maintains a collection of k clusters as the dataset evolves. When a new point is presented, it is either assigned to one of the current k clusters, or it starts a new cluster while two existing clusters are merged into one, so as the total number of clusters does not exceed the threshold k. Chen et al. [9] propose the incremental hierarchical clustering algorithm GRIN which is based on gravity theory in physics. In the *first phase*, GRIN constructs the initial clustering dendrogram, which is then flattened and its bottom levels are pruned in order to derive the so called tentative dendrogram. For each cluster, the tentative histogram keeps the centroid, the radius and the mass of the cluster. In the *second phase*, new data instances are inserted one by one and it is decided whether they belong to leaf nodes of the tentative dendrogram or are outliers. If the tentative outlier buffer exceeds some threshold, a new tentative dendrogram is reconstructed. Both [8] and [9] are approximate methods, by means that the resulting clustering after the update is not assured to be identical to the one we would obtain if we applied from scratch the static versions of the algorithms over the accumulative dataset $\mathcal{D}_t$. This is due to the fact, that the update process works upon cluster summaries rather than upon raw data; the new data at t are actually "mapped" to the closer cluster of the existing clustering (from timepoint $t - 1$).

2.3   Stream Clustering

Data streams impose new challenges for the clustering problem since "it is usually impossible to store an entire data stream or to scan it multiple times due to its tremendous volume" [14]. As a result, several methods have been proposed that first summarize the data through some summary structure and then apply clustering over these summaries instead of the original raw data. With respect to the clustering quality, these summaries might be either lossy (that is, they correspond to some approximation of the raw data) or lossless (that, is they exactly maintain the information contained in the original raw data). Agrawal et al. [2] propose the CluStream framework for clustering of evolving data streams. The clustering process is split into an online and an offline part: The online component periodically stores summary statistics (the so called, micro–clusters), whereas the offline component uses these micro–clusters for the formation of the actual clusters (the so called, macro–clusters) over a user–defined time horizon. No access to raw data is required in this method, since the clustering takes place over the microclusters, which correspond to a lossy representation of the original data. The incremental part in this case is the online component which updates the micro–clusters, whereas the clustering process is applied from scratch over these updated summaries. DenStream [7] follows the online–offline rationale of CluStream [2] but in contrast to CluStream that is specialized to spherical clusters, it can detect clusters of arbitrary shapes. In the context of their DEMON framework, Ganti et al. [12] present BIRCH+, an incremental extension of BIRCH [22]. The original BIRCH [22] first summarizes the data into subclusters and then it clusters those subclusters using some traditional clustering algorithm. The subclusters are represented very concisely through cluster features. In BIRCH+, the cluster features are maintained incrementally as updates occur, and then the clustering step takes place as in BIRCH over those (now updated) summaries. So, the incremental part is that of summary structure update, whereas clustering is then applied from scratch over the updated summary structure. The incremental version produces the same results as the static version when applied on the accumulative dataset $\mathcal{D}_t$.

2.4   High Dimensional Stream Clustering

Gao et al. [13], propose DUCStream, an incremental data stream clustering algorithm that applies the idea of dense units introduced in CLIQUE [4] to stream data. As in CLIQUE [4], the data space is split into units and a cluster is defined as a maximal set of connected dense units. Their method relies on incrementally updating, according to the update operation, the density of these units and on detecting units that change from dense to non-dense and the inverse. After the grid update phase, they identify the clusters using the original procedure of CLIQUE. DUCStream does not require access to the raw data of the past time points, but only over the summary grid structure. The incremental version produces the same results as the static version when applied to the accumulative dataset $\mathcal{D}_t$. Note that although CLIQUE is a subspace clustering algorithm, the proposed method [13] updates incrementally only the grid summary structure, whereas the clusters are discovered from scratch over the (now updated) grid. This is a clear difference to our work, where the goal is to incrementally update the existing clustering (at $t - 1$) based on the dataset updates at t, so as to finally derive the new clustering at t. Agrawal et al. [3] extend the idea of CluStream [2] to high dimensional data streams by proposing HPStream, a method for projected data stream clustering. A summary structure, the so called fading cluster structure, is proposed which comprises a condensed representation of the statistics of the points inside a cluster and can be updated effectively as the data stream proceeds. The input to the algorithm includes the current cluster structure and the relevant set of dimensions associated with each cluster. When a new point arrives, it is assigned to the closest cluster structure or if this violates the limiting radius criteria, a new cluster is created and thus some old cluster should be deleted in order for the total number of clusters to not exceed the maximum number k. In each case, the cluster structure and the relevant dimensions for each cluster are dynamically updated. Although HPStream is a subspace clustering method and we propose an incremental subspace clustering method in this work, there are core differences between the two approaches and their scopes. In particular, HPStream is targeted to stream data and thus works upon summaries and provides an approximation solution to the clustering problem. On the other hand, our INCPREDECON method works upon dynamic data, requires access to raw data (although this access is restricted to only a subset of the original dataset) and provides exact solution to the clustering problem (i.e., we obtain the same results with those obtained by applying the static PREDECON over the acumulated dataset $\mathcal{D}_t$).

2.5   Contributions

None of the existing methods can be applied to the scenario of massive, high dimensional databases that are updated over time like in a data warehouse environment. In this work, we propose an incremental version of the density based subspace preference clustering algorithm PreDeCon [6] which comprises a first step towards an integrated approach to the above listed challenges. We choose the algorithm PreDeCon because it already addresses the problem of high dimensional data (for static scenarios) and it relies on a well-known and established clustering model. Let us note that we do not discuss nor evaluate benefits and limitations of different cluster models in this paper but solely propose concepts to adapt an existing model to the various challenges of today's data.

The methods for finding subspace clusters in data streams mentioned above are to some degree related to the incremental subspace clustering in data warehouses. Both methodologies aim at providing the user with up-to-date information on subspace clusters very quickly in a dynamic, high dimensional environment. However, data streams impose different requirements on clustering algorithms and the entire data mining process. In particular, in a data warehouse, the clustering algorithm has access to all points currently in the database and not necessarily only to the most recently inserted points or to summaries of the raw data as for stream data. In addition, when clustering stream data, the algorithm for updating the summaries is restricted to sequential access to newly inserted objects and the clustering is then re-computed on the summary information only. This restriction does not apply to algorithms for incremental clustering in a data warehouse environment. Our solutions are therefore different from the data stream clustering context in these two aspects.

3   The Algorithm PreDeCon

PreDeCon [6] adapts the concept of density based clusters, introduced in DB-SCAN [11], to the context of subspace clustering. The notion of *subspace preferences* for each point defines which dimensions are relevant to cluster the point. Roughly speaking, a dimension is relevant to cluster a point if its neighborhood along this dimension has a small variance. Intuitively, a *subspace preference cluster* is a density connected set of points associated with a similar subspace preference vector.

Let $\mathcal{D}$ be a database of d-dimensional points ($\mathcal{D} \subseteq R^d$), where the set of attributes is denoted by $\mathcal{A} = \{A_1, A_2, \ldots, A_d\}$, and $dist : R^d \times R^d \to R$ is a metric distance function between points in $\mathcal{D}$. Let $\mathcal{N}_\varepsilon(p)$ be the ε-neighborhood of $p \in \mathcal{D}$, i.e., $\mathcal{N}_\varepsilon(p)$ contains all points $q \in \mathcal{D}$ with $dist(p, q) \le \varepsilon$. The variance of $\mathcal{N}_\varepsilon(p)$ along an attribute $A_i \in \mathcal{A}$ is denoted by $\mathrm{VAR}_{A_i}(\mathcal{N}_\varepsilon(p))$. Attribute A_i is considered a *preferable (relevant) dimension* for p if the variance with respect to A_i in its neighborhood is smaller than a user-defined threshold δ, i.e., $\mathrm{VAR}_{A_i} \le \delta$. All preferable attributes of p are accumulated in the so-called *subspace preference vector*. This d-dimensional vector $\bar{\mathbf{w}}_p = (w_1, w_2, \ldots, w_d)$ is defined such that $w_i = 1$ if attribute A_i is irrelevant, i.e., $\mathrm{VAR}_{A_i}(\mathcal{N}_\varepsilon(p)) > \delta$ and

$w_i = \kappa$ ($\kappa \gg 1$) if A_i is relevant, i.e., $\mathrm{VAR}_{A_i}(\mathcal{N}_\varepsilon(p)) \leq \delta$. The subspace preference vector of points defines the *preference weighted similarity* function associated with a point p, $dist_p(p,q) = \sqrt{\sum_{i=1}^d w_i \cdot (\pi_{A_i}(p) - \pi_{A_i}(q))^2}$, where w_i is the i-th component of $\bar{\mathbf{w}}_p$. Using the preference weighted similarity, the preferable attributes are weighted considerably lower than the irrelevant ones. This distance is not symmetric. A symmetric distance is defined by the *general preference similarity*, $dist_{pref}(p,q) = \max\{dist_p(p,q), dist_q(q,p)\}$. The *preference weighted $\varepsilon-$neighborhood* of a point p contains all points of $\mathcal{D}$ that are within a preference weighted distance ε from p: $\mathcal{N}_\varepsilon^{\bar{\mathbf{w}}_o}(o) = \{x \in \mathcal{D} \mid dist_{pref}(o,x) \leq \varepsilon\}$.

Based on these concepts, the classical definitions of density-based clustering have been derived:

Definition 1 (preference weighted core points [6]). *A point $o \in \mathcal{D}$ is called preference weighted core point w.r.t. ε, μ, δ, and λ (denoted by $\mathrm{CORE}_{den}^{pref}(o)$), if i) the preference dimensionality of its ε-neighborhood is at most λ and ii) its preference weighted ε-neighborhood contains at least μ points.*

Definition 2 (direct preference reachability [6]). *A point $p \in \mathcal{D}$ is directly preference reachable from a point $q \in \mathcal{D}$ w.r.t. ε, μ, δ, and λ (denoted by $\mathrm{DIRREACH}_{den}^{pref}(q,p)$), if q is a preference weighted core point, the subspace preference dimensionality of $\mathcal{N}_\varepsilon(p)$ is at most λ, and $p \in \mathcal{N}_\varepsilon^{\bar{\mathbf{w}}_q}(q)$.*

Definition 3 (preference reachability [6]). *A point $p \in \mathcal{D}$ is preference reachable from a point $q \in \mathcal{D}$ w.r.t. ε, μ, δ, and λ (denoted by $\mathrm{REACH}_{den}^{pref}(q,p)$), if there is a chain of points $p_1, \ldots, p_n$ such that $p_1 = q, p_n = p$ and p_{i+1} is directly preference reachable from p_i.*

Definition 4 (preference connectivity [6]). *A point $p \in \mathcal{D}$ is preference connected to a point $q \in \mathcal{D}$, if there is a point $o \in \mathcal{D}$ such that both p and q are preference reachable from o.*

Definition 5 (subspace preference cluster [6]). *A non-empty subset $\mathcal{C} \subseteq \mathcal{D}$ is called a subspace preference cluster w.r.t. ε, μ, δ, and λ, if all points in $\mathcal{C}$ are preference connected and $\mathcal{C}$ is maximal w.r.t. preference reachability.*

As DBSCAN, PREDECON determines a cluster uniquely by any of its preference weighted core points. As far as such a point is detected, the associated cluster is defined as the set of all points that are preference reachable from it.

4   Incremental PreDeCon

Let $\mathcal{D}$ be the accumulated data set until the time point $t - 1$ and let ζ be the corresponding clustering at $t - 1$ (built upon data set $\mathcal{D}$). Let $\mathcal{U}$ be a set of *update operations* (insertions of new points). Let $\mathcal{D}^*$ be the newly accumulated data set at time slot t, which is the result of applying $\mathcal{U}$ over $\mathcal{D}$, i.e., $\mathcal{D}^* = \mathcal{D} \cup \mathcal{U}$. The goal of incremental PREDECON is to update the so far built clustering ζ (at timepoint

$t - 1$) based on the update set $\mathcal{U}$ (at timepoint t) and thus, to derive the valid clustering ζ^* for time point t. The key observation is that the preference weighted core member property of an object might change due to the update. As a result, the existing clustering might change too, e.g., new clusters might arise, old clusters might be abolished or merged into a new cluster and so on. The challenge is to exploit the old clustering ζ at $t - 1$ (both clusters and subspaces where these clusters exist) and to adjust only that part of it which is affected by the update set $\mathcal{U}$ at time point t. Due to the density based nature of the algorithm, such an adjustment is expected (although not ensured in general) to be restricted to some (local) part of the clustering instead of the whole clustering.

We consider a dynamic environment where data are coming sequentially either as: (i) *single updates* ($|\mathcal{U}|=1$), e.g., in streams, or as (ii) *batch updates* ($|\mathcal{U}| = m > 1$), e.g., in data warehouses where updates are collected and periodically propagated. In case of *single updates*, each update is treated independently. In case of *batch updates*, the idea is to treat the effects of all these updates together instead of treating each update independently. The rationale is that the batch might contain updates that are related to each other (e.g., one update might correspond to an object that belongs to the neighborhood of another object which is also updated). This is common in many applications, e.g., news data: when a story arises usually within a small time interval there exists a burst of news articles all referring to this story.

5   Dealing with Single Updates

Due to the density based nature of PREDECON, a preference weighted cluster is uniquely defined by one of its preference weighted core points. The key idea for the incremental version is to check whether the update operation affects the preference weighted core member property of some point. If a non-core point becomes core, new density connections might be *established*. On the other hand, if a core point becomes non-core, some density connections might be *abolished*. There is also another case in PREDECON, when a core point remains core but under different preferences. Such a change might cause either the establishment of new connections or the abolishment of existing ones.

5.1   Effect on the Core Member Property

The insertion of a point p *directly affects* the points that are in the ε-neighborhood of p, i.e., all those points $q \in \mathcal{D} : dist(p, q) \leq \varepsilon$. In particular, the neighborhood of q, $\mathcal{N}_\varepsilon(q)$, might be affected, since the newly inserted object p is now a member of this neighborhood. Since $\mathcal{N}_\varepsilon(q)$ might change, the variance of $\mathcal{N}_\varepsilon(q)$ along some dimension $A_i \in A$ might also change causing A_i to turn into a preferable or non-preferable dimension. This might change the subspace preference dimensionality of q, $\mathrm{PDIM}(\mathcal{N}_\varepsilon(q))$. Also, the subspace preference vector of q, $\bar{\mathbf{w}}_q$, might change; this in turn, might result in changes in the preference ε-neighborhood of q, $\mathcal{N}_\varepsilon^{\bar{\mathbf{w}}}(q)_q$. As a result, the core member property of q might be affected. According to Def. 1, two conditions should be fulfilled in order for a point q to be core: In terms of condition 1, the preference dimensionality of q must contain at most λ dimensions (i.e., $\text{PDIM}(\mathcal{N}_\varepsilon(q)) \leq \lambda$). In terms of condition 2, the preference weighted ε-neighborhood of q should contain at least μ points.

Let p be the new point, and let $\mathcal{D}^* = \mathcal{D} \cup \{p\}$ be the new data set after the insertion of p. The addition of p might affect the core member property of any object $q \in \mathcal{N}_\varepsilon(p)$. In particular, since $\mathcal{N}_\varepsilon(q)$ changes, the variance along some attribute $A_i \in A$, i.e., $\text{VAR}_{A_i}(\mathcal{N}_\varepsilon(q))$ might also change. (i) If A_i was a non-preferable dimension (that is, $\text{VAR}_{A_i}(\mathcal{N}_\varepsilon(q)) > \delta$), it might either remain non-preferable (if still $\text{VAR}_{A_i}(\mathcal{N}_\varepsilon(q)) > \delta$) or it might become preferable (if now $\text{VAR}_{A_i}(\mathcal{N}_\varepsilon(q)) \leq \delta$). (ii) If A_i was a preferable dimension, it might either remain preferable (if still $\text{VAR}_{A_i}(\mathcal{N}_\varepsilon(q)) \leq \delta$) or it might become non-preferable (if now $\text{VAR}_{A_i}(\mathcal{N}_\varepsilon(q)) > \delta$). A change in the preference of A_i might result in changes in the subspace preference vector of q, $\bar{\mathbf{w}}_q$, since some dimension might swap from preferable to non preferable and vice versa. Thus, we can have more or less preferable dimensions comparing to the previous state (*quantitative differences*) or we can have the same dimensionality but under different preferred dimensions (*qualitative differences*). A change in $\bar{\mathbf{w}}_q$, might cause changes in both $\text{PDIM}(\mathcal{N}_\varepsilon(q))$ and in $\mathcal{N}_\varepsilon^{\bar{\mathbf{w}}_q}(q)$.

If the subspace preference dimensionality of q, $\text{PDIM}(\mathcal{N}_\varepsilon(q))$, changes, the first condition of Definition 1 (referring to dimensionality) might be violated. In particular, if $|\text{PDIM}(\mathcal{N}_\varepsilon(q))| > \lambda$, the point q cannot be core. So, if q was a core point, it now looses this property (*core → noncore*), whereas if it was non-core it still remains non-core. This is the first condition to be checked, and it is quantitative since it is based on the number of preferred dimensions (whether they exceed δ or not). If after the insertion of p, this condition holds (that is, $|\text{PDIM}(\mathcal{N}_\varepsilon(q))| \leq \lambda$), the second condition of Definition 1 (preferred neighborhood size) is to check assessing whether q is core after the update. (i) If q was a core point, and now $|\mathcal{N}_\varepsilon^{\bar{\mathbf{w}}}(q)_q| < \mu$, then q loses its core member property (*core → noncore*). Otherwise, it remains core. (ii) If q was not a core point, and now $|\mathcal{N}_\varepsilon^{\bar{\mathbf{w}}}(q)_q| \geq \mu$ then q becomes core (*noncore → core*). Otherwise, it remains non core. (iii) There is also another case of change for q, where it still remains core (*core → core*) but under different preferences (this might happen e.g., when there are qualitative changes in $\bar{\mathbf{w}}_q$). Note that, although q might remain core its neighborhood might change due to different preferred dimensions.

Note again that the objects with a changed core member property are all located in $\mathcal{N}_\varepsilon(p)$, since such a change is due to the insertion of p.

5.2   Affected Objects

So far, we referred to the objects in $\mathcal{N}_\varepsilon(p)$ that are *directly affected* by the insertion of p and we discussed when and how their core member property might change. Note however, that a change in the core member property of an object q might cause changes in the objects that are preference reachable from q (*indirectly affected*). If q was a core point before the insertion and it becomes non-core after the insertion, then any density connectivity that relied on q is destroyed. On the other hand, if q was a non-core point before the insertion and it turns into core after the insertion, then some new density connectivity based on q might arise.

We denote by $\text{AFFECTED}_\mathcal{D}(p)$ the set of points in $\mathcal{D}$ that might be affected after the insertion of p. This set contains both directly affected points (those located in $\mathcal{N}_\varepsilon(p)$, which might change their core member property after the update) and indirectly affected objects (those that are density reachable by some point in $\mathcal{N}_\varepsilon(p)$, which might change their cluster membership after the update).

Definition 6 (Affected objects). *Let $\mathcal{D}$ be a data set and let $\mathcal{D}^* = \mathcal{D} \cup \{p\}$ be the new data set after the insertion of object p. We define the set of objects in $\mathcal{D}$ affected by the insertion of p as:*
$$\text{AFFECTED}_\mathcal{D}(p) = \mathcal{N}_\varepsilon(p) \cup \{q | \exists o \in \mathcal{N}_\varepsilon(p) : \text{REACH}_{\text{den}}^{\text{pref}}(o, q) \text{ in } \mathcal{D}^*\}$$

The update of p might cause changes in the cluster membership of only some objects $q \in \text{AFFECTED}_\mathcal{D}(p)$. A naive solution would be to reapply the static PREDECON over this set in order to obtain the new clustering for the set of affected data. This way however, although one would restrict reclustering over only this subset of the data, one actually ignores any old clustering information for this set and build it from scratch. Our solution is based on the observation that any changes in $\text{AFFECTED}_\mathcal{D}(p)$, are exclusively initiated by objects that change their core member property, i.e., those in $\mathcal{N}_\varepsilon(p)$. So, instead of examining all objects in $\text{AFFECTED}_\mathcal{D}(p)$, we can start searching from objects in $\mathcal{N}_\varepsilon(p)$ and discover the rest of the affected objects on the road (those objects would belong to $\text{AFFECTED}_\mathcal{D}(p)$ though). Note also that there is no need to examine each $q \in \mathcal{N}_\varepsilon(p)$ since some objects might have not changed their core member property so related density connections from the previous clustering would be still valid. So, we need to examine only those objects in $\mathcal{N}_\varepsilon(p)$ that change their core member property after the insertion of p, instead of all objects in $\mathcal{N}_\varepsilon(p)$, so as to avoid rediscovering density connections. As already described, a possible change in the core member property of an object after the insertion of p falls into one of the following cases: (i) core $\rightarrow$ non-core, (ii) non-core $\rightarrow$ core and, (iii) core $\rightarrow$ core but under different preferences.

When the core member property of a point $q \in \mathcal{N}_\varepsilon(p)$ changes, we should consider as seed points for the update any core point $q' \in \mathcal{N}_\varepsilon(q)$. That is, the update process starts from core points in the neighborhood of the objects with changed core member property (which, in turn are all located in $\mathcal{N}_\varepsilon(p)$).

Definition 7 (Seed objects for the update). *Let $\mathcal{D}$ be a data set and let $\mathcal{D}^* = \mathcal{D} \cup \{p\}$ be the new data set after the insertion of p. We define the seed objects for the update as:*
$$\text{UPDSEED} = \{q \text{ is core in } \mathcal{D}^* | \exists q' : q \in \mathcal{N}_\varepsilon(q') \text{ and } q' \text{ changes its core property}\}$$

5.3   Updating the clustering

After the insertion of a new object p, new density connections might be established whereas existing connections might be abolished or modified. We can

```
algorithm INCPREDECON(D, U, ε, μ, λ, δ)
  for each  p ∈ U do
    1. D* = D ∪ p;
    2. compute the subspace preference vector w̄_p;
    // update preferred dimensionality and check core member property in N_ε(p)
    3. for each q ∈ N_ε(p) do
    4.    update w̄_q;
    5.    check changes in the core member property of q and if change exists, add q to AFFECTED;
    6. compute UPDSEED based on AFFECTED
    7. for each q ∈ UPDSEED do
    8.    expandCluster(D*, UPDSEED, q, ε, μ, λ);
  end;
```

Fig. 1. Pseudo code of the algorithm INCPREDECON

detect these changes starting with the seed objects in UPDSEED. As in PREDE-CON, the cluster is expanded starting from objects in UPDSEED and considering the results of the so far built clustering. The pseudo code of the algorithm is displayed in Figure 1. The existing database D, the update set U and the PRE-DECON parameters (namely, the distance threshold ε, the neighborhood size threshold μ and the dimensionality threshold λ) are the input to the algorithm. The updated clustering ζ^* is the output of the algorithm.

The algorithm works as follows: After the insertion of a point p (line 1), its subspace preference vector is computed (line 2), and its neighborhood $N_\varepsilon(p)$ is updated (lines 3–6). In particular, for each object $q \in N_\varepsilon(p)$ (line 3), we first update its subspace preference vector (line 4) and then check for any changes in the core member property of q (line 5). If the core member property of q is found to be affected, q is added to the AFFECTED set. After the AFFECTED set is computed, we derive the seed objects for the update (line 6). Based on these objects, the reorganization of the old clustering starts, which involves some call to the *expandCluster*() function of PREDECON. This is a generic solution that works on every effect caused by the update of p. Of course, there are simpler cases where we can deal with the update without invoking the *expandCluster*() procedure of PREDECON. For example, if the update of p does not affect the core member property of its neighborhood and its neighborhood belongs to exactly one cluster before the update, then p is also added to this cluster (absorption). However there are many such special cases, since, as already stated, the update of p might both destroy old density connections and create new density connections depending on the changes in the core member property of its neighborhood. The proposed method is lossless, that is the incrementally updated model ζ^* at t (which is based on the clustering model ζ at $t-1$ and on the update set U at t) is identical to the one we would obtain if we applied from scratch the traditional PREDECON over the accumulated data set D^* at time point t.

6   Dealing with Batch Updates

We now consider the case of batch updates where m ($m > 1$) points are inserted at each time point. The rationale behind this alternative is that it is possible for the batch data to be related to each other instead of independent (Consider for example an earthquake in some place and Twitter response to such an event; a flow of tweets would appear referring to that event.). Hence, instead of updating the clustering for each operation independently (as in the single update case, c.f. Section 5.3), one should consider the accumulative effect of all these batch operations on the clustering and treat them together. This way, the objects that might be affected by more than one single update operations are examined only once. Otherwise, such objects should be examined after each single operation (multiple checks).

The algorithm is similar to the algorithm for the single update case (c.f. Figure 1). The key difference is that instead of inserting objects one by one, examining the side effects of each single insert and updating the clustering based on each single insert, we now insert the whole batch (all m objects), we examine how the database is affected by the whole batch and we update the clustering model based on the whole batch. In more detail, we first insert the m objects of the batch in the existing database. Then, we continue the same rationale as with the single update case: First, we update the subspace preference vector of each of the inserted objects in the batch. Next, we check for objects with affected core member property. Recall (Section 5.1) that any objects with affected core member property lie in the neighborhood of some inserted point. Note also, that the points of the batch are all inserted into the database, so these operations also consider the newly inserted points. The AFFECTED set now contains the objects that might be affected due to the whole batch of points. Based on the AFFECTED set, the UPDSEED set is constructed which also refers to the whole batch. The points in $UpdSeed$ serve as the starting points for the cluster reorganization.

The benefit of the batch method is that some computations might take place only once, instead of after each single update as is the case for the single update method. For example, let p be an object in the database which is part of the neighborhood of both points p_1, p_2 in the batch. According to the single update case, this object should be examined twice, i.e., after each single insert, for any changes in its core member property. According to the batch update case though, this object should be examined only once.

7   Experiments

Since INCPREDECON computes the same results as PREDECON, we compare their performances in terms of efficiency. For massive data sets, the bottleneck of PREDECON and INCPREDECON is the number of range queries in arbitrary (possibly 2^d different) subspaces that cannot be supported by index structures. We report the speed-up factor defined as the ratio of the cost of PREDECON (applied to the accumulative data set $\mathcal{D}^*$) and the cost of INCPREDECON (applied to the initial data set $\mathcal{D}$ plus the updates $\mathcal{U}$).

We used a synthetic data generated according to a cluster template that describes the population of the corresponding cluster, the generating distribution, and range of dimension values for each dimension. In addition, we report a case

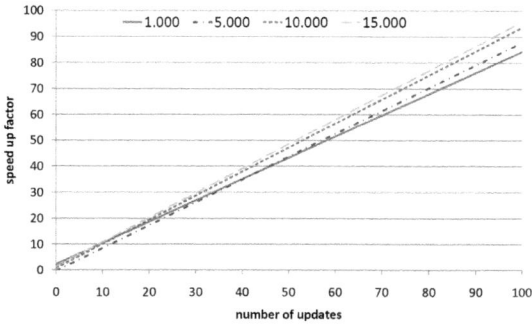

Fig. 2. Speed-up factors w.r.t. data set size

study of incrementally keeping track of clusters in real-world data. Let us again note that we do not compare different clustering models here, since the main focus of our paper is to provide a solution for incremental density-based subspace clustering in a high dimensional, dynamic environment.

7.1 Experiments on Single Updates

Varying the data set population. Four synthetic data sets of varying size between 1.000 and 15.000 objects were generated. From each data set, 100 objects were randomly extracted and used as the update set. The number of required range queries was computed after each insertion.

Figure 2 displays the speed-up factors w.r.t. the different data set sizes. IN-CPREDECON outperforms PREDECON with the speed-up factors of 2–100. As expected, with increasing number of updates the gain for INCPREDECON is higher. Analogously, the bigger the data set population is, the greater are the benefits of using INCPREDECON instead of PREDECON.

Varying the number of generated clusters. Five data sets with varying number of clusters but constant dimensionality and fixed population of each cluster were used next. From each data set, 100 objects were randomly extracted and used as the update set. Figure 3 displays the speed-up factors for all data sets. Again, INCPREDECON outperforms PREDECON for all datasets with the speed-up factors increasing with the number of updates and lying in the range [1–100]. Comparing the different data sets, however, we cannot draw some clear conclusion regarding whether more generated clusters result in greater gainings for INCPREDECON or no. This is intuitive since we generate random updates that do not necessarily correlate with the cluster structure. Thus, the number of clusters does not have a significant impact on the performance.

Varying the number of dimensions. Five data sets with varying dimensions but constant number of clusters and fixed population of each cluster were used next. From each data set, 100 objects were randomly extracted and used as the update set. Figure 4 displays the speed-up factors for all data sets. Again,

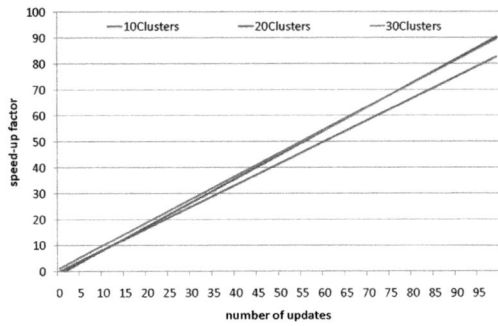

Fig. 3. Speed-up factors w.r.t. the number of generated clusters

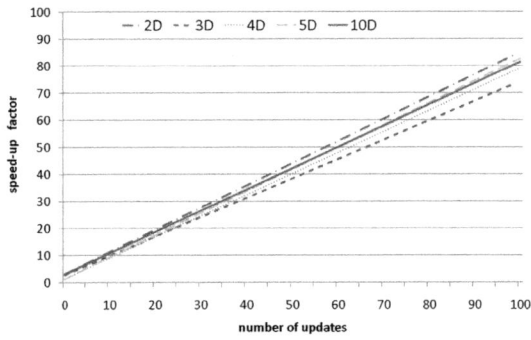

Fig. 4. Speed-up factors w.r.t. data dimensionality

INCPREDECON outperforms PREDECON with the speed up factors lying in the range 2–100. A comparison of the different data sets remains inconclusive w.r.t. whether or not more dimensions result in greater gain for INCPREDECON. This was expected since the dimensionality of the data should not have a different impact on the performances (in terms of required range queries) of PREDECON and INCPREDECON.

7.2 Experiments on Batch Updates

100 *random updates* were performed in a batch way (with batch sizes of 5, 10, 15, 20 updates) on a data set of 1.000 objects. Figure 5 displays the speed-up factors for the different batch sizes (the single update case is also partially depicted). INCPREDECON outperforms PREDECON for all different batch sizes. The highest gain exists for the single update case. As the batch size increases, the gain decreases.

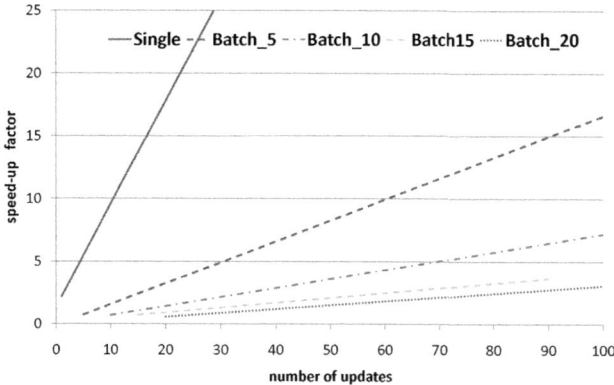

Fig. 5. Speed-up factors for random batch updates

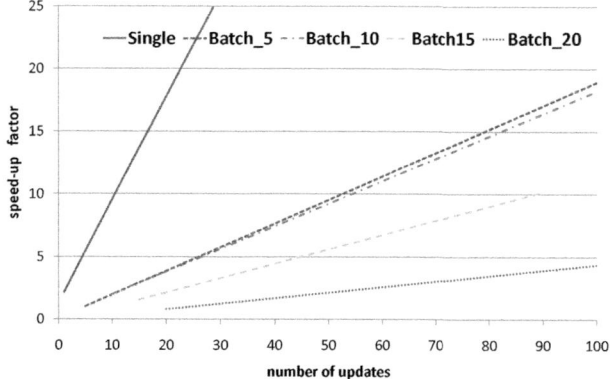

Fig. 6. Speed-up factors for "local" batch updates

The gain is expected to be even higher when the updates are not random but reflect the clustering structure. To verify this, we used an update set of objects extracted from 2 clusters in the generated data set (50 objects per cluster). As expected, the speed-up factors (cf. Figure 6) are higher compared to the random update case (cf. Figure 5).

The experiments showed the benefit of INCPREDECON versus PREDECON. The gain was very high for the single update case, whereas for the batch case the larger the batch size was, the lower the gain was. For example, in Figures 7, and 8 we can see the actual number of range queries required by PREDECON and INCPREDECON for two synthetic data sets. It can be observed that in the single update case (denoted by batch size = 1 in these figures), INCPREDECON (right bar) requires considerably less number of range queries comparing to PREDECON (left bar). As the batch size increases however, the gainings for

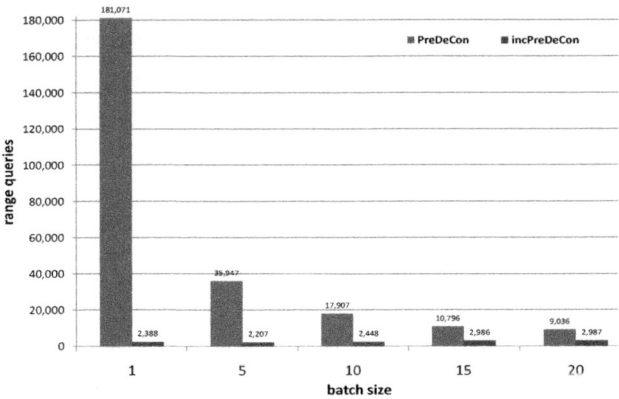

Fig. 7. Range queries for PREDECON and INCPREDECON (data set size 5.000)

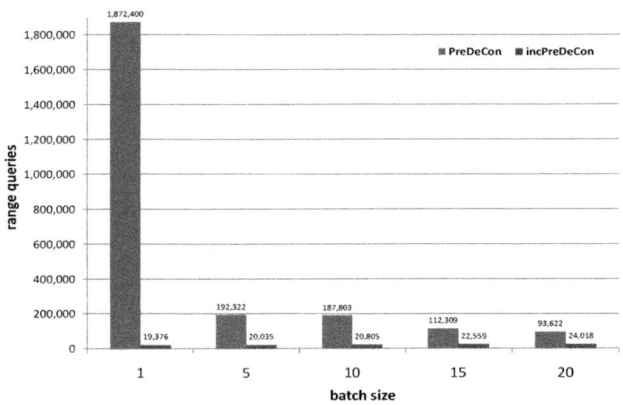

Fig. 8. Range queries for PREDECON and INCPREDECON (data set size 10.000)

INCPREDECON are decreased. Note that we run random updates in these experiments (Figure 7 and Figure 8). Greater savings are expected for "local updates", i.e., updates that correspond to a specific subcluster/ area of the population, since the batch method performs better when the update set contains related updates (recall our previous discussion on Figure 5 and Figure 6).

7.3   A Case Study on Real-World Data

We applied the original PREDECON on a small sample of 1,000 objects of the Bavarian newborn screening data, added batches of 200 objects and updated the cluster structure using INCPREDECON. Sample results from different time slots are sketched in Figure 9. Cluster 1 representing newborns suffering PKU refines the set of relevant attributes over time. This indicates that the attributes that

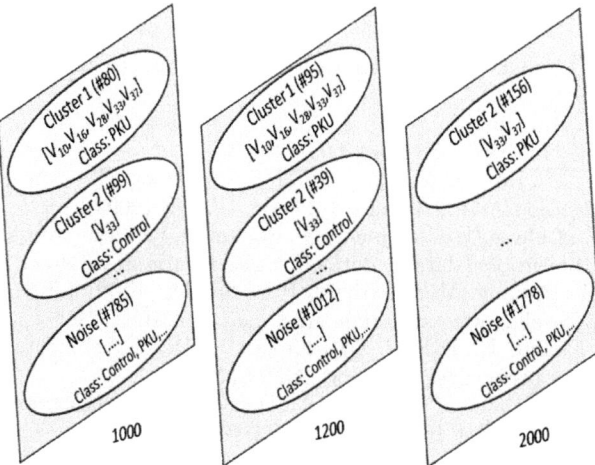

Fig. 9. Clusters on Bavarian newborn screening data evolving over time

have been evaluated as relevant for that cluster in the beginning might be false positives or might be relevant only for a subset of cluster members. The latter might indicate an interesting subtype of the PKU disease. In any case, these results might trigger the medical doctors to initiate further investigations on this issue. Cluster 2 representing a subset of the control group (healthy newborns) disappears over time since the clear subspace structure is absorbed by full dimensional noise. In fact, the attribute that is preferred by members of Cluster 2 at the beginning turns out to be not discriminative later on. Let us note that this phenomenon could not have been found by a full dimensional clustering method (e.g. by DBSCAN [11]) because the disappearance of that cluster is only possible when considering relevant projections of the feature space.

8   Conclusions

In this paper, we presented the incremental density based subspace clustering algorithm incPreDeCon. The algorithm can handle several prevalent challenges posed by today's data, including massive, dynamic, and high dimensional data sets. The update strategy, exploits the density based nature of clusters and, thus. manages to restructure only that part of the old clustering that is affected by the update. Both a single and a batch update method have been proposed. Our experimental results demonstrate the efficiency of the proposed method against the static application of PreDeCon. As future work, we plan to examine subspace clustering methods over data streams where access to raw data is usually not provided for efficiency reasons.

Acknowledgments. Irene Ntoutsi is supported by an Alexander von Humboldt Foundation fellowship for postdocs (http://www.humboldt-foundation.de/).

References

1. Aggarwal, C.C.: On change diagnosis in evolving data streams. IEEE TKDE 17(5), 587–600 (2005)
2. Aggarwal, C.C., Han, J., Wang, J., Yu, P.S.: A framework for clustering evolving data streams. In: Proc. VLDB (2003)
3. Aggarwal, C.C., Han, J., Wang, J., Yu, P.S.: A framework for projected clustering of high dimensional data streams. In: Proc. VLDB (2004)
4. Agrawal, R., Gehrke, J., Gunopulos, D., Raghavan, P.: Automatic subspace clustering of high dimensional data for data mining applications. In: Proc. SIGMOD (1998)
5. Ankerst, M., Breunig, M.M., Kriegel, H.P., Sander, J.: OPTICS: Ordering points to identify the clustering structure. In: Proc. SIGMOD (1999)
6. Böhm, C., Kailing, K., Kriegel, H.P., Kröger, P.: Density connected clustering with local subspace preferences. In: Proc. ICDM (2004)
7. Cao, F., Ester, M., Qian, W., Zhou, A.: Density-based clustering over an evolving data stream with noise. In: Proc. SDM (2006)
8. Charikar, M., Chekuri, C., Feder, T., Motwani, R.: Incremental clustering and dynamic information retrieval. SICOMP 33(6), 1417–1440 (2004)
9. Chen, C.Y., Hwang, S.C., Oyang, Y.J.: An incremental hierarchical data clustering algorithm based on gravity theory. In: Chen, M.-S., Yu, P.S., Liu, B. (eds.) PAKDD 2002. LNCS (LNAI), vol. 2336, p. 237. Springer, Heidelberg (2002)
10. Ester, M., Kriegel, H.P., Sander, J., Wimmer, M., Xu, X.: Incremental clustering for mining in a data warehousing environment. In: Proc. VLDB (1998)
11. Ester, M., Kriegel, H.P., Sander, J., Xu, X.: A density-based algorithm for discovering clusters in large spatial databases with noise. In: Proc. KDD (1996)
12. Ganti, V., Gehrke, J., Ramakrishnan, R.: DEMON: Mining and monitoring evolving data. IEEE TKDE 13(1), 50–63 (2001)
13. Gao, J., Li, J., Zhang, Z., Tan, P.N.: An incremental data stream clustering algorithm based on dense units detection. In: Ho, T.-B., Cheung, D., Liu, H. (eds.) PAKDD 2005. LNCS (LNAI), vol. 3518, pp. 420–425. Springer, Heidelberg (2005)
14. Garofalakis, M., Gehrke, J., Rastogi, R.: Querying and mining data streams: you only get one look. A tutorial. In: Proc. SIGMOD (2002)
15. Guha, S., Meyerson, A., Mishra, N., Motwani, R., O'Callaghan, L.: Clustering data streams: Theory and practice. IEEE TKDE 15(3), 515–528 (2003)
16. Jain, A.K., Murty, M.N., Flynn, P.J.: Data clustering: A review. ACM CSUR 31(3), 264–323 (1999)
17. Kriegel, H.P., Kröger, P., Gotlibovich, I.: Incremental OPTICS: efficient computation of updates in a hierarchical cluster ordering. In: Proc. DaWaK (2003)
18. Kriegel, H.P., Kröger, P., Ntoutsi, I., Zimek, A.: Towards subspace clustering on dynamic data: an incremental version of PreDeCon. In: Stream KDD 2010 (2010)
19. Kriegel, H.P., Kröger, P., Zimek, A.: Clustering high dimensional data: A survey on subspace clustering, pattern-based clustering, and correlation clustering. IEEE TKDD 3(1), 1–58 (2009)
20. Liebl, B., Nennstiel-Ratzel, U., von Kries, R., Fingerhut, R., Olgemöller, B., Zapf, A., Roscher, A.A.: Very high compliance in an expanded MS-MS-based newborn screening program despite written parental consent. Preventive Medicine 34(2), 127–131 (2002)
21. Spiliopoulou, M., Ntoutsi, I., Theodoridis, Y., Schult, R.: MONIC: modeling and monitoring cluster transitions. In: Proc. KDD (2006)
22. Zhang, T., Ramakrishnan, R., Livny, M.: BIRCH: An efficient data clustering method for very large databases. In: Proc. SIGMOD, pp. 103–114 (1996)

Hierarchical Clustering for Real-Time Stream Data with Noise

Philipp Kranen, Felix Reidl, Fernando Sanchez Villaamil, and Thomas Seidl

Data Management and Data Exploration Group, RWTH Aachen University, Germany
{kranen,reidl,sanchez,seidl}@cs.rwth-aachen.de

Abstract. In stream data mining, stream clustering algorithms provide summaries of the relevant data objects that arrived in the stream. The model size of the clustering, i.e. the granularity, is usually determined by the speed (data per time) of the data stream. For varying streams, e.g. daytime or seasonal changes in the amount of data, most algorithms have to heavily restrict their model size such that they can handle the minimal time allowance. Recently the first anytime stream clustering algorithm has been proposed that flexibly uses all available time and dynamically adapts its model size. However, the method exhibits several drawbacks, as no noise detection is performed, since every point is treated equally, and new concepts can only emerge within existing ones. In this paper we propose the LiarTree algorithm, which is capable of anytime clustering and at the same time robust against noise and novelty to deal with arbitrary data streams.

1 Introduction

There has been a significant amount of research on data stream mining in the past decade and the clustering problem on data streams has been frequently motivated and addressed in the literature. Recently the ClusTree algorithm has been proposed in [3] as the first anytime algorithm for stream clustering. It automatically self-adapts its model size to the speed of the data stream. Anytime in this context means that the algorithm can process an incoming stream data item at any speed, i.e. at any time allowance, without any parameterization by the user. However, the algorithm does not perform any noise detection, but treats each point equally. Moreover, it has limited capabilities to detect novel concepts, since new clusters can only be created within existing ones. In this paper we build upon the work in [3] and maintain its advantages of logarithmic time complexity and self-adaptive model size. We extend it to explicitly handle noise and improve its capabilities to detect novel concepts. While we improve the approach and add new functionality, it stays an anytime algorithm that is interruptible at any time to react to varying stream rates.

Due to a lack of space we will not repeat the motivation for stream clustering and anytime algorithms here. Neither can we recapitulate related work, especially the ClusTree presented in [3]. However, we stress that a good understanding of the work in [3] is indispensable for understanding the remainder of this paper.

J.B. Cushing, J. French, and S. Bowers (Eds.): SSDBM 2011, LNCS 6809, pp. 405–413, 2011.

We refer to [3] for motivation, related work and, most importantly, the ClusTree algorithm as a prerequisite for the following.

2   The LiarTree

In this section we describe the structure and working of our novel LiarTree. In the previously presented ClusTree algorithm [3] the following important issues are not addressed:

- **Overlapping:** the insertion of new objects followed a straight forward depth first descent to the leaf level. No optimization was incorporated regarding possible overlapping of inner entries (clusters).
- **Noise:** no noise detection was employed, since every point was treated equal and eventually inserted at leaf level. As a consequence, no distinction between noise and newly emerging clusters was performed.

We describe in the following how we tackle these issues and remove the drawbacks of the ClusTree. Section 2.6 briefly summarizes the LiarTree algorithm and inspects its time complexity.

2.1   Structure and Overview

The LiarTree summarizes the clusters on lower levels in the inner entries of the hierarchy to guide the insertion of newly arriving objects. As a structural difference to the ClusTree, every inner node of the LiarTree contains one additional entry which is called the noise buffer.

Definition 1. *LiarTree.* *For $m \leq k \leq M$ a LiarTree node has the structure $node = \{e_1, \ldots, e_k, CF_{nb}^{(t)}\}$, where $e_i = \{CF^{(t)}, CF_b^{(t)}\}$, $i = 1 \ldots k$ are entries as in the ClusTree and $CF_{nb}^{(t)}$ is a time weighted cluster feature that buffers noise points. The amount of available memory yields a maximal height (size) of the LiarTree.*

The noise buffer consists of a single CF which does not have a subtree underneath itself. We describe the usage of the noise buffer in Section 2.3.

Algorithm 1 illustrates the flow of the LiarTree algorithm for an object x that arrives on the stream. The variables store the current node, the hitchhiker (h) and a boolean flag indicating whether we encourage a split in the current subtree (details below). After the initialization (lines 1 to 1) the procedure enters a loop that determines the insertion of x as follows: first the exponential decay is applied to the current node in line 1. If nothing special happens, i.e. if none of the if-statements is true, the closest entry for x is determined (line 1) and the object descends into the corresponding subtree (line 1). As in the ClusTree, the buffer of the current entry is taken along as a hitchhiker (line 1) and a hitchhiker is buffered if it has a different closest entry (lines 1 to 1). Being an anytime algorithm the insertion stops if no more time is available, buffering

Algorithm 1. Process object (x)

```
1  currentNode = root; encSplit = false;
2  h = empty; // h is the hitchhiker
3  while (true) do                              /* terminates at leaf level latest */
4  |   update time stamp for currentNode;
5  |   if (currentNode is a liar) then
6  |   |   liarProc(currentNode, x); break;
7  |   end if
8  |   e_x = calcClosestEntry(currentNode, x, encSplit);
9  |   e_h = calcClosestEntry(currentNode, h, encSplit);
10 |   if (e_x ≠ e_h) then
11 |   |   put hitchhiker into corresponding buffer;
12 |   end if
13 |   if (x is marked as noise) then
14 |   |   noiseProc(currentNode, x, encSplit); break;
15 |   end if
16 |   if (currentNode is a leaf node) then
17 |   |   leafProc(currentNode, x, h, encSplit); break;
18 |   end if
19 |   add object and hitchhiker to e_x;
20 |   if (time is up) then
21 |   |   put x and h into e_x's buffer; break;
22 |   end if
23 |   add e_x's buffer to h;
24 |   currentNode = e_x.child;
25 end while
```

x and h in the current entry's buffer (line 1). The issues listed in Section 2 are solved in the procedures *calcClosestEntry* (line 1), *liarProc* (line 1) and *noiseProc* (line 1). We detail these methods in the following subsection and start by describing how we descend and reduce overlapping of clusters using the procedure *calcClosestEntry*.

2.2   Descent and Overlap Reduction

The main task in inserting an object is to determine the next subtree to descend into, i.e. finding the closest entry. Besides determining the closest entry, the algorithm checks whether the object is classified as noise w.r.t. the current node and sets an *encSplit* flag, if a split is encouraged in the corresponding subtree.

First we check whether the current node contains an irrelevant entry. This is done as in [3], i.e. an entry e is irrelevant if it is empty (unused) or if its weight $n_e^{(t)}$ does not exceed one point per snapshot (cf. [3]). In contrast to [3], where such entries are only used to avoid split propagation, we explicitly check for irrelevant entries already during descent to actively encourage a split on lower levels, because a split below a node that contains an irrelevant entry does not

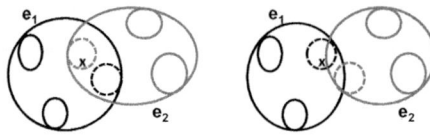

Fig. 1. Look ahead and reorganization

cause an increase of the tree height, but yields a better usage of the available memory by avoiding unused entries. In case of a leaf node we return the irrelevant entry as the one for insertion, for an inner node we set the *encSplit* flag.

Second we calculate the noise probability for the insertion object and mark it as noise if the probability exceeds a given threshold. This *noiseThreshold* constitutes a parameter of our algorithm and we evaluate it in Section 3.

Definition 2. *Noise probability.* *For a node node and an object o, the noise probability of o w.r.t. node is* $np(o) = \min_{e_i \in node} \{\{dist(o, \mu_{e_i})/r_{e_i}\} \cup \{1\}\}$ *where* e_i *are the entries of node,* r_{e_i} *the corresponding radius (standard deviation in case of cluster features) and* $dist(o, \mu_{e_i})$ *the euclidean distance from the object to the mean* μ_{e_i}.

Finally we determine the entry for further insertion. If the current node is a leaf node we return the entry that has the smallest distance to the insertion object. For an inner node we perform a local look ahead to avoid overlapping, i.e. we take the second closest entry e_2 into account and check whether it overlaps with the closest entry e_1. Figure 1 illustrates an example.

If an overlap occurs, we perform a local look ahead and find the closest entries e_1* and e_2* in the child nodes of candidates e_1 and e_2 (dashed circles in Figure 1 left). Next we calculate the radii of e_1 and e_2 if we would swap e_1* and e_2*. If they decrease, we perform the swapping and update the cluster features on the one level above (Figure 1 right). The closest entry that is returned is the one containing the closest child entry, i.e. e_1 in the example.

The closest entry is calculated both for the insertion object and for the hitchhiker (if any). If the two have different closest entries, the hitchhiker is stored in the buffer CF of its closest entry and the insertion objects continues alone (cf. Algorithm 1 line 1).

2.3   Noise

From the previous we know whether the current object has been marked as noise with respect to the current node. If so, the noise procedure is called. In this procedure noise items are added to the current noise buffer and it is regularly checked whether the aggregated noise within the buffer is no longer noise but a novel concept. Therefore, the identified object is first added to the noise buffer of the current node. To check whether a noise buffer has become a cluster, we calculate for the current node the average of its entries' weights $n^{(t)}$, their average density and the density of the noise buffer.

Definition 3. *Density*. *The density $\rho_e = n_e^{(t)}/V_e$ of an entry e is calculated as the ratio between its weighted number of points $n_e^{(t)}$ and the volume V_e that it encloses. The volume for d dimensions and a radius r is calculated using the formula for d-spheres, i.e. $V_e = C_d \cdot r^d$ with $C_d = \pi^{d/2}/\Gamma(\frac{d}{2}+1)$ where Γ is the gamma function.*

Having a representative weight and density for both the entries and the noise buffer, we can compare them to decide whether a new cluster emerged. Our intuition is, that a cluster that forms on the current level should be comparable to the existing ones in both aspects. Yet, a significantly higher density should also allow the formation of a new cluster, while a larger number of points that are not densely clustered are further on considered noise. To realize both criteria we multiply the density of the noise buffer with a sigmoid function, that takes the weights into account, before we compare it to the average density of the node's entries. As the sigmoid function we use the Gompertz function [2]

$$gompertz(n_{nb}, n_{avg}) = e^{-b(e^{-c \cdot n_{nb}})}$$

where we set the parameters b (offset) and c (slope) such that the result is close to zero ($t_0 = 10^{-4}$) if n_{nb} is 2 and close to one ($t_1 = 0.97$) if $n_{nb} = n_{avg}$ by

$$b = \frac{\ln(t_0)^{\frac{1}{1.0-(2.0/n_{avg})}}}{\ln(t_1)^{\frac{2}{n_{avg}-2}}} \qquad c = -\frac{1}{n_{avg}} \cdot \ln(-\frac{\ln(t_1)}{b})$$

Definition 4. *Noise-to-cluster event*. *For a node $node = (e_1, \ldots, e_k, CF_{nb}^{(t)})$ with average weight $n_{avg} = \frac{1}{k}\sum n_{ei}^{(t)}$ and average density $\rho_{avg} = \frac{1}{k}\sum \rho_{ei}$ the noise buffer $CF_{nb}^{(t)}$ becomes a new entry, if*

$$gompertz(n_{nb}^{(t)}, n_{avg}) \cdot \rho_n \geq \rho_{avg}$$

We check whether the noise buffer has become a cluster by now, if the encourage split flag is set to true. Note that a single inner node on the previous path with an irrelevant entry, i.e. old or empty, suffices for the encourage split flag to be true. Moreover, the exponential decay (cf. [3]) regularly yields outdated clusters. Hence, a noise buffer is likely to be checked.

    If the noise buffer has been classified as a new cluster, we create a new entry from it and insert this entry into the current node. Additionally we create a new empty node, which is flagged as *liar*, and direct the pointer of the new entry to this node. Figure 2 a-b) illustrate this noise to cluster event.

2.4   Novelty

In [3] new nodes were only created at the leaf level, such that the tree grew bottom up and was always balanced. The LiarTree allows noise buffers to transform to new clusters, i.e. we get new entries and, more importantly, new nodes

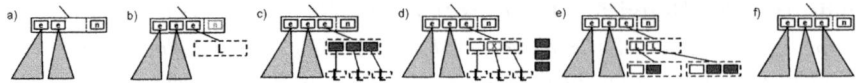

Fig. 2. The liar concept: a noise buffer can become a new cluster and the subtree below it grows top down, step by step by one node per object

---

Algorithm 2. liarProc $(liarNode, x)$
// refines the model to reflect novel concepts

---

1 create three new entries with dim dimensions $e_{new}[\,]$;
2 **for** $(d = 1$ to $dim)$ **do**
3      $e_{new}[d \bmod 3].LS[d] = (e_{parent}.LS[d])/3 + \text{offset}_A[d]$;
4      $e_{new}[(d+1) \bmod 3].LS[d] = (e_{parent}.LS[d])/3 + \text{offset}_B[d]$;
5      $e_{new}[(d+2) \bmod 3].LS[d] = (e_{parent}.LS[d])/3 + \text{offset}_C[d]$;

6      $e_{new}[d \bmod 3].SS[d] = F[d] + (3/e_{parent}.N) \cdot (e_{new}[d \bmod 3].LS[d])^2$;
7      $e_{new}[(d+1)\bmod 3].SS[d] = F[d] + (3/e_{parent}.N) \cdot (e_{new}[(d+1)\bmod 3].LS[d])^2$;
8      $e_{new}[(d+2)\bmod 3].SS[d] = F[d] + (3/e_{parent}.N) \cdot (e_{new}[(d+2)\bmod 3].LS[d])^2$;
9 **end for**

10 insert x into the closest of the new entries;
11 **if** *(liarNode is a liar root)* **then**
12   insert new entries into $liarNode$;
13 **else**
14   remove e_{parent} in parent node;
15   insert new entries into parent node;
16   split parent node (stop split at liar root);
17 **end if**

18 **if** *(non-empty liar nodes reach leaf level)* **then**
19   remove all liar flags in correspond. subtree ;
20 **else**
21   create three new empty liar nodes under $e_{new}[\,]$;
22 **end if**

---

within the tree. To avoid getting an increasingly unbalanced tree through noise-to-cluster events, we treat nodes and subtrees that represent novelty differently. The main idea is to let the subtrees underneath newly emerged clusters (entries) grow top down step by step with each new object that is inserted into the subtree until their leaves are on the same height as the regular tree leaves. We call leaf nodes that belong to such a subtree *liar nodes*, the root is called *liar root*. When we end up in a liar node during descend (cf. Algorithm 1), we call the liar procedure which is listed in Algorithm 2.

Definition 5. *Liar node.* *A liar node is a node that contains no entry. A liar root is an inner node of the liar tree that has only liar nodes as leafs in its corresponding subtree and no other liar root as ancestor.*

Figure 2 illustrates the liar concept, we will refer to the image when we describe the single steps. A liar node is always empty, since it has been created as an empty node underneath the entry e_{parent} that is pointing to it. Initially the liar root is created by a noise-to-cluster event (cf. Figure 2 b)). To let the subtree under e_{parent} grow in a top down manner, we have to create additional new entries e_i (cf. solid (red) entries in Figure 2). Their cluster features CF_{e_i} have to fit the CF summary of e_{parent}, i.e. their weights, linear and quadratic sums have to sum up to the same values. We create three new entries (since a fanout of three was shown to be optimal in [3]) and assign each a third of the weight from e_{parent}. We displace the new means from the parent's mean by adding three different offsets to its mean (a third of its linear sum, cf. lines 2 to 2). The offsets are calculated per dimension under the constraint that the new entries have positive variances. We set one offset to zero, i.e. offset$_A = 0$. For this special case, the remaining two offsets can be determined using the weight n_e^t and variance $\sigma_e^2[i]$ of e_{parent} per dimension as follows

$$\text{offset}_B[i] = \sqrt{\frac{1}{6} \cdot \left(1 - \left(\frac{1}{3}\right)^4\right) \cdot (n_e^t) \cdot \sigma_e^2[i]}, \qquad \text{offset}_C[i] = -\text{offset}_B[i]$$

The zero offset in the first dimension is assigned to the first new entry, in the second dimension to the second entry, and so forth using modulo counting (cf. lines 2 to 2). If we would not do so, the resulting clusters would lay on a line, not representing the parent cluster well. The squared sums of the three new entries are calculated in lines 2 to 2. The term $F[d]$ can be calculated per dimension as

$$F[d] = \frac{n_e^t}{3} \cdot \left(\frac{\sigma_e[d]}{3}\right)^4$$

Having three new entries that fit the CF summary of e_{parent}, we insert the object into the closest of these and add the new entries to the corresponding subtree (lines 2 to 2). If the current node is a liar root, we simply insert the entries (cf. Figure 2 c)). Otherwise we replace the old parent entry with the three new entries (cf. Figure 2 d)). We do so, because e_{parent} is itself also an artificially created entry. Since we have new data, i.e. new evidence, that belongs to this entry, we take this opportunity to detail the part of the data space and remove the former coarser representation. After that, overfull nodes are split (cf. Figure 2 d-e)). If an overflow occurs in the liar root, we split it and create a new liar root above, containing two entries that summarize the two nodes resulting from the split (cf. Figure 2 e)). The new liar root is then put in the place of the old liar root, whereby the height of the subtree increased by 1 and it grew top down (cf. Figure 2 e)).

In the last block we check whether the non empty leaves of the liar subtree already reach the leaf level. In that case we remove all liar flags in the subtree, such that it becomes a regular part of the tree (cf. line 2 and Figure 2 f)). If the subtree does not yet have full height, we create three new empty liar nodes (line 2), one beneath each newly created entry (cf. Figure 2 c)).

2.5   Insertion and Drift

Once the insertion object reaches a regular leaf, it is inserted using the leaf proce-dure (cf. algorithm 1 line 1). If there is no time left, the object and its hitchhiker are inserted such that no overflow, and hence no split, occurs. Otherwise, the hitchhiker is inserted first and, if a split is encouraged, the insertion of the hitch-hiker can also yield an overflowing node. This is in contrast to the ClusTree, where a hitchhiker is merged to the closest entry to delay splits. In the LiarTree we explicitly encourage splits to make better use of the available memory (cf. Definition 1). After inserting the object we check whether an overflow occurred, split the node and propagate the split.

Three properties of the LiarTree help to effectively track drifting clusters. The first property is the aging, which is realized through the exponential decay of leaf and inner entries as in the ClusTree (cf. [3]), a proof of invariance can be found in [3]). The second property is the fine granularity of the model. Since new objects can be placed in smaller and better fitting recent clusters, older clusters are less likely to be affected through updates, which gradually decreases their weight and they eventually disappear. The third property stems from the novel liar concept, which separates points that first resemble noise and allows for transition to new clusters later on. These transitions are more frequent on levels close to the leaves, where cluster movements are captured by this process.

2.6   Summary

To insert a new object, the closest entry in the current node is calculated. While doing this, a local look ahead is performed to possibly improve the clustering quality by reduction of overlap through local reorganization. If an object is classified as noise, it is added to the current node's noise buffer. Noise buffers can become new clusters (entries) if they are comparable to the existing clusters on their level. Subtrees below newly emerged clusters grow top down through the liar concept until their leaves reach the regular leaf level.

Obviously the LiarTree algorithm has time complexity logarithmic in its model size, i.e. the number of entries at leaf level, since the tree is balanced (logarithmic height), the loop has only one iteration per level (cf. Alg. 1) and any procedure is maximally called once followed directly by a **break** statement.

3   Experiments

We compare our performance to the ClusTree algorithm [3] and to the well known CluStream algorithm from [1] using synthetic data as in [3]. To compare to the CluStream approach we used a maximal tree height of 7 and allowed CluStream to maintain 2000 micro clusters. We calculate precision and recall using a Monte Carlo approach, i.e. for the recall we generate points inside the ground truth and check whether these are included in the found clustering, for the precision we reverse this process, i.e. we generate points inside the found clustering and check whether they are inside the ground truth. Figure 3 shows

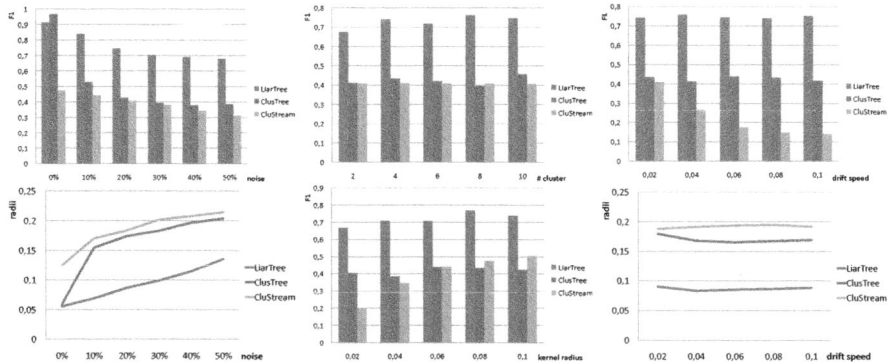

Fig. 3. Left: F1 measure and resulting radii for LiarTree, ClusTree and CluStream for different noise levels. Middle: Varying the data stream's number of clusters and their radius. Rigth: Varying the drift speed for LiarTree, ClusTree and CluStream

the resulting F1 measure and the resulting average radii of the clusters for the three approaches. In the left graphs we see that the LiarTree outperforms both competing approaches in the presence of noise, proving its novel concepts to be effective. Varying the parameters of the data stream (cf. remaining graphs) does not impair the dominance of the LiarTree.

4 Conclusions

In this paper we presented a novel algorithm for anytime stream clustering called LiarTree, which automatically adapts its model size to the stream speed. It consists of a tree structure that represents detailed information in its leaf nodes and coarser summaries in its inner nodes. The LiarTree avoids overlapping through local look ahead and reorganization and incorporates explicit noise handling on all levels of the hierarchy. It allows the transition from local noise buffers to new entries (micro clusters) and grows novel subtrees top down using its liar concept, which makes it robust against noise and changes in the distribution of the underlying stream.

Acknowledgments. This work has been supported by the UMIC Research Centre, RWTH Aachen University, Germany.

References

1. Aggarwal, C.C., Han, J., Wang, J., Yu, P.S.: A framework for clustering evolving data streams. In: VLDB, pp. 81–92 (2003)
2. Bowers, N.L., Gerber, H.U., Hickman, J.C., Jones, D.A., Nesbitt, C.J.: Actuarial Mathematics. Society of Actuaries, Itasca (1997)
3. Kranen, P., Assent, I., Baldauf, C., Seidl, T.: Self-adaptive anytime stream clustering. In: IEEE ICDM, pp. 249–258 (2009)

Energy Proportionality and Performance in Data Parallel Computing Clusters

Jinoh Kim, Jerry Chou, and Doron Rotem

Lawrence Berkeley National Laboratory
University of California, Berkeley, CA 94720, USA
{jinohkim,jchou,d_rotem}@lbl.gov

Abstract. Energy consumption in datacenters has recently become a major concern due to the rising operational costs and scalability issues. Recent solutions to this problem propose the principle of *energy proportionality*, i.e., the amount of energy consumed by the server nodes must be proportional to the amount of work performed. For data parallelism and fault tolerance purposes, most common file systems used in MapReduce-type clusters maintain a set of replicas for each data block. A *covering set* is a group of nodes that together contain at least one replica of the data blocks needed for performing computing tasks. In this work, we develop and analyze algorithms to maintain energy proportionality by discovering a covering set that minimizes energy consumption while placing the remaining nodes in low-power standby mode. Our algorithms can also discover covering sets in *heterogeneous* computing environments. In order to allow more data parallelism, we generalize our algorithms so that it can discover k-covering sets, i.e., a set of nodes that contain at least k replicas of the data blocks. Our experimental results show that we can achieve substantial energy saving without significant performance loss in diverse cluster configurations and working environments.

Keywords: Energy Management, Data Parallel Computing, Covering Subset, Node Heterogeneity.

1 Introduction

Energy consumption in scientific and commercial datacenters has increased dramatically with the introduction of high-performance, power-hungry components, such as multicore processors, high capacity memories, and high rotational speed disks. Therefore, the mounting costs of energy in datacenters has recently become a major concern. It is now estimated by EPA that in 2011 datacenters will consume up to 3% of the total energy in the U.S., while their energy consumption is doubling every 5 years [19]. Despite the technological progress and the amount of capital invested, there are significant inefficiencies in datacenters with server utilization measured at around 6% [17]. In this paper, we focus on optimizing energy consumption of compute clusters in datacenters, such as MapReduce clusters [10] often used in scientific computation [11]. The key idea is

J.B. Cushing, J. French, and S. Bowers (Eds.): SSDBM 2011, LNCS 6809, pp. 414–431, 2011.

to achieve this by placing underutilized components in lower power consumption states (i.e., standby mode).

Optimizing energy consumption in datacenters introduces several challenges. As pointed out in [21,13,4], *heterogeneity* of cluster nodes may be inevitable due to gradual replacement or addition of hardware over time. The replaced or added hardware should be "brand-new" rather than the same as the old one. Cluster heterogeneity can also be a result of a design choice. For example, the authors of [7] presented a hybrid datacenter model with two-class nodes that have different performance capabilities and power requirements for energy efficiency. In a recent work [21], heterogeneity in a MapReduce cluster was considered for job scheduling and performance improvement. There are several recent research efforts dealing with energy management for MapReduce clusters [16,15], but heterogeneity in such clusters has not been considered yet. In this paper, we examine how energy consumption can be further optimized by taking into account the different power requirements of the nodes in the cluster.

Another important requirement for energy management is *energy proportionality*, i.e., the ability to adjust energy consumption in proportion to the given workload. As mentioned in [2], server systems consume a substantial amount of energy even in idle mode (over 50% of the peak), although it could be ideally zero. Thus, a datacenter cluster still needs to consume a great deal of energy even under a very low load (e.g., at midnight), since the cluster nodes require substantial power even when no real work is done. Energy-proportionality can be a great benefit in conserving energy especially in clusters with a high degree of load variation, such as the one described in [6] where variations of over a factor of three between peak loads and light loads have been observed. This paper focuses on those two challenges, cluster heterogeneity and energy proportionality in data parallel computing clusters.

One known approach for cluster energy saving is achieved by powering on/off nodes in response to the current workload. For example, we could use cluster nodes in part to handle light loads, and save energy by deactivating the rest of the nodes not in use. In this work, we study the problem of determining which nodes should be activated or deactivated whenever it is determined that workload characteristics have changed.

More specifically, this work focuses on identifying a set of nodes that minimizes energy costs while satisfying immediate data availability for a data set required in computing. This is important since the cost of demand-based power state transitions of nodes for missing data blocks is significant in terms of both energy and performance due to the long latency needed to transition back from standby to active mode. For example, dehibernating (transitioning from standby to active mode) may require 129W for a duration of 100 seconds [15], for a node consuming 114W in idle mode. In a heterogeneous setting, such power requirements can be different from one node to another. To address this, we establish a power consumption profile for each node, and use this information in locating an optimal node set. In this paper, we refer to a group of nodes that together contain at least one replica of the data blocks needed for performing computing tasks as a CS (*covering subset*).

For high performance computing, the degree of data availability has a critical role in determining the degree of data parallelism. To consider this, we extend our node discovery algorithms to guarantee a certain degree of data availability. In its simplest form, our node discovery algorithm searches for a node set holding a single replica of the data. However, we may need a node set that has more than a single replica for each data item for certain situations. For example, for satisfying performance dictated by service level requirements we may need to activate a node set containing *two* replicas for supporting intermediate loads, rather than using a node set with a single replica.

Our key contributions are summarized as follows:

- We provide mathematical analysis of minimal CS size under the assumption of a uniform data layout as a function of the number of data blocks. We also show the validity of the theoretical model by simulation.
- We present node set discovery algorithms that find an energy-optimized node set with data availability for all required data items, for homogeneous and heterogeneous settings.
- We extend our discovery algorithms to identify a node set with any required degree of data availability, as a means of energy-proportional cluster reconfiguration.
- We present our evaluation results with respect to energy consumption and performance with a rich set of parameter settings. The results show that our techniques can achieve substantial energy saving without significant performance loss in most light workload environments. Also, we show that our power-aware technique can exploit heterogeneity successfully, yielding greater energy saving.

The paper is organized as follows. We first briefly introduce several closely related studies in the next section. In Sections 3 and 4 we present our algorithms for node set discovery for data availability and energy proportionality. In Sections 5 and 6, evaluation setup and results are presented with a rich set of parameters in diverse working environments. We finally conclude our presentation in Section 7. Additional analysis and experimental results are also available in our extended report [14].

2   Related Work

The initial work for MapReduce cluster energy management was performed in [16] based on covering subset (CS). In that work, the CS nodes are manually determined, and one replica for each data item is then placed in one of the CS nodes. Under a light load, it would be possible save energy by running the cluster with only the CS nodes activated. To enable this, the authors modified the existing replication algorithm, such that the CS nodes contain a replica of each data item. Failure of CS nodes was not considered, and as a result, any single node failure can make this scheme ineffective. Also, there was no notion of energy proportionality with gradual adjustment; rather the cluster is in either full performance mode with the entire set of nodes activated or in energy mode with only the CS nodes activated.

AIS (All-in Strategy) [15] is a different approach. AIS runs given jobs employing the entire set of nodes in the cluster to complete them as quickly as possible. Upon completion of the jobs, the entire set of nodes are deactivated to save energy until the next run. This makes sense since data parallel clusters are often used for batch-oriented computations [9]. One potential drawback can be that even with small (batched) jobs, AIS still needs to wake up the entire cluster, possibly wasting energy. Both studies (static CS and AIS) did not consider cluster heterogeneity, as we do in this work.

Rabbit [1] provides an interesting data placement algorithm for energy proportionality in MapReduce clusters. The key idea is to place data items in a skewed way across the nodes in the cluster. More specifically, node k needs to store b/k data items, where b is the total number of data items. Thus, a lower-indexed node has a greater number of data items, and it makes it possible to deactivate a higher-indexed node safely without losing data availability. Energy proportionality is also provided by allowing one-by-one node deactivation. Our approach provides energy management for clusters with the existing data layout, while Rabbit introduces its own method of data placement for energy management. Rabbit also does not consider possibility of cluster heterogeneity.

Cardosa et al. considered energy saving in a VM (Virtual Machine)-based MapReduce cluster [5]. Their approach is to place VMs in a timely balanced way, and find a way to minimize the number of nodes to be utilized, so as to maximize the number of nodes that can be idle. Subsequently, idle nodes can be considered as candidates for deactivation to save energy. One essential assumption in this work, that may not be practical, is the availability of a tool for accurate running time estimation for VMs.

3   Node Set Discovery Algorithms

In this section, we present our node discovery algorithms for a set of nodes that minimizes energy consumption subject to data availability constraints. As in [16], we refer to this node set as CS (Covering Subset). We assume that the data set statistics for the next round of computation is readily available and therefore discover the CS based on that information. This leads to a slightly different definition of CS as compared with the definition in [16]. The CS used here is not a static node set, rather it is discovered on demand based on a given list of data blocks required for computation. Thus, our CS must contain a replica for *required* data items instead of the *entire* set of data blocks in the cluster. Since data parallel computing platforms are often used for batch-style processing [9,15], the data set can be available for the next operational time window.

In this section, we first present a basic algorithm for node discovery that searches a minimal number of nodes for data availability, and then extend it

Table 1. Notations

Symbol	Description
N	Cluster node set
CS	CS node set ($CS \subseteq N$)
NCS	non-CS node set ($NCS \subseteq N$)
n	Cluster size
b	Number of data blocks
r	Replication factor
f	Fraction of low-power nodes
$P^{(i)}/T^{(i)}$	Idle power/time
$P^{(a)}/T^{(a)}$	Active power/time
$P^{(p)}/T^{(p)}$	Peak power/time
$P^{(s)}/T^{(s)}$	Standby power/time
$P^{(u)}/T^{(u)}$	Activating power/time
$P^{(d)}/T^{(d)}$	Deactivating power/time

with an energy metric for heterogeneous settings. Table 1 summarizes notations we used in this paper.

3.1   A Basic Method for CS Discovery

By definition, CS maintains at least one replica of the required data blocks. Locating such a set is NP-complete as it can be reduced to the well known *set cover* problem [8], as described in the following proposition.

Proposition 1. *A minimum CS discovery problem $CS(B, S)$ with B required blocks and a set of servers S is NP-complete, the reduction is from a minimum set cover problem $SC(U, F)$, where U is a universe of elements and F is a family of subsets of U.*

Proof. We omit the proof since it is trivial.

Figure 1 plots the size of CS for a cluster with size $n = 1024$ under two replicated environments with $r = 3$ and $r = 5$, as a function of the number of required data blocks. As the number of data blocks increases, the CS size also increases. For example, with n data blocks, the CS size ranges 20–30% of the cluster for the two replication settings. This implies that it would be possible to have energy saving of up to 70–80% in this setting. The CS size grows to 60–80% of the cluster for the case where the number of data blocks is $32n$, which is $\sim 2TB$ with the default data block size in MapReduce [10] and Hadoop [12].

We briefly discuss the theoretical analysis for the minimal CS size as a function of the number of data blocks in a uniform data distribution.

Assume that r copies of each data block are uniformly distributed on n nodes with each node holding at most one of the r copies. As previously defined, CS is a node set that contains at least one replica of each of the given b data items.

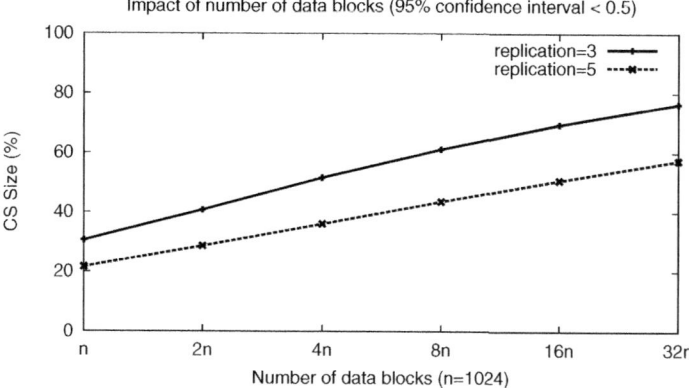

Fig. 1. CS size with respect to the number of data blocks

Lemma 1. *Let P be the probability that a randomly selected set of m nodes out of n nodes is CS. Then, P is equal to $\left(1 - \prod_{i=0}^{m-1}\left(1 - \frac{r}{n-i}\right)\right)^b$.*

Proof. The total number of ways for selecting r nodes from the available n nodes to hold the r replicas is $\binom{n}{r}$. From these possible selections, exactly $\binom{n-m}{r}$ do not place a copy in the randomly selected m nodes. We can then calculate the probability that the selected m nodes do not have any replica of a data item d_1 as $P' = \frac{\binom{n-m}{r}}{\binom{n}{r}} = \frac{(n-m)!(n-r)!}{n!(n-m-r)!} = \prod_{i=0}^{m-1}(1 - \frac{r}{n-i})$. Due to the fact that the r replicas for each of the b data items are placed independently, we get $P = (1 - P')^b$, or $P = (1 - \prod_{i=0}^{m-1}(1 - \frac{r}{n-i}))^b$.

Theorem 1. *The minimal m such that we can expect at least one CS from any given uniform data layout satisfies:* $\binom{n}{m}\left(1 - \prod_{i=0}^{m-1}\left(1 - \frac{r}{n-i}\right)\right)^b \geq 1$.

Proof. Let $M = \{M_1, M_2, \cdots, M_\ell\}$ be the collection of all sets of size m selected from n nodes. Thus $\ell = \binom{n}{m}$. By Lemma 1, we know that the probability of each M_i to be a CS is P. Let X_i be a random variable where,

$$X_i = \begin{cases} 1 & \text{if } M_i \text{ is a CS,} \\ 0 & \text{otherwise.} \end{cases}$$

Then, the expected value of X_i, $E(X_i)$, is equal to P. The expected number of CS is thus,

$$\sum_{i=0}^{\ell} E(X_i) = \binom{n}{m} P$$

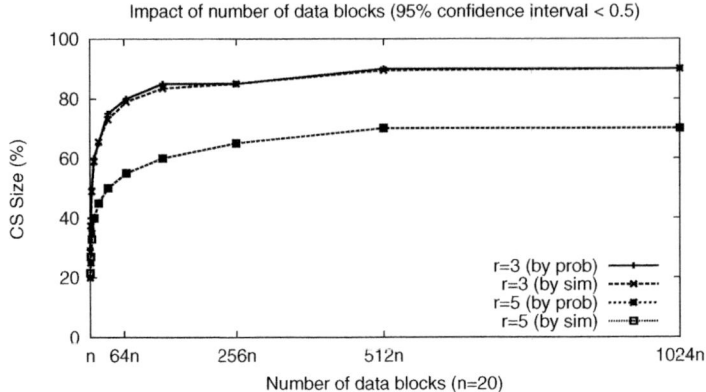

Fig. 2. Minimal CS size

Note that this is true even though the X_i's are not independent. Therefore, the minimal m that ensures existence of at least one CS must satisfy $\binom{n}{m}P \geq 1$.

Figure 2 shows the minimal CS size as a function of the number of data blocks in a small system with $n = 20$. The figure compares the analytical results based on our probabilistic model and simulation results, and we can see that they agree with each other. Also, the sub-linear shape of CS size increase over the number of blocks agrees with the mathematical work studied in [20]. Note that we used rack-unaware replication for simulation to assume the equivalent setting.

As described above, the problem of our node set discovery is simply mapped to the set cover problem, and the solution is to locate a set with the minimal size covering the data items in question. However, in a heterogeneous environment where nodes may have different power metrics, locating a minimal-size set would not be sufficient. We present a *power-aware* discovery algorithm as a solution for identifying an optimal node set in a heterogeneous cluster next.

3.2   Power-aware Discovery for Heterogeneous Clusters

Let us illustrate a heterogeneous cluster with a realistic example. Suppose there are 20 nodes in a cluster with 10 Xeons and 10 Atoms with power profiles as in Table 2. We can see that Xeons consume ten times more energy than Atoms. In such an environment, a CS with two Xeon nodes as a minimal subset may require a greater power level than a CS with ten Atom nodes. The former power requirement is $2 \cdot 315W + 8 \cdot 18W + 10 \cdot 2W = 794W$ at peak, while the latter only requires $10 \cdot 33.8W + 10 \cdot 18W = 518W$. At the idle state, the former requires 683W and the latter does 436W.

However, any technique that naively selects low-power nodes for CS discovery may not work that well. For example, in the above example, if Xeons consume only half watts than that in the table, i.e., $P^{(p)} = 315/2W = 157.5W$ and $P^{(s)} = 18/2W = 9W$, where $P^{(p)}$ stands for peak power and $P^{(s)}$ does standby power, then the power requirement for a CS with two Xeons becomes $2 \cdot 157.5W + 8 \cdot 9W + 10 \cdot 2W = 407W$, which is smaller than the energy requirement for a CS with ten Atom nodes. Hence, we need a more sophisticated approach to locate an optimal CS in heterogeneous settings, as discussed next.

Formally, CS power requirement is $P_{CS}^{(a)} + P_{NCS}^{(s)}$, where $P_{CS}^{(a)}$ is power for CS in active state and $P_{NCS}^{(s)}$ is power for non-CS nodes in standby. The energy consumption (E) for a given period of time (T) is then simply $E = (P_{CS}^{(a)} + P_{NCS}^{(s)}) \times T$. If we assume that T is fixed, our objective in identifying CS is to minimize $P_{CS}^{(a)} + P_{NCS}^{(s)}$. In other words, what we want to do here is to discover nodes for CS whose aggregated energy consumption can be minimized during time period T. This can be rewritten as follows for power P:

$$
\begin{aligned}
P &= P_{CS}^{(a)} + P_{NCS}^{(s)} \\
&= \sum_{x \in CS} P_x^{(a)} + \sum_{y \in NCS} P_y^{(s)} \\
&= \sum_{x \in CS} \left(P_x^{(a)} + P_x^{(s)} - P_x^{(s)} \right) + \sum_{y \in NCS} P_y^{(s)} \\
&= \sum_{x \in CS} \left(P_x^{(a)} - P_x^{(s)} \right) + \sum_{x \in CS} P_x^{(s)} + \sum_{y \in NCS} P_y^{(s)} \\
&= \sum_{x \in CS} \left(P_x^{(a)} - P_x^{(s)} \right) + \sum_{y \in N} P_y^{(s)} \quad\quad (1)
\end{aligned}
$$

Since the second part in Equation 1 is a constant, we can then map the node set discovery problem in a heterogeneous setting to a *weighted set cover* problem with an energy metric $(P_i^{(a)} - P_i^{(s)})$ as the weight associated with each node i. More precisely, the goal of the node set discovery problem can be cast as follows. Let G be the set of all possible covering subsets for a required set of data blocks. For covering subset $g \in G$, we define its weight $w(g)$ as the sum of weights of its nodes, i.e.,:

$$
w(g) = \sum_{x \in g} \left(P_x^{(a)} - P_x^{(s)} \right) \quad\quad (2)
$$

Then, our goal is to find a covering subset q, such that $w(q) \leq w(g)$ for all $g \in G$.

Proposition 2. *A minimum CS discovery problem $CS(B, S)$ in a heterogeneous setting is NP-complete, and it can be reduced to a minimum weighted set cover problem $WSC(U, F)$, where U is a universe and F is a family of subsets of U.*

Proof. As in Proposition 1, given a CS problem $CS(B, S)$, we can construct a corresponding set cover problem $SC(U, F)$, where for each set $f_k \in F$, we set its weight to $P_k^{(a)} - P_k^{(s)}$. Let $C \subset F$ be the minimum weighted set cover of $SC(U, F)$. Define $C' = \{s_i | u_i \in C\}$, then it is easy to see C' is also the minimum weighted set of nodes covering all blocks in B. Reversely, weighted set cover can

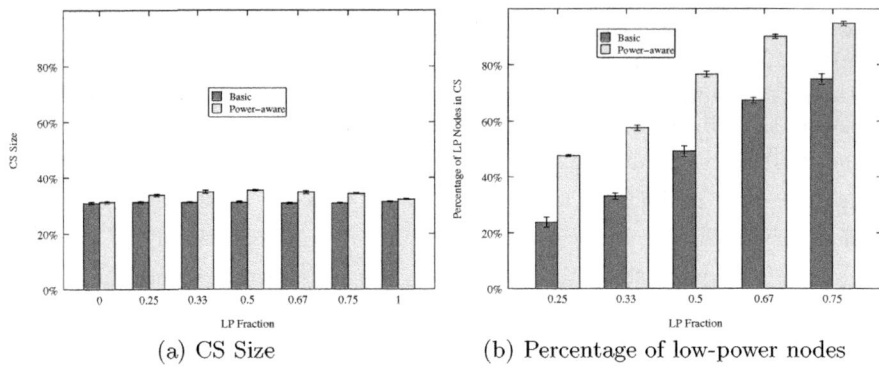

(a) CS Size   (b) Percentage of low-power nodes

Fig. 3. Comparison of basic and power-aware CS discovery

be reduced to the heterogeneous CS discovery problem, and the reduction is in polynomial time.

For an active node, its power consumption can scale from idle to peak based on workloads. That is, $P^{(a)}$ can vary over time depending on jobs running on the node. Thus, it is difficult to estimate $P_i^{(a)}$ for a given time period. In this work, we simply chose the mean between these two extreme values, $P_i^{(\bar{a})} = (P_i^{(i)} + P_i^{(p)})/2$, and use this for weight w_i for node i. However, this can be replaced with any other relevant measure.

Figure 3 compares the power-aware CS discovery algorithm with the basic CS algorithm. As above, we considered two classes of nodes, low-power (LP) and high-power (HP), based on Table 2. The two figures show CS size (Figure 3(a)) and percentage of LP nodes (Figure 3(b)) in the resulted CS, as a function of fraction of LP nodes in the cluster. In this experiment, we set the number of data blocks $b = n$ and replication factor $r = 3$. We can see that the power-aware algorithm yields a slightly bigger set for CS, but not that significant (the max gap is smaller than 4%). Figure 3(b) shows the power-aware algorithm takes a greater number of LP nodes for CS. Even with 0.25 for LP fraction, around 50% of nodes in the CS are LP nodes, while it is 25% with the basic algorithm. This power-optimized CS technique can significantly reduce energy consumption over the basic CS technique in heterogeneous settings, as we will show in Section 6.1.

3.3 Incremental CS Reorganization for Node Failure

Here, we briefly discuss the issue of CS reorganization in case of cluster configuration changes due to node failure. We assume that a new CS set is constructed periodically or on demand. Thus, any configuration change can be accounted at every construction time. However, there may be node failures, and as a result, some data blocks can be unavailable from the CS set. To deal with such failure cases, it is possible to reorganize CS incrementally by adding some nodes to keep the CS effective. Upon detection of any failure that affects the CS set, we can

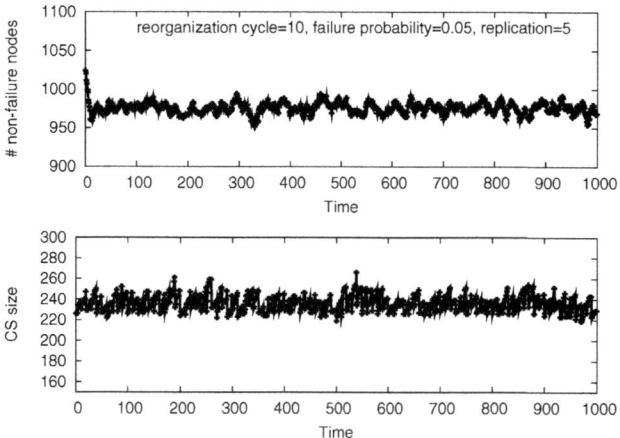

Fig. 4. CS reorganization under node failures

perform the CS discovery algorithm with inputs of the missing data blocks from the CS set and a set of non-CS nodes (i.e., NCS). The resulting set can then be added to the CS set. The incremented set may not be optimal, but still effective with required data availability. At the end of the time window for which the current CS is effective, a full reorganization is initiated to find an optimal node set for the new set of data blocks.

Figure 4 shows an example of CS reorganization over time under a node failure environment. We assumed that node failure probability is 0.005 for each node at every time unit. Probabilistically, at each time unit around 5 nodes suffer a failure in a cluster with $n = 1024$. Thus, at each time step, there would be an incremental reorganization if any CS node suffers a failure. We assume that a failed node is recovered after a deterministic amount of time (10 time units), and that a full reorganization takes place at every 10 time units. In the figure, the upper plot shows the number of nodes that do not experience failure, while the bottom plot shows CS size changes over time. In the upper plot, we can see that nodes fail and recover back, and the CS size varies accordingly in the bottom plot. As shown in the figure, CS size varies up and down over time with incremental reorganizations (increasing CS size) and full reorganizations (minimizing CS size) from the bottom one.

In this section, we have discussed node set discovery for CS that provides a single replica availability for data blocks in requirement. This can be extended to guarantee a higher degree of data availability, e.g., two replicas for each required data block. In the next section, we discuss how we can achieve this, and show how this idea can be used to provide energy proportionality in a cluster.

4   Multi-level Node Set Discovery

Here we discuss how it is possible to provide energy proportionality in this framework. In [15], the authors considered several strategies for node deactivation for non-CS nodes to support the CS approach. By deactivating (and activating) nodes one by one according to the current load, it is possible to get energy proportionality, but as the authors indicated, there may be load inequality between nodes because the number of replicas for each data block may be different for a certain time. For example, if we deactivate one node (and all the other nodes are active), there will remain $r-1$ blocks for the data blocks kept in that node, while the other blocks are maintained based on replication factor (r). This implies a possibility of load imbalance. For these types of complications, we do not rely on a node selection strategy for achieving energy proportionality. Instead, we propose a multi-level CS discovery that gives different degrees of data availability based on performance requirements for the given workload.

In our multi-level CS approach, different CS levels provide different degrees of data availability. For example, a CS set in level 2 in our framework gives 2-replica availability for the required data blocks (we call it *CS-2*). Therefore, there can be a series of CS sets from *CS-1* to *CS-r* (usually equivalent to n). In this section, we describe how we can discover such CS sets for a certain degree of data availability.

The problem of identifying *CS-k* can be mapped to the *set multicover* problem with coverage factor k, where k denotes the minimal number of times each object in question appears in the resulting set.

Proposition 3. *The CS-k(B,S) problem is NP-complete, the reduction is from the set multicover problem $SMC(U, F, k)$, where U is a universe, F is a family of subsets of U, and a required coverage factor k.*

Proof. The reduction algorithm is the same as proof 2. Since there is a one-to-one mapping between the block $b_i \in B$ and the element $u_i \in U$, any element that is covered k times in $SMC(U, F, k)$ also appears k times in the result set of *CS-k(B,S)*, and vice versa. Also, the reduction remains in polynomial time.

In [3], the authors presented an $O(k|U||F|)$ time greedy heuristic for the $SMC(U, F, k)$ problem with an approximation factor of $(1 + \ln a)$ from optimal where $a = \max_i |F_i|$.

The greedy heuristic makes a selection of a new set in each iteration. The selected set must include the max number of elements that have not been covered k times yet. We employ this greedy heuristic for our multi-level CS discovery.

Figure 5 shows the CS size compared to the cluster size, as a function of the number of data blocks in two replicated environments ($r = 3$ and $r = 5$). As shown in the figure, *CS-1* and *CS-2* have different sizes. For example with $b = 4n$ and $r = 3$, the CS size is around 50% and 80% of the cluster for *CS-1* and *CS-2*, respectively. From those sets, we can select the one with a desired data availability while considering the (expected) workload. By doing so, our multi-level CS technique can be used for achieving energy-proportionality in the cluster.

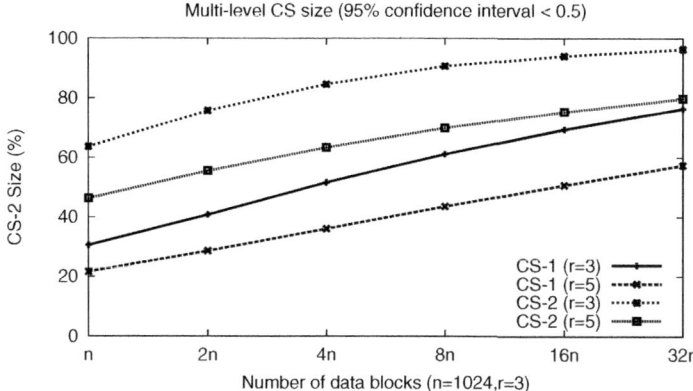

Fig. 5. Multi-level CS size

Table 2. Power model: SPECpower results from [7] and node hibernation costs from [15]

Platform	$P^{(i)}$	$P^{(p)}$	$P^{(s)}$	$T^{(d)}$	$T^{(u)}$	MaxThread	Capacity
HP (Xeon)	259.5W	315.0W	18.0W	11s	100s	8	1
LP (Atom)	25.6W	33.8W	2.0W	11s	100s	4	0.36

5   Evaluation Methodologies

For evaluation, we developed a simulator based on OMNeT++ [18] providing a discrete event simulation framework. Our simulator performs with power measures from [7,15] shown in Table 2. In the table, *MaxThread* is the max number of threads that can be concurrently run in the node, and *Capacity* refers to processing capacity. Thus in the table, we can see that an Atom node can accommodate 4 concurrent tasks at max, and its processing capacity is 0.36 of that of a Xeon. For example, if a Xeon node can run 100 instructions in a unit time, an Atom node can perform 36 instructions for that moment.

We conducted experiments extensively with a diverse set of parameters summarized in Table 3. We assume data placement follows the basic MapReduce replication properties (hence, almost close to a uniform data layout). We then inject a series of jobs to the simulator based on job arrival rate (λ). We assume λ follows an exponential distribution. Since we are more interested in light loads for energy saving, we use $\lambda = 0.5$ by default in our experiments.

Each job requires χ parallel tasks, and the processing time is defined by τ and node capacity. The task processing time (τ) consists of computation time (c) and additional time for networking for data transfer (d), and is described as $\tau = distribution(c, d)$. We assume that the computation time is deterministically calculated based on c and node capacity by the equation of *computation_time* = $c/Capacity$, whereas the data transfer time is determined by a probabilistic

Table 3. Parameters

Symbol	Description	Default value
n	Cluster size	1,024
b	Number of data blocks	$16n$
r	Replication factor	3
f	Fraction of low-power nodes	0.5
ξ	Number of jobs	1000
λ	Job arrival rate	0.5
χ	Number of tasks	n
τ	Task processing time	$Normal(300, 0.1)$

distribution. Since no previous work identified distributions for data transfer time in data parallel computing clusters, we employed two distribution models, normal and exponential, in this study. For example, $\tau = Normal(300, 0.1)$ implies $300s$ for computation time, and a positive random value (v) from a normal distribution $Normal(0, 0.1)$ determines the additional time for data transfer, with an equation of $download\_time = v \times c/Capacity$. Thus, the overall task completion time is $(1 + v) \times c/Capacity$. For exponential distributions, we randomly choose a value from the given exponential distribution with $mean = d \times c/Capacity$, and we use the chosen value as the data transfer time.

We compare the following techniques in terms of energy consumption and average turnaround time: (1) NPS without reconfiguration (hence no energy management); (2) AIS (All-in Strategy); (3) Basic CS; and (4) Power-aware CS. NPS fully utilizes nodes in the cluster, and nodes are in idle after jobs are completed. AIS also utilizes entire nodes for jobs, but keeps the cluster deactivated as soon as completing jobs until the next job comes. The Basic CS technique constructs CS dynamically without considerations of node heterogeneity, while the power-aware CS technique takes heterogeneity into account in construction.

Initially, the entire cluster is "on" for NPS and AIS, while only CS nodes selected by each algorithm are active for our *CS*-based techniques. After completing all injected jobs (i.e., ξ), we measured aggregated energy consumption and average turnaround time for each technique, and compared the measured results. We repeated experiments and provide 95% confidence intervals for statistical consideration.

6   Experimental Results

6.1   Impact of Fraction of Low-power Nodes

In this experiment, we explore the impact of the LP fraction in the cluster. We moved LP fraction from 0 to 1 for this. By definition, the two extremes (i.e. $f = 0$ and $f = 1$) refer to homogeneous settings (i.e., $f = 0$ for all high-power node setting and $f = 1$ for all low-power node setting), while the others mixes both classes of nodes based on the fraction.

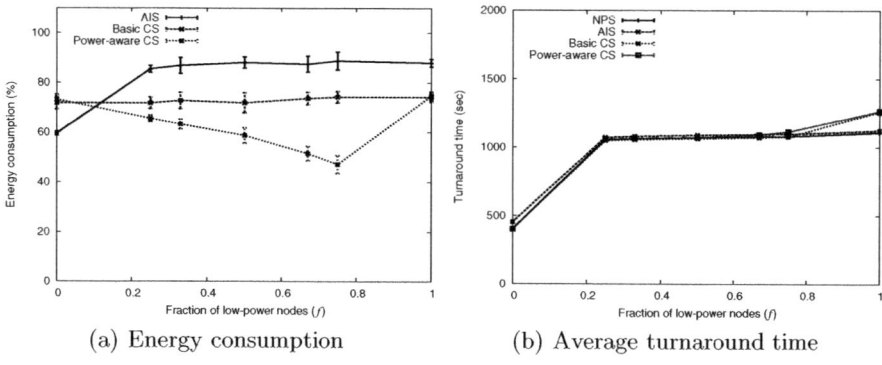

(a) Energy consumption   (b) Average turnaround time

Fig. 6. Impact of fraction of low-power nodes in the cluster

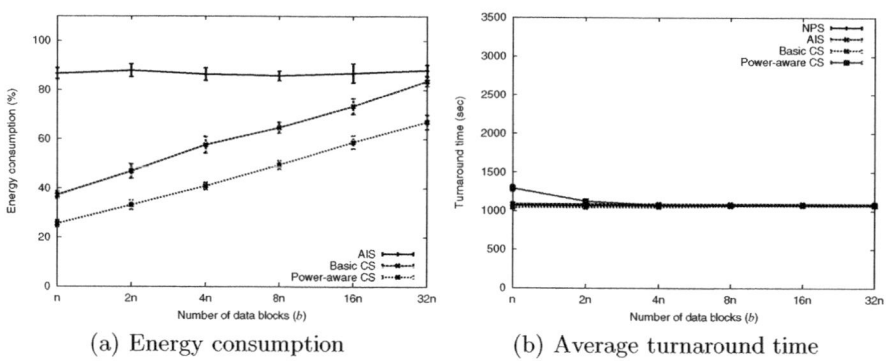

(a) Energy consumption   (b) Average turnaround time

Fig. 7. Impact of the number of data blocks

As shown in Figure 6, the power-aware technique yields the same results as the basic CS technique for the both extremes, showing around 30% energy saving. However, we can see that the power-aware technique further improves energy saving in any heterogeneous setting. With $f = 0.75$, the power-aware technique improves energy saving over 50%, as shown in Figure 6(a). For turnaround time, no significant deviations for the techniques were observed, as in Figure 6(b). This indicates that our power-aware technique enables to improve energy saving with little performance loss by exploiting cluster heterogeneity.

6.2   Impact of the Number of Data Blocks

Next, we investigate the impact of the number of data blocks since the CS size has strong correlation with this parameter. To see this, we used a diverse set of the number of data blocks, from $b = n$ (i.e., 64GB) to $b = 32n$ (i.e., 2TB). Figure 7 shows the results with respect to both energy and performance. We can see linear increases of energy consumption as the number of blocks increases for CS techniques, since a greater number of data blocks results in a larger

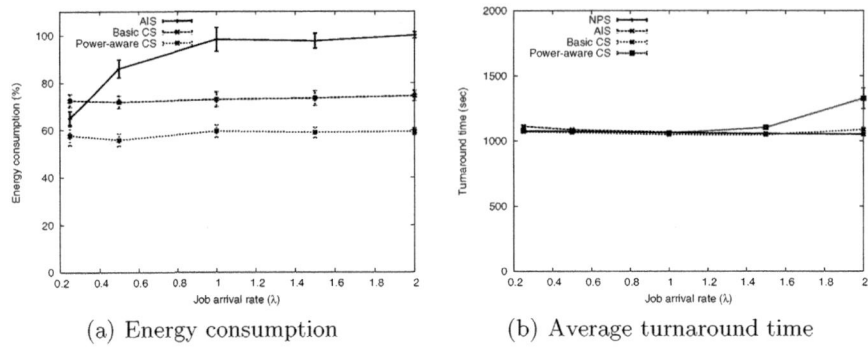

(a) Energy consumption   (b) Average turnaround time

Fig. 8. Impact of job arrival rate

CS. environment. The basic technique shows 30–60% energy saving, while the power-aware yields 40–70% saving with no noticeable performance degradation.

6.3   Impact of Job Arrival Rate

By default, we used job arrival rate $\lambda = 0.5$, since we are interested more in light load environments. In this experiment, we discuss the experimental results under varied job arrival rates. We employed a multiple set of job arrival rates from $\lambda = 0.25$ (for light load) to $\lambda = 2$ (for heavy load) in this experiment.

Figure 8 shows energy and performance as a function of λ. We see no significant changes for our CS techniques, except that power-aware CS somewhat degraded in a heavy workload environment $\lambda = 2$. Interestingly, AIS could save energy with very small job arrival rates. This is because very light loads can help to reduce the number of cluster power transition and can lengthen deactivation period at the same time. In this experiment, AIS yielded around 30% energy saving when $\lambda = 0.25$. However, little energy saving has been observed with greater job arrival rates than that with AIS.

6.4   Impact of Data Transfer Distributions

In this experiment, we employ several distribution models to consider data transfer times in the cluster. As mentioned, we consider normal and exponential distributions. For the normal distribution, we used three standard deviation values, 0.05, 0.1, and 0.25. For the exponential distribution, we used the same values as above but for the mean for the distribution. Again, these values are used to determine data transfer time according to the given distribution model, as described in Section 5.

Figure 9 shows the results. In the figure, $Norm(\sigma)$ represents a normal distribution with standard distribution σ, while $Exp(\mu)$ is for an exponential distribution with mean μ. For a diverse set of distribution models, we can see that our CS-based techniques consistently save energy around 30% for Basic and 40% for power-aware CS, but without any significant performance loss.

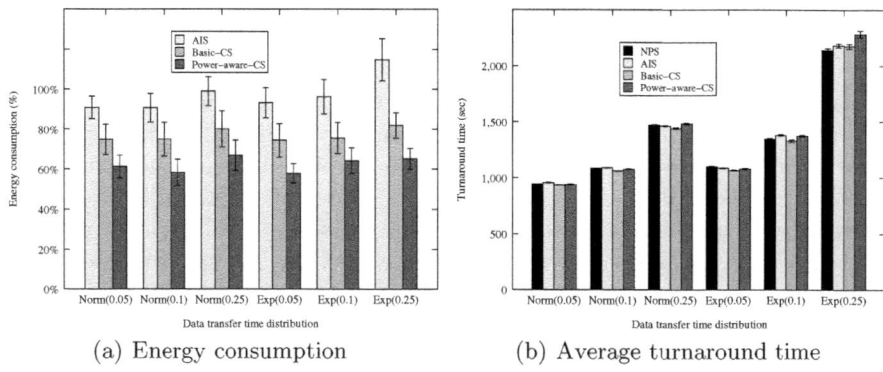

(a) Energy consumption   (b) Average turnaround time

Fig. 9. Impact of data transfer distribution

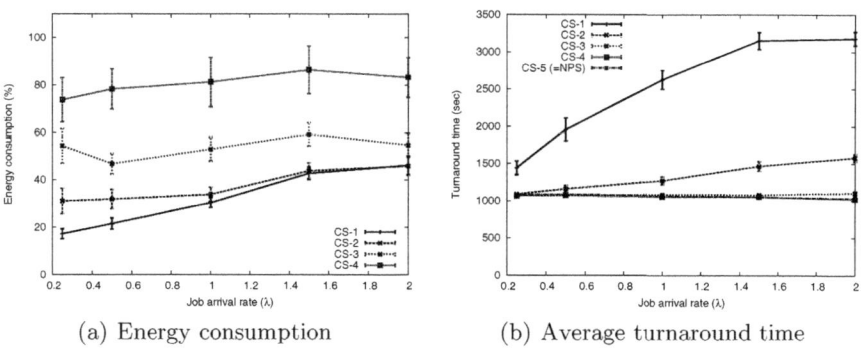

(a) Energy consumption   (b) Average turnaround time

Fig. 10. Impact of CS level

6.5 Evaluation of Multi-level CS

Finally, we present the impact of multi-level CS sets. To see the impact more clearly, we used a greater replication factor $r = 5$ and a smaller data blocks $b = n$, in this experiment. There can thus be four CS sets from *CS-1* to *CS-4* in addition to the entire cluster. For those CS sets, we varied λ to see how the CS sets respond to different loads.

Figure 10 shows the results. From the figure, we can see that each CS level gives a different degree of energy saving. For $\lambda = 1$, even with *CS-4*, it saves 20% of energy compared to NPS on average. The figure also shows that *CS-3* achieves 50% energy saving in the same setting, while *CS-2* and *CS-1* further increase saving to 70%. With respect to performance, we can see that a lower level CS shows a greater turnaround time. Thus, any appropriate CS can be chosen based on load intensity to maximize energy saving with performance guarantees.

7   Conclusions

Energy consumption in commercial and scientific datacenters has recently become a major concern due to the rising operational costs and scalability issues. For data parallelism and fault tolerance purposes, most common file systems used in MapReduce-type clusters maintain a set of replicas for each data block. Our basic idea in this work is to identify a subset of nodes, called a covering subset, that can provide a required degree of data availability for a given set of data blocks. In this work, we developed algorithms to maintain energy proportionality by discovering a covering set that minimizes energy consumption while placing the remaining nodes in low-power standby mode. In particular, we consider heterogeneity in determining a power-optimized covering set. For evaluation, we conducted experiments with a variety of parameters, such as job arrival rate and data transfer distribution. The experimental results show that power management based on our covering set algorithms can significantly reduce energy consumption, up to 70% compared to a non-power saving configuration, with little performance loss. In particular, the experimental results show that our algorithms can enhance energy saving in a heterogeneous environment by considering power metrics of individual nodes in the construction of a covering set. The results also show that our extended algorithm can be used to provide a coarse-grained level of energy proportionality based on covering sets with different degrees of data availability (thus providing different degrees of data parallelism). In the future we plan to also work on efficient scheduling algorithms for activating/deactivating nodes based on anticipatory analysis of future workloads.

Acknowledgments

This work was supported by the Director, Office of Science, Office of Advanced Scientific Computing Research, of the U.S. Department of Energy under Contract No. DE-AC02-05CH11231.

References

1. Amur, H., Cipar, J., Gupta, V., Ganger, G.R., Kozuch, M.A., Schwan, K.: Robust and flexible power-proportional storage. In: Proceedings of the 1st ACM Symposium on Cloud computing, SoCC 2010, pp. 217–228 (2010)
2. Barroso, L.A., Holzle, U.: The case for energy-proportional computing. Computer 40, 33–37 (2007)
3. Berman, P., DasGupta, B., Sontag, E.: Randomized approximation algorithms for set multicover problems with applications to reverse engineering of protein and gene networks. Discrete Appl. Math. 155(6-7), 733–749 (2007)
4. Bianchini, R., Rajamony, R.: Power and energy management for server systems. Computer 37(11), 68–74 (2004)
5. Cardosa, M., Singh, A., Pucha, H., Chandra, A.: Exploiting spatio-temporal trade-offs for energy efficient MapReduce in the cloud. Technical Report TR 10-008, University of Minnesota (April 2010)

6. Chase, J.S., Anderson, D.C., Thakar, P.N., Vahdat, A.M., Doyle, R.P.: Managing and server resources in hosting centers. In: SOSP 2001: Proceedings of the Eighteenth ACM Symposium on Operating Systems Principles, pp. 103–116 (2001)

7. Chun, B.-G., Iannaccone, G., Iannaccone, G., Katz, R., Lee, G., Niccolini, L.: An case for hybrid datacenters. SIGOPS Oper. Syst. Rev. 44(1), 76–80 (2010)

8. Chvàtal, V.: A greedy heuristic for the set-covering problem. Mathematics of Operations Research 4, 233–235 (1979)

9. Condie, T., Conway, N., Alvaro, P., Hellerstein, J.M., Elmeleegy, K., Sears, R.: MapReduce online. In: Proceedings of the 7th USENIX Conference on Networked Systems Design and Implementation, NSDI 2010, pp. 21–21 (2010)

10. Dean, J., Ghemawat, S.: MapReduce: simplified data processing on large clusters. In: OSDI 2004: Proceedings of the 6th Conference on Symposium on Opearting Systems Design & Implementation, pp. 10–10 (2004)

11. Gunarathne, T., Wu, T.-L., Qiu, J., Fox, G.: MapReduce in the clouds for science. In: CloudCom, pp. 565–572 (2010)

12. Hadoop: http://hadoop.apache.org/

13. Heath, T., Diniz, B., Carrera, E.V., Meira Jr., W., Bianchini, R.: Energy conservation in heterogeneous server clusters. In: PPoPP 2005, pp. 186–195 (2005)

14. Kim, J., Chou, J., Rotem, D.: Energy proportionality and performance in data parallel computing clusters. Technical Report LBNL-4533E, Lawrence Berkeley National Laboratory (April 2011)

15. Lang, W., Patel, J.M.: Energy management for MapReduce clusters. In: VLDB 2010 (2010)

16. Leverich, J., Kozyrakis, C.: On the (in)efficiency of Hadoop clusters. SIGOPS Oper. Syst. Rev. 44(1), 61–65 (2010)

17. http://www.mckinsey.com/clientservice/bto/pointofview/pdf/revolutionizing_data_center_efficiency.pdf

18. OMNeT++ Network Simulation Framework, http://www.omnetpp.org/

19. http://www.federalnewsradio.com/pdfs/epadatacenterreporttocongress-august2007.pdf

20. Vercellis, C.: A probabilistic analysis of the set covering problem. In: Annals of Operations Research, 255–271 (1984)

21. Zaharia, M., Konwinski, A., Joseph, A.D., Katz, R.H., Stoica, I.: Improving MapReduce performance in heterogeneous environments. In: OSDI, pp. 29–42 (2008)

Privacy Preserving Group Linkage

Fengjun Li[1], Yuxin Chen[1], Bo Luo[1], Dongwon Lee[2], and Peng Liu[2,*]

[1] Department of EECS, University of Kansas
[2] College of IST, The Pennsylvania State University

Abstract. The problem of privacy preserving record linkage is to find the intersection of records from two parties, while not revealing any private records to each other. Recently, group linkage has been introduced to measure the similarity of groups of records [19]. When we extend the traditional privacy preserving record linkage methods to group linkage measurement, group membership privacy becomes vulnerable – record identity could be discovered from unlinked groups. In this paper, we introduce threshold privacy preserving group linkage (TPPGL) schemes, in which both parties only learn whether or not the groups are linked. Therefore, our approach is secure under *group membership inference attacks*. In experiments, we show that using the proposed TPPGL schemes, group membership privacy is well protected against inference attacks with a reasonable overhead.

Keywords: Group linkage, privacy, secure multi-party computation.

1 Introduction

Record linkage (RL), also known as the *merge-purge* [12] or *object identity* [24] problem, is one of the key tasks in data cleaning [10] and integration [9]. Its goal is to identify related records that are associated with the same entity from multiple databases. When we extend the concept of "records" to *"groups of records"*, it becomes the group linkage (GL) problem [19], which is to determine if two or more *groups of records* are associated with the same entity.

RL and GL problems occur frequently in inter-database operations, in which privacy is a major concern, especially in the presence of sensitive data. In both record and group linkage, data owners need to reveal identifiable attributes to others for record-level comparison. However, in many cases, data owners are not willing to disclose any attributes unless the records are proven to be related. Here we present two GL examples, in which private attributes should not be revealed.

Example 1: As an international coordination to combat against gang violence, law enforcement units from different countries collaborate and share information.

---

* Bo Luo was partially supported by NSF OIA-1028098 and University of Kansas GRF-2301420. Dongwon Lee was partially supported by NSF DUE-0817376 and NSF DUE-0937891. Peng Liu was supported by AFOSR FA9550-07-1-0527 (MURI), ARO W911NF-09-1-0525 (MURI), NSF CNS-0905131, and NSF CNS-0916469.

J.B. Cushing, J. French, and S. Bowers (Eds.): SSDBM 2011, LNCS 6809, pp. 432–450, 2011.

Two countries will only share data when they confirm that they both possess information about the same gang group, which is represented as a set of records of gang members. Two gangs are regarded as the same when a large number of their members' records match.   □

In this example, each party holds groups (i.e. gangs) of records (i.e. members) that are identified by primary keys (i.e. names). Two records "match" only if they have identical primary keys. This scenario represents *privacy preserving group linkage with exact matching* (PPGLE) problem, in which the similarity between two inter-group members takes value from $\{0,1\}$.

Example 2: Two intelligence agencies (e.g. FBI and CIA) each obtains several pieces of intelligence documents. They would like to share the pieces if they are about the same case. Hence, the agencies need to verify that the similarity of their pieces are "very similar", in a way that does not reveal the document content in the verification process.   □

In this case, each record (e.g. a document) is represented as a vector in a term space shared by both participants. Record-level similarity is measured by a similarity function $sim(\mathbf{r}, \mathbf{s})$, and takes value in $[0, 1]$. If the similarity between two group members is smaller than a preset *record-level cut-off*, it is considered as "noise" and set to 0. The group-level similarity is defined as a function of record-level similarity (e.g. $sum()$). This scenario represents *privacy preserving group linkage with approximate matching* (PPGLA) problem.

   Privacy preserving *group* linkage (PPGL) extends privacy preserving *record* linkage (PPRL) such that participants hold groups instead of records. However, directly applying PPRL solutions to group linkage problems will suffer from *group membership inference attacks*. In such an attack, adversaries participate in the protocol with forged groups so that they can learn the group formation information of other parties, even though their groups are determined to be different in the end. To tackle this problem, we propose *threshold privacy preserving group linkage* (TPPGL) protocols for both exact matching (TPPGLE) and approximate matching (TPPGLA). The group similarity is no longer revealed to participating parties. Instead, only the result that the similarity is above or below a preset threshold is notified. In this way, private information about group membership is protected against inference attacks.

2   Problem Statement

Group linkage considers the problem of matching groups of records from multiple parities. For ease of presentation, hereafter, we use "Alice" and "Bob" to represent the two participants. In this scenario, Alice and Bob each holds a set of groups, identified as $\mathcal{R} = \{R_1, ..., R_u\}$ and $\mathcal{S} = \{S_1, ..., S_v\}$, respectively. Two groups are considered similar (i.e. linked) if and only if $SIM(R_i, S_j) \geq \theta$, where $SIM()$ is an arbitrary group similarity function and θ is a pre-negotiated threshold. In this paper, we follow the original group linkage definition [19] to use Jaccard similarity [15] as the group-level similarity measurement (see Section 3.2 for details). In real-world applications, the number of elements in groups is usually small (e.g. tens), while the number of groups tends to be large (e.g. hundreds).

In the case of *exact matching*, a record (i.e., group member) is identified by a primary key, e.g. $R = \{r_1, ..., r_m\}$ and $S = \{s_1, ..., s_n\}$, where r_p and s_q are primary keys. Two records are regarded similar if and only if their primary keys are identical, i.e. $sim(r_p, s_q) = 1$, iff $r_i = s_j$. Note that, for ease of presentation, we use R instead of R_i to denote a group.

In the case of *approximate matching*, each record is represented as a vector in a shared vector space: $R = \{\mathbf{r}_1, ..., \mathbf{r}_m\}$, and $S = \{\mathbf{s}_1, ..., \mathbf{s}_n\}$, where $\mathbf{r}_p$ and $\mathbf{s}_q$ are vectors. Two records are regarded similar if and only if their similarity is greater than a pre-negotiated record-level cut-off ρ, i.e., $sim(\mathbf{r}_p, \mathbf{s}_q) > \rho$. Here, $sim()$ is an arbitrary vector-space similarity function (e.g. cosine similarity or Euclidean distance). In this paper, we adopt the cosine similarity that is a popular similarity measure for text documents: $sim(\mathbf{r}_p, \mathbf{s}_q) = (\mathbf{r}_p \cdot \mathbf{s}_q)/(|\mathbf{r}_p||\mathbf{s}_q|)$.

Privacy preserving group linkage (PPGL): Both Alice and Bob follow a protocol to match two groups R from Alice and S from Bob. In the end, they learn $|R|$, $|S|$, and the group-level similarity $SIM(R, S)$, based on which they decide whether or not to share R and S. When exact matching or approximate matching is employed at the record level, the problem is further denoted as **PPGLE** or **PPGLA**, respectively.

The PPGL problem could be solved by existing privacy preserving set intersection protocols [1,4,11]. However, in the case where Bob holds a large number of groups with overlapping records, and Alice needs to check each of her groups against all Bob's groups, the existing solutions suffer from *group membership inference* problem, as shown in the following example.

Example 3. As shown in Figure 1 (a), Alice has a group of four records, where each record is identified by a primary key of last names. Bob has a set of three groups. Alice checks her group against Bob's three groups (using primary keys for exact matching), but fails to find a similar group (assume Jaccard similarity [15]

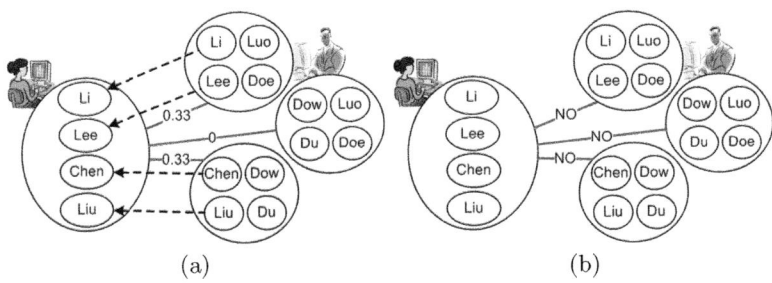

(a)   (b)

Fig. 1. (a) Privacy preserving group similarity with inference problem; (b) privacy preserving group linkage without inference problem

is used at group-level and θ is set to 0.5). In the three comparisons, both Alice and Bob learn the other's group sizes and the three similarities. Bob could easily infer the records in Alice's group via a simple derivation: (1) Alice's group should not have "Dow", "Du", "Luo", or "Doe" since its similarity with B_2 is 0; (2) Alice's group should have "Li" and "Lee" since its similarity with B_1 is 0.33 (which means the intersection size is 2); and (3) Alice's group should have "Chen" and "Liu" since its similarity with B_3 is also 0.33. Therefore, Alice's group is {Li, Lee, Chen, Liu} since the group size is known as 4.   □

In this example, Bob does not learn the content of Alice's records (i.e. attributes other than the primary keys), since the comparisons are assumed privacy preserving. However, the *group membership privacy*, i.e. the identities of group members, is disclosed. In the example, Bob infers Alice's group members by providing partially overlapped groups. An adversary may intensionally manipulate his groups in a way that group-wise similarity is always below the threshold so that he does not need to share his groups, but he is able to infer the other party's group membership privacy. To tackle such a problem, we need to develop a secure protocol that only provides a verdict of "yes" or "no" for each group-wise comparison. Hence, Bob only learns three "no"s in the above example (as shown in Figure 2 (b)), and the inference attack becomes impossible.

Threshold privacy preserving group linkage (TPPGL): Alice and Bob negotiate a threshold θ, and then follow a protocol to match two groups R and S. In the end, they only learn $|R|$, $|S|$, and a boolean result B, where $B=$true when $SIM(R, S) \geq \theta$, and $B=$false otherwise. When exact matching or approximate matching is employed at the record level, the problem is further denoted as **TPPGLE** or **TPPGLA**, respectively.

3   Preliminaries

3.1   Cryptographic Primitives

The protocols proposed in the paper adopt two special classes of cryptography algorithms for secure computation: commutative and homomorphic encryption.

Commutative Encryption. An encryption algorithm has the *commutative* property when we encrypt a message twice (with two different keys) and the resulting ciphertext is independent of the order of encryptions. Mathematically, an encryption scheme $E()$ is commutative if and only if, for any two keys e_1 and e_2 and any message m: (1) $E_{e_1}(E_{e_2}(m)) = E_{e_2}(E_{e_1}(m))$; (2) Encryption key e_i and its corresponding decryption key d_i are computable in polynomial time; and (3) $E_{e_i}()$ has the same value range.

The commutative property applies to the decryption phase too. If a message is encrypted with keys e_1 and e_2, then it can be recovered by either decrypting the cipher using d_1, followed by decryption using d_2; or decrypting using d_2, followed by d_1. Here, d_i is the corresponding secret key of e_i. Several encryption algorithms are commutative, e.g. Pohlig-Hellman, ECC, etc. In this work, we adopt

SRA encryption scheme, which is essentially RSA, except that the encryption exponent e is kept private.

Homomorphic Encryption. Homomorphic encryption represents a group of semantically-secure public/private key encryption methods, in which certain algebraic operations on plaintexts can be performed with cipher. Mathematically, given a homomorphic encryption scheme $E()$, ciphertexts $E(x)$ and $E(y)$, we are able to compute $E(x \star y)$ without decryption, i.e. without knowing the plaintext or private keys. $\star$ represents an arithmetic operation such as addition or multiplication.

Well-known homomorphic encryption schemes include: RSA, El Gamal [5], Paillier [20], Naccache-Stern [18], Boneh-Goh-Nissim [2], and etc. The Paillier cryptosystem [20,21] is additively homomorphic; the El Gamal [5] cryptosystem is multiplicatively homomorphic; and the Boneh-Goh-Nissim cryptosystem approach [2] supports one multiplication between unlimited number of additions. A more recent approach provides full support of both addition and multiplication at higher computation costs [6,25]. We omit further mathematical details in this paper, since they are out of our scope.

3.2 Related Work

The problem of privacy preserving group linkage originates from secure two-party computation and group linkage (which succeeds record linkage). We briefly summarize the literature in these areas.

Group linkage. The record linkage or merge-purge problem has been intensively studied in database, data mining, and statistics communities [27,3]. Group linkage [19] extends the scenario to take groups of records into consideration. Group-wise similarity [19] is calculated based on record-level similarity. When exact matching is enforced at the record level, *Jaccard similarity*[15] is employed at the group level: similarity of two groups (R and S) is defined as: $SIM(R, S) = |R \cap S| / |R \cup S|$. When approximate matching is applied at the record level, *bipartite matching similarity* is employed at the group level [19]. For two groups of records $R = \{\mathbf{r}_1, ..., \mathbf{r}_m\}$ and $S = \{\mathbf{s}_1, ..., \mathbf{s}_n\}$, $BM_{sim,\rho}$ is the normalized weight of M:

$$BM_{sim,\rho}(S, R) = \Big(\sum_{(\mathbf{r}_i, \mathbf{s}_j) \in M} sim(\mathbf{r}_i, \mathbf{s}_j) \Big) \Big/ (|R| + |S| - |M|)$$

where M indicates the maximum weight matching in the bipartite graph ($\mathcal{N} = R \cup S, \mathcal{E} = R \times S$). It contains all the edges whose weight is greater than ρ, i.e. $(\mathbf{r}_i, \mathbf{s}_j) \in M$ iff. $sim(\mathbf{r}_i, \mathbf{s}_j) \geq \rho$.

Privacy preserving record linkage. The original problem of *secure two/multi-party computation* was introduced in [28]. In this problem, multiple parties compute the value of a public function on private variables, without revealing the values of the variables to each other. Zero-knowledge proof [8] addresses the

Protocol 1. AES approach for set intersection [1]

Data: Alice has a group $R = \{r_1, ..., r_m\}$; and Bob has a group $S = \{s_1, ..., s_n\}$.
Result: They both learn the size of intersection: $|R \cap S|$, and nothing else.
 1: Both Alice and Bob apply hash function to their group elements to obtain: $h(R) = \{h(r_1), ..., h(r_m)\}$ and $h(S) = \{h(s_1), ..., h(s_n)\}$.
 2: Both Alice and Bob encrypt their hashed group elements to obtain: $E_r(h(R)) = \{E_r(h(r_1)), ..., E_r(h(r_m))\}$ and $E_s(h(S)) = \{E_s(h(s_1)), ..., E_s(h(s_n))\}$.
 3: Alice and Bob exchange their group, with group elements reordered.
 4: Bob encrypts what he got from Alice to obtain: $E_s(E_r(h(\mathcal{R})))$ – $\{E_s(E_r(h(r_1))), ..., E_s(E_r(h(r_n)))\}$, and return to Alice.
 5: Alice encrypts what she got from Bob in Step 3, to obtain: $E_r(E_s(h(S)))$.
 6: Alice finds out the size of intersection of the encrypted groups, $E_s(E_r(h(S)))$ (step 4) and $E_r(E_s(h(R)))$ (step 5), and shares with Bob.

problem of proving the veracity of a statement to other parties without revealing anything else. They are the earliest ancestors of privacy preserving multi-party computing. Privacy preserving record linkage with exact matching is very similar to privacy preserving set intersection: to identify the intersection of two sets of records without revealing private records. Surveys could be found at [26,11]. Among the more popular approaches, solutions based on homomorphic encryption (e.g. [4,16]) or commutative encryption (e.g. [1]) require higher computational overhead. Sanitization-based approaches (e.g. [14]) modify sensitive data so that they are not identifiable among others, but they are not suitable when there are a small number of records or the participants require perfect privacy protection. A hybrid approach [13] combines sanitization and crypto-based techniques to provide a balance among privacy, accuracy and computation (cost). In the context of approximate matching, there have been proposals on privacy preserving similar document detection (e.g. [22,17]).

3.3 Privacy Preserving Group Linkage: Baseline Solutions

Privacy preserving record linkage protocols match related records shared by two parties, without revealing any private records. This requires encrypting records so that computations (or comparisons) can be conducted on the ciphertexts. Agrawal et al. proposed a commutative encryption based solution in [1], which we refer as the AES protocol, and Freedman et al. presented a homomorphic encryption based scheme in [4], which we refer as the FNP protocol. We briefly introduce the protocols in Protocol 1 and 2. For more details, please refer to their papers [4,1].

These protocols serve as the baseline approaches for PPGLE, in which group-wise similarities are revealed to participants, but record information (for both shared and private records) is kept private. However, as we have described in Section 2, such solutions suffer from group membership inference attacks.

Protocol 2. FNP approach for set intersection size [4]

Data: Alice has a group $R = \{r_1, ..., r_m\}$; and Bob has a group $S = \{s_1, ..., s_n\}$.
Result: They both learn the size of intersection: $|R \cap S|$, and nothing else.

1: Alice creates keys for homomorphic encryption and publishes her public key.
2: Alice constructs $R(x) = \prod(x - r_i)$, and computes all the coefficients α_u that $R(x) = \sum_{u=0}^{u=m} \alpha_u x^u$. Therefore, the m degree polynomial $R(x)$ has roots $\{r_1, ..., r_m\}$.
3: Alice encrypts the coefficients and sends them ($\{\mathsf{E}(\alpha_0), \mathsf{E}(\alpha_1), ..., \mathsf{E}(\alpha_m)\}$) to Bob.
4: For each s_j, Bob evaluates the polynomial (without decryption) to get $\mathsf{E}(R(s_j))$.
5: Bob chooses a random value γ, and a pre-negotiated spacial value ν. For each $\mathsf{E}(R(s_j))$, he further computes $\mathsf{E}(\gamma * R(s_j) + \nu)$.
6: Bob permutes his set of $\mathsf{E}(\gamma * R(s_j) + \nu)$, and return them to Alice.
7: Alice decrypts all $\mathsf{E}(\gamma * R(s_j) + \nu)$. For each $s_j \in S \cap R$, she gets ν; otherwise, she gets a random value. Alice counts the number of ν values, and output.

---

4 TPPGL with Exact Matching

4.1 TPPGLE Using Commutative Encryption

In threshold privacy preserving group linkage, Alice and Bob first negotiate a group-level threshold θ: two groups are regarded similar if and only if $SIM(R, S) \geq \theta$. With exact matching at record-level, $sim(r_i, s_j) \in \{0, 1\}$. Let k be the minimum number of identical records from two groups for them to be linked, we have: $SIM(R, S) = \dfrac{k}{|R| + |S| - k} \geq \theta$. Note that we employ Jaccard similarity at group level.

If $|R| = m, |S| = n$, we have $k = \lceil (m + n)\theta/(1 + \theta) \rceil$; i.e. k is the smallest integer that is greater than or equal to $(m+n)\theta/(1+\theta)$. Therefore, the TPPGLE problem is to securely compare the actual intersection size ($|R \cap S|$) with k in a way that $|R \cap S|$ **should not be revealed to either Alice or Bob**[1]. Please note that although group sizes are revealed, it is acceptable since they could not be used to infer record identity. Similarly, privacy preserving record linkage solutions also share the number of records among participants.

To extend the AES scheme to tackle the TPPGLE problem, we enumerate all k-combinations of Alice's and Bob's elements and compare the k-combinations in privacy preserving manner. If at least one of the pairwise comparisons of k-combinations yields a positive result, we conclude that k or more elements from two groups match, so that the Jaccard similarity of the two groups has reached the threshold, and Alice and Bob should share the groups. The detailed process is shown in Protocol 3.

Figure 2 gives an example of Protocol 3. In this example, Alice and Bob each hold groups of records identified by names. Figure 2 (a) shows Alice's group of four records and Bob's group of three records. A threshold $\theta = 0.7$ is pre-negotiated. In step 1, both Alice and Bob get $k = \lceil 0.7(7)/(1 + 0.7) \rceil = 3$. In step

---

[1] This requirement makes the problem different from the well-know Yao's millionaire problem [28].

Protocol 3. K-combination approach for TPPGLE

Data: Alice has a group $R = \{r_1, ..., r_m\}$; and Bob has a group $S = \{s_1, ..., s_n\}$. They negotiate a similarity threshold θ.

Result: Both Alice and Bob learn whether or not the similarity between R and S is greater than θ, i.e. if $SIM(R, S) > \theta$, and nothing else; especially, not $SIM(R, S)$.

1: Alice and Bob both compute $k = \lceil (m + n)\theta/(1 + \theta) \rceil$, i.e. the smallest integer that is not less than $(m + n)\theta/(1 + \theta)$.

2: Alice and Bob each gets all the k-combinations of her/his own elements. There are C_k^m k-combinations from Alice and C_k^n k-combinations from Bob.

3: For each k-combination, sort the elements using a pre-negotiated order, and serialize them into a string, with a special separator between elements.

4: Both Alice and Bob follow the AES approach (Protocol 1) to find the intersection of the k-combinations. If at least one k-combination is found in the intersection, the two groups are matched, i.e. $SIM(R, S)$ is guaranteed to be greater than θ.

(a)   (b)

Fig. 2. (a) Alice and Bob's groups; (b) privacy preserving set intersection of k-combinations extracted from groups

2, Alice gets C_3^4 3-combinations from her records and Bob gets C_3^3 3-combinations from his records. In step 3, Alice sorts the elements in each 3-combination in ascending order, serializes the primary keys into a string, with "&" as the separator. As shown in Figure 2 (b), Alice and Bob continue to use AES approach to find the intersections of the strings, in a privacy preserving manner. In this example, one intersection is found, which means the two groups are considered to be matched at the threshold of 0.7. On the other hand, if we replace Alice's record "J.Doe" with another record "Z.Doe", follow the above procedures, then, none of the strings serialized from 3-combinations would match. In this case, two groups are not linked, and both Alice and Bob learn nothing about other's group.

In this approach, we avoid homomorphic encryption, which requires heavy computation. However, when Alice generates C_k^m k-combinations and Bob generates C_k^n k-combinations, the value of C_k^m and C_k^n could be too large to manipulate. Therefore, this approach is preferable when $k = \lceil (m + n)\theta/(1 + \theta) \rceil$ is (1) very small, or (2) very close to m and n. In real-world applications, k is usually close to m and n.

4.2   TPPGLE Using Homomorphic Encryption

In FNP approach, Alice and Bob pre-negotiate a special value ν, which represents a matching record. After decryption, Alice counts the number of ν values, which represents the number of records in $R \cap S$. In TPPGLE, this number should not be revealed to either party. Instead, they should only learn a Boolean value: $|R \cap S| > k$. To tackle this problem, we modify FNP approach starting from step 6. Before permuting $\mathsf{Enc}(\gamma * R(s_j) + \nu)$, Bob injects a random number (k_b) of $\mathsf{Enc}(\nu)$ elements into the result set. We assume there is $k' = |R \cap S|$, which generates k' number of $\mathsf{Enc}(\nu)$ elements in the original $\mathsf{Enc}(\gamma * R(s_j) + \nu)$ set. After the random injection, Alice decrypts the polluted set to obtain $(k_b + k')$ ν values. She has no knowledge about either k_b or k', as long as k_b is selected from a good range.

Now the problem is converted to Yao's Millionaire Problem: Alice knows $k_b + k'$ while Bob knows $k_b + k$ (the new threshold), and they want to compare two values without leaking them to each other. In our settings, $k_b + k' \ll N, k_b + k \ll N^2$. We may assume that the product of $(k' - k)$ and a random number r' is much less than N. In our solition, Bob first generates $\mathsf{Enc}(k' - k)$ from $\mathsf{Enc}(k_b + k')$ (obtained from Alice) and $\mathsf{Enc}(k_b + k)$. Hence, we are to find out whether $k' - k > 0$ or not, without revealing the actual value. Bob further randomizes the result with two positive random numbers $\gamma' \ll N$ and $\nu' < \gamma'$, obtaining $\mathsf{Enc}(\gamma' \times (k' - k) + \nu')$. Then Alice gets the cipher from Bob and decrypts it. Based on previous assumption, we may infer that $(\gamma' \times (k' - k) + \nu') > 0$ iff. $(k' - k) > 0$, and vice versa. Meanwhile, due to the cryptographic properties of Paillier cryptosystem, the decryption result should be the least positive residues of plain text modulus N. Hereby, if $\gamma' \times (k' - k) + \nu' < N/2$, we have $k' - k > 0$ and thus the two groups are linked. Otherwise, if $N/2 < \gamma' \times (k' - k) + \nu' < N$, the two groups are not linked.

5   TPPGL with Approximate Matching

In previous section, group members (records) are identified by primary keys. Therefore, two records are either "identical" or "different". In this section, we consider the problem with approximate matching. In this scenario, Alice and Bob pre-negotiate a vector space, and represent their records as vectors in this space. Since our research is more focused on group-level linkage, we adopt a simple cosine similarity function, which employs private scalar product [7], for document-level vector-space similarity.

First, in Protocol 5, we revisit the privacy preserving inner-product (scalar product) approach presented in [7]. In the protocol, z represents the dimensionality of the vector space, while μ denotes the normalized modulus of space vectors.

---

[2] k is no larger than the group size. In our assumptions, typical group size is small (e.g. tens). On the other hand, in our experiments, N is the product of two 256-bit prime numbers.

Protocol 4. Homomorphic encryption approach for TPPGLE

Data: Alice has a group $R = \{r_1, ..., r_m\}$; and Bob has a group $S = \{s_1, ..., s_n\}$. They negotiate a similarity threshold θ.

Result: Both Alice and Bob learn whether or not the similarity between R and S is greater than θ, i.e. if $SIM(R, S) > \theta$, and nothing else; especially, not $SIM(R, S)$.

1: Alice creates keys for homomorphic encryption and publishes her public key.
2: Alice constructs $R(x) = \prod(x - r_i)$, and computes all the coefficients α_u that $R(x) = \sum_{u=0}^{u=m} \alpha_u x^u$. Therefore, the m degree polynomial $R(x)$ has roots $\{r_1, ..., r_m\}$.
3: Alice encrypts the coefficients and sends them $(\{\mathsf{E}(\alpha_0), \mathsf{E}(\alpha_1), ..., \mathsf{E}(\alpha_m))$ to Bob.
4: For each s_j, Bob evaluates the polynomial (without decryption) to get $\mathsf{E}(R(s_j))$.
5: Bob chooses a random value γ, and a pre-negotiated spacial value ν. For each $\mathsf{E}(R(s_j))$, he further computes $\mathsf{E}(\gamma * R(s_j) + \nu)$.
6: Bob gets a random number k_b. He injects k_b number of $\mathsf{E}(\nu)$ values into the set he obtained from the previous step. Meanwhile, he also injects random number of random values into this set.
7: Bob permutes his polluted set of $\mathsf{Enc}(\gamma * R(s_j) + \nu)$, and returns them to Alice.
8: Alice decrypts all items in the polluted set. She then count number of ν values.
9: Assume $k' = |R \cap S|$, Alice now knows $k_b + k'$, but not k_b; Bob knows k_b, and thus $k_b + k$. Neither of them knows k'.
10: Alice encrypts $k_b + k'$, and sends it to Bob.
11: Bob gets $\mathsf{E}(k_b + k')$. With the homomorphic properties of $\mathsf{E}()$, he calculates $\mathsf{E}((k_b + k') - (k_b + k)) = \mathsf{E}(k' - k)$.
12: Bob creates two random numbers $\gamma' \ll N$ and $\nu' < \gamma'$. Bob randomizes $\mathsf{E}(k' - k)$ to $\mathsf{E}(\gamma' \times (k' - k) + \nu')$.
13: Bob return $\mathsf{Enc}(\gamma' \times (k' - k) + \nu')$ to Alice. Alice decrypts it to m, output "Yes" if $m < N/2$, or "No" if $m > N/2$.

We assume Alice has a group: $R = \{\mathbf{r}_1, \mathbf{r}_2, ..., \mathbf{r}_m\}$, and Bob has a group $S = \{\mathbf{s}_1, \mathbf{s}_2, ...\mathbf{s}_n\}$. In simple PPGLA, we conduct pairwise comparison between Alice's and Bob's vectors, and all $sim(\mathbf{r}_i, \mathbf{s}_j)$ are counted towards group similarity. Being simple in calculation, however, this approach does not provide best result, since many $sim(\mathbf{r}_i, \mathbf{s}_j)$ values are very small that they should be considered as "noise". Therefore, a better solution is to have a record-level "cut-off" ρ such that: edge is created in the bipartite graph iff. the similarity between two vertexes is larger than ρ (i.e. $sim(\mathbf{r}_i, \mathbf{s}_j) > \rho$). On the other hand, also to eliminate noises, we only consider an unlabeled bipartite graph – when two records are linked, we use "1", instead of $sim(\mathbf{r}_i, \mathbf{s}_j)$ in group-level similarities. With a binary bipartite graph, group-wise similarity becomes: $BM_{sim,\rho}(R, S) = \dfrac{k}{|R| + |S| - k}$, where $k = |M|$. To get $BM_{sim,\rho}(R, S) > \theta$, we need to have: $k > (m+n)\theta/(1+\theta)$.

Therefore, Alice and Bob need to pre-compute k_{min} based on θ, then securely compute k, and compare k with k_{min}. However, this approach is again flawed – an advisory could break the protocol by faking his groups. Let us assume that Alice and Bob each has a group of three members, while only $\mathbf{r}_1$ and $\mathbf{s}_1$ match ($sim(\mathbf{r}_1, \mathbf{s}_1) > \rho$). Bob could fake a group with (repeated) members: $\{\mathbf{s}_1, \mathbf{s}_1', \mathbf{s}_1''\}$,

---

Protocol 5. Privacy preserving inner-product [7].

Data: In a shared vector space $\mathbb{Z}_\mu^z, \mu < \sqrt{N/2z}$, Alice has her vector $\mathbf{r}$; and Bob has his vector $\mathbf{s}$.

Result: Alice and Bob both learn the inner-product $\mathbf{r} \cdot \mathbf{s}$.

1: Alice creates keys for homomorphic encryption and publishes her public key.
2: Alice encrypts each r_i in $\mathbf{r} = [r_1, ..., r_z]$ to obtain $\mathsf{Enc}(\mathbf{r}) = [\mathsf{E}(r_1), \mathsf{E}(r_2), ..., \mathsf{E}(r_z)]$.
3: Alice sends $\mathsf{E}(\mathbf{r})$ to Bob.
4: With the homomorphic properties of $\mathsf{E}()$, Bob computes the inner-product (without decryption): $\mathsf{E}(\mathbf{r} \cdot \mathbf{s}) = \mathsf{E}(r_1 s_1 + r_2 s_2 + ... + r_z s_z)$.
5: Bob sends $\mathsf{E}(\mathbf{r} \cdot \mathbf{s})$ back to Alice, Alice decrypts and publishes the result.

---

in which $\mathbf{s}_i'$ is a slightly modified version of $\mathbf{s}_i$. In this way, the manufactured group is highly likely to be linked with R. To tackle this problem, we measure the "degree of participation" from Alice and Bob, instead of using the total number of linked records. In other words, we count the number of elements from Alice that are linked to at least one element from Bob (m'), and vice versa. If we obtain m' and n' from the count, we then compare $min(m', n')$ with k_{min} to make the decision.

To implement such operations in a privacy preserving manner, we present TPPGLA in Protocol 6 (on the last page). In the protocol, Alice and Bob will use an encrypted similarity matrix M to store the intermediate results. The content of M should be private throughout the group linkage procedure. Bob first generates $\mathsf{E}(\mathbf{r}_i \cdot \mathbf{s}_j)$ with the private-preserving scalar product protocol, and subtract them by the record-level cut-off ρ. In the matrix M, each positive value (in plaintext) indicate a link at the record level (or an edge in the bipartite marching graph). If a row i has at least one positive value, it indicates that Alice's record $\mathbf{s}_i$ has participated in the linkage (i.e. linked with at least one record from Bob). To measure m' is to count number of rows that have at least one positive value, and to measure n' is to count number of columns that has at least one positive element.

To conduct such operations securely, Bob randomizes each pairwise record similarity into a encrypted "Boolean" value, with meaningful information only in the sign digit of the plain text. Before sending the cipher back to Alice, Bob injects two groups of positive and negative values with random sizes into each row and column, expanding the size of M into a larger range. If Alice counts all the positive values in row i, of M, she cannot infer whether $\mathbf{r}_i$ shares any similar records in Bob's set (i.e. the number of links between $\mathbf{r}_i$ and Bob's items, c_{ri}), if she doesn't how many positive values Bob has injected.

Further more, to protect m' and n' from been learned by either party, Bob performs another injection-permutation operation after Alice returns (encrypted) sum of each row. He injects c_{br} non-zero values into the set of c_{ri}, and make sure that Alice can only learn $m' + c_{br}$ instead of m'. Similarly, Alice can obtain $n' + c_{br}$, where n' denotes the number of shared items from Bob.

In approximate matching, if Alice shares m' records with Bob, and Bob shares n' records with Alice, the maximum bipartite matching cardinality is

Protocol 6. $TPPGLA$ with record-level cut-off

Data: Alice has a group $R = \{r_1, ..., r_m\}$; Bob has a group $S = \{s_1, ..., s_n\}$. They negotiate a record-level cut-off threshold ρ and a similarity threshold θ.

Result: Alice and Bob both learn if similarity between $\mathcal{R}$ and $\mathcal{S}$ is greater than θ, i.e. if $BM_{sim}(\mathcal{R}, \mathcal{S}) > \theta$, and nothing else; especially, not $BM_{sim,\rho}(\mathcal{R}, \mathcal{S})$.

1: Alice creates keys for homomorphic encryption and publishes her public key.

2: Alice and Bob negotiate a shared space $\mathbb{Z}_\mu^z$. They represent their vectors r_i and s_j in the space. All vectors are normalized to $\mu < \sqrt{N/2z}$.

3: Alice and Bob both compute $k = \lceil (m+n)\theta/(\rho+\theta) \rceil$.

4: For each pair r_i, s_j, they follow protocol 5 to compute $\mathsf{E}(r_i \cdot s_j)$. Instead of sending $\mathsf{E}(r_i \cdot s_j)$ to Alice, Bob chooses a random value $\gamma > 0$ to compute $\mathsf{E}(\gamma * (r_i \cdot s_j - \rho))$. The result set forms a $m \times n$ matrix $M = (\mathsf{E}(\gamma * (r_i \cdot s_j - \rho)))_{m \times n}$.

5: Bob creates two random vectors c_{b+} and c_{b-}, where $c_{b+} = (c_{b1+}, c_{b2+}, ..., c_{bm+})$ and $c_{b-} = (c_b - c_{b1+}, c_b - c_{b2+}, ..., c_b - c_{bm+})$. For each row of M, Bob injected c_{bi+} encrypted random positive values and c_{bi-} encrypted random negative values and gets $M_r = [M]_{m \times (n+c_b)}$. He then permutes each row of M_r and sends it to Alice.

6: Alice decrypts M_r into V_r. She counts the number of $\nu < N/2$ for each row, and obtains $c_{ri} + c_{bi+}$, where c_{ri} denotes the number of records in $\mathcal{S}$ that are supposed to be similar with r_i. Alice now knows $c_{ri} + c_{bi+}$ and Bob knows c_{bi+}. Neither of them knows whether r_i is similar with any record in S.

7: Alice encrypts $[c_{r1} + c_{b1+}, c_{r2} + c_{b2+}, ..., c_{rm} + c_{bm+}]$ to $[\mathsf{E}(c_{r1} + c_{b1+}), \mathsf{E}(c_{r2} + c_{b2+}), ..., \mathsf{E}(c_{rm} + c_{bm+})]$ and sends them to Bob.

8: Bob creates two positive random numbers γ and υ. He randomizes $\mathsf{E}(c_{ri} + c_{bi+} - c_{bi+})$ and gets $[\mathsf{Enc}(\gamma * c_{r1} + \upsilon), \mathsf{E}(\gamma * c_{r2} + \upsilon), ... \mathsf{Enc}(\gamma * c_{rm} + \upsilon)]$.

9: Bob creates a random number c_{br} of different random integers $0 < d_{inject} < m$. He injects c_{br} number of $\mathsf{Enc}(d_{inject})$ values into $[\mathsf{E}(\gamma * c_{r1} + \upsilon), \mathsf{E}(\gamma * c_{r2} + \upsilon), ... \mathsf{Enc}(\gamma * c_{rm} + \upsilon)]$, permutes the set and sends the result to Alice.

10: Alice decrypts all items in the polluted set. Assume $\mathcal{R}_{m'}$ is the largest subset of $\mathcal{R}_m$, $\forall r_i \in \mathcal{R}_{m'}, \exists s_j \in \mathcal{S}$, s.t. $sim(r_i, s_j) > \rho$. She then count number of non-zero values and gets $m' + c_{br}$.

11: Similarly, Alice learns $n' + c_{br}$ if we conduct the above operations in columns.

12: Now Alice knows $m' + c_{br}$ and $n' + c_{br}$, and Bob knows $k + c_{br}$. They can proceed with protocol 4 from step 9 to compare $min(m', n')$ with k. If intersection threshold k is smaller than $min(m', n')$, the $SIM(R, S)$ is guaranteed to be greater than θ.

$min(m', n')$. Alice and Bob both compute the group intersection threshold k; now Alice knows $\{m' + c_{br}, n' + c_{br}\}$ and Bob knows $k' + c_{br}$, and they want to learn if $k' < min(m', n')$. They can follow Protocol 4 from step 9 to get the final decision.

6 Security Analysis

6.1 Attacker Models

The goal of the proposed TPPGL protocols is to guarantee that (1) in the protocol, each party learns only the fact whether the groups are similar or not; and (2) no content or similarity measurement at the record level is disclosed to any party; no similarity measurement at the group-level (other than (1)) is disclosed to any party. Please note that in privacy preserving record linkage, it is convention that numbers of records from all participants are revealed. In our scenario, it is also acceptable that both parties disclose the sizes of their groups. Unlike group similarity information, group size information cannot be used to infer record identities.

In secure two-party computation problems, communication between the parties is usually assumed to be authenticated and encrypted. Therefore, our protocol is secure against *outsider adversaries* that passively monitor the network traffic but have no access to the inputs or the outputs. Hence, we further consider two types of insider adversaries, *semi-honest* (a.k.a. *honest-but-curious*) and *augmented semi-honest* adversaries, in our attacker model: (1). Semi-honest model describes a passive insider attack: both parties are assumed to properly follow protocol (so that they are "honest"); meanwhile, each party keeps all the accessible intermediate results and outputs, and tries to infer private data (so that they are "curious"). (2). In augmented semi-honest attacker model, the adversary can further change the input and output of one party, without tampering the protocol, to affect the views of the others'.

Due to space limitations, we only evaluate the **correctness** and **security** of Protocol 3, i.e. TPPGL with commutative encryption. With the same methodology, we can extend our proofs to all other protocols presented in the paper.

6.2   TPPGLE Using Commutative Encryption

First, we classify the two parties as *client* and *server* based on their roles in the protocol. The client (e.g. Alice) initiates the secure computation and gets the output, and the server (e.g. Bob) responds to the inputs from the client. The protocol is correct if it evaluates the similarity function with high probability.

STATEMENT 1. *In Protocol 3, assuming there are no hash collisions, the client learns a Boolean output, where 1 for $SIM(R,S) > \theta$, and 0 otherwise.*
PROOF. For any $k \leq min(|R|, |S|)$, assume the hash function h has no collisions on $R^k \cup S^k$. Since E_s and E_r are bijective and commutative, we have
   $v \in R^k \cap S^k$ iff $v \in R^k$ and $E_r(E_s(h(v))) = E_s(E_r(h(v)))$,
which means the same concatenated set of elements is constructed by both the client and the server. Since k is calculated from θ, we have $SIM(R,S) > \theta$ iff $\exists v \in R^k \cap S^k$. □

The security of the protocol is to preserve the privacy of the data of both client and server. Then, we have

STATEMENT 2. *TPPGLE-Commutative is secure if both parties are semi-honest or augmented semi-honest. From the protocol, the client learns the Boolean output and the size $|S|$, and the server only learns the size $|R|$.*

PROOF. We use the similar proof methodology as in [1]: it assumes a simulation using the knowledge that the client (and the server) is supposed to have according to the protocol, and the client (and the server) should not be able to distinguish the simulated view and the real view.

First, let us construct the simulator for the server. The server receives no output from the protocol, but $C_k^{|R|}$ encrypted messages from the client at step 4. Each message is the hash of the concatenation of k elements from set R, encrypted by commutative key E_r. The simulator generates k random values $z_i \in F$, where F is the message space, and then concatenates them in a random sequence. Assume the hash $h(||z_1||...||z_k||)$ is uniformly distributed ("$||$" denotes concatenation), the real view and the simulated view for the server are indistinguishable.

Then, let us construct the simulator for the client. The simulator will use R, $|S|$, and $R^k \cap S^k$. To simulate the view for the client, the simulator selects a commutative key E_h to encrypt $||z_1||...||z_k||$ for $z_i \in R \cap S, 1 \leq i \leq k$. Then the simulator generates $|S| - |S \cap R|$ random k-concatenations, and encrypts them with E_h. In real view, the $|S|$ concatenations are all encrypted by E_s. Since E_h and E_s are randomly chosen from the same key space, their distributions are identical. For the client, the real view and the simulated view are indistinguishable. □

7   Experiments

We perform our experiments on three data sets, which were adopted in [23]. As summarized in Table 1, two data sets, *co-author network* (shortly AN) and *paper citation network* (shortly CN), are extracted from academia search system Arnetminer, and the last one, *movie network* (shortly MN), is crawled from Wikipedia category "English-language films". AN represents author names as vertices and the coauthor relationships as edges, while in CN, the vertices are a set of 2,329,760 papers and the edges denote the citation relationships between the papers. Since both AN and CN are homogeneous networks, we treat each 1-neighborhood subgraph (e.g. an author and all the co-authors) as a group, and use author name and citation name as key attributes for exact matching. MN is a heterogeneous network with 142,426 relationships between the heterogeneous nodes of films, director, actors, and writers. In our experiments, we treat a heterogeneous subnet (of a selected number of nodes) as a "group", and extract the label from each node to form the content of "records". Textual similarity (e.g. cosine similarity with TF/IDF weighting) between two labels is calculated as pairwise record-level similarity.

Then, we quantitatively evaluate the performance of the proposed TPPGL protocols (Protocol 3, 4 and 6) on the three data sets. Since our focus is on the viability and performance of these approaches in privacy preserving group linkage, we measure the *end-to-end execution time* under different group sizes and thresholds θ to assess the efficiency of each protocol. To meet the computational requirements of the cryptographic operations, we implement our own Big Integer class in C# under .NET framework 3.5.

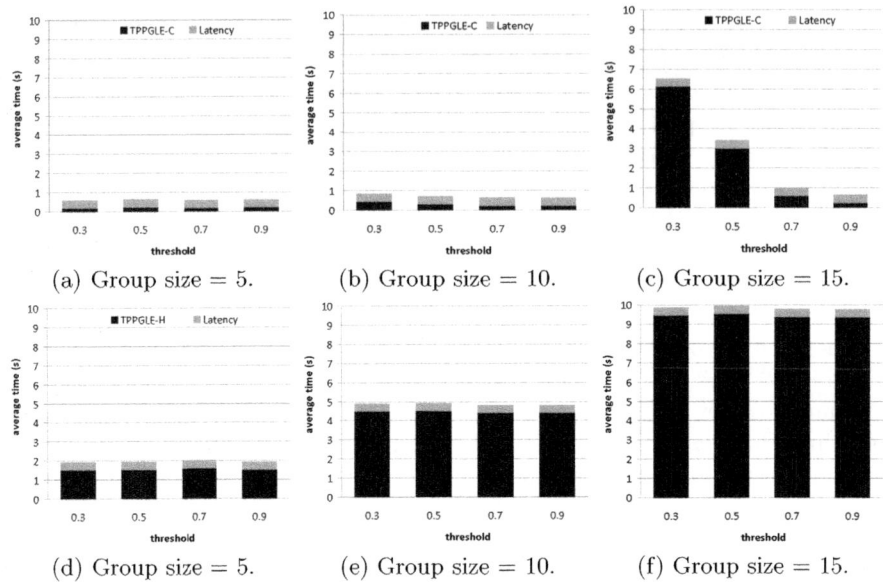

(a) Group size = 5.   (b) Group size = 10.   (c) Group size = 15.

(d) Group size = 5.   (e) Group size = 10.   (f) Group size = 15.

Fig. 3. Experimental results of TPPGLE

Table 1. A summary of three data sets

Data Set	Key	Record
AN	Author name	Authors and coauthors.
CN	Paper name	Paper and citations.
MN	-	Attributes (labels) of actors, writers, singers, etc.

7.1 TPPGLE

To evaluate the performance of TPPGLE, we first generate synthetic groups from AN and CN data sets. To form a group, we randomly pick a seed node, and follow its edges to add more members to the group. For instance, when a seed (3, J. Doe(15), 203) is selected, the next node 203 should be added to the group. We evaluate the protocol under different group sizes (e.g. each group has 5, 10, and 15 records) and different thresholds (e.g. $\theta \in \{0.3, 0.5, 0.7, 0.9\}$). For each θ, we generate 50 pairs of linked groups, and 50 pairs of unrelated groups.

Following the TPPGLE protocol using commutative encryption, we first hash each record (of Alice and Bob) into a 160-bit value, and then encrypt it with a commutative encryption function. Here, we adopt the famous SRA scheme. The average end-to-end execution time for Protocol 3 under different group sizes are shown in Figures 3(a)-(c). In each figure, the lower portion (denoted as TPPGLE-C) of the tacked bars represents the average execution time under different thresholds, and the upper portion represents network latency delay, which is estimated as the average one-hop network latency (i.e. 100ms) times the

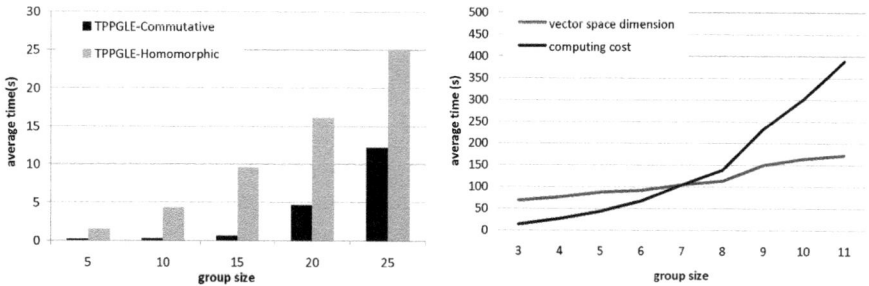

(a)TPPGLE with difference group sizes; (b) TPPGLA with difference group sizes.

Fig. 4. Computation cost of TPPGL protocols

rounds for data exchange between Alice and Bob. From the results, we see that the performance of Protocol 3 highly depends on the preset similarity threshold, especially when group size becomes large. It is easy to understand: a larger similarity threshold θ, which means k is closer to the group size, introduces less computation overhead due to the combination calculations. In real group linkage cases, it is also reasonable to select a large similarity threshold since there is no need to link two groups that are not similar.

Then, we use the same group settings to evaluate the performance of the TPPGLE protocol using homomorphic encryption (TPPGLE-H). We adopt the Paillier encryption scheme to encrypt the coefficients used in Protocol 4. The average end-to-end execution time under different group sizes are shown in Figures 3(e)-(g). The results show that the performance does not change much under different thresholds, but is greatly affected by different group sizes.

Therefore, we run another experiment to evaluate the efficiency of two TPP-GLE protocols over larger groups. We set the threshold to $\theta = 0.7$, and generate 100 pairs of groups with different sizes (5, 10, 15, 20, and 25). Figure 4(a) shows the average end-to-end execution time for both commutative encryption based approach and homomorphic encryption based approach. Apparently, when the group size increases, computation cost of the TPPGLE-H protocol shows a linear increasing trend, while in the TPPGLE-C protocol it grows almost exponentially. This is because in TPPGLE-H protocol, the computational complexity increases along with the degree of the group's polynomial representation, and thus increases linearly with the group size. In TPPGLE-C protocol, its computational complexity depends on the k-combination function: $Cost(n) = C_k^n \times \mathbf{E}(*)$. For a given k, the computation cost increases with n in a manner slower than exponential but faster than polynomial.

7.2 TPPGLA

We evaluate the validity and efficiency of TPPGLA protocol (Protocol 6) on the movie network data set. To form the group, we randomly extract a subset of

454 F. Li et al.

1000 records from MN, and calculate pairwise record-level similarities between the records. For a given record-level cut-off ρ (a preset value negotiated between Alice and Bob), we divide the records into two parts, according to whether record pairs are similar or not.

We first select k pairs of different records from the similar set as input, where k is the set-intersection threshold, and apply Protocol 6 on the k pairs. The output is "Yes", which verify the validity of the protocol. Then, we select random group pairs from MN to evaluate the efficiency of the protocol, and show the computation cost of TPPGLA under different group sizes (from 3 to 11) in Figure 4(b). From the results, we see that the computation cost increases greatly with the group size. This is because the computation cost is proportionate to group size $m \times n$ and the dimensionality of vector space, while the latter also increases with the group size. Overall, TPPGLA introduces a comparably large overhead in end-to-end execution time , however, it is the price we pay for extreme cases with strong needs for privacy protection.

8 Conclusion and Future Works

In this paper, we have presented privacy preserving group linkage, in which groups of records from two parties are compared in a way that no record content is revealed. Simple PPGL (in which both parties learn the group similarity) suffers from the group membership inference problem, which could be employed to learn the member records of the other party's groups, even though the groups are not linked. To tackle the problem, we propose threshold privacy preserving group linkage, in which both parties only learn the verdict on whether the two groups are matched or not, instead of the value of group similarity. We implemented and tested TPPGL protocols for both exact matching and approximate matching scenarios. From the experiments, we can see that TPPGL pays a price in computation in order to protect the participants' privacy.

Although our approach demonstrates strong privacy protection, the computation overhead is relatively high. In its current form, the approaches are suitable for exchanging highly sensitive information. Our future work is to further explore cryptographic methods to reduce the overall computation of our approaches. Meanwhile, we are also optimizing our existing implementations, and planning to test it over large datasets.

References

1. Agrawal, R., Evfimievski, A., Srikant, R.: Information sharing across private databases. In: SIGMOD (2003)
2. Boneh, D., Goh, E.J., Nissim, K.: Evaluating 2-DNF formulas on ciphertexts. In: Kilian, J. (ed.) TCC 2005. LNCS, vol. 3378, pp. 325–341. Springer, Heidelberg (2005)

3. Elmagarmid, A.K., Ipeirotis, P.G., Verykios, V.S.: Duplicate record detection: A survey. IEEE TKDE, 19, 1–16 (2007)
4. Freedman, M.J., Nissim, K., Pinkas, B.: Efficient private matching and set intersection. In: Cachin, C., Camenisch, J.L. (eds.) EUROCRYPT 2004. LNCS, vol. 3027, pp. 1–19. Springer, Heidelberg (2004)
5. El Gamal, T.: A public key cryptosystem and a signature scheme based on discrete logarithms. In: Blakely, G.R., Chaum, D. (eds.) CRYPTO 1984. LNCS, vol. 196, pp. 10–18. Springer, Heidelberg (1985)
6. Gentry, C.: Fully homomorphic encryption using ideal lattices. In: STOC, pp. 169–178. ACM, New York (2009)
7. Goethals, B., Laur, S., Lipmaa, H., Mielikäinen, T.: On private scalar product computation for privacy-preserving data mining. In: Park, C.-s., Chee, S. (eds.) ICISC 2004. LNCS, vol. 3506, pp. 104–120. Springer, Heidelberg (2005)
8. Goldwasser, S., Micali, S., Rackoff, C.: The knowledge complexity of interactive proof-systems. In: STOC 1985, pp. 291–304 (1985)
9. Guha, S.: Merging the results of approximate match operations. In: VLDB, pp. 636–647 (2004)
10. hai Do, H., Rahm, E.: Coma - a system for flexible combination of schema matching approaches. In: VLDB, pp. 610–621 (2002)
11. Hall, R., Fienberg, S.E.: Privacy-preserving record linkage. In: Domingo-Ferrer, J., Magkos, E. (eds.) PSD 2010. LNCS, vol. 6344, pp. 269–283. Springer, Heidelberg (2010)
12. Hernández, M.A., Stolfo, S.J.: The merge/purge problem for large databases. In: SIGMOD (1995)
13. Inan, A., Kantarcioglu, M., Bertino, E., Scannapieco, M.: A hybrid approach to private record linkage. In: ICDE, pp. 496–505 (2008)
14. Inan, A., Kantarcioglu, M., Ghinita, G., Bertino, E.: Private record matching using differential privacy. In: EDBT, pp. 123–134 (2010)
15. Jaccard, P.: Étude comparative de la distribution florale dans une portion des alpes et des jura. Bulletin del la Société Vaudoise des Sciences Naturelles 37, 547–579 (1901)
16. Kissner, L., Song, D.: Privacy-preserving set operations. In: Shoup, V. (ed.) CRYPTO 2005. LNCS, vol. 3621, pp. 241–257. Springer, Heidelberg (2005)
17. Murugesan, M., Jiang, W., Clifton, C., Si, L., Vaidya, J.: Efficient privacy-preserving similar document detection. The VLDB Journal 19, 457–475 (2010) 10.1007/s00778-009-0175-9
18. Naccache, D., Stern, J.: A new public key cryptosystem based on higher residues. In: CCS, pp. 59–66. ACM, New York (1998)
19. On, B.-W., Koudas, N., Lee, D., Srivastava, D.: Group linkage. In: IEEE ICDE, Istanbul, Turkey, pp. 496–505 (April 2007)
20. Paillier, P.: Public-key cryptosystems based on composite degree residuosity classes. In: Stern, J. (ed.) EUROCRYPT 1999. LNCS, vol. 1592, pp. 223–238. Springer, Heidelberg (1999)
21. Paillier, P., Pointcheval, D.: Efficient public-key cryptosystems provably secure against active adversaries. In: Lam, K.-Y., Okamoto, E., Xing, C. (eds.) ASIACRYPT 1999. LNCS, vol. 1716, pp. 165–179. Springer, Heidelberg (1999)
22. Scannapieco, M., Figotin, I., Bertino, E., Elmagarmid, A.K.: Privacy preserving schema and data matching. In: SIGMOD, pp. 653–664. ACM, New York (2007)

23. Tang, J., Sun, J., Wang, C., Yang, Z.: Social influence analysis in large-scale networks. In: KDD, pp. 807–816 (2009)
24. Tejada, S., Knoblock, C.A.: Learning domain-independent string transformation weights for high accuracy object identification. In: ACM SIGKDD, pp. 350–359 (2002)
25. van Dijk, M., Gentry, C., Halevi, S., Vaikuntanathan, V.: Fully homomorphic encryption over the integers. In: Gilbert, H. (ed.) EUROCRYPT 2010. LNCS, vol. 6110, pp. 24–43. Springer, Heidelberg (2010)
26. Verykios, V.S., Bertino, E., Fovino, I.N., Provenza, L.P., Saygin, Y., Theodoridis, Y.: State-of-the-art in privacy preserving data mining. SIGMOD Rec. 33(1), 50–57 (2004)
27. Winkler, W.E.: Overview of record linkage and current research directions. Technical report, Bureau of the Census (2006)
28. Yao, A.C.: Protocols for secure computations. In: SFCS 1982: Proceedings of the 23rd Annual Symposium on Foundations of Computer Science, pp. 160–164. IEEE Computer Society, Washington, DC, USA (1982)

Dynamic Anonymization for Marginal Publication

Xianmang He*, Yanghua Xiao, Yujia Li, Qing Wang, Wei Wang, and Baile Shi

Fudan University, Shanghai 200433, China
{071021057,shawyh,wangqing,wangwei1,bshi}@fudan.edu.cn

Abstract. Marginal publication is one of important techniques to help researchers to improve the understanding about correlation between published attributes. However, without careful treatment, it's of high risk of privacy leakage for marginal publications. Solution like ANGEL has been available to eliminate such risks of privacy leakage. But, unfortunately, query accuracy has been paid as the cost for the privacy-safety of ANGEL. To improve the data utility of marginal publication while ensuring privacy-safety, we propose a new technique called dynamic anonymization. We present the detail of the technique and theoretical properties of the proposed approach. Extensive experiments on real data show that our technique allows highly effective data analysis, while offering strong privacy guarantees.

Keywords: privacy preservation, marginal publication, dynamic anonymization, m-invariance.

1 Introduction

In recent year, we have witnessed the tremendous growth of the demand to publish personal data, which posed great challenges for protecting the privacy in these data. For example, medical records of patients may be released by a hospital to aid the medical study. Suppose that a hospital wants to publish records of Table 1, called microdata (T). Since attribute *Disease* is sensitive, we need to ensure that no adversary can accurately infer the disease of any patient from the published data. For this purpose, unique identifiers of patients, such as *Name* should be anonymized or excluded from the published data. However, it is still possible for the privacy leakage if adversaries have certain background knowledge about patients. For example, if an adversary knows that Bob is of age 20, Zipcode 12k and Sex M, s/he can infer that Bob's disease is bronchitis since the combination of Age, Zipcode and Sex uniquely identify each patient in Table 1. The attribute set that uniquely identify each record in a table is usually referred to as a quasi-identifier (QI for short) of the table.

---

* This work was supported in part by the National Natural Science Foundation of China (No.61003001, No.61033010 and NO.90818023) and Specialized Research Fund for the Doctoral Program of Higher Education (No.20100071120032).

J.B. Cushing, J. French, and S. Bowers (Eds.): SSDBM 2011, LNCS 6809, pp. 451–460, 2011.

Table 1. Microdata

Table 2. Generalization T^*

	Age	Zip	Sex	Disease
Bob	20	12k	M	bronchitis
Alex	19	20k	M	flu
Jane	20	13k	F	pneumonia
Lily	24	16k	F	gastritis
Jame	29	21k	F	flu
Linda	34	24k	F	gastritis
Sarah	39	19k	M	bronchitis
Mary	45	14k	M	flu
Andy	34	21k	F	pneumonia

GID	Age	Zip	Sex	Disease
1	[19-20]	[12k-20k]	M	bronchitis
1	[19-20]	[12k-20k]	M	flu
2	[20-24]	[13k-16k]	F	pneumonia
2	[20-24]	[13k-16k]	F	gastritis
3	[29-34]	[21k-24k]	F	flu
3	[29-34]	[21k-24k]	F	gastritis
4	[34-45]	[14k-21k]	*	bronchitis
4	[34-45]	[14k-21k]	*	flu
4	[34-45]	[14k-21k]	*	pneumonia

Table 3. Marginal $\langle Zip, Disease \rangle$

Table 4. GT

Table 5. BT

Zip	Disease
[12k-13k]	bronchitis
[12k-13k]	pneumonia
[14k-16k]	gastritis
[14k-16k]	flu
[19-20k]	flu
[19-20k]	bronchitis
[21k-24k]	gastritis
[21k-24k]	flu
[21k-24k]	pneumonia

GID	Zip	Batch-ID
1	[12k-13k]	1
1	[12k-13k]	2
2	[14k-16k]	4
2	[14k-16k]	2
3	[19k-20k]	1
3	[19k-20k]	4
4	[21k-24k]	3
4	[21k-24k]	3
4	[21k-24k]	4

Batch-ID	Disease	Count
1	bronchitis	1
1	flu	1
2	pneumonia	1
2	gastritis	1
3	flu	1
3	gastritis	1
4	bronchitis	1
4	flu	1
4	pneumonia	1

To protect privacy against attack guided by background knowledge, generalization has been widely used in privets anonymization solutions [1, 2, 3, 4, 5]. In a typical generalization solution, tuples are first divided into subsets (each subset is referred to as a QI-group). Then, QI-values of each QI-group are generalized into less specific forms so that tuples in the same QI-group cannot be distinguished from each other by their respective QI-values. As an example, we generalize Table 1 into Table 2 such that there exists at least two records in each QI-group. After generalization, the age(=20) of Bob has been replaced by an interval [19-20]. As a result, even if an adversary has the exact QI values of Bob, s/he can not exactly figure out the tuple of Bob from the first QI-group.

Motivation 1: Privacy leakage of marginal publication. Privacy preservation of generalization comes at the cost of information loss. Furthermore, generalization generally loses less information when the number of QI attributes is smaller [6]. Hence, to enhance the understanding about the underlying correlations among attributes, the publisher may further release a refined generalization of the projection on attributes of interest. This approach is referred to as *marginal publication*. For example, a researcher may request refined correlations of Zipcode and Disease several weeks later after the publication of Table 2. To satisfy the request, the publisher further publish Table 3, which is a more accurate generalization of $\langle Zipcode, Disease \rangle$ compared to that in Table 2, hence capturing the correlations between Zipcode and Disease better.

However, it is of possible risk of privacy leakage in solutions of marginal publication. Continue the above example. Suppose an adversary knows Bob's QI-values. Then, by Table 2 s/he infers that Bob's disease is in the set {bronchitis,

flu}. By Table 3, s/he infers that Bob has contracted the disease either pneumo-
nia or bronchitis. By combining the above knowledge, the adversary makes sure
that Bob have contracted bronchitis.

Motivation 2: Information loss of existing solutions. To overcome the
privacy leakage of marginal publication, Tao et al. [5] propose an anonymization
technique ANGEL (illustrated in Example 1), which releases each marginal with
strong privacy guarantees. Many QI-groups of the anonymized table released by
ANGEL may contain a large number of sensitive values. The number of these
values in the worst case is quadratic to the number of tuples in the QI-group.
As a result, there will exist significant average error when answering aggregate
queries. To give a clear explanation, assume that a researcher wants to derive
an estimation for the following query:

*Select Count(*) From Table GT and BT Where ZipCode ∈ [12k, 24k] And
Disease='Penumonia'.*

By estimating from Table 4 and 5, we can only get an approximate answer 4,
which is much larger than the actual query result 2(see Table 1). Then, we may
wonder whether there exists an approach that can protect privacy for marginal
publication while ensuring the data utility of the published data. This issue is
addressed in this paper.

Example 1. Suppose that the publisher need to release a marginal containing
⟨*Zipcode, Disease*⟩. If the privacy principle is 2-unique, the parameter k of AN-
GEL will be 2. After running ANGEL under this parameter, the result will be
two tables GT (shown in Table 4 which is 2-anonymity) and BT (shown in Table
5 which is 2-unique).

Related Work. Although improving the data utility of marginal publication
is desired, rare works can be found to solve this problem. We give a brief review on the previous works about marginal publication. It was shown in that
when the set of marginals overlap with each other in an arbitrarily complex
manner, evaluating the privacy risk is NP-hard [7, 8]. The work of [7], on the
other hand, is applicable only if all the marginals to be published form a decom-
posable graph. The method in [8] requires that, except the first marginal, no
subsequent marginal released can have the sensitive attribute. The work of [8]
shows that, checking whether a set of marginals violates k-anonymity is a computationally hard problem. The method in the paper [9] requires that, except the
first marginal, no subsequent marginal released can have the sensitive attribute.
For example, after publishing Table 3(Marginal ⟨*Zip, Disease*⟩), the publisher
immediately loses the option of releasing any marginal which contains the attribute Disease. This is a severe drawback since the sensitive attribute is very
important for data analysis. The work that is closest to ours is ANGEL that
is proposed by Tao et al. [5]. ANGEL can release any marginals with strong
privacy guarantees, which however comes at the cost of information loss. Please
refer to Section 1 for details.

Contributions and paper organization. To reduce the information loss of marginal publication, we propose a dynamic anonymization technique, whose effectiveness is verified by extensive experiments. We systematically explored the theoretic properties of marginal publication, and proved that the generalization principle m-invariance can be employed to ensure the privacy safety of marginal publication.

The rest of the paper is organized as follows. In Section 2, we give the preliminary concepts and formalize the problem addressed in this paper. In Section 3, we present the dynamic anonymization technique as our major solution. In Section 4, experimental results are evaluated. Finally, the paper is concluded in Section 5.

2   Problem Definition

In this section, we will formalize the problem addressed in this paper and explore some theoretic property of marginal publication.

Marginal Publication. Let T be a microdata table, which has d QI-attributes $A_1, ..., A_d$, and a sensitive attribute (SA) S. We consider that S is categorical, and every QI-attribute $A_i (1 \leq i \leq d)$ can be either numerical or categorical. For each tuple $t \in T, t.A_i (1 \leq i \leq d)$ denotes its value on A_i, and $t.A_s$ represents its SA value. We first give the fundamental concepts. A *QI-group* of T is a subset of the tuples in T. A *partition of T* is a set of disjoint QI-groups whose union equals T.

Now, we will formalize the key concepts in marginal publication. In the following texts, without loss of generality, we assume that all marginals released contain the sensitive attribute. A marginal published without the sensitive attribute is worthless for the QI-conscious adversary.

Definition 1 (Marginal). *Marginal M_j is a generalized version of certain projection on microdata T. The correspondent schema of the projection is referred to as the schema of the marginal. A trivial marginal is the marginal that contains no QI-attributes. Any other marginals are non-trivial.*

Given a microdata T, the number of its possible non-trivial marginals can be quantified by the following lemma. In following texts, without explicit statement, a marginal is always non-trivial.

Lemma 1. *There are $2^d - 1$ different non-trivial marginals, where d is the number of QI-attributes.*

Given multiple marginals of a microdata, it is possible for an adversary to infer privacy of a victim by combining knowledge obtained from different marginals. This implies that *the intersection of the sensitive sets obtained from different marginals must cover sufficiently large number of values so that the adversaries can not accurately infer the sensitive information of a victim.* One principle achieving this objective is m-invariance. We first give the definition of m-invariance and two basic concepts to define m-invariance: signature and m-unique. The privacy guarantee of m-invariance is established by Lemma 3 in the paper [4].

Definition 2 (Signature, m-Unique [4, 10]). *Let P be a partition of T, and t be a tuple in a QI-group $G \in P$. The signature of t in P is the set of distinct sensitive values in G. An anonymized version T^* is m-unique, if T^* is generated from a partition, where each QI-group contains at least m tuples, each with a different sensitive value.*

Definition 3 (m-Invariance [4, 10]). *A set S of partitions is m-invariant if (1) Each partition in S is m-unique; (2) For any partitions $P_1, P_2 \in S$, and any tuple $t \in T$, t has the same signature in P_1 and P_2.*

In this paper, we adopt the normalized certainty penalty (NCP [3]) to measure the information loss. Now, we are ready to give the formal definition about the problem that will be addressed in this paper.

Definition 4 (Problem Definition). *Given a table T and an integer m, we need to anonymize it to be a set of marginals $M_j(1 \leq j \leq r)$ such that (1)Existence: these marginals are m-invariant; (2)Optimality: and the information loss measured by NCP is minimized.*

Existence of m-Invariant marginals. Given a table T and an integer m, is it possible to generate a set of marginals $M_j(1 \leq j \leq r)$ that is m-invariant? The answer is positive. We will show in Theorem 1 that if a table T is *m-eligible*, there exists a set of marginals $M_r(1 \leq j \leq r)$ that is m-invariant. A table T is *m-eligible* if it has at least one m-unique generalization. Then, to determine the existence of m−invariant marginals for a table, we only need to find a sufficient and necessary condition to characterize m-eligibility of a table, which is given in Theorem 2.

Theorem 1. *If a table T is m-eligible, then there exists a set of marginals $\{M_1, M_2, \cdots, M_r\}$ that is m-invariant.*

Theorem 2. *A table T is m-eligible, if and only if the number of tuples that have the same sensitive attribute values is at most $\frac{|T|}{m}$, where $|T|$ is the number of tuples in table T.*

3   Dynamic Anonymization Technique

In this section, we will present our the detail of our solution: dynamic anonymization, which contains three steps: partition, assign and decomposition. Each step will be elaborated in following texts.

The Partitioning Step. The partitioning step aims to partition tuples of T into disjoint sub-tables T_i such that each T_i is m-eligible. The detailed procedure is presented in Figure 1. Initially, S contains T itself (line 1); then, each $G \in S$ is divided into two generalizable subsets G_1 and G_2 such that $G_1 \cup G_2 = G$, $G_1 \cap G_2 = \emptyset$ (line 5-7). Then for each new subset, we check whether $G_1(G_2)$ satisfies m-eligible (line 8). If both are generalizable, we remove G from S, and add G_1, G_2 to S; otherwise G is retained in S. The attempts to partition G are tried k times and tuples of G are randomly shuffled for each time (line 3-4). Our experimental results show that most of G can be partitioned into two m-eligible sub-tables by up to $k = 5$ tries. The algorithm stops when no sub-tables in S can be further partitioned.

In the above procedure, the way that we partition G into two subsets G_1 and G_2 is influential on the information loss of the resulting solution. To reduce information loss, we *distribute tuples sharing the same or quite similar QI-attributes into the same sub-tables.* For this purpose, we artificially construct two tuples $t_1, t_2 \in G$ with each attribute taking the maximal/minimal value of the corresponding domains, and then insert them G_1 and G_2 separately (line 6). After this step, for each tuple $w \in G$ we compute $\Delta_1 = NCP(G_1 \cup w) - NCP(G_1)$ and $\Delta_2 = NCP(G_2 \cup w) - NCP(G_2)$, and add tuple w to the group that leads to lower penalty (line 7). After successfully partitioning G, remove the artificial tuples from G_1 and G_2 (line 8).

Input: A microdata T, integers k and m
Output: A set S consisting of sub-tables of T;
/* the parameter k is number of rounds to partition G*/
1. $S = \{T\}$;
2. While($\exists G \in S$ that has not been partitioned)
3.   For $i = 1$ to k
4.   Randomly shuffle the tuples of G;
5.   Set $G_1 = G_2 = \emptyset$;
6.   Add tuple t_1 (t_2) of extremely maximal (minimal) value to G_1 (G_2);
7.   For any tuple w
               compute Δ_1 and Δ_2.
               If($\Delta_1 < \Delta_2$) then Add w to G_1, else add w to G_2;
8.   If both G_1 and G_2 are m−eligible
               remove G from S, and add $G_1 - \{t_1\}, G_2 - \{t_2\}$ to S, **break**;
9.Return S;

Fig. 1. The partitioning step

The Assigning Step. After the partitioning step, we enter into the assigning step, which is accomplished by the assign algorithm proposed in paper [4]. Given a set of sub-tables T_i passed from the previous phase, the assigning step is to divide each T_i into buckets such that each bucket constitutes a bucketization. The concepts about bucket and bucketization are given by following definitions.

Definition 5 (Bucket [10],Bucketization). *Given T and T^*, a bucket B is a set of tuples in T whose signatures in T^* are identical. The signature of a bucket B is the set of sensitive values that appear in B. A bucketization U is a set of disjoint buckets, such that the union of all buckets equals T.*

The Decomposition Step. In real applications, a publisher may want to release a set of marginals $\{M_1, M_2, \cdots, M_r\}$ that overlap with each other in an arbitrary manner. To help publishers accomplish this, we use the third step: decomposition step to produce a set of marginals $\{M_1, M_2, \cdots, M_r\}$ that are m-invariant. Depending on marginals of different attribute sets, the bucketization U is decomposed differently. Each decomposition of U is a partition of the microdata T. All the partitions constitute an m-invariant set while offering strong privacy guarantees.

Definition 6 (Decomposition [10]). *Let B be a bucket with signature K. A decomposition of B contains $\frac{|B|}{|K|}$ disjoint QI-groups whose union is B, and all of them have the same signature K.*

The decomposition algorithm runs iteratively and maintains a set bukSet of buckets. Let $B \in U$ be a bucket with a signature containing $s \geq m$ sensitive values $\{v_1, v_2, ..., v_s\}$. The decomposition phase starts by initializing a set bukSet$= \{B\}$. Then, we recursively decompose each bucket B_i in bukSet that contains more than s tuples into two buckets B_1 and B_2 until each bucket in bukSet contain exactly s tuples. The final bukSet is returned as the QI-groups for generalization. The resulting decomposition is guaranteed to be m-invariant, which is stated in Theorem 3.

Now, we elaborate the detailed procedure to decompose a single bucket $B \in U$ with the signature $K = \{ v_1, v_2, \cdots, v_s \}$ into B_1, B_2. Suppose the schema of marginal $M_j (1 \leq j \leq r)$ is $\langle A_1, A_2, ..., A_t \rangle$. We first organize B into s groups such that the i-th $(1 \leq i \leq s)$ group denoted by Q_i contains only the tuples with the sensitive value v_i. Then, by one attribute A_i, we can sort the tuples in each group into the ascending order of their A_i values. After sorting by A_i, we assign the first $\frac{|Q_i|}{2}(1 \leq i \leq s)$ tuples to B_1, and the remaining tuples to B_2. In this way, we get a decomposition of B by A_i. Similarly, we can get another $t - 1$ decompositions by A_j with $j \neq i$. Among all the t decompositions, we pick the one that minimizes the sum of $NCP(B_1)$ and $NCP(B_2)$ as the final decomposition.

Theorem 3. *Given the bucketization U, the marginals $M_j(1 \leq j \leq r)$ produced by the decomposing algorithm are m-invariant.*

Since our marginals $M_j(1 \leq j \leq r)$ enforce m-invariance, we can guarantee the privacy preservation of marginals produced by above decomposition, which is given in following corollary. The computation of above decomposition is also efficient enough (see Theorem 4).

Corollary 1. *A QI-conscious adversary has at most $\frac{1}{m}$ confidence in inferring the sensitive value of any individual in $M_j(1 \leq j \leq r)$, even if s/he is allowed to obtain all versions of $M_j(1 \leq j \leq r)$.*

Theorem 4. *For a single bucket B, the decomposition algorithm can be accomplished in $O(|B| \cdot log^2(|B|) \cdot t)$, where t is the size of attributes of the required marginal, and $|B|$ is the cardinality of bucket B.*

4   Experimental Evaluation

In this section, we experimentally evaluate the effectiveness and efficiency of the proposed technique. We utilize a real data set CENSUS (http://ipums.org) that is widely used in the related literatures. We examine five marginals M_1, M_2 ... M_5, whose dimensionalities are 2, 3, ..., 6, respectively. Specifically, M_1 includes attributes Age and Occupation, M_2 contains attributes of M_1 and Gender(for simplicity, we denote $M_2 = M_1 \cup \{Gender\}$), $M_3 = M_2 \cup \{Education\}$, $M_4 = M_3 \cup \{Marital\}$, and $M_5 = M_4 \cup \{Race\}$. We run all experiments on a PC with 1.9 GHz CPU and 1 GB memory. For comparisons, the results of the state-of-art approach: ANGEL will also be given.

Utility of the Published Data. Since data utility of ANGEL can't be measured by NCP, for the fairness of comparison, we evaluate the utility of the published data by summarizing the accuracy of answering various aggregate queries on the data, and has been widely accepted in the literature [3, 4, 10, 11]. Specifically, each query has the form: *select count(*) from M_j where $A_1 \in b_1$ and $A_2 \in b_2$ and $\cdots$ and $A_w = b_w$*, where w is query dimensionality. $A_1, ..., A_{w-1}$ are $w - 1$ arbitrary distinct QI-attributes in M_j, but A_w is always Occupation. Each $b_i(1 \leq i \leq w)$ is a random interval in the domain of A_i. The generation of $b_1, \cdots, b_w$ is governed by a real number $s \in [0, 1]$, which determines the length of range $b_i(1 \leq i \leq w)$ as $\lfloor |A_i| \cdot s^{1/w} \rfloor$. We derive the estimated answer of a query using the approach explained in [11]. The accuracy of an estimation is measured by its relative error, which is measure by $|act - est|/act$ where act and est denote the actual and estimated results, respectively.

We conduct the first set of experiments to explore the influence of m on data utility. Towards this, we vary m from 4 to 10. Figure 2 plots the error as a function m. Compared to ANGEL that produces about 50%-200% average error, the published data produced by our algorithm is significantly more useful. It is quite impressive to see that the error of our algorithm is consistently below 11 % despite of the growth of m. Figure 3 shows the error as a function of $M_i(2 \leq i \leq 5)$(ANGEL is omitted due to its high query error). We can see that the accuracy increases as the number of QI-attributes of the marginal decreases, which can be attributed to the fact that M_i is more accurate than $M_{i+1}(2 \leq i \leq 4)$.

We evaluate the influence of parameter s on the query error. The result is shown in Figure 4. Evidently, the accuracy is improved for larger s, which can be naturally explained since larger s implies larger query intervals, which in return reduce the query error.

Efficiency. We evaluate the overhead of performing marginal publication. Figure 5 shows the cost of computing marginals $(M_1, \cdots, M_5)$ for m varying from 4 to 10. We can see that the cost drops as m grows. However, the advantages of our method in data utility do not come for free. From figure 5, we can see that the time cost of our algorithm is 115% to 143% of ANGEL, however, which is acceptable especially for those cases where query accuracy is the critical concern.

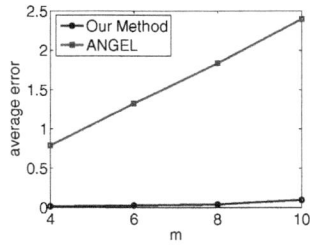

Fig. 2. Accuracy vs. $m(M_5)$

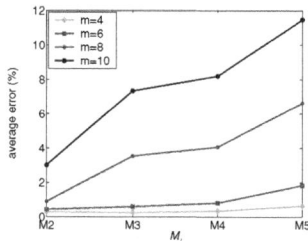

Fig. 3. Accuracy vs. $M_i(s = 0.1)$

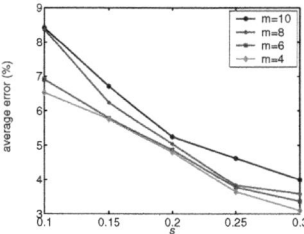

Fig. 4. Accuracy vs. $s(M_5)$

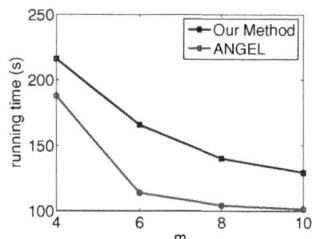

Fig. 5. Running time

5   Conclusion

In this paper, we systematically investigate characteristics of marginal publications. We propose a technique called dynamic anonymization to produce a set of anonymized marginals for a given schema of marginals. As verified by extensive experiments, the marginals produced by our approach not only guarantees the privacy safety of published data but also allows high actuary of query estimation.

References

1. Sweeney, L.: k-anonymity: a model for protecting privacy. Int. J. Uncertain. Fuzziness Knowl. -Based Syst. 10(5), 557–570 (2002)
2. Samarati, P.: Protecting respondents' identities in microdata release. IEEE Trans. on Knowl. and Data Eng. 13(6), 1010–1027 (2001)
3. Xu, J., Wang, W., Pei, J., Wang, X., Shi, B., Fu, A.W.-C.: Utility-based anonymization using local recoding. In: KDD 2006, New York, pp. 785–790 (2006)
4. Xiao, X., Tao, Y.: M-invariance: towards privacy preserving re-publication of dynamic datasets. In: SIGMOD 2007, pp. 689–700. ACM, New York (2007)
5. Tao, Y., Chen, H., Xiao, X., Zhou, S., Zhang, D.: Angel: Enhancing the utility of generalization for privacy preserving publication. TKDE, 1073–1087 (2009)
6. Aggarwal, C.C.: On k-anonymity and the curse of dimensionality. In: VLDB 2005, pp. 901–909. VLDB Endowment (2005)

7. Kifer, D., Gehrke, J.: Injecting utility into anonymized datasets. In: SIGMOD 2006, New York, NY, USA, pp. 217–228 (2006)
8. Yao, C., Wang, X.S., Jajodia, S.: Checking for k-anonymity violation by views. In: VLDB 2005, pp. 910–921. VLDB Endowment (2005)
9. Wang, K., Fung, B.C.M.: Anonymizing sequential releases. In: KDD 2006, New York, NY, USA, pp. 414–423 (2006)
10. Xiao, X., Tao, Y.: Dynamic anonymization: accurate statistical analysis with privacy preservation. In: SIGMOD 2008, pp. 107–120 (2008)
11. Zhang, Q., Koudas, N., Srivastava, D., Yu, T.: Aggregate query answering on anonymized tables. In: ICDE 2006, pp. 116–125 (2007)

Pantheon: Exascale File System Search for Scientific Computing

Joseph L. Naps, Mohamed F. Mokbel, and David H.C. Du

Department of Computer Science and Engineering, University of Minnesota,
Minneapolis, MN, USA
{naps,mokbel,du}@cs.umn.edu

Abstract. Modern scientific computing generates petabytes of data in
billions of files that must be managed. These files are often organized, by
name, in a hierarchical directory tree common to most file systems. As
the scale of data has increased, this has proven to be a poor method of file
organization. Recent tools have allowed for users to navigate files based
on file metadata attributes to provide more meaningful organization. In
order to search this metadata, it is often stored on separate metadata
servers. This solution has drawbacks though due to the multi-tiered architecture of many large scale storage solutions. As data is moved between
various tiers of storage and/or modified, the overhead incurred for maintaining consistency between these tiers and the metadata server becomes
very large. As scientific systems continue to push towards exascale, this
problem will become more pronounced. A simpler option is to bypass
the overhead of the metadata server and use the metadata storage inherent to the file system. This approach currently has few tools to perform
operations at a large scale though. This paper introduces the prototype
for Pantheon, a file system search tool designed to use the metadata
storage within the file system itself, bypassing the overhead from metadata servers. Pantheon is also designed with the scientific community's
push towards exascale computing in mind. Pantheon combines hierarchical partitioning, query optimization, and indexing to perform efficient
metadata searches over large scale file systems.

1 Introduction

The amount of data generated by scientific computing has grown at an extremely
rapid pace. This data typically consists of experimental files that can be gigabytes
in size and potentially number in the billions. Tools for managing these files are
built upon the assumption of a hierarchical directory tree structure in which files
are organized. Data within this tree are organized based on directory and file
names. Thousands of tools, such as the POSIX API, have been developed for
working with data within this hierarchical tree structure.

The POSIX API allows for navigation of this hierarchical structure by allowing users to traverse this directory tree. While the POSIX API is sufficient
for directory tree navigation, its ability to search for specific files within the
directory tree is limited. Within the confines of the POSIX API, there are three

J.B. Cushing, J. French, and S. Bowers (Eds.): SSDBM 2011, LNCS 6809, pp. 461–469, 2011.
© Springer-Verlag Berlin Heidelberg 2011

basic operations that one is able to use to search a directory hierarchy for desired information: *grep*, *ls*, and *find*. Each one of these operations searches for data in their own way, and come with their own requirements and limitations. The *grep* operation performs a naïve brute force search of the contents of files within the file system. This approach presents an obvious problem, namely its lack of scalability. A single *grep* search would need to be performed over gigabytes, or even terabytes of information. As *grep* is not a realistic solution for data at current scale, it clearly will not be a solution for future scale. A better possible route is to use the POSIX operation *ls*, that is instead based on file names. The *ls* operation simply lists all files that are within a given directory. In order to facilitate a more efficient file search, scientists used *ls* in conjunction with meaningful file names. These files names would contain information such as the name of experiments, when such experiments were run, and parameters for the experiment. By using such names, along with the wild-card(*) operator, one would perform a search for desired information based on file names. This solution also had its own problems. First, this technique is dependent on a consistent application of conventions between file names. Even something such as parameters being in different orders could prevent such a search from returning the needed information. Second, as experiments grow in complexity, more parameters must be maintained, resulting in long file names that are difficult to remember and work with. The POSIX operation *find* allows navigation of files via metadata informaiton, a much more attractive file search option than either *grep* or *ls*. Metadata represents information about the file, as opposed to information within the file. Such information includes items such as file owner, file size, and time of last modification. Unfortunately, the *find* is not sufficient for the large scale searches needed for the scientific computing community.

To solve the limitations imposed by simple POSIX commands, research began to develop full featured metadata search tools at an enterprise (e.g. Google [1], Microsoft [2], Apple [3], Kazeon [4]) as well as the academic level (e.g. Spyglass [5]). These tools added to the richness of current metadata searching capabilities, but also possessed their own limitations. Enterprise solutions typically index their data by using a standard database management system. This stores all metadata information as flat rows, thus losing the information that can be inferred from the directory hierarchy itself. Also, the need for a single metadata server can cause scalability problems in large, distributed computing systems. Relationships between files based on their location in the hierarchy are lost. Spyglass [5] exploited these hierarchical structure relationships, but at the expense of losing the query optimization and indexing powers of a database management system.

In this paper, we present the prototype of the Pantheon system. Pantheon is a file system search tool designed for the large scale systems used within the scientific computing community. Pantheon combines the query optimization and indexing strategies of a database management system with the ability to exploit the relationships between files based on locality used in current file system search. For our initial prototype, we focused on the effects of basic database style query optimization and indexing when implemented over a more tailored file system partitioning scheme. To this end, we implemented a detailed partitioning algorithm with simple query optimization and indexing built on top.

Pantheon's core is separated into three primary components: partitioning, query optimizer, and indexing. The partitioning component is responsible for separating the directory hierarchy into disjoint partitions. We present a general partitioning algorithm, but any custom algorithm may be used. Partitioning is needed in order to avoid a system-wide bottleneck. Without partitioning, all searches would be forced to go through a single set of indexes. This would create an obvious bottleneck that would severely limit the scalability of the system. Query optimization is a well known technique from database management systems [6]. The Pantheon optimizer collects statistics on a per-partition basis, and evaluates query plans using a basic cost model. This strategy selects predicates that will prune the largest number of possible files from the result. This simple technique results in a significant performance boost over picking predicates at random. The indexing component maintains B^+-Tress and hash tables, also on a per-partition basis. A single index is kept for every metadata attribute that can be searched. Taking this approach gives Pantheon two distinctive advantages. First, this indexing method ties in very well with our query optimization. Second, the use of individual indexes allows Pantheon to quickly adapt should attributes be added to the system, as could be the case with extended attributes.

The rest of the paper is organized as follows. Section 2 discusses work related to Pantheon. Section 3 gives a high level overview of the Pantheon system. Section 4 details the partitioning system used in Pantheon. Section 5 discusses the Pantheon query optimizaer and interface. Section 6 gives an overview of the indexing system used in Pantheon. Section 7 looks at the experimental evaluation of the Pantheon system. The paper is concluded with Section 8.

2   Related Work

At an enterprise level, numerous products have been developed allowing for metadata search [1–4]. At an academic level, the closest work to Pantheon is Spyglass [5]. Spyglass uses a technique known as hierarchical partitioning [7], which is based on the idea that files that are close to each other within the directory tree tend to be searched together often. Pantheon presents an algorithm that exapnds upon this idea in two primary ways. First, many modern large scale systems use storage architectures that involve multiple tiers of storage. In such systems, data is moved between multiple layers of storage in a dynamic fashion. From the standpoint of the file system, this results in sudden changes to the directory hierarchy that must be accounted for. Pantheon's partitioning algorithm is able to adapt to data being migrated into the directory hierarchy. Second, Pantheon monitors query patterns and allows for the partition structure to be changed based on changes to query loads. For indexing, Spyglass uses a single KD-Tree [8] built over each partition. This approach to indexing has several drawbacks. First, using a multi-dimensional index limits the performance scalability of the system if the number of attributes being indexed were to grow very large. By splitting attributes into multiple indexes Pantheon is able to adapt in the case that additional attributes are introduced to the system more gracefully. Second, having a single index per partition means that Spyglass is unable to take advantage of the attribute distribution of a partition. Spyglass also lacks any form of a query optimizer. Using query optimization, in conjunction with a richer set of indexing structures, Pantheon is able to make intelligent decisions when dealing with queries to the file system.

3   System Architecture

Figure 1 gives the architecture for the Pantheon system. Partitions exist over the storage layer of the system, and within each partition we have the query optimizer, where distribution statistics are stored, as well as the partition indexes. The figure also gives the basic flow of a query within the Pantheon system. The query begins at the partition map. This map simply determines which partitions must be accessed in order to respond to the query. Each required partition produces the result of the query, which is then passed back to the user.

Pantheon also uses a modular design in its operations. Each of the three primary components are totally independent of one another. Only basic interfaces must remain constant. We believe this to be important for two primary reasons. First, by modularizing the design, we give scientists the ability to quickly add custom features based on their individual needs. Such examples of this could include a custom partitioning module or a different indexing structure that may be more suited to their data and querying properties. Second, a modular design will make it easier for additional components to be added to the system to adapt to new storage and architecture paradigms.

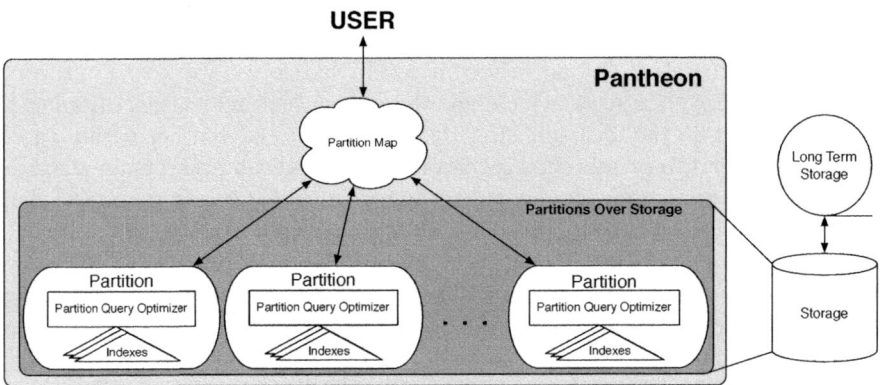

Fig. 1. Pantheon System Architecture

4   Partitioning

The partitioner is the heart of the Pantheon system. Without the partitioner, we would be forced to construct a single set of indexes over the data space that we wish to search. This will create a massive bottleneck that would slow down all aspects of the system, and severely limit system scalability. Similar techniques can be seen in distributed file systems [9–12].

A common pattern found in studies on metadata [5, 13–15] is that of spatial locality of metadata. Spatial locality is the general concept that files that are located close to one another in the directory hierarchy tend to have significantly more similarities in metadata values and tend to be queried together more often. This is typically the result of how files tend to be organized by users within the directory hierarchy. So, files that are owned by a user u will tend to reside close to one another in the directory tree, i.e. they will tend to reside in u's home directory. When lookinging possible algorithms for partitioning our directory tree we explored works that looked into disk page based tree partitioning [16]. The general idea of our partitioning algorithm is as follows. We begin with the root of the directory tree R and proceed to find all leaves of the directory tree. From this point we place each leaf into its own partition and mark it as being processed. We then proceed up the tree processing all interior nodes such that all of their children have been marked as processed. If the interior, parent, node is able to be merged into a partition with all of its children, we merge them. In the event that there is not enough room, we create a new partition with only this interior node. Following this step, the interior node is marked as processed. This work continues all the way up the tree until we get to the root node. For more specifics about this process, refer to the pseudocode presented in Algorithm 1.

Initially, the entire directory tree must be partitioned in this manner. Since this process may take some time, it is run as a background process so that normal system function may continue.

5   Query Optimizer

Query optimization is a well studied problem in the field of database management systems. Database systems typically use a cost based model [6] to estimate optimal query plans. The gain in bringing the idea of the query optimizer from databases to file systems is significant. Query optimization research is on of the primary reasons that database systems have been able to perform so well in real world environments.

Formally, the job of Pantheon's query optimizer is a follows: Given a query Q and a series of predicates $P_1, \ldots, P_n$, the query optimizer finds a plan for evaluation of these predicates that is efficient. Using the query optimizer, indexes are used to prune the possible result set. From there a scan can be performed over this pruned data space. If done properly, this pruned data space will be significantly smaller than the original. This results in a scan that can be done very quickly. More so, this scan can be performed as a pipelined process that is done as results are being returned from the index.

Algorithm 1. Pantheon Partitioning Algorithm

1: Pantheon-Tree-Partition(T)
2:    Input: A tree rooted at T
3:    Output: A mapping from nodes in T to partitions
4: **while** There are nodes in T not yet processed **do**
5:    Choose a node P that is a leaf or one where all children have been processed
6:    **if** P is a leaf node **then**
7:      Create a new partition C containing node P
8:    **else**
9:      Let $P_1, \ldots, P_n$ be the children of P
10:     Let $C_1, \ldots, C_n$ be the partitions that contain $P_1, \ldots, P_n$.
11:     **if** Node P and the contents of the partitions $C_1, \ldots, C_n$ can be merged into a single partition **then**
12:       Merge P and the contents of $C_1, \ldots, C_n$ into a new partition C, discarding $C_1, \ldots, C_n$.
13:     **else**
14:       Create a new partition C containing only P
15:     **end if**
16:   **end if**
17: **end while**

For Pantheon's query optimization, the decisions is based primarily on the selectivity of a given predicate. The selectivity represents how effective that predicate is at reducing the overall data set. A predicate will low selectivity percentage will prune more extraneous items, while a predicate with high selectivity percentage will prune less such values. This distinction is important, as not choosing the proper index when evaluation a query can lead to a significant decrease in query response time.

To track the selectivity, we need to keep basic statistics about the files on a per partition basis. For each attribute within a partition, we construct a histogram. Given an input value, these histograms quickly return an estimate as to the percentage of values that will satisfy that query.

6 Indexing

The initial indexing implementation uses simple and well known indexing structures as a baseline evaluator. We use multiple single dimensional indexes over the attributes in conjunction with query optimization. In the event that a new attribute is added to the file system, we simply construct an index over that new attribute, and continue operation as normal.

The metadata attributes being indexed are those typically found in a standard file system including: file mode, file owner, file group, time of last access, time of last modification, time of last status change, and file size. These attributes are then separated into two groups. One group represents those attributes for which range queries make sense. This includes all of the time attributes as well as file size. The remaining attributes are those where only equivalence queries make

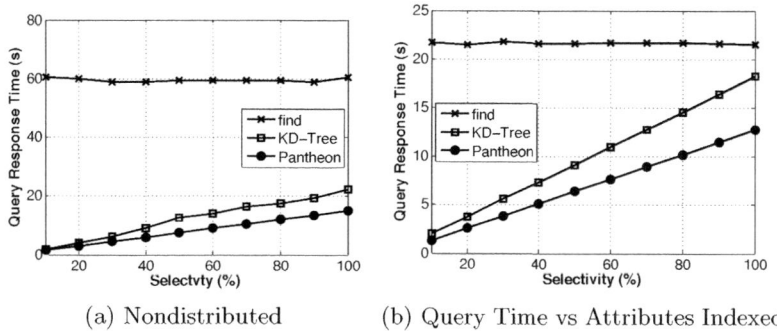

(a) Nondistributed   (b) Query Time vs Attributes Indexed

Fig. 2. Query Response Time vs Selectivity

sense. These include file mode, owner, and group. Each attribute that has been deemed an equivalence attribute is indexing using hash table. Each attribute that will be searched over a range is indexed using a B^+-Tree. These indexes were chosen due to the fact that each handle their own respective query types very well.

7   Experimental Evaluation

Experimental evaluation is meant to provide us a baseline for which future work may be compared. There are two other techniques that we test Pantheon against. The first is the POSIX operation *find*. This is simply to show Pantheon's viability over the naïve method. The second is testing Pantheon's processing over that of a KD-Tree. This is the indexing used by the Spyglass system, and provides a good competitor to examine Pantheon's strengths and weaknesses.

Pantheon is implemented as a FUSE module [17] within a Linux environment over an ext4 file system. For the default partition cap size we used 100,000. This was the same cap used in [5] and we see no reason to change this for experiments where partition size is held constant. The default selectivity used is 20% unless otherwise noted. The default number of attributes indexed was 8.

Experimentation was done over two different system configurations. The first of which is refered to as the nondistributed configuration. This was done on a single node system consisting of a dual-core 3 GHz Pentium 4 with 3.4 GB of RAM. The distributed tests were done on a 128 node cluster. Each node in the cluster consisted of two processors at 2.6 GHz with 4 GB of RAM.

In Figure 2 we see the effect on query response time when we vary the selectivity of a single query predicate. First, It shows that *find* is not any competition to Pantheon. As such, it will not be considered in future experiments. Second, it shows that using selectivity as a metric for the query optimizer is a good idea. In both cases we see that if Pantheon evaluates based on the most selective index, there is an improvement in query response time over that of a KD-Tree.

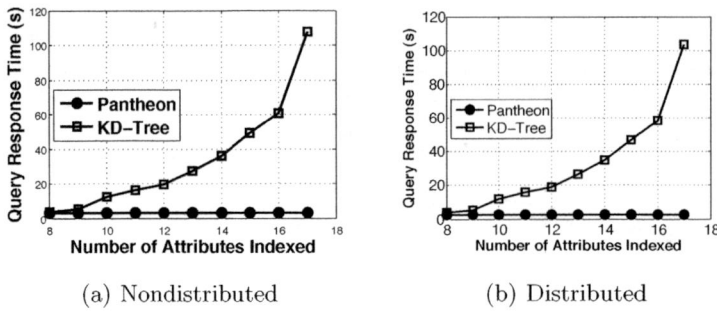

(a) Nondistributed   (b) Distributed

Fig. 3. Query Response Time vs Number of Attributes Indexed

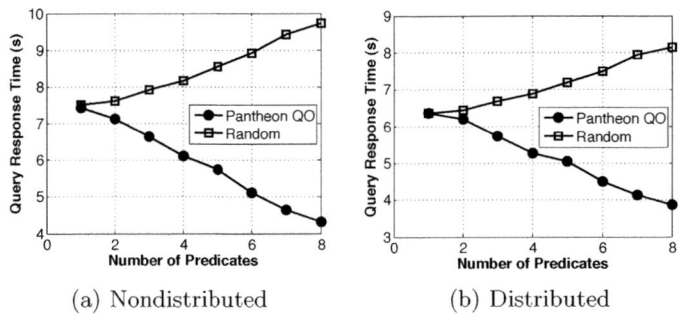

(a) Nondistributed   (b) Distributed

Fig. 4. Query Response Time vs Number of Predicates

Figure 3 relates query response time to the number of attributes being indexed. Here is where Pantheon shows significant improvement over a KD-Tree based approach. As the number of dimensions increases without the partition size changing, the performance of the KD-Tree suffers greatly. Due to the fact that Pantheon indexes attributes separately, it does not show any noticeable change as the number of attributes increases.

Figure 4 displays how the Pantheon query optimizer is able to improve query performance as the number of predicates increases. Here, we generated query predicates for random attributes with random values. These results strengthen the case for using query optimization. If predicates are chosen at random, we see is significant increase in the overall time needed to response to queries.

8   Conclusion

Here we have presented the foundational work for the Pantheon indexing system. Pantheon represents a combination of ideas from both file system search and database management systems. Using these ideas Pantheon plays on the strength of each of the two fields to accomplish its goal. We have shown through experimentation that Panteon is either competitive or outperforms current file system indexing strategies. We intend to use this prototype as a test bed for future work in aspects of partitioning, query optimization, and indexing within the context of file system search.

References

1. Inc., G.: Google enterprise, http://www.google.com/enterprise
2. Inc., M.: Enterprise search from microsoft, http://www.microsoft.com/enterprisesearch
3. Apple, Spotlight server: Stop searching, start finding, http://www.apple.com/server/macosx/features/spotlight
4. Kazeon: Kazeon Search the enterprise, http://www.kazeon.com
5. Leung, A., Shao, M., Bisson, T., Pasupathy, S., Miller, E.: Spyglass: Fast, scalable metadata search for large-scale storage systems. In: Proccedings of the 7th Conference on File and Storage Technologies, pp. 153–166. USENIX Association (2009)
6. Selinger, P., Astrahan, M., Chamberlin, D., Lorie, R., Price, T.: Access path selection in a relational database management system. In: Proceedings of the 1979 ACM SIGMOD International Conference on Management of Data, pp. 23–34. ACM, New York (1979)
7. Weil, S., Pollack, K., Brandt, S., Miller, E.: Dynamic metadata management for petabyte-scale file systems. In: Proceedings of the 2004 ACM/IEEE Conference on Supercomputing, p. 4. IEEE Computer Society, Los Alamitos (2004)
8. Bentley, J.: Multidimensional binary search trees used for associative searching. Communications of the ACM 18(9), 509–517 (1975)
9. Ousterhout, J., Cherenson, A., Douglis, F., Nelson, M., Welch, B.: The Sprite network operating system. Computer 21(2), 23–36 (1988)
10. Weil, S., Brandt, S., Miller, E., Long, D., Maltzahn, C.: Ceph: A scalable, high-performance distributed file system. In: Proceedings of the 7th Symposium on Operating Systems Design and Implementation, p. 320. USENIX Association (2006)
11. Pawlowski, B., Juszczak, C., Staubach, P., Smith, C., Lebel, D., Hitz, D.: NFS version 3 design and implementation. In: Proceedings of the Summer 1994 USENIX Technical Conference, pp. 137–151 (1994)
12. Morris, J., Satyanarayanan, M., Conner, M., Howard, J., Rosenthal, D., Smith, F.: Andrew: A distributed personal computing environment. Communications of the ACM 29(3), 201 (1986)
13. Agrawal, N., Bolosky, W., Douceur, J., Lorch, J.: A five-year study of file-system metadata. ACM Transactions on Storage (TOS) 3(3), 9 (2007)
14. Douceur, J., Bolosky, W.: A large-scale study of file-system contents. ACM SIGMETRICS Performance Evaluation Review 27(1), 70 (1999)
15. Leung, A., Pasupathy, S., Goodson, G., Miller, E.: Measurement and analysis of large-scale network file system workloads. In: USENIX 2008 Annual Technical Conference, pp. 213–226. USENIX Association (2008)
16. Diwan, A., Rane, S., Seshadri, S., Sudarshan, S.: Clustering techniques for minimizing external path length. In: Proceedings of the International Conference on Very Large Data Bases, Citeseer, pp. 342–353 (1996)
17. FUSE, File system in user space, http://fuse.sourceforge.net

Massive-Scale RDF Processing Using Compressed Bitmap Indexes

Kamesh Madduri and Kesheng Wu

Lawrence Berkeley National Laboratory, Berkeley CA 94720, USA
{KMadduri,KWu}@lbl.gov

Abstract. The Resource Description Framework (RDF) is a popular data model for representing linked data sets arising from the web, as well as large scientific data repositories such as UniProt. RDF data intrinsically represents a labeled and directed multi-graph. SPARQL is a query language for RDF that expresses subgraph pattern-finding queries on this implicit multigraph in a SQL-like syntax. SPARQL queries generate complex intermediate join queries; to compute these joins efficiently, this paper presents a new strategy based on bitmap indexes. We store the RDF data in column-oriented compressed bitmap structures, along with two dictionaries. We find that our bitmap index-based query evaluation approach is up to an order of magnitude faster the state-of-the-art system RDF-3X, for a variety of SPARQL queries on gigascale RDF data sets.

Keywords: semantic data, RDF, SPARQL query optimization, compressed bitmap indexes, large-scale data analysis.

1 Introduction

The Resource Description Framework (RDF) was devised by the W3C consortium as part of the grand vision of a semantic web[1]. RDF is now a widely-used standard for representing collections of linked data [3]. It is well-suited for modeling network data such as socio-economic relations and biological networks [11,13]. It is also very useful for integrating data from dynamic and heterogeneous sources, in cases where defining a schema beforehand might be difficult. Such flexibility is key to its wide use. However, the same flexibility also makes it difficult to answer queries quickly. In this work, we propose a new strategy using bitmap indexes to accelerate query processing.

A record in the RDF data model is a triple of the form ⟨subject, predicate, object⟩. If these records are stored in a data management system as a three-column table, then all queries except a few trivial ones would require self-joins, and this would be inefficient in practice. The most commonly used query language on RDF data is called SPARQL [10]. To speed up the SPARQL query answering process, there have been a number of research efforts based on modifying existing data base systems and developing specialized RDF processing

---

[1] More information about RDF can be found at http://www.w3.org/RDF/

J.B. Cushing, J. French, and S. Bowers (Eds.): SSDBM 2011, LNCS 6809, pp. 470–479, 2011.

systems. For example, popular commercial database systems (DBMS) such as ORACLE have added support for RDF [7]. A number of research database management systems have also been applied to RDF data [1, 12]. Special-purpose RDF storage systems include Virtuoso RDF[2], Jena[3], and hyperGraphDB[4].

The most common indexing techniques in database systems are variants of B-Trees or bitmap indexes. The techniques for indexing RDF data generally follow these two prototypical methods as well. Among existing B-Tree indexing methods, RDF-3X is one of the best performers in terms of SPARQL query processing speed [8].

Two recent bitmap indexing methods, BitMat and RDFJoin, have demonstrated performance on par with RDF-3X [2, 5]. The BitMat index creates a 3D bit-cube with the three dimensions being subject, predicate, and object. This cube is compressed and loaded into memory before answering any queries. This technique has been shown to be quite efficient, but due to its reliance on the whole bit-cube to be in memory, it is difficult to scale to larger datasets.

The RDFJoin technique breaks the 3D bit-cube used by BitMat into six separate bit matrices. Each of these bit matrices can be regarded as a separate bitmap index, and therefore can be used independently from each other. Thus, the RDFJoin approach is more flexible and can be applied to larger datasets [6].

Our work significantly improves on the above bitmap index-based strategies. We create a space-efficient representation of the RDF data (discussed in Section 2). By utilizing a compute-efficient bitmap compression technique and carefully engineering the query evaluation procedure (Section 3), we dramatically reduce the query processing time compared to the state-of-the-art RDF-3X processing system.

2   Bitmap Index Construction

We first explain the data structures used in our work. We describe them as bitmap indexes here, because each of them consists of a set of key values and a set of compressed bitmaps, similar to the bitmap indexes used in database systems [9, 14]. However, the key difference is that each bitmap may not necessarily correspond to an RDF record (or a row), as in database systems.

For RDF data, one can construct the following sets of bitmap indexes:

Column Indexes. The first set of three bitmap indexes are for three columns of the RDF data. In each of these indexes, the key values are the distinct values of subjects, predicates, or objects, and each bitmap represents which record (i.e., row) the value appears in. This is the standard bitmap index used in existing database systems [9, 14].

Unlike conventional bitmap indexes, our indexes for subject and object share the same dictionary. This strategy is taken from the RDFJoin approach [6]. It

[2] http://www.openlinksw.com/dataspace/dav/wiki/Main/VOSRDF
[3] http://openjena.org/
[4] http://www.hypergraphdb.org/

eliminates one dictionary from the three bitmap indexes, and allows the self-join operations to be computed using integer keys instead of string keys. This is a trick used implicitly in many RDF systems.

Composite Indexes. We can create three composite indexes, each with two columns as keys. The keys are composite values of predicate-subject, predicate-object, and subject-object. This ordering of the composite values follows the common practice of RDF systems. As in normal bitmap indexes, each composite key is associated with a bitmap. However, unlike the normal bitmap index where a bitmap is used to indicate which rows have the particular combination of values, our bitmap records values the other column has. For example, in a composite index for predicates and subjects, each bitmap represents what values the objects have.

In a normal bitmap index, there are many columns not specified by the index key. Therefore, it is useful for the bitmap to point to rows containing the specified key values, so that any arbitrary combination of columns may be accessed. However, in the RDF data, there are only three columns. If the index key contains information about two of the three columns already, directly encoding the information about the third column in the index removes the need to go back the data table and is a more direct way of constructing an index data structure.

To effectively encode the values of the third column in a bitmap, we use a bitmap that is as long as the number of distinct values of the column. In the example of a predicate-subject index, each bitmap has as many bits as the number of distinct values in objects. Since we use a unified subject-object dictionary, the bitmap has as many bits as the number of entries in the dictionary. To make it possible to add new records without regenerating all bitmap indexes, our dictionary assigns a fixed integer to each known string value. A new string value will thus receive the next available integer. When performing bitwise logical operations, we automatically extend the shorter input bitmap with additional 0 bits. This allows us to avoid updating existing bitmaps in an index, which can reduce the amount of work needed to update the indexes when new records are introduced in a RDF data set.

Join Indexes. A normal join index represents a cross-product of two tables based on an equality join condition. Because the selection conditions in SPARQL are always expressed as triples, the join operations also take on some special properties, which we can take advantage of when constructing the join indexes. Note that for SPARQL queries, joins are typically across properties. Thus, the most commonly-used join indexes for RDF data would map two property identifiers to a corresponding bitmap, and there can be three such indexes based on the positions of the variable. In the current version of our RDF processing system, we chose not to use construct join indexes due to the observation that most of the test queries could be solved efficiently with just composite indexes. We will investigate use of join indexes for query answering in future work.

3   Query Evaluation and Optimization

SPARQL is a query language that expresses conjunctions and disjunctions of triple patterns. Each conjunction, denoted by a dot in SPARQL syntax, nominally corresponds to a join. A SPARQL query can also be viewed as a graph pattern-matching problem: the RDF data represents a directed multigraph, and the query corresponds to a specific pattern in this graph, with the possible degrees of freedom expressed via wildcards and variables. Figure 1 gives an example SPARQL query. Informally, this query corresponds to the question "produce a list of all scientists born in a city in Switzerland who have/had a doctoral advisor born in a German city". This query is expressed with six triple patterns, and each triple pattern can either have a variable or a literal in the three possible positions. The goal of the query processor is to determine all possible variable bindings that satisfy the specified triple patterns.

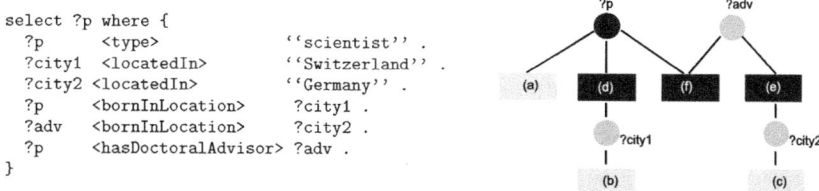

```
select ?p where {
    ?p        <type>                ''scientist'' .
    ?city1   <locatedIn>           ''Switzerland'' .
    ?city2   <locatedIn>           ''Germany'' .
    ?p        <bornInLocation>      ?city1 .
    ?adv      <bornInLocation>      ?city2 .
    ?p        <hasDoctoralAdvisor>  ?adv .
}
```

Fig. 1. An example SPARQL query (left) and a graph representation of the query triple patterns (right). The rectangular nodes in the graph represent triple patterns. The labels (a)-(f) correspond to the ordering of the patterns in the query.

To understand how the constructed bitmap indexes can be utilized to answer a SPARQL query, consider the "graph" representation of the query shown in Figure 1. Each triple pattern is shown as a rectangular node. Two triple nodes are connected via query variables they may share, and these variables are represented using circular nodes. Further, the triple patterns are colored based on the number of variable positions in the pattern. The light blue-colored blocks have one variable and one literal in their pattern, whereas the dark blue blocks represent patterns with two variables. Similarly, the dark brown circular node represents the output variable, and the nodes in light brown color are other variables in the query. Such a query graph succinctly captures the query constraints, and forms the basis for a possible query evaluation approach.

For query evaluation, consider representing each variable using a bitmap. For instance, the variable p can be initialized to a bitmap of size n_{SO} (where n_{SO} is the cardinality of the combined subject-object dictionary), with all subject bits set to 1. Observe that triple patterns that have only one variable in them can be resolved by composite index (in our case, PSIndex and POIndex) lookups. For instance, the key corresponding to predicate $<type>$ and object "scientist" can be determined using dictionary lookups, and then a bit vector corresponding to all possible subjects that satisfy the particular condition can be obtained with a single composite index lookup. Performing a conjunction just translates to performing a bitmap logical "AND" operation with the initialized bitmap. Similarly, we can initialize and update bitmaps corresponding to *city1* and *city2* in the figure. The other triples (d), (e), and (f) have two variables in their pattern, and so we are required to perform joins. The bit vectors give us a sorted list of index values, and so we employ the nested loop merge join to determine the final binding.

The key primitives in our query processing system are dictionary lookups, composite index lookups, bit vector AND and OR operations, and nested loop merge joins after decompressing bit vectors. We were able to express all the benchmark queries, some with as many 15 triple patterns, compactly using these primitives. The query required less than 20 lines of source code in almost all cases. In most cases, the lookup operations are performed with integer identifiers of the string values obtained through the dictionaries. These integer identifiers are directly used as indices into arrays of bitmaps in the bitmap indexes.

4   Experimental Evaluation and Analysis

4.1   Experimental Setup

Data sets. We choose a variety of data sets and test instances to test our new query evaluation scheme. First, we experiment with synthetic data sets of different sizes using the Lehigh University Benchmark suite LUBM [4]. LUBM is a popular benchmark for evaluating triple-stores, with a recommended list of fourteen queries that stress different aspects related to query optimizations. We use large subsets of two datasets: the Billion Triples Challenge[5] data and the UniProt[6] collection. The Billion Triples dataset encapsulates public domain web crawl information, whereas UniProt is a proteomics repository. We implement three sample queries for each of these datasets. Both these datasets are significantly more complex, noisy, and heterogeneous compared to LUBM. We use queries recommended by the UniProt RDF data publishers, and ones similar to prior RDF-3X query instances. We also present query results with the Yago [13] dataset, which is comprised of facts extracted from Wikipedia. This dataset contains about 40 million triples, and the number of distinct predicates (337,000) is significantly higher than LUBM. We use here a set of queries that were previously used to evaluate RDF-3X.

Test Systems and Software. Our primary test machine data5 is a Linux workstation with a quad-core Intel Xeon processor with a clock speed of 2.67 GHz, 8 MB L2 cache, and 8 GB RAM. The disk system used to store the test data is a software RAID concatenating two 1TB SATA disks in RAID0 configuration. The second test machine named euclid is a shared resource at NERSC[7]. It is a Sunfire x4640 SMP with eight 6-core Opteron 2.6 GHz processors and 512 GB of shared memory. On this system, the test files are stored on a GPFS file system

---

[5] http://challenge.semanticweb.org/

[6] http://www.uniprot.org/downloads

[7] More information about NERSC and euclid can be found at http://www.nersc.gov

Table 1. Data, Index, and Database sizes in GB for different data sets

Data set # triples	LUBM 1M	LUBM 50M	LUBM 500M	Yago 40M	UniProt 220M	BTC 626M
Raw data	0.125	6.27	62.30	3.56	30.58	65.19
FastBit Dictionaries	0.032	0.79	8.22	1.30	3.05	2.48
FastBit Indexes	0.016	1.59	15.41	1.20	6.30	15.03
RDF-3X DB	0.058	2.83	33.84	2.75	—	—

shared by thousands of users. Therefore, we may expect more fluctuations in I/O system performance.

We use FastBit v1.2.2 for implementing our bitmap index-based RDF data processing approach. We built the codes using the GNU C++ compiler v4.4.3 on data5 and the PGI C++ compiler v10.8 on euclid. For parsing the data, we use the Raptor RDF parser utility (v2.0.0).

There are numerous production and prototype research triple-stores available for comparison, a majority of which are freely available online[8]. In this paper, we chose to compare our bitmap index strategies against version 0.3.6 of RDF-3X [8]. RDF-3X is a production-quality RDF-store widely used by the research community for performance studies, and prior work shows that it is significantly faster, sometimes by up to two orders of magnitude, than alternatives such as MonetDB and Jena-TDB. We also experimented with the bitmap indexing-based approach BitMat [2], but found that RDF-3X consistently outperforms BitMat for a variety of queries, including high selectivity queries.

For all the queries, we present cold cache performance results, which correspond to the first run of the query, as well as "warm cache" numbers, which are an average of ten consecutive runs, excluding the first.

4.2   Results and Discussion

Index construction and sizes. Table 1 lists the sizes of the FastBit dictionaries and indexes after the construction phase. We observe that the cumulative sum of the dictionary and index sizes is substantially lower than the raw data size for all the data sets. As a point of comparison, we present the size of the RDF-3X B-tree indexes (which internally stores six compressed replicas of the triples compactly) for these datasets. For the data sets studied, our approach requires slightly lower disk space than RDF-3X.

The dictionary and index construction times range from 20 seconds on data5 for the 1M triple LUBM data set, to nearly four hours for the BTC 626M triple data set on euclid. These index construction times were comparable to RDF-3X's construction times.

LUBM Query Performance. We next evaluate query performance of our bitmap index-based approach for LUBM data sets of different sizes. In Table 2,

---

[8] Please see http://semanticweb.org/wiki/Tools for a list of tools.

Table 2. LUBM benchmark SPARQL query evaluation times (in milliseconds) for a 50 million triple data set on data5-sata

	Q1	Q2	Q3	Q4	Q5	Q6	Q7
Cold caches							
FastBit	0.30	1320	1.26	0.65	0.34	139	0.643
RDF-3X	0.43	572	2.9	0.75	2.1	4150	4.62
Warm caches							
FastBit	0.167	1311	0.92	0.40	0.19	135	0.46
RDF-3X	0.31	544	0.193	0.70	1.95	4021	1.52
Speedup	1.86×	0.42×	0.21×	1.75×	10.26×	29.8×	3.30×

	Q8	Q9	Q10	Q11	Q12	Q13	Q14
Cold caches							
FastBit	7.85	9457	0.313	0.263	2.61	0.36	636
RDF-3X	55.6	1431	1.65	0.41	17.2	3.9	14190
Warm caches							
FastBit	6.34	9288	0.179	0.148	2.34	0.34	467
RDF-3X	50.4	1369	0.336	0.35	7.44	1.7	13770
Speedup	7.95×	0.15×	1.87×	2.36×	3.17×	5.0×	29.5×

we compare the cold and warm caches performance of queries for the LUBM-50M data set. We do not observe a substantial difference between the cold and warm cache times (i.e., the difference is not as pronounced as RDF-3X), which may indicate that the indexes may be already cached in main memory and I/O activity is minimal. Overall, we observe that our strategy outperforms RDF-3X by a significant margin in the warm cache case, particularly for the 5M dataset. Studying the queries individually, we observe that the speedup is higher for simple two or three triple pattern queries (such as queries 5, 10, and 14). The results for the slightly more complex queries (queries 2, 8, 9) are mixed: RDF-3X is faster on queries 2 and 9, whereas our bitmap index-based approach is faster for query 8. We surmise that this may be because we picked a non-optimal join ordering when executing queries 2 and 9. Table 3 presents performance results for the same set of queries on a 500M data set, but on the euclid system. Interestingly, RDF-3X query 2 performance is significantly slower, and our test harness times out for this particular instance. Another trend apparent on investigating the relative performance is that the average speedup remains the same as the data size increases.

Performance on Multi-pattern Complex SPARQL Queries. We next present results for three additional fixed-size datasets. In Table 4, we summarize performance achieved for sample queries on the large-scale UniProt and Billion Triple data sets. The compressed bitmap composite indexes are very useful in case of the UniProt queries, where there are several triple patterns sharing the same join variable. They help prune the tuple space significantly, and the query execution times are comparable to previously-reported RDF-3X numbers.

Table 3. LUBM benchmark SPARQL query evaluation times (in milliseconds) for a 500 million triple data set on euclid

	Q1	**Q2**	**Q3**	**Q4**	**Q5**	**Q6**	**Q7**
Cold caches							
FastBit	1.92	17481	17.2	3.2	0.62	2560	1.75
RDF-3X	1.58	—	85.9	199.6	1.25	91300	560.6
Warm caches							
FastBit	1.73	8344	7.81	1.21	0.41	2344	1.21
RDF-3X	0.875	—	0.984	2.344	1.11	80039	3.47
Speedup	0.51×	—	0.13×	2.71×	2.68×	34.1×	2.87×

	Q8	**Q9**	**Q10**	**Q11**	**Q12**	**Q13**	**Q14**
Cold caches							
FastBit	9.43	278.1	1.27	2.44	15.7	5.32	11231
RDF-3X	204.2	41.2	870.1	30.2	1051.12	15082.3	—
Warm caches							
FastBit	7.23	140.4	0.38	1.52	11.51	2.53	10682
RDF-3X	71.8	28.1	1.01	0.94	124.2	18.5	—
Speedup	9.93×	0.2×	2.66×	0.62×	10.79×	7.31×	—

Table 4. UniProt and Billion Triple datasets SPARQL query evaluation times (in milliseconds) on euclid

	UniProt			Billion Triples		
	Q1	**Q2**	**Q3**	**Q1**	**Q2**	**Q3**
Warm caches time	1.71	262	30.4	12.35	443.42	378.21

Table 5. FastBit query evaluation performance improvement achieved (geometric mean of individual query speedup) over RDF-3X for various data sets. [†] Speedup on euclid.

	LUBM-5M	**LUBM-50M**	**LUBM-500M**[†]	**YAGO**
Speedup	12.96×	2.62×	2.81×	1.38×

Table 5 gives the overall performance improvement achieved for queries using FastBit versus RDF-3X, when taking the geometric mean of the execution times into consideration. We observe that FastBit outperforms RDF-3X for both LUBM data sets of various sizes, as well as queries on the Yago data set.

5   Conclusions and Future Work

This paper presents the novel use of compressed bitmaps to accelerate SPARQL queries on large-scale RDF repositories. Our experiments show that we can efficiently process queries with as many as 10 to 15 triple patterns, and query execution times compare very favorably to the current state-of-the-art results. Bitmap indexes are space-efficient, and bitvector operations provide an intuitive mechanism for expressing and solving ad-hoc queries. The set union and intersection operations that are extensively used in SPARQL query processing are extremely fast when mapped to bitvector operations.

We plan to extend and optimize our RDF data processing system in future work. We will speed up data ingestion by exploiting parallel I/O capabilities and distributed memory parallelization. Our current dictionary and index creation schemes provision for incremental updates to the data. We intend to study the cost of updates, both fine-grained as well as batched updates. We do not yet support a full SPARQL query parser, and the join ordering steps in our query optimization scheme can be automated; we plan to research these problems in future work.

Acknowledgments

This work is supported by the U.S. Department of Energy under Contract No. DE-AC02-05CH11231. We thank Amrutha Venkatesha for implementing and optimizing the FastBit-based LUBM query code.

References

1. Abadi, D.J., Marcus, A., Madden, S.R., Hollenbach, K.: Scalable semantic web data management using vertical partitioning. In: Proc. 33rd Int'l. Conference on Very Large Data Bases (VLDB 2007), pp. 411–422 (2007)
2. Atre, M., Chaoji, V., Zaki, M.J., Hendler, J.A.: Matrix "bit" loaded: a scalable lightweight join query processor for RDF data. In: Proc. 19th Int'l. World Wide Web Conference (WWW), pp. 41–50 (2010)
3. Bizer, C., Heath, T., Berners-Lee, T.: Linked data – the story so far. Int'l. J. Semantic Web Inf. Syst. 5(3), 1–22 (2009)
4. Guo, Y., Pan, Z., Heflin, J.: LUBM: A benchmark for OWL knowledge base systems. Web Semant. 3, 158–182 (2005)
5. McGlothlin, J.P., Khan, L.: Efficient RDF data management including provenance and uncertainty. In: Proc.14th Int'l. Database Engineering & Applications Symposium (IDEAS 2010), pp. 193–198 (2010)
6. McGlothlin, J.P., Khan, L.R.: RDFJoin: A scalable data model for persistence and efficient querying of RDF datasets. Tech. Rep. UTDCS-08-09, Univ. of Texas at Dallas (2008)
7. Murray, C.: RDF data model in Oracle. Tech. Rep. B19307-01, Oracle (2005)
8. Neumann, T., Weikum, G.: RDF-3X: a RISC-style engine for RDF. In: Proc. VLDB Endow., vol. 1, pp. 647–659 (August 2008)
9. O'Neil, P.: Model 204 architecture and performance. In: Proc. of HPTS , vol 359. LNCS, pp. 40–59 (1987)

10. Prud'Hommeaux, E., Seaborne, A.: SPARQL query language for RDF. In: World Wide Web Consortium. Recommendation REC-rdf-sparql-query-20080115 (January 2008)
11. Redaschi, N.: Uniprot in RDF: Tackling data integration and distributed annotation with the semantic web. In: Proc. 3rd Int'l. Biocuration Conf. (2009)
12. Sidirourgos, L., Goncalves, R., Kersten, M., Nes, N., Manegold, S.: Column-store support for RDF data management: not all swans are white. In: Proc. VLDB Endow., vol. 1, pp. 1553–1563 (August 2008)
13. Suchanek, F.M., Kasneci, G., Weikum, G.: YAGO: A large ontology from Wikipedia and WordNet. Web Semant. 6, 203–217 (2008)
14. Wu, K., Otoo, E., Shoshani, A.: Optimizing bitmap indices with efficient compression. ACM TODS 31(1), 1–38 (2006)

Database-as-a-Service for Long-Tail Science

Bill Howe, Garret Cole, Emad Souroush,
Paraschos Koutris, Alicia Key, Nodira Khoussainova, and Leilani Battle

University of Washington, Seattle, WA
{billhowe,gbc3,soroush,pkoutris,akey7,nodira,leibatt}@cs.washington.edu

Abstract. Database technology remains underused in science, especially in the long tail — the small labs and individual researchers that collectively produce the majority of scientific output. These researchers increasingly require iterative, ad hoc analysis over ad hoc databases but cannot individually invest in the computational and intellectual infrastructure required for state-of-the-art solutions.

We describe a new "delivery vector" for database technology called SQLShare that emphasizes ad hoc integration, query, sharing, and visualization over pre-defined schemas. To empower non-experts to write complex queries, we synthesize example queries from the data itself and explore limited English hints to augment the process. We integrate collaborative visualization via a web-based service called VizDeck that uses automated visualization techniques with a card game metaphor to allow creation of interactive visual dashboards in seconds with zero programming.

We present data on the initial uptake and usage of the system and report preliminary results testingout new features with the datasets collected during the initial pilot deployment. We conclude that the SQLShare system and associated services have the potential to increase uptake of relational database technology in the long tail of science.

1 Introduction

Relational database technology remains remarkably underused in science, especially in the *long tail* — the large number of relatively small labs and individual researchers who, in contrast to "big science" projects [22,25,30], do not have access to dedicated IT staff or resources yet collectively produce the bulk of scientific knowledge [5,24]. The problem persists despite a number of prominent success stories [7,18,19] and an intuitive correspondence between exploratory hypothesis testing and the ad hoc query answering that is the "core competency" of an RDBMS. Some ascribe this underuse to a mismatch between scientific data and the models and languages of commercial database systems [9,17]. Our experience (which we describe throughout this paper) is that standard relational data models and languages can manage and manipulate a significant range of scientific datasets. We find that the key barriers to adoption lie elsewhere:

1. *Setup.* Local deployments of database software require too much knowledge of hardware, networking, security, and OS details.

J.B. Cushing, J. French, and S. Bowers (Eds.): SSDBM 2011, LNCS 6809, pp. 480–489, 2011.
© Springer-Verlag Berlin Heidelberg 2011

2. *Schemas.* The initial investment required for database design and loading can be prohibitive. Developing a definitive database schema for a project at the frontier of research, where knowledge is undergoing sometimes daily revision, is a challenge even for database experts. Moreover, the corpus of data for a given project or lab accretes over time, with many versions and variants of the same information and little explicit documentation about connections between datasets and sensible ways to query them.

3. *SQL.* Although we corroborate earlier findings on the utility of SQL for exploratory scientific Q&A [30], we find that scientists need help writing non-trivial SQL statements.

4. *Sharing.* Databases too often ensconce one's data behind several layers of security, APIs, and applications, which complicate sharing relative to the typical status quo: transmitting flat files.

5. *Visualization.* While SQL is appropriate for assembling tabular results, our collaborators report that continually exporting data for use with external visualization tools makes databases unattractive for exploratory, iterative, and collaborative visual analytics.

As a result of these "S4V" challenges, spreadsheets and ASCII files remain the most popular tools for data management in the long tail. But as data volumes continue to explode, cut-and-paste manipulation of spreadsheets cannot scale, and the relatively cumbersome development cycle of scripts and workflows for ad hoc, iterative data manipulation becomes the bottleneck to scientific discovery and a fundamental barrier to those without programming experience.

Having encountered this problem in multiple domains and at multiple scales, we have released a cloud-based relational data sharing and analysis platform called SQLShare [20] that allows users to upload their data and immediately query it using SQL — no schema design, no reformatting, no DBAs. These queries can be named, associated with metadata, saved as views, and shared with collaborators. Beyond the basic upload, query, and share capabilities, we explore techniques to partially automate difficult tasks through the use of statistical methods that exploit the shared corpus of data, saved views, metadata, and usage logs.

In this paper, we present preliminary results using SQLShare as a platform to solve the S4V problems, informed by the following observations:

- We find that cloud platforms drastically reduce the effort required to erect and operate a production-quality database server (Section 2).
- We find that up-front schema design is not only prohibitively difficult, but unnecessary for many analysis tasks and even potentially harmful — the "early binding" to a particular fixed model of the world can cause non-conforming data to be overlooked or rejected. In response, we postpone or ignore up-front schema design, favoring the "natural" schema gleaned from the filenames and column headers in the source files (Section 2).
- We find that when researchers have access to high-quality example queries, pertinent to their own data, they are able to self-train and become productive SQL programmers. The SQL problem then reduces to providing such

examples. We address this problem by deriving "starter queries" — auto-matically — using the statistical properties of the data itself (as opposed to the logs, the schema, or user input that we cannot assume access to.) More-over, we analyze the corpus of user-provided free-text metadata to exploit correlations between English tokens and SQL idioms (Section 3).

– We find that streamlining the creation and use of views is sufficient to facil-itate data sharing and collaborative analysis (Section 2).
– We find the awkward export-import-visualize cycle can be avoided by eager pre-generation of "good" visualizations directly from the data (Section 4).

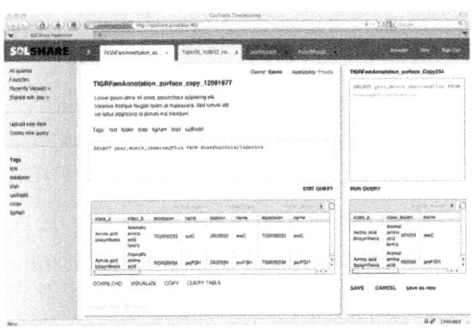

Number of uploaded datasets	772
Number of non-trivial views	267
Number of queries executed	3980
Number of users	51
Max datasets, any user	192
Total size of all data	16.5 GB
Size of largest table, in rows	1.1M

Fig. 1. Screenshot of SQLShare

Fig. 2. Early usage for the SQLShare system during a 4-month pilot period. We use data collected during the pi-lot to perform preliminary evaluation of advanced features.

2   SQLShare: Basic Usage and Architecture

The SQLShare platform is currently implemented as a set of services over a relational database backend; we rely on the scalability and performance of the underlying database. The two database platforms we use currently are Microsoft SQL Azure [28] and Microsoft SQL Server hosted on Amazon's EC2 platform.

The server system consists of a REST API managing access to database and enforcing SQLShare semantics when they differ from conventional databases. The "flagship" SQLShare web client (Figure 1) exercises the API to provide upload, query, sharing, and download services. The following features highlight key differences between SQLShare and a conventional RDBMS.

No Schema. We do not allow *CREATE TABLE* statements or any other DDL; tables are created directly from the columns found in the uploaded files. Just as a user may place any file on his or her filesystem, we intend for users to put any table into the SQLShare "tablesystem." By identifying patterns in the data (keys, union relationships, join relationships, attribute synonyms) and exposing them to users through views, we can superimpose a form of schema post hoc, incrementally — a *schema later* approach (Section 3).

Tolerance for Structural Inconsistency. Files with missing column headers, columns with non-homogeneous types, and rows with irregular numbers of columns are all tolerated. We find that data need not be pre-cleaned for many tasks (e.g., counting records), and that SQL is an adequate language for many data cleaning tasks.

Append-Only. We claim that science data should never be destructively updated. We therefore do not support tuple-level updates; errors can be logically replaced by uploading a new version of the dataset. This approach allows aggressive caching and materialization to improve performance.

Simplified Views. Despite their utility, we find views to be underused in practice. We hypothesize that the solution may be as simple as avoiding the awkward CREATE VIEW syntax. View creation in SQLShare is a side effect of querying — the current results can be saved by simply typing a name. This simple UI adjustment appears sufficient to encourage users to create views (Table 2).

Metadata. Users can attach free-text metadata to datasets; we use these metadata to support keyword search and to inform simple query recommendations by mining correlations between English tokens and SQL idioms (Section 3).

Unifying Views and Tables. Both logical views and physical tables are presented to the user as a single entity: the *dataset*. By erasing the distinction, we reserve the ability to choose when views should be materialized for performance reasons. Since there are no destructive updates, we can cache view results as aggressively as space will allow. However, since datasets can be versioned, the semantics of views must be well-defined and presented to the user carefully. We find both snapshot semantics and refresh semantics to be relevant, depending on the use case. Currently we support only refresh semantics.

3   Automatic Starter Queries

When we first engage a new potential user in our current SQLShare prototype, we ask them to provide us with 1) their data, and 2) a set of questions, in English, for which they need answers. This approach, informed by Jim Gray's "20 questions" requirements-gathering methodology for working with scientists [18], has been remarkably successful. Once the system was seeded with these examples, the scientists were able to use them to derive their own queries and become productive with SQL. The power of examples should not be surprising: Many public databases include a set of example queries as part of their documentation [15,30]. We adopt the term *starter query* to refer to a *database-specific* example query, as opposed to examples that merely illustrate general SQL syntax. In our initial deployment of SQLShare, starter queries were provided by database experts, usually translated from English questions posed by the researchers. In this section we show preliminary results in generating a set of starter example queries from a set of tables by analyzing their statistical properties only — no schema, no query workload, and no user input is assumed to exist.

Our approach is to 1) define a set of heuristics that characterize "good" example queries, 2) formalize these heuristics into features we can extract from the data, 3) develop algorithms to compute or approximate these features from the data efficiently, 4) use examples of "starter queries" from existing databases to train a model on the relative weights of these features, and 5) evaluate the model on a holdout test set. In this context, we are given just the data itself: In contrast to existing query recommendation approaches, we cannot assume access to a query log [21] , a schema [31], or user preferences [2].

We are exploring heuristics for four operators: union, join, select, and group by. In these preliminary results, we describe our model for joins only. Consider the following heuristics for predicting whether two columns will join: (1) A foreign key between two columns suggests a join. (2) Two columns that have the same active domain but different sizes suggest a 1:N foreign key and a good join candidate. For example, a fact table has a large cardinality and a dimension table has a low cardinality, but the join attribute in each table will have a similar active domain. (3) Two columns with a high similarity suggest a join. (4) If two columns have the same active domain, and that active domain has a large number of distinct values, then there is evidence in favor of a join.

Join heuristics 1-4 above all involve reasoning about the union and intersection of the column values and their active domains, as well as their relative cardinalities. We cannot predict the effectiveness of each of these heuristics a priori, so we train a model on existing datasets to determine the relative influence fo each. For each pair of columns x, y in the database, we extract each feature in Table 1 for both set and bag semantics.

Table 1. Features used to predict joinability of two columns, calculated for set and bag semantics

Join card. estimate	$\frac{|x|_b|y|_b}{max(|x|,|y|)}$
max/min card.	$max/min(|x|, |y|)$
card. difference	$abs(|x| - |y|)$
intersection card.	$|x \cap y|$
union card.	$|x \cup y|$
Jaccard	$\frac{|x \cap y|}{|x \cup y|}$

Table 2. Number of occurrences of the top 8 pairs of co-occurring tokens between English and SQL, for 4 users

	u_1	u_2	u_3	u_4	all
(join,join)	5	6	3	2	4
(thaps@phaeo,join)	-	-	1	-	18
(counts,join)	7	7	5	6	13
(flavodoxin,readid)	1	-	-	-	16
(tigr,gene)	7	-	-	4	16
(cog@kog,gene)	7	-	-	6	18

Preliminary Results We train a decision tree on the Sloan Digital Sky Survey logs (SDSS) [30], and then evaluate it on the queries collected as examples from the pilot period of SQLShare. To determine the class of a specific instance, we traverse the ADTree and sum the contributions of all paths that evaluate to true. The sign of this sum indicates the class.

For the SDSS database, we have the database, the query logs, and a set of curated example queries created by the database designers to help train users in writing SQL. We use the log to train the model, under the assumption that the joins that appear in the logs will exemplify the characteristics of "good" joins we would want to include in the starter queries.

For the SQLShare dataset, we use *the same model learned on the SDSS data* and see if it can be used to predict the example queries written for users. The key hypothesis is that the relative importance of each of these generic features in determining whether a join will appear in an example query do not vary significantly across schemas and databases. In this experiment, we find that the model classifies 10 out of 13 joins correctly, achieving 86.0% recall. To measure precision, we tested a random set of 33 non-joining pairs. The model classified 33 out of 37 of these candidates correctly, achieving 86.4% precision.

We observe that the model encoded several intuitive and non-intuitive heuristics. For example, the model found, unsurprisingly, that the Jaccard similarity of the active domains of two columns is a good predictor of joinability. But the tree also learned that similar columns with high cardinalities were even more likely to be used in a join. In low-similarity conditions, the model learned that very high numbers of distinct values in one or both tables suggests a join may be appropriate even if the Jaccard similarity is low. Overall, this simple model performed well even on a completly unrelated schema.

To improve this score, we are also exploring how to exploit the English queries provided by users as part of the 20 questions methodology. Natural language interfaces to databases typically require a fixed schema and a very clean training set, neither of which we have. However, we hypothesize that term co-ocurrence between metadata descriptions of queries and the SQL syntax provides a signal that can be incorporated to our ranking function for starter queries. To test this hypothesis using the full set of queries saved in SQLShare, we first pruned examples with empty or automatically generated descriptions, as well as all descriptions that included the word "test." Second, we tokenized the SQL and English descriptions into both single words and pairs of adjacent words. Finally, we computed the top k pairs of terms using a modified tf-idf measure. The support for these co-occurences appear in the cells of Table 2. We see that some term pairs refer to structural elements of the SQL (`join,join`). This rules may help improve our example queries for all users by helping us prune the search space. Other frequent co-occurring terms are schema dependent, which may help personalize query recommendations, or help determine which science domain is relevant to the user.

4   VizDeck: Semi-automatic Visualization Dashboards

VizDeck is a web-based visualization client for SQLShare that uses a card game metaphor to assist users in creating interactive visual dashboard applications without programming. VizDeck generates a "hand" of ranked visualizations and UI widgets, and the user plays these "cards" into a dashboard template, where they are syncronized into a coherent web application. By manipulating the hand dealt — playing one's "good" cards and discarding unwanted cards — the system learns statistically which visualizations are appropriate for a given dataset, improving the quality of the hand dealt for future users.

Figure 4 shows a pair of screenshots from the vizdeck interface. VizDeck operates on datasets retrieved from SQLShare; users issue raw SQL or select from a list of public datasets. After retrieving the data, VizDeck displays 1) a *dashboard canvas* (Figure 3 (left)) and 2) a *hand of vizlets* (Figure 3 (right)). A vizlet is any interactive visual element — scatter plots and bar charts are vizlets, but drop down boxes are also vizlets.

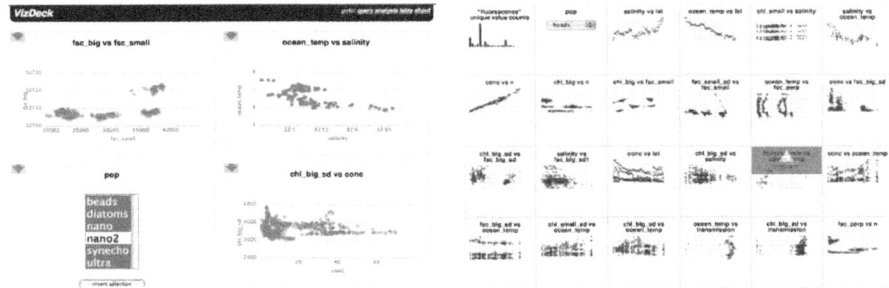

Fig. 3. Screenshots from the VizDeck application. (left) A VizDeck dashboard with three scatter plots and one multi-select box for filtering data. (right) A "hand" of vizlets that can be "played" into the dashboard canvas. The green arrow appears when hovering over a vizlet, allowing promotion to the dashboard with a click.

By interacting with this ranked grid, a user can *promote* a vizlet to the dashboard or *discard* it. Once promoted, a vizlet may be *demoted* back to the grid. Promoted vizlets respond to brushing and linking effects; items selected in one vizlet control the items displayed in the other vizlets. Multiple filtering widgets are interpreted as a conjunctive expression, while multiple selections in a single widget (as in Figure 3 (left)) are interpreted as a disjunction. By promoting visualizations and widgets, simple interactive dashboards can be constructed with a few clicks. Crucially, the user can *see* the elements they are adding to the dashboard before they add them — we hypothesize that this "knowledge in the world" [29] will translate into more efficient dashboard construction with less training relative to other visualization and mashup systems [11,23]. The user study to test this hypothesis is planned for future work.

Ranking. VizDeck analyzes the results of a query to produce the ranked list of vizlets heuristically. For example, a scatter plot is only sensible for a pair of numeric columns, and bar charts with too many bars are difficult to read [26].

As part of ongoing work, we are incorporating user input into the ranking function. We interpret each promote action as a "vote" for that vizlet, and each discard action as a vote against that vizlet, then assign each vizlet a score and train a linear model to predict this score from an appropriate feature vector. We are currently collecting data to evaluate our preliminary model.

Preliminary Results. A potential concern about our approach is that the number of potential vizlets is either too large (as to be overwhelming) or too small (making a ranking-based approach unnecessary). To test the feasibility of the approach, we applied VizDeck to all public datasets in the SQLShare system.

Figure 4 (left) shows the elapsed time to both retrieve data and generate (but not render) the vizlets for each public query, where each query is assigned a number. Most queries (over 70%) return in less than one second, and 92% return in less than 8.6 seconds, the target response time for page loads on the web [8]. Figure 4 (right) shows the number of vizlets generated by each query. Most queries (73 out of 118) generated a pageful of vizlets or less (30 vizlets fit on a page at typical resolutions). Of the remaining 45 queries, 29 return more than 90 vizlets, suggesting that a ranking-based approach is warranted.

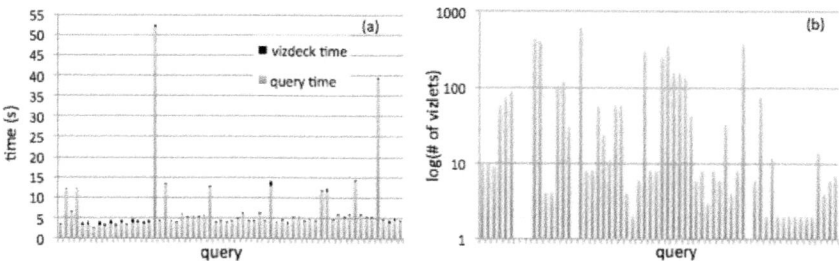

Fig. 4. Two experiments using VizDeck with the public queries in the SQLShare system. At left, we see the elapsed time for executing the query and generating the vizlets. The total time is dominated by query execution rather than VizDeck analysis. At right, we see the number of vizlets generated for each query.

5   Related Work

Other database-as-a-service platforms either do not support full SQL [4,14] or provide a conventional schema-oriented database interface [3,28].

The VizDeck system builds on seminal work on automatic visualization of relational data using heuristics related to visual perception and presentation conventions [26]. More recent work on intelligent user interfaces attempts to infer the user's task from behavior and use the information to recommend visuaizations [16]. Dork et al. derive coordinated visualizations from web-based data sources [11]. Mashup models have been studied in the database community [12,1], but do not consider visualization ensembles and assume a pre-existing repository of mashup components.

Query recommendation systems proposed in the literature rely on information that we cannot assume access to in an ad hoc database scenario: a query log [21], a well-defined schema [31], or user history and preferences [2]. The concept of dataspaces [13] is relevant to our work; we consider SQLShare an example of a (relational) Dataspace Support Platform. The Octopus project [10] provides a tool to integrate ad hoc data extracted from the web, but does not attempt

488   B. Howe et al.

to derive SQL queries from the data itself. The generation of starter queries is
related to work on schema mapping and matching [6,27]: both problems involve
measuring the similarity of columns.

References

1. Abiteboul, S., Greenshpan, O., Milo, T., Polyzotis, N.: Matchup: Autocompletion
   for mashups. In: ICDE, pp. 1479–1482 (2009)
2. Akbarnejad, J., Chatzopoulou, G., Eirinaki, M., Koshy, S., Mittal, S., On, D.,
   Polyzotis, N., Varman, J.S.V.: Sql querie recommendations. PVLDB 3(2) (2010)
3. Amazon Relational Database Service (RDS), http://www.amazon.com/rds/
4. Amazon SimpleDB, http://www.amazon.com/simpledb/
5. Anderson, C.: The long tail. Wired 12(10) (2004)
6. Bernstein, P.A., Melnik, S.: Model management 2.0: manipulating richer mappings.
   In: SIGMOD Conference, pp. 1–12 (2007)
7. Boeckmann, B., Bairoch, A., Apweiler, R., Blatter, M.C., Estreicher, A., Gasteiger,
   E., Martin, M.J., Michoud, K., O'Donovan, C., Phan, I., Pilbout, S., Schneider, M.:
   The SWISS-PROT protein knowledgebase and its supplement TrEMBL in 2003.
   Nucleic Acids Research 31(1), 365–370 (2003)
8. Bouch, A., Kuchinsky, A., Bhatti, N.: Quality is in the eye of the beholder: meeting
   users' requirements for internet quality of service. In: Proceedings of the SIGCHI
   Conference on Human Factors in Computing Systems, CHI 2000, pp. 297–304.
   ACM, New York (2000)
9. Brown, P.G.: Overview of scidb: large scale array storage, processing and analysis.
   In: Proceedings of the 2010 International Conference on Management of Data,
   SIGMOD 2010, pp. 963–968. ACM, New York (2010)
10. Cafarella, M.J., Halevy, A.Y., Khoussainova, N.: Data integration for the relational
    web. PVLDB 2(1) (2009)
11. Dörk, M., Carpendale, S., Collins, C., Williamson, C.: Visgets: Coordinated visu-
    alizations for web-based information exploration and discovery. IEEE Transactions
    on Visualization and Computer Graphics 14, 1205–1212 (2008)
12. Elmeleegy, H., Ivan, A., Akkiraju, R., Goodwin, R.: Mashup advisor: A recommen-
    dation tool for mashup development. In: ICWS 2008: Proceedings of the 2008 IEEE
    International Conference on Web Services, pp. 337–344. IEEE Computer Society,
    Washington, DC, USA (2008)
13. Franklin, M.J., Halevy, A.Y., Maier, D.: From databases to dataspaces: A new
    abstraction for information management. SIGMOD Record 34(4) (December 2005)
14. Google fusion tables, http://www.google.com/fusiontables
15. Gene ontology, http://www.geneontology.org/
16. Gotz, D., Wen, Z.: Behavior-driven visualization recommendation. In: Proceedings
    of the 13th International Conference on Intelligent User Interfaces, IUI 2009, pp.
    315–324. ACM, New York (2009)
17. Graves, M., Bergeman, E.R., Lawrence, C.B.: Graph database systems for ge-
    nomics. IEEE Eng. Medicine Biol. Special Issue on Managing Data for the Human
    Genome Project 11(6) (1995)
18. Gray, J., Liu, D.T., Nieto-Santisteban, M.A., Szalay, A.S., DeWitt, D.J., Heber,
    G.: Scientific data management in the coming decade. In: CoRR abs/cs/0502008
    (2005)

19. Heber, G., Gray, J.: Supporting finite element analysis with a relational database backend; part 1: There is life beyond files. Technical report, Microsoft MSR-TR-2005-49 (April 2005)
20. Howe, B.: Sqlshare: Database-as-a-service for long tail science, http://escience.washington.edu/sqlshare
21. Khoussainova, N., Kwon, Y., Balazinska, M., Suciu, D.: Snipsuggest: A context-aware sql autocomplete system. In: Proc. of the 37th VLDB Conf. (2011)
22. Large Hadron Collider (LHC), http://lhc.web.cern.ch
23. Lin, J., Wong, J., Nichols, J., Cypher, A., Lau, T.A.: End-user programming of mashups with vegemite. In: Proceedings of the 13th International Conference on Intelligent User Interfaces, IUI 2009, pp. 97–106. ACM, New York (2009)
24. Big science and long-tail science. Term attributed to Jim Downing, http://wwmm.ch.cam.ac.uk/blogs/murrayrust/?p=938
25. Large Synoptic Survey Telescope, http://www.lsst.org/
26. Mackinlay, J.: Automating the design of graphical presentations of relational information. ACM Transactions on Graphics 5, 110–141 (1986)
27. Madhavan, J., Bernstein, P.A., Rahm, E.: Generic schema matching with cupid. In: VLDB (2001)
28. Microsoft SQL Azure, http://www.microsoft.com/windowsazure/sqlazure/
29. Norman, D.: The design of everyday things. Doubleday, New York (1990)
30. Sloan Digital Sky Survey, http://cas.sdss.org
31. Yang, D.X., Procopiuc, C.M.: Summarizing relational databases. In: Proc. VLDB Endowment, vol. 2(1), pp. 634–645 (2009)

Data Scientists, Data Management and Data Policy

Sylvia Spengler

National Science Foundation, 4201 Wilson Boulevard,
Arlington, Virginia, USA
sspengle@nsf.gov

US science agencies have or will soon have a requirement that externally funded projects have "data management plans." Projects with a large budget or a tradition of data access and repositories do not see the impact as significant. However, the impact of the requirement can be particularly challenging for single investigators and small collaborations, especially in multidisciplinary research. These data represent what is known as Dark Data (Heidorn, 2008) in the long tail of science, where the data sets may be relatively small and the funding and expertise for handling also small. But just developing tools or putting computer scientists with the investigators is not sufficient.

The challenges exist at multiple levels: in the data themselves, where many different formats and systems are used; social and cultural issues of reward, recognition, rights and protections; in the persistent issues of investment and valuation. This panel will explore the challenges in this shotgun wedding, illustrate some success stories and explore ways that agencies can support the necessary changes.

Reference

Heidorn, B.P.: Shedding light on the dark data in the long tail of science. Library Trends 57, 280–289 (2008)

J.B. Cushing, J. French, and S. Bowers (Eds.): SSDBM 2011, LNCS 6809, p. 490, 2011.
© Springer-Verlag Berlin Heidelberg 2011

Context-Aware Parameter Estimation
for Forecast Models in the Energy Domain

Lars Dannecker[1], Robert Schulze[1], Matthias Böhm[2],
Wolfgang Lehner[2], and Gregor Hackenbroich[1]

[1] SAP Research Dresden, Chemnitzer Str. 48, 01187 Dresden, Germany
{lars.dannecker,robert.schulze}@sap.com
[2] Technische Universität Dresden, Database Technology Group
Nöthnitzer Str. 46, 01187 Dresden, Germany
{matthias.boehm,wolfgang.lehner}@tu-dresden.de

Abstract. Continuous balancing of energy demand and supply is a fundamental prerequisite for the stability and efficiency of energy grids. This balancing task requires accurate forecasts of future electricity consumption and production at any point in time. For this purpose, database systems need to be able to rapidly process forecasting queries and to provide accurate results in short time frames. However, time series from the electricity domain pose the challenge that measurements are constantly appended to the time series. Using a naive maintenance approach for such evolving time series would mean a re-estimation of the employed mathematical forecast model from scratch for each new measurement, which is very time consuming. We speed-up the forecast model maintenance by exploiting the particularities of electricity time series to reuse previously employed forecast models and their parameter combinations. These parameter combinations and information about the context in which they were valid are stored in a repository. We compare the current context with contexts from the repository to retrieve parameter combinations that were valid in similar contexts as starting points for further optimization. An evaluation shows that our approach improves the maintenance process especially for complex models by providing more accurate forecasts in less time than comparable estimation methods.

Keywords: Forecasting, Energy, Maintenance, Parameter Estimation.

1 Introduction

In the energy domain, the balancing of energy demand and supply is of utmost importance. Especially, the integration of more renewable energy sources (RES) poses additional requirements to this balancing task. The reason is that RES highly depend on exogenous influences (e.g., weather) and thus, their energy output cannot be planned like traditional energy sources. In addition, RES cannot be stored efficiently and must be used when available.

Several research projects such as MIRACLE [1], and MeRegio [2] address the issues of real-time energy balancing and improved utilization of RES. Current

J.B. Cushing, J. French, and S. Bowers (Eds.): SSDBM 2011, LNCS 6809, pp. 491–508, 2011.
© Springer-Verlag Berlin Heidelberg 2011

approaches in this area have in common that they require accurate predictions of energy demand and energy supply from RES at each point in time. For this purpose, mathematical models so called forecast models are used to model the behavior and characteristics of historic energy demand and supply time series. The most important classes of forecast models are: autoregressive models [3], exponential smoothing models [4] and models that apply machine learning [5]. Forecast models from all classes employ several parameters to express different aspects of the time series such as seasonal patterns or the current energy output. To exactly describe the past behavior of the time series, these parameters are estimated on a training data set by minimizing the forecast error (i.e., the difference between actual and predicted values) that is measured in terms of an error metric like the Mean Square Error (MSE) [3] or the (Symmetric) Mean Average Percentage Error ((S)MAPE) [6]. The so created forecast model instances are used to predict future values up to a defined horizon (e.g., one day). To allow efficient forecast calculations as well as the transparent reuse of forecast models, forecasting increasingly gains direct support by database systems. Besides research prototypes such as the Fa system [7] or the Forecast Model Index [8], forecasting has also been integrated into commercial products like Oracle OLAP DML [9] or Microsoft SQL Server Data Mining Extension [10]. However, the forecasting process remains inherently expensive, due to a large number of simulations involved in the parameter estimation process that can increase exponentially with the number of parameters when using a naive estimation approach.

Todays optimization algorithms used for parameter estimation can be divided into two classes: (1) Algorithms that need derivable objective functions and (2) algorithms that can be used with arbitrary objective functions. We focus on algorithms of the second class, because they can be used with any forecast models and error metrics and are hence more general. We can further classify algorithms of this class into local and global optimization algorithms. Global optimization algorithms such as Simulated Annealing [11] consider the whole solution space to find the globally optimal solution at the cost of slow convergence speed. In contrast, local optimization algorithms such as Nelder-Mead simplex search [12] follow a directed approach and converge faster at the risk of starving into local optima. These algorithms also highly depend on the provision of good starting points. Due to the limitations of local and global optimization algorithms, we enhance the parameter estimation by an approach that exploits knowledge about the time series context to store and reuse prior parameter combinations as starting points. This paper makes the following contributions:

First, we outline our general forecast model maintenance approach in Section 2.
Second, we introduce our Context-Aware Forecast Model Repository (CFMR) and define basic operations to preserve old parameter combinations in Section 3.
Third, we describe how to retrieve parameter combinations efficiently from the CFMR and how to revise them to yield the final parameters in Section 4.
Fourth, we evaluate our approach and demonstrate its advantages over comparable parameter estimation strategies in Section 5.
Finally, we present related work in Section 6 and conclude the paper in Section 7.

2   Context-Aware Forecast Model Maintenance

The core idea underlying our approach is to store previously used forecast models and in particular their parameter combinations in conjunction with information about the time series context during which they were valid (i.e., produced accurate forecasts) into a *Context-Aware Forecast Model Repository* (CFMR). Assuming that similar contexts lead to similar model parameters, we retrieve beneficial starting values for further optimization from the repository by comparing the current time series context with previous contexts stored in the CFMR.

The term *context* has been coined in machine learning to describe the conglomerate of background processes and influence factors, which drives the temporal development of data streams [13]. Regarding electricity demand and supply time series, we identify influences from meteorological (e.g., weather), calendar (e.g., seasonal cycles, public holidays) and economic (e.g., local law) factors. While each of these factors takes an individual state at each point in time, they form in their entirety a specific time series context. Since electricity demand and supply time series are regularly updated with new measurements, the state of the influence factors changes over time, and hence the entire context. Such *context drifts* can modify the behavior and characteristics of the subjacent time series in unanticipated ways. We adopt the classification of Zliobaite [14] and distinguish three types of context drift based on their duration and number of re-occurrences. Figure 1 illustrates the different types of context drifts:

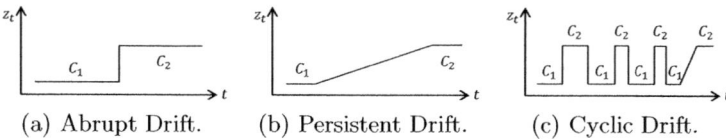

(a) Abrupt Drift.   (b) Persistent Drift.   (c) Cyclic Drift.

Fig. 1. Different Types of Context Drift

Abrupt Drift. A new context abruptly replaces the current context and causes a disruptive change within a short time frame. Example: power outages.
Persistent Drift. The old context slowly transforms into a new context that establishes permanently. Examples: gradual production changes, wind turbine aging.
Cyclic Drift. Old contexts re-appear and alternate. Examples: seasonal patterns (daily, weekly, yearly season), public holidays.

Context drifts can decrease the forecast accuracy, if the forecast model is not able to reflect the new situation and adapt to the changed time series characteristics. The reason is that changing contexts often lead to changing optimal forecast model parameters. Figure 2(a) illustrates the development of such optimal values for three example parameters of the electricity-tailored forecast model *EGRV* [15]. It can be seen that the optimal parameters greatly fluctuate over time, which we ascribe to context drift. In addition, the stationary parameter search space for a single parameter often exhibits multiple local minima for some

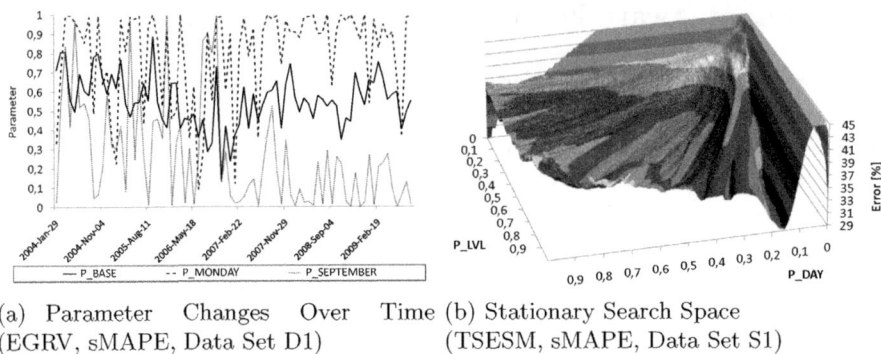

(a) Parameter Changes Over Time
(EGRV, sMAPE, Data Set D1)

(b) Stationary Search Space
(TSESM, sMAPE, Data Set S1)

Fig. 2. Illustration: Temporal and Stationary Parameter Spaces

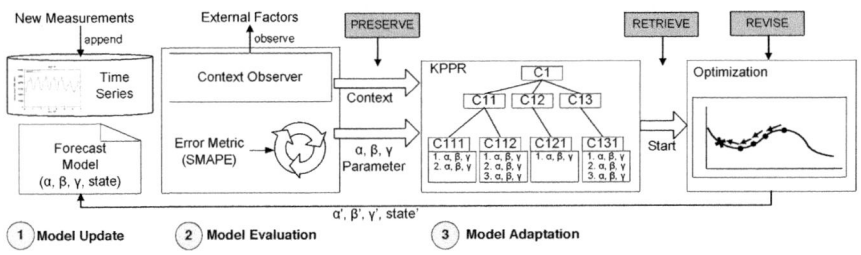

Fig. 3. Context-Aware Forecast Model Maintenance Process

error metrics as shown in Figure 2(b), which greatly decreases the probability of finding global optima using pure local or global searching. This clearly motivates a context-aware adaptation strategy of forecast model parameters to modified contexts from suitable starting points. In the following, we describe our general approach to maintain continuously accurate forecasts in the face of context drift.

Figure 3 illustrates the general forecasting process. First, new measurements are simply appended to the time series and the state of the forecast model is incrementally updated. This step is computational inexpensive and therefore not in the focus of the paper. Second, we continuously observe the forecast accuracy and the development of the current context using different model evaluation techniques. The recognition of context drifts requires enhanced model evaluation capabilities since changing contexts can increase the forecast error at arbitrary times. Evaluation techniques that regularly trigger model adaptations after fixed intervals turn out to be insufficient due to the difficulty of defining suitable adaptation intervals. If the interval is too long, context drifts and increasing errors occurring between adaptations are missed and lead to worse forecasts. Conversely, too short intervals may trigger unnecessary adaptations. We overcome these problems by a threshold-based model evaluation strategy that continuously checks the current forecast error against a defined threshold to ensure a maximal forecast error. The forecast model is adapted as soon as the forecast error surpasses the threshold. While this strategy guarantees that a maximal forecast error is not exceeded, it shares the major drawback with fixed interval model adaptation as it also depends on the definition of suitable thresholds. To this end, we propose an ensemble strategy that combines several individual evaluation strategies. Compared to using only one evaluation technique, such combinations of multiple maintenance strategies reduce the dependence on single adaptation criteria and make it easier to determine suitable values for them.

Third, our model adaptation approach is inspired by an artificial intelligence technique called case-based reasoning (CBR) [16]. The idea of CBR is to solve new problems from solutions of previously encountered problems similarly to the way humans reason by preserving, retrieving and revising previous experiences in the face of new problems. We apply the CBR paradigm to forecast model adaptation by *preserving* old parameter combinations and information about their context (i.e., when they produced accurate forecasts) in a Context-Aware Forecast Model Repository (CFMR) to solve later adaptation problems (Figure 3: context=$C1,C12,...$; parameters=α, β, γ). Upon triggering model adaptations, we first *retrieve* promising parameter combinations from the CFMR that were valid in a context similar to the new context. These parameter combinations are *revised* by using them as input for optimization algorithms. This approach is not limited to the energy domain, CBR-based techniques can be applied to arbitrary forecasting applications where similar models are periodically reused.

3   Preserving Forecast Models Using Time Series Context

The *Context-Aware Forecast Model Repository* (CFMR) allows to store and retrieve previous parameter combinations based on context information. When a forecast model is invalidated, the CFMR is searched for parameter combinations that produced accurate forecasts in similar past contexts. The retrieved parameters then serve as starting points for subsequent local optimization. Consider for example a cloudy and rainy day. The meteorological state influences the context and hence, the shape of the time series. Provided the weather conditions change to a similar state on a later occasion, we can search the CFMR for parameter combinations that were valid during such weather conditions.

3.1   Model History Tree

The CFMR is organized as a binary search tree named *Model History Tree*:

Definition 1 (Model History Tree). *A **model history tree** mht, defined over the similarity attributes $a_1, \ldots, a_n$, a maximum leaf node capacity c_{max} and a parameter vector size $V \in \mathbb{N}$, is a decision tree whose nodes are either decision nodes or leaf nodes.*

- *Decision nodes contain a splitting attribute $\dot{a}_i \in \{a_1, \ldots, a_n\}$, a splitting value $\dot{s} \in \dot{a}_i$ and references to the left and right successor nodes. Splitting*

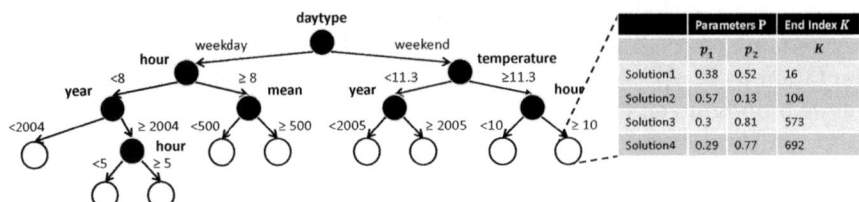

Fig. 4. Example Model History Tree

attributes are different influence factors (compare Section 2) of the time series and divide the stored parameter combinations into several classes.
- Leaf nodes *contain a list* $[a_i]$ *of similarity attribute values, at most* $c \leq c_{max}$ *parameter vectors* $p = [p_i | i = 1, \ldots, P]$, *and for each vector an end index* K *representing the last time the parameters were used.*

Figure 4 shows an exemplary model history tree, built over the similarity attributes *daytype, temperature, hour, year* and *mean*. The highlighted leaf node stores four parameter vectors, each of which contains two parameter combinations. The tree essentially forms a recursive partitioning of the parameter space into a set of disjoint subspaces whereas splitting attributes can be thought of as $(n-1)$-dimensional, axially parallel hyperplanes. At each decision node, the tree branches the parameter space into parameter combinations with attribute values smaller than the splitting attribute and those with attribute values greater or equal than the splitting attribute. Leaf nodes store the actual parameter combinations along with the corresponding end indices.

We generally distinguish numerical, nominal and cyclic similarity attributes a_i, which can take values within a domain $[a_i^{min}, a_i^{max}]$. Cyclic attributes are numerical or nominal attributes, whose instance values repeat every c indexes, i.e., $a_i = a_{i+c}$. Accordingly, they are typically connected with seasonal cycles. Table 1 presents an example selection of possible splitting attributes.

The similarity attributes guide the search in the CFMR for promising parameter combinations. That way, the parameter space is initially restricted and the majority of old parameter vectors can quickly be excluded from further

Table 1. Similarity Attributes for Electricity Demand and Supply

		Numerical	Nominal	Cyclic	a_i^{min}	a_i^{max}	Example
Temporal	Year		✓		2000	2020	2005
	Month		✓	✓	1	12	Apr (4)
	Day		✓	✓	1	7	Tue (2)
	Special Day		✓		0	1	False (0)
Exogenous	Temperature	✓			-30	40	27.4
	Wind Speed	✓	✓		0	30	15 m/s
	Electricity Price	✓			0	100	70.38 €/MWh
Statistical	Mean $\bar{z}$	✓			0	40000	12435.5 MW
	Variance σ^2	✓			0	10000	1719.6 MW²

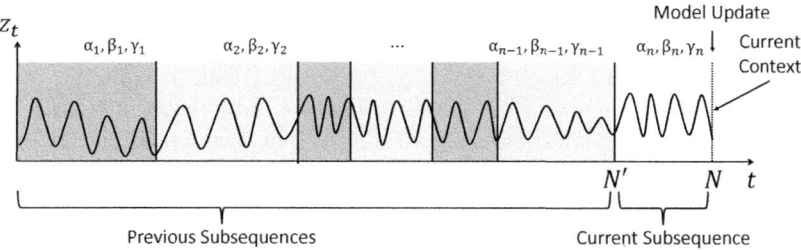

Fig. 5. Time Series Subsequences

processing. However, searching the tree may still result in a large number of results. To this end, we further restrict the solutions by comparing the shapes of corresponding past time series subsequences with the most current subsequence. For each parameter combination we save the time series index K the parameters were used the last time and compare the similarity between the most recent and past subsequences. The parameter combinations with the highest similarity score are finally chosen as starting values for subsequent optimization.

Figure 5 illustrates the subsequences of the time series. The parameter combinations $(\alpha_i, \beta_i, \gamma_i)$ were used to forecast during the corresponding subsequence. The time frame of the current subsequence is marked with indices from N' to N.

3.2 Inserting Models into the Model History Tree

Algorithm 1 defines how parameter combinations are inserted into the Model History Tree. The algorithm first traverses from the root node to the leaf node

---

Algorithm 1. mhtInsert().

---

input : $currContext$, $paramComb$, $endIndex$
if $treeSize() = 0$ **then** makeLeafNode()   // *Create tree if necessary*
$curr\_node \longleftarrow$ getRoot()   // *Traverse to leaf node;*
repeat
    // *i is index of splitting attribute*/;*
    if $curr\_node.\dot{s}_i < currContext[i]$ **then** $curr\_node \longleftarrow curr\_node$.left();
    else $curr\_node \longleftarrow curr\_node$.right();
until $isLeafNode(curr\_node)$;
$curr\_node$.add($currContext$, $paramComb$, $endIndex$);
if $nodeSize(curr\_node) > c_{max}$ **then**
    $(\dot{a}_i, \dot{s}_i) \longleftarrow$ computeSplittingAttributeAndValue();
    $curr\_node$.setSplittingAttributeAndValue($\dot{a}_i, \dot{s}_i$);
    $(left, right) \longleftarrow$ makeLeafNodes();
    $curr\_node$.setSuccessors($left, right$);
    foreach $context$, $paramComb$, $endIndex$ in $curr\_node$ **do**
        if $context[i] < \dot{s}_i$ **then** $left$.add($context$, $paramComb$, $endIndex$);
        else $right$.add($context$, $paramComb$, $endIndex$);
    makeDecisionNode($curr\_node$);

---

that represents the most similar context by comparing the similarity attributes. Afterwards, the new vector is added to the node. If the number of stored vectors c exceeds the maximum node capacity c_{max}, the leaf node is split into two leaf nodes and subsequently converted into a decision node with references to the new successors (divide-and-conquer strategy). The new splitting attribute $\dot{a}_i$ and split value (cut-point) $\dot{s}_i \in \dot{a}_i$ is determined from the models stored in the node. We could potentially rotate the splitting attribute and use the average over all values in the node. This strategy however suffers from the fact that for some attributes we cannot assign unambiguous values (e.g., the attribute values 'hour' 4, 6, 7, 10 yield an average of 6.75, which has no corresponding hour attribute). Thus, using the simple average of all values, especially of nominal or cyclic splitting attributes, is not possible in all cases. For this reason, we use the median which works on numerical as well as nominal and cyclic attributes. The basic idea behind the median is to choose a central value that partitions the possible values $[a_i]$ for attribute a with $i = N', \ldots, N$ in even halves. A prerequisite for the median is to first sort the attribute values in ascending order, leading to a list $[a_j]$ with $j \in \{N', \ldots, N\} \wedge a_j \le a_{j+1}$. For attributes that do not have a natural order, we apply an artificial order. However, such attributes are very seldom in the energy domain.

Definition 2 (Median). *Provided a_{min} and a_{max} are the minimum and maximum attribute values in the ordered list $[a_j]$, the **median** $\tilde{a}$ over $[a_j]$ is defined as follows:*

$$
\tilde{a} = \begin{cases} a_{min + \frac{max - min}{2}} & , \text{ if } N' - N \text{ even} \\ \frac{1}{2}\left(a_{min + \frac{max - min - 1}{2}} + a_{min + \frac{max - min - 1}{2} + 1}\right) & , \text{ else.} \end{cases} \tag{1}
$$

Example 1. Consider the numerical attribute *temperature* with $a_{i=7} = $ '13.5', $a_{i=8} = $ '12.3', $a_{i=9} = $ '15.6' and $a_{i=10} = $ '14.2'. Sorting in ascending ordering gives $[a_{i=8/j=7}, a_{i=7/j=8}, a_{i=10/j=9}, a_{i=9/j=10}]$. Because $N' - N = 10 - 7 = 3$ is uneven, we apply the second formula on the ordered list $[a_j]$ and obtain:

$$
\tilde{a} = \frac{1}{2}(a_{j=7+\frac{10-7-1}{2}} + a_{j=7+\frac{10-7-1}{2}+1}) = \frac{1}{2}(a_{j=8} + a_{j=9}) = \frac{1}{2}(a_{i=7} + a_{i=10}) = \text{'13.25'}
$$

Using the median as defined above to partition the parameter combinations ensures that selecting any of the attributes as splitting attribute results in the same number of models in both successors. However, the median does not distinguish homogeneously and heterogeneously spread attributes (compare Figure 6) and thus, cannot be used to choose an appropriate splitting attribute. We can assume that attributes with higher density towards the ends constitute better spliting attributes, as the median value separates both halves more clearly. For this reason, we additionally apply the *(Percental) Inter-quartile Range* ((P)IQR) as a measure of dispersion within attributes values and therefore as a measure for the suitability of the attribute as splitting attribute.

Definition 3 (Inter-quartile Range / Percental Inter-quartile Range).
*The **inter-quartile range** (IQR), defined over the list of attribute instances*
$[a_i | i = 1, \ldots, N]$, *denotes the average of the first and third quartiles:*

$$IQR = \frac{\tilde{a}^3 - \tilde{a}^1}{2}$$

with $\tilde{a}^1 = \tilde{l}$ and $l = \{a_i \leq \tilde{a}\}$ 1st quartile (median of left half)

with $\tilde{a}^3 = \tilde{r}$ and $r = \{a_i \geq \tilde{a}\}$ 3rd quartile (median of right half)

To ensures that attributes with a homogenous distribution, but large total range and thus large IQR, are not preferred over those with a heterogeneous distribution and a small range the IQR is normalized by the total range of attribute values leading to: $PIQR = \frac{IQR}{2(a_N - a_1)}$.

The attribute with the highest PIQR-value, i.e., the one with the lowest dispersion, is chosen as splitting attribute.

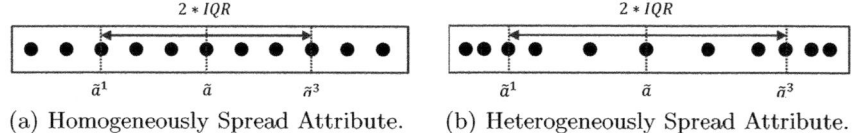

(a) Homogeneously Spread Attribute.   (b) Heterogeneously Spread Attribute.

Fig. 6. Attribute Spread

Balancing of the Model History Tree

Although the median and PIQR guarantee an equal distribution of parameter combinations for single nodes, model history trees can still degenerate to imbalanced, list-like trees when new parameter combinations are always added to one side of the splitting value. The reason is that the employed median heuristics makes *local* decisions only, i.e., the models in the node considered for splitting represent only a small subregion of the whole tree. In order to keep the tree globally balanced, we supplement it with a *global* balancing strategy which is based on measuring the heights of subtrees:

Definition 4 (Node Height / Balanced Node)
*The **height** $h(n)$ of a node n is defined as*

$$h(n) = \begin{cases} 0 & \text{, if } n \text{ is a leaf node} \\ 1 + \max(h(n.left, n.right))) & \text{, if } n \text{ is a decision node.} \end{cases}$$

*If a node n is **B(max)-balanced** if its balance factor $B(n)$ meets the following criteria: $B(n) = |h(n.left) - h(n.right)| \leq B_{max}$.*

$B(max)$-balanced nodes possess the property that the heights of their left and right subtrees differ at most by the predefined maximal balance factor B_{max} (e.g., $B_{max} = 3$). Global balance of the tree can then be assured by regularly checking the balancing condition. In case of imbalance the model history tree is regenerated, which means to re-build the tree from scratch, although other balancing strategies such as AVL-tree-like rotation [17] are conceivable in future work. When the tree is regenerated, the splitting decisions made at upper intermediate nodes is based on all models below that node. As we employ the median as splitting rule, it is guaranteed that the resulting tree is balanced again.

4   Retrieving and Revising Parameters from the CFMR

After introducing and defining the model history tree of the CFMR to preserve forecast model parameter combinations, we now show how to use the CFMR to quickly retrieve promising parameter combinations and conduct further revision.

4.1   Retrieving Forecast Models

Algorithm 2 defines how forecast model parameter combinations are retrieved from the model history tree. We provide the current context $currContext$ (i.e., the state of the similarity attributes at time N) and an auxiliary variable $best$. The algorithm is an adapted k-nearest neighbor search and based on the principle of backtracking, which means that it first descents to a leaf node and gradually adds solutions on its way back to the root. The k most similar models with respect to the current situation are obtained using the following process:

Algorithm 2. mhtRetrieve().

---

input : $currContext, currNode, best$

if $isLeafNode(currNode)$ **then**
 foreach $(context, paramComb, endIndex)$ in $currNode$ **do**
  $dist \longleftarrow getEuclidDist(currContext, context)$
  if $dist < getMaximumDist(best)$ **then**
   $best.update(\ context, paramComb, endIndex, dist\ )$
   $distanceComputation(paramComb, endIndex)$
else
 if $currNode.\dot{s}_i < currContext[i]$ **then**
  $best, maxDist \leftarrow mhtRetrieve(currContext, currNode.left, best)$
  if $bobTest(currContext,\ currNode,\ maxDist)$ **then**
   $best, maxDist \leftarrow mhtRetrieve(currContext, currNode.right, best)$
 else
  $best, maxDist \leftarrow mhtRetrieve(currContext, currNode.right, best)$
  if $bobTest(currContext,\ currNode,\ maxDist)$ **then**
   $best, maxDist \leftarrow mhtRetrieve(currContext, currNode.left, best)$
return $best$

---

1. Traverse from the root to the leaf node that corresponds best to the provided situation (repeatly execute second if-branch).

2. Compute the Euclidian distance at the leaf node between the provided context $currContext(v)$ and any old context(w) stored in the leaf node. The Euclidean distance yields small distances for models which agree in important attributes (v_i/w_i represent the single attributes of context1 and context2).

$$getEuclidDist(context_1(v), context_2(w)) = \sqrt{\sum_{i=l}^{n}(v_i - w_i)^2}$$

Save results in *best*, calculate subsequence similarity for each intermediate result.

3. Ascend to the root node. Perform a *ball-overlap-bounds* (bob) test at each intermediate node to evaluate the existence of additional solutions in opposite branches by checking for points closer than the worst point in *best*. Test by intersecting a n-dimensional ball with radius $getMaximumDist(best)$ and the splitting hyperplane: $getMaximumDist(best) \geq |currContext[i] - node.\dot{s}_i|$. The bob-test evaluates whether the ball around the worst intermediate result overlaps the hyperplane. If true, descend into other branch to search for better solutions.

Example 2. Figure 7 illustrates the execution of Algorithm 2. Attribute a_1 is non-cyclic and attribute a_2 is cyclic.

1. Descent to leaf node corresponding best to provided context and compute the Euclidean distance to all models in the node(O and P). Save results to *best*. Start subsequence similarity calculation.
2. Ascent and perform bob-test at predecessor. Negative — continue ascent.
3. Perform bob-test at next node. Positive bob-test. Find R as nearest neighbor. Update *best*. Start subsequence similarity calculation.
4. Perform bob-test at root. Negative — algorithm finishes. Result: R.

In the worst case, the k-nn-search evaluates all nodes in the tree. While this misbehavior is very unlikely, we avoid long runtimes by further processing the intermediate results in parallel (subsequence similarity, optimization) and using them as temporary parameter combinations for forecasting. This procedure is feasible because even the first intermediate results were found in a node that at least is a good approximation of the current situation. Later results are only accepted if they improve upon the currently known worst intermediate result.

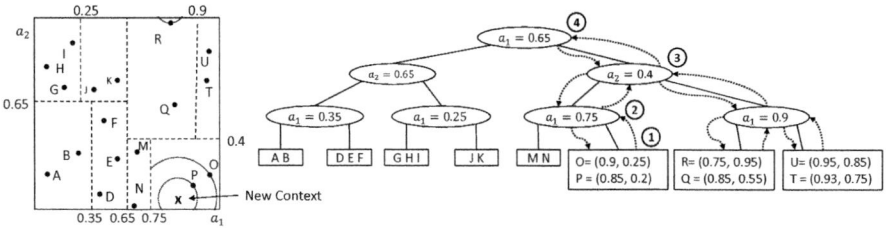

Fig. 7. Example: Model Retrieval in a 2-Dimensional Model History Tree

4.2   Comparing Time Series Similarity

In addition to comparing the contexts, we also compare the shapes of the most current time series values and past subsequences corresponding to found parameter combination candidates to identify the most promising parameter combinations. A high degree of coincidence between current and past load curves is regarded as sign for the similarity of the corresponding contexts. The distance between recent and past subsequences is expressed in terms of the Pearson cross-correlation coefficient $R_{zz'}(\tau)$:

$$R_{zz'}(\tau) = \frac{\sum\limits_{i=1}^{N-\tau}(z_i - \bar{z})(z'_{i+\tau} - \bar{z'})}{\sqrt{\sigma_z^2 \sigma_{z'}^2}}.$$

High cross-correlation values entail strong similarity of the involved subsequences. Other distance measures such as Dynamic Time Warping [18] can be used as well. The main side condition imposed by the aforementioned definition, is that both subsequences have to be equally long. To ensure equally long time series, we chose a fixed length of the subsequence from a corresponding point of the time series and hence obtain as *past* and *current* subsequences. However, the cross-correlation will be low for phase-shifted subsequences (i.e., different start/end indices), even though their shapes are similar. We overcome this difficulty by shifting one sequence past the other through the specification of an lag τ. The lag τ is denoted with respect to the ending indexes K (end of former situation) and N (end of current situation) and the (known) period s of a seasonal cycle: $\tau = |K \bmod s - N \bmod s|$. This lag specification aligns the subsequences by cutting the outer values of both sequences. Altogether, the cross-correlation can be applied to time series which changes at most its amplitude, offset and level over time, but not its periods. This makes it a good choice for electricity demand time series, which possess comparatively stable seasonal cycles. As a result the similarity search provides the parameter combinations that yield the most similar subsequences.

4.3   Revising Forecast Models

A re-estimation of forecast model parameters serves as further refinement of the retrieved parameter combinations. We concurrently perform a local (e.g., Nelder-Mead) and a global optimization (e.g., Simulated Annealing). The starting points for the local search are the parameter combination candidates provided by the CFMR. Due to the continuous adaptation of the forecast model to drifting context, we assume that the parameters changed gradually only with respect to the old and current situation. For this reason, we find the global optimal parameter combination with high probability close to the provided starting point. However, it is still advisable to check for regions not covered by local search. The employed global search runs in parallel to the local search, because it is independent of starting values. We continue the global search even after the local search found its optimal solution. Thus, we consider all areas of the solution space. Due to the long run time of the global search, the search runs asynchronously in the background, while the solution found by the local search is used as an intermediate parameter combination. If the global search finds a better solution, we use this solution as an additional starting point for the global optimization.

5 Evaluation

In this evaluation we proof the claims of our approach and show that with the help of the CFMR we can increase the parameter estimation efficiency by means of delivering more accurate forecasts in a shorter time frame compared to other approaches. Our evaluation compares the accuracy and the time necessary to gain the accuracy and is based on two forecast models and three electricity data sets from different parts of the European electricity market.

Data Set D1: National Grid Electricity Demand from National Grid (publicly available [19]). Electricity demand of the United Kingdom. Measures used: INDO, January 1st 2002 to December 31st 2009, $30min$ resolution.

Data Set D2: EnBW MeRegio Household Energy Demand from MIRACLE partner EnBW. Energy demand from 86 anonymized customers — We used customers 7 (D2a, more predictable behavior) and 40 (D2b, hardly predictable behavior). Measures used: November 1st 2009 to June 30th 2010, $1h$ resolution.

Data Set S1: CRES Photo-Voltaic Energy Supply from MIRACLE partner CRES. Supply of a 22kW photovoltaic panel. Measures used: January 11th 2008 to December 16th 2008, $1min$ resolution aggregated to $30min$ resolution.

The evaluation was conducted on a AMD Athlon 4850e with 4 GB RAM, Microsoft Windows 7 64bit and Microsoft Visual Studio C++ 2010.

For our evaluation we employed two forecast models tailor-made for the forecasting of energy demand and supply. The first model is Triple Seasonal Exponential Smoothing (TSESM) [20]. TSESM involves five forecast model parameters with values from 0 to 1, which lead to a five-dimensional solution space for the parameter estimation. The second model is named EGRV model and defines a separate model for each hour of the day to avoid the explicit modeling of the complex daily season [15]. In addition, different influences such as the current day and month are included as separate variables. Thus, one hourly model involves around 31 parameters, depending on the incorporated influences.

To contrast our approach, we used the following four common local and global parameter estimation approaches:

- *Monte-Carlo*: Iteratively evaluate random solutions.
- *Simulated Annealing*: Global search without starting values.
- *Random-Restart Nelder-Mead*: Iterated local searching with starting values obtained by random search.
- *Single Nelder-Mead*: Local search with current parameters as starting values

The forecast models were optimized for one-step ahead forecasts and using the sMAPE error metric [6]. We traced the lowest sMAPE obtained during optimization every two seconds and averaged over four runs per approach. For EGRV,

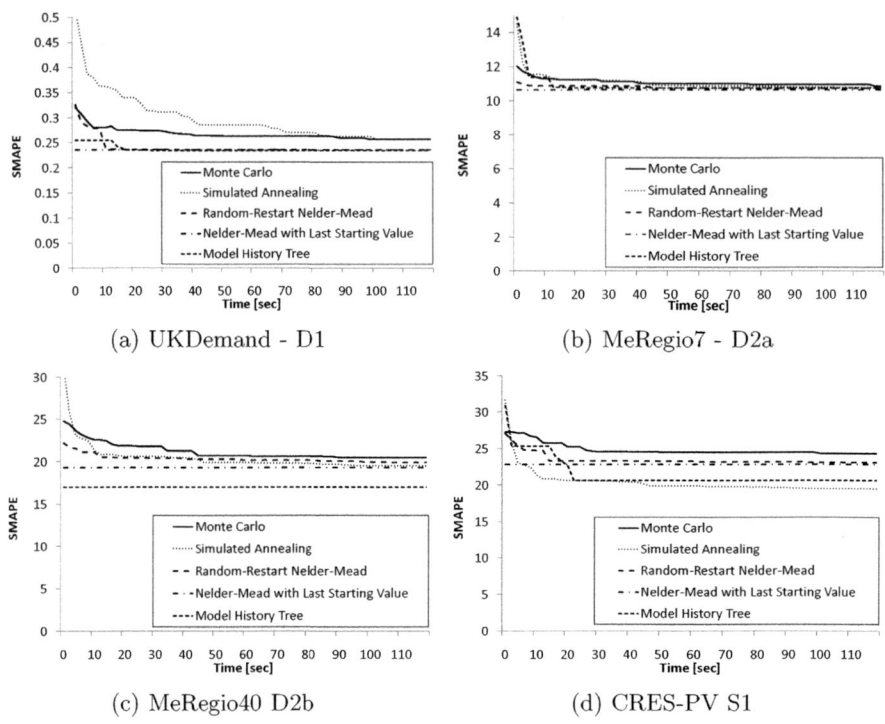

Fig. 8. Time vs. Accuracy - Triple Seasonal Exponential Smoothing

the random-restart Nelder-Mead strategy was budgeted at five minutes, because unsuitable starting points led to a very slow convergence. The presented results illustrate a single point in time only and are based on threshold-based model adaptations. In detail, we used data-set specific error thresholds/sliding windows sizes of 6%/12 D1,20%(30%)/3 D2a(D2b) and 30%/2 S1, which lead to about 100 models stored into the tree at the time the model was re-estimated. We repeated our evaluation at other points in time and achieved similar results.

Figure 8 illustrates the results for the Triple Seasonal Exponential Smoothing model. We observe that our approach in general quickly reaches good accuracies on all data sets. However, the subsequent global search does not find better parameter combinations. The reason could be that the parameters found by local searching are very good and with a high probability already the global optimal solution. For the datasets D1 and D2b our approach achieved the best results regarding accuracy and time for the entire test period. There, the single simulated annealing approach achieved the worst results for data set D1 and the Monte-Carlo approach performed worse for data set D2b. For data set D2a and S1 our approach also achieved good results but performed not as well as other approaches. Regarding data set D2a our approach had a good start, but both local search approaches that involve the Nelder-Mead algorithm achieved better results at the start. Simulated annealing and the Monte-Carlo approach performed worse at the start. With further progression all approaches achieved similar results and differ by less than 0.5% SMAPE. However, our approach performed worst on this data set. For dataset S1, our approach converged slower than all approaches except the Monte-Carlo sampling, but at the end it achieved the second best result. Only simulated annealing performed better by less than a half percent. We blame the results to the sequential execution of the local searches which are occasionally supplied unfavorable starting values in the beginning and the best starting values in the end. All other approaches differ by a more significant amount of two or more percent. Overall we can state that for the triple seasonal exponential smoothing our approach achieved good results on all data sets. In two cases it performed worse than other approaches, but however the other approaches have a larger divergence concerning their results, e.g. simulated annealing performed worse for data set D1 and best for data set S1. In contrast, our approach constantly achieved very good results, which leads to the educated guess that it is useable for all data sets from the energy domain without prior evaluation. We furthermore observed only small overhead from using the tree. Depending on the data set, the context computation, model insertion and model retrieval took together always less than 4 msec. We also tested scalability of these operations for trees with up to 20,000 models and obtained a joint worst case insertion and access time of less than 0.6 sec, which is negligible in comparison to the overall re-estimation time.

Figure 9 illustrates the results for the EGRV forecast model. In general, due to a much higher number of parameters, these models are more challenging to update than the previously discussed seasonal exponential smoothing models. In addition, we can observe that the differences in the reached maximal accuracy between the best and worst strategy are larger than for smoothing models which means that the choice of a good re-estimation strategy is hence more critical. Our approach achieved the best results for all evaluated data sets by means of both accuracy and time. All strategies obtained improvements particularly quickly within the first minute of execution, but with further progression they were not able to reach the accuracy of our approach. For data set D1 and D2a the accuracy gap between our approach and its competitors is comparatively large. For data set D2b the local search strategies at the end achieved similar but slightly worse results. In contrast, the produced accuracies of the global search strategies were far off. Regarding the supply data set S1 four out of five approaches converged to a similar result and except for Nelder-Mead with Last Starting Value the difference in accuracy is rather small. The results produced by Simulated Annealing are also at least as good or better than the results found by Monte-Carlo, which confirms our choice of simulated annealing as our global coverage strategy. In addition the random-restart Nelder-Mead strategy shows only slow convergence in average, but it often finds good results after some minutes. This demonstrates the need to start optimization from suitable starting parameters. Again, the run-time overhead for inserting and retrieving models from the CFMR depends on the data set, but was less than 5 msec in the worst case. Further experiments with 20,000 EGRV models in the tree still show access

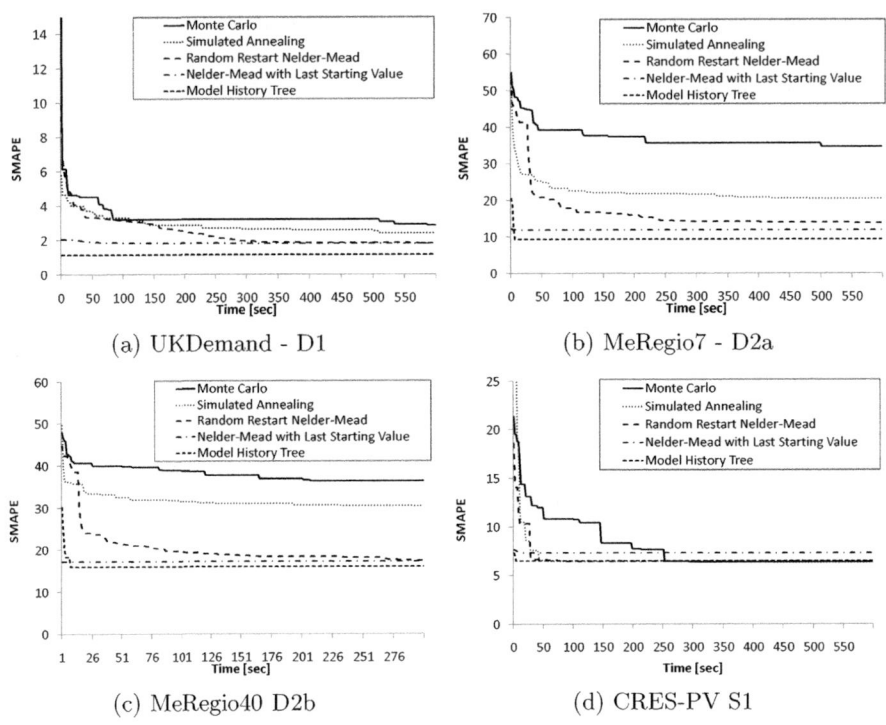

Fig. 9. Time vs. Accuracy - EGRV

times of less than 1.1 sec which suggests that the tree scales also for large history bases. Overall, our approach achieved better results when used in conjunction with a more complex forecast model. For all data sets the CFMR outperformed all other estimation approaches, which makes it the most suitable approach for models that involve a large number of parameters. An interesting effect we observed during evaluation was the steadily improving capability of the tree to provide parameters that were already optimal without further optimization. For later stages of the demand data sets, optimization was fully redundant and improved the result only on special days such as Christmas. However, to ensure the optimality of the result, the parallel optimization should still be conducted.

6   Related Work

Approaches that use aspects similar to our approach exist in various domains: Kohara et al. described a system that uses prior knowledge and information about events manually extracted from newspapers in conjunction with neural networks to improve stock market predictions [21]. In the domain of conceptual modeling Becker et al. proposed the reuse of knowledge from other modelers or former projects by using context-based modeling that combines reuse mechanisms like aggregation, restriction and specialization of previous models. Context-base modeling exploits the context of a conceptual model to for example restrict available constructs and their relations [22]. Breitman et al. proposed a similar solution that stores conceptual models created by expert designers in a repository that is used by less experienced designers to later create new conceptual models [23]. Luan Ou et al. introduced an approach for process models in the business intelligence domain. They store process models in a model base and use CBR and rule-based reasoning techniques to quickly find the right process for a given task. For new data mining tasks they retrieve the most similar model from the model base and present it to the user for confirmation. The similarity measures base on domain knowledge about the process models [24]. Compared to our approach, all presented solutions utilize similar basic ideas, especially, (1) the creation of a case base for later reuse and (2) the usage of context or domain knowledge to efficiently find suitable solutions. However, they are applicable in their specific domain only and cannot directly be applied to the forecasting domain. Our approach is the first adaptation of CBR that exploits the context of time series, for which reason it involves specific aspects like the utilization of a decision tree, a similarity measurement and a subsequent optimization.

7   Conclusion

In this paper, we presented a novel parameter estimation approach that exploits the context of a time series to quickly find starting parameters for further optimization. There, we used our Context-Aware Forecast Model Repository (CFMR) to store the parameter combinations in conjunction to their associated context. Besides basic definitions, we described the preservation, retrieval and revision of forecast models within our repository. Our evaluation on four datasets showed that our solution provides an efficient way to estimate parameters, especially when dealing with complex forecast models. In most cases we were superior to all evaluated competitors in providing more accurate forecasts in less time. There is plenty of future work, which includes the context-aware estimation of parameters, an experimental investigation of influences between different context components and a even better parallelization of our approach.

Acknowledgment

The work presented in this paper has been carried out in the MIRACLE project funded by the EU under the grant agreement number 248195.

References

1. MIRACLE Project (2011), http://www.miracle-project.eu
2. MeRegio Project (2011), http://www.meregio.de/en/
3. Box, G.E.P., Jenkins, G.M., Reinsel, G.C.: Time Series Analysis: Forecasting and Control. John Wiley & Sons Inc., Chichester (2008)

4. Winters, P.R.: Forecasting sales by exponentially weighted moving averages. Management Science, 324–342 (April 1960)
5. Bunnoon, P., Chalermyanont, K., Limsakul, C.: A computing model of artificial intelligent approaches to mid-term load forecasting: a state-of-the-art- survey for the researcher. Int. Journal of Engineering and Technology 2(1), 94–100 (2010)
6. Hyndman, R.J.: Another look at forecast-accuracy metrics for intermittent demand. Foresight: The International Journal of Applied Forecasting 4, 43–46 (2006)
7. Duan, S., Babu, S.: Processing forecasting queries. In: VLDB (2007)
8. Fischer, U., Rosenthal, F., Boehm, M., Lehner, W.: Indexing forecast models for matching and maintenance. In: IDEAS. Dresden University of Technology (2010)
9. Oracle test: Oracle - Driving Strategic Planning with Predictive Modeling (2008)
10. Microsoft: SQL Server 2008 - Predictive Analysis with SQL Server (2008)
11. Kirkpatrick, S., Gelatt, C.D., Vecchi, M.P.: Optimization by simulated annealing. Science. New Series 220(4598), 671–680 (1983)
12. Nelder, J., Mead, R.: A simplex method for function minimization. The Computer Journal 7(4), 308–313 (1965)
13. Widmer, G., Kubat, M.: Learning in the presence of concept drift and hidden contexts. Machine Learning 23(69) (1996)
14. Zliobaite, I.: Learning under concept drift: An overview. Technical report, Vilnius University (2009)
15. Ramanathan, R., Engle, R., Granger, C.W., Vahid-Araghi, F., Brace, C.: Short-run forecasts of electricity loads and peaks. International Journal of Forecasting 13(2), 161–174 (1997)
16. Aamodt, A., Plaza, E.: Case-based reasoning: foundational issues, methodological variations, and system approaches. AI Communications 7(1), 39–59 (1994)
17. Sedgewick, R.: Algorithms. Addison-Wesley, Reading (1988)
18. Berndt, J., Donald, J.C.: Using Dynamic Time Warping to Find Patterns in Time Series. Technical report, Stern School of Business, New York University (1994)
19. Nationalgrid UK.: Metered half-hourly electricity demands (2010), http://www.nationalgrid.com/uk/Electricity/Data/Demand+Data/
20. Taylor, J.W.: Triple seasonal methods for short-term electricity demand forecasting. European Journal of Operational Research 204, 139–152 (2009)
21. Kohara, K., Ishikawa, T., Fukuhara, Y., Nakamura, Y.: Stock price prediction using prior knowledge and neural networks. Intelligent Systems in Accounting, Finance & Management 6(1), 11–22 (1998)
22. Becker, J., Janiesch, C., Pfeiffer, D.: Towards more Reuse in Conceptual Modeling - A Combined Approach using Contexts. In: CAiSE Forum (2007)
23. Breitman, K.K., Barbosa, S.D.J., Casanova, M.A., Furtado, A.L., Hinchey, M.G.: Using analogy to promote conceptual modeling reuse. In: ISoLA, pp. 111–122 (2007)
24. Ou, L., Peng, H.: XML and knowledge based process model reuse and management in business intelligence system. In: Shen, H.T., Li, J., Li, M., Ni, J., Wang, W. (eds.) APWeb Workshops 2006. LNCS, vol. 3842, pp. 117–121. Springer, Heidelberg (2006)

Implementing a General Spatial Indexing Library for Relational Databases of Large Numerical Simulations

Gerard Lemson[1], Tamás Budavári[2], and Alexander Szalay[2]

[1] MPA
lemson@mpa-garching.mpg.de
[2] JHU
{budavari,szalay}@jhu.edu

Abstract. Large multi-terabyte numerical simulations of different physical systems consist of billions of particles or grid points and hundreds to thousands of snapshots. Increasingly these data sets are stored in large object-relational databases. Most statistical analyses involve extracting various spatio-temporal subsets. Existing built-in spatial indexes in commercial systems lack essential features required for many applications in the physical sciences. We describe a library that we have implemented in several languages and platforms (Java/Oracle, C#/SQL Server) based on generic space-filling curves, implemented as plug-ins. The index provides a mapping of higher dimensional space into the standard linear B-tree index of any relational database. The architecture allows intersections with different geometric primitives. The library has been used for cosmological N-body simulations and isotropic turbulence, providing sub-second response time over datasets exceeding several tens of terabytes. The library can also address complex space-time challenges, like temporal look-back into past light-cones of cosmological simulations.

Keywords: spatial indexing, numerical simulations, relational databases.

1 Introduction

Astronomical data is doubling every year. Much of the growth in observational data is due to the emergence of ever larger detectors, CCD mosaic cameras. The raw images from optical surveys in astronomy are starting to reach the PB/year data rates. The resulting object catalogs are placed in large databases, a trend started with the Sloan Digital Sky Survey (SDSS)[17,18]. These databases are usually publicly available both through open SQL interfaces, and various web services. There is a standardization effort under way, to define a set of atomic, or core services, that can be used to build a system that federates all astronomy data into a Virtual Observatory[1]. The astronomy community has been very quick in embracing SQL as a new way to access and analyze these observational

---

[1] See http://www.ivoa.net

J.B. Cushing, J. French, and S. Bowers (Eds.): SSDBM 2011, LNCS 6809, pp. 509–526, 2011.
© Springer-Verlag Berlin Heidelberg 2011

data sets. Sharable user databases placed server side emerged as a convenient collaborative environment[12].

However, not only observational data are growing exponentially. In order to interpret the observations, astronomers have traditionally been running large numerical simulations on all astronomical scales, from the formation of black holes to stellar evolution, the formation of galaxies and the large-scale structure of the Universe. Traditionally the simulations were analyzed as they were ran. At most a few (up to about 100) checkpoints were saved for subsequent analyses. As simulations grow in size, writing the checkpoints becomes harder and harder, not to mention their long-term archival and public access. Over the last ten years the typical cardinality of state-of-the-art cosmological simulations has grown from about 100 million to 300 billion. These simulations are run on the world's largest supercomputers, accessible only to a small fraction of the astronomy community. Yet, there is a growing need by astronomers to be able to compare observations to first-principle simulations. It is difficult today to publish a paper related to statistical analyses of large extragalactic surveys without comparing the results to large simulations. There is a set of reference simulations, like the Millennium[15,3,1], Aquarius[16], Via Lactea-2[4], Coyote[8], and many others, that represent the state of the art.

As a result, there is a growing need and pressure not only to make the data from these simulations public, but also accessible. The expectations for data access and interfaces are now set by the standards of the large on-line databases, like SDSS. Traditionally, the public interface to these simulations was through downloading the snapshot files from a central location to the users and analyzing it at their local facility. As the simulation sizes are growing very rapidly (for the Millennium XXL simulation[1] a single snapshot is 7 TB) this is no more feasible. Furthermore, each simulation code has its own proprietary data format. The users need to learn the subtle details of each simulation, customize I/O libraries, and have non-trivial amounts of storage and processing at their location in order to even get started.

Consequently, there is a substantial ongoing effort to bring a good fraction of the available reference simulations on-line, in the form of publicly queryable relational or object-relational databases, and the International Virtual Observatory Alliance is in the process of defining service interfaces[2].

There are several different types of data in these simulations. In CFD we have the data on a structured grid, with the physical parameters, like velocity, density, pressure, dissipation, magnetic fields (for MHD), etc. For particle based simulations we have discrete particles at the lowest level, representing the dark matter, which are then aggregated into friends-of-friends groups, then to halos and subhalos and finally into observable galaxies. These higher level objects are the observable ones, but by linking them to the individual dark matter particles we can build up the so-called merger-trees, the formation history of halos and galaxies over cosmic time.

---

[2] http://www.ivoa.net/theory

The typical queries astronomers want to run against the data are related to various spatial and temporal extractions of a set of galaxies, halos or dark matter particles. Some of the search patterns are quite simple, like finding all objects in a given spherical region of space at a given time, others can be much more complicated, like building up a virtual observation of a past light-cone (see Section 4.2).

The scale of the largest simulations is growing extremely fast. The currently ongoing Silver River will have 50 billion dark matter particles. Even if only the minimal information is stored (particle label, position and velocity), a single snapshot is 3.2TB. The current plan is to generate 800 snapshots, for a total of 2.6PB of output data. Over the years we found that open source databases have yet to scale to such data sizes, and as a result most production sites have used commercial database platforms when it came to a large scale deployment. However, it is also clear that the spatial libraries and features offered by these platforms are far from satisfactory for the kinds of spatio-temporal queries mentioned above.

In this paper we describe a framework for indexing and searching such databases based on space filling curves, that is capable of representing these search patterns, has excellent performance, and has been implemented and used extensively in two of the most popular commercial database environments, Oracle and Microsoft SQL Server, in deployments reaching tens of terabytes.

2   Methodology

2.1   The Problem: Fast Spatial Querying of Points in a Box

Consider the results from a large-scale cosmological N-Body simulation stored in a relational database, containing tens of billions of particles. These particles are represented as points, stored in a table with positions over a finite, cubic subvolume L^3 of 3D space. The coordinates of the points are floating point numbers, stored in columns (X, Y, Z). In general, the points also have a (discrete) time coordinate T and a velocity vector. This is not important for our main argument, but the fact that e.g. Oracle Spatial (version 11g2) can not mix ordinary and spatial columns in a single index makes that solution not relevant.

Typical spatial queries require finding points (at a given time) that are in some way near to each other in space, or are located inside a region of a certain shape. For the simplest applications in astronomy the subvolumes will be boxes, spheres or cones, although our framework is able to deal with more complex shapes as well. See Fig. 1 for an illustration in 2 dimensions. An important feature of many of these simulations is that they assume periodic boundary conditions. We want this fact to be reflected in our query results, see Fig.1-b. If a query volume crosses a boundary, we want to retrieve also the points that we would find if the box were truly replicated.

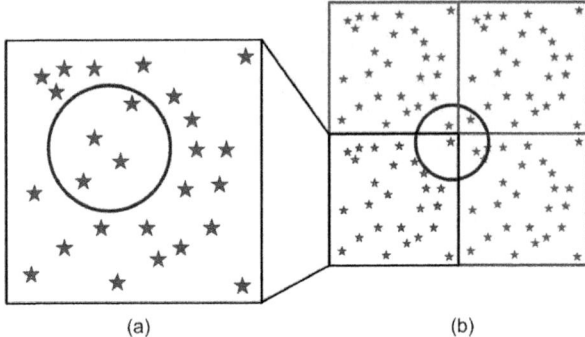

Fig. 1. (a) shows a typical spatial query. Objects are searched which fall in a query shape, here a circle. (b) shows a spatial query volume that crosses the boundary of the simulation volume (black boundary). Due to the periodic boundary conditions the query should also retrieve objects in the replicated boxes (red boundaries).

2.2 Space-Filling Curves and Octrees

The naive implementation of these spatial queries scans the complete table, checking for each point whether it falls in the subvolume. For typical tables with billions of rows that algorithm is too inefficient. Standard database indexes built on existing columns, such as B-Trees, are also useless, since they do not preserve three-dimensional coherence.

Our solution to this problem depends on a discretization of the spatial coordinates, following classical techniques, based on octrees and space-filling curves [2,6,7,10,13,14]. Space filling curves have the property that they map a set of cells in a higher dimensional volume to the unit interval $[0, 1]$. The space filling curves have the property that points close to each other in the embedding space are also typically close to each other along the curve. This property is extremely important for accessing large spatial data sets stored on external storage (hard disks). Due to advances in the density of modern hard disk drives, sequential I/O performance has become increasingly faster, while random seeks have remained almost constant over the last ten years. As a result, sequential I/O patterns are essential to deal with data sets measured in terabytes or larger.

In our approach we create a hierarchical partitioning of the L^3 volume of the simulation into a set of regular grid cells, organized into an octree of a predetermined depth. Each point is assigned to the leaf node on the octree it lies in. The tree nodes are labeled by the mapping provided by the particular space filling curve used (z-index, Peano, Hilbert, etc). This label, stored as an unsigned integer is represented as a column in the database. Since points in the same cell will have the same address, a standard B-tree index defined on this address alone will already speed up queries for points in the same cell.

As illustrated in Fig.2, space (assumed square) is subdivided in 2^D child cells, these cells are again subdivided and so on until an preset recursion level B is reached. The total number of cells in D-dimensional space will be 2^{BD}. Integer

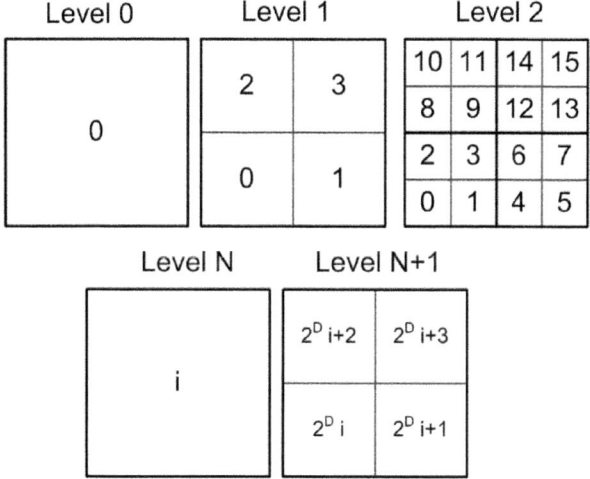

Fig. 2. A simulation box is recursively subdivided into $2^{B \times D}$ cells where B is the maximum recursion level. At each level indexes are assigned to cells based on the index of their parent cells as indicated in the bottom row. This assures that child cells are consecutively numbered and provides an ordering of the cells along a space-filling curve, in this example z z-curve.

indexes are assigned to the cells recursively as well. A cell at level N with index i will be subdivided in 2^D cells at level $N+1$ with indexes in $[2^D \times i, 2^D \times (i+1)-1]$. Clearly if a grid cell high in the hierarchy is contained in the volume, so will all of its children. Moreover, all these children will have consecutive index values, which implies that points organized according to this index can be retrieved very efficiently. This fact is an important factor in our query algorithm.

Precisely *how* the child indexes are subdivided over the child cells is a further detail that has some consequences for the performance. There are many different space filling curves known[13]. The algorithm in Fig.2 uses a constant orientation of the cells, corresponding to a Z-Curve, for our implementation we use a Hilbert curve ([13], Ch. 2).

The advantage of the Hilbert curve is that two cells adjacent on the index are also adjacent in the grid. The various families of space-filling curves have slightly different statistical properties, which have been quantified using their correlation function[11], finding the Hilbert curve to have the best clustering properties for spatio-temporal indexes such as our implementation (see also [5]). Nevertheless, our library is implemented in such a way that different space filling curves can be easily added as plug-ins.

2.3 The Query Algorithm

Our algorithm for finding points in a given query shape makes use of the nested hierarchy of grid cells. As illustrated in Fig. 3 it recurses into the simulation volume at the highest level, keeping boxes that are completely contained and recursing down into the children of boxes that partially intersect, or completely contain the query volume. Thus we iterate until some maximum resolution level which should be no larger than the level at which the points are stored in the database. At the end of the recursion we are left with a collection of boxes that are fully contained within the query volume, and a number with a partial overlap.

In pseudo code this can be written as follows:

```
1.   initialize FULL list
2.   initialize PARTIAL list
3.   set currentlevel = 0
4.   add SIMULATION_BOX to PARTIAL
5.   while currentlevel < MAX_LEVEL do
6.     initialize TEMP_PARTIAL list
7.     foreach box on PARTIAL do
8.       if(QUERY_VOLUME contains box) then
9.         add box to FULL
10.      else if(QUERY_VOLUME intersects box) then
11.        add children of box to TEMP_PARTIAL
12.    set PARTIAL = TEMP_PARTIAL
13.    currentlevel++
```

Points in boxes in the FULL list are all contained in the volume, points in the PARTIAL need to be checked individually whether they lie in the query volume. To find the points in these boxes we need to translate each box to a range query over the index column. At the deepest recursion level boxes have a single address, at higher levels they correspond to consecutive ranges as explained above.

To minimize the number of disjoint ranges, a simplification step may be taken that joins consecutive ranges from neighboring cells together. For the PARTIAL lists this simplification may join ranges that are not direct neighbors, across cells not in the PARTIAL lists. The maximum fraction of cells thus included is parametrized by a tolerance, which we have set to $1/3$, but needs further experimenting.

The algorithm above does not yet take into account the periodic boundary conditions. To do so, we modify steps 3. and 4. in the pseudo-code above. The first step now is to check whether the query volume is completely contained in the simulation box. If so, we continue as before. If not, we iteratively replicate the box in the directions in which the query volume extends beyond the boundaries of the box. This inverse recursion stops when a replicated box is created that completely contains the query volume. In the example in Fig. 1-b this is the outer box consisting of 4 copies of the original box as children.

This box now takes the role of the simulation box at the start of the recursion down, its children consisting of translated copies of the original simulation. These translation vectors must be taken into account when returning the final results, so that point positions can be suitably adjusted. For more details see the implementation notes in the next section.

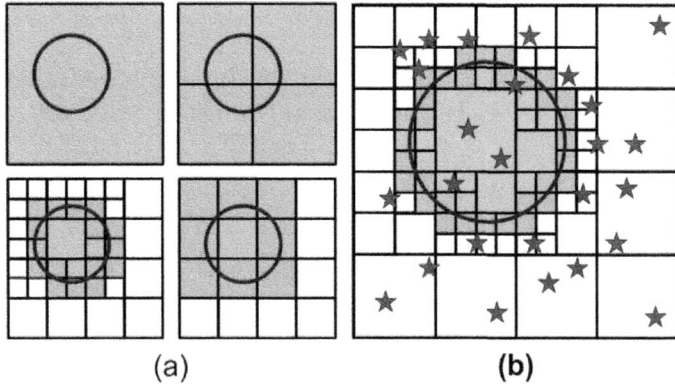

Fig. 3. The recursion from the simulation box down to leaf nodes is shown. The goal is to find an approximation to the query volume with boxes drawn from different levels in the hierarchy. The left side panel displays 4 recursion steps, in clock-wise direction from level 0 to 3. Pink cells indicate those that overlap the query region, but are not contained in it. Blue cells are fully contained and white cells are outside the volume. As soon as a cell is colored blue or white, the recursion stops. On the right panel the final result is shown, at level 4, with stars indicating the points that are the target of the query. Note, that we find blue cells from different levels and that the remaining pink cells will trace the boundary of the volume with arbitrary precision if the maximum recursion level is increased.

3   Query Shapes

Scientific questions on N-Body simulations often translate to a series of geometric constraints on the distribution of points. To accommodate a wide variety of studies, we formally introduce *query shapes* that users can create, store and manage within the same RDBMS, where the data reside. We implement routines and custom data types that make these shapes first-class citizens in the database and enable efficient spatial searching by connecting them to the relational engine's B-tree indexing mechanism. An expandable framework is implemented that defines the fundamental geometric primitives. From these simple shapes, the users can build more complex, custom query shapes that possess the same functional qualities.

3.1   Geometric Primitives

The usual suspects for the most basic building blocks in 3D include boxes, spheres and ellipsoids, cylinders, cones and frustums. Our collection of primitives is currently limited to the most basic ones but continuously growing with the demand of every new project. Among these, the most important is the axis-parallel box that plays a major role in the organization of the data. For example, a simulation has a bounding box, which in turn is recursively divided into smaller boxes down to the grid cells. This Box is also central to the space filling curve and the indexing.

The primitives implement a simple, yet, sufficiently versatile interface that is primarily used to describe their topological relation to other 3D objects. It is formally defined as

```
public interface IShape
{
    TopoPoint Contains(Point p);
    TopoShape GetTopo(Box box);
    Box GetBoundingBox();
}
```

The simplest is the containment test of a given point. Although this is conceptually a yes or no problem, it becomes a three-valued logic in the presence of numerical imprecisions. A point can be inside or outside of a shape, or so close to its boundary that a decisive answer is impossible. The TopoPoint enum captures this distinction and replaces a boolean return type. It becomes important when checking the containment of a grid cell. One can quickly test the vertices of a Box against a shape to draw conclusions about their relation to the given shape. For example, if all vertices of a box are inside a convex but one is outside, we know there is a partial overlap. If the outside point close to the boundary within the numerical imprecision, a full coverage can be established.

Similarly, the topological relation of shapes are formally enumerated in the definition of the TopoShape. They can be identical, touching, intersecting, disjoint, or contain each other. This interface is sufficient to enable the use of spatial indexing. As described in sec. 2.3, an IShape is intersected with successively more refined axis-aligned boxes (the octree nodes), using the TopoShape relation to determine which boxes are considered further. These boxes are mapped to a range on the space filling curve.

In addition to the normal TopoShape relations, we also find it useful to distinguish certain cases as *inconclusive*. This can also occur when an approximate algorithm is used for overlap calculations instead of a possibly more expensive exact test. When this happens for spatial indexing, we can still proceed assuming the worst case scenario and let the precise containment test decide for every point separately. This only costs us in performance but not in accuracy.

3.2   Building Composites

Our library supports the building of composite shapes from the primitives in a recursive manner. We implement the Boolean algebra via binary intersection, union and difference operations that are generic composite shapes. Figure 4 illustrates these composites on different primitives along with the results of their queries.

These generic composites implement the same IShape interface as the primitives. They do so by recursively using its member's identical interface. For example, a point is inside a union of two shapes, if it is inside any of them. Or a

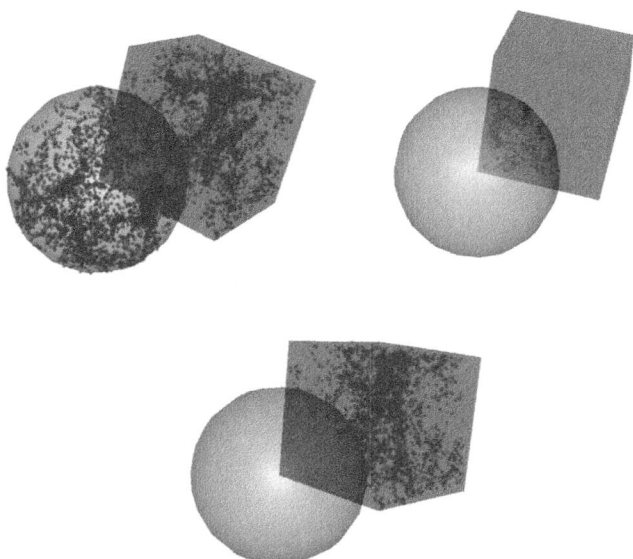

Fig. 4. Generic composite shapes implement Boolean operators to enable advanced querying. We show the union, intersection and difference of a box and a sphere. A composite of composites is defined recursively.

box is fully covered by an intersection if it is covered by all/both shapes. And so on recursively all the way down the binary tree to the primitives.

With the composite shapes there is only a practical limit to the queries that users can execute. We note, however, that even the most complicated query shapes will translate to a set of ranges on the space filling curve, hence the pre-filtering using the B-tree of the RDBMS will be unaffected. Performance penalty is only seen on the exact geometrical containment tests on a tiny subset of the data.

We have defined a simple grammar for representing composite query shapes as strings to facilitate their use in SQL. Examples are shown in section 5.1 below.

4   Representing Physics

A cosmological simulation evolves a model of the universe from early times to the present. It produces outputs at discrete points in time, called snapshots. In our database we need to have explicit representations of these simulations. We need to enumerate the simulation domain, the particles describing the matter in the universe, and the cosmological background within which they evolved. We also need some of the numerical artifacts such as the discrete time steps and the (periodic) boundary conditions. Here we briefly describe how these are represented in the software and describe some of the functions.

4.1   Simulations and Cosmology

A Simulation is a rectangular PeriodicBox, often a cube of size L. It contains particles which are represented as 3D Point objects with coordinates (X, Y, Z). We assume the simulation's box to be subdivided into an octree grid with maximum depth B, with 2^{3B} cells. The cells are indexed using a space filling curve. The type of curve is used is optional, but we have mainly tested the Hilbert curve, which we will use from now on. We will refer to the cell's index as PHKey. The simulation box can calculate for each Point what cell is it in.

A simulation is run in a specific cosmological background, represented by a Cosmology. The cosmology determines how internal units are translated to physical ones, what the physical time is corresponding to a particular discrete output snapshot, and can transform from time to redshift and to different possible cosmological distance measures. These features are especially important for light-cone calculations, see next section.

4.2   Light Cones

A typical simulation produces a large number of output snapshots at different time steps. Under the assumption of periodic boundary conditions, one can mimic real observations in the universe in so-called "light-cones". Due to the finite speed of light looking far in space requires to look back in time as well. In order to create such a light-cone observation we use snapshots at progressively earlier times, as well as replicated volumes. Fig. 5 shows such a query configuration. Here the yellow and blue parts of the cone represent different snapshots

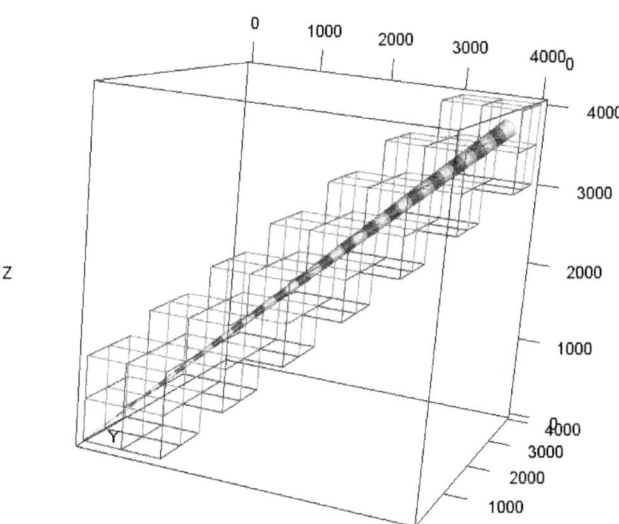

Fig. 5. A light-cone traversing a large number of replicated volumes. These have to be "observed" at different consecutive times, indicated by the yellow and blue cone segments.

in time, while the small cubes show the replicated simulation volumes. Furthermore, for a more precise calculation the positions at a look-back time in between snapshots could be properly interpolated, using the velocities.

Fig. 6 shows how the individual segments roughly cover the fundamental simulation box.

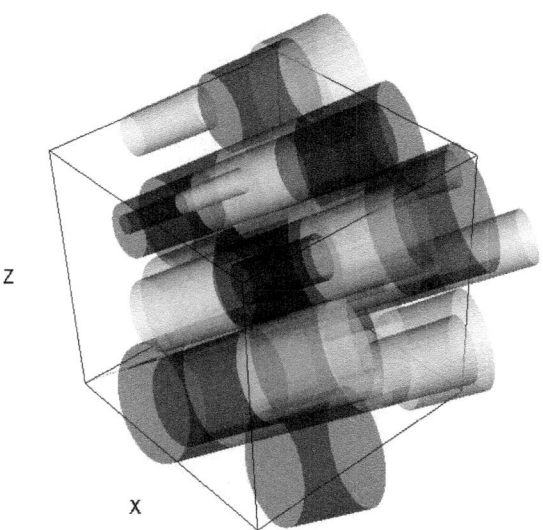

Fig. 6. The wrapped covering of a light-cone such as the one in Fig. 5 inside the fundamental box itself. Care must be taken that the simulation box is covered as completely as possible, with possibly minimal overlap of different segments.

A light cone *is a* `Cone` in the periodically replicated simulation box, where the depth of the cone represents not just a distance form an observer, but also a different time of the simulation. A `LightCone` is modeled therefore as a collection of cone *segments*, one corresponding to each snapshot that the light cone intersects, with minimum and maximum depth defined by the corresponding snapshot interval.

To calculate the sequence of segments we need a listing of the snapshots for a simulation and a translation of cosmic time to distance in the box. The snapshots will generally be stored in the database. The translation of time to distance is implemented using a Cosmology class [19].

A light-cone observation is now a spatial query using the collection of segments, but since the query needs to take into account the time for each segment, a LightCone is not a composite shape as defined above. The segments themselves are a primitive, `ConeSegment`, built from a `Cone` with an extra property, `minDepth`. Instead each segment is treated separately, leading to a typical cover as in Fig. 7.

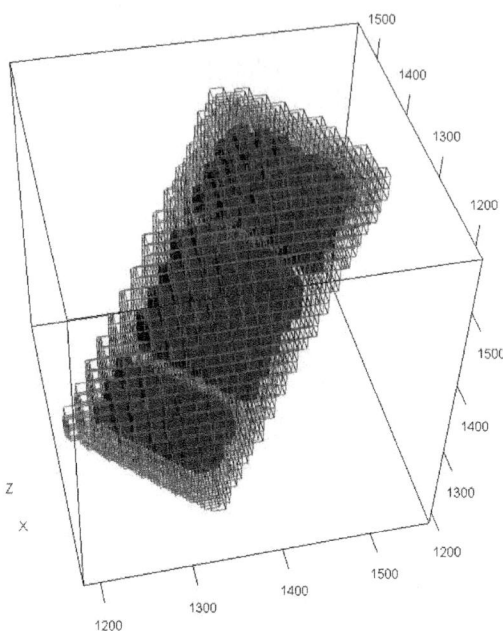

Fig. 7. Covers for consecutive cone segments from a light-cone. The red cells are completely contained in a segment, the partial cells are green. Each segment corresponds to a different time, hence we can not combine the segments into a composite shape, and partial cells appear on the separators between them.

5   The Implementation

Our first choice of programming language was C# for several reasons. Besides the rapid development cycles of the .NET programming model, special consideration was given to its integration with Microsoft's SQL Server. The RDBMS is tightly coupled with the runtime, which enables elegant (and fast) implementation of advanced extensions. Our solution decidedly follows a pattern that is straightforward to port to other architectures, operating systems or database engines. Recently we have completed a Java port on the Linux platform to enable similar functionalities in Oracle.

A modular design has been followed from the beginning of the project to provide programming interfaces at all levels. Dynamic libraries contain the core routines to deal with the 3D geometry and query shapes. On top of that there is an assembly for space filling curves and their related functionalities. Building on these packages a lightweight wrapper of SQL routines is added. Here we focus on the SQL extension that provides high-level access to the shapes and the indexes, which facilitate the efficient queries.

5.1   New Types in SQL

Our SQL Server API makes extensive use of User-Defined Types (hereafter UDTs) written in C#. These custom types can encapsulate the objects of our libraries, and directly expose the relevant methods to SQL users. While the shape primitives such as `Sphere`, `Box`, `Cone`, `ConeSegment` are directly made available for the advanced users, a high-level shape UDT is also provided for convenience. To create a sphere object in SQL with center $(X, Y, Z) = (1, 2, 3)$ and radius 10, one can simply write

```
DECLARE Sphere @s = Sphere::New(1,2,3,10)
```

or alternatively access it though the generic `Shape` type that wraps an `IShape` instance, see Section 3, as

```
DECLARE Shape @s = Shape::NewSphere(1,2,3,10)
```

The `Shape` UDT also provides access to composite shapes and implements the Boolean operations. For example, the union of two shapes is another shape that is constructed as

```
DECLARE Shape @u = Shape::Union(@s1,@s2)
```

A flexible string representation is also introduced to describe the shapes in a human readable format. A sphere is defined as `SPHERE[1,2,3,10]`, a box would be `BOX[1,1,1,2,2,2]` or their union as `UNION[BOX[..],SPHERE[..]]`; see the project website for the full grammar. Using the strings we can create a union by just an assignment,

```
DECLARE Shape @u = 'UNION[SPHERE[1,2,3,10],SPHERE[1,2,5,9]]'
```

All shape UDTs implement the method `ContainsPoint(@x,@y,@z)` just like the underlying objects implement it as part of the `IShape` interface. It returns 1 if the given point is contained in the shape, and 0 otherwise. Also these shapes do not take into account periodic boundary conditions.

Other UDTs are specific to the simulations and their cosmological hypotheses. A `Simulation` is primarily defined by the box of its volume and the cosmological model. The `Cosmology` UDT captures the physics. It is instantiated with model parameters, e.g., Ω, Ω_{baryon}, Ω_Λ and the Hubble constant. It has methods for calculus in space-time, e.g., the co-moving distance of an object at given redshift. Functions like these are necessary to determine the apparent properties if objects, when comparing to real observations. In addition, we also define the resolution of the space filling curves.

5.2   New Spatial Index in SQL

A query shape is approximated by the union of cells along the space filling curve, see Section 2. Such covers are returned by custom SQL routines as a table. Each row represents an interval along the space filling curve. A User-Defined Function (hereafter UDF) is introduced to make this more convenient `fSimulationCover(@sim Simulation, @query Shape)`. The function returns a table of the following columns:

```
KeyMin bigint, KeyMax bigint, FullOnly bit,
ShiftX real, ShiftY real, ShiftZ real
```

The `KeyMin` and `KeyMax` columns define the range of grid cells. The value of `FullOnly` indicates whether the cells in the range are known to be completely contained in the shape (1) or not (0). The `Shift*` columns represent a translation vector to be applied to the resulting points and allows periodic boundary conditions to be taken into account as described in sec. 2.3. Particles found in a cell in the range should be translated by this vector before they can be assumed to be in the query volume. To quickly select some particles within a shape @shp from a simulation @sim, one could just use the inner cover, `FullOnly=1`. The following query runs exceptionally fast and only uses the B-tree index on the PHKey to fetch the results.

```
SELECT p.ID           --  Return the particle IDs
FROM Spatial3D.fSimulationCover(@sim,@shp) AS c
    INNER JOIN Particles AS p
        ON p.PHKey BETWEEN c.KeyMin AND c.KeyMax
WHERE c.FullOnly = 1              --  Inner cover
```

To select *all* the particles we have to include the cells on the boundary, the PARTIAL cells from the pseudo-code in sec. 2.3. These are given by `FullOnly=0`, and to perform the precise geometry cut we need to include the shape's `ContainsPoint()` method

```
SELECT p.ID           --  Return the particle IDs
FROM Spatial3D.fSimulationCover(@sim,@shp) AS c
    INNER JOIN Particles AS p
        ON p.PHKey BETWEEN c.KeyMin AND c.KeyMax
WHERE c.FullOnly = 0              --  Boundary cover
    AND @shp.ContainsPoint( p.X+c.ShiftX,
                            p.Y+c.ShiftY,
                            p.Z+c.ShiftZ ) = 1
```

We note the use of the shift vector when testing the particles' positions for containment.

The division in the FULL and PARTIAL lists of key ranges is currently being investigated. The potential drawback to this approach is that the boundary volume can become fragmented and, hence, require lots of short intervals in the PARTIAL list to be represented. This could slow down the query more than the speed obtained by the reduction in the number of explicit particle containment tests. Getting the best possible performance is a delicate balancing act that seems to depend on the actual problem. Often, however, whichever method is selected, the queries run very fast. The performance of our solution is discussed later in Section 6.

5.3   Remote Deployment through SQL

One of the most attractive features of Microsoft's SQL Server has been its extensibility. Since the release in 2005 the engine hosts a Common Language Runtime

(CLR) and stores managed code as part of the database. The .NET assemblies are created from the DLLs of the intermediate language. We encapsulate all required libraries into SQL scripts using hexadecimal representation. This way we can deploy the spatial infrastructure across any SQL connection even remotely. Our software package also comes with batch scripts that execute SQL installers with parameterized target, which enables seamless deployment to a cluster of databases. While we can easily deploy our solution to all production databases, the first use cases such as the Millennium Database use a dedicated database called fSpatial3D where all the code was deployed once. Users explicitly reference the routines in that database in all queries, which simplifies the update process of the library and allows multiple version to coexist.

5.4   Application to the Millennium Database

The Spatial3D library was first deployed in the "Millennium Database"[3][9]. That database contains data products derived from the Millennium simulations [15,3]. The Millennium simulation contained 10 billion particles in a box of 685 Mpc on the side. Altogether 64 snapshots were calculated and stored, for a total of O(20) TB. The database contains currently only derived products such as halo and galaxy catalogs, the tables contain O(700-1000) million objects each.

For spatial queries the relevant columns are (snapnum,phkey,x,y,z) and these have been combined in a B-Tree index. The Spatial3D library is loaded in a separate database, fSpatial3D and to have access to the UDTs, users have to either connect directly to that database, or start all queries that make use of these types explicitly with a use fSpatial3D statement.

The schema spatial3d contains the UDTs along with various cover functions, and sims holds several utilities for the different simulations stored in the database. For example, parameters of the Millennium simulation are conveniently returned by the following function,

```
CREATE FUNCTION sims.Millennium()
RETURNS spatial3d.Simulation AS
BEGIN
RETURN spatial3d.Simulation::New('BOX[0,0,0,500,500,500]',
        spatial3d.Cosmology::New(0.25,0.75,0,0.73,0.04,-1),8)
END
```

Queries typically look for co-located galaxies, e.g., in spheres of 10Mpc radii around the 10 most massive groups. The following command selects objects from Guo database's Millennium table using separate friends-of-friends collections in a table FoF. This real-life example is written using a Common Table Expression (CTE) to dynamically create set of query shapes around the groups.

---

[3] See http://www.mpa-garching.mpg.de/millennium and in particular http://www.g-vo.org/Millennium

```
WITH QueryShapes(FoFID,Sph) AS
(
    SELECT TOP 10 FoFID,
           spatial3d.Shape::NewSphere(x,y,z,10)
    FROM MField.dbo.FoF
    WHERE snapnum = 63
    ORDER BY m_tophat200 DESC
)
SELECT DISTINCT s.FoFID, g.GalaxyID, g.x+r.ShiftX AS x,
                                     g.y+r.ShiftY AS y,
                                     g.z+r.ShiftZ AS z
FROM QueryShapes s
    CROSS APPLY spatial3d.fSimulationCover
                      (sims.Millennium(), s.Sph, 8) r
    INNER JOIN Guo2010a.dbo.MR g
        ON g.PHKey BETWEEN r.KeyMin and r.KeyMax
WHERE g.snapnum = 63                    -- Specific snapshot
  AND ( (r.fullonly=1)                  -- Inner cover
        OR
        (r.fullonly=0                   -- Boundary cover
        AND s.sph.ContainsPoint( g.x+r.ShiftX,
                                 g.y+r.ShiftY,
                                 g.z+r.ShiftZ ) = 1)
      )
```

The query returns in a few seconds delivering about 85,000 galaxies out of more than 1 billion objects in the database and renders a web page with the new results. Most of this time is actually spent in transferring the results over the web to the browser.

6   Discussion

The Spatial3D library presented in this paper has so far been successfully deployed in MS SQLServer databases containing very large simulations of cosmological systems as well as homogeneous turbulence. SQLServer does not have native support for spatial queries in three and four dimensional spaces, hence building our own library was the only option. This was greatly facilitated by the Common Language Runtime, which provides an almost seamless integration of complex procedural code and the T-SQL query language.

The performance has been satisfactory. Queries for objects in a rectangular box in a table with 600 million objects see response times < 0.1 sec, allowing support for an interactive graphical "click-and-query" interface. It is possible to tune the performance in a number of ways. Space filling curves other than the Hilbert curve can be plugged in to the code without great problems. The simplification step, in which cell ranges are merged, sometimes across outer cells can be modified using a tolerance parameter, decreasing the number of seeks on the disk drives. The desired recursion depth down to which a query shape should be approximated can be explicitly set. The optimal choice of these settings depends on the hardware characteristics of the system under consideration, and the details of the typical query workload.

The main issue with our framework is that it is external to the main database engine. The reliance on user defined functions backed by procedural code makes it difficult for the database engine to incorporate statistics on the spatial distribution of the points in the design of optimal query plans. Particularly when columns from outside the B-Tree index with spatial columns are requested, the default query plans can be very inefficient. It is often possible however to fix them with simple changes to the query, such as substituting a forced LEFT OUTER JOIN for an INNER JOIN.

A smaller issue, at least for SQLServer is the lack of support for table-valued methods or polymorphism of function arguments. This makes for a slightly less elegant interface than a proper object-oriented approach would offer.

An initial port to Java for deployment in Oracle (version 11g2) has been undertaken. Oracle's Spatial solution has support for 3D queries, but turns out to be unsuitable for time dependent simulations, where a time coordinate must be constrained together with the spatial columns. The required mixed indexes are not supported. Furthermore, periodic boundary conditions need separate coding, as do light-cone queries. The integration of Java in Oracle is not as elegant as the CLR solution in SQLServer. Nevertheless initial tests indicate that performance is comparable, showing that the main time is spent in the disk I/O rather than the in-memory calls to the library.

We will publish our solution to the community under a BSD license. We plan to extend it with more query shape primitives, such as frustums, cylinders and ellipsoids and to querying for more general shapes than point objects.

Notes and Comments. We thank Volker Springel for providing us with the C-version of the Hilbert library. AS and TB are supported by a grant from the Gordon and Betty Moore Foundation, and NSF grants ITR-AST-0428325, OCI-104114 and OCI-106256. The Millennium Simulation databases used in this paper and the web application providing online access to them were constructed as part of the activities of the German Astrophysical Virtual Observatory.

References

1. Angulo, R., Springel, V., White, S.D.M., et al.: (in preparation 2011)
2. Bayer, R.: The Universal B-Tree for Multidimensional Indexing: General Concepts. In: World-Wide Computing and Its Applications 1997 (WWCA 1997), Tsukuba, Japan, pp. 198–209 (1997)
3. Boylan-Kolchin, M., Springel, V., White, S.D.M., et al.: Resolving cosmic structure formation with the Millennium-II Simulation. Monthly Notices of the Royal Astronomical Society 398, 1150–1164 (2009)
4. Diemand, J., Kuhlen, M., Madau, P., et al.: Clumps and streams in the local dark matter distribution. Nature 454, 735–738 (2008)
5. Faloutsos, C., Rong, Y.: DOT: A Spatial Access Method Using Fractals. In: Proceedings of the Seventh International Conference on Data Engineering, Kobe, Japan, April 8-12, pp. 152–159 (1991)

526   G. Lemson, T. Budavári, and A. Szalay

6. Jagadish, H.V.: Linear Clustering of Objects with Multiple Atributes. In: Proceedings of the 1990 ACM SIGMOD International Conference on Management of Data, Atlantic City, NJ, May 23-25, pp. 332–342 (1990)
7. Kamel, I., Faloutsos, C.: Hilbert R-tree: An Improved R-tree using Fractals. In: Proceedings of the 20th International Conference on Very Large Data Bases (VLDB 1994), Santiago de Chile, Chile, September 12-15, pp. 500–509 (1994)
8. Lawrence, E., Heitmann, K., White, M., et al.: The Coyote Universe III. Simulation Suite and Precision Emulator for the Nonlinear Power Spectrum, Astrophysical Journal 713, 1322–1331 (2010)
9. Lemson, G., and the Virgo Consortium: Halo and Galaxy Formation Histories from the Millennium Simulation: Public release of a VO-oriented and SQL-queryable database for studying the evolution of galaxies, in the eprint arXiv:astro-ph/0608019 (2006)
10. Markl, V.: MISTRAL: Processing Relational Queries using a Multidimensional Access Technique Ph. D. Thesis, TU München (1999), http://mistral.informatik.tu-muenchen.de/results/publications/Mar99.pdf
11. Moon, B., Jagadish, H.V., Faloutsos, C., Saltz, J.H.: Analysis of the Clustering Properties of Hilbert Space-filling Curve. IEEE Transactions on Knowledge and Data Engineering 13(1) (2001)
12. O'Mullane, W., Li, N., Nieto-Santisteban, M.A., Thakar, A., Szalay, A.S., Gray, J.: Batch is back: CasJobs, serving multi-TB data on the Web, Microsoft Tech Report MSR-TR-2005-19 (2005)
13. Sagan, H.: Space-Filling Curves. Springer, Heidelberg (1994)
14. Samet, H.: Foundations of Multidimensional and Metric Data Structures. Morgan-Kaufmann, San Francisco (2006) ISBN 0-12-369446-9
15. Springel, V., White, S.D.M., Jenkins, A., et al.: Simulations of the formation, evolution and clustering of galaxies and quasars. Nature 435, 629–636 (2005)
16. Springel, V., Wang, J., Vogelsberger, M., et al.: The Aquarius Project: the sub-haloes of galactic haloes. Monthly Notices of the Royal Astronomical Society 391, 1685–1711 (2008)
17. Szalay, A., Gray, J., Thakar, A., Kuntz, P., Malik, T., Raddick, J., Stoughton, C., Vandenberg, J.: The SDSS SkyServer Public Access to the Sloan Digital Sky Server Data. In: Proc SIGMOD 2002 Conference, pp. 570–581 (2002)
18. Szalay, A.S., Kunszt, P., Thakar, A., Gray, J., Slutz, D., Brunner, R.: Designing and Mining Multi-Terabyte Astronomy Archives: The Sloan Digital Sky Survey. In: Proc. SIGMOD 2000 Conference, pp. 451–462 (2000)
19. Taghizadeh-Popp, M.: CfunBASE: A Cosmological Functions Library for Astronomical Databases Publications of the Astronomical Society of the Pacific, vol. 122, pp. 976–989 (2010)

Histogram and Other Aggregate Queries in Wireless Sensor Networks⋆

Khaled Ammar and Mario A. Nascimento

Department of Computing Science,
University of Alberta, Edmonton, Canada
{kammar,mn}@cs.ualberta.ca

Abstract. Wireless Sensor Networks (WSNs) are typically used to collect values of some phenomena in a monitored area. In many applications, users are interested in statistical summaries of the observed data, e.g., histograms reflecting the distribution of the collected values. In this paper we propose two main contributions: (1) an efficient algorithm for answering *Histogram* queries in a WSN, and (2) how to efficiently use the obtained histogram to also process other types of aggregate queries. Our experimental results show that our proposed solutions are able to substantially extend the lifespan of the WSN.

1 Introduction

A typical Wireless Sensor Network (WSN) consists of nodes, equipped with sensors, distributed in an area and connected, via a tree-like topology, to a base station. Typically, WSN nodes have limited resources in terms of power, CPU and memory. The base station is a full-fledged computer system with light limitations on memory, CPU, or bandwidth. Battery lifetime is considered the most important resource in WSN nodes as the required power for transmission is significantly higher than the required power for data processing in a WSN node [7]. Thus, it is vital that query processing algorithms for WSN are energy-efficient.

WSNs are typically used to observe some phenomena about a monitored area and are becoming common in many applications domains [4,2]. Simple aggregate queries like *Max*, *Average* and *Sum* are sufficient for a large number of applications where a high-level (summary) view of the data suffices, e.g., when looking for abnormal behavior. However, more complex aggregates queries such as *Histogram* provide a broader picture and are mandatory for many applications. For example, in the Electronic Nose project [2], any single value is not important by itself, but, the distribution of the sensor values is used as a chemical signature to classify the material as being safe or unsafe.

We assume that a WSN has N nodes $s_j \in S$ $(1 \leq j \leq N)$ spread in a monitored area. Each node s_j in S periodically measures a value v_j. In fact, every value has an associated timestamp, however in order to lighten the notation

---

⋆ This research was partially supported by NSERC Canada.

J.B. Cushing, J. French, and S. Bowers (Eds.): SSDBM 2011, LNCS 6809, pp. 527–536, 2011.
© Springer-Verlag Berlin Heidelberg 2011

we do not denote it unless necessary. Nodes are connected to the base station by a routing tree, where the base station is the root and, they can reach the base station by multi-hop routing through other nodes. We assume that the connection between nodes are reliable (no link failure), and we focus on the data aggregation problem only. Users are connected to the WSN through the base station where they can submit queries and collect the respective answers.

We define a *Histogram* query as: $Q = (Lb, Ub, b_1, b_2, b_3, ...b_B, epoch)$, where *epoch* is the time lapse between any two consecutive histogram answers. The lower and upper bound values of the measured phenomena are Lb and Ub. Each b_i is one of B bins in the *Histogram* query and it is defined as $b_i = [Lb_i, Ub_i[\; \forall 1 \leq i < B$ and $b_B = [Lb_B, Ub_B]$. Furthermore, $Ub_i \leq Lb_j \; \forall i < j$ and $\bigcup_{1 \leq i \leq B}\{b_i\} = [Lb, Ub]$ and $Lb_1 = Lb$ and $Ub_B = Ub$. The answer for a *Histogram* query is $H = (h_1, h_2, ..., h_B)$, where $h_i = |\; \{(s_j, v_j) \mid Lb_i \leq v_j < Ub_i, s_j \in S\} \;|$. Naturally, a sensor's value v_j may change at any epoch and so does the query answer.

This paper presents two main contributions. The first is an efficient distributed algorithm to answer *Histogram* queries in WSNs. This algorithm requires less than half amount of energy used by the classical TAG algorithm [6]. Our second contribution is to show how to answer other aggregate queries using histogram at the cost of a very small overhead on the WSN.

The rest of this paper is organized as follows: Section 2 reviews how TAG answers a *Histogram* query and details our proposed algorithm. How to compute approximate as well as exact answers for other aggregate queries using a histogram as a starting point is discussed in Section 3. Section 4 presents our experiments and Section 5 discusses briefly the related work . Finally, Section 6 concludes the paper and presents a few future directions for further research.

2   In-Network Algorithms for *Histogram* Queries

A straightforward technique to build a histogram is to periodically gather all values from all sensors at the base station and then build a histogram. The classical TAG algorithm decreases the number of required messages extensively comparing to the straightforward technique [6]. The authors define a model to answer aggregate queries in WSNs using an in-network approach. The process can be visualized as a routing tree where the base station is the root and nodes send their aggregate answer as messages up the tree towards that root. This process continues until the base station receives all aggregate answers and can construct the query answer.

TAG works for a *Histogram* query as follows. Each sensor should send exactly one message on every epoch but the message sizes (in bits) are different depending on the node type. The size of a message from a leaf node is $\log_2 B + \log_2 N$ bits (bin id + number of values in a bin), whereas the size of a message from a non-leaf node depends on the values distribution on the histogram and is bounded by $O(\log_2 N \times B)$ bits.

2.1   Histogram Incremental Updates (HIU) Algorithm

A value does not change a histogram answer if its change was within its current bin's lower and upper bounds. This histogram property motivates us to look closer into the histogram construction process. Instead of sending its data every epoch, a sensor can build an update message based on the previous round's data. In our algorithm, sensors receive incremental histogram updates, merge them together and then forward to their parents, and so forth. The process continues until the histogram in the base station is updated.

We assume that each sensor node's data consists of two objects: its value v_j and the histogram summarizing its subtree H_j. In-node caching is an essential component in the HIU algorithm. Thus, each node also caches its value and its subtree's histogram from the previous round in $\tilde{v}_j$ and $\tilde{H}_j$, respectively.

The HIU algorithm works as follows. Every round, each node s_j caches its value and its histogram by copying v_j into $\tilde{v}_j$ and H_j into $\tilde{H}_j$. Then, each leaf node updates its histogram H_j based on its value v_j while an intermediate node updates its histogram H_j based on its value and also based on the other messages received from its children. If the updated histogram in H_j is different than cached histogram $\tilde{H}_j$ in any node, this node should send a single message. There are three message types in HIU: (1) A leaf node may send its value only. An intermediate node sends either (2) its updated histogram H_j, or (3) an update message U_j summarizing the difference between H_j and $\tilde{H}_j$, whichever smaller bitwise. A more complex compression could be implemented for this function, e.g., [9,10]. A detailed discussion about compression algorithms in WSN, however, is beyond the scope of this paper.

The update message U_j is a set of pairs (k, u_k), where k is a bin id and $u_k = h_k - \tilde{h}_k \ \forall h_k \in H_j$ and $\tilde{h}_k \in \tilde{H}_j$. Recall that a histogram $H = (h_1, h_2, ..., h_B)$. An update value, u_k, could be positive or negative but cannot equal *zero*. If an update value equals *zero* because this bin was not changed, it is automatically removed from the update message.

Received update-messages in any non-leaf node may cancel each other in which case nothing is sent forward. For example, consider an intermediate node C that has two subtrees, A and B. Subtree A has x nodes where their value moved from bin b_k to b_l. On the other hand, the subtree B has $x - 1$ nodes where their values moved from bin b_l to b_k. If these two updates are merged together, then subtree C has only one value moved from b_k to b_l. Moreover, if node C's value moved from b_l to b_k, then C should not send any update to its parent at all.

3   Other Aggregate Queries

A histogram provides a broad picture for values in the WSN and is a starting point for more statistical analysis. Occasionally, a user might like to know more specific information (e.g. *Max* or *Average*) about those values represented by the histogram. We can compute approximate and exact answers for several aggregate queries using a previously obtained histogram in the base station.

The approximate solutions have bounded accuracy levels [1] and the exact solutions can be computed with very low extra overhead on the WSN.

Communication devices in some WSN mandate the sensor to send messages of fixed size only [8]. In this case, sending less information will not decrease the energy consumption because all buckets should have the same size. These idle bytes can be used to send extra information, with no extra cost, in order to facilitate computing the exact answer. We use this strategy to compute exact answers for *Max, Min, Sum*, and *Average* queries.

While obtaining a histogram answer, each node can collect required information and aggregate them to facilitate computing the exact answer in the base station. The only change to the previously described HIU algorithm is that, if an exact answer is required, leaf nodes values should be sent even if $H_j = \tilde{H}_j$. The kind of required information depends on the required aggregate query. For example, in case of *Max* (or *Min*) queries, all intermediate nodes who construct a histogram, H_j, should also report information about the maximum (or minimum) value in their subtree.

Because the base station (and all intermediate nodes) knows the exact answer for *Count*, it can compute the exact result for *Average* if the exact *Sum* is available. The exact answer for *Sum* can be computed if each intermediate node sends the total sum of its subtree while leaf nodes send their own values only.

4   Performance Evaluation

For our simulations we implemented TAG and HIU assuming both of them are using a Shortest Path (logical) Tree (SPT) for the underlying routing tree. We make the following assumptions about the required storage: (1) an observed/sensed node value consumes 2 bytes, (2) a complete histogram size depends on the number of bins, i.e., it requires $2 \times B$ bytes, where B is the number of the bins in the histogram, and (3) a pair in an update message requires 3 bytes, 1 for the bin id and 2 for the update value.

We investigate our algorithms with respect to five parameters (Radio range R, histogram size in terms of number of bins B, average amount of change in sensor's value δ, the probability that a sensor's value change ρ, and number of nodes in the WSN N). Table 1 has a list of all tested values for all parameters. While testing one parameter, we use the default value (denoted in bold) of all other parameters.

We used two datasets, a synthetic and a real one. Due to space constraints we do not show the results obtained using the real dataset[1]. Nonetheless, the results we obtained using the real dataset are available in an extended version of this paper [1].

Our synthetic dataset consists of N nodes uniformly distributed in an area of $200m \times 200m$. The values of all sensors are initialized uniformly between 1 and 2^{16}. In each round, a sensor's value could change with a probability ρ. In case of change, a sensor value is increased by an exponential random variable (equally

---

[1] http://www.select.cs.cmu.edu/data/labapp3/index.html

Table 1. Studied parameters and their values (default values in **Bold**)

Parameter	Used Values
R (WSN node's radio range)	20, **30**, 40, 50, 60
B (Histogram size in terms of number of bins)	5, 10, **20**, 40, 60
δ (Average amount of change)	1%, 25%, **50%**, 75%, 100%
ρ (Probability of change)	1%, 25%, **50%**, 75%, 100%
N (Number of Sensors)	1000, 2000, **3000**, 4000, 5000

likely to be negative or positive). The average of the exponential random variable is δ% of 2^{16}. We assume that all sensors capable of sensing values between 0 and 2^{16} only. If a value exceeds that range in either direction, it is assumed to be either 0 or 2^{16}, respectively. All Figures show the average values obtained over 20 runs. During each run, the sensor locations are randomly distributed and the base station is randomly selected among one of the sensors. In order to ensure a fair comparison, both TAG and HIU use exactly the same setup.

Since the main typical goal within the realm of WSN research is the minimize energy consumption we use network lifetime as the performance indicator. Network lifetime is counted in number of rounds until the first node dies. In all our experiments we assume that each battery's initial budget is $30mJ$. Energy consumption is calculated after [5], i.e., $E_t = S + t \times b \times d^2$ and $E_r = r \times b$, where $S = 50$ nJ is the setup cost to send any message, $t = 10$ pJ and $r = 50$ nJ are the required amount of energy to send or receive one bit for one meter, respectively. The message size in bits is b, while the euclidean distance (in meters) between the sender and the receiver is d.

4.1   *Histogram* Queries

Figure 1 shows the HIU and TAG performance when changing δ, ρ, R and B. Because TAG performance does not depend on changes in sensors' values, a network using TAG algorithm died after about 2700 rounds regardless of the change probability (ρ) or amount of change per round (δ). Figures 1(a) and 1(b) show that HIU performs better when the changes of nodes' values happen less frequently because it caches the result of the previous round and send updates only, if required. For higher update frequencies (ρ) or update with large changes (δ), HIU performance becomes stable. The reason for that is two fold. First, HIU selects whether to send an update message or a histogram message. This arbitration guarantees a worst case scenario if more than half of the bins are changed. Second, because the node's histogram is constructed in network, many of these changes are not forwarded as they can cancel each other in the early stages of the routing tree.

Figure 1(c) shows that both TAG and HIU perform better when the radio range is small. This seems to contradict the following intuition: the smaller the radio range the more hops are required from leaf nodes to reach the base station, the more messages and then the shorter network lifetime. In reality, each node sends a message to reach all the other nodes within its range regardless of the

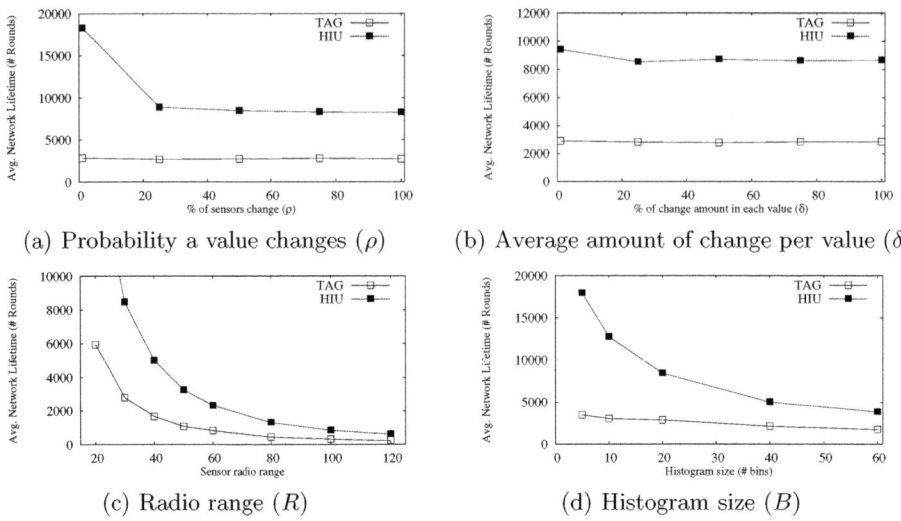

(a) Probability a value changes (ρ)   (b) Average amount of change per value (δ)

(c) Radio range (R)   (d) Histogram size (B)

Fig. 1. Network Lifetime analysis

real distance between the sender and the receiver. The larger the radio range the larger the energy consumed, because energy consumption is based on how far a message can reach and is not based on the euclidean distance between the sender and receiver. The Figure shows that even though the performance of both HIU and TAG is better when the radio range is smaller, HIU multiplies the network lifetime three or four times comparing to TAG.

Figure 1(d) is an evidence that HIU can still multiply the network lifetime by at least a factor of two as the number of bins increases. A larger number of bins means a higher probability that the number of changed bins increases and then HIU performs deteriorates. However, TAG requires all intermediate nodes to send their partial state regardless of number of bins, i.e., TAG also performs worse when increasing number of bins.

More experiments on [1] show that HIU can scale efficiently and handle WSNs with large number of nodes better than TAG. We basically increase the network density with N. HIU has the same performance regardless of the number of sensors in the field. TAG's performance decreased dramatically because the more sensors in the field the higher probability of occupying all histogram bins. Recall that TAG sends the bin's count if the bin is occupied by one or more values. On the other hand, because of our encoding, the values distribution does not influence HIU performance. The key factor is the how frequent values change and by how much.

4.2 Exact Answers for Other Aggregate Queries

Regardless of the algorithm used to construct a histogram in the base station, a histogram allows computing approximate answers for several other aggregate queries without any overhead (as discussed in Section 3).

Because the main target of our experiment now is to study the HIU overhead cost for computing an exact answer, we now use the average amount of bytes sent per sensor per round as our performance indicator. Every round, the total number of sent bytes from all nodes during all previous rounds are calculated and then divided by number of sensors. For brevity we constrain our presentation to a particular type of query, namely *Max* queries.

Based on [6], every sensor should, due to in-network aggregation, send exactly 2 bytes to collect the maximum value using TAG. HIU collects the *Max* information while constructing the histogram. HIU's performance depends on the amount of changes in the network because it uses in-network caching and send data to update this cache. Initially HIU requires more bytes to be sent, but as time goes, the average total number of sent bytes per round is decreased and eventually reaches a steady state. Recall that the first round in HIU consumes a large amount of energy due to sending the largest amount of bytes comparing to other rounds because there is no cached information.

Figure 2 shows the amortized analysis for TAG and HIU algorithms in computing the exact *Max* using four parameters: δ, ρ, R and B. The main goal is to show that HIU can outperform TAG in the long run, after a few rounds.

Figure 2(a) shows the influence of change probability on HIU. If the probability is 100% then HIU needs one extra byte from each sensor (on average) per round. As the probability gets smaller, the overhead decreases. The figure shows that lower values of ρ leads to a smaller HIU cost but TAG's performance stay the same. If the probability is 1% only, not shown in the figure, HIU outperforms TAG by about 1.8 bytes which means 90% less bytes than TAG. It is worth mentioning that HIU's cost includes, also, constructing an accurate histogram in the base station while TAG (in this experiment) computes the maximum value only. The histogram in the base station offers computing approximate answers for many other aggregate queries. This means, if the target is computing the *Max* query only, then HIU is better only if sensors change their values not very often ($\rho \leq 40\%$).

In Figure 2(b) we assume that $\rho = 50\%$ and investigate the influence of the amount of change (δ). If δ is very small (1%) HIU will outperform TAG after 3 rounds only. If δ is very large (100%), HIU ties with and slightly outperforms TAG. Recall that a sensor can sense a specific range of values. If the value is bigger than the maximum value, a sensor will report its maximum limit. If the average amount of change is 100% then there is a high probability that all sensors end up detecting only the maximum or minimum limits because the change could be positive or negative. It is clear that, in the long term, the amount of change does not have a significant influence on the results in Figure 2(b). Regardless of the value of the amount of change δ, the average number of bytes is very close to 2.05. The exception for this conclusion is when δ is very small, e.g., $\delta = 1\%$ because changing a sensor value by 1% on average will unlikely change its bin in the histogram (assuming the bin's width is resonable) and then unlikely to cause a sensor to send any data.

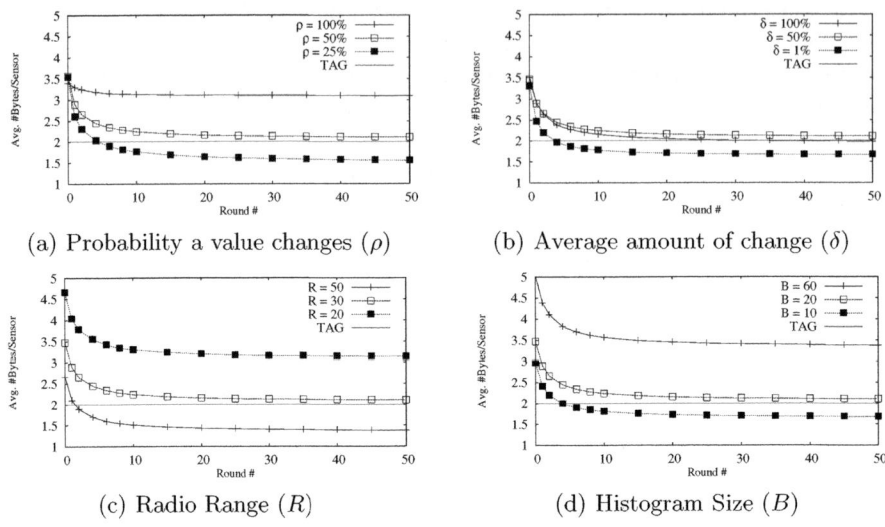

Fig. 2. Cost of running exact-Max (X-axis is number of rounds)

HIU performance depends on the bin size (number of bins) because the number of values in smaller bins is more likely to change every epoch. Although the error bound of all approximate answers get worse when bin size increase, the HIU algorithms perform better while computing exact *Max*. Figure 2(d) shows that decreasing number of bins can make HIU outperforms TAG very early, round = 3, even if the probability of change and amount of change are both 50%. TAG outperforms HIU when δ or ρ equals 50% but if we decrease number of bins, e.g. B=10, HIU wil outperform TAG. The major fraction of the HIU cost is paid to construct the histogram. Decreasing histogram size decreases the histogram overhead but increases the *Max* overhead $(log(Ub_i - Lb_i))$. However, this overhead is already very small comparing to the *Histogram* cost.

The sensor's radio range influences the logical tree structure. A short radio range requires the WSN to build a logical tree with larger depth than a long radio range. Increasing the average number of hops for sensors to reach the base station does not have any influence on TAG because every sensor will send a single message of fixed size (2 bytes) any way. In HIU, the message size is usually bigger than TAG. It also depends on the values and sensors distribution. The shorter the radio range, the more the number of hops which requires HIU to send more bytes. Figure 2(c) shows that increasing the radio range makes HIU's total cost less than TAG's total cost after only 3 rounds.

Even though a graph is not shown we also investigated the influence on the number of nodes. Network density depends on the number of sensors and it has no influence on the TAG algorithm to compute *Max*. In all cases, each sensor should report its value. In the case of HIU, the more sensors available in the area the more opportunities to save and decrease the amount of sent messages.

5   Related Work

There has been not much work done in the literature to construct a histogram of WSN values since Madden et. al. proposed TAG algorithm in 2002 [6]. Chow et.al. proposed an algorithm to construct a spatio-temporal histogram [3]. The main idea is to construct an approximate spatio histogram that is updated with every time any sensor reading reaches the base station. This approximate *Histogram* is used for location monitoring. The authors proposed a basic and adaptive approach to construct an approximate histogram in the base station. The main idea is collecting values at the base station and then construct the histogram. The energy saving comes from constructing an efficient approximate histogram instead of an exact one. Since our algorithm construct an exact histogram using an in-network algorithm and we do not require all values to be sent to the base station, we did not compare their approach with HIU.

6   Conclusions and Future Work

In this paper we proposed a new algorithm (HIU) that uses in-network aggregation and in-node caching to reduce the energy consumption for constructing a *Histogram* query. Obtaining a histogram in the base station helps in computing bounded approximate answers for other aggregate queries. Moreover, we proposed algorithms that use HIU to compute exact answers for these aggregate queries. HIU outperforms the TAG algorithm (current state-of-the-art to answer a *Histogram* query) multiplying the network lifetime, on average, about three times. In the long term, HIU outperforms TAG's algorithm in computing the *Max* query if the amount and/or probability of changes in observed values are low. Nonetheless, we have shown that a small histogram size can compute an exact answer for a *Max* query less expensively than TAG.

We would like to develop a cost model for computing the cost of *Histogram* queries. Moreover, in our work, we assume that network communication between sensors is perfect with no losses. In real world applications, this is not very realistic. In-node caching can be useful to help reduce the impact of any network failure. How to make HIU able to reduce the impact of any network failure is another issue that we would like to address in the future.

References

1. Ammar, K., Nascimento, M.A.: Histogram and other aggregate queries in wireless sensor networks. Technical Report TR 11-03, Department of Computing Science, University of Alberta (2011)
2. Burl, M., Sisk, B., Vaid, T., Lewis, N.: Classification performance of carbon black-polymer composite vapor detector arrays as a function of array size and detector composition. Sensors and Actuators B: Chemical 87(1), 130–149 (2002)
3. Chow, C.Y., Mokbel, M.F., He, T.: Aggregate location monitoring for wireless sensor networks: A histogram-based approach. In: Proc. of MDM, pp. 82–91 (2009)

4. Collins, S., et al.: New opportunities in ecological sensing using wireless sensor networks. Frontiers in Ecology and the Environment 4(8), 402–407 (2006)
5. Coman, A., Sander, J., Nascimento, M.A.: Adaptive processing of historical spatial range queries in peer-to-peer sensor networks. Distrib. Parallel Databases J. 22(2), 133–163 (2007)
6. Madden, S., Franklin, M.J., Hellerstein, J.M., Hong, W.: Tag: a tiny aggregation service for ad-hoc sensor networks. SIGOPS Oper. Syst. Rev. 36(SI), 131–146 (2002)
7. Madden, S., Franklin, M.J., Hellerstein, J.M., Hong, W.: The design of an acquisitional query processor for sensor networks. In: Proc. of ACM SIGMOD, pp. 491–502 (2003)
8. Pinedo-Frausto, E.D., Garcia-Macias, J.A.: An experimental analysis of zigbee networks. In: Proc. of LCN, pp. 723–729 (2008)
9. Pradhan, S., Kusuma, J., Ramchandran, K.: Distributed compression in a dense microsensor network. IEEE Signal Processing Magazine 19(2), 51–60 (2002)
10. Ying, B., Liu, W., Liu, Y., Yang, H., Wang, H.: Energy-efficient node-level compression arbitration for wireless sensor networks. In: Proc. of the ICACT, pp. 564–568 (2009)

Efficient In-Database Maintenance
of ARIMA Models

Frank Rosenthal and Wolfgang Lehner

Dresden University of Technology,
Database Technology Group
{frank.rosenthal,wolfgang.lehner}@tu-dresden.de

Abstract. Forecasting is an important analysis task and there is a need of integrating time series models and estimation methods in database systems. The main issue is the computationally expensive maintenance of model parameters when new data is inserted. In this paper, we examine how an important class of time series models, the *AutoRegressive Integrated Moving Average* (ARIMA) models, can be maintained with respect to inserts. Therefore, we propose a novel approach, *on-demand* estimation, for the efficient maintenance of maximum likelihood estimates from numerically implemented estimators. We present an extensive experimental evaluation on both real and synthetic data, which shows that our approach yields a substantial speedup while sacrificing only a limited amount of predictive accuracy.

Keywords: AutoRegressive Integrated Moving Average Models, Parameter Estimation, Integrated Forecasting.

1 Introduction

Time series can be encountered in many domains like business (e.g. sales or inventory), industry (e.g. yield rates or power consumption) and science (e.g. sunspot activity or ambient temperature). One important *time series analysis* task is *forecasting*, which can be achieved by creating a *model* of the time series. A suitable model represents the process that generated the time series in question and reproduces the dynamics of the time series, e.g. deterministic or stochastic trend or seasonal patterns. Forecasts of the time series can then be derived from the model.

A widely used class of time series models are the *AutoRegressive Integrated Moving Average* (ARIMA) models [2] that are able to model a wide range of real-world time series from various domains [6,8]. A key task in the creation of time series models is *parameter estimation*, which is the determination of values of the model parameters that provide an optimal fit to a time series. Optimality can be determined through different optimization criterion, e.g. a least squares approach. However, the best parameter values can usually be obtained through the *maximum likelihood* approach that uses the model specific *likelihood function* as optimization criterion [2]. The maximization of this function is often implemented through computationally expensive numerical procedures, e.g.

J.B. Cushing, J. French, and S. Bowers (Eds.): SSDBM 2011, LNCS 6809, pp. 537–545, 2011.
© Springer-Verlag Berlin Heidelberg 2011

Quasi-Newton algorithms, where the parameter values are iteratively adjusted until the likelihood becomes maximal.

There has been a rising research interest in integrating forecasting functionality in database management systems [3,5,1] and there are also first practical implementations, e.g. in the analysis services in SQL Server 2008 [9]. Integrated forecasting offers the key advantage of reusing models to answer repeated or similar queries [4]. The expensive parameter estimation is performed only once and its costs amortize through using the model for answering many queries.

What has not been considered so far is the issue of *maintenance*, i.e. how to update parameter estimates when real time passes and new data becomes available. Simply ignoring the new time series elements will cause the model to become outdated and not reflect the true process any more. *Reestimation* of the parameters after every update guarantees that model parameters are always based on all available information, but it is also the most expensive possible maintenance procedure. A simple approach to maintenance is *periodic reestimation*, where parameters are reestimated after a certain number of new tuples have been inserted. This is less expensive, but omitting estimation without any consideration of the impact of the new tuples on the estimated parameters can still increase the forecast error arbitrarily [10].

In this paper, we propose *on-demand* estimation, a novel approach for efficiently maintaining maximum likelihood parameter estimates acquired through numerical optimization. We focus our presentation on ARIMA models, but the approach is in principal applicable to any estimation scheme using numerical optimization.

To make maintenance more efficient, on-demand estimation involves constructing a *boundary synopsis* around a found optimum on the likelihood function. The boundary helps to decide whether updates will change the optimum and thereby the estimated parameters significantly. Note that we can decide on the impact of the update without actually estimating the parameters anew. If the update will not change the optimum we can skip the expensive reestimation and we perform it otherwise.

2   On-Demand Estimation

Parameter estimation is the task to find parameter values ζ that provide the optimal fit to a given time series $x_{1:t}$. Updates to a time series expand it, e.g. an update x_t to a time series $x_{1:t}$ yields $x_{1:t+1}$. Parameter estimation using numerical procedure involves the evaluation of a function on each element of the time series at least once, but normally several times. Hence, estimating parameters is computationally expensive and becomes more expensive with each update.

We focus our discussion on parameter estimation for ARIMA(p, d, q) models. The structural parameters p, d and q are set a priori and can not be estimated. Only p and q have an influence on the likelihood function, while d controls the preprocessing step *differencing* (not relevant for this paper). We employ one

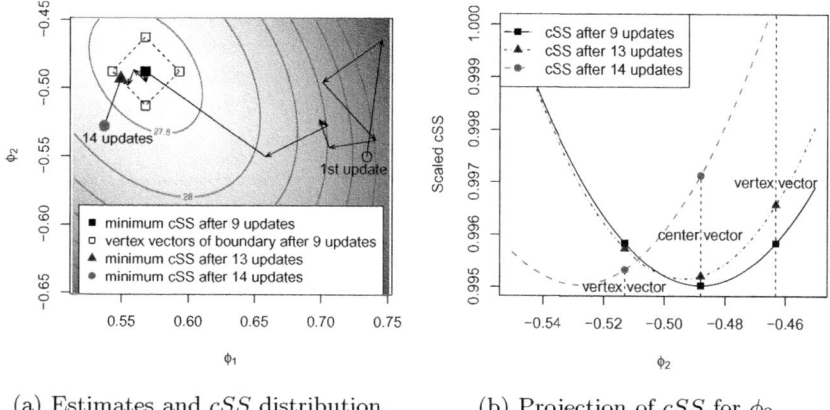

(a) Estimates and cSS distribution.   (b) Projection of cSS for ϕ_2.

Fig. 1. Example evolution of `ARIMA`$(2, 0, 0)$ estimates

variant of the likelihood function of `ARIMA` models, the *conditional likelihood* [2]. Maximization of it can be realized by minimizing the *conditional sum of squares* $cSS(\phi, \theta)$, where ϕ and θ form the parameter vector ζ to be estimated:

$$cSS(\phi, \theta) = \sum_{i=p}^{t}(a_i(\phi, \theta, x_{1:t}))^2 = \sum_{i=p}^{t}\left(x_i - \sum_{j=1}^{p}\phi_j x_{i-j} - \sum_{k=1}^{q}\theta_k a_{i-k}(\phi, \theta, x_{1:t}) \right)^2$$

An $a_i(\phi, \theta, x_{1:t})$ term is the difference of x_i, the prediction from the autoregressive part of the model that uses a linear combination of the past $x_{i-p:i-1}$ and the prediction from the moving average part that uses a linear combination of the past $a_{i-q:i-1}$. Hence, the cSS can be interpreted as the sum of the *one-step-ahead prediction errors*. An important property of the cSS is that we can incrementally maintain it for fixed parameter values (ϕ, θ): $cSS_{t+1}(\phi, \theta) = cSS_t(\phi, \theta) + (a_{t+1}(\phi, \theta, x_{t+1}))^2$. Note that incremental maintenance of the objective function for fixed parameter values can be derived for many other parameter estimation tasks, e.g. for any numerical least squares estimator. Therefore, our approach could be applied there as well.

Overview. The basic idea of on demand estimation is to decide whether an update x_t, to a time series $x_{1:t-1}$ is influential enough to change the location of the optimum on the objective function significantly and hence justifies the use of estimation on the expanded time series $x_{1:t}$ to get the current parameter estimates. Hence, for `ARIMA` models we need to assess how heavily an update changes the distribution of the cSS.

We introduce our approach using an example. Figure 1(a) shows how the estimates of the parameters ϕ_1 and ϕ_2 of an `AR`(2) process change after successive updates. The contour lines represent the cSS distribution after nine updates were applied and the filled square marks the minimum cSS after these nine updates. We see how nearly all earlier estimates varied heavily after updates, which

---

Algorithm 1. On-Demand Estimation

---
cv: center vector; vv_i: vertex vector; tol: tolerance

```
 1: for each update tuple x_t do
 2:    if exists(syn) then
 3:       cSS_cv,t ← cSS_cv,t-1 + calculate_cSS(cv, x_t)
 4:       for each vv in syn do
 5:          cSS_vv,t ← cSS_vv,t-1 + calculate_cSS(vv, x_t)
 6:          if cSS_vv,t < cSS_cv,t then
 7:             syn ← build_synopsis(estimate_parameters(x_1:t), tol)
 8:          end if
 9:       end for
10:    else
11:       syn ← build_synopsis(estimate_parameters(x_1:t), tol)
12:    end if
13: end for
```

---

requires reestimation. However, after the ninth update, the next four updates result only in small changes.

To decide on the impact of the next update we construct a boundary synopsis that consists of the four *vertex vectors* vv (empty squares in Figure 1(a)) around the *center vector* cv (filled square). The center vector is the minimum of the cSS found by the last estimation, while the vertex vectors represent acceptable tolerance in the parameter estimates. Figure 1(b) shows a projection of the cSS depending only on ϕ_2 with $\phi_1 \approx 0.57$, i.e. a vertical cut through the cSS from top to bottom. The continuous line with squares is a scaled version of the cSS after nine updates and with the center vector at the minimum. After each of the next four updates, estimates change only slightly as can be seen in Figure 1(b) where the cSS after the four updates (dash-dotted line with triangles) still has its minimum inside of the boundaries, although the true minimum is not at the cv any more. A significant change happens after the next update. We can see in Figure 1(b) that the minimum of the dashed line with circles is now outside the boundary. When we look at the left vertex vector, we notice that cSS is now smaller than at the center vector. This is our indication that the minimum has shifted and that we must apply estimation to find the new minimum. This approach is efficient since we can maintain the cSS incrementally at all parameter vectors, requiring to evaluate the objective function only on the update and adding it to the stored cSS.

Formal Definition. Algorithm 1 formalizes these considerations. On the first run of on-demand estimation we have no synopsis yet. It stores center and vertex vectors as well as the cSS at these vectors. The synopsis is constructed from a parameter estimate yielded by *estimate_parameters*$(x_{1:t})$ and a user defined tolerance *tol* (line 11).

For any successive update, we update $cSS_{cv,t}$ for the center vector (line 3). Since cSS can be updated incrementally, we need to evaluate the objective function *calculate_cSS* for a parameter vector (cv and the vv) only on the update

x_t. We also update $cSS_{vv,t}$ for each vertex vector (line 5) and stop if we find a vertex vector where $cSS_{vv,t} < cSS_{cv,t}$, i.e. when we have an indication of a significant change. We then proceed as in the initial construction of the synopsis. If we update all vertex vectors without detecting a significant change, we can stop processing the update, since the old estimate is still good with respect to the tolerance.

Form of the Boundary. We propose two boundary structures: *Hypercube* and *Simplex*. While the hypercube offers absolute uniform coverage of all dimensions of the parameter space, the simplex is the structure with the smallest possible number of vertices to bound the center vector in all dimensions [11]. The hypercube can simply be constructed from a new parameter estimate, i.e. the d-dimensional center vector $cv = (cv_1, \ldots, cv_d)$ and tolerance *tol* by creating a pair of new vectors $vp_i = \{(cv_1, \ldots, cv_i + tol, \ldots, cv_d), (cv_1, \ldots, cv_i - tol, \ldots, cv_d)\}$ for each dimension $i = \{1, \ldots, d\}$. The boundary is the union $\bigcup_{i=1}^{d}\{vp_i\}$ and consists of $2d$ vectors.

A simplex in d dimensions is composed of $d + 1$ vectors. For example, in two dimensions the simplex is a triangle. The boundary is looser than with the hypercube, since we have fewer vertices at which to check for a shift. A simplex can be constructed from a center vector $cv = (cv_1, \ldots, cv_d)$ by creating a new vector v_i for each dimension $i = \{1, \ldots, d\}$: $v_i = \{(cv_1, \ldots, cv_i + tol * o, \ldots, cv_d)\}$, where $o = 1$ if $cv_i > 0$ and $o = -1$ if $cv_i < 0$. This yields d vectors with an orientation o away from the coordinate origin. The last vector is acquired by scaling cv in the direction of the origin in a way that the new vector v_{d+1} has a distance of *tol* to cv. The simplex is made up of $\bigcup_{i=1}^{d+1}\{v_i\}$.

Adaptation. We now introduce an approach for the hypercube that adapts the boundary to the variance of the parameter estimates. We achieve this by increasing the tolerance for parameters that vary heavier than others, i.e. by constructing vertex vectors that are further away from the center vector, and vice versa. An adapted hypercube can be constructed from a center vector $cv = (cv_1, \ldots, cv_d)$, tolerance *tol* and a vector of scale factors f_i with $i = 1, \ldots, d$ by creating a pair of new vectors $vp_i = \{(cv_1, \ldots, cv_i + f_i * tol, \ldots, cv_d), (cv_1, \ldots, cv_i - f_i * tol, \ldots, cv_d)\}$ for each dimension. To determine the f_i, we calculate the empirical variance Var_i per dimension i from the elements of the parameter estimates $cv_{1,i}, \ldots, cv_{m,i}$ in that dimension i. The scale factor f_i is defined as:

$$f_i = d * \frac{Var_i(cv_{1,i}, \ldots, cv_{m,i})}{\sum_{j=1}^{d} Var_j(cv_{1,j}, \ldots, cv_{m,j})}$$

Hence, the scale factor f_i is a measure of the variance in dimension i relative to the overall variance of the estimates. We normalize the fraction by multiplying with d and gain a factor f_i that is larger than one if the variance is above average and vice versa. If all dimensions vary equal, all scale factors are $f_i = 1$.

Internal Estimation. Internal estimation is a heuristic that seeks to improve on the base strategy of using the center point as a fixed parameter estimate. We introduce it for the hypercube. The basic idea is to shift our parameter estimate from the center vector cv in the likely direction of the true cSS minimum. Our shifted estimate ζ is determined as:

$$\zeta = cv + \sum_{i=1}^{2d} \frac{d_i}{\sum_{i=1}^{2d} d_i} (vv_i - cv)$$

where $d_i = (cSS_{vv_i} - cSS_{cv})^{-1}$, i.e. the reciprocal of the cSS difference between vertex and center. This gives the most weight to vertices that have a relative low cSS which is an indication that the true estimate lies in the direction of that vertex. We normalize the d_i to sum to one and multiply the resulting weights with the directional vectors $(vv_i - cv)$. The weighted sum of all $2d$ directional vectors gives the final estimate.

3   Evaluation

We implemented all presented maintenance approaches in the R programming language. Our evaluation consists of a series of simulations to evaluate run time behavior and accuracy. A simulation run is composed of an initial batch estimation on a starting window $x_{1:S}$ of the time series at hand. S is set large enough to give a first parameter estimate (e.g. $S = p + 1$ for ARIMA($p,0,0$)). We then expand the time series to $x_{1:S+1}$, $x_{1:S+2}$ and so on until the end of the series and we call the maintenance procedure after each expansion.

The *baseline* approach consists in using estimation for each update. We also report results for *periodic estimation* for specific rates of estimation. In the figures, periodic x means that after each x tuples parameters were estimated. We measured accuracy and runtime behavior and we present the latter as *speedup* in terms of the average estimation time in comparison to the baseline.

We used both real and synthetic data for our evaluation. The synthetic set is composed of several simulated time series of different length that behave according to given ARIMA models of different order and parameterization. The first real dataset are the 645 yearly time series from the M3 competition [8], an extensive empirical evaluation of the predictive performance of a large number of forecasting methods. The second real data set is derived from the half hourly measured electricity demand in Great Britain, which is published by *National Grid* [7]. We used the *Initial Demand Outturn* of 2008 and aggregated it to the sum per day yielding a 347 element time series with seasonal patterns per week as well as over the year.

Synthetic Data. To evaluate the loss in accuracy caused by skipping estimation we present the deviation of estimates for periodic and on-demand estimation with respect to the estimates yielded by the baseline. As can be seen from Figure 2(a), the mean deviation, as root mean squared error, lies well inside the bounds of the used tolerance. We can see that the Hypercube estimators have tighter bounds. Note that the internal estimation (IE), the adaptation (A) and the

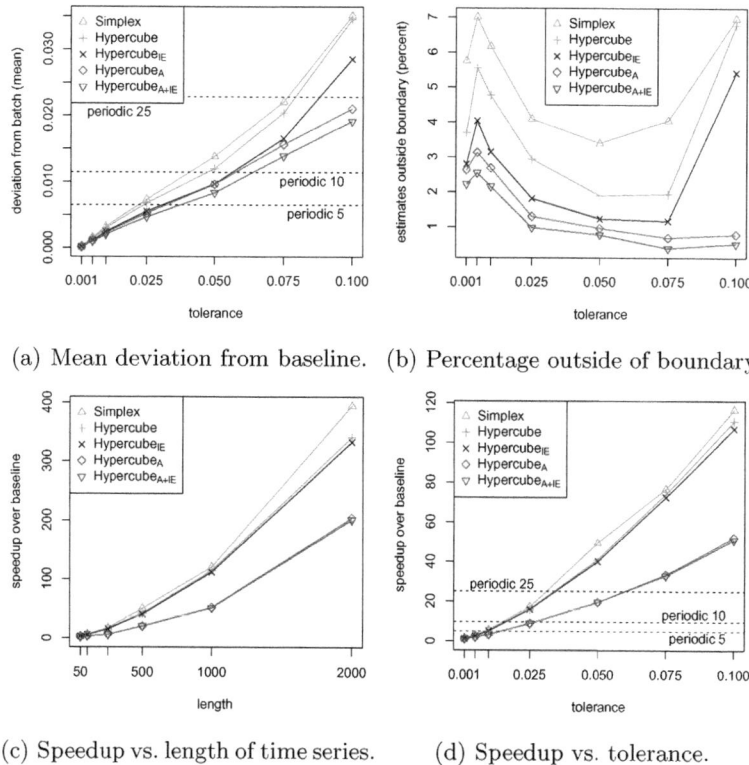

(a) Mean deviation from baseline.   (b) Percentage outside of boundary.

(c) Speedup vs. length of time series.   (d) Speedup vs. tolerance.

Fig. 2. Experimental Results on Synthetic Data

combination (A+IE) of these two modifications each reduce the mean deviation. Because we can only check cSS values at the vertex vectors, the boundary is not strict. Figure 2(b) shows the percentage of estimates that varied larger than the set tolerance. The ratio is relatively large for very small tolerances because the covered parameter space is equally small. The percentage of boundary violations sinks with increasing tol, but rises again for those approaches that do not adapt the boundary.

We see from Figure 2(c) where $tol = 0.05$ that the speedup increases as expected with the length of the overall simulated time series. The reason is that the benefit for each skipped update is larger for larger time series. Speedup also increases with increasing tolerance (Figure 2(d), $length = 500$), whereby we see an speedup of about 50 for the Simplex and about 40 for the static Hypercube estimators at $tol = 0.05$. Looking at Figures 2(a) and 2(d) conjointly, we see that for a given tolerance, e.g. $tol = 0.05$, the variants of on-demand estimation in comparison to periodic estimation can offer a smaller error, e.g. adapted Hypercube vs. periodic 25, a greater speedup, e.g. Simplex vs. periodic 10, or both e.g. adapted Hypercube with internal estimation vs. periodic 10.

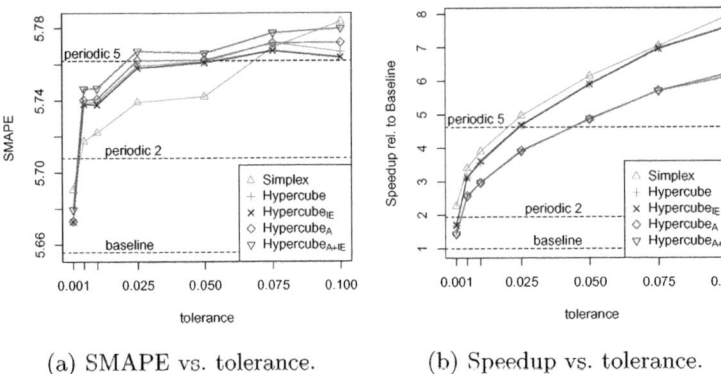

(a) SMAPE vs. tolerance.   (b) Speedup vs. tolerance.

Fig. 3. Experimental Results on M3 Competition Data

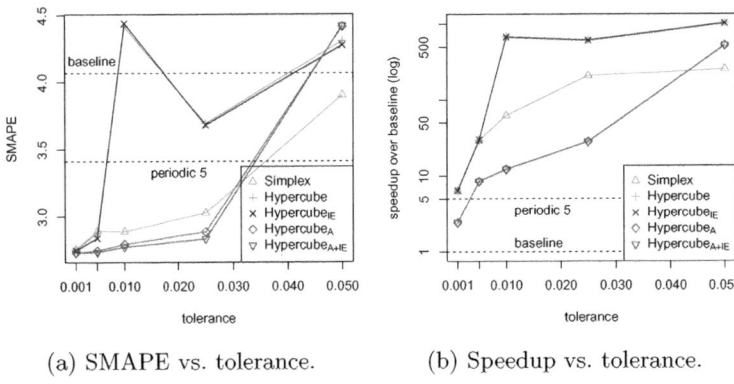

(a) SMAPE vs. tolerance.   (b) Speedup vs. tolerance.

Fig. 4. Experimental Results on Energy Demand Data

Real Data. We used the time series from the M3 competition to test the robustness of our approach under adverse conditions. These series are very short with a minimum length of only 28 elements. Hence, there is little that can be gained from omitting estimation and the estimates on such short time series may vary strongly with each update. However, there are still inserts that do not change the parameters significantly and on-demand estimation identifies these updates. As can be seen form Figure 3(b), even on this adverse data set we are never slower than the baseline and can even achieve speedups, e.g. a speedup of about 6 for the Hypercube approach with $tol = 0.05$.

We evaluated accuracy using a series of one-step-ahead forecasts $\hat{x}_{t+1}$ from the models and measuring the error to the real value x_{t+1} using the *symmetric mean absolute percentage error* SMAPE $= 100 * t^{-1} \sum_{i=1}^{t} (|x_t - \hat{x}_t|)(|x_t| + |\hat{x}_t|)^{-1}$ (range: 0% to 100%). We can see from Figure 3(a) that on-demand and periodic estimation yield a slightly worse but acceptable SMAPE than the baseline. The variation between approaches is small, but for larger tolerances the speedup of on-demand estimation is higher than that of periodic estimation with only slightly greater error.

Figure 4(a) shows the SMAPE of the energy dataset. Note that the SMAPE for nearly all of our approaches is lower than the SMAPE of the baseline. This can occur in real world data, where skipping an update might decrease the forecast error. We can also see an outlier for the two static Hypercube approaches at tolerance $tol = 0.01$ which does not show up in the adaptive approaches that perform better. The speedup on this data set is huge for all our approaches e.g. about 250 for the Simplex at $tol = 0.025$ (Figure 4(b), logarithmic scale). Periodic estimation has a smaller speedup and greater error.

4   Conclusion

We proposed on-demand estimation, a maintenance strategy that uses expensive estimation only when necessary and we presented several on-demand estimators. With these estimators we achieved a considerable speedup over the baseline approach on synthetic and real data with only little impact on predictive accuracy. Our approaches also outperform periodic reestimation by offering either a smaller error or a greater speedup.

References

1. Agarwal, D., Chen, D., ji Lin, L., Shanmugasundaram, J., Vee, E.: Forecasting high-dimensional data. In: SIGMOD, pp. 1003–1012 (2010)
2. Box, G., Jenkins, G., Reinsel, G.: Time Series Analysis: Forecasting and Control, 4th edn. Wiley, Chichester (2008)
3. Duan, S., Babu, S.: Processing Forecasting Queries. In: VLDB, pp. 711–722 (2007)
4. Fischer, U., Rosenthal, F., Boehm, M., Lehner, W.: Indexing forecast models for matching and maintenance. In: IDEAS, pp. 26–31 (2010)
5. Ge, T., Zdonik, S.B.: A skip-list approach for efficiently processing forecasting queries. PVLDB 1(1), 984–995 (2008)
6. Gooijera, J.G.D., Hyndman, R.J.: 25 years of time series forecasting. International Journal of Forecasting 22(3), 443–473 (2006)
7. Grid, N.: Demand data,
   http://www.nationalgrid.com/uk/Electricity/Data/Demand+Data
8. Makridakis, S., Hibon, M.: The M3-Competition: results, conclusions and implications. International Journal of Forecasting 16(4), 451–476 (2000)
9. Microsoft: PredictTimeSeries - Microsoft SQL Server Books (2008),
   http://msdn.microsoft.com/en-us/library/ms132167.aspx
10. Tashman, L.J.: Out-of-sample tests of forecasting accuracy: an analysis and review. International Journal of Forecasting 16(4), 437–450 (2000)
11. Weisstein, E.W.: Simplex, http://mathworld.wolfram.com/Simplex.html

Recipes for Baking Black Forest Databases

Building and Querying Black Hole Merger Trees from Cosmological Simulations

Julio López, Colin Degraf, Tiziana DiMatteo, Bin Fu, Eugene Fink, and Garth Gibson

Computer Science and Physics departments, Carnegie Mellon University

Abstract. Large-scale N-body simulations play an important role in advancing our understanding of the formation and evolution of large structures in the universe. These computations require a large number of particles, in the order of 10-100 of billions, to realistically model phenomena such as the formation of galaxies. Among these particles, black holes play a dominant role on the formation of these structure. The properties of the black holes need to be assembled in merger tree histories to model the process where two or more black holes merge to form a larger one. In the past, these analyses have been carried out with custom approaches that no longer scale to the size of black hole datasets produced by current cosmological simulations. We present algorithms and strategies to store, in relational databases (RDBMS), a forest of black hole merger trees. We implemented this approach and present results with datasets containing 0.5 billion time series records belonging to over 2 million black holes. We demonstrate that this is a feasible approach to support interactive analysis and enables flexible exploration of black hole forest datasets.

1 Introduction

The analysis of simulation-produced black hole datasets is vital to advance our understanding of the effect that black holes have in the formation and evolution of large-scale structures in the universe. Increasingly larger and more detailed cosmological simulations are being used to gain insight on the evolution of massive black holes (Sec. 2). The simulations store the data in a format that is not readily searchable or easy to analyze. Purpose-specific custom tools have often been preferred over standard relational database management systems (RDBMS) for the analysis of datasets in computational sciences (Sec. 3). The assumption has been that the overhead incurred by the database will be prohibitive. Previous studies of black holes have used custom tools. However, this approach is inflexible as these tools often need to be re-developed for carrying out new studies and answering new questions. As part of our goal of reducing the time to science, we developed an approach that leverages RDBMS to analyze black hole datasets (Sec. 4). This approach enables fast, easy and flexible data analysis. A major benefit of the database approach is that now the astrophysicists are able to interactively ask ad-hoc questions about the data and test hypotheses by writing relatively simple queries and processing scripts. We present: (1) A set of algorithms and approaches for processing, building and querying black hole merger tree datasets. (2) A compact

J.B. Cushing, J. French, and S. Bowers (Eds.): SSDBM 2011, LNCS 6809, pp. 546–554, 2011.

database representation of the merger trees. (3) An evaluation of the feasibility and relative performance of the presented approaches. Our evaluation (Sec. 5) shows that it is feasible to support the analysis of current black hole datasets using a database approach. An extended version of the results presented here is also available [13].

2 Motivation: Black Holes and the Structures in the Universe

Black holes play an important role in the process by which structures, such as galaxies, are organized in the universe. To understand these phenomena, large-scale cosmological numerical simulations are used. They cover a vast dynamic range of spatial and time scales with an extremely large number of particles, in excess of 10^{10} in principle.

Black Hole Datasets. The simulations produce three types of datasets: snapshots, group membership and black holes. Snapshots contain complete information for all the particles in the simulation at a given time step. In recent simulations, snapshots require close to 100 TB of storage. The group files contain the membership of particles to groups, such as dark matter halos. The black hole files contain the black hole data with high temporal resolution. They contain two main types of records. (1) Black hole property records contain the id, simulation time, mass, and other properties. (2) Merger events records indicate when a pair of black holes merge with one another and contain the ids and masses of the two black holes, as well as the time when the event occurred. A *black hole merger tree* comprises the set of merger event records along with the detailed property records for the black holes involved in the mergers.

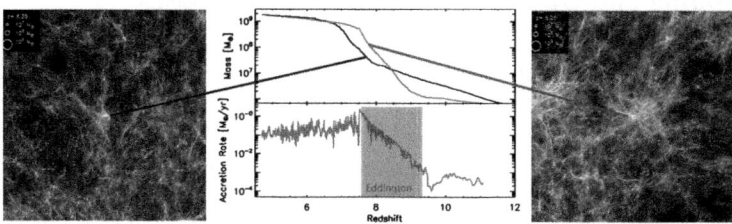

Fig. 1. Sample black holes. This figure shows the gas distribution around two large black holes and their respective light curves and accretion rate history for the most massive one.

Analysis of Black Hole Datasets. Recent observations imply that black holes with billion solar masses are already assembled when the universe is only 800 million years old. An objective of the analyses of simulation-generated black hole datasets is to explain the formation of these objects. There are two types of analyses we want to perform on black hole datasets. The first type requires queries based on a specific redshift (i.e., simulation time), often selecting a subset according to their mass and growth rate. These analyses aim to characterize the properties of black hole properties that exist at a specific time, including the number and density of black holes as a function of mass [6] or

luminosity [4], how they cluster and the correlation between black holes and the galaxies in which they are found [3,6,5]. The second type of analyses requires processing the
detailed growth history of individual black holes. An example is shown in Fig. 1. These
histories help us understand how black holes grow, the relative importance of black hole
mergers vs. gas accretion.

3   Background and Related Work

Database techniques have been adopted to manage and analyze datasets in a variety of
science fields such as medical imaging [2], bioinformatics [15] and seismology [16].
In astronomy, RDBMS have been used to manage the catalogs of digital telescope sky
surveys such as the Sloan Sky Digital Survey (SSDS) [1,9]. Database techniques have
been used in observational astronomy to perform anomaly detection [10] among others, and data-intensive approaches have been used for spatial clustering [7,11]. RDBMS
have not been as widely used for the analysis of cosmological simulations, in part due to
the challenge posed by the massive multi-terabyte datasets generated by these simulations. The German Astrophysical Virtual Observatory (GAVO) has led in this aspect by
storing the Millenium Run dataset in an RDBMS and enabling queries to the database
through a web interface [12]. GAVO researchers proposed a database representation for
querying the merger trees of galactic halos. We are using RDBMS to support interactive analysis of cosmological simulation datasets. We present techniques for building
and querying the merger trees of black holes, along with a compact database representation for these trees.

4   Building and Querying Black Forest Databases

Database Design. To support the queries needed for the analysis of BH datasets, we
transform the the simulation output into RDBMS tables. The database comprises two
main tables as shown in Fig. 2: *BlackHoles* (BH), *MergerEvents* (ME). Querying this
database consists of two steps: (1) building the merger tree from the ME table to obtain
the ids of the black holes in the tree; (2) querying the BH table to retrieve the associated
history for the black holes. The input for a query is the id of a black hole of interest
(qbhid). The desired output for step 1 is the ids of all the black holes in the same
merger tree as qbhid. Notice that the ME records do not have explicit links to other ME
records that belong to the same merger tree. The approaches for building and querying
the merger trees are presented below.

Approach 1: Recursive DB Queries. Given a qbhid, this approach finds the root of the
tree by repeatedly querying the ME table. Once the root is found, it recursively queries
the ME table for each of the root's children (left, right) as shown in the BuildTree
procedure. This simple approach works well when only a small number of merger trees
are being queried and the resulting trees have few records.

Approach 2: In-Memory Queries. This approach consists in using a single database
query for loading all the records from the ME table into a set in memory (MESet) and
then looking up in MESet the events that belong to a tree. The algorithm is the following.

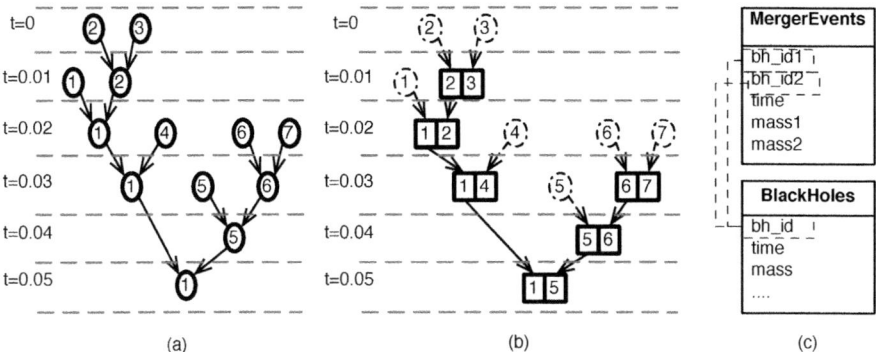

Fig. 2. (a) Black hole merger tree. Leaf nodes (at the top) correspond to black holes. Interior nodes correspond to black holes that merge. (b) DB representation: only the interior nodes of the tree, i.e., merger events, are stored, the dashed circles, corresponding to the leaf nodes, are not explicitly stored. (c) Basic schema for main tables in the black holes database: mergerevents (ME) and blackholes (BH).

Given a query qbhid, add it to a queue pq of pending black holes. For each element bh in the queue, fetch from MESet the records that match bh (i.e., r.bh1 = bh). For each matching record r, add the corresponding r.bh2 to the pq queue. Repeat this process until every element of pq has been processed (i.e., the end of the queue is reached). At the end of the procedure, pq contains the ids belonging to the corresponding tree.

Approach 3: In-Memory Forest Queries. This approach builds on the previous one. The basic idea is to build all the merger trees in the dataset with a single scan of the ME table, instead of building a single tree as in the previous approach. This approach incurs extra work to build all the trees. However, this cost is amortized when a large number of queries need to be processed. This approach is based on the Union-Find algorithm [8] and adjusted to handle the peculiarities of the merger events representation. The process is described in the procedure BuildForestInMemory.

Approach 4: ForestDB. The ForestDB approach builds on the techniques used in the In-Memory Forest approach. The basic idea is to build the black hole forest in the same way as in the in-memory case. Then tag each tree with an identifier (tid). The forest can be written back into a table in the database that we will call merger events forest (MF). This is done as a one-time pre-processing step. The schema for this table is the same as the ME's schema (see Fig. 2), with the addition of the tid field. Two conceptual steps are performed at query time to extract a merger tree for a given qbhid. First, search the MF table for a record matching qbhid. The tid field can be obtained from the record found in this step. Second, retrieve from the MF table all the records that have the same tid. These two steps can be combined in a single SQL query. Moreover, the detailed history for the black holes in the tree can be retrieved from the BH table using a single query that uses tid as the selection criteria and joins the MF and BH tables. Indices on the bh1, bh2 and tid fields are required to speed up these queries. Alternatively, the

---

Procedure BuildTree(bhroot, ctime): *Recursively build a merger tree rooted at bhroot*

   // *Find all the records that have the bh1 field = bhroot*
1 **type** TreeNode {id, time, left, right }
2 TreeNode node = NULL, pnode = NULL
3 qresult = SELECT bh2, time FROM ME WHERE bh1 = bhroot AND time ≤ ctime
   ORDER BY time DESC
4 **for** (bh2, time) **in** qresult **do**
5   node = new TreeNode(id, time)
6   node.right = BuildTree(bh2, time)
7   **if** pnode is not null **then**
8   pnode.left = node // *set left child for previous node in the result*
9   pnode = node
10 **return** node // *node is the latest event (tree root), it may be null*

Procedure BuildForestInMemory(db)

---

   input   : DB with the ME table
   output : A forest containing all the merge trees in ME
1 cursor = SELECT bh1, bh2, time FROM ME // *Scan over all ME records*
2 **for** (bh1, bh2, time) **in** cursor **do**
3   node = new TreeNode(bh1, time, bh2)
4   bh2Map.put(bh2, node) // *Map from bh2 to this node*
5   bh1Map.addToList(bh1, node) // *Map from bh1 to a node list*
6 **for** node **in** bh2Map **do**
7   node.right = bh1Map.get(node.bh2) // *Create link for right-side child, it may be null*
8 forest = emptySet()
9 **for** lst **in** bh1Map **do**
10   sortbytime(lst)
11   createLinkOnBh1(lst) // *Create links from lst[n-1].left to lst[n]*
12   findRootAndAddToForest(lst, forest)
13 **return** (forest, bh1Map, bh2Map)

---

indices on bh1 and bh2 can be replaced by an additional auxiliary indexed table to map from bhid to tid.

The MF table only stores the membership of the merger event records to a particular tree. Notice that the MF table does not explicitly store the tree structure, i.e., the parent-child relationships. Also, the MF table only stores the internal nodes of the merger tree. The leaves are not explicitly stored. Instead the relevant data (such as the leaf's bhid) is stored in the parent node. This makes for a more compact representation as it requires fewer records in the MF table.

5   Evaluation

We implemented the approaches described above using Python and SQLite. Our evaluation aims to characterize the relative performance of these approaches and determine the feasibility of using RDBMSs in the analysis of black holes datasets. For this purpose, we ran a set of experiments using a dataset produced by the largest published cosmology simulation to date.

Workload. The dataset was produced by a cosmological simulation using the GADGET-3 [14] parallel program. The simulation contained 66 billion particles. At the end of the simulation, there are 2.4 million black holes. The size of the resulting black holes dataset is 84 GB. The black hole history table contains 420 million records corresponding to 3.4 million unique black holes and 1 million merge events. Figure 3 shows the distribution of tree sizes in number of merger events in the ME table. The storage requirements for the tables and associated indexes is shown in the table in Figure 3.

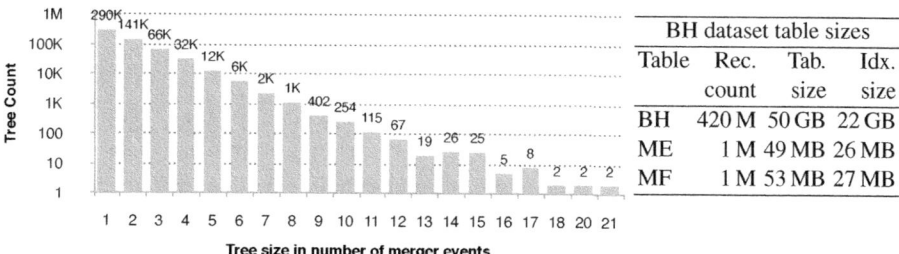

	BH dataset table sizes		
Table	Rec.	Tab.	Idx.
	count	size	size
BH	420 M	50 GB	22 GB
ME	1 M	49 MB	26 MB
MF	1 M	53 MB	27 MB

Fig. 3. Left: Distribution of tree sizes in the black holes dataset. The X axis is the size of a merger tree measured as the number of events in a tree. The Y axis is the number of trees of that size in $\log_{10}$ scale. Right: Sizes of tables and indexes in the BH database.

Performance. To characterize the performance of the developed approaches, we conducted a series of micro benchmark experiments that correspond to the steps involved in answering queries for the detailed time history of merger trees. The experiments were run on a server host with 2 GHz CPUs, 24 GB of memory and a SATA disk.

Building Merger Trees. The first set of micro benchmark experiments corresponds to the steps needed to build the merger trees for a set of query black holes (qbhs). We compared three of the approaches explained in Sec. 4: (a) Recursive DB – RDB, (b) In-memory – IM, and (c) Forest DB – FDB. The In-memory Forest approach was only used to build the tables for FDB. For these experiments we selected black holes (*qbhs*) that belonged to merger trees in the ME table. We timed the process of satisfying a request to build one or more merger trees specified by the requested qbhs. The processing time includes the time required to issue and execute the database query, retrieve and postprocess the result to build the trees.

In the first experiment, we kept the tree size fixed at 5 and varied the number of black holes for which a tree is requested (number of qbhs). The results for the different approaches are shown in Fig. 4. The X axis is the qbh count varying from 1 to 10K. The Y axis shows the processing time (seconds) in log scale. For qbh counts less than 1K, both the RDB and FDB approaches are faster than the In-Memory approach. The RDB approach is not as expensive as we originally thought for small queries, either in the

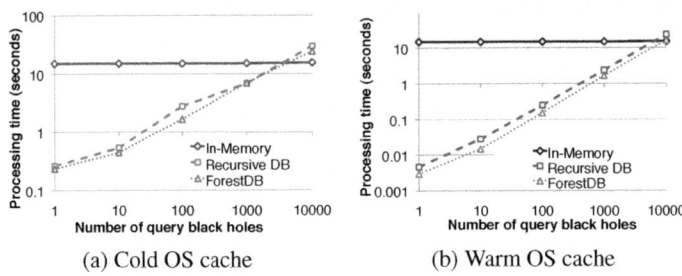

(a) Cold OS cache   (b) Warm OS cache

Fig. 4. Running time to obtain the merger trees for the different approaches. These results correspond to a tree of size 5. The X axis is the number of trees being queried at once in a batch. The Y axis is the elapsed time in seconds (log scale) to retrieve the corresponding records from the ME table. The cases with cold (a) and warm (b) OS caches are shown.

number of qbhs or the requested tree size. It was surprising to find out that for the cold OS cache setup (Fig. 4a), the processing time for RDB and FDB does not differ significantly. For the warm OS cache, there is a (constant in log scale) difference between RDB and FDB. The IM approach pays upfront a relatively large cost of 15 seconds to load the entire ME table, then the processing cost per requested qbh is negligible, and thus can be amortized for a large number of qbhs.

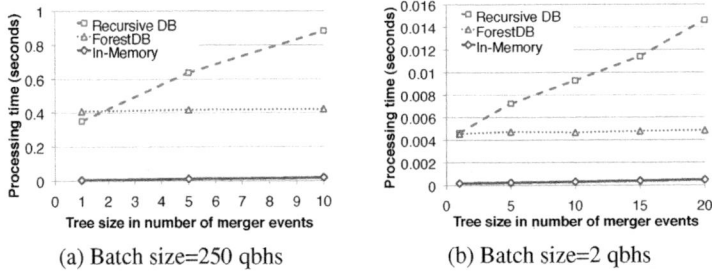

(a) Batch size=250 qbhs   (b) Batch size=2 qbhs

Fig. 5. Processing time for building the merger trees using various approaches. This experiment was performed with a warm OS cache and a cold DB cache. The X axis is the size of the resulting tree; (a) and (b) show the time to process 250 qbhs and 2 qbhs per request respectively. The Y axis is the elapsed time to build the number of trees of each size.

Figure 5 shows the effect of the merger tree size on the request processing time. In this experiment the requests were grouped by tree sizes (X axis = 1, 5, 10, 15, 20). This experiment was performed with a warm OS cache and cold database cache. The initial load time for the IM approach is not included in the processing time shown in the graphs, only the time to build the tree in memory. The running time for the RDB approach increases as the trees get larger. This is due to the larger number of queries to the ME table needed to process each tree in the recursive approach. The FDB approach requires a single query to the ME table per requested tree.

Retrieving the Time History for Merger Trees. In the second set of experiments, we retrieved the detailed time history for a set of trees retrieved in the previous step. This entails retrieving from the BH table all the records for the corresponding BH in a given merger tree. For each tree size (1, 5, 10, 15), we retrieved the BH histories for 100 trees of that size. Figure 6a shows the elapsed time in seconds to retrieve the detail records from the BH table. The times are shown for an unsorted indexed BH table and a BH table sorted by the black hole id. As expected for this query pattern, sorting by the BH id is beneficial. Figure 6b shows the elapsed time according to the number of records that were retrieved from the BH table. Each data point corresponds to a merger tree that resulted in retrieving the number of BH records shown in the X axis. The Y axis is the elapsed time in seconds for the unsorted and sorted BH tables.

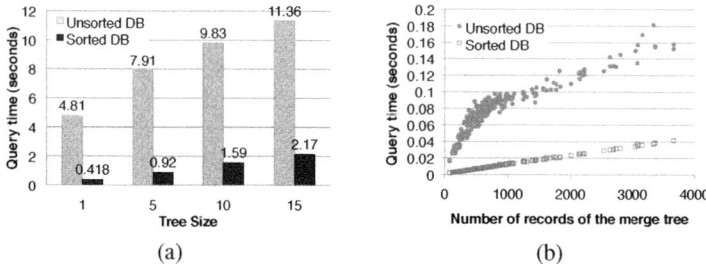

(a)   (b)

Fig. 6. Time to retrieve the detail BH history from the BH table for merger trees of various sizes. The running times for queries to sorted and unsorted BH tables are shown. Figure (a) shows the elapsed time grouped by tree size. Figure (b) shows the same data grouped by the number of BH records comprising the merger trees.

6 Conclusion

Rapid, flexible analysis of black hole datasets is key to enable advances in astrophysics. We presented a set of algorithms for processing these data using a database approach. The database approach is not only flexible, but also exhibits good performance to support interactive analysis.

Acknowledgments. This research was sponsored in part by the National Science Foundation (NSF), under award CCF-1019104; the Gordon and Betty Moore Foundation, in the eScience project; and the support of the companies of the PDL Consortium. The computations were supported by an allocation of advanced computing resources supported by the NSF and the TeraGrid Advanced Support Program. Computations were performed at the Carnegie Mellon Cloud Cluster (wiki.pdl.cmu.edu/opencloudwiki) and the National Institute for Computational Sciences.

References

1. Abazajian, et al.: 7-th Data Release of the Sloan Digital Sky Survey. ApJS 182 (2009)
2. Cohen, S., Guzman, D.E.: SQL.CT: Providing data management for visual exploration of ct datasets. In: SSDBM: Scientific and Statistical Database Management (2006)
3. Colberg, J.M., di Matteo, T.: Supermassive black holes and their environments. Monthly Notices of the Royal Astronomical Society (NMRAS) 387, 1163–1178 (2008)
4. DeGraf, C., et al.: Faint-end quasar luminosity functions from cosmological hydrodynamic simulations. Monthly Notices of the Royal Astronomical Society (NMRAS) 402 (2010)
5. DeGraf, C., et al.: Quasar Clustering in Cosmological Hydrodynamic Simulations: Evidence for mergers. ArXiv e-prints (2010)
6. Di Matteo, T., et al.: Direct Cosmological Simulations of the Growth of Black Holes and Galaxies. Astrophysical Journal (ApJ) 676, 33–53 (2008)
7. Fu, Ren, López, Fink, Gibson: Discfinder: A data-intensive scalable cluster finder for astrophysics. In: High Performance Distributed Computing (HPDC) (2010)
8. Galler, Fischer: An improved equivalence algorithm. Comm. ACM 7(5) (1964)
9. Ivanova, et al. : MonetDB/SQL meets skyserver: the challenges of a scientific database. In: Scientific and Statistical Database Management (SSDBM), p. 13 (2007)
10. Kaustav Das, J.S., Neill, D.: Anomaly pattern detection in categorical datasets. In: Knowledge Discovery and Data Mining (KDD) (2008)
11. Kwon, Y., Nunley, D., Gardner, J.P., Balazinska, M., Howe, B., Loebman, S.: Scalable clustering algorithm for N-body simulations in a shared-nothing cluster. In: Gertz, M., Ludäscher, B. (eds.) SSDBM 2010. LNCS, vol. 6187, pp. 132–150. Springer, Heidelberg (2010)
12. Lemson, Springel: Cosmological simulations in a relational database: Modelling and storing merger trees. In: Astronomical Data Analysis Software and Systems (2006)
13. Lopez, et al.: Recipes for baking black forest databases. Tech. Rep. CMU-PDL-11-104, Carnegie Mellon, PDL (2011)
14. Springel, V.: The cosmological simulation code GADGET-2. Monthly Notices of the Royal Astronomical Society (NMRAS) 364, 1105–1134 (2005)
15. Xu, W., Ozer, S., Gutell, R.R.: Covariant evolutionary event analysis for base interaction prediction using a relational database management system for RNA. In: Winslett, M. (ed.) SSDBM 2009. LNCS, vol. 5566, pp. 200–216. Springer, Heidelberg (2009)
16. Yang, Y.-S., et al.: Isee: Internet-based simulation for earthquake engineering - Part I: Database approach. Earthquake Engineering & Structural Dynamics 36, 2291–2306 (2007)

CrowdLabs: Social Analysis and Visualization for the Sciences

Phillip Mates, Emanuele Santos, Juliana Freire, and Cláudio T. Silva

SCI Institute, University of Utah, USA

Abstract. Managing and understanding the growing volumes of scientific data is one of the most challenging issues scientists face today. As analyses get more complex and large interdisciplinary groups need to work together, knowledge sharing becomes essential to support effective scientific data exploration. While science portals and visualization Web sites have provided a first step towards this goal, by aggregating data from different sources and providing a set of pre-designed analyses and visualizations, they have important limitations. Often, these sites are built manually and are not flexible enough to support the vast heterogeneity of data sources, analysis techniques, data products, and the needs of different user communities. In this paper we describe CrowdLabs, a system that adopts the model used by social Web sites, allowing users to share not only data but also computational pipelines. The shared repository opens up many new opportunities for knowledge sharing and re-use, exposing scientists to tasks that provide examples of sophisticated uses of algorithms they would not have access to otherwise. CrowdLabs combines a set of usable tools and a scalable infrastructure to provide a rich collaborative environment for scientists, taking into account the requirements of computational scientists, such as accessing high-performance computers and manipulating large amounts of data.

Keywords: Computational Sciences, Cyberinfrastructure, Visualization.

1 Introduction

The infrastructure to design and conduct scientific experiments has not kept pace with our collective ability to gather data. This has led to an unprecedented situation: data analysis and visualization are now the bottleneck to discovery. This problem is compounded as interdisciplinary groups collaborate and need to perform a wide range of analyses targeted to multiple audiences.

We posit that by facilitating the *social analysis of scientific data*, we can overcome many of these challenges. When users share their analyses and visualizations, they can benefit from the collective wisdom: by querying analysis specifications which make sophisticated use of tools, along with data products and their provenance, users can learn by example from the reasoning and/or analysis strategies of experts; expedite their scientific training in disciplinary and inter-disciplinary settings; and potentially reduce the time lag between data acquisition and scientific insight.

In this paper, we describe *CrowdLabs*, a system that adopts the model used by social Web sites and integrates a set of usable tools and a scalable infrastructure to provide a

J.B. Cushing, J. French, and S. Bowers (Eds.): SSDBM 2011, LNCS 6809, pp. 555–564, 2011.

rich collaborative environment for scientists. Similar to social Web sites, CrowdLabs aims to foster collaboration, but unlike these sites, it was designed to support the needs of computational scientists, including the ability to access high-performance computers and manipulate large volumes of data. By providing mechanisms that simplify the publishing and use of analysis pipelines, it allows IT personnel and end users to collaboratively construct and refine portals. Thus, CrowdLabs lowers the barriers for the use of scientific analyses and enables broader audiences to contribute insights to the scientific exploration process, without the high costs incurred by traditional portals. In addition, it supports a more dynamic environment where new exploratory analyses can be added on-the-fly.

Another important feature of CrowdLabs is the support for provenance [5, 8]. Publishing scientific results together with their provenance—the details of how the results were obtained—not only makes the results more transparent, but it also enables others to reproduce and validate the results. CrowdLabs leverages provenance information (*e.g.*, workflow/pipeline specifications, libraries, packages, users, datasets and results) to provide a richer sharing experience: users can search and query this information. In addition, provenance is made accessible through the Web site and an API. This allows users to connect results published in an article or wiki page to the pipelines and data served by CrowdLabs, greatly simplifying the creation of provenance-rich publications.

The remainder of the paper is organized as follows. We review related work in Section 2. In Section 3, we describe the main components of CrowdLabs. Information on deploying CrowdLabs at www.crowdlabs.org is given in Section 4. In Section 5, we describe the different ways of sharing content and of making reproducible documents using CrowdLabs. We conclude in Section 6, where we outline directions for improvements and future work.

2   Related Work

While there have been several efforts focused on sharing scientific data, relatively little work has gone into sharing analysis and visualization specifications (pipelines). To this end, closely related to our approach is myExperiment [13], a collaborative environment for sharing pipelines and other digital objects. myExperiment supports versioning of pipelines and can execute certain types of pipelines. However, because its focus is on pipelines that integrate bioinformatics-related Web services, myExperiment does not support data- and compute-intensive pipelines.

Recently, a number of sites have come online which aim to support social analysis and visualization of small, tabular data. Tableau Public[1] provides infrastructure for users to publish interactive visualizations on the Web. Many Eyes [18] and Swivel [17] (no longer available) are public social data analysis Web sites, where users can upload data, create visualizations of that data, and leave comments on either visualizations or datasets. Built upon Many Eyes is Many Eyes Wikified, which is based on Dashiki [11], a wiki-based Web site for collaboratively building visualization dashboards.

The ability to run and interact with compute-intensive pipelines or simulation jobs is rapidly becoming essential for most scientists. The HUBzero Platform for

---

[1] http://www.tableausoftware.com/public

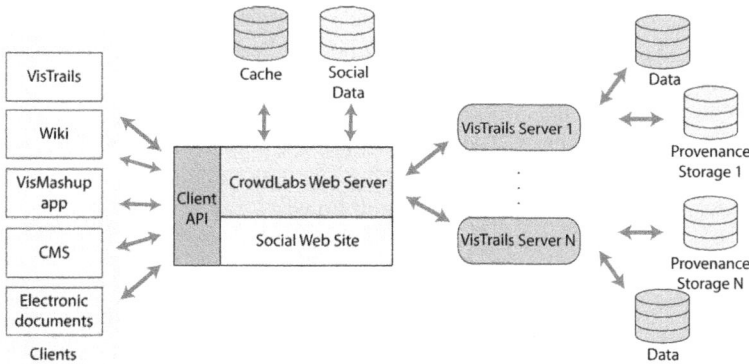

Fig. 1. CrowdLabs system architecture

Scientific Collaboration [12] allows researchers to access and share scientific simulation and modeling tools. It was created to support nanoHUB.org [16], an online community for the Network for Computational Nanotechnology (NCN). To publish a tool, developers have to connect to a special workspace machine, where they compile their code and use NCN's open source toolkit Rapture to create friendly GUIs. End users access the resulting tools using an ordinary Web browser and launch simulation runs on the national Grid infrastructure using virtual machines, without having to download or compile any code. Similar to HUBzero, CrowdLabs enables the use of HPC resources and is flexible enough to be deployed on the cloud. But the use of the pipeline computation model allows it to provide additional functionality: besides the support for provenance and queries over this information [15], it is possible to deploy customized (and easy-to-use) Web applications [14].

3 System Overview

The CrowdLabs infrastructure is general and can be integrated with any workflow management system that runs in server mode and exposes an API. In this paper we describe how the CrowdLabs infrastructure is used with the VisTrails [2, 9] and VisMashup [14] systems. VisTrails is an open-source provenance management and scientific workflow system that was designed to support the scientific discovery process. VisTrails provides unique support for data analysis and visualization, a comprehensive provenance infrastructure, and a user-centered design. The system combines and substantially extends useful features of visualization and scientific workflow systems. For more details about VisTrails, please refer to [2, 9]. VisTrails was modified to provide access to workflow provenance in a client-server mode.

The VisMashup [14] system simplifies the creation, maintenance, and use of customized, workflow-based applications (or mashups). It supports the tasks required in the construction of custom applications: from querying and mining workflow collections and their provenance (for finding relevant workflows and parameters that should be exposed in the application) to automatic generation of the application and associated user interface, that can be deployed as a desktop or Web application. For use within CrowdLabs, VisMashup was extended to support a Web-based user interface for interacting with workflows.

The CrowdLabs architecture consists of two main components: *CrowdLabs Web Server*, which provides the *Social Web Site*, and a *Client API* and the *VisTrails Server*, which handles workflow-related tasks. The CrowdLabs architecture is depicted in Figure 1 and a description of each component follows.

3.1   CrowdLabs Web Server

Client API. CrowdLabs provides a Web-based interface for sharing workflows and provenance. The system includes a repository of workflow results (e.g., visualizations), datasets, and libraries, and while the CrowdLabs Web site provides a useful platform for sharing and collaboration, the social and provenance data can be useful in other contexts. In order to expose CrowdLabs resources to a diverse set of clients, the site employs a RESTful HTTP API [6]. This API identifies visualization and social resources, providing uniform resource identifiers (URIs) for clients to retrieve, add, update and delete them.

The data analysis and visualization resources defined by the system are vistrails, workflows, vismashups, packages, and datasets, while the social resources include profiles, projects, groups, and blogs. Adhering to the RESTful architecture, each of these resources has various different representations associated with them. Visualization resources might include provenance data, meta-data (modules, documentation, *etc.*), application files (vistrails, data files), as well as visualization results. Social resources might include blog posts, discussion topics, and notices, along with ratings, tags, and comments, which are linked to the visualization resources. For example, for a client to access the XML representation corresponding to the workflow with id 117, they would use the URI *http://www.crowdlabs.org/vistrails/workflows/get_xml/117/*.

The RESTful API enables basic CrowdLabs functionality to be integrated into the VisTrails desktop application and other extensions. Users can login, add, and update vistrails and datasets to CrowdLabs through the *Web Repository Options* dialog from within the VisTrails desktop application. The LaTeX extension described in Section 5 also uses the RESTful API for embedding workflows in PDF documents.

Following the example of myExperiment's Google Gadget and Facebook App [13], providing an API encourages developers to extend functionality and creates an open development environment.

Social Web Site. To foster user interaction, CrowdLabs incorporates a social Web site that is based around user-created content and social networking tools. Users can make friends, join groups, write blogs, and create projects, topics, and wikis. In addition, they can add, edit, and delete VisTrails related data such as vistrails, workflows, vismashups, packages, and datasets. Tied to each of these VisTrails related objects are ratings, tags, comments, and projects. This social data not only encourages an environment of user discussions and interaction, but enables the use of crowd sourcing to find good-quality visualizations through user ratings, better categorize datasets and workflows by user tagging, and troubleshoot problems through comments and discussion topics. We also let users share their work off-site by providing syntax to embed interactive vismashups on Web sites as well as static visualization results on the Web, wikis, and within LaTeX documents (see Section 5).

Cache. CrowdLabs is set up to generate content dynamically. This creates potential efficiency issues, since some workflows can take a long time to run. It is important to avoid delays when presenting pages to users, otherwise they can get discouraged and avoid using the site. We use different forms of caching to speed up common operations.

In the CrowdLabs Web server there are two caches: the results cache and the provenance cache. The results cache is used to store images and other files generated by workflows and vismashups. When there is a request for a workflow execution result, for instance when a vismashup run, the system first checks if that workflow has been executed before. If so, it uses the files already in the cache, otherwise it will forward the request to the proper VisTrails server instance.

The provenance cache stores information about the workflows and vistrails uploaded to the CrowdLabs Web site. This information currently includes the named workflows in a vistrail, who created them and the packages and modules referenced in the workflows together with their documentation. The system takes advantage of the fact that VisTrails change-based provenance model records information about workflow evolution [9]: all the workflows are versioned.

3.2   VisTrails Server

The *VisTrails server* is one of the most important components of CrowdLabs. It provides the link between the workflow provenance and the rest of the system. The VisTrails server is a multi-threaded server that uses the XML-RPC protocol to answer client requests. The most common requests are: execute a workflow or a vismashup, add or remove a vistrail or a vismashup from the database, get the packages and modules used in a vistrail or workflow and other information associated with the workflows.

The CrowdLabs Web server communicates with VisTrails server instances via XML-RPC requests, enabling communication with multiple remote VisTrails servers. Scientific teams can thus host their own VisTrails servers as a way to meet their computing and data storage needs.

Another key feature of the VisTrails server is that it maintains its own cache (separate from CrowdLabs results cache) for keeping the results of executed workflows or vismashups. When both components are in the same machine, CrowdLabs can be configured to use the VisTrails server cache to avoid redundant storage. The VisTrails server also has the ability to start and communicate with other VisTrails server instances using the same XML-RPC protocol. This allows the creation of clusters of servers that work transparently with the rest of CrowdLabs.

4   Deploying CrowdLabs

Depending on the particular application, it is possible to use different deployment configurations for CrowdLabs (see Figure 2). Here we will describe the current system deployment at www.crowdlabs.org and explain some of its key capabilities. We encourage readers to access the site, but bear in mind that it is constantly under development.

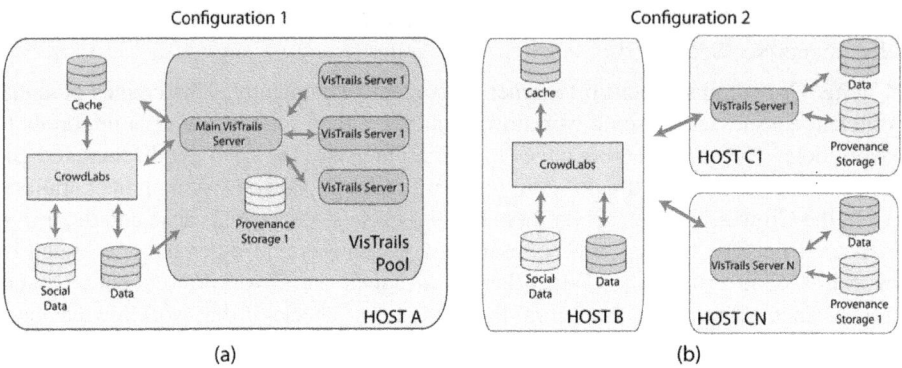

Fig. 2. Different configurations of deploying CrowdLabs. (a) All the components are located on the same machine. (b) VisTrails servers execute on dedicated machines.

System configuration. CrowdLabs is currently deployed as shown in Figure 2(a). The core system and the four instances of the VisMashup server share a 8-core Intel Xeon 2.66 GHz machine with 24 GBs of RAM running Linux. The CrowdLabs webserver cache and social data along with VisMashup workflow specifications and provenance are stored in MySQL databases. The CrowdLabs Web site is implemented using the Python Web framework Django and the VisTrails server is implemented in Python.

Projects and servers. It is possible that users would like to organize their content into different projects. An example is the ALPS [1] project, which contains all the vistrails, vismashups, workflows and the information they use together with the discussions and blog posts created about them. This allows for defining different levels of visibility: Groups can have discussions and upload workflows that are only visible to the people involved in the project. The ability to selectively disclose information for people outside the group is extremely important for scientists, who may work for many years before deciding to release certain types of data. Another advantage is that a project can have its own dedicated VisTrails server. This creates the possibility of having specialized servers for different types of workflows. For example, for the ALPS project, we are in the process of deploying a VisTrails server on one of their machines so users on the CrowdLabs Web site can execute workflows and vismashups that access all the required resources on ALPS's file servers for running the simulations. These different servers allow the system to grow in functionality without compromising the overall performance.

5   Sharing Content and Supporting Reproducible Research

An important motivation for us to create CrowdLabs was to make it easier for scientists to publish provenance-rich, reproducible results. While it is widely accepted that scientific publications should include detailed provenance so that others can both reproducte and validate the results, in practice, doing so is challenging, both for authors and reviewers. Even when authors provide data sets and computer code, reviewers must configure their systems so that they can compile and run the code; they must also navigate between code, data and text, identify important parameters and manually enter the values specified in the text. Recently, the renewed interest on this subject in different communities has led to different scientific publishing approaches (see [7] for an overview). CrowdLabs simplifies the process of packaging workflows and results for publication. Authors can create documents whose digital artifacts (*e.g.,* figures) include a deep caption: detailed provenance information which contains the specification of the computational process (or workflow) and associated parameters used to produce the artifact. CrowdLabs supports the publication on wikis, other Web sites, and scientific documents. Readers need not install any special software and can interact with the results through a Web interface. Next, we briefly describe the different mechanisms that CrowdLabs supports for sharing content.

Interactive versus static content. CrowdLabs supports the generation of static content (e.g., images, animations, tables, XML pages) as well as interactive ones. Static content is generated directly from the workflow specification. In particular, we use VisTrails' flexible spreadsheet infrastructure to generate the different types of output. Anything that can be displayed in a spreadsheet cell can be re-routed through CrowdLabs. Due to limitations of current browsers, it is hard to provide fully interactive content, in particular, for 3-D visualizations. Instead, we use the capabilities of VisMashup to allow for interactive widgets to be placed next to an output on a browser (see Figure 3). As the user modifies the exposed inputs, the system computes the resulting visualization, and makes it available.

Publishing on the Web. To publish workflows onto other Web sites, the CrowdLabs site provides embeddable HTML of either a static image linking back to the workflow page or an embeddable vismashup object. CrowdLabs also integrates with Wikis by extending the wiki markup language. Users can embed either static visualizations or VisMashups onto Wiki pages by including `<vistrail />` or `<vismashup />` tags (provided through the *Embed this Workflow in a Wiki* link present on the workflow pages of CrowdLabs Web site). When using these tags on a CrowdLabs-enabled Wiki, they are replaced with images generated from XML-RPC execution requests submitted to a VisTrails server. Not only does this create an easy way to share workflow results, it provides versioning of scientific results to Wiki technology lacking such possibilities.

Publishing scientific documents. We believe that one of the most interesting applications of a system like CrowdLabs is the impact that it can have on printed media. Often we see published scientific results and wonder how they were generated or would like to compute new results using different data. Currently this is nearly impossible. Every time an image is cut and pasted from a workflow or visualization tool to a paper, most of the lineage information is lost. In CrowdLabs, we advocate a direct linkage from images and workflows results presented in documents to their provenance. To make this possible, CrowdLabs uses a technique analogous to the one used to extend the wiki. We defined a `\vistrail` command that takes in the information necessary to identify and execute the workflow and include the images produced in-place. A LaTeX style file parses the information inside the command, sends it to a python script that builds and makes a HTTP request to CrowdLabs. The images are then downloaded and included as a hyperlinked regular `\includegraphics` command.

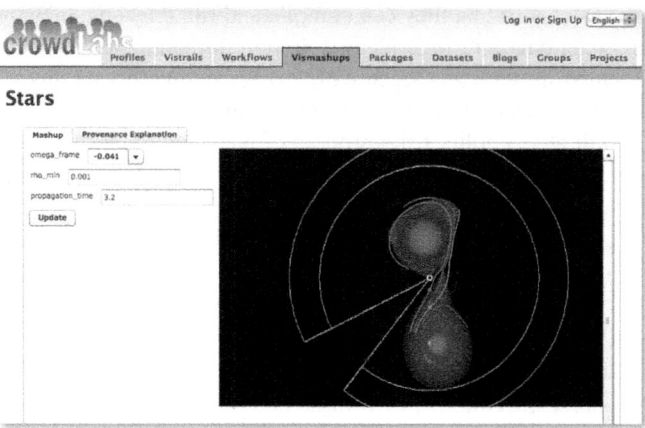

Fig. 3. Interacting with the visualization of a binary star system simulation using VisMashup. Users change parameters on the left and see the resulting visualization on the right. Available at http://www.crowdlabs.org/vistrails/medleys/details/5/.

Sharing content from other tools. The techniques presented here are easily extensible to any other system that supports provenance and is capable of producing result images or files. For instance, ParaView running with the VisTrails Provenance Explorer plugin [3] could be easily extended to to share visualizations on CrowdLabs.

6   Conclusions and Future Work

The CrowdLabs system builds on infrastructure that our group has been working on since early 2005, and it provides the "last mile" to the scientists. For its most basic use, it does not require any installation of tools in the user's machine, and we see this as an enormous advantage. The barrier of entry is quite small, and it is possible for users to perform a wide range of data analysis and visualization tasks without ever having to install any tools. Besides being deployed on www.crowdlabs.org and on the VisTrails Wiki (www.vistrails.org), the system is also being used at the CMOP [4] site at the Oregon Health & Science University and the ALPS [1] site at ETH in Zurich. We believe many small research groups that do not have all the resources and infrastructure needed for data analysis and visualization tasks can benefit from CrowdLabs. In order to accommodate a growing community, we expect the need for the following new functionality:

Provenance querying and analytics. By mining the data in the CrowdLabs provenance repository, we will be able discover of patterns that can potentially simplify the notoriously hard and time-consuming process of designing and refining scientific workflows [10]. Also useful are advanced querying capabilities that allow users to better explore the workflow, provenance and data.

Improved Web-enabled interfaces and graphics. Although it is possible to use CrowdLabs completely from a Web browser, some advanced functionality, such as interaction with 3D visualization, is not currently supported. One of the big challenges is that Web 3-D graphics are not standardized at this moment, creating a major obstacle in supporting high-end visualization over the Web. Due to some data being remote, we believe that we will need to also add streaming and multi-resolution techniques to our data analysis and visualization workflows.

System improvements. There are a number of system improvements that are needed, including improved scalability in terms of the size of provenance information and data; and a more sophisticated security model.

Acknowledgments. Our research has been funded by the National Science Foundation (grants IIS-0905385, IIS-0746500, ATM-0835821, IIS-0844546, CNS-0751152, IIS-0713637, OCE-0424602, IIS-0534628, CNS-0514485, IIS-0513692, CNS-0524096, CCF-0401498, OISE-0405402, CCF-0528201, CNS-0551724), the DoE SciDAC (VACET and SDM centers), and IBM Faculty Awards (2005, 2006, 2007, and 2008).

References

1. The ALPS project, `http://alps.comp-phys.org`
2. Bavoil, L., Callahan, S., Crossno, P., Freire, J., Scheidegger, C., Silva, C., Vo, H.: VisTrails: Enabling Interactive Multiple-View Visualizations. In: IEEE Visualization 2005, pp. 135–142 (2005)
3. Callahan, S.P., Freire, J., Scheidegger, C.E., Silva, C.T., Vo, H.T.: Towards provenance-enabling paraView. In: Freire, J., Koop, D., Moreau, L. (eds.) IPAW 2008. LNCS, vol. 5272, pp. 120–127. Springer, Heidelberg (2008)
4. NSF Center for Coastal Margin Observation and Prediction (CMOP), `http://www.stccmop.org`
5. Davidson, S.B., Freire, J.: Provenance and scientific workflows: challenges and opportunities. In: SIGMOD, pp. 1345–1350 (2008)
6. Fielding, R.T.: Architectural Styles and the Design of Network-based Software Architectures. Ph.D. thesis, University of California, Irvine (2000)
7. Fomel, S., Claerbout, J.: Guest editors' introduction: Reproducible research. Computing in Science Engineering 11(1), 5–7 (2009)
8. Freire, J., Koop, D., Santos, E., Silva, C.T.: Provenance for computational tasks: A survey. Computing in Science & Engineering 10(3), 11–21 (2008)
9. Freire, J.-L., Silva, C.T., Callahan, S.P., Santos, E., Scheidegger, C.E., Vo, H.T.: Managing rapidly-evolving scientific workflows. In: Moreau, L., Foster, I. (eds.) IPAW 2006. LNCS, vol. 4145, pp. 10–18. Springer, Heidelberg (2006)
10. Koop, D., Scheidegger, C.E., Callahan, S.P., Freire, J., Silva, C.T.: Viscomplete: Automating suggestions for visualization pipelines. IEEE TVCG 14(6), 1691–1698 (2008)
11. McKeon, M.: Harnessing the Web Information Ecosystem with Wiki-based Visualization Dashboards. IEEE TVCG 15(6), 1081–1088 (2009)
12. McLennan, M., Kennell, R.: HUBzero: A Platform for Dissemination and Collaboration in Computational Science and Engineering. Computing in Science & Engineering 12(2), 48–53 (2010)

13. Roure, D.D., Goble, C., Stevens, R.: The design and realisation of the virtual research environment for social sharing of workflows. Future Generation Computer Systems 25(5), 561–567 (2009)
14. Santos, E., Lins, L., Ahrens, J., Freire, J., Silva, C.: VisMashup: Streamlining the Creation of Custom Visualization Applications. IEEE TVCG 15(6), 1539–1546 (2009)
15. Scheidegger, C., Koop, D., Vo, H., Freire, J., Silva, C.: Querying and creating visualizations by analogy. IEEE TVCG 13(6), 1560–1567 (2007)
16. Strachan, A., Klimeck, G., Lundstrom, M.: Cyber-Enabled Simulations in Nanoscale Science and Engineering. Computing in Science & Engineering 12(2) (March/April 2010)
17. Swivel, http://www.swivel.com
18. Viegas, F.B., Wattenberg, M., van Ham, F., Kriss, J., McKeon, M.: ManyEyes: A site for visualization at internet scale. IEEE TVCG 13(6), 1121–1128 (2007)

Heidi Visualization of R-tree Structures over High Dimensional Data

Shraddha Agrawal, Soujanya Vadapalli, and Kamalakar Karlapalem

International Institute of Information Technology, Hyderabad, India
shraddha@students.iiit.ac.in, soujanya@iiit.ac.in,
kamal@iiit.ac.in

Abstract. High dimensional index structures are used to efficiently answer range queries in large databases. Visualization of such index structures helps in: (a) *visualization of the data set* in a hierarchical format of the index structure, (b) *"explorative querying"* on the data set, similar to explorative browsing on the web, (c) *index structure diagnostics*: visualizing the structure along with its performance statistics enables the user to make changes to structure for better performance. To the best of our knowledge, there is no such visualization for high dimensional index structures.

1 Heidi Visualization of R-Trees

R-Tree visualization for high dimensional data is addressed by [2], which uses Parallel co-ordinates and Star co-ordinates for hyperbox details (number of children and region bounding it). Heidi [5] is a system to visualize high dimensional data clusters. Heidi is a 2-D matrix oriented data visualization, the rows and columns along the matrix reflect the data points. The data points are grouped and ordered according to the clusters provided. Heidi displays a $n \times n$ image, each pixel and its color denotes the closeness of a pair of points in various subspaces.

In this paper, Heidi is extended to visualize index structures by changing the point-ordering as per the index structure hierarchy. The visualization would reflect on the index structure characteristics. R-Tree [3] structure is a hierarchical structure defined over the points; points grouped into Minimum Bounding Boxes (MBBs) and MBBs are again grouped into larger MBBs. A multi-dimensional rectangle is referred to as a Minimum Bounding Box (MBB[1]) and it bounds a set of objects that are located within its boundaries. R*-Trees [1] is a variant of R-Tree and objective of the R*-Trees is to minimize the area covered by MBB, overlap between MBBs, MBB margins and storage utilization.

The MBBs are like the clusters having groups of points within a bounding box and Heidi brings forth the various subspace overlaps between the MBBs. The hierarchy of the index structure is restored by grouping points based on the MBBs to which they belong and then ordering the MBBs with respect to the corresponding parent MBBs and so on. In this work, Heidi is customized for the R-Tree and R*-Tree structures; though, other structures also could be viewed if given in a specific hierarchical format to the system.

---

[1] For 2-D data, a MBB is referred to as Minimum Bounding Rectangle (MBR).

J.B. Cushing, J. French, and S. Bowers (Eds.): SSDBM 2011, LNCS 6809, pp. 565–567, 2011.

2   Data Analytics for R-Tree and R*-Tree Index Structures

In Heidi visualization, the pixel color reflects the closeness between a pair of points in various subspaces. The color histogram of the Heidi visualization image gives subspace data analytics. The histogram is obtained for each cluster block; this reflects on the extent of an MBB-MBB overlap and the subspace in which it is maximal. As dimensionality of the data increases, there is a high likelihood of subspace overlaps; points within a MBB can share kNN relationships with points in another MBB in a lower subspace. The extent of kNN relationships reflect on the prominence of the subspace.

The following analytics are presented to the user along with the visualization: (i) Number Of MBBs, (ii) depth of each MBB in the hierarchy (length of the path from the root to the MBB), (iii) MBB density (calculated as the fraction of points present in the MBB) and (iv) Inter MBB overlaps (statistics for each subspace and in which subspace is the overlap maximum).

3   Explorative Querying

The most common operation performed over an R-Tree index is a range query, which finds all objects that a query region intersects. The problem of finding kNNs from R-trees has been introduced by Roussopoulos [4]. Given a point range-query (mentioned by a point in the data set and the k to compute the kNNs), the query result is displayed in the Heidi visualization. The points satisfying the query are highlighted in red. The spread of the red color in the matrix helps the user in understanding which MBBs are checked for the query. If the MBBs are present in different branches of the structure, it implies that there is scope for improvement in re-organizing the structure. If the query result falls within one MBB, then the data is structured properly. The query points highlighted could be selected (by clicking) to perform a new range query. This process of generating queries on the fly by observing the data is termed "explorative querying"; where in the user has no specific query in mind, but after looking at the visualization and the patterns, the user tries to identify interesting sets of points satisfying a query criteria.

Future work includes building a *R-Tree Diagnostics Tool* which overlays the R-Tree performance statistics with the Heidi Visualization; visually aiding the user to identify frequent page faults and MBB overlaps to amend the structure iteratively. An intuitive interactive user interface needs to be built to realize the concept of "explorative querying" over the data set (performance enhanced with a R-Tree).

References

1. Beckmann, N., Kriegel, H., Schneider, R., Seeger, B.: The r*-tree: An efficient and robust access method for points and rectangles. In: SIGMOD 1990, pp. 322–331. ACM Press, New York (1990)
2. Gimnez, A., Rosenbaum, R., Hlawitschka, M., Hamann, B.: Using r-trees for interactive visualization of large multidimensional datasets. In: Proceedings of the 6th International Symposium on Visual Computing (November 2010)

3. Guttman, A.: R-trees: A dynamic index structure for spatial searching. In: SIGMOD 1984, pp. 47–57. ACM Press, New York (1984)
4. Roussopoulos, N., Kelley, S., Vincent, F.: Nearest neighbor queries. In: SIGMOD, pp. 71–79 (1995)
5. Vadapalli, S., Karlapalem, K.: Heidi matrix: Nearest neighbor driven high dimensional data visualization. In: VAKD, SIGKDD (2009)

Towards Efficient and Precise Queries over Ten Million Asteroid Trajectory Models

Yusra AlSayyad, K. Simon Krughoff, Bill Howe, Andrew J. Connolly,
Magdalena Balazinska, and Lynne Jones

University of Washington, Seattle, WA
{yusra@astro,krughoff@astro,billhowe@cs,ajc@astro,magda@cs,
ljones@astro}.washington.edu

1 Introduction

The new generation of telescopes under construction return to the same area
of the sky with sufficient frequency to enable tracking of moving objects such
as asteroids, near-earth objects, and comets [4,5]. To detect these moving objects, one image may be subtracted from another (separated by several days or
weeks) to differentiate variable and moving sources from the dense background
of stars and galaxies. Moving sources may then be identified by querying against
a database of expected positions of known asteroids. At a high-level, this task
maps onto executing the query: "Return all known asteroids that are expected
to be located within a given region at a given time." We consider the problem
of querying for asteroids in a specified interval in space and time, specifically as
applied to populating the simulations of the data flow from the Large Synoptic
Survey Telescope (LSST).

Spatio-temporal databases have been well studied[3], however this problem
introduces new challenges. (1) *Number of Objects*: A characteristic Solar System
model[1] contains over 11 million asteroids[2]. (2) *Complex Trajectory Models*: An
asteroid's trajectory can be calculated precisely[2] at any given point in time, but
evaluation of the model describing its position is prohibitively expensive at query
time. (3) *Accuracy Constraints*: The LSST simulated images require accuracy
within 5 milliarcseconds[3] (mas), precluding the use of coarse statistical approximation methods. Additional requirements include 30-second query response time
for a 9.6 sq. degree circular search area and a 10TB storage limit.

Within this context, we develop and evaluate two approaches. (1) Model each
trajectory with a *bounding envelope*, use a spatial index to reduce the search
space, and evaluate the exact positions of the asteroids at the given epoch using
the ephemeris calculation code. (2) Model each trajectory by a set of positions
sampled frequently enough to *interpolate* their positions within a given accuracy

---

[1] Representing the asteroids potentially observable by LSST and Pan-STARRS.

[2] From its orbital elements - six variables that completely describe an orbit.

[3] For comparison, 1 milliarcsecond (mas) corresponds to the apparent diameter of a
dime about 3700 kilometers away.

J.B. Cushing, J. French, and S. Bowers (Eds.): SSDBM 2011, LNCS 6809, pp. 568–570, 2011.
© Springer-Verlag Berlin Heidelberg 2011

threshold at runtime. We implement all solutions in a relational database and evaluate them on a dataset of asteroid trajectories. Please refer to our technical report for details[1].

2 Evaluation

We find that the bounding envelope method provides exact accuracy and tunable storage requirements, but at the cost of lengthy query times. Total response time is extended by the necessary calculation of exact positions within the returned envelopes. The interpolation method can drastically reduce query time, but at the expense of either accuracy or storage space. With third degree polynomial interpolation, a sampling rate meeting the accuracy requirements exceeds our available storage. Table 1 summarizes a subset of the storage, accuracy and query time tradeoffs for "worst-case" searches in the densest region of the sky (returns 20K objects per 9.6 sq. deg. search area). This dense region forms a narrow strip across the sky. Typical searches in less dense regions (at least 20° from the ecliptic) return <1000 objects in under 2s for both methods.

Table 1. Preliminary Evaluation Summary

		Storage[a]	Accuracy	Response Time	
			Error	Query[b]	Evaluate positions[c]
		(billion rows)	(mas)	(s)	(s)
Bounding Envelope	Biweekly[d]	45	exact	20	+15
	Daily[e]	80	exact	20	+4
Interpolation	Inaccurate	200	<1000	3	-
	Accurate	900	<5	3	-

[a] Storage required to support queries over a 10yr range.
[b] Bounding env. uses MS Spatial Index; interpolation a hierarchical triangular mesh.
[c] Time to evaluate 20K exact positions with ephemeris calculation code.
[d] Store daily bounding envelopes and biweekly orbital elements.
[e] Store daily bounding envelopes and daily orbital elements.

The challenge of predicting the positions of sources extends beyond the question of tracking asteroids. In general, our techniques are applicable to a class of spatio-temporal trajectory search problems, where the true positions of the objects can be predicted by the evaluations of complex, often non-linear models that are extremely accurate but computationally expensive.

Acknowledgments. This work is funded by the NSF Cluster Exploratory (CluE) grant (IIS-0844580) and NSF CAREER award (IIS-0845397).

References

1. AlSayyad, Y., Krughoff, K.S., Howe, B., Connolly, A.J., Balazinska, M., Jones, L.: Technical Report UW-CSE-11-04-03, University of Washington (April 2011)
2. Grav, T., Jedicke, R., Denneau, L., Holman, M.J., Spahr, T.: Pan-STARRS Moving Object Processing System Team. In: The Pan-STARRS Synthetic Solar System Model and its Applications. Bulletin of the American Astronomical Society, vol. 38, p. 807 (December 2007)
3. Güting, R., Schneider, M.: Moving Objects Databases. Morgan Kaufmann, San Francisco (2005)
4. Kaiser, N., et al.: Pan-STARRS: A Large Synoptic Survey Telescope Array. In: Society of Photo-Optical Instrumentation Engineers (SPIE) Conference, vol. 4836, pp. 154–164 (December 2002)
5. LSST Science Collaborations. LSST Science Book, V2.0. ArXiv (December 2009)

Keyword Search Support for Automating Scientific Workflow Composition

David Chiu[1], Travis Hall[1], Farhana Kabir[1], and Gagan Agrawal[2]

[1] School of Engineering and Computer Science, Washington State University
[2] Dept. of Computer Science and Engineering, Ohio State University

1 Introduction

Keyword search is indispensable for locating relevant documents the web. Yet, at the same time, we have also grown aware of its limitations. It is often difficult to reach esoteric information obscured deep within various domains. For example, consider an earth science student who needs to identify how much a certain area in a nearby park has eroded since 1940 for a school project. Certainly, if this exact erosion information had previously been published onto a web page, a search engine could probably locate it effortlessly. But due to the exactness of this query, the likelihood that such a web page exists is slim. However, avenues for obtaining this information exist in the form of scientific workflows, which can be implemented using web service composition.

The maturation of the semantic web has prompted, if not already produced, the mass sharing of scientific web services and data sets. Keyword-based search engines like *seekda!* (http://webservices.seekda.com) and *GeoPortal* (http://www.geowebportal.org) are sites where users can publish and search data sets and web services. Woogle [2], a similarity search engine for services, can aid users in identifying relevant services for composition. *myExperiment* [3] is an online social networking community for users to share, customize, and execute scientific workflow plans. However, workflow planning (and web service composition) still requires intrinsic knowledge from end-users in terms of which data sets and services to use and domain-level expertise.

We describe our system, *Auspice*, which can perform on-demand, automatic workflow planning given a set of keywords. Web services and data sets are indexed on their user-defined tags and metadata. Given this index and a search query, Auspice not only returns previously defined workflows, but also identifies and composes any known services together with appropriate data sets to derive the information sought by the query. This strategy is in contrast to previously mentioned works which search predefined plans and services. We further propose and evaluate a relevance model that ranks the retrieved composed plans.

2 System Overview

We have developed keyword querying support for generating service-based workflow plans on the fly. These plans may or may not be previously developed by other users, and we have created an IR-based retrieval model which attempts to capture each workflow's relevance to the submitted keywords. Particularly, the workflows are ranked according

J.B. Cushing, J. French, and S. Bowers (Eds.): SSDBM 2011, LNCS 6809, pp. 571–572, 2011.
© Springer-Verlag Berlin Heidelberg 2011

to a score computed as a function of the number of scientific concepts relative to the query. In our system, scientific web services and data sets are assumed to represent (or derive) certain virtual objects simply referred to as *concepts*. In a previous effort, we proposed an ontology to capture these concept derivation relationships [1]. This recursive concept derivation is analogous to composing services. Given a targeted domain concept, c, it may be derived using a data set, d, or a web service operation s, whose input parameters, $(x_1, \ldots, x_p)$ may again be substantiable by respective scientific concepts $(c(x_1), \ldots, c(x_p))$.

This chaining process continues until a terminal element (either a service without input or a data file) has been reached on all paths. One can envision then, that when given a target concept along with attributes like date, location, etc., that applying this technique can yield a set of workflow plans by considering all derivation paths from the originating concept. Our system is depicted in Figure 1. Auspice maps a set of keywords to concepts within a predefined domain ontology, which captures the concept derivation relationships. Once the set of concepts has been identified, it is sent to the workflow planner, which com-

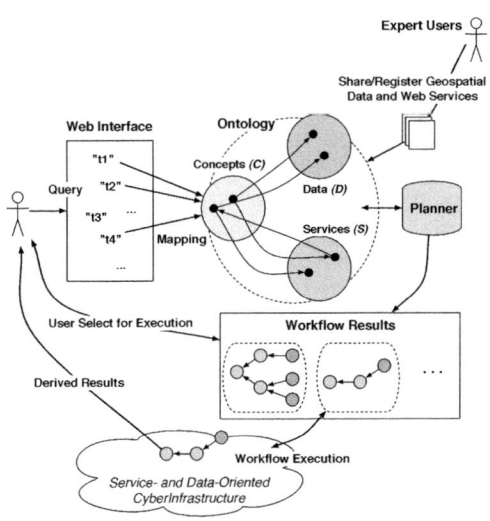

Fig. 1. Auspice Interface and Concept

poses services together with data files automatically and returns a ranked list of workflow candidates to the user. The user can then select a suitable workflow plan to execute over the service-enabled compute environments.

An evaluation over a geoinformatics case study has been conducted using our IR metrics. Our collaborators from Department of Geodetic Sciences at Ohio State University performed a blind relevance feedback over six queries, and we are observing agreeable precision values: 78%, 77.3%, and 76.2% for the Top 3, 5, and 10 retrieved workflows respectively. We will present our methods and preliminary results in detail in the poster.

References

1. Chiu, D., Agrawal, G.: Enabling ad hoc queries over low-level scientific data sets. In: Winslett, M. (ed.) SSDBM 2009. LNCS, vol. 5566, pp. 218–236. Springer, Heidelberg (2009)
2. Dong, X., Halevy, A., Madhavan, J., Nemes, E., Zhang, J.: Similarity search for web services. In: Proc. of VLDB, pp. 372–383 (2004)
3. Goble, C.A., et al.: Myexperiment: a repository and social network for the sharing of bioinformatics workflows. Nucl. Acids Res., 429 (May 2010)

FastQuery: A General Indexing and Querying System for Scientific Data⋆

Jerry Chou, Kesheng Wu, and Prabhat

Lawrence Berkeley National Laboratory
University of California, Berkeley, CA 94720, USA
{jchou,kwu,prabhat}@lbl.gov

Introduction. Modern scientific applications are producing vast amounts of data[2]. In many cases, the essential information needed for understanding scientific processes is stored in a relatively small number of data records. Efficiently locating the interesting data records is indispensable to the overall analysis procedure. The data structures most effective at performing these search operations is known as indices [2]. This work implements a flexible way of applying such an index to a range of different scientific data.

Instead of requiring the users to place their data into centralized data management system (DBMS), we propose to build indices alongside of the existing data outside of DBMS. To demonstrate the effectiveness of this approach, we have chosen to use an indexing software called FastBit [3]. In a number of earlier studies, FastBit has been demonstrated to support scientific applications well. We have worked in the past on applying this indexing software to a small class of well-defined HDF5 files; the resulting system was called HDF5-FastQuery [1]. FastQuery is a complete redesign of HDF5-FastQuery to support a generic I/O layer. Furthermore, it has significantly improved usability and performance as we describe next.

FastQuery Architecture. The overall FastQuery system is illustrated in Fig. (a). This generic design allow arbitrary array data to be mapped into the relational data model supported by FastBit indexing system. The core of this design is a dynamic variable table that groups arrays and subarrays with the same number of elements into a logical table required by the relational data model. On top of this data structure, we implement the query processing and index building functions. To insulate the naming schemes used by scientific data formats and naming scheme acceptable to FastBit, we implement a parser which is responsible for understanding the subarray syntax.

To separate FastQuery from any specific scientific format, we abstracted out the required I/O operations into a File I/O interface. This allows us to implement a handful of functions to adapt FastQuery into any array-based scientific data format.

Performance Measurements. To illustrate the effectiveness of the new FastQuery design, in Fig. (b), we show its performanance with a set of synthetic data on a workstation

---

⋆ This work was supported by the Director, Office of Science, Office of Advanced Scientific Computing Research, of the U.S. Department of Energy under Contract No. DE-AC02-05CH11231. The authors would also like to thank Allen Sanderson, John Shalf, Quincey Koziol and Wes Bethel for the helpful discussions leading up to the design and specification of FastQuery API.

J.B. Cushing, J. French, and S. Bowers (Eds.): SSDBM 2011, LNCS 6809, pp. 573–574, 2011.

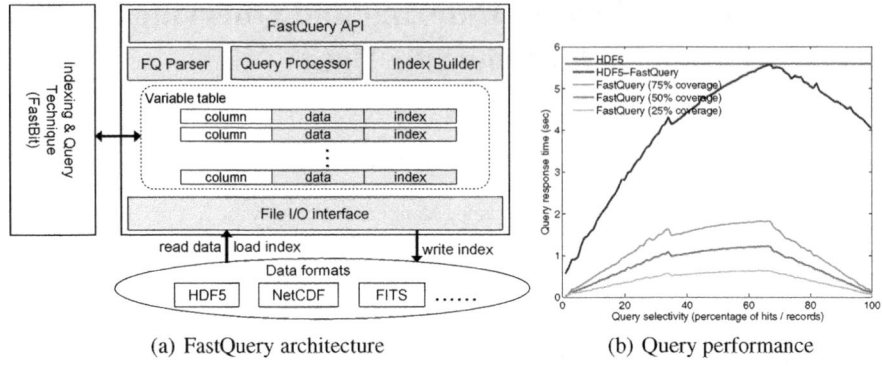

(a) FastQuery architecture   (b) Query performance

with 2.67 GHz Intel processor, 4 GB of memory and a commodity SATA disk. The set of queries use the same variable, but have different range conditions so that they select different number of records. With FastQuery, we further select the subarray that covers a specified percentage which we call coverage in Fig. (b).

With 75% coverage, the new implementation of FastQuery is at least 2.5 times faster than the previous version of HDF5-FastQuery and is on average 5.5 times faster. With 25% coverage, the new implementation is on average 13 times faster.

Conclusions. We have designed and implemented FastQuery, a general indexing and querying system for scientific data. FastQuery utilizes the FastBit bitmap indexing technology to support semantic query on common data format, such as HDF5 and NetCDF. We significantly extended our previous work, HDF5-FastQuery, by addressing the usability, applicability and flexibility issues of the indexing and querying system. Through the evaluations, we demonstrated that the FastQuery implementation is significantly more efficient than the previous version.

References

1. Gosink, L., Shalf, J., Stockinger, K., Wu, K., Bethel, W.: HDF5-FastQuery: Accelerating complex queries on HDF datasets using fast bitmap indices. In: SSDBM, pp. 149–158 (2006)
2. Shoshani, A., Rotem, D. (eds.): Scientific Data Management: Challenges, Technology, and Deployment, ch. 6. Chapman & Hall/CRC Press (2010)
3. Wu, K., Ahern, S., Bethel, E.W., Chen, J., Childs, H., Cormier-Michel, E., Geddes, C., Gu, J., Hagen, H., Hamann, B., Koegler, W., Lauret, J., Meredith, J., Messmer, P., Otoo, E., Perevoztchikov, V., Poskanzer, A., Prabhat, O.R., Shoshani, A., Sim, A., Stockinger, K., Weber, G., Zhang., W.-M.: FastBit: Interactively searching massive data. In: SciDAC (2009)

Retrieving Accurate Estimates to OLAP Queries over Uncertain and Imprecise Multidimensional Data Streams

Alfredo Cuzzocrea

ICAR-CNR and University of Calabria
I-87936, Cosenza, Italy
`cuzzocrea@si.deis.unical.it`

Abstract. In this paper, we introduce a novel framework for estimating OLAP queries over uncertain and imprecise multidimensional data streams, along with three relevant research contributions: (i) a probabilistic data stream model, which describes both precise and imprecise multidimensional data stream readings in terms of nice confidence-interval-based Probability Distribution Functions (PDF); (ii) a possible-world semantics for uncertain and imprecise multidimensional data streams, which is based on an innovative data-driven approach that exploits "natural" features of OLAP data, such as the presence of clusters and high correlations; (iii) an innovative approach for providing theoretically-founded estimates to OLAP queries over uncertain and imprecise multidimensional data streams that exploits the well-recognized probabilistic estimators theory.

1 Introduction

Modern data stream applications and systems are more and more characterized by the presence of uncertainty and imprecision that make the problem of dealing with uncertain and imprecise data streams a leading research challenge. This issue has recently attracted a great deal of attention from both the academic and industrial research community, as confirmed by several research efforts done in this context [4,9,5,2]. Uncertain and imprecise data streams arise in a plethora of actual application scenarios ranging from environmental sensor networks to logistic networks and telecommunication systems, and so forth.

While some recent papers have tackled the problem of efficiently representing, querying and mining uncertain and imprecise data streams [4,9,5,2], to the best of our knowledge, there does not exist in the literature research initiatives that deal with the problem of efficiently OLAPing [6] uncertain and imprecise multidimensional data streams, with explicit emphasis over multidimensionality of data [1,11,10,7]. In order to fulfill this relevant gap, we first introduce the problem of estimating OLAP queries over uncertain and imprecise multidimensional data streams, which can be reasonably considered as the first research attempt towards the definition of OLAP tools over uncertain and imprecise multidimensional data streams exposing complete OLAP functionalities, such as on-the-fly

J.B. Cushing, J. French, and S. Bowers (Eds.): SSDBM 2011, LNCS 6809, pp. 575–576, 2011.

data summarization and indexing. In particular, we propose a framework that is able of effectively and efficiently provide theoretically-founded estimates to OLAP queries over uncertain and imprecise multidimensional data streams, as a first step towards building more complex OLAP tools.

The framework for OLAPing uncertain and imprecise multidimensional data streams proposed in our research builds on some previous results that have been provided by recent efforts. Particularly, [3], which focuses on static data, introduces a nice Probability Distribution Function (PDF) [8]-based model that allows us to capture the uncertainty of OLAP measures. Furthermore, imprecision of OLAP data with respect to OLAP hierarchies available in the multidimensional data stream model is meaningfully captured by means of the so-called possible-world semantics [3]. This semantics allows us to evaluate OLAP queries over uncertain and imprecise static data, while also ensuring some well-founded theoretical properties, namely consistency, faithfulness and correlation-preservation [3]. The possible-world semantics [3] is exploited and significantly extended in our research, and specialized to the more challenging issue of dealing with uncertain and imprecise multidimensional data streams, along which several original and innovative research contributions.

References

1. Cuzzocrea, A., et al.: Improving OLAP analysis of multidimensional data streams via efficient compression techniques. In: Intelligent Techniques for Warehousing and Mining Sensor Network Data. IGI Global (2009)
2. Jin, C., et al.: Sliding-window top-k queries on uncertain streams. PVLDB 1(1) (2008)
3. Burdick, D., et al.: OLAP over uncertain and imprecise data. In: VLDB (2005)
4. Cormode, G., et al.: Sketching probabilistic data streams. In: ACM SIGMOD (2007)
5. Cormode, G., et al.: Exponentially decayed aggregates on data streams. In: IEEE ICDE (2008)
6. Gray, J., et al.: Data cube: A relational aggregation operator generalizing group-by, cross-tab, and sub-totals. Data Mining and Knowledge Discovery 1(1) (1997)
7. Han, J., et al.: Stream cube: An architecture for multi-dimensional analysis of data streams. Distributed and Parallel Databases 18(2) (2005)
8. Papoulis, A.: Probability, Random Variables, and Stochastic Processes, 2nd edn. McGraw-Hill, New York (1994)
9. Jayram, T.S., et al.: Estimating statistical aggregates on probabilistic data streams. In: ACM PODS (2007)
10. Chen, Y., et al.: Multi-dimensional regression analysis of time-series data streams. In: VLDB (2002)
11. Cai, Y.D., et al.: MAIDS: Mining alarming incidents from data streams. In: ACM SIGMOD (2004)

Hybrid Data-Flow Graphs for Procedural Domain-Specific Query Languages

Bernhard Jaecksch[1], Franz Faerber[1], Frank Rosenthal[2], and Wolfgang Lehner[2]

[1] SAP AG, Dietmar-Hopp-Allee 16, 69190 Walldorf, Germany
{b.jaecksch,franz.faerber,wolfgang.lehner}@sap.com
[2] TU Dresden, Institute for System Architecture,
Database Technology Group, 01062 Dresden, Germany
frank.rosenthal@tu-dresden.de

Abstract. Domain-specific query languages (DSQL) let users express custom business logic. Relational databases provide a limited set of options to execute business logic. Usually, stored procedures or a series of queries with some glue code. Both methods have drawbacks and often business logic is still executed on application side transferring large amounts of data between application and database, which is expensive. We translate a DSQL into a hybrid data-flow execution plan, containing relational operators mixed with procedural ones. A cost model is used to drive the translation towards an optimal mixture of relational and procedural plan operators.

1   Introduction

Relational databases provide with SQL a standardized and powerful query language. Although, SQL can be considered as a domain-specific language (DSL), its scope is broad and generic. To keep the user within confined boundaries of specific application domains a domain-specific query language (DSQL) is better suited to the task. Stored procedures incorporate procedural and declarative elements and can be used to express business logic. But they are pre-compiled into native C programs with embedded SQL commands. Hence, execution is driven by procedural C-code interspersed with SQL statements. This makes it difficult to optimize the entire procedure.

2   Contribution

We propose a mechanism to translate a procedural DSQL into the data-flow execution model of the underlying database. Normally, a database execution plan contains relational operators and the graph describes the execution in a declarative way. To overcome the mismatch between an imperative language and the declarative plan, we introduce a hybrid execution plan incorporating both aspects in one plan. Different, to typical stored procedure translation, we translate into a data-flow graph interspersed with procedural elements.

J.B. Cushing, J. French, and S. Bowers (Eds.): SSDBM 2011, LNCS 6809, pp. 577–578, 2011.
© Springer-Verlag Berlin Heidelberg 2011

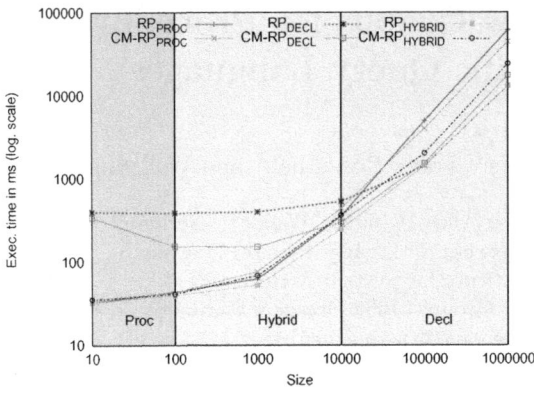

Translation Type	Percent
Declarative	58%
Hybrid	38%
Procedural	4%

Fig. 1. Actual execution times vs. cost model (CM) predictions for procedural (PROC), declarative (DECL) and hybrid (HYBR)

Fig. 2. Translation types of different DSQL scripts

The goal is to translate into a plan with as few procedural elements as possible. Because we express procedural logic by declarative means, we benefit from well-known optimizations and the data-flow graph representation allows the easy exploitation of parallelism. Although, there are many cases where procedural statements can be expressed entirely by relational operators, for others it is impossible. In other cases, both translation variants are possible and we provide a cost model that drives the translation process towards an optimal plan in terms of execution time. We identified procedural patterns that can be expressed in a purely declarative way using a combination of relational operators.

Figure 1 shows evaluation results for an example script translated into a hybrid plan based on our cost model and depending on the input size. The prediction by the cost model are included as well. The shaded areas specify the type of plan that was selected. Our setting is an industrial setup, where an existing DSQL for business planning is used and most complex data processing is done on application side. We see a demand, that the database layer has to provide means to handle business logic from various domains in order to process the tasks closer to the data. To provide an insight, how how typical customer scripts can be translated, we did a survey of custom scripts from over 50 customers, classifying them into translatable to a purely declarative plan, a hybrid plan or only procedural. The results shown in Table 2 suggest that a large percentage of typical scripts can be translated into a hybrid plan with only relational operators.

3   Conclusion

We proposed a translation of a domain-specific query language (DSQL) into a data-flow execution graph that contains relational and procedural operators. Furthermore, we devised a cost model that guides the selection between procedural and relational operators to find an optimal hybrid plan based on the size of the input data.

Scalable and Automated Workflow in Mining Large-Scale Severe-Storm Simulations

Lei Jiang[1,3], Gabrielle Allen[1,3,*], and Qin Chen[2,3]

[1] Department of Computer Science, Louisiana State University
[2] Department of Civil and Environmental Engineering, Louisiana State University
[3] Center for Computation and Technology, Louisiana State University
{ljiang,gallen}@cct.lsu.edu, qchen@lsu.edu

Abstract. The simulation of large-scale complex systems, such as modeling the effects of hurricanes or storms in coastal environments, typically requires a large amount of computing resources in addition to data storage capacity. To make an efficient prediction of the potential storm surge height for an incoming hurricane, surrogate models, which are computationally cheap and can reach a comparable level of accuracy with simulations, are desired. In this paper, we present a scalable and automated workflow for surrogate modeling with hurricane-related simulation data.

Keywords: automated workflow, scalability, severe-storm simulation, data mining, surrogate model.

As today the cyberinfrastructure development keeps progressing in many research organizations across the world, the abundance of high-performance computing resources brings new possibilities to domain scientists and engineers in terms of large-scale parallel simulations for complex phenomena. A typical application scenario with high significance comes from severe storms such as hurricanes. A 6-day storm surge simulation can take more than 2,000 CPU hours, which means a 35-hour run on a cluster with 64 cores. Deterministic physics-based simulations are the primary way as a guide for decision makers but become time-consuming and inflexible for real-time predictions. Simulation data, especially for a set of simulations with designed input parameter space, can be analyzed and knowledge can be extracted from them. As an effort to simulation data mining [1], we focus on surrogate modeling from storm simulation data, which leads to lightweight models mimicking the behavior of simulations on points of interest (POI).

In the paper, we first describe the surrogate modeling approaches with large-scale simulation data, where each simulation generates values on multiple POIs. In this way, to make predictions on a target variable at a point or location in the simulation, we use functional data analysis for storm simulations with designed parameter space in two circumstances: the prediction of maximum storm surge height as scalar response and the time-series prediction of surge profile

---

* Corresponding author: Gabrielle Allen (216 Johnston Hall, Baton Rouge, LA 70803).

J.B. Cushing, J. French, and S. Bowers (Eds.): SSDBM 2011, LNCS 6809, pp. 579–580, 2011.

as functional response. It is needed to train a different model for each point of interest. Also, for complex systems with dynamics on the response surface at a single point, it is desired to dig out more information. Spatio-temporal causal links [2] can be constructed for the same simulation output variable between locations or across variables. As links are contingent upon the simulation input, like hurricane tracks in our scenario, a granger causality test is performed ahead of regression in order to detect the links that tend to be invariant. Such a modeling framework can thereby supply confidence level and measurements of uncertainty to time-critical predictions.

A scalable and automated workflow is then important to facilitate the data mining process. Our workflow combines parallel simulation, distributed data archive, and high-performance data mining in the same framework. We exploit task-level parallelism to achieve scalability. Two modes in data processing are involved: *i)* a piece of data from each simulation is needed for modeling and those from multiple simulations are assembled for a modeling task, namely *Task Assembling*; and *ii)* the processing can be separately performed for each simulation and finally the results are to be reduced for generalization (Map-Reduce [3]). Then, in the implementation, several components are included: simulation data archive [4], parameter space, pattern space, data mining job pool and model base. Besides scalability, it also ensures that modeling process is automatically handled with continuously increasing data in the archive.

In the demonstration at the SSDBM 2011 conference, we show the workflow performance as well as some surrogate modeling results. Future directions of the work include creating a generic workflow for more data sources and the surrogate models themselves can also be further elaborated.

Acknowledgments

The authors acknowledge the contributions of Kelin Hu (Louisiana State University). This work is funded by NSF/EPSCoR EPS-0701491 (CyberTools) and NSF/EPSCoR EPS-1010640 (Coastal Hazards Collaboratory). Computing resources were provided by the Louisiana Optical Network Initiative (LONI).

References

1. Brady, T.F., Yellig, E.: Simulation data mining: a new form of computer simulation output. In: Proc. of the Winter Simulation Conference, pp. 285–289 (2005)
2. Lozano, A., Li, H., et al.: Spatial-temporal causal modeling for climate change attribution. In: Proc. of the 16th ACM SIGKDD Conference on Knowledge Discovery and Data Mining, KDD 2009 (2009)
3. Dean, J., Ghemawat, S.: MapReduce: simplified data processing on large clusters. In: Proc. of USENIX Symposium on Operation Systems Design and Implementation, OSDI 2004 (2004)
4. Bhagawaty, H., Jiang, L., et al.: Design, Implementation and Use of a Simulation Data Archive for Coastal Science. In: Proc. of Int'l ACM Symposium on High-Perf. Parallel and Distributed Computing (HPDC 2010) Workshops, pp. 651–657 (2010)

Accurate Cost Estimation Using Distribution-Based Cardinality Estimates for Multi-dimensional Queries

Andranik Khachatryan and Klemens Böhm

Institute for Program Structures and Data Organization (IPD)
Karlsruhe Institute of Technology (KIT), Germany
{khachatryan,klemens.boehm}@kit.edu

1 Introduction

Cardinality estimation is crucial for query optimization. The optimizer uses cardinality estimates to compute the query-plan costs. Histograms are one of the most popular data structures used for cardinality estimation [2]. Because histograms compress the data set, the cardinality estimates issued are not exact. We model these estimates as random variables, and denote the cardinality of query q by $card(q)$. For a query-plan with a cost function $v(\cdot)$, the plan cost is:

$$cost = E[v(card(q))] \qquad (1)$$

It is the expected value of the cost function applied to the random variable $card(q)$. Conventionally, the optimizers assume the cost is close to linear and use the approximation

$$cost \approx v(E[card(q)]) \qquad (2)$$

The reason why (2) is easier to use is that it requires estimating the expected value of the cardinality; in contrast, (1) needs the probability distribution of cardinalities to compute the expected cost. However, non-linear costs are commonplace, meaning that (2) often does not yield accurate cost estimates. In order to overcome this we propose estimating the cardinality distribution and use (1). This is challenging because we need to start issuing an accurate cardinality distribution instead of just the expected value. Ideally, we would like to use existing data structures such as histograms for this purpose. To this end, we propose a method which uses the previously executed query results (query feedback) as a sample and builds the cardinality distribution from this sample. As the underlying data structure we use STHoles [1] which is a multi-dimensional self-tuning histogram.

2 The STHoles Histogram

STHoles organizes the muldi-dimensional data space into rectangular buckets, like an R+ tree. The buckets cannot partially overlap but can be nested. The histogram estimates the number of tuples in a query using the uniformity assumption. For the query q and histogram S,

$$n(q) = \sum_{b \in S} n(b) \cdot \frac{vol(b \cap q)}{vol(b)} \qquad (3)$$

J.B. Cushing, J. French, and S. Bowers (Eds.): SSDBM 2011, LNCS 6809, pp. 581–582, 2011.

where $n(\cdot)$ is the number of tuples, $vol(\cdot)$ is the volume of a region, and $b \in S$ are the buckets.

3   The Sample-Based Method

In order to approximate the random variable $card(q)$, we use the past query execution results as a sample. The cumulative distribution of the random variable X is approximated by

$$F_m(z) = \frac{1}{m} \sum_{i=1}^{m} I(x_i \leq z) \tag{4}$$

where $\{x_1,\ldots,x_m\}$ is the sample and $I(P)$ is the indicator function. Let the histogram bucket b enclose the query q, with child buckets $\{b_1,\ldots,b_m\}$. We use the already observed selectivities within the region of b to approximate the distribution of selectivities. These are a) the selectivities of child buckets $\{sel(b_1),\ldots,sel(b_m)\}$, b) the selectivity which corresponds to the region covered by b, excluding child buckets: $s = (n(b) - \sum n(b_i))/(vol(b) - \sum vol(b_i))$.

Using $\{s, sel(b_1),\ldots,sel(b_m)\}$ as a sample, we approximate the cumulative distribution of selectivities inside the bucket. We weight the bucket selectivities with the volume of the intersection, to reflect the fact that larger intersections should have more weight. For the selectivity s this is $(1 - \sum vol(b_i)/vol(b))$, for a child bucket b_i this is $vol(b_i)/vol(b)$. The formula for the cardinality distribution becomes:

$$Pr(sel(q) \leq x) = (1 - \sum_{i=1}^{m} \frac{vol(b_i)}{vol(b)}) \cdot I(s \leq x) + \sum_{i=1}^{m} \frac{vol(b_i)}{vol(b)} \cdot I(sel(b_i) \leq x) \tag{5}$$

Given a cost function $v(\cdot)$, the query-plan cost is:

$$cost = \sum_{x=0}^{M} v(x) Pr(card(q) = x)$$

where M is the maximal possible value of the cardinality, and

$$Pr(card(q) = x) = Pr(sel(q) = x \cdot vol(q)) = Pr(sel(q) \leq x \cdot vol(q)) - Pr(sel(q) \leq x \cdot vol(q) - 1) \tag{6}$$

The sample-based method enables us to obtain cardinality distributions instead of point cardinalities. The method is computationally inexpensive; moreover, it uses only the information already contained in a histogram.

References

1. Bruno, N., Chaudhuri, S., Gravano, L.: STHoles: a multidimensional workload-aware histogram. SIGMOD Record (2001)
2. Ioannidis, Y.: The history of histograms (abridged). In: VLDB (2003)

Session-Based Browsing for More Effective Query Reuse[*]

Nodira Khoussainova, YongChul Kwon, Wei-Ting Liao,
Magdalena Balazinska, Wolfgang Gatterbauer, and Dan Suciu

Department of Computer Science and Engineering
University of Washington, Seattle, WA, USA
{nodira,yongchul,liaowt,magda,gatter,suciu}@cs.washington.edu

1 Introduction

Scientists today are able to generate and collect data at an unprecedented scale [1]. Afterwards, scientists analyze and explore these datasets. Composing SQL queries, however, is a significant challenge for scientists because most are not database experts.

In this work, we leverage the collaborative environment that many scientists work in, which often includes a shared database with many scientists asking queries over it. As such, we utilize the collective knowledge of all the users by providing new users with access to a log of past queries, which can be used as starting points for writing new queries. However, navigating a large log of queries can be difficult and overwhelming.

In this paper, we introduce the *Smart Query Browser* (SQB) system. SQB supports efficient retrieval of relevant queries using what we call *session-based browsing*. We also show results from a user study where we investigated whether SQB speeds up the query formulation by supporting better query reuse. [1]

2 SQB Overview

To start, SQB provides keyword search over a query log. Instead of simply listing all matching queries, it presents the results as a *set of query sessions*. A query session, as introduced in our previous work [3], is a set of queries written by a user to achieve a single task. For example, an astronomer who wants to find, in the Sloan Digital Sky Survey database [5], all the stars of a certain brightness in the r-band within 2 arc minutes of a known star, is likely to write multiple SQL queries before completing this task.

SQB thus allows users to view each result query in the context of the task that it aimed to complete. With this approach, SQB helps the user to more rapidly identify relevant queries because the user can decide on the relevance of entire sessions. It also helps users see how simple queries evolved into more complex ones. Finally, query sessions enable users to discriminate between high-quality and low-quality queries: queries that appear near the end of a session tend to be of higher quality because the author has spent time to edit and improve the query.

---

[*] This work was partially supported by NSF CAREER award IIS-0845397 and IIS-0627585.
[1] We invite the reader to read our technical report for more details on SQB and the user study [4].

J.B. Cushing, J. French, and S. Bowers (Eds.): SSDBM 2011, LNCS 6809, pp. 583–585, 2011.

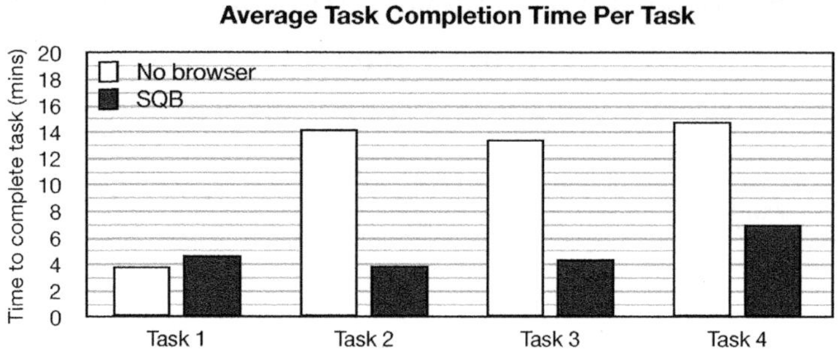

Fig. 1. Mean task completion time per interface, grouped by task

3 Evaluation

We performed a user study with 16 participants to investigate whether SQB speeds up the query formulation through query reuse. We find that, on average, SQB allows users to complete their tasks 2.3 times faster compared to having no access to a query browser.

The evaluation dataset consists of all SQL queries written by students in an undergraduate database class, offered at the University of Washington in 2008. These queries were logged as students worked on nine different problems for an assignment that used the IMDB database [2]. For the user study, we used a subset of this query log with 492 queries. Each participant in the user study was asked to translate four English sentence questions into four SQL queries.

Figure 1 presents the average completion time per task across the users. Note that a smaller completion time is better. We see that the SQB interface greatly outperforms the interface with no-browser for three of the tasks. Task 1 is a select-from-where query that can be written easily, and thus there is no benefit from SQB. In contrast, Task 4 is both difficult to write (i.e. requires a GROUP BY and either TOP or a NOT EXISTS subquery) but is not similar to any past query. The most similar query requires the user to make structural changes to the query before achieving the goal. Despite this heavy editing, SQB still helps users complete the task in less than half the time compared to no browser. The queries for Tasks 2 and 3 are also complex, requiring a GROUP BY and a self-join, respectively. However, there are similar queries in the query log. Therefore, we see a more than 3-fold improvement in speed with SQB.

4 Conclusion

We presented SQB, a tool for browsing through past SQL queries. The key insight behind SQB's design is the concept of query sessions. We showed that query sessions help speed up query composition by organizing queries in a large repository in a manner that facilitates the identification of relevant, high-quality queries to use as example.

References

1. The Fourth Paradigm: Data-Intensive Scientific Discovery (2009)
2. IMDB course assignment, http://www.cs.washington.edu/education/courses/cse444/08au/project/project1/project1.html
3. Khoussainova, N., Kwon, Y., Balazinska, M., Suciu, D.: SnipSuggest: Context-aware Auto-completion for SQL. In: Proc. VLDB Endow., vol. 4, pp. 22–33 (2010)
4. Khoussainova, N., Kwon, Y., Liao, W.-T., Balazinska, M., Gatterbauer, W., Suciu, D.: SQB: Session-based Query Browsing for More Effective Query Reuse. Technical Report 2011-04-02, Department of Computer Science and Engineering, University of Washington (2011)
5. Sloan Digital Sky Survey, http://www.sdss.org/

The ETLMR MapReduce-Based ETL Framework

Xiufeng Liu, Christian Thomsen, and Torben Bach Pedersen

Dept. of Computer Science, Aalborg University
{xiliu,chr,tbp}@cs.aau.dk

Abstract. This paper presents *ETLMR*, a parallel Extract–Transform–Load (ETL) programming framework based on MapReduce. It has built-in support for high-level ETL-specific constructs including star schemas, snowflake schemas, and slowly changing dimensions (SCDs). ETLMR gives both high programming productivity and high ETL scalability.

There is an ever-increasing demand for ETL tools to process very large amounts of data efficiently. Parallelization is a key technology to achieve the needed performance. Therefore, the "cloud computing" technology *MapReduce* [2] offering flexibility and scalability is interesting to apply to ETL parallelization. However, MapReduce is a general framework and lacks direct support for high-level ETL-specific constructs such as star schemas, snowflake schemas and SCDs. It is thus still very complex to implement a parallel ETL program with MapReduce and the ETL programmer's productivity is low. This paper presents the parallel ETL framework *ETLMR* [1] which directly supports high-level ETL constructs on MapReduce. The complexity of MapReduce is hidden from the user who only has to specify the transformations to apply and declarations of source data and destination tables. This makes it very easy to develop a highly scalable ETL program as only few lines of code and configuration are required.

ETLMR uses and extends the framework pygrametl [3] for code-based ETL and further uses Disco [4] as MapReduce platform. Fig 1 shows the architecture. An ETL flow in ETLMR comprises two sequential phases: dimension processing and fact processing, each of which runs as a separate job on the MapReduce platform. A job consists of a number of MapReduce instances (or tasks) running in parallel on several nodes in a cluster. An instance reads data from input files in a distributed file system (DFS) and processes the data and inserts it into dimension tables and/or fact tables in the DW.

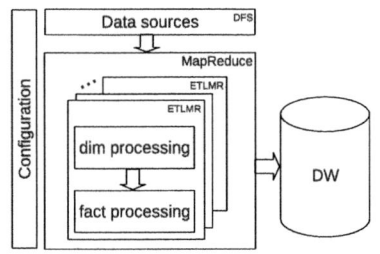

Fig. 1. The ETLMR framework

Configuration. ETLMR facilitates parallel ETL implementation by using a configuration file which defines dimension tables and fact tables. An object is

J.B. Cushing, J. French, and S. Bowers (Eds.): SSDBM 2011, LNCS 6809, pp. 586–588, 2011.

created for each of the target tables in a single statement where table name, attribute names, etc. are given. ETLMR supports different types of dimensions including slowly changing dimensions and snowflaked (i.e., normalized) dimensions. All the dimension classes have a common interface offering dimension operations such as *insert*, *lookup*, etc. In addition, the configuration file specifies the transformations to apply to the data and the number of map and reduce tasks to use (i.e., the level of parallelization). It is thus extremely easy to scale up/scale down an ETLMR-based program by only changing these numbers.

Dimension processing. During the dimension processing, ETLMR uses the (parallel instances of) MapReduce's *map* function to apply user-defined transformations to the source data. Further, the map function is used to find (i.e., project) the attributes that are relevant for the different dimensions. These subsets of the data are then processed by the (parallel instances of) MapReduce's *reduce* function to be inserted into the dimension tables. ETLMR offers several methods for doing this. The simplest method is *one dimension one task (ODOT)*. With ODOT, there is one, and only one, reduce instance for each dimension table. This makes it easy to avoid duplicated data, duplicated key values, etc. as all insertions to a given dimension table is done by one instance. Obviously, this method does, however, not scale well. Another supported method is *one dimension all tasks (ODAT)* in which all reduce instances (of which there can be any number) process data for all dimensions. This can, however, lead to problems with duplicated values (one reducer cannot see what another is about to insert) and duplicated key values. To remedy this, ETLMR, features the novel technique *post-fixing* where such problematic data autmatically is corrected when all data has been processed. Further, ETLMR offers dimension processing methods specialized for snowflaked dimensions. By processing the participating dimension tables in safe orders which respect foreign key relationships between tables, post-fixing can be avoided with these methods. Finally, ETLMR offers "offline" dimensions. With these, ETLMR does not communicate directly with the DW DBMS but instead stores data locally in the nodes. This leads to better performance.

Fact processing. In the fact processing, ETLMR looks up dimension keys in the (now processed) dimensions, does aggregation of fact data if needed, and bulk-loads the fact data into the DW. Several tasks do this in parallel.

Scalability. We have evaluated the scalability of the different processing methods on realistically sized datasets. The experiments [1] show that ETLMR achieves a nearly linear speedup in the number of tasks and compares favourably with other MapReduce data warehousing tools. To load a simple snowflake schema required 14 statements in ETLMR. In the MapReduce data processing frameworks Hive and Pig, the same schema required 23 and 40 statements, respectively. For a more complex schema with SCDs, ETLMR still only requires 14 statements, while Pig and Hive are not able to handle SCDs at all.

References

1. http://www.cs.aau.dk/~xiliu/etlmr/ as of (April 13 ,2011)
2. Dean, J., Ghemawat, S.: MapReduce: A Flexible Data Processing Tool. CACM 53(1), 72–77 (2010)
3. Thomsen, C., Pedersen, T.B.: pygrametl: A Powerful Programming Framework for Extract-Transform-Load Programmers. In: Proc. of DOLAP, pp. 49–56 (2009)
4. http://www.discoproject.org as of (April 13 ,2011)

Top-k Similarity Search on Uncertain Trajectories [*]

Chunyang Ma[1], Hua Lu[2], Lidan Shou[1], Gang Chen[1], and Shujie Chen[1]

[1] Department of Computer Science, Zhejiang University, China
mcy@cs.zju.edu.cn, {should,cg}@zju.edu.cn, pzbcsj@gmail.com
[2] Department of Computer Science, Aalborg University, Denmark
luhua@cs.aau.dk

Abstract. Similarity search on uncertain trajectories has a wide range of applications. To the best of our knowledge, no previous work has addressed similarity search on uncertain trajectories. In this paper, we provides a complete set of techniques for spatiotemporal similarity search on uncertain trajectories.

1 Introduction

Large amounts of trajectory data have been and are continuously being collected. In real applications the trajectories often are uncertain due to various factors, e.g., hardware limits and privacy concerns.

Motivated in a way akin to a broad range of applications on certain trajectories, e.g., traffic control and movement pattern mining, similarity search is also of interest and importance on uncertain trajectories. A key issue in the uncertain context is an appropriate distance function to measure the dissimilarity between two uncertain trajectories. The majority of existing metrics such as Discrete Time Warping, Longest Common Subsequences, etc. are focused on certain time series analysis only. Euclidean distance is not a reliable indicator of similarity in uncertain context [5]. The only technique combining both spatial distance and probabilities together [4,3,1] suffers from the problem that users have to specify parameters of ranking function to get the desirable results.

We in this paper propose a novel and effective metric, *p-distance*, to measure the dissimilarity between two uncertain trajectories. Based on the p-distance metric, we define a top-k similarity query (KSQ) on uncertain trajectories. To facilitate processing KSQs, we propose a novel index structure called *UTgrid* to manage all uncertain trajectories. The results of an extensive experimental study show that UTgrid based KSQ algorithms are efficient and scalable.

2 Problem Statement

In a 2D space, an uncertain trajectory X is represented as $\{(R_1, t_1), \ldots, (R_n, t_n), pdf\}$, where R_i is the uncertain range at time t_i, associated with the corresponding pdf_i.

We define the *p-distance* of an uncertain trajectory as the quantity of all other trajectories in the database that may be nearer than it to a query trajectory Q during the query time interval.

---

[*] This work is supported in part by the National Science Foundation of China (NSFC Grant No. 60803003, 60970124).

J.B. Cushing, J. French, and S. Bowers (Eds.): SSDBM 2011, LNCS 6809, pp. 589–591, 2011.

Given a query trajectory Q and an integer K, a top-k similarity query (KSQ) retrieves from the database k trajectories with the smallest p-distances with respect to the query trajectory. KSQs give users considerable convenience in that they do not need to specify any ranking functions for the trajectories.

3   Processing KSQs

To process KSQs efficiently we design UTgrid, a practical index for spatiotemporal *uncertain* *trajectories* based on *grid* partitioning techniques. In particular, UTgrid first partitions the spatial dimensions (the space domain) into a set of non-overlapping cells for spatial comparison, and then creates a temporal index within each spatial partition (cell) to help temporal elimination.

We also develop detailed algorithms for processing KSQs by exploiting the UTgrid.

4   Experiments

All the experiments are run on a desktop PC with a 2.66GHz CPU and 4GB RAM. The page size is 4096 bytes. We use a real dataset, the Geolife GPS trajectory dataset [2].

We compare UTgrid with two simple approaches, namely *BoundPruning* and sequential scan. The BoundPruning method uses only the boundaries of uncertain ranges of trajectories but no probability information to prune unqualified trajectories. We vary the value of k from 1 to 80 to see its impact on the query performance. As shown in Figure 1 UTgrid always outperforms both other techniques in terms of pruning power, total query processing time and the number of disk page accesses (I/O).

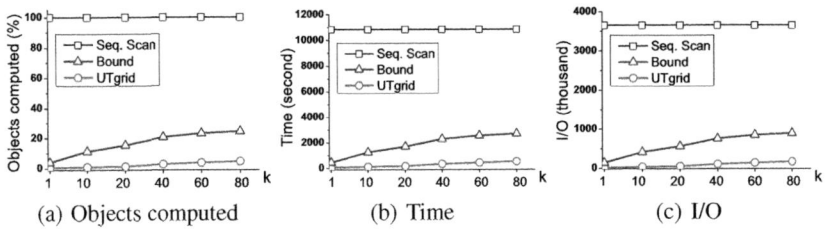

(a) Objects computed   (b) Time   (c) I/O

Fig. 1. Effect of k on real datasets

5   Conclusion

In this paper we address top-k similarity search on uncertain trajectories. We introduce *p-distance* to measure the dissimilarity between two uncertain trajectories and define top-k similarity query (KSQ) based on it. We design *UTgrid* for indexing uncertain trajectories and develop query processing algorithms. We also conduct an extensive experimental study to demonstrate the efficiency of our proposals.

References

1. Aßfalg, J., Kriegel, H.-P., Kröger, P., Renz, M.: Probabilistic similarity search for uncertain time series. In: Winslett, M. (ed.) SSDBM 2009. LNCS, vol. 5566, pp. 435–443. Springer, Heidelberg (2009)
2. Geolife gps trajectories, http://research.microsoft.com/en-us/downloads/b16d359d-d164-469e-9fd4-daa38f2b2e13/default.aspx
3. Lian, X., Chen, L., Yu, J.X.: Pattern matching over cloaked time series. In: ICDE, pp. 1462–1464 (2008)
4. Yeh, M.-Y., Wu, K.-L., Yu, P.S., Chen, M.-S.: Proud: a probabilistic approach to processing similarity queries over uncertain data streams. In: EDBT, pp. 684–695 (2009)
5. Yuen, S.M., Tao, Y., Xiao, X., Pei, J., Zhang, D.: Superseding nearest neighbor search on uncertain spatial databases. In: TKDE, pp. 1041–1055 (2010)

Fast and Accurate Trajectory Streams Clustering

ICAR-CNR
`masciari@icar.cnr.it`

Abstract. Trajectory data streams are huge amounts of data pertaining to time and position of moving objects. They are continuously generated by different sources exploiting a wide variety of technologies (e.g., RFID tags, GPS, GSM networks). Mining such amounts of data is challenging, since the possibility to extract useful information from this peculiar kind of data is crucial in many application scenarios such as vehicle traffic management, hand-off in cellular networks, supply chain management. Moreover, spatial data streams poses interesting challenges both for their proper definition and acquisition, thus making the mining process harder than for classical point data. In this paper, we address the problem of trajectory data streams clustering, that revealed really challenging as we deal with data (trajectories) for which the order of elements is relevant.

1 Introduction

Data Clustering is one of the challenging mining techniques exploited in the knowledge discovery process[3]. Clustering huge amounts of data is a difficult task since the goal is to find a suitable partition in a unsupervised way (i.e. without any prior knowledge) trying to maximize the similarity of objects belonging to the same cluster and minimizing the similarity among objects in different clusters. Many different clustering techniques have been defined in order to solve the problem from different perspective, i.e. partition based clustering (e.g. *K-means*[6]), density based clustering (e.g. *DBScan*[1]), hierarchical methods (e.g. *BIRCH*[10]) and grid-based methods (e.g. *STING* [9]). In this paper we deal with trajectory data streams that collect data pertaining to time and the position of moving objects (or groups of objects). Trajectory data carry information about actual positions and timestamps of moving objects at a detail level often unnecessary. Indeed, many proposals split the search space in regions having the suitable granularity and represent them as areas tagged by an annotated symbol. The sequence of regions define the trajectory traveled by a given object. Based on the above representation of trajectory data (i.e. region based instead of a sequence of multidimensional points) mining steps are performed based on proper techniques. Thus, regioning is a common assumption in trajectory data mining [4,2]. Since, in many application scenarios we need to work on the original data points, we propose in this work an approach that works directly on the original two dimensional trajectory representation. Moreover, few proposal

J.B. Cushing, J. French, and S. Bowers (Eds.): SSDBM 2011, LNCS 6809, pp. 592–593, 2011.
© Springer-Verlag Berlin Heidelberg 2011

address the trajectory clustering in an incremental way, thus many approaches are not suitable for datastreams except the one presented in [5]. As trajectories flow into the system we perform a data pre-elaboration based on a proper filtering of the multidimensional points based on *Lifting Schemes*[8]. The aim of lifting is to represent a spatial signal (i.e. the whole trajectory) using a shorter sequence by a proper filtering step that allow prediction and update of proper coefficients. We define a clustering strategy based on multidimensional Fourier Analysis in order to catch "structural" dissimilarities between trajectories. The basic intuition exploited is that a trajectory has a "natural" interpretation as a time series (namely, a discrete-time signal), in which numeric values summarize some relevant features of the elements belonging to the trajectory. In a sense, the analysis of the way the signal shapes differ can be interpreted as the detection of different locations crossed by the trajectories. Moreover, the analysis of the frequencies of common signal shapes can be seen as repeated crossing of the same location. In this context, the proposed approach is an efficient technique, which can satisfactorily evaluate how much two trajectories are similar w.r.t. the structural features previously discussed. Indeed, the exploitation of Non Separable Fourier Transforms (in particular we use Discrete Fourier Transform - *DFT* [7]) allows to abstract from minor details which, in most application contexts, should not affect the similarity estimation (e.g., multiple occurrences of the same location due to simple traffic problems). Thus, the comparison is less sensitive to minor mismatches. Moreover, a frequency based approach allows to estimate the similarity through simple measures (e.g., vector distances) which are computationally less expensive than techniques based on the direct comparison of the original trajectory structures.

References

1. Ester, M., Kriegel, H.P., Sander, J., Xu, X.: A density-based algorithm for discovering clusters in large spatial databases with noise. In: KDD 1996 (1996)
2. Giannotti, F., Nanni, M., Pinelli, F., Pedreschi, D.: Trajectory pattern mining. In: KDD 2007, pp. 330–339 (2007)
3. Han, J., Kamber, M.: Data Mining: Concepts and Techniques. Morgan Kaufmann, San Francisco (2000)
4. Lee, J.G., Han, J., Li, X., Gonzalez, H.: TraClass: trajectory classification using hierarchical region-based and trajectory-based clustering. PVLDB 1(1) (2008)
5. Li, Z., Lee, J.-G., Li, X., Han, J.: Incremental clustering for trajectories. In: Kitagawa, H., Ishikawa, Y., Li, Q., Watanabe, C. (eds.) DASFAA 2010. LNCS, vol. 5982, pp. 32–46. Springer, Heidelberg (2010)
6. Lloyd, S.: Least squares quantization in pcm. IEEE TOIT 28 (1982)
7. Press, W.H., et al.: Numerical Recipes in C++. Cambridge University Press, Cambridge (2001)
8. Taubman, D., Secker, A.: Lifting-based invertible motion adaptive transform (limat) framework for highly scalable video compression. IEEE Transactions on Image Processing 12(12), 1530–1542 (2003)
9. Wang, W., Yang, J., Muntz, R.R.: Sting: A statistical information grid approach to spatial data mining. In: VLDB 1997, pp. 186–195 (1997)
10. Zhang, T., Ramakrishnan, R., Livny, M.: Birch: An efficient data clustering method for very large databases. In: SIGMOD 1996, pp. 103–114 (1996)

Data-Driven Multidimensional Design for OLAP

Oscar Romero and Alberto Abelló

Universitat Politècnica de Catalunya, Barcelona, Spain
{aabello,oromero}@essi.upc.edu

Abstract. OLAP is a popular technology to query scientific and statistical databases, but their success heavily depends on a proper design of the underlying multidimensional (MD) databases (i.e., based on the *fact / dimension* paradigm). Relevantly, different approaches to automatically identify *facts* are nowadays available, but all MD design methods rely on discovering functional dependencies (FDs) to identify *dimensions*. However, an unbound FD search generates a combinatorial explosion and accordingly, these methods produce MD schemas with too many dimensions whose meaning has not been analyzed in advance. On the contrary, i) we use the available ontological knowledge to drive the FD search and avoid the combinatorial explosion and ii) only propose dimensions of interest for analysts by performing a statistical study of data.

1 Introduction

i) Our approach avoids generating too much results by mixing data mining and OLAP technologies. The purpose of this work is to demonstrate the feasibility and benefits of performing a statistical study of data to filter and prioritize the dimensional concepts found in the sources for a given fact, so that the designer can focus on these to decide and define his/her requirements for an OLAP application. ii) Furthermore, we tackle the usual assumption that a RDBMS is the most common kind of data sources we may find, by benefiting from a conceptual formalization of the domain (in our case, an OWL DL ontology) to avoid a combinatorial explosion of the statistical study.

Eventually, our approach identifies, for each fact, all the dimensional concepts and uses statistical evidences to filter out those of no relevance for data analysis.

2 Sketched Idea

Essentially, instead of blindly looking for FDs, our approach only tests combination of concepts likely to be interesting dimensional concepts for a given fact and its measures. We address the reader to [1] for further details on how to exploit the ontological knowledge available and the well-known FD theory to generate multi-concept FDs in a smart way.

Here, we extend our previous work with a statistical study to filter out those combinations of interest for the user. In [2] we can see how to perform an analysis of variance (ANOVA). This is a test designed to decide whether the difference in

J.B. Cushing, J. French, and S. Bowers (Eds.): SSDBM 2011, LNCS 6809, pp. 594–595, 2011.
© Springer-Verlag Berlin Heidelberg 2011

the means of several samplings are due to differences in the populations or can be reasonably attributed to chance fluctuations alone. We propose to measure the importance when an attribute *partitions* a fact measure. Based on this objective evidence, we should choose the dimensional attributes based on the gain of entropy on partitioning each measure of interest. In our ANOVA tests, the hypothesis of "no difference" in the population of the different subsets is the *null hypothesis*. If this is rejected in our statistical analysis with a given confidence level, we will propose this attribute (or set of attributes) as an ***Interesting Dimension*** (**ID** from here on).

The* interesting Dimension *Function: This function is called whenever a combination of attributes is likely to be an ID (see [1]). Up to this step everything has been verified at the conceptual level. Then, we verify whether this combination of dimensional concepts is interesting to analyze a given measure by querying data. Prior to perform the statistical analysis, we first disregard candidate IDs with too many instances, since the end-user will be overwhelmed by the amount of values. Indeed, statisticians consider that useful categorical variables should have, at most, some tens of values. Relevantly, in case of querying a RDBMS, this pruning rule disregards combinations by just querying the catalog. Those combinations satisfying this rule are still candidates to be an ID, and we verify it by performing a one-way ANOVA test over data, as explained in [2], with the following query:

```
SELECT (SUM(gr.s)/(#distinct-1))/(SUM(POWER(A-grAvg,2))/(#tuples-#distinct)) AS fFisher
FROM t, (SELECT attrSet AS id, avg(A) AS grAvg, POWER(AVG(A)-(SELECT avg(A) FROM t),2) AS s
         FROM t WHERE joinConds GROUP BY attrSet) gr
WHERE attrSet=gr.id;
```

Where *attrSet* are the attributes conforming the *feasible ID* to be verified, *t* the table or tables containing those attributes (*join conditions* should be added if necessary), *#distinct* is the number of different values for *setAttr* and *#tuples* the number of tuples in the fact table. Then, the credibility of the null hypothesis is obtained by placing the result of this query in a Fisher distribution with the corresponding degrees of freedom.

However, this is not enough to decide whether this combination of attributes is an ID or not, because we could detect an ID due to the influence of another ID. Thus, once we detect an ID, we perform a two-way ANOVA test involving it and any other ID detected before (by means of a similar query) to discard the possibility that this is an ID just because another one is and there is some relationship between them (e.g., a multivalued dependency). Importantly, our approach can be used for other kind of data sources, as we would only need to adapt these SQL queries to the available data sources technology.

References

1. Abelló, A., Romero, O.: Using Ontologies to Discover Fact IDs. In: ACM 13th International Workshop on Data Warehousing and OLAP, pp. 3–10. ACM, New York (2010)
2. Wonnacott, T.H., Wonnacott, R.J.: Introductory Statistics. Wiley & Sons, Chichester (1990)

An Adaptive Outlier Detection Technique for Data Streams[*]

Shiblee Sadik and Le Gruenwald

School of Computer Science, University of Oklahoma, Norman, OK 73071

Abstract. This work presents an adaptive outlier detection technique for data streams, called Automatic Outlier Detection for Data Streams (A-ODDS), which identifies outliers with respect to all the received data points (global context) as well as temporally close data points (local context) where local context are selected based on time and change of data distribution.

1 Introduction

An outlier is a data point which is significantly different from other data points [1]. Although outliers are interesting to the user, a handful of techniques are available for data streams, which are adopted from existing outlier detection techniques for regular data with ad-hoc modifications. A number of those techniques use sliding window and detect outliers inside the window [2]; but an outlier for a particular window may appear as an inlier in another window; hence the notion of outlier in a data stream window is not very concrete. Auto-regression based techniques construct a model and compute a metric for each data point [3] where a data point is an outlier if the corresponding metric is beyond a certain cut-off limit. However finding a proper auto-regression model and cut-off limit is a not a trivial task and requires expert knowledge. Statistics based techniques [1] assume a fixed data distribution while data streams have varying distribution. In this work we present A-ODDS to detect outliers based on the deviations of a data point with respect to global and local contexts.

2 The Proposed Technique: A-ODDS and Experimental Results

Our approach is based on two deviation factors for the global and local contexts, called Global Deviation Factor (GDF) and Local Deviation Factor (LDF), respectively. GDF represents the deviation of a data point with respect to the entire history data points; and LDF represents the deviation of a data point with respect to the recent data points; both deviation factors are calculated from neighbor density.

GDF of a data point is the relative distance from the average neighbor density of the entire history data points to its neighbor density; and LDF of a data point is the relative distance from the average neighbor density of the recent data points to its neighbor density. A data point is identified as an outlier if either its GDF or LDF goes beyond three standard deviations away from its respective average. The choice of

---
[*] This work has been supported in part by the NASA under the grants No. NNG05GA30G.

J.B. Cushing, J. French, and S. Bowers (Eds.): SSDBM 2011, LNCS 6809, pp. 596–597, 2011.

three standard deviation dispersion ensures a significant dispersion of a data point from other data points [1] and does not require the user to select cut-off limits.

Our local context selection scheme for LDF is based on two intuitive criteria: first, data points in local context have to be temporally close and second, they have to follow similar distribution. We choose data points in between two consecutive concept drifts as local context as they are close temporally and expected to follow similar distribution; hence LDF finds outliers non-conformist to the recent trend. GDF and LDF use the dynamically adaptive data distribution function for neighbor density computation that we presented in [5].

We conducted simulation experiments using a real dataset collected from California Irrigation Management [4] to compare A-ODDS with the three existing algorithms: auto-regression based algorithm ART [3], sliding window based algorithm ODTS [2] and distance-based outlier detection algorithm DBOD-DS [5], in terms of outlier detection accuracy (Jaccard Coefficient (JC) [2]) and execution time. On average, as shown in Table 1, A-ODDS gives the best accuracy among all the techniques; however, when measuring execution time, while A-ODDS requires less time than DBOS-DS, it takes more time than ART and ODTS.

Table 1. Average JC and Execution Time

	Average accuracy (JC)	Average execution time (ms)
A-ODDS	0.7095	1.3656
DBOD-DS	0.1585	1.8485
ODTS	0.1467	0.0405
ART	0.1373	0.3005

3 Conclusions and Future Work

This paper presents an overview of A-ODDS and its accuracy and efficacy compared to existing algorithms. For future work, we will perform extensive empirical studies and extend it to multi-dimensional and multiple heterogeneous data streams.

References

1. Barnett, V., Lewis, T.: Outliers in Statistical Data. Wiley Series in Probability and Mathematical Statistics. John Wiley & Sons Inc., Chichester (1994)
2. Basu, S., Meckesheimer, M.: Automatic outlier detection for time series: an application to sensor data. Knowledge Information System (2007)
3. Curiac, D., Banias, O., Dragan, F., Volosencu, C., Dranga, O.: Malicious Node Detection in Wireless Sensor Networks Using an Autoregression Technique. In: ICNS 2007 (2007)
4. California Irrigation Management Information System. web-link,
   http://wwwcimis.water.ca.gov/cimis/welcome.jsp
   (accessed January 2010)
5. Sadik, S., Gruenwald, L.: DBOD-DS: Distance Based Outlier Detection for Data Streams. In: Bringas, P.G., Hameurlain, A., Quirchmayr, G. (eds.) DEXA 2010. LNCS, vol. 6261, pp. 122–136. Springer, Heidelberg (2010)

Power-Aware DBMS: Potential and Challenges

Yi-cheng Tu[1], Xiaorui Wang[2], and Zichen Xu[1]

[1] Department of Computer Science and Engineering
4202 E. Fowler Ave., ENB118
Tampa, FL 33620
[2] Department of Electrical Engineering and Computer Science
The University of Tennessee
Knoxville, TN 37996 Univ. of Tennessee
{ytu,zxu5}@mail.usf.edu, xwang@eecs.utk.edu

Abstract. Energy consumption has become a first-class optimization goal in computing system design and implementation. Database systems, being a major consumer of computing resources (thus energy) in modern data centers, also face the challenges of going green. In this position paper, we describe our vision on this new direction of database system research, and report the results of our recent work on this topic. We describe our ideas on the key issues in designing a power-aware DBMS and sketch our solutions to such issues. Specifically, we believe that the ability for the DBMS to dynamically adjust various knobs to satisfy energy (and performance) goals is the main technical challenge in this paradigm. To address that challenge, we propose dynamic modeling and tuning techniques based on formal feedback control theory. Our preliminary data clearly show that the energy savings can be significant.

Keywords: Power-aware DBMS, feedback control, energy profile identification, system modeling, power cost estimation.

1 Introduction

The steep increase of energy consumption of computers have made power management a critical issue in system design and implementation. As shown in our previous work [2], a basic design of P-DBMS introduces multiple control knobs that enable real-time adjustment of system behavior. A typical database system bears many uncertainties in its workload and environment therefore the problem of DBMS control for energy-saving purposes cannot be mapped into a conventional optimization problem. Our proposal to tackle this problem is to view it as an optimal control problem and utilize rigorous control-theoretical analysis and system design techniques to accomplish energy saving and performance goals. Our control-based solution, in contrast to *ad hoc* heuristics that are widely used in solving similar tuning problems, has the advantage of providing guaranteed performance and resistance to system/environmental dynamics.

J.B. Cushing, J. French, and S. Bowers (Eds.): SSDBM 2011, LNCS 6809, pp. 598–599, 2011.

2   Overview of P-DBMS Design

Our vision of building an energy-efficient DBMS is to enhance current DBMS components with energy-related functionalities, rather than building these components from scratch. This allows us to minimize the impacts on the current DBMS architecture that is well-designed for performance-driven query processing. The design goal of the system is as follows:

PROBLEM 1. *Given a performance bound, the power consumption of the database system should be minimized.*

The above design goal reflects the idea that performance is the most critical issue, while saving energy cost is a best-effort requirement based on a tolerance level of performance degradation.

3   Feedback Control for Power Optimization

Traditionally, solutions to adaptive power management problems, as well as those in self-tuning databases, heavily rely on heuristics. Recently, however, feedback control theory has been successfully applied to power control in servers and database tuning [1]. The benefit of having control theory as a theoretical foundation is that we can have (1) standard approaches to choosing the right control parameters so that exhaustive iterations of tuning and testing are avoided; (2) theoretically guaranteed control performance such as accuracy, stability, short settling time, and small overshoot; and (3) quantitative control analysis when the system is suffering unpredictable workload variations. This rigorous design methodology is in sharp contrast to heuristic-based adaptive solutions that rely on extensive empirical evaluation and manual tuning.

Intuitively, the power mode of hardware should be set to a level such that the system performance converges to the tolerance bound mentioned in PROBLEM 1. This is because making the performance better than the bound is not necessary and also implies less energy saving. We are developing a *feedback control loop* to satisfy the runtime power and performance requirements set in PROBLEM 1.

4   Conclusions

In this paper, we elaborated on the issues in building such a PDBMS system and how such issues can be resolved. Our idea was to utilize formal feedback control theory to achieve effective modeling and system control. We conclude that power-aware DBMS is a meaningful and interesting approach for tackling the problem of energy-efficient database systems.

References

1. Lightstone, S., et al.: Control theory: a foundational technique for self managing databases. In: ICDE Workshops, pp. 395–403 (2007)
2. Xu, Z., Tu, Y., Wang, X.: Exploring power- performance tradeoffs in database systems. In: Proceedings of ICDE (2010)

Author Index

Abelló, Alberto   594
Agrawal, Gagan   571
Agrawal, Shraddha   565
Allen, Gabrielle   579
AlSayyad, Yusra   568
Ammar, Khaled   527
Assent, Ira   150
Augsten, Nikolaus   274
Aung, Htoo Htet   369

Bach Pedersen, Torben   586
Balazinska, Magdalena   568, 583
Battle, Leilani   480
Bernecker, Thomas   37
Böhm, Klemens   351, 581
Böhm, Matthias   491
Brown, Paul   1
Budavári, Tamás   509

Chen, Gang   589
Chen, Keke   332
Chen, Lei   168, 312
Chen, Qin   579
Chen, Shujie   589
Chen, Yuxin   432
Cheng, Hong   255
Chin Jr., George   189
Chiu, David   571
Chou, Jerry   414, 573
Cohen-Boulakia, Sarah   73
Cole, Garret   480
Connolly, Andrew J.   568
Critchlow, Terence   189, 253
Cuzzocrea, Alfredo   575

Dannecker, Lars   491
Degraf, Colin   546
Denise, Alain   73
Dey, Saumen C.   225
DiMatteo, Tiziana   546
Du, David H.C.   461

Faerber, Franz   577
Fink, Eugene   546
Freire, Juliana   555

Frew, James   244
Fu, Bin   546

Gándara, Aída   189
Gatterbauer, Wolfgang   583
Gibson, Garth   546
Gruenwald, Le   596
Günnemann, Stephan   150
Guo, Shumin   332

Hackenbroich, Gregor   491
Hall, Travis   571
Hamel, Sylvie   73
He, Xianmang   451
Howe, Bill   480, 568

Ivanescu, Anca-Maria   150

Jaecksch, Bernhard   577
Janée, Greg   244
Ji, Shengyue   17
Jiang, Lei   579
Jin, Wei   293, 322
Johnson, Theodore   129
Jones, Lynne   568

Kabir, Farhana   571
Karlapalem, Kamalakar   565
Kazimianec, Michail   274
Key, Alicia   480
Khachatryan, Andranik   351, 581
Khoussainova, Nodira   480, 583
Kim, Jinoh   414
Köhler, Sven   207
Kopper, Jonida   351
Koutris, Paraschos   480
Kranen, Philipp   405
Kremer, Hardy   150
Kriegel, Hans-Peter   37, 387
Kröger, Peer   387
Krughoff, K. Simon   568
Kwon, YongChul   583

Lee, Dongwon   432
Lehner, Wolfgang   491, 537, 577

Lemson, Gerard   509
Li, Chen   17
Li, Fengjun   432
Li, Yujia   451
Lian, Xiang   168
Liao, Wei-Ting   583
Lin, Xuemin   312
Liu, Peng   432
Liu, Ruilin   109
Liu, Xiufeng   586
López, Julio   546
Lu, Hua   589
Ludäscher, Bertram   207, 225
Luo, Bo   432

Ma, Chunyang   589
Madduri, Kamesh   470
Maier, David   55
Mamoulis, Nikos   37
Masciari, Elio   592
Mates, Phillip   555
McPhillips, Timothy   207
Megler, V.M.   55
Mokbel, Mohamed F.   461
Müller, Emmanuel   351

Naps, Joseph L.   461
Nascimento, Mario A.   527
Ntoutsi, Irene   387

Patil, Manish   91
Peng, Peng   312
Pinheiro da Silva, Paulo   189
Poliakov, Alex   1
Prabhat,   573

Qiao, Miao   255

Raman, Suchi   1
Reidl, Felix   405
Renz, Matthias   37
Riddle, Sean   207
Romero, Oscar   594
Rosenthal, Frank   537, 577
Rotem, Doron   414

Sadik, Shiblee   596
Sanchez Villaamil, Fernando   405
Santos, Emanuele   555
Schulze, Robert   491
Seidl, Thomas   150, 405
Shah, Rahul   91
Shi, Baile   451
Shkapenyuk, Vladislav   129
Shou, Lidan   589
Silva, Cláudio T.   555
Sivaramakrishnan, Chandrika   189
Slaughter, Peter   244
Souroush, Emad   480
Spengler, Sylvia   490
Stonebraker, Michael   1
Suciu, Dan   583
Szalay, Alexander   509

Tan, Kian-Lee   369
Thankachan, Sharma V.   91
Theodoratos, Dimitri   109
Thomsen, Christian   586
Tian, Fengguang   332
Tu, Yi-cheng   598

Vadapalli, Soujanya   565

Wang, Hui (Wendy)   109
Wang, Qing   451
Wang, Wei   451
Wang, Xiaorui   598
White, Signe   189
Wu, Kesheng   470, 573
Wu, Xiaoying   109

Xiao, Yanghua   451
Xu, Huiqi   332
Xu, Zichen   598

Yang, Jiong   293, 322
Yu, Jeffrey Xu   255

Zhao, Dongyan   312
Zimek, Arthur   387
Zinn, Daniel   207, 225
Zou, Lei   312
Zuefle, Andreas   37